W9-AOP-002

The Beginning of the
Age of Dinosaurs

The Beginning of the Age of Dinosaurs

Faunal change across the Triassic–Jurassic boundary

Edited by

Kevin Padian

Department of Paleontology
University of California, Berkeley

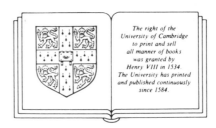

The right of the
University of Cambridge
to print and sell
all manner of books
was granted by
Henry VIII in 1534.
The University has printed
and published continuously
since 1584.

Cambridge University Press
Cambridge
New York   Port Chester
Melbourne   Sydney

Published by the Press Syndicate of the University of Cambridge
The Pitt Building, Trumpington Street, Cambridge CB2 1RP
40 West 20th Street, New York, NY 10011, USA
10 Stamford Road, Oakleigh, Melbourne 3166, Australia

First published 1986
First paperback edition 1988
Reprinted 1990

Printed in the United States of America

Library of Congress Cataloging-in-Publication Data
The Beginning of the age of dinosaurs.
Papers from the Symposium on Faunal Change Across
the Triassic–Jurassic Boundary, held Oct. 31, 1984,
in conjunction with the 44th Annual Meeting of the
Society of Vertebrate Paleontology at Berkeley, Calif.
1. Vertebrates, Fossil – Congresses.   2. Paleontology
– Triassic – Congresses.   3. Paleontology – Jurassic –
Congresses.   4. Evolution – Congresses.   I. Padian,
Kevin.   II. Symposium on Faunal Change Across the
Triassic–Jurassic Boundary (1984 : Berkeley, Calif.)
III. Society of Vertebrate Paleontology. Meeting
(44th : 1984 : Berkeley, Calif.)
QE841.B39   1986   567   86–9692

British Library Cataloguing-in-Publication Data
The Beginning of the age of dinosaurs : faunal
change across the Triassic – Jurassic boundary.
1. Vertebrates, Fossil   2. Geology,
Stratigraphic – Mesozoic
I. Padian, Kevin
566   QE841

ISBN-0-521-30328-1 hardback
ISBN 0-521-36779-4 paperback

The participants in the Triassic–Jurassic symposium and the authors of the papers in this volume would like to dedicate our efforts in this behalf to three men who have contributed significantly to the knowledge and understanding of vertebrate evolution in the early Mesozoic Era, and to the advancement of so many aspects of lower vertebrate paleontology.

Charles L. Camp

Joseph T. Gregory

Samuel P. Welles

In memory of the late Charles Camp and with sincerest appreciation and respect to Drs. Gregory and Welles, the authors hope that the works presented in this volume may reflect the influence and stimulus that you have had on all of us, and will continue to have for a long time to come.

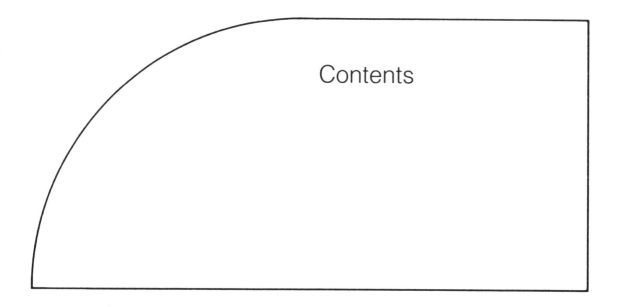

Contents

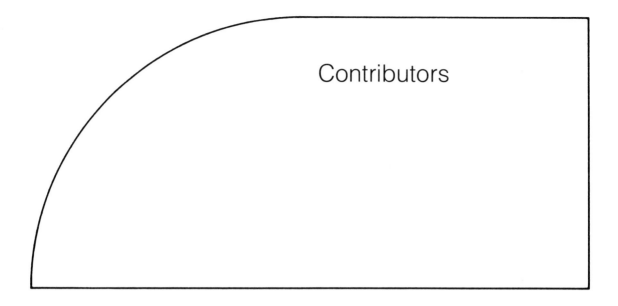

Contributors

Sidney R. Ash
Department of Geology and Geography
Weber State College
Ogden, UT 84408

John Attridge
Birkbeck College
University of London
London WC1E 7HX, England

Donald Baird
Department of Geology
Princeton University
Princeton, NJ 08544

Michael J. Benton
Department of Geology
The Queen's University
Belfast BT7 1NN, Northern Ireland

José F. Bonaparte
CONICET
Museo Argentino de Ciencias Naturales
Av. Angel Gallardo 470
1405 Buenos Aires, Argentina

Kenneth Carpenter
Academy of Natural Sciences
19th & The Parkway
Philadelphia, PA 19103

Sankar Chatterjee
The Museum
Texas Tech University
Lubbock, TX 79409-4499

James M. Clark
Department of Anatomy
University of Chicago
1025 East 57th St., Chicago IL 60637

William A. Clemens
Department of Paleontology
University of California
Berkeley, CA 94720

Edwin H. Colbert
Department of Geology
Museum of Northern Arizona
Flagstaff, AZ 86001

A. W. Crompton
Museum of Comparative Zoology
Harvard University
Cambridge, MA 02138

K. H. Cui
Institute of Vertebrate Paleontology and
   Paleoanthropology
P.O. Box 643
Beijing, Peoples' Republic of China

David E. Fastovsky
Department of Geology and Geophysics
University of Wisconsin
Madison, WI 53706

R. M. Frank
Unité de Recherche INSERM W157
Faculté de Chirurgie Dentaire
67000 Strasbourg, France

Eugene S. Gaffney
Department of Vertebrate Paleontology
American Museum of Natural History
Central Park West at 79th, New York, NY 10024

Peter M. Galton
Department of Biology
University of Bridgeport
Bridgeport, CT 06601

Hartmut Haubold
Department of Geological Sciences and
    Geiseltalmuseum
Martin-Luther-University
402 Halle (Saale), German Democratic Republic

J. Hemmerlé
Unité de Recherche INSERM W157
Faculté de Chirurgie Dentaire
67000 Strasbourg, France

R. A. Long
Museum of Paleontology
University of California
Berkeley, CA 94720

Amy R. McCune
Section of Ecology and Systematics
Cornell University
Ithaca, NY 14853-0239

Phillip A. Murry
Department of Physical Sciences
Tarleton State University
Stephensville, TX 76402

Paul E. Olsen
Department of Geology
Lamont-Dougherty Geophysical Observatory
Palisades, NY 10964

Kevin Padian
Department of Paleontology
University of California
Berkeley, CA 94720

J. Michael Parrish
University of Colorado Museum
Campus Box 315
Boulder, CO 80309

Bobb Schaeffer
Department of Vertebrate Paleontology
American Museum of Natural History
Central Park West at 79th, New York, NY 10024

Denise Sigogneau-Russell
Institut de Paléontologie
8 Rue de Buffon
75005 Paris, France

Hans-Dieter Sues
Museum of Comparative Zoology
Harvard University
Cambridge, MA 02138

A. L. Sun
Institute of Vertebrate Paleontology and
    Paleoanthropology
P.O. Box 643
Beijing, Peoples' Republic of China

Laurie R. Walter
Department of Biological Sciences
Chicago State University
95th Street at King Drive, Chicago, IL 60628

Samuel P. Welles
Museum of Paleontology
University of California
Berkeley, CA 94720

J. M. Zawiskie
Department of Geology
Wayne State University
Detroit, MI 48202

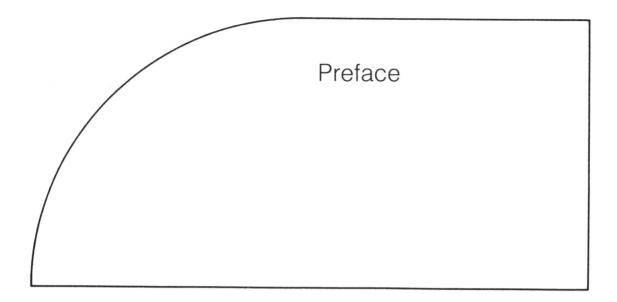

Preface

The Symposium on Faunal Change across the Triassic–Jurassic Boundary was held on October 31, 1984, in conjunction with the 44th Annual Meeting of the Society of Vertebrate Paleontology (SVP) at Berkeley, California. Some 200 of the 400 members attending the SVP meeting arrived a day early to attend this special symposium, a mark of the revival of interest within the past decade in how the "Age of Dinosaurs" began. Most of the contents of this volume were presented during the course of the day by over two dozen contributors, who in the ensuing months edited and honed their manuscripts to what is now a comprehensive summary of the state of knowledge of what happened among terrestrial vertebrates across the Triassic–Jurassic boundary.

The general theme of this volume is faunal change in all its various aspects. Some chapters focus on sweeping trends; others on changes in faunas or adaptations; and others on individual animals, their systematics, functional morphology, or stratigraphic importance. Yet all are related to the general theme of evolutionary change in early Mesozoic faunas that witnessed the rise of the dinosaurs, crocodiles, mammals, lepidosaurs, pterosaurs, and turtles, and the eclipse of many groups of diapsids and synapsids that dominated the Triassic and even the Late Permian. This book could not have been assembled ten years ago, because it was not generally recognized that there even was an Early Jurassic record of terrestrial tetrapods. Even now, the question of the age assignments of many supposed "Early Jurassic" horizons is strongly contested, and much evidence of these healthy controversies can be found in these pages. Other issues of phylogenetic relationships, functional morphology, and even basic anatomy of many of these fossil groups have been advanced during this time, and the chapters in this book reflect some of these advancements.

The dedication of this volume should surprise no one but those to whom it is dedicated. It is, we hope, a fitting tribute to the years of work on Triassic and Jurassic vertebrates by Joe Gregory and Sam Welles and to the legacy of the late Charles L. Camp. These three men have, more than any others, been responsible for the great collections of lower vertebrates in the Museum of Paleontology of the University of California, the largest west of the Mississippi, as well as for so many advances in our knowledge of this field. Though both Dr. Gregory and Dr. Welles participated actively in the assembly of this symposium and its volume of proceedings – indeed, how could it have been held without them? – their colleagues kept them in the dark about the dedication, which was suggested independently by many of them and applauded by all. To these gentlemen, our sincerest appreciation and respect.

I would like to thank all of the contributors to the symposium and this volume for their participation, care in preparation of their manuscripts, and patience with the reviewing and editing process. Their willingness to meet strict deadlines and their general spirit of cooperation and support for the project are greatly appreciated.

Each manuscript was peer-reviewed by at least two readers and closely edited for style, attention to reviewers' comments, and coordination with other manuscripts. The reviewers were uniformly constructive and helpful, and frequently offered new information and insights that greatly enriched the volume. The privilege of anonymity was offered but seldom exercised by the reviewers, who included (in part) D. Baird, M. J. Benton, D. Brinkman, S. Chatterjee, J. M. Clark, W. A. Clemens, E. H. Colbert, W. P. Coombs, P. J. Currie, P. Dodson, J. A. Doyle, P. M. Galton, J. T. Gregory, H. Haubold, J. A. Hopson, N. Hotton III, J. H. Hutchison,

F. A. Jenkins Jr., J. S. McIntosh, P. A. Murry, P. E. Olsen, J. H. Ostrom, J. M. Parrish, F. E. Peterson, T. Rowe, W. A. S. Sarjeant, N. Simmons, H.-D. Sues, K. S. Thomson, D. B. Weishampel, S. P. Welles, and J. Zawiskie. For special help and suggestions in editing I especially want to thank Paul Olsen, Mike Parrish, Hans Sues, Jacques Gauthier, Joe Gregory, Russ Ciochon, Don Baird, Peter Dodson, Bill Curtis, and Susan Abrams. Above all, my greatest appreciation goes to Bill Clemens for his constant help, advice, collegiality, and support.

Financial support for this project came in part from the Host Committee of the 44th SVP Meeting, which underwrote costs of the symposium and of editing the volume. Finally, I should like to thank the editorial staff of Cambridge University Press, especially Helen Wheeler and Richard Ziemacki, for their support and encouragement.

Kevin Padian

Berkeley
June 1985

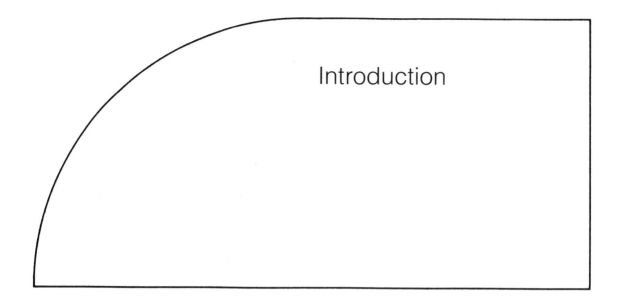

Introduction

To many paleontologists and evolutionary biologists, it may seem strange to convene a symposium about what we euphemistically call "the beginning of the Age of Dinosaurs," the interval of time encompassing the Triassic–Jurassic boundary some 200 MYA. Most attention in the past few years has been given to the *end* of the "Age of Dinosaurs," with the publication of the Alvarez theory (Alvarez et al. 1980) that an extraterrestrial bolide impact may have been largely responsible for some extinctions at the end of the Cretaceous period. There is no doubt that the end of the Cretaceous will always fascinate paleontologists and capture the imagination of the public, but extinction is only one-half of the evolutionary process; the other half is origination. There cannot be an end without a beginning, and so the circumstances under which the dinosaurs arose and replaced other animals of the Triassic faunas are of great evolutionary interest.

The beginning of the dinosaurs

To begin this book, it will be necessary to explain some concepts and circumstances. The "Age of Dinosaurs" is regarded as that period of time encompassing the Late Triassic through Late Cretaceous, or most of the Mesozoic Era in which dinosaurs dominated terrestrial faunas (about 215–65 MYA). Its "beginning," the subject of this book, roughly spans about 25 MY, from the Carnian (~215–210 MYA) and Norian (~210–200 MYA, including the "Rhaetian"; see Chapter 25) stages of the Late Triassic through the early part of the Early Jurassic (the "Liassic," containing the Hettangian, Sinemurian, Pliensbachian, and Toarcian stages). The stages are recognized on the basis of the type European geological sections, mostly marine, and have been extended to terrestrial sections there and elsewhere largely on the basis of fossil pollen remains (see Chapters 1 and 2).

Dinosaurs

The next concept to explain is the "dinosaur" itself. Sir Richard Owen erected the concept in 1841 on somewhat questionable grounds (Desmond 1979); most of the poorly known remains included at that time in the Dinosauria were not even dinosaurs. In 1888, Seeley sundered the group, which was much better known by then, into the two orders, Ornithischia and Saurischia, based on the configuration of the pubis in the respective groups, and since then many arguments have hinged on whether Dinosauria is monophyletic or polyphyletic (e.g., Bakker and Galton 1974; Charig 1976). Most workers accept monophyly on the basis of character complexes associated with upright, initially bipedal stance and parasagittal locomotion; these characters include a fully perforate acetabulum and an offset femoral head, coupled with a reduced fibula, mesotarsal ankle, and digitigrade posture. Charig (1972), in a seminal paper, explored the consequences of three broadly defined grades of archosaur locomotion associated with specific archosaurian groups and showed how the features listed above were characteristic of dinosaurs. Later work building upon Charig's showed that all of these features except the perforate acetabulum were also present in pterosaurs (Padian 1983) and in other close relatives of the dinosaurs, such as *Lagosuchus* (Bonaparte 1975); therefore, Dinosauria needs a more rigorous definition.

Recently, Gauthier (1984) has integrated such a concept into a character analysis of the major divisions of the diapsid reptiles. The Dinosauria can be defined by the fact that its common ancestor possessed a series of diagnostic characters. The vertebral column was highly regionalized, and there were at least three sacral vertebrae. Digits 4 and 5 on the hand were reduced in size and number of phalanges, and the distal end of the first metacarpal is offset so that the first digit is somewhat opposable to the others. The pelvic bones have a reduced acetabular contact, so that the acetabulum is at least partly perforated; there is a strong supraacetabular buttress and a long deep channel below the posterior ilium for the M. caudofemoralis brevis. The tibia has a crescentic cnemial crest and is transversely widened at the distal end, corresponding to its articulation with the astragalus, which has a unique ascending process. The fibula and calcaneum are reduced, and the astragalus and calcaneum together form a double roller joint against the distal tarsals. The long metatarsals are closely appressed; the foot is functionally tridactyl, and the stance digitigrade. [Details on these and other characters may be found in Gauthier (1984).] Many of these characters also apply to pterosaurs and *Lagosuchus*, among other taxa; so their strict applicability to the taxon Dinosauria is not certain. But these close relatives of dinosaurs are often difficult to interpret. Many of these relatives (e.g., *Lagosuchus*) are known from only partial skeletons, and some essential characters are not preserved; in other groups, such as the pterosaurs, the skeleton is already so highly modified that it is difficult to determine whether, for example, the ancestral pterosaurian hip socket was ever perforated, or whether the first metacarpal was offset distally. The sequence of transformation of characters is ambiguous (Padian 1984).

It is clear, however, that the characters shared by Dinosauria and these other two groups plus Ornithosuchidae set them apart from other Archosauria (Gauthier 1984). This realization is a considerable advancement over the traditional recognition of "thecodontians," unspecified archosaurs distinguished by primitive characters, as the source of all more derived archosaurian groups. In this book, the authors have generally tried to avoid the use of paraphyletic groups, that is those including an ancestor and some, but not all, descendants. Hence, "thecodonts" (nondinosaurian archosaurs), "prosauropods" (nonsauropodan sauropodomorphs), and other such terms are frequently placed in quotation marks. It is worth noting that, in strictly monophyletic terms, birds should be included among the theropod (carnivorous) dinosaurs because they are descended from some of their Mesozoic members (Bakker and Galton 1974; Ostrom 1976). So the Age of Dinosaurs may not yet be over, given the diversity of living dinosaurs.

The first dinosaurs

When are dinosaurs first found? Apart from some apocryphal footprint records from the Middle Triassic (Anisian–Ladinian; Chapter 15), both footprints and bones of dinosaurs first reliably appear in the Carnian. In many cases, it is not possible to place these dinosaur-bearing horizons more precisely within the Carnian; it requires independent data from palynology or radiometric dating to corroborate an age based on vertebrate biostratigraphy. Therefore, the limits of resolution may only be approximate for these cases. It is important, too, to distinguish between the concepts "stratigraphically oldest" and "most primitive." In either case, the best candidate is generally recognized as *Staurikosaurus*, from the "Rhynchocephalia assemblage" of the upper part of the Santa Maria Formation of Brazil. Bonaparte (1982) regards these beds as latest Middle Triassic (pre-Carnian) in age. However, even if these beds are Carnian in age, *Staurikosaurus* is too primitive a dinosaur to fit within either the Ornithischia or Saurischia, as are *Herrarasaurus* and *Ischisaurus*, two partially known dinosaurs from the lower third of the Ischigualasto Formation of Argentina. These forms, therefore, are most reasonably interpreted as both the oldest and most primitive dinosaurs. Bonaparte regards the Ischigualasto as "approximately the lower half of the Late Triassic," which approximately corresponds to the Carnian in his correlation chart (Bonaparte 1982, Fig. 1). If these beds approximate the tempo of Carnian deposition, which is not at all certain, they may be presumed to be of early Carnian age; Galton (Chapter 16) regards the fossiliferous Ischigualasto beds as middle Carnian. From the lower third of the Ischigualasto also comes "Triassolestes" preocc., renamed *Trialestes* by Bonaparte (1982); its strange mixture of dinosaurian (mesotarsal ankle) and crocodilian (elongated carpus) features led Bonaparte (1982) to reassign the type and a second specimen to the sphenosuchid crocodylomorphs, although some workers (for example, Reig 1963) have regarded the type specimen as a composite of two or more different animals. *Pisanosaurus*, the earliest known ornithischian, comes from the middle third of this formation (Bonaparte 1982).

There are other middle Carnian records of dinosaurs. Dutuit (1972) described *Azendohsaurus* from the Argana Formation in the High Atlas of Morocco; the beds, which predate the onset of rifting in that area of Pangea, were dated by fossil pollen (Cousminer and Manspeizer 1976). Dutuit regarded *Azendohsaurus* as an ornithischian, although the remains are so fragmentary and primitive that some workers have suspected that it may be a prosauropod or perhaps a dinosaur not assignable to either Ornithischia or Saurischia. The Cumnock Formation of the Newark Supergroup of North America has yielded ornithischian remains in beds dated by pollen as middle Carnian (Olsen 1984), as has the Pekin Formation, in which dinosaur tracks are also found (P. E. Olsen pers. comm.). Thus, it is clear that by this time ornithischians and saurischians had diverged from each other; there are no records of the more archaic forms of dinosaur such as those found in the earlier beds of South America.

By the late Carnian, theropod, sauropodo-morph, and ornithischian types were well estab-lished. The Petrified Forest Member of the Chinle Formation is divided into upper and lower units; the lower unit contains abundant pollen of late Carnian age, and the sparser pollen of the upper unit appears to be commensurate with the late Carnian or early Norian (see references in Chapter 12). The upper unit has yielded theropod remains referable to *Coelophysis* (see Chapter 5), while the *Placerias* quarry near St. Johns, Arizona, biostratigraphically corre-lated with the lower unit of the Petrified Forest, contains isolated dinosaur bones (R. A. Long, quoted in Chapter 9). Also in the late Carnian is the famous Ghost Ranch Quarry of the Chinle Forma-tion of New Mexico, the source of numerous com-plete skeletons referred by Colbert to *Coelophysis* (see Chapter 5) and tied biostratigraphically to the Petrified Forest Member. Several kinds of dinosaurs come from the Dockum Formation of Texas, in-cluding ornithischians and theropods (see Chapters 9–11). Though Chatterjee perhaps rightfully regards some of the Dockum to be of Norian age, inde-pendent evidence for this is still awaited, and the other authors support a late Carnian age, based on extensive fossil pollen sampling. Other late Carnian dinosaurs include occurrences in the Fundy Basin of Nova Scotia [Wolfville Formation: ornithischians, sauropodomorphs and theropods (Baird and Olsen 1983)] and ornithischian records in the Pekin For-mation (Wadesboro Basin) and New Oxford For-mation (Gettysburg Basin), all in the Newark Supergroup (Galton 1983); the last was recorded as "*Thecodontosaurus*" *gibbidens*, a "prosauropod," but may be an ornithischian (Galton 1983; P. E. Olsen pers. comm.). Olsen and Galton (1984) also noted the presence of the "prosauropod" dinosaur in the Lower Elliott Formation ("Lower Red Beds") of South Africa, and the associated vertebrate fauna of rauisuchians and capitosaurid amphibians implies that the beds are of Carnian age. In addition to these, there are several unpublished or questionable rec-ords of dinosaurs from South Africa, Scotland, In-dia, and Arizona, all apparently of late Carnian age. By the Norian, dinosaur bones and tracks become reasonably abundant in Europe, eastern North America, southern Africa, and South America, bringing the Age of Dinosaurs into full swing.

Some background to the Triassic

The Triassic was an extraordinary time in ver-tebrate history. More orders of nonmammalian tet-rapods lived during the Triassic than at any other period (Padian and Clemens 1985), and only the evolutionary explosions of the mammals and birds during the Tertiary surpass it in diversity. This is because the Triassic really contains three major faunal elements: (1) groups that survived from ear-lier periods, (2) groups that lived only during the Triassic, and (3) groups that appeared at the end of the Triassic but reached their greatest diversities in later periods. To understand the circumstances that prevailed at the Triassic–Jurassic boundary, it is nec-essary to describe these faunal elements briefly.

1. At its beginning, some 225 MYA, the Triassic was populated mainly by groups that were holdovers from the Paleozoic, having survived the great extinctions at the end of the Permian. To un-derstand the Triassic faunal changes we have to delve back in the Paleozoic to the Pennsylvanian period some 325 MYA, when the two great branches of the amniote vertebrates diverged (Carroll 1982). These two branches were the Synapsida, including the mammals and their earlier relatives, and the Diap-sida, the "typical" reptiles that number among them the ichthyosaurs, plesiosaurs, lizards, snakes, croc-odiles, pterosaurs, dinosaurs (including birds), and many other groups. From what the records of ver-tebrate history tell us, the Synapsida was the first group to exploit the terrestrial habitats: Many of them grew quite large, and they radiated into a series of carnivorous and herbivorous adaptive zones dur-ing the Permian, while the Diapsida maintained a lower profile (Romer 1966, 1968).

The largest, most diverse and dominant mem-bers of the Early Triassic faunas were various mem-bers of the Therapsida, a group of synapsids often misleadingly called the "mammal-like reptiles" be-cause they include their descendants the mammals; however, they were not "reptiles" in the same sense that the diapsids were, and it is best to think of the synapsids as a completely different lineage of am-niotes. The prominence of these nonmammalian therapsids waned during the Triassic until only a couple of lineages were left in the Late Triassic and Early Jurassic; their highest known stratigraphic re-cord is in the upper Middle Jurassic of China (Sig-ogneau–Russell and Sun, 1981). However, the synapsid lineage did not peter out entirely, because the first members of the group conventionally known as the Mammalia appeared at the very end of the Triassic. (Their history is another story; see Lille-graven, Kielan-Jaworowska, and Clemens 1979.)

In addition to these early Therapsida, other "Paleozoic holdovers" included various amphibian groups (Carroll 1977), the enigmatic Procolophonia, and some small, inconspicuous diapsids. At the be-ginning of the Triassic many of these diapsid groups were still rather undifferentiated, and for many years were lumped into a taxon called the "Eosuchia." However, as Gauthier's (1984) work has shown, by this time the Diapsida had already split into two major groups, the Lepidosauromorpha (including the lizards, snakes and sphenodontids) and the Archosauromorpha (including the crocodiles, dinosaurs, and many other groups). Of these, the archosauromorphs of the Early Triassic included the carnivorous and often sizable Proterosuchia and Erythrosuchia, as well as smaller forms such as *Euparkeria* (Ewer 1965) that appear to be very close to the derivation of the later archosaurs.

2. According to present knowledge, some nineteen groups of tetrapods, distinct enough to be separated at the family, ordinal, or subordinal level from other such groups, are indigenous to Triassic strata. They include plagiosaurs, metoposaurs, phytosaurs, aetosaurs, rauisuchids, poposaurids, lagosuchids, proterosuchians, erythrosuchians, proterochampsians, trilophosaurs, rhynchosaurs, hypsognathids, scleromochlids, erpetosuchids, diademodontids, traversodontids, chiniquidontids, and theroteinids. Many of the archosaurian groups, such as the phytosaurs, aetosaurs, rauisuchians, ornithosuchids, and their distant cousins, the rhynchosaurs and trilophosaurs, are the most diverse and abundant members of the Late Triassic (Carnian and early Norian) faunas (see Chapters 9–12, 24–6). It is not clear why so many of these groups radiated so quickly or why they are confined to the Triassic. They may have been caught in what Charig (1979) regarded as the "paleotetrapod–neotetrapod" transition between the classic Paleozoic and Mesozoic faunas, or it could be that environments were changing too fast for these groups to keep up [for one view see Tucker and Benton (1982)]. Although there is little evidence of direct competitive replacement, it does seem as if these groups ecologically replaced the Early Triassic therapsid faunas just as they were in turn replaced by the dinosaurian faunas at the end of the Triassic (for a dialogue on this compare Chapters 24 and 26).

3. Finally, toward the end of the Triassic several new groups of vertebrates began to appear in terrestrial faunas. Most were diapsids, especially archosaurs – related to the archosaurs that were then dominating the faunas, such as the phytosaurs, aetosaurs, rauisuchians, and ornithosuchids. These included crocodylomorphs, pterosaurs, and both orders of dinosaurs, as well as sphenodontid lepidosaurs, the nondiapsid chelonians, and several groups of therapsid synapsids, including tritylodontids, tritheledontids, and morganucodontids – the last generally considered the first true mammals. Even the first "salamander-like" lissamphibians left their first records. In fact, although evidence of "lizards" from this interval of time is somewhat suspect (Estes 1984), and although birds did not emerge until the Late Jurassic (with the appearance of *Archaeop-*

teryx), the origins of the modern vertebrate fauna were essentially in place by the end of the Triassic, some 200 MYA. Of course, we could not place the crocodiles, turtles, salamanders, and mammals of that time within the taxonomic bounds of their living representatives, but they would still have been quite recognizable to a modern visitor.

This tripartite succession of Triassic faunas, with its rapid evolutionary radiations and high turnover rates, is one of the most fertile areas for further study in vertebrate paleontology. Several of the authors represented in this volume address questions related to the replacements of these faunas through time (for example, Colbert, Welles, Olsen, Sues, Benton, and Zawiskie). Were these changes the result of competitive replacement, made possible by improvements in locomotion and physiology, as Charig (1984) and Bakker (1980) have argued? Or did the dinosaurs simply get a "lucky break" when the earlier and no less well-adapted beasts disappeared for completely different reasons (Benton 1984)? The data may not yet be complete enough to decide the question, and they may never be, but the lack of a firm answer has forced ever more refined inquiries and has stimulated new ways of thinking about the question.

The problem of the Triassic–Jurassic boundary

As recently as ten years ago, paleontologists had a very different picture of the Late Triassic and Early Jurassic than they have now. To put it succinctly, there were simply no known Early Jurassic terrestrial vertebrates, except for the odd dinosaur or two washed into some marine sediments in England (*Scelidosaurus*) or Germany (*Ohmdenosaurus*). This was especially disappointing because, as just noted, the Late Triassic horizons already provided enticing glimpses of the vertebrate groups that would soon take over the terrestrial faunas for the rest of the Mesozoic. The frustration of the situation is nicely epitomized in a passage in A. S. Romer's *Notes and Comments on Vertebrate Paleontology* (1968), in which Romer stated that "one of our greatest desideratas in the vertebrate fossil record, apart from better remains of the earliest tetrapods, would be a record of the Lower Jurassic." Romer noted that after the abundant faunas of the Late Triassic there was nothing until the great dinosaur faunas of the Upper Jurassic Morrison and Tendaguru Beds. He knew, as everyone did, that great evolutionary changes took place during that interval, but there was no way to grasp them, because all over the world the spreading seas erased most records of terrestrial vertebrates for the rest of the Jurassic.

In the ensuing decade, however, things began to change. Paleobotanists, such as Bruce Cornet, started to compare palynoflorules of the Newark Supergroup with those of Triassic and Jurassic formations of Europe and found unmistakably Jurassic spore and pollen assemblages in the upper portion of the Newark (Cornet 1977). Geophysicists began to date the Newark's basalt flows and came up with surprisingly young ages (Armstrong and Besancon 1970). Correlations of the basins of the Newark revealed that not all were deposited at the same time, and when rearranged the vertebrate faunas of those basins told a very different story. The lower beds were dominated by remains of nondinosaurian archosaurs, such as phytosaurs, aetosaurs, and primitive archosaurian footprints, as well as several small aquatic and aerial reptiles (Colbert 1970; Olsen 1979). In the upper beds, these faunas gave way to dinosaurs and crocodiles. There was no trace of the other nondinosaurian forms (Olsen 1980), and, while the lower beds were still recognized to be of Late Triassic age, the upper beds had been reassigned to the Lower Jurassic (Cornet, Traverse, and McDonald 1973; Olsen and Galton 1977).

Other reconsiderations followed in other places around the world, including South Africa, China, South America, and Europe. Realignments of faunas and new assignments of age led to a series of hypotheses that was perhaps most tersely and completely summarized for the first time in Paul Olsen and Peter Galton's 1977 *Science* article, which questioned the magnitude of the end-Triassic "extinction event" and proposed a sweeping revision of the assigned ages of many formations usually considered Late Triassic. According to them and the sources they consulted, most of the Newark Supergroup, the Glen Canyon Group of the southwestern United States, the Upper Stormberg Series of southern Africa, and the Lower Lufeng Series of China were properly regarded as Early Jurassic in age. These hypotheses have been discussed, tested, and revised, and continue to be subjected to intense analysis.

In essence, the revised view of Triassic–Jurassic vertebrate biostratigraphy implies not one major paradigm shift, but two. The first is the reassignment of many "Late Triassic" beds to the Early Jurassic. The second, discussed above, is the realization that in beds properly regarded as Triassic, dinosaurs are present but are usually not the most abundant members of these assemblages and are certainly not as diverse as the phytosaurs, aetosaurs, and other nondinosaurian archosaurs. For example, R. A. Long, in surveying the thousands of bones collected by Charles Camp from the Chinle Formation, believes that only a handful can be positively identified as dinosaurian, contrary to their original identifications (Chapter 9). Several other paleontologists working on Triassic faunas have reached the same conclusion about collections in other museums, including those of the University of Michigan, the University of Texas, and the American Museum (Chapter 5), as well as those of Europe (Chapter 24). This pattern of dinosaurian rarity in the Upper Triassic has been borne out in the results of recent field work in the Newark, the Chinle, and the Dockum (see Chapters 9–12 and 20).

On the other hand, some Late Triassic fossil horizons around the world show good admixtures of all these animals. The Ghost Ranch Quarry of the Chinle Formation (late Carnian) in New Mexico (Colbert 1947) may be unusual in its abundance of dinosaurs compared to the rest of the Chinle and Dockum, but the Norian beds of Germany and South Africa are rich in prosauropods and other dinosaurs (Weishampel 1984; Chapter 24), and the Los Colorados Formation of South America shows many aspects of what we might expect to be a "transitional" type of fauna between the Triassic and Jurassic (Bonaparte 1982; Olsen and Galton 1984), although its exact age and faunal unity raise questions that are not completely resolved (Chapters 20 and 25). So it is possible that during the later Norian, dinosaurs began to "take over" terrestrial faunas, but measures of living abundance are difficult to assess from fossil remains. Possible preservational bias and small samples continue to plague paleontologists studying this issue.

There are many large questions that still need to be addressed. Was the change from "Late Triassic"- to "Early Jurassic"-type faunas in lockstep around the world, or did they change in some areas before others? To what extent might the logical framework of biostratigraphic correlations be circular? What other lines of evidence can be invoked to test these hypotheses, and are these fine-tuned enough to resolve the questions further? Was there an end-Triassic vertebrate extinction "event," or not, or were there several? Olsen and Sues (Chapter 25) argue that vertebrate extinctions were quite concentrated near the very end of the Triassic – a situation made more intriguing by the presence of the approximately contemporaneous Manucowagan impact crater in Nova Scotia. Crompton and Attridge (Chapter 17) also suggest the complicity of some catastrophic agent in the sharp decline of certain herbivorous groups near the end of the Triassic. At this point, both temporal and causal connections are at a very gross level of resolution; but improvements in the understanding of taxonomic and faunal changes across the Triassic–Jurassic boundary, such as many authors report in this book, will be the grist for answering many questions about macroevolutionary patterns and processes that relate to one of the most important intervals in the history of land vertebrates.

A brief word about the organization of this book. The first introductory chapters are designed to familiarize the nonspecialist with some of the background of the beginning of the Age of Dinosaurs. We are fortunate to have the dean of American vertebrate paleontologists, Dr. E. H. Colbert, to provide us with his experienced perspective on the history of the Triassic–Jurassic question, and he has given us an elegantly written epitome of current problems in the field. Dr. Sidney R. Ash's expertise on Triassic–Jurassic paleobotany is complemented by his long-standing interests in geological and paleovertebrate changes during this time, and Dr. Ash was kind enough both to participate actively in the symposium held at Berkeley and to contribute a brief survey of the contemporaneous plants and their distributions through time and space. To both these gentlemen we express our deepest appreciation and thanks.

The remainder of the book follows an approximate stratigraphic organization, with sections on the Triassic, on changes across the boundary, and on the Jurassic, concluding with papers addressing large-scale macroevolutionary patterns and trends. The organization is not strict, and aspects of several problems are to be found in many chapters. A brief summary and prospectus ends the book. It is hoped that these contributions will synthesize and clarify the problems facing our understanding of the beginning of the Age of Dinosaurs and stimulate channels of future research and communication.

References

Alvarez, L. W., W. Alvarez, F. Asaro, and H. Michel. 1980. Extraterrestrial cause for the Cretaceous–Tertiary extinction. *Science* 208: 1095–108.

Armstrong, R. L., and J. Besancon. 1970. A Triassic time scale dilemma: K–Ar dating of Upper Triassic igneous rocks, eastern U.S.A. and Canada and post-Triassic plutons, western Idaho, U.S.A. *Eclogae Geol. Helv.* 63: 15–28.

Baird, D., and P. E. Olsen. 1983. Late Triassic herpetofauna from the Wolfville Fm. of the Minas Basin (Fundy Basin), Nova Scotia, Canada. *Geol. Soc. Am. Abstr. Progr.* 15: 122.

Bakker, R. T. 1980. Dinosaur heresy – dinosaur renaissance: Why we need endothermic archosaurs for a comprehensive theory of bioenergetic evolution. *In* Thomas, R. D. K., and E. C. Olson (eds.), *A Cold Look at the Warm-Blooded Dinosaurs.* AAAS Selected Symposium 28 (Boulder, Colorado: Westview Press), pp. 351–462.

Bakker, R. T., and P.M. Galton. 1974. Dinosaur monophyly and a new class of vertebrates. *Nature (London)* 248: 168–72.

Benton, M. J. 1984. Dinosaurs' lucky break. *Nat. Hist.* 93 (6): 54–9.

Bonaparte, J. F. 1975. Nuevos materiales de *Lagosuchus talampeyensis* Romer (Thecodontia Pseudosuchia) y su significado en el origin de los Saurisquios. Chanares inferior, Triasico Medio de Argentina. *Acta Geol. Lill.* 13(1): 5–90.

1982. Faunal replacement in the Triassic of South America. *J. Vert. Paleontol.* 2(3): 362–71.

Carroll, R. L. 1977. Patterns of amphibian evolution: an extended example of the incompleteness of the fossil record. *In* Hallam, A. (ed.), *Patterns of Evolution, as Illustrated by the Fossil Record* (Amsterdam: Elsevier Scientific), pp. 405–38.

1982. Early evolution of reptiles. *Ann. Rev. Ecol. Syst.* 13: 87–109.

Charig, A. J. 1972. The evolution of the archosaur pelvis and hindlimb: an explanation in functional terms. *In* Joysey, K. A., and T. S. Kemp, (eds.), *Studies in Vertebrate Evolution* (London: Oliver and Boyd), pp. 121–55.

1976. Dinosaur monophyly and a new class of vertebrates: a critical review, *In* Bellairs, A. d'A., and C. B. Cox (eds.), *Morphology and Biology of Reptiles* (London: Academic Press), pp. 65–104.

1979. *A New Look at the Dinosaurs* (London: Heinemann).

1984. Competition between therapsids and archosaurs during the Triassic period: a review and synthesis of current theories. *Symp. Zool. Soc. London* 52: 597–628.

Colbert, E. H. 1947. Little dinosaurs of Ghost Ranch. *Nat. Hist.* 56(9): 392–99 and 427–8.

1970. The Triassic gliding reptile *Icarosaurus. Bull. Am. Mus. Nat. Hist.* 143: 85–142.

Cornet, B. 1977. The palynostratigraphy and age of the Newark Supergroup. Ph.D. thesis, Department of Geosciences, University of Pennsylvania.

Cornet, B., A. Traverse, and N. G. McDonald. 1973. Fossil spores, pollen, fishes from Connecticut indicate Early Jurassic age for part of the Newark Group. *Science* 182: 1243–7.

Cousminer, H. L., and W. Manspeizer. 1976. Triassic pollen date, Moroccan High Atlas, and the incipient rifting of Pangea in the Middle Carnian. *Science* 191: 943.

Desmond, A. J. 1979. Designing the dinosaur: Richard Owen's response to Robert Edmond Grant. *Isis* 70: 224–34.

Dutuit, J. M. 1972. Decouverte d'un dinosaure ornithischien dans le Trias superieur de l'Atlas occidental marocain. *Compt. Rend. Acad. Sci., Paris D* 275: 2841–4.

Estes, R. 1984. *Handbuch der Palaoherpetologie*, Teil 10A, *Sauria Terrestris, Amphisbaenia* (Stuttgart: Gustav Fischer).

Ewer, R. 1965. The anatomy of the thecodont reptile *Euparkeria capensis* Broom. *Phil. Trans. Roy. Soc. London B* 248(751): 379–435.

Galton, P. M. 1983. The oldest ornithischian dinosaurs in North America from the Late Triassic of Nova Scotia, N.C., and Pa. *Geol. Soc. Am. Abstr. Progr.* 15: 122.

Gauthier, J. A. 1984. A cladistic analysis of the higher systematic categories of the Diapsida. Ph.D. Dissertation, Department of Paleontology, University of California, Berkeley.

Lillegraven, J.A., Z. Kielan-Jaworowska, and W. A. Clemens, (eds.). 1979. *Mesozoic Mammals: the First Two-Thirds of Mammalian History* (Berkeley: University of California Press).

Olsen, P. E. 1979. A new aquatic eosuchian from the Newark Supergroup (Late Triassic – Early Jurassic) of North Carolina and Virginia. *Postilla (Yale Peabody Museum)* 176: 1–14.

1980. A comparison of vertebrate assemblages from the Newark and Hartford Basins (Early Mesozoic, Newark Supergroup) of Eastern North America. *In* Jacobs, L. L. (ed.), *Aspects of Vertebrate History* (Flagstaff, Arizona: MNA Press).

1984. Comparative paleolimnology of the Newark Supergroup: a study of ecosystem evolution. Ph.D. Thesis, Dept. of Biology, Yale University.

Olsen, P. E., and P. M. Galton. 1977. Triassic–Jurassic tetrapod extinctions: are they real? *Science* 197: 983–6.

1984. A review of the reptile and amphibian assemblages from the Stormberg of southern Africa, with special emphasis on the footprints and the age of the Stormberg. *Palaeontal Afr.* 25: 87–110.

Ostrom, J. H. 1976. *Archaeopteryx* and the origin of birds. *Biol. J. Linn. Soc.* 8(2): 91–182.

Padian, K. 1983. A functional analysis of flying and walking in pterosaurs. *Paleobiology* 9(3): 218–39.

1984. The origin of pterosaurs. *In* Reif, W.-E., and F. Westphal (eds.), *Third Symposium on Mesozoic Terrestrial Ecosystems: Short Papers* (Tubingen: ATTEMPTO), pp. 163–9.

Padian, K., and W. A. Clemens. 1985. Terrestrial vertebrate diversity: episodes and insights. *In* J. W. Valentine (ed.), *Phanerozoic Diversity Factors* (Princeton, New Jersey: Princeton University Press).

Reig, O. A. 1963. La presencia de dinosaurios saurisquios en los "Estratos de Ischigualasto" (Mesotriasico superior) de las provincias de San Juan y La Rioja (Republica Argentina). *Ameghiniana* 3(1): 3–20.

Romer, A. S. 1966. *Vertebrate Paleontology*, 3rd ed. (Chicago: University of Chicago Press).

1968. *Notes and Comments on Vertebrate Paleontology* (Chicago: University of Chicago Press).

Sigogneau-Russell, D., and A.-L. Sun. 1981. A brief review of Chinese therapsids. *Geobios* 14(2): 275–9.

Tucker, M. E., and M. J. Benton. 1982. Triassic environments, climates, and reptile evolution. *Palaeogeogr., Palaeoclimatol., Palaeoecol.* 40: 361–79.

Weishampel, D. B. 1984. Trossingen: E. Fraas, F. von Huene, R. Seeman and the "Schwabische Blindwurm" *Plateosaurus. In* Reif, W.-E., and F. Westphal (eds.), *Third Symposium on Mesozoic Terrestrial Ecosystems: Short papers* (Tubingen: ATTEMPTO), pp. 249–53.

The Beginning of the Age of Dinosaurs: the time and the setting

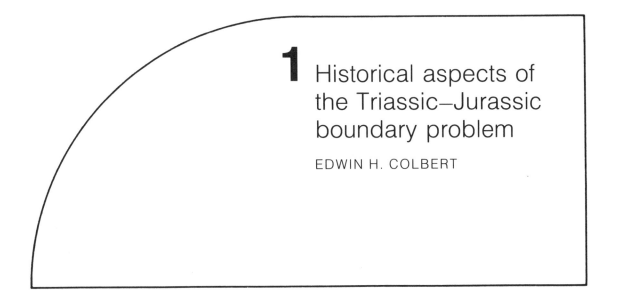

1 Historical aspects of the Triassic–Jurassic boundary problem

EDWIN H. COLBERT

Introduction

The Triassic and Jurassic systems, like the other systems of the geological column, were first studied and defined in Europe. J. G. Lehman and G. C. Füchsel defined the Buntsandstein and Muschelkalk divisions of the tripartite Triassic in 1780; L. von Buch added the Keuper in 1820; and in 1834 F. von Alberti brought these rock units together as the Triassic system. The name Jurassic was introduced by Brongniart and von Humboldt in 1795, but it remained for L. von Buch to determine the position of the Jurassic in 1837. F. A. Quenstedt, who did much of the early work on the Triassic, joined with von Buch to distinguish the Lower, Middle, and Upper Jurassic, which von Buch named the Black, Brown, and White Jura, respectively (Schmidt 1928, Brinkman 1960).

The type or Germanic facies of the Triassic is composed of Lower and Upper continental facies, separated by a Middle marine incursion, and is readily distinguishable from the type Jurassic, which at its base and throughout its extent is composed of marine sediments. At an early date in the study of stratigraphic geology, however, it became apparent that the line of demarcation between uppermost Triassic beds and the supposed lowermost Jurassic beds in Europe was not quite as clear-cut as originally had been supposed. This was because of the discovery of Rhaetic sediments, whose type facies may be found in Swabia. The name Rhaetic was established in 1861 by C. W. Gümbel and gained quick acceptance. The base of the type Rhaetic is defined by the appearance of marine shells, which, of course, sets it apart from the underlying continental Keuper. Above the shell-bearing layer are red bone beds containing the remains of small vertebrates.

Almost from the beginning, geologists have been troubled about where the Rhaetic should be placed. Among German scholars it has been regarded as an uppermost subdivision of the Triassic; French authorities have been inclined to place it (as the Infralias) at the base of the Liassic, the lowest division of the Jurassic. According to this latter view, the final or Norian stage of the Triassic would be essentially limited to the Germanic Gypskeuper, in which are found gigantic stereospondyl amphibians, numerous phytosaurs, aetosaurs, and early saurischian dinosaurs, particularly the prosauropod, *Plateosaurus*. This exuberant fauna certainly may be contrasted with the fossils of the overlying Rhaetic sediments, limited to the bones of very small tetrapods.

So it is that here, in the type area of the Triassic, the dilemma of a Triassic–Jurassic boundary made an early appearance. It is a question that is concerned with the problem of trying to establish a sharp divisional line in what is essentially a continuum of sedimentary and organic evolutionary history. This chapter will be concerned with this problem of the Triassic–Jurassic boundary so far as it is expressed by the physical development of sediments and the evolutionary development of tetrapods.

The Triassic–Jurassic boundary in eastern North America

In 1856, W. C. Redfield proposed the term *Newark Group* to apply to rocks in eastern North America previously designated as the "New Red" sandstones, the Keuper, or simply as the Triassic. As he stated: "I propose the latter designation [*Newark Group*] as a convenient name for these rocks [of New Jersey] and to those of Connecticut Valley, with which they are thoroughly identified by footprints and other fossils, and I would include also the contemporary sandstones of Virginia and North Carolina" (Redfield 1856, quoted by Wilmarth 1938). Such rocks had been investigated geologically and paleontologically at various times during the early and middle years of the nineteenth century. Thus, Redfield published a number of papers during those years on fossil fishes from the Newark beds of New Jersey, Connecticut, and Virginia. Interestingly, in two of his later contributions, he ascribed Newark fishes to the "Liassic and Oolitic periods" (Redfield 1856) and the "Liassic and Jurassic periods" (Redfield 1857). Furthermore, Edward Hitchcock and James Deane had separately made numerous studies of the dinosaur footprints of the Connecticut Valley during the 1830s to the 1850s and had each published several papers describing the tracks as having been made by giant birds. The name Dinosauria was proposed by Richard Owen in 1842; dinosaurs did not become well known until some thirty years later.

In 1904, R. S. Lull published a study of Connecticut Valley footprints in the *Memoirs of the Boston Society of Natural History*, the title of which was "Fossil Footprints of the Jura–Trias of North America." In his discussion of the age of the beds containing the footprints, he states that they are "generally conceded to be upper Triassic, equivalent to the Keuper and Rhaetic of Europe" (Lull 1904, p. 468). Within the text, Lull does not mention Jurassic or Liassic as possible ages for the Newark rocks, but from the title of the work one is led to wonder if he may have considered the Rhaetic of Jurassic affinities. Perhaps it is significant that, in his revision of "Triassic Life of the Connecticut Valley," published in 1953, he states that the Portland Formation, his topmost division of the Newark Group, is "apparently not older than the Upper Keuper" (Lull 1953, p. 17).

Generally speaking, however, the trend among most geologists and paleontologists until recent years was to regard the Newark Group as entirely of Late Triassic affinities. Within the past decade, there has been a revision of this view, based not only upon new interpretations of fossil vertebrates and other animal remains, but also and particularly upon palynological evidence, augmented by radiometric data. Accordingly, the view has shifted in favor of a Late Triassic and Early Jurassic age for the Newark Group, the division indicated by Olsen in 1980 as being at the base of the Orange Mountain Basalt, or First Watchung, in the Newark Basin (Olsen 1980a). Previously, Olsen had proposed that the Newark rocks be elevated to the status of a supergroup, based on the evidence of palynology and fossil vertebrates, with the Passaic Formation of the Newark Basin, immediately below the Orange Mountain Basalt, representing the final phase of Triassic sedimentation (Olsen 1978). In this connec-

tion, it should be mentioned that radiometric measurements of the Newark Basin basalts indicate an Early Jurassic age for them.

With this new evidence indicating that the Newark sequence in the Newark Basin contains both Late Triassic and Early Jurassic sediments, there is reason to think that the same holds true for the Triassic rocks exposed in the Connecticut Valley and to the northeast in Nova Scotia (Baird and Take 1959; Olsen and Baird 1982). Thus, Lull's remark quoted above, to the effect that the Portland Formation in Connecticut is "not older than the Upper Keuper," may indicate a foreshadowing in his mind as to the possibility of the Triassic–Jurassic boundary dividing the Newark sequence in Connecticut.

Olsen and Galton (1977) and Olsen (1980a) recognized three faunal zones within the Newark Supergroup corresponding in a general way with palynomorph zones that had been proposed by Cornet (1977). The lowest of these zones, including most of the Lower Newark beds, is considered of Late Triassic age, being assigned specifically to the Carnian and Norian stages. The middle zone, of Rhaetian age, is represented by the uppermost part of the Passaic Formation in the Newark Basin. The topmost zone, considered of Liassic or Hettangian–Sinemurian age, is represented by those Newark beds that include the Newark volcanics and related sediments.

In a review of the biostratigraphy and igneous activity in the Newark Supergroup, Olsen (1983) states that "there is absolutely no evidence of any igneous activity prior to biostratigraphic zones corresponding to the early Jurassic." As far as tetrapods are concerned, Olsen correlates the three zones with footprints, the lowest or zone 1 being the *Brachychirotherium*–"*Anchisauripus*" *milfordensis* zone, the middle or zone 2 being the *Batrachopus*–*Grallator* zone, and the highest or zone 3 being the *Anomoepus*–*Grallator* (*Eubrontes*) zone. Although these footprints, when identified, with close attention to details, are indicative of the three established zones, they also show that there was a persistence of early saurischian dinosaurs from Late Triassic into Early Jurassic years. Consequently, the end of Triassic history and the beginning of the Jurassic in eastern North America is followed closely by the beginning of igneous activity (in turn, a result of the tectonic events foreshadowing the rifting between Mauretania and eastern North America) rather than by any profound faunal changes.

The Triassic–Jurassic boundary in western North America

The Triassic–Jurassic boundary within the Newark Supergroup has thus been established, perhaps tentatively and perhaps somewhat artificially, near the initiation of vulcanism that may have marked the very preliminary stages of rifting between two segments of Pangaea. In the western part of the continent, there is no such apparent geological evidence to indicate the end of one period and the beginning of another. Here, sedimentation in many localities would seem to have proceeded continuously, or almost continuously, with the result that there have been years of confusion and argument as to the ages of the sediments involved. The problem is concerned principally with the rocks of the Glen Canyon Group above the Triassic Chinle Formation, and composed from bottom to top of the Wingate, Moenave, Kayenta, and Navajo formations. Differences of opinion range from the inclusion of the entire group within the Triassic, to inclusion entirely within the Jurassic, or to something in between. The problem is not easy, especially when one is confronted with the interfingering of formations, such as one sees in the relationships between the Kayenta and Navajo formations (Colbert and Gregory 1957).

Charles Camp, one of the pioneers in the study of Mesozoic tetrapods in the United States Southwest, realized in the 1920s and 1930s that there was a problem. In northern Arizona he could clearly distinguish the Upper Triassic Chinle Formation, named by H. E. Gregory (1917), but above this unit he encountered what for many years he called the "orange red" beds. These were the sediments that Ward (1901) had placed at the base of his "Painted Desert beds," and that had been recognized by H. E. Gregory (1917) as the "undifferentiated La Plata and McElmo of the Moenkopi Plateau." These colorful rocks subsequently were thought to be the equivalent of the Wingate Formation because they occur above the Chinle beds in an area where the Wingate is absent. Lateral tracing, however, showed that the orange red beds constitute a distinct mappable unit, named the Moenave Formation, stratigraphically higher than the Wingate and rather closely related to the Kayenta Formation, which is immediately above them.

The Wingate sandstone, now regarded as the lowermost unit of the Glen Canyon Group, had been known since the latter part of the nineteenth century, having been described by Dutton in 1885, but, until the monographic study of the uppermost Triassic and Jurassic rocks of the Navajo Country by Harshbarger, Repenning, and Irwin in 1957, there was a great deal of confusion between the true Wingate and the much later Entrada Formation. These authors were able to correlate the Wingate as known in northern Arizona and New Mexico with sediments below the Entrada at Fort Wingate, thereby solving a long misunderstood stratigraphic relationship (Harshbarger et al. 1957).

In 1955, Averitt et al. stated that "on the basis of new evidence" the United States Geological Survey recognizes the Wingate Formation to be of Triassic age, the Moenave to be questionably Triassic, the Kayenta to be questionably Jurassic, and the Navajo either to be questionably Jurassic or Jurassic without question. (As has been remarked, there is interfingering between the last two formations.)

In 1961, Lewis, Irwin, and Wilson, in a paper on the age of the Glen Canyon Group, regarded the Wingate to be Late Triassic, the Moenave and Kayenta to be questionably Late Triassic, and the Navajo to be questionably Late Triassic or Jurassic. This paper, of course, differs little from the views expressed by Averitt and his coauthors.

In 1967, Wilson and Stewart reviewed the problem and outlined some sedimentary criteria for distinguishing between Upper Triassic and basal Glen Canyon sediments, whatever the age of the latter might be. They stipulated that:

1. The uppermost occurrences of bentonitic clay are in Upper Triassic rocks; there are no bentonitic clays in the Moenave and Kayenta formations.
2. There is a color change, from gray-red-purple in the Upper Chinle to red-brown in the Moenave and Kayenta beds.
3. There is an increase in grain sizes from the Upper Chinle to the basal Glen Canyon sediments. (Wilson and Stewart 1967)

Upon the basis of such criteria, one might regard the Chinle and presumably the Wingate to be of Triassic age and the Glen Canyon beds above the Wingate to be of Jurassic age.

The age and relationships of the Kayenta Formation and its fauna have been reviewed recently (Colbert 1981).

What is the evidence of the fossil vertebrates, especially the tetrapods? Some comparisons may be made as in Table 1.1.

What is one to make of these columns in Table 1.1? At first glance the differences between the Chinle and Wingate, on the one hand, and the Moenave and Kayenta, on the other, seem reasonably clear. To the left are labyrinthodont amphibians, euryapsids, thecodonts, and dicynodonts. To the right are turtles, lizards, pterosaurs, crocodilians, tritylodonts, and mammals. Saurischian dinosaurs cut across the line, and, although *Coelophysis* is an undoubted Triassic coelurosaur, the age relationships of *Dilophosaurus* are not so unequivocal. It might be an advanced Triassic form; it might just as well be a precocious Jurassic genus.

The discovery within the past decade of a complete ornithischian skeleton in the Kayenta Formation raises some interesting questions. This dinosaur, *Scutellosaurus*, is a fabrosaurid, closely related to *Fabrosaurus* and *Lesothosaurus* from the Red Beds of the Stormberg Series in South Africa. The South African fabrosaurids would seem to constitute a Late Triassic beginning for the ornithischian dinosaurs. Should *Scutellosaurus* be viewed in the same light, or is it a persistent fabrosaurid living into Jurassic times?

The ancestral crocodilian, *Protosuchus*, is found in the Moenave Formation. Here, as in the case of the ornithischian dinosaurs, comparisons may be made with the South African Red Beds, in which are found primitive crocodilians.

Some of the particularly significant reptiles of the Kayenta Formation are the tritylodonts, which are in many respects at the threshold between reptiles and mammals. Numerous specimens representing the recently named genus, *Kayentatherium*, closely related to *Tritylodon* (again, a South African Red Bed genus), as well as to *Bienotherium* from the Lower Lufeng beds of China, have been recovered from Kayenta sediments. Because the Red Beds are considered by many authorities to be of

Triassic age, these interesting reptiles would seem to tip the age scale of the Kayenta Formation into Late Triassic time. *Bienotherium* may be either of Late Triassic or Early Jurassic age, according to how one views the age relationships of the Lower Lufeng beds (see next section). The tritylodont *Oligokyphus* also has been found in the Kayenta Formation, and in Europe this genus is definitely a Jurassic form.

Recently, a mammal, *Dinnetherium*, has been described from the Kayenta Formation by Farish Jenkins and his associates, who within the past few years have opened new vistas of Kayenta life (Jenkins, Crompton, and Downs 1983). Here, again, the age of this fossil can be argued; is it truly a Triassic mammal?

So, historically speaking, the Triassic–Jurassic boundary in the continental beds of western North America is in a state of uncertainty, despite excellent field and laboratory studies carried out in recent years. In this part of the world, the boundary prob-

Table 1.1

Chinle	Wingate	Moenave	Kayenta[a]
Amphibia			Amphibia
Labyrinthodontia			
Metoposaurus			
Reptilia			Reptilia
Eosuchia			
Tanytrachelos		Chelonia	Chelonia
Trilophosaurus			Lizards
			Sphenodontids
Thecodontia			Pterosauria
			Rhamphinion
Hesperosuchus			
Typothorax			
Desmatosuchus			
Poposaurus			
Rauisuchid			
thecodonts			
Rutiodon	*Rutiodon*		
		Crocodilia	Crocodilia
		Protosuchus	*Eopneumatosuchus*
Saurischia			Saurischia
Coelophysis			*Dilophosaurus*
Therapsida			Trackways
Placerias			Ornithischia
			Scutellosaurus
			Trackways
			Therapsida
			Kayentatherium
			Oligokyphus
			Mammalia
			Dinnetherium

[a]Some of the Kayenta identifications are made upon as yet unpublished data.

lem is still open to vigorous debate, a debate that in many of its aspects founders in a quagmire of definitions.

The Triassic–Jurassic boundary in Yunnan, China

In 1940, Bien announced the discovery of Triassic red beds in Yunnan, China, containing the remains of saurischian dinosaurs and "mammals" – these latter subsequently proving to be tritylodont therapsids. The Lufeng beds, named after its type locality, and the contained fauna were then the subjects of extensive research by Chung Chien Young, who published a considerable series of papers describing the sediments and fossils. By 1951, when Young published his monograph on the Lufeng Saurischian fauna, the Lufeng Series, as then known in Yunnan, had been divided into the Lower and Upper Lufeng Formations, a succession some thousand feet in thickness, resting upon Precambrian metamorphics and capped by Jurassic sediments. The Lufeng sediments are flat-lying and essentially continuous, although a disconformity just beneath a thin, red sandstone is taken to mark the boundary between the Lower and Upper Lufeng divisions of the Lufeng Series.

In recent years, the Lower and Upper Lufeng sequences have each been subdivided into four lesser entities, presently called "members." The tetrapods found in these several subdivisions as they comprise the Lower and Upper Lufeng beds may be compared as shown in Table 1.2.

The faunal differences between the Lower and Upper Lufeng sediments are obvious. The presence of capitosaurs, supposed thecodonts, and prosauropods in the Lower Lufeng beds could indicate a Late Triassic age, approximately equivalent to the Germanic Keuper. This is the opinion of Simmons (1965). *Lufengosaurus*, known from very complete materials, is a close counterpart of *Plateosaurus* from the Keuper of southern Germany. As for *Bienotherium*, this genus, also known from excellent fossils, is very close to *Tritylodon* itself and to *Kayentatherium* found in the Kayenta Formation. *Tritylodon* in the Red Beds and the tritylodont of the Kayenta Formation may indicate a Liassic presence for this group of therapsids in Africa and North America, and the same may be true for *Bienotherium*, espe-

Table 1.2

Lower Lufeng		Upper Lufeng
	Amphibia	
Capitosauridae		
	Reptilia	
Thecodontia		Chelonia
Dibothrosuchus		
Strigosuchus		
Pachysuchus[a]		*Plesiochelys*
Crocodilia		
Platyognathus[b]		
Microchampsa		
Saurischia		Saurischia
Lukousaurus		
Sinosaurus		Megalosauridae
Gyposaurus		
Lufengosaurus		Brachiosauridae
Yunnanosaurus		
Ornithischia		
Tatisaurus		
Tawasaurus		
Ictidosauria		
Bienotherium		
Lufengia		
Dianzhongia		
Yunnania		
Oligokyphus		

[a]A questionable phytosaur. The type specimen is now lost.
[b]Considered by Simmons to be a thecodont, although he points out the crocodylomorph nature of many of its characters.

cially as it is associated with *Oligokyphus* in the Lower Lufeng Formation – the latter genus being an undoubted Jurassic form in Europe. Recent students in China, notably Sun and Cui, favor a Jurassic age for the entire Lufeng sequence. On the other hand, the presence of thecodonts (if they are thecodonts) in the Lower Lufeng Formation may be variously regarded either as indicating a Late Triassic age or as representing an extension of this order of reptiles into Early Jurassic times, as would seem to have been the case with certain labyrinthodont amphibians in Australia (as discussed in the section on Triassic–Jurassic boundary in Australia later in this chapter). The plateosaurs of the Lower Lufeng beds certainly resemble prosauropod dinosaurs in the Upper Triassic sediments of other regions, although we know that prosauropods seem to have crossed the boundary at some places. Megalosaurs and brachiosaurs attest to the Jurassic relationships of the Upper Lufeng Formation.

Therefore, it would seem that the Triassic–Jurassic boundary is still a matter of some question in western China. The sedimentary record does not indicate a profound break within the Lufeng Series, although there is a small disconformity between the Lower and Upper Lufeng sediments that may indicate the boundary. Otherwise, the question depends upon the interpretation of certain elements in the Lower Lufeng fauna. Thus, these fossils indicate a Late Triassic, or perhaps an Early Jurassic, age. There the matter rests: a matter of personal opinion, as is so often the case in any attempt to determine this troublesome boundary within the Mesozoic System.

The Triassic–Jurassic boundary in peninsular India

Triassic and Jurassic beds with tetrapod fossils have been known in peninsular India for about a century. Indeed, the Maleri Formation of Late Triassic age and the Kota Formation of Early Jurassic age were described by King in 1881 (Sastry et al. 1977). The Maleri Formation, the object of intensive studies by P. L. Robinson, S. L. Jain, T. Roy-Chowdhury, and T. S. Kutty of the Indian Statistical Institute, has been designated to be of Carnian and early Norian age (Jain, Robinson, and Roy-Chowdhury 1964; Robinson 1967). The Kota Formation, also the object of vigorous studies by these same paleontologists, has been recognized by them to be of early Liassic age (Jain, Robinson, and Roy-Chowdhury 1962). Recently, Kutty and Roy-Chowdhury have described the Dharmaram Formation, above the Maleri, to be of late Norian and Rhaetian age (Kutty 1969; Kutty and Roy-Chowdhury 1970).

The Maleri beds contain a fauna that shows close relationships with the Germanic Keuper and with the Chinle of western North America. It has yielded the stereospondyl amphibian *Metoposaurus* and characteristic phytosaurs, as well as coelurosaurian dinosaurs, but it also contains a rhynchosaur, *Paradapedon*, that is closely related to the Brazilian genus, *Scaphonyx*. These tetrapods, therefore, reflect the position of peninsular India during Triassic time as a closely integrated segment of Gondwanaland yet open to the influx of Laurasian reptiles.

The Dharmaram Formation contains prosauropod dinosaurs.

As stated above, the Kota Formation long ago was assigned to an early Liassic position upon the basis of its fishes, such as *Lepidotes, Tetragonolepis, Paradapedium*, and *Indocoelacanthus* (Jain 1973, 1974). A number of years ago, the Indian Statistical Institute paleontologists made the surprising discovery of a giant sauropod dinosaur in the Kota Formation, at a horizon below that of some of the Liassic fishes. This dinosaur, described under the generic name of *Barapasaurus*, gives dramatic proof of the fact that the sauropods attained gigantic size very quickly, at the beginning of Jurassic time (Jain et al. 1975).

Thus, the Triassic–Jurassic boundary is indicated in peninsular India by a time span during which the prosauropod dinosaurs presumably disappeared while the sauropods appeared as full-fledged giants. It was not a long interval, but it clearly was the temporal boundary between typical Late Triassic dinosaurs and those gigantic sauropods that are so very characteristic of Jurassic history.

The Triassic–Jurassic boundary problem in South Africa

The discussion of the Triassic–Jurassic boundary problem, as seen within the sequence of Glen Canyon Group formations in southwestern North America, and particularly the comparisons that have been made between certain tetrapods in the Moenave and Kayenta formations of Arizona and the Red Beds of South Africa, leads to a consideration at this place of the South African Karroo System. The full succession, from the Dwyka Series at the base, through the overlying Ecca Series, the Beaufort Series, and, at the top, the Stormberg Series – in all comprising some 35,000 ft of sediments occupying the Great Karroo Basin – is one of the classic examples of continental sedimentation. Moreover, it is a succession of strata and faunas in which boundaries are subject to question because it is a remarkable and almost continuous record of life through millions of years of earth history. In the Karroo, the position of the boundary between the Permian and the Triassic can be argued, as can that between the

Triassic and Jurassic. Our problem, of course, is with the latter boundary as it may be present within the Stormberg Series.

The Stormberg Series consists of the Molteno Formation at the bottom, followed by the Red Beds, the Cave Sandstone, and the Drakensberg volcanics. [In recent years the South African Committee for Stratigraphy has introduced many new names to replace older stratigraphic terms; for example, the Elliot Formation for the Red Beds and the Clarens Formation for the Cave Sandstone. Indeed, the name Stormberg has been discarded, in spite of its usage since 1859. As J. N. J. Visser has recently pointed out: "Almost 30 stratigraphic names have been proposed or are partly in use for the Stormberg and Drakensberg Groups in South Africa, Botswana, Namibia, and Zimbabwe, which to any reader without a lexicon could be utterly confusing. For this reason the well-established names like Molteno Sandstone, Red Beds and Cave Sandstone are informally used . . . " (Visser 1984, p. 6). In the present discussion the older terms also will be generally used.]

Such a remarkable series of sediments, for the most part highly fossiliferous, was destined to attract the interest of paleontologists at an early date. As a result, Karroo fossils were studied during the nineteenth century by able paleontologists, among whom there may be mentioned T. H. Huxley and Harry Govier Seeley. However, it was Robert Broom who, during the early years of this century, raised a curtain that revealed to the world the richness and variety of Karroo amphibians and reptiles. Broom was followed by Sydney Haughton (who gave particular attention to the stratigraphic relationships of Karroo sediments, with special emphasis on the Stormberg Series), by van Hoepen, and by an impressive array of present-day paleontologists and stratigraphers. As a result, our knowledge of Karroo sediments and fossils is broad and detailed. Thus, South Africa has been and is a central reference point, the pivotal region, for students of Gondwana tetrapod faunas (see, especially, Second Gondwana Symposium 1970).

For many years, opinions concerning the age of the Stormberg beds paralleled ideas of the Newark Group in that the entire sequence, at least below the Drakensberg Lavas, was supposed to be of Late Triassic age (e.g., Ellenberger et al. 1964). Consequently, the fossils found in the Red Beds/Elliot and Cave Sandstone/Clarens formations were looked upon as diagnostic for Upper Triassic Gondwana faunas. This was especially true for such forms as the tritylodonts and the saurischian dinosaurs. Moreover, the discovery and recognition of small ornithischians in these two formations has been regarded by many as marking the Triassic appearance of the Order Ornithischia in the geological record.

The sedimentary structures of the Red Beds would seem to indicate a warm climate and seasonal rainfall during the time of their deposition. At some intervals, eolian sands encroached into the areas where the Red Beds were being deposited, so that there were simultaneous depositions of Red Beds and sands. This would account for the interfingering that is seen between the Red Beds and the Cave Sandstone, and particularly the intergradation down from the Cave Sandstone, which commonly forms a thick cap above the Red Beds, into the characteristic stratified layers of the Red Beds. So there are transitional beds between the Red Beds and the Cave Sandstone. Because tetrapods (and sometimes the same tetrapods) are found in the typical Red Beds, in the transitional deposits, and in the Cave Sandstone, the paleontologist is confronted with a situation where lines of demarcation are not clearly established.

Many years ago Haughton (1924) noted that there was a change from heavy-limbed dinosaurs to small, cursorial types between the lower and upper levels of the Red Beds, a change that has been confirmed in recent years by A. W. Crompton. Furthermore, Crompton found that small dinosaurs, particularly the ornithischian *Heterodontosaurus*, occur in the transitional beds just below the base of typical Cave Sandstone deposits, and below this level, still in the upper part of the Red Beds, is found *Tritylodon*, while in the lower levels of the Red Beds are large dinosaurs such as *Euskelosaurus* (A. W. Crompton pers. comm.).

A comparison between the Red Beds and the Cave Sandstone involving certain tetrapods may be made (Table 1.3) without any attempt to indicate a line of demarcation between Triassic and Jurassic forms.

In the light of the transitional sedimentation between the Red Beds and the Cave Sandstone, and in the light of the occurrences of the same or closely related reptiles in the two formations and in the transitional beds between them, where is the Triassic–Jurassic boundary to be established in South Africa? Are all of the sediments of the Stormberg Group to be considered of Triassic age? The presence of thecodonts might support this view if one defines the continental Triassic as indicated, at least in part, by the presence of thecodont reptiles, but how is one to view the presence of some advanced reptiles, such as *Tritylodon*, the ornithischian dinosaurs, and the mammal *Megazostrodon*?

In 1980, Paul Olsen indicated the lower Red Beds as being Carnian plus Norian plus Rhaetian in age, the upper Red Beds as being Hettangian, and the Cave Sandstone as being Sinemurian (Olsen 1980a,b). This would establish an internal division between the Triassic and Jurassic within the Stormberg Series, paralleling the division that Olsen sees within the Newark Group and the division advocated by some paleontologists and stratigraphers within the Glen Canyon Group.

Thus, as in the case of the North American successions, we are confronted once again with the problem of definitions. Where the boundary between the Triassic and Jurassic is to be placed depends in no little measure upon the bias of the student who is making the choice.

The Triassic–Jurassic boundary in South America

Much of our present knowledge concerning Mesozoic tetrapods in South America is due to the excellent field work and research carried on in recent years by the late Llewellyn Price and Mario Barberena in Brazil and by the late A. S. Romer and his colleagues and by José Bonaparte in Argentina. Here the Triassic faunas are of Gondwanic type, with therapsids dominating the Early and Middle Triassic assemblages, while the Late Triassic faunas show, for the first time in this part of the world, a dominating position being assumed by the archosaurs (Bonaparte 1978).

According to Bonaparte, there would seem to

be a hiatus in the preservation of sediments at the top of the Brazilian and Argentinian Triassic – in what has been designated as the Coloradian provincial age in southern South America (Bonaparte 1982). Thus, the faunas from the La Esquina and El Tranquilo beds in Argentina represent the highest Triassic tetrapod faunas known, even though these are at the top of the Coloradian, not at the very upper limit of the Triassic. The Roca Blanca Formation of Liassic age is separated from the underlying El Tranquilo Formation by an angular unconformity. There is not at the present time a difficult boundary problem, such as has been discussed for other regions.

In 1978, in a general analysis of the Late Triassic and Early Jurassic tetrapods, Bonaparte pointed out that the rhynchosaurs, all of the thecodonts, the protosuchians, the prosauropods, and the anomodonts became extinct at the end of Triassic time. (It should be mentioned here that Olsen and Galton, in their paper of 1977, which obviously was not seen by Bonaparte when he made his statement, demonstrated that upon the basis of new information the protosuchians and prosauropods do extend into the Lower Jurassic.) Various other archosaurian groups had their beginnings in the final stages of Triassic history and continued into the later Mesozoic or were of post-Triassic origins (Bonaparte 1978, pp. 388–9).

Table 1.3

Red Beds/Elliot Formation		Cave Sandstone/Clarens Formation
	Therapsida	
Lycorhinus		
Pachygenelus		*Diarthrognathus*
Tritylodon		
	Thecodontia	
Sphenosuchus		*Pedeticosaurus*
	Crocodilia	
Erythrochampsa		*Notochampsa*
	Saurischia	
		Syntarsus
Thecodontosaurus		*Thecodontosaurus*
Euskelosaurus		*Gryposaurus*
Plateosaurus		*Massospondylus*
Melanorosaurus		
(and other genera)		
	Ornithischia	
Heterodontosaurus		*Heterodontosaurus*
Lesothosaurus		*Geranosaurus*
	Mammalia	
Megazostrodon		

The Triassic–Jurassic boundary in Australia

Upper Triassic and Lower Jurassic sediments are present in eastern Australia, in which there are rare occurrences of tetrapods. These fossils have been subjected to careful scrutiny by Warren, Bartholomai, Thulborn, and other contemporary Australian paleontologists, with the result that upon the basis of even so sparse a fossil record there have arisen questions concerning the Triassic–Jurassic boundary.

Two labyrinthodont amphibians have been of particular interest: *Austropelor*, described by Longman (1941) and redescribed by Colbert (1967), and *Siderops*, recently described by Warren and Hutchinson (1983). The first of these fossils was found in the Marburg Sandstone, considered by Whitehouse (who wrote a section on the age of the Marburg in Longman's type description) to be of Jurassic affinities; but in 1955, Whitehouse revised his earlier opinion, largely because of the presence of *Austropelor* in the Marburg, and placed this horizon definitely within the Upper Triassic. The Marburg Sandstone is part of a sequence involving, from bottom to top, the Triassic Bundamba Group, the Marburg Sandstone, and the Jurassic Walloon Coal Measures, with no major breaks involved. Where should the boundary be placed?

Warren and Hutchinson (1983) described *Siderops*, a brachyopid, known from ample materials preserved in the Upper Evergreen Formation of eastern Australia, a horizon that not only is definitely of Liassic affinities, but may be regarded even as of late Liassic age. It seems evident that at least in Australia some labyrinthodonts definitely persisted into Early Jurassic time. Therefore, it is quite possible that the Marburg Sandstone with *Austrobrachyops* is a Jurassic horizon, as it was originally designated in 1941. On the basis of present evidence, it would appear that in this part of Gondwanaland there was a crossing of the Triassic–Jurassic boundary by the labyrinthodont amphibians, a group heretofore firmly regarded as having become extinct with the close of Triassic time. Obviously, there is still much to be learned about the Triassic–Jurassic boundary in Australia.

Conclusions

As has so often proved to be the case in the study of geology and paleontology, the line of separation between two periods of earth history, in this case the Triassic and Jurassic, may not be easy to determine. In various circumstances, clear stratigraphic breaks within successions that include Upper Triassic and Lower Jurassic continental sediments are not readily distinguishable. Notable instances are seen in the Newark Group and the Glen Canyon Group of eastern and southwestern North America, respectively, in the Stormberg beds of South Africa, in the Lufeng Series of Yunnan, China, and perhaps in the Bundamba to Walloon sediments of Australia. It is possible that in other situations where there does seem to be a definite stratigraphic break between sediments of Triassic and Jurassic ages, future discoveries and interpretations may add complications to the picture as it is now seen. Moreover, future discoveries in other localities may add similar complex boundary problems.

It has, therefore, sometimes been necessary for paleontologists to attempt definitions of the Triassic–Jurassic boundary upon the basis of the fossils. As far as tetrapods are concerned, the results have been equivocal and very much subject to personal bias. Does the presence of thecodont reptiles ensure the Triassic as the age of the sediments in which they occur? How are we to interpret the presence of the genus *Tritylodon* and closely related tritylodonts? Did such therapsids cross the line from the Triassic into the Jurassic, as now seems to have been the case for labyrinthodont amphibians in Australia? Are the fabrosaurid ornithischians of Triassic or Liassic age, or both? These are some of the questions with which paleontologists have struggled for many years. Indeed, the problem of Triassic or Jurassic age relationships goes back to the early days of geological studies, to the arguments as to where the Rhaetic should be placed.

So, the historical aspect of the Triassic–Jurassic boundary problem presents a story full of doubts, and it would seem that it will be a story filled with unresolved problems for many years into the future. It is the old theme of attempting to divide a continuing record into discrete segments and place them in labeled pigeonholes. Perhaps in some respects it is an exercise in futility, but it does add interest and a little bit of heat to our study of early Mesozoic history.

References

Averitt, P., J. S. Detterman, J. W. Harshbarger, C. A. Repenning, and R. F. Wilson. 1955. Revisions in correlations and nomenclature of Triassic and Jurassic formations in southwestern Utah and northern Arizona. *Bull. Am. Assoc. Petrol. Geol.* 39(12): 2515–24.

Baird, D. E., and W. F. Take. 1959. Triassic reptiles from Nova Scotia. *Geol. Soc. Am. Bull.* 70: 1565–6.

Bien, M. N. 1940. Discovery of Triassic saurischian and primitive mammalian remains at Lufeng, Yunnan. *Bull. Geol. Soc. China* 20(3–4): 225–34.

　1941. "Red Beds" of Yunnan. *Bull. Geol. Soc. China*

21(2–4): 157–98.

Bonaparte, J. F. 1978. El Mesozoico de America del Sur y sus Tetrapodos. Tucuman, *Opera Lilloana* 26: 1–596.

1982. Faunal replacement in the Triassic of South America. *J. Vert. Paleontol.* 2: 362–71.

Brinkmann, R. 1960. *Geologic Evolution of Europe* (New York: Hafner).

Colbert, E. H. 1967. A new interpretation of *Austropelor*, a supposed Jurassic labyrinthodont amphibian from Queensland. *Mem. Queensl. Mus.* 15: 35–41.

1981. A primitive ornithischian dinosaur from the Kayenta Formation of Arizona. *Mus. N. Ariz. Bull. Ser.* 53: 1–61.

Colbert, E. H., and J. T. Gregory. 1957. Correlation of continental Triassic sediments by vertebrate fossils. *In* Correlation of the Triassic Formations of North America Exclusive of Canada. J. B. Reeside, Jr., et al. *Bull. Geol. Soc. Am.* 68: 1456–67, Chart 8a.

Cornet, B. 1977. The palynostratigraphy and age of the Newark Supergroup. Ph.D. Thesis, Department of Geosciences, Pennsylvania State University. 505 pp.

Ellenberger, F., P. Ellenberger, J. Fabre, L. Ginsburg, and C. Mendrez. 1964. The Stormberg Beds of Basutoland. *Proc. 22nd Int. Geol. Congr., New Delhi* 9(9): 320–30.

Gregory, H. E. 1917. Geology of the Navajo country. *U.S. Geol. Surv. Prof. Paper* no. 93.

Harshbarger, J. S., C. A. Repenning, and J. H. Irwin. 1957. Stratigraphy of the Uppermost Triassic and the Jurassic rocks of the Navajo country. *U.S. Geol. Surv. Prof. Paper* no. 291.

Haughton, S. H. 1924. The fauna and stratigraphy of the Stormberg Series. *Ann. S. Afr. Mus.* 12: 323–497.

Jain, S. L. 1973. New specimens of Lower Jurassic holostean fishes from India. *Paleontology* 16: 149–77.

1974. *Indocoelacanthus robustus* n. gen., n. sp. (Coelacanthidae, Lower Jurassic), the first fossil coelacanth from India. *J. Paleontol.* 48: 49–62.

Jain, S. L., P. L. Robinson, and T. K. Roy-Chowdhury. 1962. A new vertebrate fauna from the Early Jurassic of the Deccan, India. *Nature (London)* 194: 755–7.

1964. A new vertebrate fauna from the Triassic of the Deccan, India. *Quar. J. Geol. Soc., London* 120: 115–24.

Jain, S. L., T. S. Kutty, T. Roy-Chowdhury, and S. Chatterjee. 1975. The sauropod dinosaur from the Lower Jurassic Kota formation of India. *Proc. Roy. Soc. London* A 188: 221–8.

Jenkins, F. A., Jr., A. W. Crompton, and W. R. Downs. 1983. Mesozoic mammals from Arizona: new evidence on mammalian evolution. *Science* 222: 1233–5.

Kutty, T. S. 1969. Some contributions to the stratigraphy of the Upper Gondwana formations of the Pranhita-Godavari Valley, central India. *J. Geol. Soc. India* 10: 33–48.

Kutty, T. S., and T. Roy-Chowdhury. 1970. The Gondwana sequence of the Pranhita-Godavari Valley, India, and its vertebrate fauna. *Second Gondwana Symposium, South Africa, 1970, Proceedings and Papers* (Pretoria: Pretoria Council for Scientific and Industrial Research), pp. 303–8.

Lewis, G. E., J. H. Irwin, and R. F. Wilson. 1961. Age of the Glen Canyon Group (Triassic and Jurassic) on the Colorado Plateau. *Geol. Soc. Am. Bull.* 72: 1437–40.

Longman, H. 1941. A Queensland fossil amphibian (*Austropelor*). *Mem. Queensl. Mus.* 12: 29–32.

Lull, R. S. 1904. Fossil footprints in the Jura-Trias of North America. *Mem. Boston Soc. Nat. Hist.* 5: pp. 461–557 and Plate 72.

1953. Triassic life of the Connecticut Valley. *Conn. Geol. Nat. Hist. Surv., Bull.* 81: 1–331.

Olsen, P. E. 1978. On the use of the term Newark for Triassic and Early Jurassic rocks of eastern North America. *1978 Newsl. Stratigr.* 7(2): 90–5.

1980a. Triassic and Jurassic formations of the Newark Basin. *In* Manspeizer, W. (ed.), *Field Studies of New Jersey Geology and Guide to Field Trips* (Newark, New Jersey: Rutgers University Press), pp. 2–39.

1980b. Fossil great lakes of the Newark Supergroup in New Jersey. *In* Manspeizer, W. (ed.), *Field Studies of New Jersey Geology and Guide to Field Trips* (Newark, New Jersey: Rutgers University Press), pp. 352–98.

1983. Relationship between biostratigraphic subdivisions and igneous activity in the Newark Supergroup. *Geol. Soc. Am. Abst.* 15: 93.

Olsen, P. E., and D. Baird. 1982. Early Jurassic vertebrate assemblages from the McCoy Brook fm. of the Fundy Group (Newark Supergroup, Nova Scotia, Ca.). *Geol. Soc. Am. Abst. Prog.* 14(1–2): 70.

Olsen, P. E., and P. M. Galton. 1977. Triassic–Jurassic tetrapod extinctions: are they real? *Science* 197: 983–5.

Redfield, W. C. 1856. On the relations of the fossil fishes of the sandstone of the Connecticut and other Atlantic States to the Liassic and Oolitic periods. *Am. J. Sci.* 22: 357–63.

1857. On the relations of the fossil fishes of the sandstone of Connecticut and other Atlantic States to the Liassic and Jurassic periods. *Am. Assoc. Advan. Sci., 10th Meeting, Albany, 1856*, pp. 180–88.

Robinson, P. L. 1967. The Indian Gondwana formations – a review. *In Gondwana Stratigraphy, IUGS Symposium, Buenos Aires, 1967* (Paris: Unesco), pp. 201–68.

Sastry, M. V. A., et al. 1977. Stratigraphic lexicon of Gondwana formations of India. *Geol. Surv. India, Miscell. Publ. No. 36.*

Schmidt, Martin. 1928. *Die Lebewelt unserer Trias.* (Ferdinand Rau, Öhringen).

Second Gondwana Symposium, South Africa. 1970. *Proceedings and Papers.*

Sigogneau-Russell, D, and Ai-Lin Sun. 1981. A brief review of Chinese synapsids. *Géobios* 14(2): 275–9.

Simmons, D. J. 1965. The non-therapsid reptiles of the Lufeng Basin, Yunnan, China. *Fieldiana: Geol.* 15(1): 1–93.

Visser, J. N. J. 1984. A review of the Stormberg Group and Drakensberg Volcanics in southern Africa. *Paleontol. Afr.* 25: 5–27.

Ward, L. F. 1901. Geology of the Little Colorado Valley. *Am. J. Sci., 4th Ser.* 12: 401–13.

Warren, A. A., and M. N. Hutchinson. 1983. The last labyrinthodont? A new brachyopoid (Amphibia, Temnospondyli) from the early Jurassic Evergreen formation of Queensland, Australia. *Phil. Trans. Roy. Soc. London* 303(1113): 1–62.

Whitehouse, F. W. 1955. The geology of the Queensland portion of the Great Australian Artesian Basin. *Queensl. Parliam. Papers, 1955* 2: 653–76.

Wilmarth, G. 1938. Lexicon of Geologic Names of the United States. *U.S. Geol. Surv. Bull., No. 896.*

Wilson, R. F., and J. H. Stewart. 1967. Correlation of Upper Triassic and Triassic (?) formations between southwestern Utah and southern Nevada. *U.S. Geol. Surv. Bull. 1244-D*, pp. D1–D20.

Young, C. C. 1951. The Lufeng saurischian fauna in China. *Palaeontol. Sin.* No. 134, New Ser. c, No. 13, pp. 1–96, Plates 1–12.

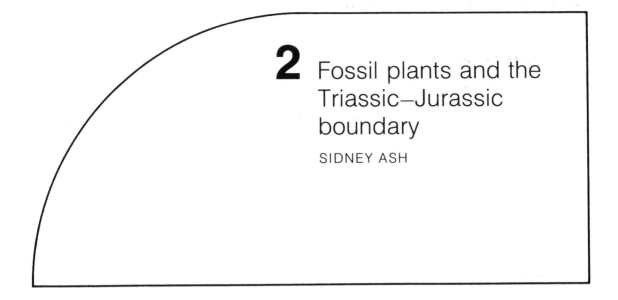

2 Fossil plants and the Triassic–Jurassic boundary

SIDNEY ASH

Introduction

In this chapter, the land floras of the Upper Triassic (Carnian–Rhaetian stages) and Lower Jurassic (Lias stage) are reviewed briefly, and apparent evolutionary trends are examined. Certain representative Upper Triassic and Lower Jurassic forms are illustrated. Although plant megafossils are discussed almost exclusively in this chapter with only occasional reference to pollen and spores (palynomorphs), it should be noted that data obtained from both types of fossils are in general agreement regarding the stratigraphic and geographic distribution of plants and their evolution. Furthermore, because palynomorphs are often widely distributed, they may complement or clarify conclusions based on other types of plant fossils.

Because this chapter is included in a volume that will be read primarily by vertebrate paleontologists, a few comments need to be made about some of the problems that paleobotanists face. These problems are caused by the fact that land plants, unlike vertebrates, have the habit of "falling apart" before death as well as afterward, such as when they lose their leaves at the end of a growing season. Thus, most plant fossils consist of the remains of just one part of a plant (for example, leaves, seeds, or spores) and only rarely are two or more of them found attached to one another. To complicate matters still further, the wood and leaves and sometimes other parts may be so similar in unrelated plants that it is difficult to distinguish them. Furthermore, the various plant parts, even parts of the same plant, are often preserved in different ways; the woody stems may be petrified, while the leaves may be compressed or just leave an impression. As a consequence, it is often difficult to relate dispersed fossilized plant parts to one another with assurance, and it is often difficult to reconstruct entire fossil

plants. Thus, it is usually necessary to give each plant part a different name even if they are actually parts of the same plant. Sometimes plant fossils are so distinctive that they can be referred to a particular family, and so they are given a "natural" generic name just as in the case of living plants. If, on the other hand, they cannot be assigned to a family, they are referred to an artificial or "form" genus. An example of a form genus is *Araucarioxylon*, a name that is applied to a type of gymnospermous wood that occurs in several families and orders.

Typically, the whole plant is not given a formal name even if most or all of its parts are known. For example, the wood of the plant that bore the leaf *Glossopteris* is referred to the form genus *Araucarioxylon*, the root is assigned to *Vertebraria*, and so forth, and the entire plant is merely called the "*Glossopteris* plant."

In spite of these difficulties, plant fossils can still be used in biostratigraphic and paleoecological studies if the right parts are preserved. Several biostratigraphic zones based on plant fossils have been devised for various parts of the geological record, including, as shown below, the Upper Triassic and Lower Jurassic.

Systematic review

In this section, the more common land plants of the Upper Triassic and Lower Jurassic are reviewed systematically by order and family. The general characteristics of the plants in each of the groups are probably known by most readers so they are not discussed here. More detailed information about plant fossils can be found in the paleobotanical books recently published by Stewart (1983) and by Taylor (1981).

During the Upper Triassic and Lower Jurassic, the land flora was fairly luxuriant and no longer impoverished as it had been during the preceding period. The floras contained a few holdovers from the Upper Paleozoic, such as some of the lycopods, horsetails, ferns, and seed ferns. In addition, the floras contained many new species and genera that are assigned to these groups and to the cycads, Voltziales, and Coniferales. They also included one new order, the Bennettitales and a number of new families. Many of the plants were very similar to modern forms and may have been directly ancestral to living species. Some certainly were species of living genera. Certain ones combine the features of several living forms, and others looked unlike any now living on earth. As a whole, the floras generally had a distinctly modern aspect. In the Southern Hemisphere, the floras were distinctive during the Upper Triassic and Lower Jurassic and contained many forms characteristic of that region. There were a number of common ferns and fern allies in both the northern and southern floras, but there were many fewer common seed plants. In general, the forms that are common to the two regions are fairly scarce.

Lycopods

Lycopod fossils occur at a surprising number of localities in Upper Triassic and Lower Jurassic rocks throughout the world (Stewart 1983). Most of them are the remains of herbaceous plants apparently closely related to living species of *Lycopodium*, *Selaginella*, and *Isoetes*. Because of this, they have commonly been referred to *Lycopodites*, *Selaginellites*, or *Isoetites* depending on their presumed relationships. In some cases, they have even been referred to living genera as in the case of the *Selaginella* (Fig. 2.1a), which was described a few years ago from the Upper Triassic of Arizona (Ash 1972). The remains (stems, cones, certain large spores) of more robust types of plants that compare with *Lepidodendron* and other of the treelike lycopods of the Carboniferous occur rarely in these same strata (Dobruskina 1974). It is apparent from the fossil evidence that herbaceous lycopods were present in small numbers during the Late Triassic and Early Jurassic and that treelike forms were also present, but they were comparatively scarce (Taylor 1981). There is no evidence that the lycopods underwent any significant evolutionary changes during this time, except possibly for a further decline in the number of treelike forms.

Horsetails

The remains of horsetails are widely distributed in Upper Triassic and Lower Jurassic strata throughout the world. Many of the fossils are the remains of herbaceous plants that resemble the living horsetail *Equisetum* rather closely. As a consequence, they have often been referred to that genus or to *Equisetites* (Figs. 2.1b, c). Spores that are very similar to those produced by *Equisetum* occur widely in the same strata. Other herbaceous horsetails that occur in the Upper Triassic and Lower Jurassic include *Phyllotheca* and *Schizoneura* (Fig. 2.1d). Treelike plants resembling the characteristic Carboniferous horsetail *Calamites* occur rather commonly in Upper Triassic and Lower Jurassic strata, but they are usually referred to the genus *Neocalamites*. Although they did not reach the size of the giant *Calamites*, some specimens of *Neocalamites* are fairly large. For example, specimens in the position of growth in the Upper Triassic of western Colorado are nearly 6 m tall and about 30 cm in diameter (Holt 1947). Apparently the horsetails did not undergo much evolutionary change during the Late Triassic and the Early Jurassic, except perhaps for a decline in the number of treelike forms.

Figure 2.1. Representative Upper Triassic and Lower Jurassic lycopod, horsetails, and ferns. **a**, *Selaginella*. **b**, **c**, *Equisetites* stems. **d**, *Schizoneura*. **e**, *Marattiopsis*. **f**, *Pseudodanaeopsis*. **g**, *Asterotheca*. **h**, *Todites*. **i**, *Osmundopsis*. Scale = 0.2 cm for (**a**), 1 cm for (**b**)–(**i**).

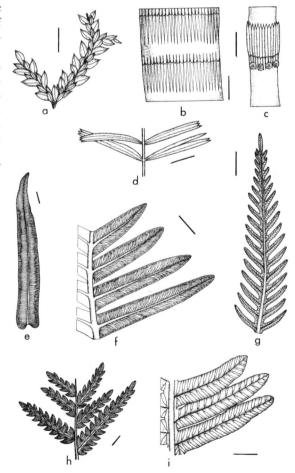

Ferns

The ferns were very diverse and abundant during the Carboniferous, but near the end of the period they began a rapid decline that continued into the Permian and resulted in the extinction of several families (Arnold 1964). Consequently, they were comparatively uncommon by the beginning of the Triassic. However, many new forms appeared during the Triassic and Early Jurassic, and the ferns began to expand once again (Arnold 1964). Some of these new forms are assigned to new families, and others are placed in older ones. In general, the ferns of the Upper Triassic and Lower Jurassic have a modern aspect, and most of them can be easily assigned to living families without question.

Marattiaceae

This family was widely distributed throughout the world during the Late Triassic and Early Jurassic (Arnold 1964). Most of the marattiaceous fossils found in Upper Triassic strata are the remains of leaves that are usually referred to the genera *Marattiopsis* (Fig. 2.1e), *Danaeopsis, Pseudodanaeopsis* (Fig. 2.1f), and *Asterotheca* (Fig. 2.1g). Some have even been referred to the living genus *Marattia* because they are so similar. The family seems to have been particularly common in the Upper Triassic (Carnian stage) of eastern North America (Fontaine 1883), Mexico (Silva-Pinada 1981), and Australia and Argentina (Herbst 1977a,b). The family seems to have declined rather abruptly later during the Upper Triassic, and only a few genera such as *Marattiopsis* survived into the Lower Jurassic.

Osmundaceae

The Osmundaceae appears to have been one of the largest families of Late Triassic and Early Jurassic ferns (Arnold 1964). Many of the floras of this age contain several species of the osmundaceous ferns *Todites* (Fig. 2.1h) and *Cladophlebis* (Fig. 2.2a). Other possible members of the family found in these floras are *Cladotheca* and *Osmundopsis* (Fig. 2.1i) (Harris 1931).

Schizaeaceae

It is uncertain when the family arose, but the oldest megafossil that can be safely assigned to it is *Klukia* (Fig. 2.2d), which first occurs in rocks of Early Jurassic age in Poland (Raciborski 1894). Stewart (1983) suggests that the family was also present during the Upper Triassic because there are reports of its characteristic spores in Carboniferous strata (Baxendale and Baxter 1977).

Matoniaceae

The leaves of the plants in this family are palmate and, in general, large and showy. *Phlebopteris* (Fig.

2.2b), which is first met in rocks of Late Triassic age (Carnian stage) at many localities in the Northern Hemisphere seems to be the oldest member of the family (Corsin and Waterlot 1979). During the Early Jurassic, the family expanded greatly, and several new genera, such as *Selenocarpus* and *Matonidium*, appeared together with some new species of *Phlebopteris* (Harris 1931).

Dipteridaceae

In this family, the leaves are also palmate and showy, and many are large. The oldest member of the family is *Dictyophyllum* (Fig. 2.2c), which first appears in the Middle Triassic of Australia (Webb 1982). Three other genera of dipteridaceous ferns were present during the Late Triassic, and the family became a significant element in the northern floras of the Early Jurassic (Oishi and Yamasita 1936; Corsin and Waterlot 1979). The most common Late Triassic and Early Jurassic members of the family are *Clathropteris* (Fig. 2.2f) and *Dictyophyllum* (which includes

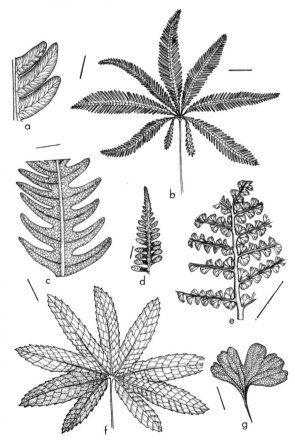

Figure 2.2. Representative Upper Triassic and Lower Jurassic ferns. **a**, *Cladophlebis*. **b**, *Phlebopteris*. **c**, *Dictyophyllum*. **d**, *Klukia*. **e**, *Gleichenites*. **f**, *Clathropteris*. **g**, *Hausmania*. Scales = 1 cm.

Thaumatopteris of many authors). A somewhat rarer genus in strata of this age is *Hausmania* (Fig. 2.2g).

Gleicheniaceae

The early history of this large family is unclear, but it appears that the oldest member of the family occurs in the Middle Triassic of Argentina (Herbst 1972). It is so close to the living fern *Gleichenia* that it has been referred to *Gleichenites* (Fig. 2.2e). Fragmentary specimens of the same genus are found at a few Late Triassic and Early Jurassic localities in the Northern Hemisphere, and the family remained insignificant until sometime later in the Mesozoic (Arnold 1964).

Seed ferns

The seed ferns of the Mesozoic are not very well defined because they have no simple characters that can be used to separate them from some other groups of seed-bearing plants (Harris 1964). Consequently, there are differences of opinion about which fossil plants should be considered seed ferns and which should be assigned elsewhere. As they are now generally recognized, they constitute a rather heterogeneous group of plants, some of which were dominants in certain lower Mesozoic floras. Typically, the seed ferns are very common in Upper Triassic and Lower Jurassic strata of the Southern Hemisphere, and, except for one family, they are somewhat rare in the Northern Hemisphere. Except for the *Glossopteris* plant, it is still not possible to reconstruct any of the plants in this order with much assurance, even though most of their parts are known.

Glossopteridaceae

This family apparently occurs only in the southern continents where it probably arose during the Permian (Schopf 1973). The family quickly reached its peak later during that period and then began to decline. It was still present in small numbers during the Upper Triassic, and seemingly became extinct before the begining of the Jurassic, at least in the southern continents (Anderson and Anderson 1983). Leaves that are somewhat similar to those usually assigned to the family have been reported from a few localities in the Upper Triassic and the Jurassic of Mexico and Central America (Delevoryas and Person 1975; Ash 1981). These fossils may be just examples of parallel evolution and may not have anything to do with the Glossopteridaceae because reproductive structures characteristic of the family do not occur outside of the southern continents. The *Glossopteris* plant is now reconstructed as a large tree (Gould and Delevoryas 1977). The most famous and common fossil included in this family is the tongue-shaped leaf *Glossopteris* (Fig. 2.3a). The

stems are assigned to *Araucarioxylon*; the roots to *Vertebraria*; the seed-bearing organs to *Scutum*, *Austroglossa*, and so forth; the dispersed seeds to *Pterygospermum*; and the pollen-bearing organs to *Glossotheca*, *Eretmonia*, and so forth (Fig. 2.3b).

Peltaspermaceae

The record of this family extends back into the Upper Permian of the northern continents where the family was still fairly abundant in the uppermost Triassic (Townrow 1960). It was also very common in the Upper Triassic of the southern continents (Anderson and Anderson 1983). Presumably the family became extinct during the Upper Triassic because it does not appear to be present anywhere in the Lower Jurassic (Dobruskina 1975). Most of the leaves are assigned to *Lepidopteris* (Fig. 2.3c) and *Scytophyllum*. The seed-bearing organ is referred to *Peltaspermum*, and the pollen-bearing organ is assigned to *Antevsia*.

Corystospermaceae

This family occurs principally in the Triassic of the southern continents and India, where it apparently dominated most of the floras of that time (Schopf 1973; Petriella 1979). The leaf is a forked, pinnate frond that is highly polymorphic. Retallack (1977)

Figure 2.3. Representative Upper Triassic and Lower Jurassic seed ferns. **a**, *Glossopteris*. **b**, *Eretmonia*. **c**, *Lepidopteris*. **d**, *Dicroidium*. **e**, *Caytonia*. **f**, *Pteruchus*. **g**, *Sagenopteris*. **h**, *Umkomasia*. Scales = 1 cm.

recognizes three genera of leaves in this family, *Dicroidium* with fifteen species and sixteen varieties, *Johnstonia* with four species and four varieties, and *Xylopteris* with four species and two varieties. The reproductive structures are much more conservative and show much less variation. The seed-bearing organs are assigned to *Umkomasia* (Fig. 2.3h), and the pollen-bearing organs are assigned to *Pteruchus* (Fig. 2.3f).

Caytoniaceae

This small but fascinating family of pteridosperms occurs throughout the world in the early Mesozoic, but is most common in Western Europe and adjacent areas (Harris 1970). Representatives of the family are first found in small numbers in the Rhaetian stage of the Upper Triassic of Sweden (Nathorst 1878) and east Greenland (Harris 1964). They also occur there in the Lower Jurassic and later became quite widespread elsewhere in the northern continents. The palmate leaves are assigned to *Sagenopteris* (Fig. 2.3g), the seed-bearing organs to *Caytonia* (Fig. 2.3e), and the pollen-bearing organs to *Caytonanthus*.

Bennettitales

This order (also called the Cycadeoidales) was an important component of most Mesozoic floras prior to the sudden expansion of the flowering plants

Figure 2.4. Representative Upper Triassic and Lower Jurassic Bennettitaleans (**a–e**) and cycad (**f**). **a**, *Zamites*. **b**, *Nilssoniopteris*. **c**, *Otozamites*. **d**, *Pterophyllum*. **e**, *Dictyozamites*. **f**, *Pseudoctenis*. Scales = 1 cm.

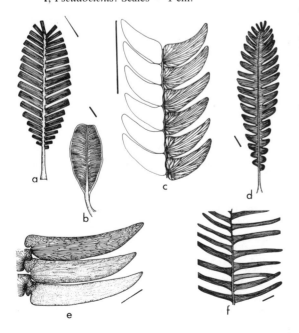

during the Middle Cretaceous. At one time, the Bennettitales were thought to be closely related to the Cycadales because the leaves and stems of both orders are strikingly similar in gross morphology. With the discovery by Thomas and Bancroft (1913) that the epidermal structures of the leaves in the two orders differed greatly, it became apparent that the two groups were only distantly related. This has been confirmed by a study of their reproductive structures and stems. The Bennettitales probably arose sometime in the Early or Middle Triassic because definite Bennettitalean fossils are very common in Upper Triassic floras in many parts of the world (Arnold 1953). These fossils include the leaves called *Zamites* (Fig. 2.4a), *Otozamites* (Fig. 2.4c), *Nilssoniopteris* (Fig. 2.4b), and *Pterophyllum* (Fig. 2.4d) and several types of reproductive structures (Harris 1973). Bennettitalean fossils became still more common in the Lower Jurassic with the appearance of many new species of those genera and of the new genus *Dictyozamites* (Fig. 2.4e), which appeared for the first time in the Lower Jurassic of Iran (Barnard 1965). Although the Bennettitales seem to have expanded somewhat in numbers and during the Late Triassic and Early Jurassic, the order apparently did not evolve very much during this time.

Cycads

The fossil record of the cycads is poorly known (Harris 1961). One of the principal reasons for this is that it is difficult or impossible to differentiate clearly cycad and pteridosperm leaves. Also if the epidermal structures are not preserved, it is usually impossible to distinguish most cycad and bennettitalean leaves. What evidence we have seems to indicate that the cycads were never very common (Arnold 1953). Recent discoveries show that the order originated in the Upper Carboniferous or Lower Permian (Mamay 1976). Cycad-like leaves occur in the Lower and Middle Triassic, and the order seems to have become fairly well established during the Upper Triassic (Norian stage). Examples of Upper Triassic cycad leaves and reproductive structures include *Bjuvia*, *Palaeocycas*, *Leptocycas*, and *Pseudoctenis* (Fig. 2.4f). *Pseudoctenis* also occurs in the Lower Jurassic in large numbers (Harris 1961). Several cycad trunks, such as *Lyssoxylon* and *Charmorgia*, have been described from the Upper Triassic, but none are known from the Lower Jurassic (see Ash 1984).

Voltziales

This extinct order is often called the "transition conifers" because the reproductive structures of the plants included in it seem to be intermediate between those of certain Paleozoic plants and of modern conifers (Miller 1977). Most of the genera in this order are based on cones and cone scales. Their foliage is poorly known, but what has been studied suggests to Miller (1977) that at least some of these plants looked something like the "Norfolk Island Pine," *Araucaria excelsa*.

Voltziaceae

This order appears to have been present during the Permian if not earlier (Stockey 1982). Characteristic Upper Triassic members found in the Northern Hemisphere include the cones and associated foliage referred to *Voltzia* and *Pseudovoltzia* (Fig. 2.5a). Another common Upper Triassic member, *Swedenborgia*, is widespread in the Northern Hemisphere from Greenland to China (Florin 1960). It also extends into the Lower Jurassic in the same region. A less common member of the order is *Aethophyllum*, which is only known from the Upper Triassic of France (Miller 1982).

Figure 2.5. Representative Upper Triassic and Lower Jurassic Voltziales (**a, b**), Taxaleas (**c**), Coniferales (**d–o**). **a**, *Pseudovoltzia*. **b**. *Hirmerella*. **c**, *Palaeotaxus*. **d**, *Mataia*. **e**, *Rissikia*. **f**, *Elatocladus*. **g**, *Schizolepis*. **h**, *Araucarites*. **i**, *Palissya*. **j**, *Stachyotaxus*. **k**, *Palissya*. **l**, *Stachyotaxus*. **m, n**, *Pagiophyllum*. **o**, *Podozamites*. All scales for (**a**)–(**l**) = 1 cm; for (**m**)–(**n**) scales = 0.2 cm.

Cheirolepidiaceae

This widespread Upper Jurassic family may have arisen earlier because its characteristic pollen grains, *Corollina* (*Classopollis* of some authors), are found in Upper Triassic rocks, although the foliage and reproductive structures are not known from this time. A common cone that is assigned to this family is *Hirmerella* (Fig. 2.5b). Most other parts of the *Hirmerella* plant, such as the leafy shoot, bark, pollen, and seed cones and the pollen, are known (Watson 1982).

Taxales

The oldest fossils that can be attributed to this family with any confidence are some leafy shoots and fruits called *Paleotaxus* (Fig. 2.5c) in the Lower Jurassic of Sweden (Florin 1958). According to Miller (1977), the family did not expand appreciably until the Middle Jurassic.

Coniferales

Details of the early Mesozoic history of the conifers are still somewhat unclear because the fossil evidence needed to clarify it is scarce and often difficult to interpret (Miller 1977). There is fairly strong evidence indicating that several living and one or more extinct families of conifers were present during the Upper Triassic. The living families found in the Upper Triassic apparently had arisen earlier in that period, and a few may be represented in the Lower Jurassic.

Podocarpaceae

This family arose during the Lower Triassic of the southern continents and was still confined to that region in the Upper Triassic (Miller 1977). *Mataia* (Fig. 2.5d) and *Rissikia* (Fig. 2.5e) from the Upper Triassic of Australia and South Africa certainly belong to the Podocarpaceae. *Mataia* also ranges into the Lower Jurassic in Australia and New Zealand (Townrow 1967).

Araucariaceae

Although a considerable number of early Mesozoic fossils have been referred to this family, the evidence for such an assignment is often weak (Miller 1977). Most of the Upper Triassic and Lower Jurassic fossils that have been assigned to the Araucariaceae are pieces of secondary wood and have usually been referred to the form genus *Araucarioxylon*. It should be noted in this regard that the assignment of fossil wood to that genus does not automatically mean that the wood actually came from a plant that is in the Araucariaceae. Similar wood also occurs in other coniferous families, and it may be a basic wood type in the order (J. A. Doyle pers. comm.). As noted above, the same type of wood occurs in the Glossopteridaceae (Gould and Delevoryas 1977). Impressions and compressions of other plant parts that might belong in the family occur in the Upper Triassic and Lower Jurassic, with cone and cone parts being placed in *Araucarites* (Fig. 2.5h) and foliage in certain species of the genera *Pagiophyllum* (Figs. 2.5m,n) and *Brachyphyllum* (Stockey 1982).

Cupressaceae
There is a little fossil evidence suggesting that this family was present in the Upper Triassic and Lower Jurassic. It includes some leafy twigs, small pieces of wood, and detached cones from the Upper Triassic of France that appear to be cupressaceous (Lemoigne 1967). Also there is foliage from the Lower Jurassic of Israel that has been assigned to this family by Chaloner and Lorch (1960), but the material is reminiscent of the Cheirolepidiaceae (Alvin 1982).

Taxodiaceae
The geological record of this family is sparse, and there is little solid evidence that it evolved until the Middle Jurassic (Miller 1977). Some of the foliage called *Elatocladus* (Fig. 2.5f) from the Lower Jurassic of Sweden and east Greenland may represent this family (Florin 1958).

Pinaceae
Evidence for this family in pre-Cretaceous rocks is somewhat weak (Miller 1977). Some small seed cones called *Compsostrobus*, some associated pollen cones containing the pollen *Alisporites*, and foliage similar to *Elatocladus* (Fig. 2.5f) in the Upper Triassic of North Carolina have been assigned to the family (Delevoryas and Hope 1973). Also some leafy shoots of the *Pityocladus* type and associated cone scales called *Schizolepis* (Fig. 2.5g) that occur in the Upper Triassic (Rhaetian stage) of Sweden have been assigned to the family by Seward (1919). Nothing that appears to belong to the Pinaceae except for the foliage called *Elatocladus* has been described from the Lower Jurassic.

Palissyaceae
This extinct family is represented at a few Upper Triassic (Rhaetian stage) and Lower Jurassic localities in Europe and east Greenland (Florin 1958). The fossils assigned to this family include the seed scales and foliage called *Stachyotaxus* (Figs. 2.5j, l) and *Palissya* (Figs. 2.5i, k).

An uncomfortably large group of unassigned coniferous fossils occur in the Upper Triassic and Lower Jurassic. These fossils cannot be easily accommodated in any generally accepted family for a variety of reasons (Miller 1977). Most of these unassigned fossils are sterile foliage of various types. Some of them compare fairly closely with those of certain living species of *Araucaria, Pinus*, and other living conifers, but in the absence of additional data, such as associated reproductive structures, they cannot be assigned to any particular family. Until a few years ago, authors rather arbitrarily and inconsistently dealt with this problem by assigning the fossils to various living and extinct groups. Then in 1969, Harris recommended that sterile conifer foliage be assigned to one of eight genera that he thinks represent most kinds of sterile conifer foliage. Most authors now follow his suggestion. The more common genera now used for sterile Upper Triassic and Lower Jurassic coniferous foliage are *Pagiophyllum* (Figs. 2.5m, n), *Brachyphyllum, Elatocladus* (Fig. 2.5f), and *Podozamites* (Fig. 2.5o). Other common unassigned coniferous fossils from the Upper Triassic and Lower Jurassic are the pollen cones *Masculostrobus* and *Pityanthus* and the seed cones *Callipitys*.

Ginkgoales
After having arisen during the Upper Carboniferous or Lower Permian, this ancient order expanded rapidly during the Triassic, and by the end of the period included at least seven genera with numerous species (Florin 1936). Certain Late Triassic and Early Jurassic fossils so closely resemble the leaf of the living *Ginkgo* (Fig. 2.6a) that they are placed in that genus or in *Ginkgoites*. Two other common Upper Triassic and Lower Jurassic genera that are placed in this family are *Baiera* (Fig. 2.6b) and *Sphenobaiera* (Dorf 1958). During the Lower Jurassic, the order seems to have expanded still further with the appearance of a number of new species. By this time, the order was worldwide in distribution.

Czekanowskiales
The fossils now placed in this order were formerly referred to the Ginkgoales. However, in 1951, Harris reported that the leaves of this group were associated with a bizarre seed-bearing structure that

Figure 2.6. Representative Upper Triassic and Lower Jurassic Ginkgoales (**a, b**) and Czekanowskiales (**c, d**). **a**, *Ginkgo*. **b**, *Baiera*. **c**, *Czekanowskia*. **d**, *Lepidostrobus*. Scales = 1 cm.

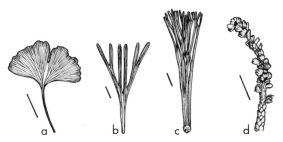

differed significantly from that found in the Ginkgoales, and a few years later they were placed in a separate family, the Czekanowskiales, by Pant (1959). The leaves placed in this family, *Czekanowskia* (Fig. 2.6c) and *Solenites*, are long and linear and are borne in bundles. The seed-bearing structure, *Leptostrobus*, is a cone with bivalved capsules (Fig. 2.6d). The relationships of this order to the Ginkgoales and to other plants seems to be remote. This order seems to be restricted to the Northern Hemisphere, where it first appears in the Upper Triassic of Siberia (Krassilov 1970). The order expanded into western Europe and adjacent areas during the Rhaetic and was present there in small numbers in the Lower Jurassic (Vakhrameev et al. 1970).

Distribution

The geographic distribution of Upper Triassic and Lower Jurassic floras has been discussed by Barnard (1973), Dobruskina (1982), and Vakhrameev et al. (1970) in connection with their comprehensive discussion of the general distribution of Mesozoic floras. Earlier Florin (1960) discussed the distribution of the conifers throughout the geological record. As these authors show, Upper Triassic and Lower Jurassic floras are fairly abundant and widely distributed in the Northern Hemisphere, particularly in Europe and adjacent areas in Russia, but rather scarce and unevenly distributed in the rest of the world. The abundance of known floras in the Northern Hemisphere is partly biased by the number of workers in the region in contrast to the situation in the rest of the world, and also, of course, it is related to the distribution of fossiliferous Upper Triassic and Lower Jurassic rocks. Barnard's predrift map shows that the floras are absent in the very high paleolatitudes (above 65°) and in a broad belt in the tropics between about 20° north and 30° south.

Several limited attempts have been made to identify floral zones in the Upper Triassic and Lower Jurassic. One of the earliest attempts was made by Harris (1931) when he recognized a zone of *Lepidopteris ottois* in the Rhaetic in western Europe and an overlying zone, the zone of *Thaumatopteris schenki*, in the lower part of the Lias. His zones have been recognized as being fairly worldwide in extent. More recently, Ash (1980) has recognized two floral zones in the Carnian stage of the Upper Triassic of North America: a zone of *Eoginkgoites* and an overlying zone of *Dinophyton*. So far, neither zone has been recognized outside of North America.

By the Late Triassic, the strong provinciality of the land floras that had existed during the late Paleozoic, particularly in North America (Read and Mamay 1964), had almost entirely disappeared, although the northern and southern floras were still quite distinct (Barnard 1973). The northern floras can be divided into a northerly or "Siberian" floristic region and a southerly or "Eurasian" floristic region (Vakhrameev 1964). Although the geographical division between them is not sharp, the differences are major. Barnard (1973) suggests that the differences may be due to some environmental cause other than latitudinal effect.

Paleoecological considerations

It is generally accepted that plant fossils, like living plants, reflect the climate under which they grew. Thus the occurrence of water-loving plants, such as lycopods, horsetails, and ferns, in the Late Triassic and Early Jurassic suggests that the climate was probably moist and warm to hot at that time. The cycads and ginkgos in these fossil floras probably lived in nearby slightly higher and drier areas. The remains of still more xeric plants, such as the conifers, do occur in those floras, but the fossils are often fragmentary, suggesting that they were transported some distance from much drier upland areas. These interpretations seem to be confirmed by the fine structures of the epidermises on the leaves. The abundance of ferns and other plants with large leaves as high as a paleolatitude of 60° indicates that a warm, moist climate extended much closer to the poles than it does nowdays.

Many of the floral niches that had been left vacant after the mass extinctions of the Permian were still being filled during the Upper Triassic, a process that continued into the Jurassic (Fredericksen 1972). Thus, a large number of new forms evolved during the Upper Triassic and Lower Jurassic. Most of them represented old groups, but some represented new ones. These resulting Mesophytic floras were transitional between the ancient or Paleophytic flora of the Paleozoic and the modern or Cenophytic flora of the Upper Cretaceous and Cenozoic (Gothan and Weyland 1954).

Discussion

The plant megafossils found in Upper Triassic and Lower Jurassic strata show little interruption at the Upper Triassic – Lower Jurassic boundary. Many of the changes that did occur involved the seed ferns. For example, the Glossopteridaceae, Peltaspermaceae, and the Corystospermaceae apparently became extinct at or near the end of the Triassic. On the other hand, another family of seed ferns, the Caytoniaceae, seems to have expanded somewhat during the Lower Jurassic after arising in the Rhaetic. One family of conifers, the Palissyaceae, also arose in the Rhaetic and expanded later in the Jurassic. Several fern families, including the Osmundaceae, Matoniaceae, and Dipteridaceae, continued to increase in numbers and variety in the

Lower Jurassic as did the Bennettitales. Possibly some new families of conifers, such as the Cupressaceae and Pinaceae, arose in the Upper Triassic, but the evidence for this is fairly weak. The Taxales and the Taxodiaceae appear to have arisen during the Lower Jurassic.

The palynomorphs extracted from the strata above and below the Triassic–Jurassic boundary in Somerset at the proposed (George et al. 1969) type section of the basal zone of the Jurassic system have been examined recently by Fisher (1981), who reported that there are no major palynological breaks that correlate with the boundary. This was also reported earlier at a number of other localities where the boundary is preserved in England (Fisher 1972), Austria, Germany (Schuurman 1979), and Greenland (Pederson and Lund 1980).

Conclusions

Plant megafossils and palynomorphs do not show a significant change at the Triassic–Jurassic boundary. In fact, a comparison of the floras on either side of the boundary shows that they are rather similar. A few families became extinct in the Upper Triassic, and a few families arose just before the end of the period in the Rhaetic. A few others originated in the Lower Jurassic. Many families that were present during the Upper Triassic expanded during the Lower Jurassic or remained unchanged. In general, the fossils show that the plants were slowly evolving with one assemblage gradually giving away to another during the Upper Triassic and Lower Jurassic and that the floras were slowly becoming more modern in aspect.

Acknowledgments

I am grateful to C. N. Miller, Jr. (Missoula, Montana) and J. A. Doyle (Davis, California) for their constructive comments about this report. Research was supported by the Earth Sciences Section of the National Science Foundation (Grant EAR–8218054).

References

Alvin, K. L. 1982. Cheirolepidiaceae: Biology, structure and paleoecology. *Rev. Palaeobot. Palynol.* 31: 71–98.

Anderson, J. A., and Anderson, H. M. 1983. Palaeoflora of southern Africa. *Molteno Formation*, Vol. 1, Parts 1 and 2. (Rotterdam: A. A. Balkema).

Arnold, C. A. 1953. Origin and relationships of the cycads. *Phytomorphology*. 3:51–65.

1964. Mesozoic and Tertiary fern evolution and distribution. *Mem Torrey Bot. Club* 21:58–66.

Ash, S. R. 1972. Late Triassic plants from the Chinle Formation in northeastern Arizona. *Palaeontology* 15:598–618.

1980. Upper Triassic floral zones of North America. *In*

Dilcher, D. L., and T. N. Taylor (eds.), *Biostratigraphy of Fossil Plants*. Dowden (Stroudsburg, Pennsylvania: Hutchinson and Ross), pp. 153–70.

1981. Glossopterid leaves from the early Mesozoic of northern Mexico and Honduras. *Paleobotanist* 28–9:201–6.

1984. A short thick cycad stem from the Upper Triassic of Petrified Forest National Park, Arizona and vicinity. *Mus. N. Ariz. Bull.* 54:17–32.

Barnard, P. D. W. 1965. Flora of the Shemshak Formation, part 1, Liassic plants from Dorud. *Riv. Ital. Paleont.* 71:1123–68.

1973. Mesozoic floras. *Palaeont. Assoc. Spec. Pap. Palaeont.* 12:175–87.

Baxendale, R. W., and Baxter, R. W. 1977. A new fertile schizaeceous fern from Middle Pennsylvanian Iowa coal balls. *Univ. Kansas Sci. Bull.* 51:283–9.

Chaloner, W. G., and Lorch J. 1960. An opposite-leaved conifer from the Jurassic of Israel. *Palaeontology* 2:236–42.

Corsin, P. and Waterlot, M. 1979. Paleobiogeography of the Dipteridaceae and Matoniaceae of the Mesozoic. *4th Intern. Gondwana Symp.* 1:51–70.

Delevoryas, T., and Hope, R. C. 1973. Fertile coniferophyte remains from the Late Triassic Deep River Basin, North Carolina. *Am. J. Bot.* 60:810–18.

Delevoryas, T. and Person, C. P. 1975. *Mexiglossa varia* gen. et sp. nov., A new genus of glossopteroid leaves from the Middle Jurassic of Oaxaca, Mexico. *Palaeontographica B* 180:82–119.

Dobruskina, I. A. 1974. Triassic lepidophytes. *Paleontol. J.* 3:384–97.

1975. The role of peltaspermacean pteridosperms in Late Permian and Triassic floras. *Paleontol. J.* 4:536–49.

1982. Triassic floras of Eurasia. *Acad. Sci. U.S.S.R. Trans.*, 365:1–196 (in Russian).

Dorf, E. 1958. The geological distribution of the Ginkgo family. *Bull. Wagner Free Inst. Sci.* 33(1):1–10.

Fisher, M. J. 1972. The Triassic palynofloral succession in England. *Geosci. Man* 4:101–9.

1981. Palynology and the Triassic/Jurassic boundary. *Rev. Palaeobot. Palynol.* 34:129–35.

Florin, R. 1936. Die Fossilen Ginkgophyten von Franz-Joseph-Land. I. Spezieller Teil. *Palaeontographica B* 81:71–173.

1958. On Jurassic Taxads and conifers from northwestern Europe and eastern Greenland. *Acta Horti Bergiana* 17:259–402.

1960. The distribution of conifer and taxad genera in time space. *Acta Horti Bergiana* 20:121–312.

Fontaine, W. M. 1883. Contributions to the knowledge of the older Mesozoic flora of Virginia. *U.S. Geol. Survey Monogr.* 6:1–144.

Frederiksen, N. O. 1972. The rise of the Mesophytic flora. *Geosci. Man* 4:17–28.

George, T. N., et al. 1969. Recommendations on stratigraphic usage. *Proc. Geol. Soc. London* 1656:139–66.

Gothan, W., and Weyland, H. 1954. *Lehrbuch der Palaobotanik* (Berlin: Akademie-Verlag).

Gould, R., and Delevoryas, T. 1977. The biology of

Glossopteris: evidence from petrified seed-bearing and pollen-bearing organs. *Alcheringa* 1:387–99.

Harris, T. M. 1931. The fossil flora of Scoresby Sound, east Greenland. Part 1. Cryptogams (Exclusive of Lycopodiales). *Meddelelser om Gronland.* 85(2):1–104.

——— 1951. The fructification of *Czekanowskia* and its allies. *Phil. Trans. Royal Soc. London B* 235:483–508.

——— 1961. The fossil cycads. *Palaeontology* 4:313–23.

——— 1964. The Yorkshire Jurassic flora. II. *Caytoniales, Cycadales and Pteridosperms* (London: British Museum) (Nat. Hist.).

——— 1969. Naming a conifer. *In* Santarau, H., et al. (eds.), *J. Sen Memorial Volume.* (Calcutta, India: Botanical Society Bengal), pp. 243–52.

——— 1970. The stem of *Caytonia. Geophytology* 1(1):23–9.

——— 1973. *The Strange Bennettitales, Nineteenth Sir Albert Charles Seward Memorial Lecture* (Lucknow, India: Birbal Sahni Institute of Paleobotany).

Herbst, R. 1972. *Gleichenites potrerillensis* n. sp. del Triasico medio de Mendoza (Argentina), con comentarios sobre las Gleicheniaceae fosiles de Argentina. *Ameghiniana.* 9(1):17–22.

——— 1977a. Sobre Marattiales (Filicopsidae) Triasicas de Argentina y Australia. Parte I. El género *Asterotheca. Ameghiniana* 14:1–18.

——— 1977b. Sobre Marattiales (Filicopsidae) Triasicas de Argentina y Australia. Parte II. Los géneros *Danaeopsis* y *Rienitsia. Ameghiniana* 14:19–32.

Holt, E. L. 1947. Upright trunks of *Neocalamites* from the Upper Triassic of western Colorado. *J. Geol.* 55:511–13.

Krassilov, V. A. 1970. Approach to the classification of Mesozoic "Ginkgoalean" plants from Siberia. *Paleobotanist* 18:12–19.

Lemoigne, Y. 1967. Paleoflore a Cupressales dans le Trias–Rhétien du Contentin. *Compt. Rend. Acad. Sci. Paris* 264:715–18.

Mamay, S. 1976. Paleozoic origin of the Cycads. *U.S. Geol. Survey. Prof. Pap. 934.*

Miller, C. N. 1977. Mesozoic conifers. *Bot. Rev.* 43:218–80.

——— 1982. Current status of Paleozoic and Mesozoic conifers. *Rev. Palaeobot. Palynol.* 37:99–114.

Nathorst, A. G. 1878. Om Floran Skanes Kolforande Bildningar. *Sver. Geol. Unk., Ser. C* 27: 1–56.

Oishi, S., and Yamasita, K. 1936. On the fossil Dipteridaceae. *J. Fac. Sci. Hokkaido Imper. Univ. Ser. 4* 3:135–84.

Pant, D. D. 1959. The classification of gymnospermous plants. *Paleobotanist* 6:65–70.

Pedersen, K. R., and Lund, J. J. 1980. Palynology of the plant-bearing Rhaetian to Hettangian Kap Stewart Formation, Scoresby Sund, east Greenland. *Rev. Paleobot. Palynol.* 31:1–16.

Petriella, B. 1979. Sinopsis de las Corystospermaceae (Corystospermales, Pteridospermophyta) de Argentina. I. Hojas. *Ameghiniana* 16:81–102.

Raciborski, M. 1894. Flora Kopalna ogniotrwalych glinek Krakowskich. Czesc. 1. Rodniowce (Archaegoniatae). *Pam. mat.-przyr. Akad. Um., Krakow* 18:142–243.

Read, C. B., and Mamay, S. 1964. Upper Paleozoic floral zones and floral provinces of the United States. *U.S. Geol. Survey, Prof. Pap. 454-K.*

Retallack, G. J. 1977. Reconstructing Triassic vegetation of eastern Australasia: a new approach for the biostratigraphy of Gondwanaland. *Alcheringa.* 1:247–77.

Schopf, J. M. 1973. The contrasting plant assemblages from Permian and Triassic deposits in southern continents. (Mem. *Can. Soc Petr. Geologists* 2: 379–97.

Schuurman, W. M. L. 1979. Aspects of Late Triassic Palynology. 3. Palynology of latest Triassic and earliest Jurassic deposits of the northern limestone Alps in Austria and southern Germany, with special reference to a palynological characterization of the Rhaetian stage in Europe. *Rev. Paleobot. Palynol.* 27:53–75.

Seward, A. C. 1919. *Fossil Plants*, Vol. 4 (Cambridge: Cambridge University Press).

Silva-Pinada, A. 1981. Asterotheca y plantas asociadas de la Formación Huazchal (Triasico Superior) del Estado de Hidalgo. *Univ. Nat. Auton. Mex. Inst. Geol. Rev.* 5:47–54.

Stewart, W. N. 1983. *Paleobotany and the Evolution of Plants* (Cambridge: Cambridge University Press).

Stockey, R. A. 1982. The Araucariaceae: an evolutionary perspective. *Rev. Paleobot. Palynol.* 37:133–54.

Taylor, T. N. 1981. *Paleobotany, An Introduction to Fossil Plant Biology* (New York: McGraw-Hill).

Thomas, H. H., and Bancroft, N. 1913. On the cuticle of some recent and fossil cycadean fronds. *Trans. Linnean Soc. Lond. Ser. B* 8:155–204.

Townrow, J. A. 1960. The Peltaspermaceae, a pteridosperm family of lower Mesozoic age. *Palaeontology* 3:333–61.

——— 1967. On *Rissikia* and *Mataia* podocarpaceous conifers from the lower Mesozoic of southern lands. *Pap. Proc. Roy. Soc. Tasmania* 101:103–6.

Vakhrameev, V. A. 1964. Jurassic and Early Cretaceous floras of Eurasia and the palaeofloristic provinces of this period. *Trans. Geol. Inst. Acad. Sci. U.S.S.R. Moscow* 102:1–261 (in Russian).

Vakhrameev, V. A., Dobruskina, I. A. Zaklinskaja, E. D., and Meyen, S. V. 1970. Palaeozoic and Mesozoic floras of Eurasia and phytogeography of this time. *Trans. Geol. Inst. Akad. Sci. USSR Moscow* 208:1–424 (in Russian).

Watson, J. 1982. The Cheirolepidiaceae: a short review. *In* Nautiyal, D. D. (ed.) *Phyta: Studies on Living and Fossil Plants* (Allahabad, India: Society of Plant Taxonomists), pp. 265–73.

Webb, J. A. 1982. Triassic species of *Dictyophyllum* from eastern Australia. *Alcheringa* 6:79–91.

II Late Triassic vertebrate taxa and faunas

3 Thoughts on the origin of the Theropoda

SAMUEL P. WELLES

The subject of this discussion is the origin of the Theropoda; therefore, the first objective must be to present my concept of the definition of this taxon. Very briefly, theropods are bipedal dinosaurs of a basically carnivorous stock. They had, as Barsbold (1983), stressed, a great potential to diversify. They had a mesotarsal ankle joint. As noted by Colbert (1964) and others, they had developed a dolichoiliac pelvis (with an enlargment of the anterior blade of the ilium) and, I might add, a tendency to increase the height of the ilium.

Sir Richard Owen (1842) coined the word Dinosauria to include the giant fossil reptile remains that were being discovered in southern England. Marsh (1881b) included in the Dinosauria the five suborders Sauropoda, Stegosauria, Ornithopoda, Theropoda, and Hallopoda, with a sixth suborder Coeluria questionably included. The term Theropoda, meaning "beast foot," was new; the group was characterized as "carnivorous, digitigrade, with prehensile claws; pubes co-ossified in front; post-pubis present; vertebrae more or less cavernous; limb bones hollow" (Marsh 1881b, p. 423). It included only the Allosauridae. Some seven years later, Seeley (1888) noted that the Dinosauria could be divided into two major groups: the Ornithischia with a "bird-like" pelvis, and the Saurischia with a "reptilian" triradiate pelvis. Seeley's terms are used in all modern classifications, but Owen's Dinosauria is often dropped, as, for example, by Romer (1966), who treated the Saurischia and Ornithischia as orders of the Subclass Archosauria; but Camp et al. (1972) retained Dinosauria as a superorder of the Infraclass Archosauria.

Whether or not we retain the term Dinosauria in our classification is a matter of judgment. I agree with Camp et al. that because the name is so widely known that it should be used if at all practical, and also because it specifies a name for a natural group. We can at least include the Theropoda as a suborder of the Order Saurischia, and in this we can find almost unanimous agreement. Most of us also agree that theropods are bipedal saurischians that first appeared in the Middle Triassic and became dominant in the Jurassic, carrying on and diversifying throughout the Cretaceous, a time span of some 170 MY.

In terms of dominance, there was a time in the Late Triassic when the great carnivorous pseudosuchian archosaurs – phytosaurs, rauisuchids, poposaurids, and others – preyed upon all else, including the labyrinthodont amphibians, the rhynchosaurs, the aetosaurs, and the therapsids. Primitive theropods were present, but it is hardly conceivable that these agile little reptiles could have competed with the carnivorous pseudosuchians, much less have been large enough to have caused their extinction. Rather, toward the end of the Triassic, the labyrinthodonts dwindled, and this great biomass was eliminated from the food chain; the last of them disappeared by the earliest Jurassic. It almost seems as though the pseudosuchians ate themselves out of existence by killing off the labyrinthodonts. They could also be blamed for the extinction of the rhynchosaurs, therapsids, aetosaurs, and other large-bodied, slow-moving reptiles. The turtles, which first appear in the Late Triassic, might have been saved by their armor, while the theropods and other bipeds were agile enough to escape. Throughout the Jurassic and Cretaceous, there can be little doubt that the theropods dominated, and were capable of preying upon or scavenging all terrestrial life.

Although the first theropods must have appeared in Middle Triassic times, we have no conclusive record of them. When we get into the Late Triassic, we have a relatively bountiful record that, however, created a very special problem during the early days of vertebrate paleontology. There were carnivorous teeth and jaws, but none was definitely associated with a postcranial skeleton. In contrast, the many cranial elements that were associated with postcranial material were of a kind called prosauropod or plateosaurid. These were moderately large bipedal reptiles, and Huene considered them quite distinct from the smaller, lightly built saurischians. He therefore separated off the latter as the Suborder Coelurosauria (Huene 1914). [It might be noted that Marsh (1881a) had long ago coined the term Coeluria for similar forms.] In contrast to his Coelurosauria, Huene grouped the great bulk of the Saurischia in his Suborder Pachypodosauria to include as three "Familien Hauptkreis" (which I would translate as infraorders): the Carnosauria, Prosauropoda, and Sauropoda. Huene considered the Prosauropoda the direct ancestors of the Sauropoda, but this concept was greatly weakened by the work of Attridge (1963) and Charig, Attridge, and Crompton (1965), who found better ancestral forms in South Africa, thus leaving the plateosaurs as an aberrant end group. However, later work has eclipsed that of Charig et al., and nearly everyone now believes that the ancestors of Sauropoda are to be found within the "Prosauropoda," so that the Sauropodomorpha, including both groups, is itself a natural group.

The great similarity between the postcranial skeletons of the Late Triassic carnivores and herbivores that had given Huene so much trouble was emphasized by Colbert (1964) when he resurrected the Suborder Paleopoda to include both kinds. His Paleopoda was of equivalent rank to the Suborders Theropoda and Sauropoda. Within his Paleopoda, he included the carnivores in the Palaeosauria and the herbivores in the Plateosauria. Unfortunately, Palaeosauria was preoccupied, and so for it Colbert (1970) substituted Teratosauria. (It should be noted here that Huene's *Teratosaurus* has since been recognized by many as a rauisuchian pseudosuchian, very distant from dinosaurs but with many convergent features, and the term "Teratosauria" in the context of dinosaurs is consequently misleading. However, much thought about the post-Triassic radiation of theropods historically depended on the recognition of the "Teratosauria" as true, large theropods of the Late Triassic, and that is how we will regard them here.)

Here, then, in the Late Triassic forms our concept of Theropoda goes back either through the "Teratosauria" or through the more lightly built forms that Huene separated as Coelurosauria. Whichever we do will satisfy the two main characteristics of the group: bipedality and a great potential for diversification.

Huene's wide separation of the Coelurosauria from the other Saurischia has been made less obvious by the later discovery of taxa that are intermediate between the Coelurosauria and the other theropoda. This has led Ostrom (1969, 1978) in his studies of *Deinonychus* and *Compsognathus* to return to the old Marshian category of Theropoda to include even the coelurosaurs. Barsbold (1977, 1983) has likewise lumped all the carnivorous bipeds into the Order Theropoda, although he did separate them into Small Theropods and Large Theropods. His 1983 classification is comprehensive and includes the bizarre Late Cretaceous forms from Mongolia. My own study of *Dilophosaurus* and related taxa (Welles 1984) led me to agree with Ostrom and Barsbold in returning to the old concept of Theropoda to include all the many diverse carnivorous bipeds.

If we accept this all-inclusive concept of the Theropoda, and that they originated in the carnivorous bipeds of the Late Triassic (Colbert's "Teratosauria"), we can now ask, what was the origin of the "Teratosauria"?

The time-honored concept is of an origin in the Late Permian from forms such as *Archosaurus*, which is the first known representative of the Proterosuchia, and also of the Archosauria. The Proterosuchia then gave rise to the Pseudosuchia, each of which was summarized in the *Handbuch der Palaeoherpetologie*, by Charig and Sues (1976) and by Krebs (1976), respectively. There was an increased trend to bipedality, and the crocodiloid tarsus of the Pseudosuchia was thought by some to have evolved into the hinged mesotarsal ankle of the theropods. Thus, the Pseudosuchia gave rise to the "Teratosauria" in the Late Triassic.

This was a nice, neat concept. However, protests were raised by those of us who did not believe that the crocodiloid tarsus of the Pseudosuchia, with its astragalus a part of the crus and its calcaneum a part of the pes, could evolve into the theropod mesotarsal tarsus. Among the dissenters has been Krebs (1963), who even more strongly later (Krebs 1976) united the Pseudosuchia and Crocodilia in the Order Suchia, denying their derivation from the known proterosuchians. Hughes (1963) also denied the possibility of evolving the dinosaur tarsus from the crocodiloid tarsus of the Pseudosuchia. Welles and Long (1974), after analyzing the theropod tarsus, came to a similar conclusion.

The crocodiloid tarsus is a highly developed and specialized structure in which a peg and socket develops between the astragalus and the calcaneum so that the two can rotate against each other around this peg (Fig. 3.1). The calcaneum develops a heel and has a convex proximal surface for the fibula. A concave spout also protrudes medially behind the peg from the astragalus. The proximal surface of the astragalus is a spiral, as is the distal end of the tibia. This tarsus was thus highly adapted to a rotation of the vertical crus upon the fixed pes, the rotation raising the crus. I contend that this is so specialized a structure, universal among Pseudosuchia and Crocodilia, that its possessors were locked into the various gaits of the crocodiles and could never develop the bipedality that is so characteristic of the theropods. I, therefore, conclude that the Pseudosuchia (= "crocodile-normal" tarsus-bearing archosaurs of many authors) was a blind alley, giving rise only to the Crocodilia, because their tarsus is so highly specialized in the wrong direction for the Theropoda.

This leaves us with the Proterosuchia and perhaps some taxa that were wrongly included in the Pseudosuchia. These would have a primitive tarsus with a tendency for the astragalus to form a cup below the tibia and to develop a distal roller against the distal tarsals: in short, a mesotarsal hinge. The calcaneum would likewise develop a concave proximal facet for the fibula and a distal roller. It would move together with the astragalus as extensions of the crus, rolling over the distal tarsals in a parasagittal vertical plane. This would satisfy our requirement of bipedality, and it is thus among the Proterosuchia with a noncrocodilian tarsus that we will find the ancestors of the Theropoda.

I should like to consider further the development of the high ilium based upon the long "dolichoiliac" type identified by Colbert. This increase in height is observed in all later theropods. The attainment of bipedality was possible with the acquisition of the mesotarsal joint by the theropods, and I believe that the increase in height of the ilium is connected with the continued development of bipedality.

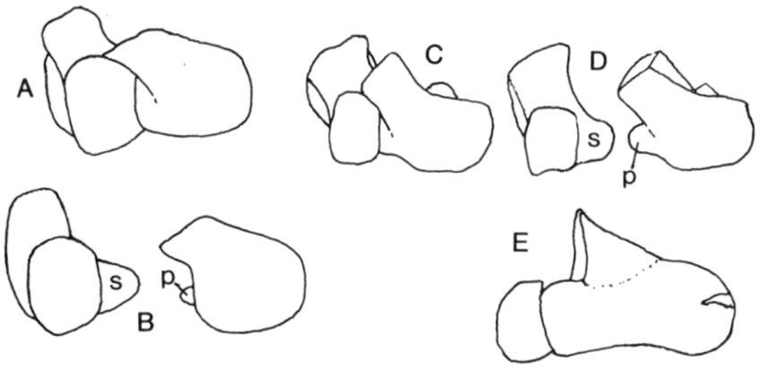

In any animal, the two primary functions of a limb are to hold up the body and to move it. In a biped, both these functions are relegated entirely to the hindlimb, and so special problems are generated. Most, if not all, previous studies of theropod pelvic musculature have been concerned with the fore-and-aft rotation of the leg. This involves the muscles that effect the propulsion and recovery strokes of the femur, and these analyses treat the leg as a pendulum swinging freely from the acetabulum.

This, to me, is the wrong approach to the study of hip muscle function. Instead, the problem should consider the foot as the fulcrum, the basis for action; the limb is a pillar, a lever, rocking in a parasagittal plane upon this base.

When standing still, the two legs of a biped support the body from the two acetabula as though it were slung between them. However, when movement begins, one foot must be lifted from the ground and that support of the sling is lost. There is now a tendency for the sling, the body, and the lifted leg to rotate downward. Thus, one of the problems of a slowly walking biped is to prevent this downward rotation, or collapse of the sling. It is therefore necessary for much of the pelvic musculature, especially those muscles originating on the iliac blade, to prevent this collapse, and to do this throughout the forward movement of the body.

This tendency to collapse is somewhat compensated by inertia, and is also countered by shifting the body and fixed acetabulum sideways in order to bring the center of gravity more nearly above the foot. The total stress is thus reduced, but there is still a considerable rotational force to be supported at the fixed acetabulum. There is also an evolutionary tendency to counter this collapse by narrowing the hips and bringing the acetabula closer together.

This need to prevent the collapse of the body and free leg around the fixed acetabulum is solved by the development of powerful muscles that originate on the iliac blade and insert on the outside of the leg. Three of these muscles are very important: the M. iliotibialis, originating along the entire top of the ilium; the M. iliofemoralis, originating on the anterior blade below the former; and the M. iliofibularis, originating behind the latter (Fig. 3.2). These muscles are ideally situated to prevent the collapse, but would have little propulsive power. The main propulsion comes from muscles originating on the posterior dorsal vertebrae (M. puboischiofemoralis) and the anterior caudals (M. caudifemoralis).

Figure 3.1. **A**, Anterior view of right tarsus of a pseudosuchian archosaur (*Ticinosuchus*); **B**, the two bones disarticulated, showing the peg on the astragalus (p) that inserts in the calcaneum, and the spout on the calcaneum (s) directed medially behind the peg; **C** and **D**, corresponding views of crocodile tarsus; **E**, anterior view of a theropod dinosaur tarsus (*Dilophosaurus*): dotted line is the suture with ascending process. **A–D** from Krebs 1976. Astragalus is to the right. Scales vary.

As body size increased, the strain on these anticollapse muscles became greater. This was countered in the larger Theropoda by an increase in the height of the ilium, affording larger areas for the origins of these muscles. A comparable increase in height of the ilium is also evident in the brachyiliac Plateosauroidea. However, this group failed to develop an anterior blade on the ilium, and so was specialized in the opposite (brachyiliac) direction from the Theropoda.

Another, and perhaps even more important, function of these anticollapse muscles is to prevent side sway of the hips while walking or running, and especially while changing direction. As support shifts from side to side or direction changes, there is a tendency for the hip to sway outward. This outward pressure becomes stronger as pace angulation diminishes and the feet are brought nearer the midline. This is countered, as in the body collapse, by the external iliac muscles, especially the M. iliotibialis.

In conclusion, I suggest that we seek the origin of the Theropoda in the Proterosuchia. They might already have been bipedal, but, if not, would have rapidly become so. I would not expect their lineage to continue through the Pseudosuchia, with their aberrant crocodiloid tarsus, or through the "Teratosauria," with their specialized ilium. They must come through dolichoiliac bipeds that had a theropod tarsus and a tendency to increase the height of the ilium. They also carried a great potential for diversification.

Figure 3.2. Diagrammatic view of the pelvic region of a theropod in anterior view, showing functions of the M. iliotibialis (hatched) and the M. iliofemoralis (stippled) in presenting collapse of the body when the other hindlimb is lifted. The M. iliofibularis is hidden.

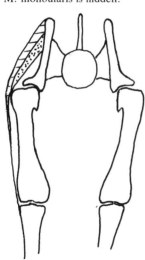

References

Attridge, J. 1963. The Upper Triassic Karroo deposits and fauna of southern Rhodesia. *S. Afr. J. Sci.* 59: 242–7.

Barsbold, R. 1977. K evolyutsii khishchnykh dinozavrov. Pp. 48–56 in Barsbold et al., 1977 (q.v.).

　　1983. Carnivorous dinosaurs from the Cretaceous of Mongolia *Trans. Joint Soviet–Mongolian Paleontol. Exped.* 19: 5–119 [in Russian].

Barsbold, R., E. I. Vorob'yeva, and B. Luvsandanzan (eds.) 1977. Fauna, flora i biostratigrafiya Mezozoya i Kainozoya Mongolii *Sovmestnaya Sov.-Mong. Nauchno-Issled. Geol. Ekxped., Tr.* 4: 1–171 [in Russian].

Camp, C. L., R. H. Nichols, B. Brajnikov, E. Fulton, and J. A. Bacskai. 1972. Bibliography of fossil vertebrates 1964–1968. *Geol. Soc. Am. Mem.* 134.

Charig, A. J., J. Attridge, and A. W. Crompton. 1965. On the origin of the sauropods and the classification of the Saurischia. *Proc. Linn. Soc. London* 176: 197–221.

Charig, A. J., and H.-D. Sues. 1976. Suborder Proterosuchia Broom 1906b. *In* Kuhn, O. (ed.), *Handbuch der Palaeoherpetologie*, Vol. 13, *Thecodontia* (Stuttgart; Gustav Fischer), pp. 11–39.

Colbert, E. H. 1964. Relationships of the saurischian dinosaurs. *Am. Mus. Novit.* 2181: 1–24.

　　1970. A saurischian dinosaur from the Triassic of Brazil. *Am. Mus. Novit.* 2401: 1–39.

Huene, F. 1914. Saurischia and Ornithischia. *Geol. Mag.* 1(6): 444–5.

Hughes, B. 1963. The earliest archosaurian reptiles. *S. Afr. J. Sci.* 59: 221–41.

Krebs, B. 1963. Bau und Funktion des Tarsus eines Pseudosuchiers aus der Trias des Monte San Giorgio (Kanton Tessin, Schweiz). *Palaeontol. Z.* 37: 88–95.

　　1976. Pseudosuchia. *In* Kuhn, O. (ed.), *Handbuch der Palaeoherpetologie*, Vol. 13, *Thecodontia.* (Stuttgart: Gustav Fischer), pp. 40–95.

Marsh, O. C. 1881a. New order of extinct Jurassic reptiles (Coeluria). *Am. J. Sci.* 21(3): 339–40.

　　1881b. Principal characters of American Jurassic dinosaurs. Part 5. *Am. J. Sci.* 21(3): 417–23.

Ostrom, J. H. 1969. Osteology of *Deinonychus antirrhopus*, an unusual theropod from the Lower Cretaceous of Montana. *Yale Univ., Peabody Mus. Nat. Hist., Bull.* 30: 1–165.

　　1978. The osteology of *Compsognathus longipes* Wagner. *Zitteliana.* 4: 73–118.

Owen, R. 1842. Report on British fossil reptiles. Part II. *Rept. Brit. Assoc. Adv. Sci., Plymouth*, 11: 60–204.

Romer, A. S. 1966. *Vertebrate Paleontology*, 3rd ed. (Chicago: University of Chicago Press).

Seeley, H. G. 1888. On the classification of the fossil animals commonly named Dinosauria. *Proc. Roy. Soc. London* 43: 165–71.

Welles, S. P. 1984. *Dilophosaurus wetherilli* (Dinosauria, Theropoda): osteology and comparisons. *Palaeontographica. A* 185: 85–180.

Welles, S. P., and R. A. Long. 1974. The tarsus of theropod dinosaurs. *Ann. S. Afr. Mus.* 64: 191–217.

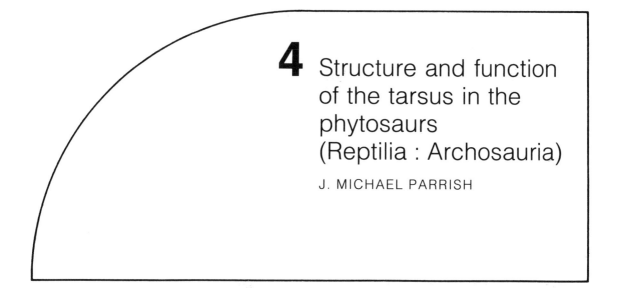

4 Structure and function of the tarsus in the phytosaurs (Reptilia : Archosauria)

J. MICHAEL PARRISH

Introduction

The phytosaurs, comprising the archosaurian Suborder Parasuchia, are distinguished by distinctive cranial morphology, with elongate rostra and dorsally placed external nares. They are restricted stratigraphically to the Upper Triassic, with the possible exception of *Mesorhinus*, which Jaekel (1910) claimed came from the Lower Triassic of Germany.

Although phytosaurs are among the most common of Late Triassic archosaurs, the morphology of their tarsus has not been well documented. Huene (1926) tentatively identified an isolated element (USNM 2160) as a right astragalus. Although the element is not clearly identifiable from the figures, it is definitely not a phytosaur astragalus. Case (1929) identified UMMP 7445 as a phytosaur astragalus, and UCMP 10604 as a phytosaur calcaneum. The figures were apparently reversed, as the former is a calcaneum, whereas the latter is an astragalus. Furthermore, both belong to rauisuchians. Camp (1930) identified an element (UCMP 27008) as a left astragalus of the phytosaur *Rutiodon adamanensis*. The element identified as an astragalus is instead the broken distal end of the tibia. Thus, none of the publications on phytosaurs prior to the 1970s correctly identified the tarsi of phytosaurs.

Chatterjee (1978) published a description of two articulated skeletons of the phytosaur *Parasuchus* from the Maleri Formation of India, including the first record of articulated phytosaur tarsi in the literature. However, his description does not provide a basis for thorough functional studies.

The purpose of this chapter is to describe the phytosaur tarsus on the basis of abundant material from the southwestern United States and to consider the implications of the structure of the tarsus and pes for locomotor abilities.

Materials and methods

Phytosaur tarsal material is fairly common in the Late Triassic formations of North America. However, because tarsals are most frequently found as isolated elements, much of the known material has been misidentified or unlabeled in museum collections. Study of two associated phytosaur skeletons, heretofore undescribed, has facilitated identification of the tarsal elements. USNM 18313 is a largely complete but disarticulated skeleton of *Rutiodon* from the Chinle Formation (Upper Triassic) of Apache County, Arizona. Both sets of proximal tarsals are known and are very well preserved. These elements will form the main basis of the description of the phytosaur tarsus. UCMP V2816/27235 is an articulated skeleton collected in 1930. It has a complete right foot with an articulated astragalus and distal tarsus, although the calcaneum is missing. Disarticulated phytosaur tarsal elements are also in the collections of the University of California Museum of Paleontology, the Texas Memorial Museum, and the Museum of Northern Arizona.

The functional implications of the structure of the phytosaur tarsus were considered by studying the morphology of the best preserved material and comparing the observed structures with those seen in other early archosaurs and in modern crocodilians and lizards. The mechanics of the tarsus were considered in light of the structure of the rest of the hindlimb.

Description

Because the phytosaur tarsus has never been fully described, I will discuss it in some detail here. As mentioned above, the best preserved, associated phytosaur tarsal material from North America is from an associated *Rutiodon* skeleton from the

Chinle Formation (USNM 18313) (Figs. 4.1–4.3), which has both proximal tarsi in articulation. The other phytosaur tarsal material from the Chinle and Dockum Formations all exhibits the basic morphology described below.

The proximal tarsi are flat, broad elements. The calcaneum consists of a broad body proximally and an elongate, posterolaterally directed tuber. The proximal face of the bone is occupied by a large,

Figure 4.1. Right proximal tarsus of *Rutiodon* (USNM 18313), dorsal view. Scale = 1 cm. A = astragalus; C = calcaneum; a = anterior intratarsal facets; p = posterior intratarsal facets; t = calcaneal tuber; ff = fibular facets; pf = perforating foramen homologue; tf = tibial facet.

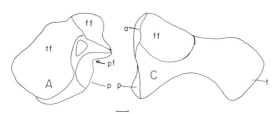

Figure 4.2. Right proximal tarsus of *Rutiodon*. USNM 18313. Ventral view. Scale = 1 cm. See Figure 4.1 for definition of labels.

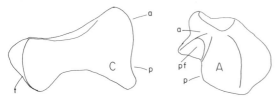

ovoid fibular facet that angles down gently to the anterior margin of the bone and is bordered along its medial and posteromedial borders by a curving groove that somewhat undercuts the edge of the fibular facet. Otherwise, the proximal surface of the bone is rather smooth and featureless, expanded medially by the posterior of the two facets for astragalocalcaneal articulation and concave posteromedially as the anteriorly expanded body thins into the tuber.

The distal end of the tuber is expanded mediolaterally and somewhat dorsoventrally. As opposed to the condition in crocodile-normal archosaurs, there is no groove on the distal face of the tuber. The lateral face of the bone is concave in proximal view. The ventral surface of the bone is rugose anteriorly, forming a gently beveled, flat surface for distal tarsal articulation. The posterior part of the ventral surface forms the ventral margin of the tuber.

The astragalus can be divided into body and neck, with the neck much narrower than the body in proximal view. On the lateral margin of the neck is a broad fibular facet, triangular in shape. This facet is angled anterolaterally. The bulk of the body of the astragalus is occupied by a broad, rectangular tibial facet that is medially horizontal, but curves gently proximally at its posterolateral extent. The posteroproximal border of the astragalar neck is occupied by a triangular concavity, whereas the anterior margin of the neck contains a rhomboidal concavity that faces anterolaterally. The posterior part of the body is occupied by a deep recess ventral to the neck that forms the remnant of the perforating foramen by the posteromedially projecting "ball" for

Figure 4.3. Right proximal tarsus of *Rutiodon*. USNM 18313. **A**, posterior view; **B**, anterior view. Scale = 1 cm. See Figure 4.1 for definition of labels.

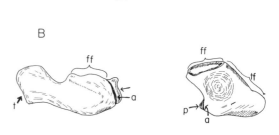

calcaneal articulation and by a broadly rounded, roughened surface underlying the body of the astragalus. Anteriorly, the body of the astragalus is occupied by a steeply sloping, posteroventrally angled surface for articulation with the distal tarsus and metatarsus. In ventral view, the body of the astragalus is broadly rounded. A thin ridge occupies the ventral surface of the neck adjacent to the anterior and posterior concavities, flaring to form a T-shaped surface below the ventral margin of the fibular facet.

The articulation between astragalus and calcaneum consists of two pairs of facets. Anteriorly, there is a flat, very gently concave surface on the ventrolateral edge of the neck of the astragalus. This facet articulates with a smooth, convex surface on the lateral edge of the anterior surface of the calcaneum. Just posterior to the proximal facet for calcaneal articulation on the astragalus is a deep semicircular notch descending anteroventrally and corresponding to the astragalar part of the perforating foramen in primitive diapsids. The calcaneum does not contribute to a perforating foramen in phytosaurs. Instead, a flat surface is seen on the medial calcaneum opposite the surface where the homologue of the perforating foramen of the astragalus would be. It separates the proximal astragalar facet from the concave posterior facet for astragalar articulation. Posteroventral to the perforating foramen homologue on the astragalus is the second calcaneal facet: a broadly rounded, posterolaterally directed ball. It articulates with a rounded but shallow socket on the posterior part of the medial margin of the calcaneum. The line of articulation between the astragalus and calcaneum in phytosaurs is sharply angled in an anterolateral direction relative to the mediolateral long axis of the astragalus. The two pairs of facets for articulation between the proximal tarsals lie in the same horizontal plane.

Chatterjee's (1978) illustrations of the *Parasuchus* tarsus from the Maleri Formation of India suggest a proximal tarsus more crocodile-like than that described here. However, my examination of photographs and casts of the Indian *Parasuchus* tarsal material indicates that it is fundamentally similar in structure to those of the North American forms. Thus, our present state of knowledge indicates a single basic morphotype for the phytosaur proximal tarsus.

The distal tarsus of phytosaurs is known in the *Parasuchus* specimens from India (ISI R 42 and 43) and from the articulated *Rutiodon tenuis* specimen from the Chinle on display at the University of California at Berkeley (UCMP V2816/27235). It consists of a large, wedge-shaped fourth distal tarsal and a smaller cuboidal third tarsal.

Complete phytosaur pedes in articulation are known from the specimens of *Parasuchus* described

by Chatterjee (1978) from the Maleri Formation of India (Fig. 4.4a). No complete North American phytosaur pes is known in articulation, although a metatarsus with a few phalanges articulated is known in UCMP V2816/27235 (Fig. 4.4b).

Five pedal digits are present, and the *Parasuchus* specimens from the Maleri Formation have a phalangeal formula of 2–3–4–5–4 (Fig. 4.4a). The metatarsus is similar in basic morphology to that of *Proterosuchus*, with a hooked fifth metatarsal articulating proximally to the other four metatarsals. The proportions of metatarsals are somewhat different than in *Proterosuchus*. The third and fourth digits are the longest and are subequal in length. The second is somewhat shorter, the fifth is shorter still, and the first metatarsal is the shortest.

In phytosaurs, the ridge projecting from the posteroventral corner of the fifth metatarsal is broader than in *Proterosuchus* and projects posterolaterally rather than posteriorly. The other four metatarsals are expanded proximally, especially on their lateral sides where they underlie their medial counterparts. In *Parasuchus*, the ungual phalanges decrease in relative size from the first to the fifth and have a distinct longitudinal groove laterally (Chatterjee 1978).

Functional analysis

In his discussion of the tarsus of *Parasuchus*, Chatterjee (1978) stated that the phytosaur tarsus was functionally crurotarsal, and of the crocodile-normal type. Thus, the main ankle joint would be between the proximal tarsi, by a ball on the astragalus articulating with a socket on the calcaneum. My examination of the material from North America

Figure 4.4. Phytosaur left pedes. Scale = 1 cm. **A**, *Parasuchus*, Indian Statistical Institute ISI 42. Traced from photographs provided by the Indian Statistical Institute. **B**, *Rutiodon tenuis*, UCMP V2816/27235. Shaded digits on the *Rutiodon* pes restored after *Parasuchus*; calcaneum is from USNM 18313, scaled to the UCMP astragalus.

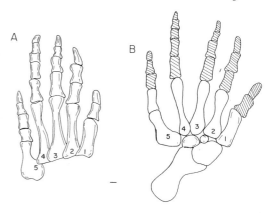

and casts of the Indian material indicates a much more restricted degree of movement between astragalus and calcaneum than is seen in archosaurs, with the main ankle joint between the proximal tarsi.

In crocodilians, the calcaneum rotates about the axis formed by the ball-and-socket joint between the proximal tarsi. In phytosaurs, although a shallow but conventional ball and socket of the crocodile-normal type comprises the posterior of the two pairs of facets between proximal tarsi, marked rotation of the calcaneum on the astragalus was prevented by the close articulation of the anterior facet pair (Fig. 4.5). Although a modest degree of plantarflexion of the calcaneum on the astragalus was apparently possible, the major movement of the pes on the crus at this joint was a twisting motion, with the calcaneum rotating on the posterior astragalar ball while sliding anteromedially along the anterior facet of the astragalus. With modest plantarflexion of the calcaneum, the neck of the astragalus locks against the body of the calcaneum just medial to the calcaneal fibular facet, preventing any further movement at the joint. Similarly, the anterior intratarsal facets abut each other when the bodies of the two bones are horizontal, preventing any dorsiflexion of the foot at the astragalocalcaneal joint.

The joint between astragalus and tibia consists of two gently saddle-shaped facets. By analogy with the ligament patterns in lizards and crocodilians (Brinkman 1979), the two bones were probably held together immovably by ligaments in life. As in crocodilians, the joint between astragalus and fibula was more flexible. It is evident from the shape of the fibular facet of the calcaneum that modest movement

of the calcaneum on the astragalus was possible. The facet is elongated somewhat anteroposteriorly, allowing for continuous contact between the facet and the fibula with pedal flexion, although the roller-like fibular facet structure seen in crocodilians, rauisuchians, and aetosaurs is not seen. The rounded aspect of the distal part of the fibula and its flat distal calcaneal facet indicates that the fibula was free to rotate on the calcaneum, allowing long-axis rotation of the crus. However, the fibular facet of the astragalus is markedly angled such that it is inclined proximomedially. The normal attitude of the fibula, and thus the crus as a whole, was a marked medial inclination (Fig. 4.7C). In crocodilians, although the fibular facet of the astragalus is inclined slightly in this manner, it lacks the extreme angulation seen in phytosaurs.

The phytosaur calcaneal tuber is different from those in crocodilians in that it is both relatively more elongate and directed strongly laterally, in contrast to the posterolateral orientation in crocodilians. In crocodilians (and, presumably, in other crocodile-normal archosaurs), the calcaneal tuber serves as a lever around which the pedal plantarflexors, notably the M. gastrocnemius, travel from the posterior side of the crus to the plantar surface of the foot. The lateral direction of the tuber allows the plantarflexors to angle down from the crus and travel anteriorly to the plantar surface of the foot. The angulation of the plantarflexors at the tuber helps them to rotate the pes around the long axis of the crus as well as to plantarflex the foot with contraction.

The limited range of flexion and extension possible at the astragalocalcaneal joint indicates that most ankle movement was achieved between the proximal and distal tarsi in phytosaurs. The joints between proximal and distal tarsi were of the simple roller type (Fig. 4.4).

Joints between distal tarsi and metatarsi, between metatarsi and phalanges, and among phalanges are all of the simple roller type, mainly permitting flexion and extension.

In considering the functional capabilities of the tarsus and pes in phytosaurs, it is also necessary to consider briefly the structure and function of the proximal limb and pelvis (Figs. 4.6–4.8). The hip joint in phytosaurs is characterized by a shallow, laterally directed acetabulum and by a poorly defined femoral head, which is not separated from the shaft of the femur by a distinct neck. This type of joint, analogous to that observed in lizards (e.g., Snyder 1954), permits long-axis rotation of the femur in the acetabulum as well as protraction and retraction of the femur. The absence of a prominent femoral head or a pronounced supraacetabular lip on the ilium suggests that significant femoral adduction [an important component of the erect "high walk"

Figure 4.5. Posterior views of articulated *Rutiodon* tarsus, showing full range of potential rotation between proximal tarsal elements. Scale = 1 cm. **A**, Fully dorsiflexed position; (1) indicates where the anterior facet pair abut, preventing further dorsiflexion (partially obscured by calcaneum). **B**, Fully plantarflexed position; (2) indicates that further plantarflexion of the calcaneum is prevented by contact between the neck of the astragalus and the fibular facet of the calcaneum.

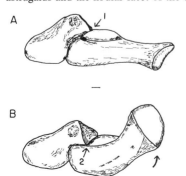

seen in crocodilians (e.g., Brinkman 1980)], was not probable in phytosaurs.

In the knee joint, a broad distal femur with small, distally directed condyles articulated with the broad, flat ends of the proximal tibia and fibula. Thus, the joint had no apparent osteological specializations that restricted movement to any extent.

Long-axis rotation about both the femur and crus were osteologically possible, whereas no structures are seen that could facilitate femoral adduction or movement of the limbs in a parasagittal plane. The absence of the latter group of specializations suggests that phytosaurs did not employ the erect gait seen in dinosaurs (e.g., Charig 1972), pterosaurs (Padian 1983), and in some "thecodontians" (e.g., Bakker 1971; Bonaparte 1983; Parrish 1983, unpubl. obs.) and crocodylomorphs (Crush 1984).

The "semiimproved" (e.g., Charig 1972; Chatterjee 1978) or "semierect" (e.g., Bakker and Galton 1974; Parrish 1983) gait in archosaurs is exemplified by that of modern crocodilians, in which two step cycles are commonly employed: one essentially a sprawling gait, the other an approximation of a parasagittal erect gait termed the "high walk." Chatterjee (1978) suggested a crocodilian, semiimproved gait for the phytosaur *Parasuchus*, citing two chief points as evidence. First, he noted expansion of the pelvis in phytosaurs relative to the plate-like condition in proterosuchians. Second, he noted that phytosaurs have a crocodile-like femur. The pelvis in phytosaurs is best represented by a three-dimensional specimen of the genus *Rutiodon* from the Dockum Formation of Crosby County, Texas, housed at the University of Michigan Museum of Paleontology (UMMP 7244) (Fig. 4.6). In this spec-

Figure 4.6. *Rutiodon* pelvis (UMMP 7224). Scale = 1 cm. **A**, Left lateral view. **B**, Anterior view of articulated pubes.

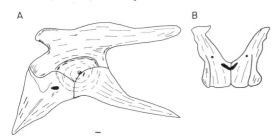

Figure 4.8. Limb movement in *Rutiodon*. **A**, Lateral view of restored right hindlimb at beginning of propulsive phase. As inferred, the propulsive phase incorporated the following rotatory movements: 1, rotation of the femur about its long axis; 2, retraction of the femur in a near-horizontal plane; 3, rotation of the crus about its long axis; 4, rotation of the pes about the long axis of the crus. **B**, Limb at end of propulsive phase, lateral view. **C**, Anterior view of right hindlimb, to show marked angling of crus when articulated with tarsus. Pes is omitted for clarity. Scale = 5 cm.

Figure 4.7. *Rutiodon* limb bones. **A,B**, Left femur. After MacGregor (1906). Scale = 1 cm. **A**, Dorsal view. **B**, Ventral view. **C,D**, Left tibia (UCMP 34236) and fibula (UCMP 70/4–7). Scale = 1 cm. **C**, Anterior view; **D**, posterior view. f = fourth trochanter, t = iliofibularis trochanter.

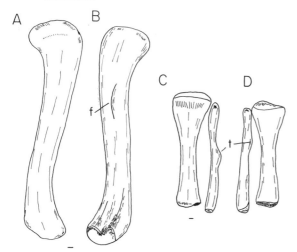

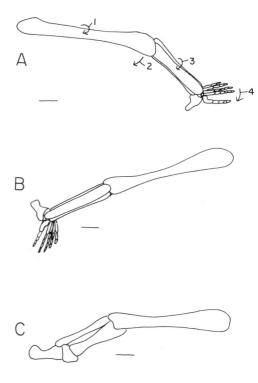

imen, the ventral pelvic elements form a continuous plate, interrupted only by a short gap between the two elements below the acetabulum. Comparison with the pelvis of the early archosaur *Proterosuchus* (e.g., Cruickshank 1972) indicates that no marked downward expansion of the ventral elements has taken place in *Rutiodon*. Thus, the origins of the femoral protractors and retractors at the distal ends of the pubis and ischium are located near the axis of femoral long-axis rotation, and could cause such rotation in addition to protracting and retracting the femur (Parrish 1983).

The phytosaur femur (Fig. 4.7) is superficially quite similar to that of crocodilians. However, significant differences are present. In phytosaurs, the distal condyles of the femur are directed distally and are poorly developed. This is the primitive condition for archosaurs, as seen in *Proterosuchus* and *Erythrosuchus* (Parrish 1983). In crocodilians, the condyles project ventrally as well as distally and are much better defined. Another point of difference is the degree of torsion seen in the femoral shaft, as defined by the angle formed by the long axes of the proximal and distal ends of the bone. Measurement of a large sample of phytosaur and crocodilian femora (Parrish 1983) demonstrated that the femora of phytosaurs display much more marked torsion than do those of crocodilians. Torsion of the shaft is favored when the femur and crus meet at a marked angle relative to the vertical plane, because this allows long-axis rotation of the femur to be translated into protraction and retraction of the crus. Torsion of the femur, coupled with the angulation of the distal end of the femur, allows flexion of the knee joint while the femur is held horizontal (Fig. 4.7).

Thus, the predominant mode of terrestrial locomotion in phytosaurs appears to have been a sprawling gait similar to those observed in lizards and inferred in the earliest archosaurs. This locomotor pattern was characterized by long-axis rotation of the femur and crus, and of the pes about the long axis of the crus (Fig. 4.7). The major ankle joint was of the mesotarsal type, between proximal and distal tarsi. However, a limited degree of twisting of the calcaneum on the astragalus was also possible, probably employed during pushoff of the pes from the substrate.

Phytosaurs have been interpreted as semiaquatic (e.g., Camp 1930; Parrish and Long 1983, and manuscript in prep.), and thus it is worth considering how the hindlimb might have been used in aquatic locomotion. The tails of *Parasuchus* and *Rutiodon* are characterized by high neural spines, and the mediolaterally compressed, dorsoventrally elongated profile of the tail suggests that it may have been the primary locomotory organ for aquatic locomotion. The horizontally positioned femora and flexible hindlimb joints in phytosaurs could have allowed the hindlimb to have been used in swimming. However, no hypertrophy of the number or length of the phalanges is seen in known phytosaur pedes, so particular aquatic specializations of the hindlimb for swimming are not seen. Chatterjee's (1978) interpretation that the tail was the main aquatic locomotory organ and that the limbs were largely passive in swimming seems to be a reasonable interpretation of the morphological evidence.

Trackways referred to the Phytosauria

Baird (1957) tentatively suggested that the footprint taxon *Apatopus* from the Late Triassic of the Newark Supergroup of the eastern United States (Olsen 1980) had phytosaurian affinities (Fig. 4.9). Furthermore, the stratigraphic ranges of phytosaurs and *Apatopus* (late Carnian–Rhaetian) are similar in the Newark Supergroup (Olsen 1980). However, as will be shown below, the morphology of the *Apatopus* tracks is inconsistent with the known pedal morphology of phytosaurs.

Baird (1957) cited similarities in gait pattern between *Apatopus* tracks and those of a modern alligator using the "high walk" and suggested that the animal responsible for *Apatopus* had crocodile-like limb morphology and function. He cited similarities between MacGregor's (1906) restorations of the manus and pes of *Mystriosuchus* as supportive evidence for the assignment.

Figure 4.9. Phytosaur pedal morphology and *Apatopus* footprints. Right pes in dorsal view. **A**, Hypothetical *Rutiodon* pes print based on the composite foot shown in Fig. 4.4B. Dashed circular line indicates possible pad formed by bodies of astragalus and calcaneum. **B**, *Apatopus* footprint from Milford, New Jersey. **B** is after Baird (1957). Scale = 1 cm.

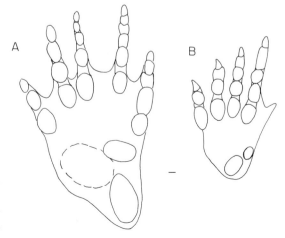

However, comparisons of *Apatopus* and the articulated phytosaur material from India and North America do not support Baird's view. *Apatopus* pes impressions are notable (but not unique) among trackways attributed to archosaurs in that they display a fourth digit that is longer than the third (Fig. 4.9). This condition is primitive for diapsids and is displayed in the archosaurs *Proterosuchus* (Cruickshank 1972) and *Erythrosuchus* (Cruickshank 1978), as well as in other nonarchosaurian Triassic diapsids such as rhynchosaurs (e.g., Benton 1983) and *Trilophosaurus* (Gregory 1945). In the articulated *Parasuchus* pedes from India, a more typical archosaurian pedal pattern is seen, with the third digit somewhat longer than the fourth (Fig. 4.4A). No articulated pedes with complete digits are known from North America, so the differences observed in digital length between *Apatopus* and *Parasuchus* do not entirely preclude the possibility that the *Apatopus* trackmaker could have been of one of a North American form displaying different digital proportions.

Other morphological characters that vary between *Apatopus* and known phytosaur material involve the relative size and position of the fifth metatarsal and the calcaneum. In *Apatopus*, both the fifth metatarsal and the calcaneum are represented by small, ovoid pads, in close proximity to one another. Figure 4.9A represents an inferred phytosaur track as reconstructed from the restored *Rutiodon* foot. The sturdy proximal part of the fifth metatarsal of *Rutiodon* makes a much larger, mediolaterally elongate pad impression than the nearly circular impression seen for the fifth metatarsal in *Apatopus*. The relatively massive, posterolaterally directed calcaneal tuber in *Rutiodon* makes a large, ovoid pad impression as well. Even with the tarsus positioned such that the calcaneal tuber is somewhat posteriorly directed, the pad that would be made by the tuber is posterior and slightly lateral to the fifth metatarsal pad, as opposed to the posteromedial position in *Apatopus*.

Although comparisons are made between a restored *Rutiodon* pes and *Apatopus*, the parts of the foot that reflect the greatest differences between the two (i.e., the fifth metatarsal and the tarsus) are well known from the Chinle Formation. UCMP V2816/27235 has articulated metatarsals and an astragalus, whereas both proximal tarsi in articulation are known from USNM 18313.

The fact that *Apatopus* retains the primitive diapsid digital proportions eliminates the possibility that it could be the trackway of most well known, comparably sized Late Triassic quadrupedal archosaurs, such as aetosaurs or rauisuchians. No archosaurs with this type of foot are known beyond the Middle Triassic (Parrish 1983). However, two other groups of large Late Triassic diapsids retain a relatively elongate fourth digit. Trilophosaurids are known from the Dockum (Gregory 1945) and Chinle (Jacobs and Murry 1980; R. A. Long pers. comm.). However, the aberrant manus of *Trilophosaurus* effectively eliminates the described species *T. buettneri* as a candidate for the *Apatopus* trackmaker.

Rhynchosaurs are known from the Late Triassic Wolfville (Baird 1963) and Dockum (Chapter 9) formations of North America, as well as from numerous other Late Triassic formations worldwide (Benton 1983). Rhynchosaur mani are not commonly known in articulation. The rhynchosaur manus is frequently restored with the fourth digit longer than the third (e.g., Woodward, 1907). However, in *Hyperodapedon* (Benton 1983, Fig. 31g), the phalangeal series of the fourth digit is shorter than that of the third digit, suggesting a symmetrical manus not unlike that seen in *Apatopus*. Thus, the possibility that *Apatopus* could have been made by a rhynchosaur cannot be discounted at this point. Because the *Apatopus*-type foot is primitive for archosaurs, the possibility remains that it was made by a large archosaur that retained this condition and had a symmetrical manus as well. The forelimbs of early archosaurs are not generally well known, so it is unclear how widespread the symmetrical manus is throughout the basal archosaurs.

If the hypothesis presented here regarding a sprawling gait in phytosaurs is correct, it also offers an explanation why true trackways of phytosaurs have not been previously recognized. If the power stroke of the hindlimb included rotation of the pes about the long axis of the crus with limb retraction, such pedal rotation could obscure any clear pedal impresssion that might be made by a stationary foot. Thus, phytosaur trackways might be preserved but not clearly identifiable, owing to the lack of clear digital impressions. However, it should be noted that in modern lizards, such as the Komodo dragon, clear pedal impressions are made because rotation of the pes about the long axis of the crus takes place after the pes has pushed off the ground in the propulsive stroke (Padian and Olsen 1984; Parrish, unpubl. obs.).

Phylogenetic implications of phytosaur tarsal structure

Several recent studies (e.g., Cruickshank 1979; Thulborn 1980; Brinkman 1981; Bonaparte 1983; Chatterjee 1982) have presented phylogenies of the "Thecodontia" and other archosaurian groups based on variations in the morphology of the distal tarsus. The consensus of some of these studies (e.g., Brinkman 1981; Chatterjee 1982; Bonaparte 1983) is that there are two divergent lineages within Archosauria, with one derived group displaying "crocodile-normal" tarsi and the other displaying "crocodile-reverse" tarsi. Each of these functional crurotarsal joint types can be derived from the basic diapsid tarsal type as seen in *Proterosuchus*, in which two pairs of concave–convex facets are seen between astragalus and calcaneum. Although the two facet pairs in *Proterosuchus* and some other early diapsids prevent any movement between the two elements, the dorsal of the two pairs can be shown to be homologous to the facets in the crocodile-reverse tarsi, whereas the ventral pair correspond to the facets participating in crocodile-normal tarsi (e.g., Parrish 1983, and manuscript in prep.).

In phytosaurs, both of the tarsal facets present in *Proterosuchus* are well developed. However, the ventral facet pair has developed into a ball-and-socket joint of the crocodile-normal type. Although the joint is shallow and the range of movement at the joint is limited, the presence of the crocodile-normal joint may nonetheless be considered a derived character shared by the phytosaurs and the crocodile-normal archosaurs. However, the retention of the well-developed dorsal facet pair in phytosaurs, which is not present in any of the crocodile-normal forms, indicates that the Phytosauria may be the sister group of the remaining crocodile-normal taxa.

Any phylogeny based primarily on a single character, such as the proximal tarsus, must be regarded as very tentative, and subject to revision in the face of other characters. A phylogenetic analysis of the "Thecodontia" as a whole, employing other characters, is underway (Parrish, manuscript in prep.).

Conclusions

The tarsus of the phytosaurs, previously incompletely known, is described on the basis of abundant material from the Late Triassic of the southwestern United States. Articulation between the astragalus and calcaneum is by two pairs of articular facets, similar to the configuration in *Proterosuchus*. Although limited movement was possible between the astragalus and calcaneum, the predominant ankle movement was between proximal and distal tarsi. The morphology of the remainder of the hindlimb and pelvis in phytosaurs is consistent with a sprawling gait, as observed in modern lacertilians and early diapsids such as *Proterosuchus*. The ichnotaxon *Apatopus*, tentatively identified as a phytosaur trackway (Baird 1957), is morphologically inconsistent with the articulated pedes of the phytosaurs *Parasuchus* and *Rutiodon*. However, pedal morphology of trilophosaurids and rhynchosaurs is consistent with that of *Apatopus*. The partial development of a crocodile-normal crurotarsal joint in phytosaurs, coupled with their retention of the prim-

itive archosaur condition of two sets of articular facets between astragalus and calcaneum, suggests that phytosaurs may be the sister group of the crocodile-normal archosaurs.

Abbreviations of cited repositories

UCMP, University of California Museum of Paleontology, Berkeley

UMMP, University of Michigan Museum of Paleontology, Ann Arbor

USNM, United States National Museum, Smithsonian Institution

Acknowledgments

For loan of specimens, I would like to thank Dr. N. H. Hotton of the U.S. National Museum and Dr. Kevin Padian of the University of California Museum of Paleontology. Robert Long of the UCMP kindly granted me access to material he had under study. I would also like to thank Dr. Wann Langston, Jr. of the Texas Memorial Museum and Mike Morales of the Museum of Northern Arizona for permission to study material in their institutions. Dr. Tappan Roy-Chowdhury of the Indian Statistical Institute was kind enough to send me photographs and casts of the Maleri *Parasuchus* specimen. This work has benefited from my discussions with Robert Bakker, Donald Brinkman, Sankar Chatterjee, Robert Long, and Paul Olsen. Thanks are also extended to Donald Baird, Donald Brinkman, Kevin Padian, and John Zawiskie for their careful reviews of the manuscript.

References

Bakker, R. T. 1971. Dinosaur physiology and the origin of mammals. *Evolution* 25: 636–58.

Bakker, R. T. and P. M. Galton. 1974. Dinosaur monophyly and a new class of vertebrates. *Nature (London)* 248: 168–72.

Baird, D. 1957. Triassic reptile footprint faunules from Milford, New Jersey. *Bull. Mus. Comp. Zool.* 117:449–520.

1963. Rhynchosaurs in the Late Triassic of Nova Scotia. *Geol. Soc. Amer., Spec. Pap.* 73:107.

Benton, M. J. 1983. The Triassic reptile *Hyperdapedon* from Elgin: functional morphology and relationships. *Phil. Trans. Roy. Soc. London B* 302:605–720.

Bonaparte, J. F. 1981. Classification of the Thecodontia *Géobios Mem. Spec.* 6: 99–112.

1983. Classification of the Thecodontia. *Géobios Mem. Spec.* 6:49–55.

Brinkman, D. 1979. The structural and functional evolution of the diapsid tarsus. Ph.D. dissertation, McGill University.

1980. The hindlimb step cycle of *Caiman sclerops* and the mechanics of the crocodile tarsus and metatarsus. *Can. J. Zool.* 58:2187–200.

Camp, C. L. 1930. A study of the phytosaurs, with a description of new material from western North America. *Mem. Univ. Calif.* 10:1–174.

Case, E. C. 1929. Description of the skull of a new form of phytosaur, with notes on the characters of described North American phytosaurs. *Mich. Univ. Mus. Pal. Mem.* 2:1–56.

Charig, A. J. 1972. The archosaur pelvis and hindlimb: an explanation in functional terms. *In* Joysey, K. A., and T. S. Kemp (eds.), *Studies in Vertebrate Evolution* (New York: Winchester Press)r, pp. 121–56.

Chatterjee, S. 1978. A primitive parasuchid (phytosaur) reptile from the Upper Triassic Maleri Formation of India. *Palaeontology* 21:83–127.

Cruickshank, A. R. I. 1972. The proterosuchian thecodonts, *In* Joysey, K. A., and T. S. Kemp (eds.), *Studies in Vertebrate Evolution* (Oliver and Boyd, Edinburgh), pp. 89–119.

  1978. The pes of *Erythrosuchus africanus* Broom. *Zool. J. Linn. Soc. London* 62:166–77.

  1979. The ankle joint in some early archosaurs. *S. Afr. J. Sci.* 75:168–78.

Crush, P. J. 1984. A Late Upper Triassic sphenosuchid crocodilian from Wales. *Paleontology*, 27:131–57.

Gregory, J. T. 1945. Osteology and relationships of *Trilophosaurus*, University of Texas Publ. No. 4401 pp. 273–359.

Huene, F. F. von. 1926. Notes on the age of the continental Triassic beds in North America, with remarks on some fossil vertebrates. *Proc. U.S. Nat. Mus.* 69:1–10.

Jacobs, L. L., and P. A. Murry. 1980. The vertebrate community of the Triassic Chinle Formation near St. John's, Arizona. *In* Jacobs, L. L. (ed.), *Aspects of Vertebrate History* (Flagstaff, Arizona: Museum Northern Arizona Press), pp. 55–72.

Jaekel, O. 1910. Uber einen neuen belodonten aus der Buntsandstein der Bernberg. *Gessel. Naturf. Freunde Berlin, Sitzungber.* 1910:197–229.

MacGregor, J. H. 1906. The Phytosauria, with especial reference to *Mystriosuchus* and *Rhytiodon*. *Mem. Am. Mus. Nat. Hist.* 9:27–100.

Olsen, P. E. 1980. A comparison of vertebrate assemblages from the Newark and Hartford Basins (Early Mesozoic, Newark Supergroup) of eastern North America. *In* Jacobs, L. L. (ed.), *Aspects of Vertebrate History* (Flagstaff, Arizona: Museum of Northern Arizona Press), pp. 35–54.

Padian, K. 1983. A functional analysis of flying and walking in pterosaurs. *Paleobiology* 9:218–39.

Padian, K., and P. E. Olsen. 1984. Footprints of the Komodo Monitor and the trackways of fossil reptiles. *Copeia* 1984:662–71.

Parrish, J. M. 1983. Locomotor adaptations in the hindlimb and pelvis of the Thecodontia (Reptilia : Archosauria). Ph.D. Thesis, Dept. of Anatomy, University of Chicago.

Parrish, J. M., and R. A. Long, 1983. Vertebrate paleoecology of the Late Triassic Chinle Formation, Petrified Forest and vicinity, Arizona. *Geol. Soc. Am. Abst. Prog.* 15:285.

Snyder, R. C. 1954. The anatomy and function of the pelvic girdle and hindlimb in lizard locomotion. *Am. J. Anat.* 95:1–46.

Thulborn, R. A. 1980. The ankle joints of archosaurs. *Alcheringa* 4(3–4):241–61.

Woodward, A. S. 1907. On *Rhynchosaurus artceps* (Owen). *Rep. Br. Assoc. Advan. Sci.* 1906:293–329.

5 On the type material of *Coelophysis* Cope (Saurischia : Theropoda) and a new specimen from the Petrified Forest of Arizona (Late Triassic: Chinle Formation)

KEVIN PADIAN

Introduction

During the summer of 1982, a field party sponsored by the Museum of Paleontology of the University of California (UCMP) began a paleontological survey of the Chinle Formation (Petrified Forest Member) in the Petrified Forest National Park, Arizona (Chapter 12). Near the southwest border of the north end of the park, not far from Lacey Point, and stratigraphically in the "Upper Unit" characterized by a fauna containing the aetosaur *Typothorax* and phytosaurs of the *Rutiodon tenuis* complex (Camp 1930), a series of three low-sloping hills was virtually covered with vertebrate bone (UCMP locality V82250). Most of the material collected was variously dissociated surface float of several vertebrate taxa, but reasonably complete material of three vertebrates was collected: a metoposaurid amphibian, a crocodylomorph archosaur, and a small theropod dinosaur. Only the pelvis and hindlimbs of the theropod (UCMP 129618) remained, but these have proved sufficient to compare the specimen to *Coelophysis* (Cope 1889). In order to justify the comparison and the diagnosis of the material, the type material of *Coelophysis* must be briefly considered before describing the new specimen, which is the first theropod from the Petrified Forest (late Carnian–early Norian) and the most complete dinosaurian material yet known from the Triassic of Arizona.

The type material

In 1887, Cope published two papers on the partial remains of a small theropod dinosaur recovered from what he believed to be Triassic strata of New Mexico. In the first of these papers (Cope 1887a), he referred the remains to Marsh's genus *Coelurus*, known from the Upper Jurassic Morrison Formation of Como Bluff, Wyoming. At that time Cope claimed to have "nearly all parts of the skeleton, excepting jaws and teeth," but this assessment seems to have been overly optimistic. He described, but did not illustrate, two species. *Coelurus longicollis* was based on cervical, dorsal, and caudal vertebrae (one each) and a femur. A bit larger than Marsh's *C. fragilis*, Cope estimated it at about the size of a greyhound. The second species, *C. bauri*, was smaller and was designated by a cervical vertebra, a sacrum, and the distal end of a femur.

In his second paper (Cope 1887b), a review of North American Triassic vertebrates, Cope repeated his description of these taxa but considered them distinct from *Coelurus* because the cervicals were amphicoelous. Instead, he referred the New Mexican material to Meyer's European form *Tanystropheus*, based on the long, slender caudal vertebrae of the latter. [These were actually cervicals; *Tanystropheus* was not recognized as a giraffe-necked marine reptile until much later (see Wild 1973)]. Cope distinguished a third species among his material, even smaller than the first two, and named it *T. willistoni* on the basis of a partial acetabular border and a single dorsal centrum. Besides naming this new species, Cope also referred some material, presumably from the same initial collection, to *T. longicollis* and *T. bauri*. Again, he did not illustrate the material, designate museum numbers or types, and kept the locality or localities private.

Colbert (1964) recovered the only available piece of locality information: a label from collector D. Baldwin indicating that all the material came from "about 400 feet below gypsum stratum 'Arroyo Seco' Rio Arriba Co., New Mexico." Colbert identified this stratum as belonging to the Petrified Forest Member of the Chinle Formation, about 25 miles east of Gallina, New Mexico, and probably from the Ghost Ranch locality later worked by field crews from the American Museum of Natural History (AMNH) in 1947 and 1948. However, other data may suggest several different collections or that the material may have been gathered from several localities. For example, the distinctness of preservation of Cope's material from the later AMNH collection must be underscored, because the latter is much better preserved and more complete. Also, Baldwin's label states that only one tooth was collected, yet there are several referred to *Coelophysis* under AMNH 2733 (these are actually phytosaur teeth). Unfortunately, the question may never be resolved. Cope (1889) referred to his material only once more, and that was to remove it from *Tanystropheus*: "I have recently learned that the reputed vertebrae of the latter genus possesses [*sic*] no complete neural canal, so that the position in the skeleton of these elements, on which the genus was founded, becomes problematical" (Cope 1889). *Coelophysis* was the new generic name for his three species.

Table 5.1. *Inventory of AMNH* Coelophysis *material, with identifications by previous authors (arranged roughly in order of description)*

AMNH #	Cope 1887a species ID	Cope 1887a element ID	Cope 1887b species ID	Cope 1887b element ID	Huene 1906 species ID	Huene 1906 element ID	Huene 1906 figure #	Huene 1915 species ID	Huene 1915 element ID	Huene 1915 fig. #
2701	C.1.*	cerv. v.	C.1.	?3rd cerv. v.	C.1.	Epistropheus	X:2	C.1.	Epistropheus	28
2715	C.1.*	dors. v.	C.1.	dors. v.	C.1.	dors. v.	11, X:3	C.b.	dors. v.	46[a]
2702	C.1.*	caud. v.	C.1.	caud. v.	C.1.	mid-caud. v.	X:4	C.1.	mid-caud. v.	32
2704	C.1.*	femur	C.1.	femur	C.1.	1. femur	19, XI:1	C.1.	r. femur	40[b]
2717	C.b.*	cerv. v.	C.b.	cerv. v.				C.b.	3rd or 4th cerv. v.	42
2722	C.b.*	sacrum	C.b.	sacrum	C.b.	sacrum	XII:1	C.w.	sacr. & pubic process	54[c]
2725	C.b.*	dist. femur	C.b.	dist. femur						
2735			C.1.	caudal centrum	C.1.	post. caudal v.	X:5	C.w. or C.b.	incompl. caudal v.	58
2708			C.1.	ilium #1				C.b.	r. ilium	48[d]
2705			C.1.	ilium #2	C.1.	r. ilium	14-17, X:9-10	C.1.	r. ilium	38
2706			C.1.	pubis	C.1.	r. pubis	18, XI:2	C.1.	r. pubis	39
?2716			C.1.	ischium						--[e]
2707			C.1.	phalanx	C.1.	ph1, dII, ?1.h.	X:7	C.1.	ph.1, 1gst. digit	36
2703			C.1.	ungual	C.1.	ungual	13, X:6	C.1.	ungual	37
2720			C.b.	3rd cerv. v.				C.b.	4th or 5th cerv. v.	43
2723			C.b.	dors. v.	C.b.	ant. dors. v.	XI:3	C.w.	dorsal v.	53[f]
2724			C.b.	head of pubis				C.sp.	head of r. pubis	61[g]
2719			C.b.	head of ischium				C.b.	prox. 1. ischium	49
2718			C.b.	dist. ischium				C.b.	dist. 1. ischium	50[h]
2721			C.b.	head of tibia	C.b.	r. tibia		C.1.	same, or 1g. mt.	41[i]
2726			C.w.*	partial ilium				C.w.	r. ilium (part)	60[j]
2727			C.w.*	dors. centrum				?C.w.	caud. or sm. dors. v.	57[k]
2730					C.1.	dist. 1. mcIII	12, X:8	C.1.	dist. end of 1g. mc.	34[l]
2744					C.b.	mid-caud. v.	XI:4	C.w.	anterior caud. v.	55[m]
2739								?C.1.	dors. v.	29[n]
2749								?C.1.	dors. v.	30
2729								?C.1.	?dist. end of ischium	31
2731								C.1.	artic. end of scapula	33[o]
2728								C.b. or C.1.	dist. end of sm. mc.	35[p]
2751								?C.b.	9th or 10th cerv. v.	44
2752								C.b.	ant. dors. v.	45
2750								C.b.	sacrum	47
2745								?C.b.	prox. fibula	51
2736								C.w.	upper arch of cerv. v.	52
2734								C.w.	mid-caud. v.	56
2740								C.w.	dist. end of mt.	64
2738								C.w.	pubic & ischial frags.	62-3[q]
2737								C.w.	prox. r. humerus	59
2732	C.sp.	five vert. frags.								
2742	C.sp.	misc. bone frags.								
2743	C.sp.	one vert. & six frags.								
2746	C.sp.	five caud. v. frags.	(as yet undescribed material)							
2747	C.sp.	misc. foot frags.								
2748	C.sp.	misc. vert. frags.								
2753	C.sp.	vert. & bone frags.								

Friedrich von Huene was the next to consider Cope's *Coelophysis*. In the course of preparing a review of non-European Triassic dinosaurs, Huene (1906) requested a set of casts of the specimens from H. F. Osborn, and so became the first to illustrate at least part of the material. (Some could not be found or identified.) This included all but the ischium of *C. longicollis*, but only a dorsal vertebra and the sacrum of *C. bauri*, and nothing of *C. willistoni*. Using Cope's measurements and descriptions, Huene was fairly successful in sorting out the type and referred material, although he provided no museum numbers. He believed that the sacrum had three vertebrae, not four as Cope had stated: The first was a dorsal. Later, however Huene (1915) accepted Cope's original view, and Huene referred as much identifiable material as possible to Cope's three species, supplementing the first with many drawings. Unfortunately, Huene still provided no specimen numbers and designated no type species; but more unfortunately, he reassigned much of the material from one species to another – including some type material of *C. longicollis* and *C. bauri*. Huene's reassignments were evidently based on size differences; but as Colbert (1964) pointed out, the *Coelophysis* material probably all represents ontogenetic stages of a single species. In most recent publications, only *C. bauri*, the type species arbitrarily designated by Hay (1930), is used. However, for the sake of historical preservation, Cope's material should be identified as far as possible by his original names, and the type material indicated as such. In Table 5.1, I have attempted to do this by giving the AMNH catalog numbers of all material in Cope's collection and by summarizing the identification and illustration of this material.

It should be noted that the museum numbers in Table 5.1 were only given to these specimens in 1973. Before 1973, all material was listed under the numbers AMNH 2701–2708, with several bones in each lot. Thanks to Charlotte Holton of the Department of Vertebrate Paleontology at the AMNH, records of the original cataloging of Cope's material have been preserved; these may have some bearing on the identification of certain type and referred elements. The original and present listings are compared in Table 5.2.

Other material referred to Coelophysis

Case (1922, 1927, 1932) reported vertebrate material from the Upper Triassic Dockum Formation of West Texas that he referred to *Coelophysis*. This initially included "a posterior cranial region and a string of cervical and anterior dorsal vertebrae." He redescribed the vertebrae in 1927 after further preparation revealed that their neural spines were unusually high. Finally, in 1932, he referred a series of small caudal vertebrae to *Coelophysis*. However, probably none of this material belongs to *Coelophysis* or to any dinosaur; the same applies to the other isolated remains (ilium, femur, vertebrae, and

Notes to Table 5.1 (cont.)

*Type.

[a]Huene (1915) noted that this is part of the type of *C.l.*, but reassigned it.

[b]The identification as a right femur is correct.

[c]Huene (1915) described this as including the sacrum of *C.w.* (!) plus the pubic process of an ilium of *C.b.* crushed together.

[d]Huene (1915) referred this to *C.b.*, noting Cope's original assignment to *C.l.*

[e]This bone has never been surely identified since Cope's (1887b) original mention.

[f]Huene (1915) noted that this was part of the type of *C.b.* (!)

[g]Actually is from the left pubis.

[h]Actually is the distal right ischium.

[i]Huene (1915) believed this to be lost.

[j]Huene (1915) believed this to be lost.

[k]Huene (1915) recognized this was Cope's type, but doubted its identification. Huene (1906) believed it lost.

[l]May be a distal matatarsal.

[m]Huene (1915) reassigned this material from *C.b.*

[n]Formerly placed with the pelvis in 2706; 2739 was originally assigned to a fragment that turned out to be part of another element (see footnote *q*).

[o]Actually the proximal left pubis.

[p]May be a distal metatarsal.

[q]The specimens figured in Huene's (1915) Figures 62 and 63 in fact fit together as part of a left ischium, including the distal end. Originally given the numbers 2738 and 2739, they are now both 2738, and 2739 has been reassigned (see footnote *n*).

teeth) described in Case's papers. Huene (1932) created a new taxon, *Spinosuchus caseanus*, for the long-spined vertebral column, and tentatively referred the basicranium (which may be rauisuchian) to it. Case's other material is archosaurian, but not clearly dinosaurian.

American Museum excavations at Ghost Ranch in 1947 and 1948 obtained numerous skeletons of a small theropod referred by Colbert (1947) to *Coelophysis*. Colbert (1964) compared these remains to a cast of the long-destroyed *Podokesaurus* from the Connecticut Valley (Newark Supergroup, eastern North America), and concluded that the latter should be synonymized with *Coelophysis*. This is possible, although *Podokesaurus* is very difficult to reconstruct with confidence, as Colbert made clear. The general resemblances between the interpretable remains of both taxa are most likely only primitive for theropods. Recent research has moved the *Podokesaurus*-bearing beds (probably Portland Formation) from the Late Triassic to the Early Jurassic (Olsen and Galton 1977); so the two genera are not as close in time as originally thought. This temporal shift also applies to the bone casts of another small theropod from the Connecticut Valley (probably also Portland Formation) previously referred to *Coelophysis* (Colbert and Baird 1958). I take these remains to be two tibiae and fibulae, presumably of a theropod but otherwise undiagnostic.

The Petrified Forest material

The University of California Museum of Paleontology specimen UCMP 129618 consists of most bones of the pelvic region and hindlimb of a small theropod, collected not far from Lacey Point, in the southwestern portion of the northern end of Petrified Forest National Park, Arizona (V82250). The locality was discovered by Mrs. Ann Preston, who with her husband Dr. Robert Preston was exploring the Park for caves and prehistoric carvings suggesting astronomical observations by prehistoric Indians. The UCMP field party returned with her to the site and gathered a rich collection of bone and bone fragments lying over the surface of three low, contiguous mudstone hills. Two partial skeletons, including the theropod and crocodylomorph, were eventually jacketed and returned to Berkeley, along with other archosaur remains (phytosaur, aetosaur, and ornithischian), metoposaurian bones, and nearby collections of gastropods and unionid bivalves.

Vertebrae

Vertebral remains are confined to a few centra of typical theropod aspect (Fig. 5.1). There is one complete dorsal centrum 45 mm long, of which the anterior face has a diameter of 29 mm and the posterior face 27 mm; their dorsal borders are slightly flattened. Diameter at midcentrum is 13 mm, and the neural canal is 5 mm wide. Two halves of dorsal centra have end diameters of 27 and 26 mm, and may together represent the following dorsal. The sacral vertebrae are fused end to end; this fusion is stronger than the bone at midshaft, and so there are no complete sacrals, and only two pairs of fused sacral ends. These are 25 mm in diameter. They represent either three or four of the sacrals, of which there were presumably five in dinosaurs plesiomorphically (Gauthier 1984). Finally, there is an anterior caudal centrum 33 mm long with end diameters of 26 and 29 mm. The centrum slopes posteriorly so that the posterior face is lower than the anterior. The neural canal is quite deep, and part of the posterior base of the left transverse process is preserved. There is no visible chevron facet, so I presume this vertebra is from the anterior caudal series.

Table 5.2. *Original and new AMNH catalog numbers of* Coelophysis *material*

Former number	New numbers	Original taxon[a]	Notes
2701	2701–2708, 2715, 2716	*C. l.**	2709–2714 are aetosaurian
2702	2717–2725	*C. b.**	2702A is now 2722
2703	2726–2727	*C. w.**	
2704	2749–2753	*C. sp.*	
2705	2716 (part), 2734–2738, 2740, 2742–2743	*C. w.*	2741 is archosaurian scutes
2706	2728–2732, 2739	*C. l.*	2733 is phytosaur teeth
2707	2746–2748	(*C. sp.*)	⎱ Fragmentary material,
2708	2716 (part)	(*C. sp.*)	⎰ not described

[a]*b.* = *bauri*; *l.* = *longicollis*; *w.* = *willistoni*; * = type material.

Pelvis

Only the anterior and posterior extremities of the left ilium remain; the latter shows the mediolateral arch for the posterior exit of the M. caudofemoralis brevis. Two other pieces preserve the contact between the ilium and the pubis and between the ilium and the ischium. The distal ends of the left pubis and ischium are preserved, along with fragments of the shaft.

The right pelvis (Fig. 5.2) is far more complete. The acetabular region was found complete, though fragmented, in situ, and over fifty pieces were eventually assembled to form the ilium and circumacetabular region. Curiously, the right astragalocalcaneum was diagenetically fused to pieces of the medial surface of the right ilium, just above the acetabulum. Because it could not be removed without destroying the iliac surface and because it is more poorly preserved than the left astragalocalcaneum, it has been left as found.

The dorsal ridge of the ilium as preserved is 165 mm long. The anterior extent of the iliac blade cannot be determined, but, as preserved, it reaches as far as the anterior extent of the suture between the ilium and pubis, about 20 mm in front of the acetabulum. The height of the ilium is 80 mm, as measured from the pubic and ischiadic peduncles. Its posterior process is incomplete at 113 mm posterior to the approximate center of the acetabulum, or about 76 mm posterior to its departure from the circumacetabular region. The anterior extreme of the pubic peduncle and the posterior extreme of the ischiadic peduncle are separated by 88 mm.

Despite its fragmentation, there are several salient features in the ilium. First, the margins along the dorsal and ventral borders of its posterior process are thickened, suggesting a slight hollow on the lateral face of the ilium, the origin of the MM. iliofemoralis and puboischiofemoralis internus of archosaurs. The lateral surface of the ilium is concave in all other theropods (e.g., *Dilophosaurus, Compsognathus, Deinonychus, Ornithomimus, Tyrannosaurus, Archaeopteryx*, and all birds). Second, the supraacetabular crest is prominent, as in all dinosaurs plesiomorphically (Gauthier 1984). It proceeds laterally from the supraacetabular buttress of the ilium and then dips ventrally, to form a distinct "hood" with a pronounced mediolateral arch over the dorsal margin of the acetabulum. The exact shape of the interior (medial) margin of this arch

Figure 5.2. *Coelophysis bauri* (UCMP 129618). Right pelvis in lateral (above) and medial (below) views. Scale divisions = 1 cm. Abbreviations: acet, acetabulum; anti, antitrochanter; il, ilium; is, ischium; is ped, ischial peduncle of ilium; n, notch in supraacetabular crest for passage of M. caudofemoralis; ob for, border of obturator foramen; p, pubis; pu ped, pubic peduncle of ilium; pv arch, posteroventral arch (= "brevis fossa") for passage of M. caudofemoralis brevis; r att, sacral rib attachments; s, sutures between pelvic bones; sac, supraacetabular crest. Sutures have been emphasized; inner arch of acetabulum is shown by a dashed line in the medial view.

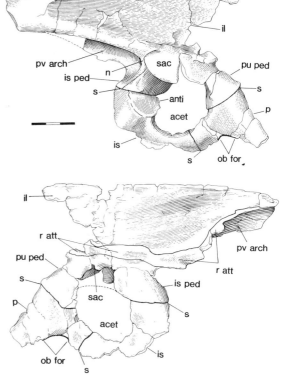

Figure 5.1. *Coelophysis bauri* (UCMP 129618). A–C, dorsal vertebrae in dorsal (A), lateral (B), and anterior (C) views; D–E, sacral vertebrae, fused at ends, broken at midcentrum, in lateral view; F–G, anterior caudal vertebra in left lateral (F) and posterior (G) views. Scale divisions = 1 cm.

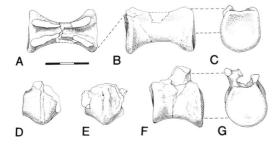

cannot be ascertained because some pieces are missing, but it is at least as deep as the supraacetabular crest. The "open" acetabulum usually thought characteristic of dinosaurs is, in fact, not completely open; the medial margin still extends about 25–35 percent into the acetabulum in all dinosaurs except large sauropods, which have secondarily evolved a columnar stance (Gauthier 1984).

Anteriorly, the supraacetabular crest appears to taper smoothly into the pubic peduncle of the ilium; posteriorly, it appears incomplete, but comparison with *Dilophosaurus* and other early theropods shows that there is a real notch between the ischiadic peduncle and the supraacetabular crest (Welles 1984). This notch presumably reflects the origin of the M. caudofemoralis brevis from the femur, inasmuch as the divided, arched ventral border of the posterior process of the ilium begins here. This feature, the "brevis fossa" of some authors, will henceforth be referred to as the "posteroventral arch" of the ilium, following Welles (1984).

The posteroventral arch is synapomorphic of dinosaurs (Gauthier 1984), but is most pronounced in theropods. At its anterior extreme, near the supraacetabular notch, it faces laterally; but as it proceeds posteriorly, rising toward the dorsal edge of the ilium, the internal wall of the arch becomes shallower and the external wall ("posterolateral shelf" of Welles) becomes deeper, so that the posterior half of the arch faces medially. Anteriorly, the internal wall of the arch slopes downward to contact the sacral ribs just above and behind the acetabulum. There were at least three attachments in the acetabular region, and the suggestion of room for two more behind it (Fig. 5.2).

The bone surrounding the acetabulum, especially the anterior, dorsal, and posterior borders, is more robust than anywhere else in the pelvis. The dorsal border thickens as discussed above; the pubic and ischiadic peduncles thicken distally, and the pubis and ischium thicken proximally to contact them. The acetabular face of the ischiadic peduncle (antitrochanter) is more oblique than that of the pubic peduncle; its characteristic configuration is conspicuous in Cope's specimens of *Coelophysis*, as in all theropods. Its junction with the proximal end of the ischium is a broad, anterolaterally oriented notch reminiscent of the shape of a glenoid fossa. The pubis, by contrast, meets head-on with the pubic peduncle of the ilium.

The shape of the acetabulum is an ovoid, inclined anterodorsally; the inner dorsal border, as noted, is incomplete in this specimen, although its configuration can be approximated. The maximum and minimum diameters of the oval approach 50 and 30 mm, respectively. The suture between the pubis

and the ischium is an anteroventrally sloping contact, situated well in the anterior half of the ventral acetabular margin. The three pelvic bones are solidly fused. The ventral plate of bone connecting the left and right pelves in *Syntarsus* (Raath 1969) is not present.

Based on comparison to Colbert's reconstructions and unpublished figures of the Ghost Ranch specimens, as well as to Raath's (1969) figures of *Syntarsus*, the right pubis of the specimen at hand (Fig. 5.3) appears to lack between 10 and 30 mm at a break about one-fourth of the way down the shaft; its preserved length is 200 mm. Most of the dorsal border of the obturator foramen may still be observed; it was quite large, as in *Syntarsus*. Distally, the pubes arch downward, twist medially, and form

Figure 5.3. *Coelophysis bauri* (UCMP 129618). Ischia (**A–D**) and pubes (**E–I**). **A**, Distal left ischium, lateral view; **B**, medial view of **A**, with contact area between ischia hatched; **C**, distal view of ischia; **D**, cross section of ischia at level d in **A**; **E**, right pubis (all but proximal end), lateral view; **F**, left pubis, medial view of distal end; **G**, anterior view of distal pubes, suture reconstructed; **H**, distal view of **G**; **I**, cross section of pubes at level i in **F**. Scale divisions = 1 cm. Abbreviations: s, suture between pubes; z, contact zone between distal ends of pubes.

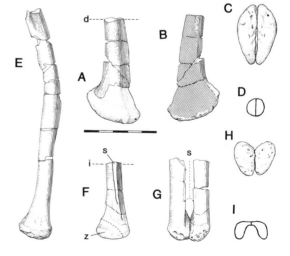

a medial apron of bone 3 mm thick. The apron extends at least 95 mm along the distal part of the pubis, then ends 25 mm before the tip, where the distal ends form a broad medial contact; they are roughly heart-shaped in distal profile. The pubic foot characteristic of later theropods is almost completely undeveloped: dorsally there is only a slight inflection at the tip, and ventrally only a very slight expansion, a condition that is plesiomorphic for theropods.

The ischia (Fig. 5.3) are far less complete; only the immediate acetabular region and 70 mm of both distal ends are preserved. The expansion of the distal ischia is greater than that of the pubes, and the midline contact continues over the entire medial surfaces of both halves. The ischia are torpedo-shaped in distal profile. The distal expansion of the pubis is greater than that of the ischium in all carnosaurs, deinonychosaurs, birds, and ornithomimoids (*sensu lato*), but primitively in theropods the reverse is true. Unfortunately, the transition in theropods is not well understood; *Compsognathus* is advanced in this respect and *Ornitholestes* is also reconstructed in this way. *Segisaurus* apparently retains the primitive condition seen in *Coelophysis*, *Syntarsus*, and *Dilophosaurus*.

Hindlimb

Femora, tibiae, and fibulae are preserved essentially complete, although their shafts are broken and distorted. No gaps are present in the tibiae, left femur, and left fibula; the right fibula and femur are irreparable in only one place. The left femur is 245 mm; the right, at 235 mm, may be lacking some 10 mm. The left tibia is 255 mm, the left fibula 250 mm; the right tibia measures 270 mm and the right fibula 265 mm. The variation between left and right sides is not uncommon in individual fossils, and diagenetic distortion may be largely responsible. The 5-mm difference between respective tibial and fibular lengths is explained by the form of the astragalocalcaneum with which they articulate: the surface for reception of the fibula is 5 mm higher than that for the tibia. When correctly assembled, the proximal ends of tibia and fibula are nearly level at the knee, with the fibula extending between the lateral and fibular condyles of the femur. In the following discussion, then, the orientation of the femur will be regarded as roughly horizontal, and that of the tibia as roughly vertical.

The femur (Fig. 5.4) has a distinct head, offset as in all dinosaurs and pterosaurs from the long axis of the shaft, which is not sigmoid as in many other archosaurs (see below). The theropod condition can be defined by several features. There are two grooves, one running mediolaterally along the proximal end of the femur (gr1), and the second seen in ventral view separating the head from the shaft (gr2). At the proximal end of the second groove is a pronounced tuberosity (t) on the proximoventral face of the head. On the dorsal face of the head, a slight ridge (r) runs almost parallel to the groove on the ventral side. The prominence, usually called the greater trochanter (gt), is prominent on the lateral side of the proximal shaft. The lesser or anterior trochanter (lt) begins just below the head on the dorsal surface of the shaft. It has a knurled config-

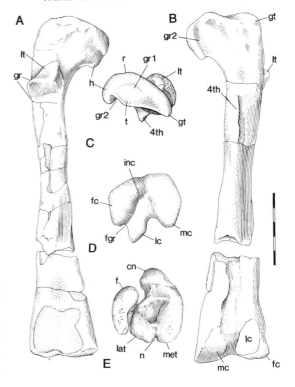

Figure 5.4. *Coelophysis bauri* (UCMP 129618). **A–D**, right femur in dorsal (**A**), ventral (**B**), proximal (**C**), and distal (**D**) views, drawn from a composite of left and right femora. **C** and **D** are oriented as if in erect posture. **E**, Left tibia and fibula in proximal view. Scale divisions = 1 cm. Abbreviations: cn, cnemial crest; f, fibula; fc, fibular condyle; fgr, fibular groove; gr, groove lateral to lesser trochanter; gr1, gr2, grooves on proximal and ventral sides of femoral head; gt, greater trochanter; h, head of femur; inc, incisura; lat, lateral tuberosity of proximal tibia; lc, lateral condyle of distal femur; lt, lesser trochanter; mc, medial condyle of distal femur; met, medial tuberosity of proximal tibia; n, notch; r, ridge on femoral head; t, proximal tuber on head of femur, which articulates with the antitrochanter; 4th, fourth trochanter.

uration, distinctly not shelflike as in later theropods. Its proximal area rises to an indented apex just below the level of the femoral head. A groove runs laterally from the lesser trochanter to produce a slight ridge along the side of the shaft. The fourth trochanter begins on the ventral surface at a level just distal to the lesser trochanter on the dorsal face. It runs parallel to the lateral ridge for about 40 mm. The greater trochanter is homologous to the "obturator ridge" of birds (Howard 1929).

The shaft of the femur is bowed dorsoventrally, and the distal end, corresponding to about 30 percent of the femoral length, curves slightly laterally. This dinosaurian configuration, also seen in pterosaurs, is distinct from the "sigmoid" femur of crocodiles and primitive archosaurs. The "sigmoid" femur is S-shaped because it is curved in both dorsoventral and mediolateral planes. The plane of the distal condyles (of which the lateral is larger than the medial) is offset 90° from the transverse plane of the femoral head; but in pterosaurs and dinosaurs, there is no S-shaped curvature, because the shaft is not curved mediolaterally. The medial distal condyle is larger than the lateral, and the femoral head is offset only about 45° from the plane of the distal condyles. I believe that sigmoid curvature has erroneously been attributed to many early dinosaurs, as well as to pterosaurs, because of a failure to recognize that the shaft is bowed dorsoventrally and that the head is actually directed dorsomedially into the acetabulum at a 45° angle in posterior view (see Fig. 5.4C), not horizontally.

In the present specimen, both the anterior and posterior intermuscular lines characteristic of birds (Howard 1929) are prominent on the femoral shaft.

The distal end of the femur is approximately as wide as the proximal end. Viewed distally with the shaft horizontal, the medial and lateral condyles (terminology of Howard 1929) are of approximately equal width (about 22 mm) and are separated by a pronounced incisura in the center of the distal face. The medial condyle (36 mm) is shallower than the lateral (43 mm), which is developed as a ventrolateral tuber diagnostic of all dinosaurs including birds, and is also present (although smaller) in pterosaurs and *Lagosuchus*. It descends ventrally from the middle of the distal face for about 30 mm from the incisura and tapers to a blunt edge below the fibular condyle. Ventrally, the lateral condyle forms a ridge running proximally for about 15 mm, terminating in a hooklike notch reminiscent of the shape of the femoral head. The ventral channel between the lateral condyle and the medial condyle is deep and broad and extends proximally for an indeterminate length. There is also a slight groove on the ventral face between the lateral and fibular con-

dyles. This is where the fibula articulates with the femur.

The tibia (Figs. 5.4 and 5.5) is characteristically theropod in the shape of the head, the fibular crest, the "twisted" configuration, and the distal articulation with the astragalocalcaneum. The proximal face of the tibia is triangular, with a shallow notch separating two posterior tuberosities and the anterior cnemial crest (cn) twisting anterolaterally in front of the fibula. The proximal face of the tibia is higher medially than laterally, and it slopes posteriorly. The tibia receives the medial condyle of the femur, and the fibula articulates with the lateral and fibular condyles, of which the former fits between tibia and fibula. The fibular crest (f cr) increases

Figure 5.5. *Coelophysis bauri* (UCMP 129618). A–C, Left tibia and fibula in posterior (**A**), anterior (**B**), and distal (**C**) views. **D–F**, Left astragalus and calcaneum in proximal (**D**), anterior (**E**), and posterior (**F**) views. Scale divisions = 1 cm. Abbreviations: as, astragalus; asc, ascending process of astragalus; ca, calcaneum; em, emarginaton on tibia for reception of ascending process; f cr, fibular crest; fl, flexor attachment; lig, ligamental attachment; s, suture between astragalus and ascending process; x, accessory articulation of tibia and astragalus. Other abbreviations as in Fig. 5.4.

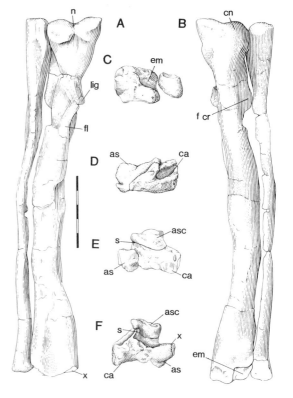

gradually as it descends from the proximal end for about 40 mm, then drops suddenly for 5 mm and ends. Opposite the fibular crest on the lateral side of the shaft is a prominent ligamental attachment (lig), and posterior to this is the flexor attachment (fl). All of these basal theropod features are retained essentially unmodified in birds (Howard 1929).

The distal end of the tibia is rectangular in outline, but has a complex topography. Viewed anteriorly, there is a right triangular emargination (em) with two corners represented by the lateral and medial extremes of the distal end; the third corner is about 20 mm up the lateral face, and the hypotenuse descending to the medial extremity is concave. This emargination reflects the articulation of the ascending process of the astragalus. There is also, as Welles (1984) notes for *Dilophosaurus*, a shallow notch (x) near the posteromedial corner of the tibia for reception of a corresponding prominence (x) on the astragalus. The "twisted" conformation of the theropod tibia often described amounts to a statement that the tibia is broader anteroposteriorly at its proximal end, and broader transversely at its distal end. The proximal expansion is slightly exaggerated by the curvature of the cnemial crest. This conformation is present in the type material of *Coelophysis*, and in the present specimen, as in all theropods.

The fibula is a narrow, straplike bone 12–18 mm broad and about 6 mm thick. The proximal end of the fibula is comma-shaped, with its thickest part anterior and its tail wrapping around the posterolateral tuberosity of the tibia. The medial face of the fibula also thickens just below its proximal end, as it hugs the lateral surface of the tibial tuberosity. The fibula has a shallow medial groove for the fibular crest of the tibia. The distal end of the fibula is subtriangular and fits snugly into an obliquely oriented transverse depression in the astragalocalcaneum.

The right astragalocalcaneum, as noted, was fused diagenetically to the inner wall of the right ilium and has been left in place to preserve the clarity of the features of the ilium. Only the distal portion is clearly preserved; the ascending process (asc) appears to have been lost before the tarsus and ilium were fused. Comparison of the two tarsi shows that the astragalar portion of the left tarsus lacks about 5 mm of its mediodistal border. All other details have been gleaned from the left member. The astragalus and calcaneum (Fig. 5.5) are completely fused to each other with no discernible trace of a suture (note the break in the left specimen, which is not the suture). However, there appears to be a suture (s) between the astragalus and its ascending process, which Welles (1983) first recognized in *Dilophosaurus*. It is interesting to note that in *Dilopho-saurus* fusion of the astragalus and calcaneum is not as complete as that between the astragalus and its ascending process, whereas in UCMP 129618 the reverse is true. Ontogenetic and phylogenetic patterns, however, must be elucidated by a larger sample. The ascending process in UCMP 129618 is more robust than in other theropod specimens, but it is difficult to tell if this appearance is merely the result of diagenetic distortion and mineralization.

As noted, the facet for articulation of the fibula is about 5 mm higher than that for the tibia. The fibular facet slopes laterally, and its transverse axis is somewhat oblique. The tibial facet is oriented transversely and has little discernible slope.

Distal tarsals 3 and 4 and metatarsals III and IV are preserved as discrete elements (Fig. 5.6). Distal tarsal 3 is fused to metatarsal III and is complete except for the anterior tip which is chipped off. Otherwise, it conforms exactly to the outline of the proximal metatarsal. Distal tarsal 4 also follows the general shape of its proximal metatarsal. The proximal faces of both distal tarsal 4 and metatarsal IV are concave, with slight ridges along their edges (except the medial edge, which is depressed). The anterior face of distal tarsal 4 is convex; the medial and lateral faces are concave; and the posterior end is developed into a cuboid tuber. Medially, distal tarsal 4 articulates with distal tarsal 3 and metatarsal III. A notch (n) on the posterolateral face apparently

Figure 5.6. *Coelophysis bauri* (UCMP 129618). Right distal tarsals (dt) and metatarsals. **A**, Proximal view; **B**, distal tarsal 4 in proximal, lateral, and distal views; **C**, metatarsals III and IV in anterior (center) and lateral views. Scale divisions = 1 cm. cap = metatarsal cap; n = notch in distal tarsal 4; Vf, facet on metatarsal IV for reception of metatarsal V.

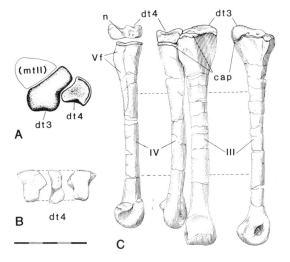

allowed passage of a flexor tendon. Metatarsal V would have been located distal to this groove on the posterolateral surface of metatarsal IV. Distal tarsal 4 thins by 2 mm anteriorly and laterally, but this reduction is compensated by a corresponding thickening of the proximal end of metatarsal IV – principally of the metatarsal cap, which is up to 3 mm thick in front and only 1 mm thick behind.

Metatarsal III is long and straight, with a shaft of nearly uniform diameter. Its length is 150 mm, of which the proximal 30 mm is expanded and deepened obliquely, and the distal 20 mm is transversely expanded. The proximal half of the metatarsal looks as if it had been twisted 45° so that its anterior face is displaced laterally and its posterior face displaced medially. The proximal face is long and rectangular, with a triangular posterior expansion. The lateral corner of this expansion is inflected for reception of the fourth distal tarsal and metatarsal; the opposite side, which articulates with metatarsal II, is straighter with only a slight lower expansion. As noted, distal tarsal 3 is fused to the metatarsal and conforms to its outline. It is 2.5 mm thick anteriorly, and it thickens posteriorly to 12.5 mm to form part of the tuberosity just mentioned. The metatarsal cap is 2 mm thick anteriorly, but is obscured posteriorly by overhang of the distal tarsal.

The distal end of metatarsal III is an expanded condyle 20 mm wide, marked by deep collateral ligament fossae on its lateral and medial faces. The radius of the condyle, measured from these fossae, is 10 mm. The surface of the distal end is slightly convex dorsally (with the medial side slightly higher than the lateral), almost imperceptibly concave distally, and markedly concave ventrally. There is a slight depression on the ventral face just proximal to the condyle.

Metatarsal IV is bowed medially, so that its distal end points slightly laterally. Its length is 133 mm, of which the proximal 13 mm are slightly expanded and the distal 20 mm are formed by the condyle, which has a radius of 10 mm measured as above. The medial face of the proximal end is concave along 20 mm where it articulates snugly with metatarsal III. There is a slight ridge on either side of this concavity. The next 60 mm are flattened slightly for the continuation of this contact. The posterior face of metatarsal IV is directed slightly laterally, is flattened for almost its entire length, and is slightly concave for the proximal 45 mm. This surface received the fifth metatarsal, which is not preserved here. A small triangular chip is missing from the posteromedial side of the proximal end where metatarsals III and IV articulated; otherwise the bone is complete.

In anterior view, the distal end of metatarsal IV is as narrow as the shaft (10 mm); its condyle is

slightly convex except ventrally, where a marked concavity is exaggerated by the characteristic lateral spur that proceeds posterolaterally from the lower portion of the lateral condylar face. A groove runs posterodistally along the upper face of the spur, which extends 12.5 mm from the body of the condyle. There is no collateral ligament fossa on the lateral side of the metatarsal.

Six pedal phalanges (Fig. 5.7) were preserved wholly or in part; in addition, there is one proximal and one distal portion of the unguals. The largest phalanx lacks its distal half and is chipped along one proximal edge. By its size and shape, I take it for phalanx 1 of digit III because its proximal width of 20 mm corresponds to the width of the distal metatarsal III, as does its convex, unkeeled proximal surface. Its dorsal process is only slightly developed, corresponding to the slight concavity of distal meta-

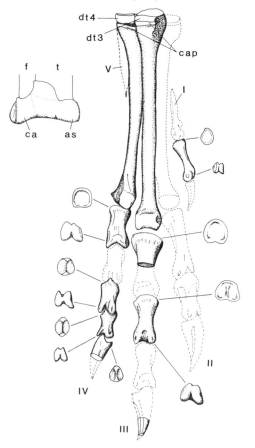

Figure 5.7. *Coelophysis bauri* (UCMP 129618). Right ankle and foot in anterior view. The astragalus articulates with digits II and III, and the calcaneum with digit IV. Missing phalanges reconstructed in dotted lines, after Raath's (1969) *Syntarsus*. Scale divisions = 1 cm. ca, calcaneum; as, astragalus; f, fibula; t, tibia; cap, metatarsal cap; dt3, dt4, distal tarsals 3 and 4.

tarsal III. Its ventral face is flattened. At its midshaft break, it is 16 mm long, 12 mm across, and 13 mm high.

I take the second largest phalanx for the second of the same digit because of its size and straightness. It is 37 mm long, 16 mm high, and 18 mm across at its proximal end, 11–12 mm high and 14 mm across at its distal end, and its minimum diameter at midshaft is 10 mm. Its proximal end has pronounced dorsal and ventral processes and a very slight vertical keel on the proximal face. The distal end is a double condyle of which the medial condyle is 1 mm longer and deeper.

The first, third, and fourth phalanges from the fourth toe are preserved. The first is 30 mm long, with a concave proximal face; most of the proximal border has been chipped off. The shaft is inflected so that its distal end curves slightly medially, and the lateral distal condyle consequently extends farther than the medial. The medial face of the phalanx, including its distal condyle, is somewhat ventromedially depressed, and the medial condyle is slightly lower than the lateral. This phalanx is about 50 percent larger than the comparable phalanx in *Syntarsus*.

The third phalanx is 22 mm long. Its proximal face is vertically keeled, and it has long dorsal and ventral processes that extend over its adjacent phalanx. Like the first phalanx, its shaft is slightly inflected medially, and its medial face is ventrolaterally depressed; also, its medial distal condyle is lower than the lateral. There is a rough depression on the medial face near the proximal end.

The fourth phalanx is 19 mm long, but otherwise just like the third. The proximal ungual fragment fits it well. All three phalanges preserved have deep collateral ligament fossae and a shallow pit on their dorsal faces just behind the distal condyles. This seems to be a widespread feature in theropods.

Finally, there is a long, thin, curved, lightly constructed phalanx that clearly was not weight bearing. Its proximal face is round and concave, and its distal condyles are close together, as they are typically in antungual phalanges. This phalanx is clearly the first of the first digit, and its length of 29 mm is about 50 percent larger than the comparable phalanx in *Syntarsus* (Raath 1969).

The pelvis and hindlimbs are reconstructed in Figure 5.8.

Referral of the Petrified Forest specimen to Coelophysis

Cope's original material of *Coelophysis* is relatively scrappy, except for some vertebral centra, a pubis (AMNH 2706), the acetabular region of one ilium (2705), a sacral vertebral series (2722), a distal right ischium (2718), several phalanges, and a poorly

preserved right femur (2704). Insofar as these elements in the Petrified Forest material can be compared to the type material, they correspond exactly in both size and detail to the largest pieces in Cope's collection, and could almost be taken for the same individuals were it not for the known discrepancy in locality. The structure of the peduncles and acetabular region of the ilium, the shape and length of the pubis, the undeveloped pubic foot, the larger ischial foot, and the details of articulation of the left and right pubes and ischia are some of the salient features that are indistinguishable between the two collections. Other preserved structures, such as the vertebral centra and phalanges, are equally indistinguishable but not especially different in this respect from those of other theropods primitively.

Figure 5.8. *Coelophysis bauri*, reconstruction of pelvis and hindlimb from UCMP 129618. Shown as foot touches the ground during walking, at beginning of propulsive phase. Femur is drawn slightly forward of its actual articulation with the acetabulum in order to show position of the head with respect to the antitrochanter. Missing portions restored from *Syntarsus* (Raath 1969). Scale divisions = 5 cm.

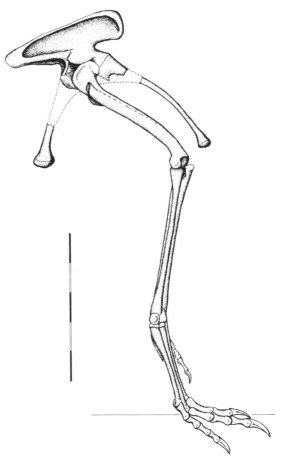

The problem is that *Coelophysis* has never been diagnosed with respect to its peculiar synapomorphies and the synapomorphies it may share with other theropods. Cope showed that it was a theropod, Huene showed that it was a very primitive theropod and differentiated it from other primitive saurischians, and Case confused the issue by incorrectly referring other material to it. None of these authors, to my knowledge, ever diagnosed the taxon. Colbert (1964) was the first to do so, in comparing *Podokesaurus* to *Coelophysis*. This diagnosis must be used for comparing the Petrified Forest material to the AMNH material.

At the outset it must be noted that Colbert's diagnosis was based mostly on the new Ghost Ranch specimens, not strictly on the type material of *Coelophysis*. Of twenty-four diagnostic characters listed, only one-third are preserved in the type material. However, because these features did not appear to differ between the type and Ghost Ranch material, Colbert regarded the synonymy of the two collections to be justifiable. This is important to note because there are differences between the Ghost Ranch and Petrified Forest specimens that may require taxonomic reevaluation when full descriptions of the former are published.

Colbert's diagnostic characters visible in the type material relate to the light build, size, amphicoelous vertebrae, long ilium, subequal pubic and femoral lengths, flat distally thickened pubis, rodlike proximally expanded ischia, and "curved" femur with large external trochanter. However, these are plesiomorphic characteristics of theropods: *Syntarsus*, *Procompsognathus*, *Podokesaurus*, *Segisaurus*, *Dilophosaurus*, and *Ceratosaurus* all also have similar dental and vertebral counts, forelimb–hindlimb ratios, and phalangeal formulas (as far as preserved material indicates). In fact, I can find no characteristics that can be proposed as apomorphies of the type material of *Coelophysis* or of the Petrified Forest material. It remains to be seen whether there are any in the Ghost Ranch material. As such, *Coelophysis* and its referred specimens stand as perfectly plesiomorphic theropods, insignificantly different from *Procompsognathus* in all respects (Ostrom 1981), except perhaps for the relatively long metatarsals of the latter.

The question is whether a taxon with no synapomorphies can be diagnosed. Because synapomorphies (shared derived characters) are only "derived" relative to a more restrictive or inclusive set of taxa, the answer would seem to be yes, as long as reference is made to a phylogenetic analysis in which other inclusive and restrictive synapomorphic levels are specified (Gauthier 1984). In such a context, a taxon with no synapomorphies of its own may

be diagnosed by a combination of character states both plesiomorphic at more restrictive (i.e., further derived) levels and synapomorphic compared to more general (i.e., primitive) levels (Table 5.3). [Gauthier (1984) has given the name *metataxon* to all such taxa with no apomorphies of their own.] For example, in all theropods, of which *Coelophysis* and *Procompsognathus* are among the most plesiomorphic, the foot is reduced to three functional digits, and the astragalus and calcaneum are fused to each other. An ascending process of the astragalus is present in *Procompsognathus* and in the Petrified Forest specimen, as well as in all other theropods. On the other hand, the astragalus and calcaneum are not fused to the tibia and fibula in the Petrified Forest material, in the Ghost Ranch material, or in *Dilophosaurus*, but they are in *Syntarsus* and most other theropods. Unfortunately, the type material of *Coelophysis* is too incomplete even for characterizations of this sort to be meaningful. Samples are too small to determine whether many such features are merely ontogenetic, a problem that could be approached by comparing fusions in postcranial elements. However, if the Petrified Forest material can be referred to *Coelophysis*, the diagnosis (which also applies to the level Theropoda) may be advanced as follows.

Systematic paleontology
Monophyletic hierarchy
Amniota: Diapsida: Archosauria: Dinosauria: Saurischia: Theropoda.

Coelophysis
Plesion for the Theropoda, a metataxon *senso* Gauthier (1984).

Diagnosis
As for the Theropoda primitively. Adapted in part from Colbert's (1964) diagnosis of *Coelophysis*, from Padian (1982), and from Gauthier's (1984) survey of diapsid phylogeny. Five sacral vertebrae (Gauthier 1984). Long low ilium with posterior process developed into a laterally twisting arch with the posterior end of the arch facing medially and the anterior end laterally, beginning just posterior to the acetabulum. Supraacetabular crest wide and with a low overhang, notched posteriorly. Antitrochanter facing anterolaterally, with its adjacent ischial surface directed dorsolaterally (Gauthier 1984). Pubis as long as femur. Ischium slightly shorter than pubis and femur; left and right halves with a broad medial contact extending most of their length; the broad distal expansion larger than the pubic "foot." Femur with distinct head characterized by transverse proximal groove, ventral groove separating head from shaft

matched by a slight ridge on dorsal surface; femoral shaft bowed dorsoventrally with distal end curved outward; well-developed subterminal medial, lateral, and fibular condyles; lesser trochanter located distal to the femoral head and knoblike; fourth trochanter a long ventral ridge opposite the lesser trochanter. Tibia slightly longer than femur, with expanded, laterally curving cnemial crest, and two posterior condyles; fibular crest on lateral face; distal end indented for reception of astragalus. Fibula reduced, straplike, with comma-shaped proximal end, and distal end oriented anterolaterally in calcaneal portion of astragalocalcaneum. Astragalus and calcaneum fused to each other at some point in ontogeny. Calcaneum reduced; ascending process of astragalus fitting over the anterior face of the tibia; distal face of astragalocalcaneum consists of two rounded condyles. Distal tarsals 2 and 4 concave proximally; distal tarsal 3 convex proximally. Metatarsals I and V reduced, the first with two reduced

phalanges, the fifth with none. Functionally tridactyl pes with a spur on the lateral side of the distal end of metatarsal IV; phalangeal formula 2–3–4–5–0. Digit III longest; II and IV subequal; I and V reduced.

If the Ghost Ranch material is referred to *Coelophysis*, other synapomorphies basal to theropods, many of which were listed by Colbert (1964), may be added. The forelimb is about half the length of the hindlimb; the manus has several distinct carpals and four digits, of which the fourth is quite reduced; a fifth metacarpal may persist. The scapula retains only a slight distal expansion and is not straplike as in later carnosaurs and coelurosaurs. The skulls of *Coelophysis* and *Procompsognathus* are similarly low and long, with many compressed, serrated teeth (Colbert lists twenty-six above and twenty-five below) and a pronounced spacing between premaxilla and maxilla. All these are primitive theropod characters.

Table 5.3. *Some synapomorphies of the most primitive Theropoda* (Coelophysis) *compared to character state transformations in later theropods*[a]

Basal Theropoda	Later transformations
Scapula more than three times the length of the coracoid	Scapula straplike, not expanded distally
Four digits on hand (fourth reduced)	Three digits (two in tyrannosaurids)
Posteroventral arch of ilium; orientation of acetabulum and antitrochanter (see text)	Backward rotation of pubis; synsacral fusion; incorporation of more vertebrae into sacrum
Pubis and femur nearly equal in length; ischium slightly shorter	Pubis longer than or equal to femur; femur longer than or equal to ischium
Ischial "foot" expanded; pubic "foot" not	Pubic "foot" greatly expanded
Configuration of femoral head (see text); lesser trochanter raised, knoblike, distal to head	Neck of femoral head more slender; lesser trochanter shelflike, migrating proximally
Anterolaterally expanded cnemial crest on tibia; fibular crest; fibula straplike and greatly reduced	Cnemial crest further divided into internal and external crests; further reduction of both fibula and fibular crest
Astragalus and calcaneum fused; ascending process of astragalus is a separate ossification	Astragalus, calcaneum, and ascending process fused to each other and to the tibia and fibula
Foot reduced to three functional digits; digit I reduced and rotated medially; digit V absent except for metatarsal	Digit I fully rotated to a posterior position; digit V completely lost
Distal tarsal 4 concave, 3 convex proximally; distal tarsal 3 fused to metatarsal III (distal tarsal 2 fails to ossify in pterosaurs and dinosaurs).	Distal tarsals 3 and 4 fused to each other; distal tarsals 3–4 and metatarsals II–IV all fused into a tarsometatarsus

[a]This partial list above omits basal dinosaurian characters of a fully regionalized vertebral column, five sacral vertebrae, offset femoral head, hooded, open acetabulum, tuber on the lateral distal condyle of the femur, reduced fibula, etc.

Type species

Coelophysis bauri was designated by Hay (1930). Colbert's (1964) judgment that all *Coelophysis* material described by Cope represents growth stages of a single species is accepted as probable. Welles (1984) has erected *Longosaurus* for the material originally referred to *C. longicollis*; however, the absence of more complete material inhibits considerations of ontogenetic variation, preservation, sexual dimorphism, and individual variation, and I prefer to reserve judgment on the new taxon pending more complete, differentiable material.

Diagnosis

As for the genus, and for the Theropoda.

Range

Late Triassic; possibly Early Jurassic. Southwestern and perhaps eastern United States, Germany, South Africa, and South America.

Referred taxa

As far as is known, *Coelophysis* lacks many synapomorphies of other early theropods, for example, the frontal crests of *Dilophosaurus*, the nasal crest of *Ceratosaurus*, and possibly the fused pelvis and ankle of *Syntarsus*. *Podokesaurus* and *Procompsognathus*, as well as *Avipes* and several other partial theropod finds are virtually indistinguishable from *Coelophysis*, except for particulars of size and proportions that may be unreliable in such a small and incompletely preserved sample. Hence, all taxa possessing the basal theropod synapomorphies listed above but lacking any of their own are potentially referrable. These include *Avipes* (Huene 1932), "*Triassolestes*" (Reig 1963), *Podokesaurus*, and perhaps others. Material referred to *Halticosaurus* (Huene 1932) may or may not be theropod, is greatly in need of reevaluation, and will not be further discussed here.

Comments

Gauthier (1984) redefined the theropod group Ceratosauria, including within it the taxa *Ceratosaurus, Syntarsus, Coelophysis, Segisaurus, Sarcosaurus*, and several unassigned specimens, including UCMP 129618. Diagnosis of the group was not provided, although several potential synapomorphies were cited. Gauthier and T. Rowe are preparing a full phylogenetic analysis of the Ceratosauria (n. comb.) and its place among the Theropoda, so the hypothesized level of generality of some synapomorphies proposed here may be subject to modification in the light of fuller analysis of character distributions and currently unpublished specimens.

Discussion

The genus *Coelophysis* presents the taxonomist with a peculiar, although not unique, problem. It was named on scrappy material that was never adequately diagnosed or compared with other taxa. Other material has been referred to the genus, often with little justification. Phylogenetic analysis requires unique synapomorphies in order to characterize taxa, yet even referred material of *Coelophysis* is unable to suggest any that do not characterize all theropods primitively. Can the genus *Coelophysis* have any integrity? I have argued above that it can, as long as it is characterized by the presence of certain synapomorphies and the absence of others. Phylogenetic analysis and classification are flexible enough to recognize such metataxa (*senso* Gauthier) as the closest thing to an ancestral species discernible in the fossil record. The next question is how to treat material that would be referred to such a plesiomorphic taxon, especially when the type specimens are so incomplete.

Historically, Colbert's (1947) referral of the Ghost Ranch material (which includes many complete skeletons) to *Coelophysis* made Cope's old genus immediately familiar; however, it has obscured the fact that the type material is too incomplete to provide much information that can decide whether or not to refer any new material to *Coelophysis*. Cope's *Coelophysis* is simply the rather undistinguished remains of a primitive theropod, whose partial details match those of many genera diagnosed on other skeletal features that are not preserved in Cope's material. The Petrified Forest material, as far as it goes, is indistinguishable from (or identical to) Cope's, and so is referred solely on grounds primitive to all theropods. The Ghost Ranch material may satisfy the same criteria, but this question is under Dr. Colbert's purview. It happens that Cope's best preserved material appears to be more comparable in size and build to the Petrified Forest material than to the smaller, more gracile Ghost Ranch specimens. Colbert (1964) may well be correct that Cope's three species are merely ontogenetic stages of the same species, but it is very difficult to demonstrate that all three collections are of the same genus and species.

Phylogenetically, description of the Petrified Forest material invites comparisons with other archosaurs, including later theropods and of course birds. As the foregoing diagnosis shows, some plesiomorphic characteristics of theropods seen in *Coelophysis* have been transformed in more derived theropods, whereas many others remain essentially unmodified even in birds. The latter include the posteroventral arch of the ilium, the orientation of the acetabulum and antitrochanter, the femoral head and distal end, the shape and articulation of the proximal tibia and fibula, the cnemial crest, the fibular crest, the configuration and processes of the femur and tibia, the articulation of the ankle (of which the fused avian elements may be separable into their primitive theropod components in ontogenetic series), the tridactyl foot, and all details of its phalangeal formulas and proportions. The morphology of the pelvis and hindlimbs supports the immediate ancestry of birds among Mesozoic coelurosaurian dinosaurs (Ostrom 1976; Padian 1982; Gauthier 1984), whereas no details controvert it. Those who claim that the similarities are convergent (e.g., Tarsitano and Hecht 1980) must logically deduce this by demonstrating that another monophyletic taxon is closer to birds. The assertion that the similarities are functionally or adaptively convergent is logically and empirically insufficient, because pterosaurs, which were similarly bipedal and very closely related to dinosaurs, do not share the above synapomorphies of Mesozoic theropods and modern birds, although they do share many synapomorphies with these taxa that reflect their close relationship (Padian 1983a,b, 1984; Gauthier 1984). The Petrified Forest material, therefore, provides insight into an early stage of theropod evolution, retaining many primitive features, but with many others already locked in place that have persisted in the only living group of dinosaurs, the birds.

Acknowledgments

I am most grateful to E. H. Colbert, J. A. Gauthier, J. H. Ostrom, J. M. Parrish, T. Rowe, and S. P. Welles for helpful discussions, critical review of the manuscript, and access to specimens and manuscripts. Charlotte Holton of the American Museum of Natural History was especially diligent and cooperative in straightening out the curatorial history of Cope's specimens. The fieldwork of R. A. Long, J. M. Parrish, J. R. Bolt, K. Ballew, J. Elzea, J. Woodcock, and Ann Preston is greatly appreciated, as is the cooperation of the officials and staff of the Petrified Forest National Park, in particular former Superintendent Roger K. Rector and Chief Ranger Chris Andress. Kyoko Kishi prepared the specimen, and Jaime Pat Lufkin drew all figures except Figures 5.6A, 5.7, and 5.8. This work was supported by a grant from the donors of the Petroleum Research Fund of the American Chemical Society (No. 13577-G2), and by the Museum of Paleontology of the University of California.

References

Camp, C. L. 1930. A study of the phytosaurs, with descriptions of new material from Western North America. *Mem. Univ. Calif.* 10: 1–161.

Case, E. C. 1922. *New Reptiles and Stegocephalians from the Upper Triassic of Western Texas.* (Carnegie Institute, Washington, D.C.), Publ. No. 321, pp. 1–84.

1927. The vertebral column of *Coelophysis* Cope. *Contrib. Mus. Geol. Univ. Mich.* 2(10): 209–22.

1932. On the caudal region of *Coelophysis* sp. and on some new or little known forms from the Upper Triassic of West Texas. *Contrib. Mus. Paleontol. Univ. Mich.* 4(3): 81–91.

Colbert, E. H. 1947. Little dinosaurs of Ghost Ranch. *Nat. Hist.* 56(9): 392–9 and 427–8.

1964. The Triassic dinosaur genera *Podokesaurus* and *Coelophysis*. *Am. Mus. Novit.* 2168: 1–12.

Colbert, E. H., and D. Baird. 1958. Coelurosaur bone casts from the Connecticut Valley Triassic. *Am. Mus. Novit.* 1901: 1–11.

Cope, E. D. 1887a. The dinosaurian genus *Coelurus*. *Am. Nat.* 21: 367–9.

1887b. A contribution to the history of the Vertebrata of the Trias of North America. *Proc. Am. Phil. Soc.* 24(126): 209–28.

1889. On a new genus of Triassic Dinosauria. *Am. Nat.* 23: 626.

Gauthier, J. A. 1984. A cladistic analysis of the higher systematic categories of the Diapsida. Ph.D. Thesis, Department of Paleontology, University of California, Berkeley.

Hay, O. P. 1930. *Second Bibliography and Catalogue of the Fossil Vertebrata of North America*, Vol. II (Washington, D.C., Carnegie Institute), Publ. No. 390.

Howard, H. 1929. The avifauna of Emeryville Shellmound. *Univ. Calif. Publ. Zool.* 32(2): 301–94.

Huene, F. von. 1906. Ueber die Dinosaurier der Aussereuropaischen Trias. *Geol. Pal. Abh.* 12: 99–156.

1915. On reptiles of the New Mexican Trias in the Cope collection. *Bull. Am. Mus. Nat. Hist.* 34: 485–507.

1932. Die fossile Reptilordnung Saurischia. *Monogr. Geol. Pal.* (1) 4, pts. 1 and 2.

Olsen, P. E., and P. M. Galton. 1977. Triassic–Jurassic tetrapod extinctions: are they real? *Science* 197: 983–6.

Ostrom, J. H. 1976. *Archaeopteryx* and the origin of birds. *Biol. J. Linn. Soc.* 8: 91–182.

1981. *Procompsognathus*: theropod or thecodont? *Palaeontographica A* 175: 179–95.

Padian, K. 1982. Macroevolution and the origin of major adaptations: vertebrate flight as a paradigm for the analysis of patterns. *3rd N. Am. Paleontol. Conv. Proc.* 2: 387–92.

1983a. Osteology and functional morphology of *Dimorphodon macronyx* (Buckland) (Pterosauria: Rhamphorhynchoidea) based on new material in the Yale Peabody Museum. *Postilla* 189: 1–44.

1983b. A functional study of flying and walking in pterosaurs. *Paleobiology* 9: 218–39.

1984. The origin of pterosaurs, *In* Reif, W.-E. and F. Westphal (eds.), *Proceedings, Third Symposium on Mesozoic Terrestrial Ecosystems* (Tuebingen: ATTEMPTO), pp. 163–8.

Raath, M. A. 1969. A new coelurosaurian dinosaur from the Forest Sandstone of Rhodesia. *Arnoldia (Rhodesia)* 28(4): 1–25.

Reig, O. A. 1963. La presencia de dinosaurios saurisquios en los "Estratos de Ischigualasto" (Mesotriasico Superior) de las provincias de San Juan y La Rioja (Republica Argentina). *Ameghiniana* 3(1): 3–20.

Tarsitano, S., and M. K. Hecht. 1980. A reconsideration of the reptilian relationships of *Archaeopteryx*. *Zool. J. Linn. Soc.* 69(2): 149–82.

Welles, S. P. 1983. Two centers of ossification in a theropod astragalus. *J. Paleontol.* 57: 401.

1984. *Dilophosaurus wetherilli* (Dinosauria, Theropoda): osteology and comparisons. *Palaeontographica* 185A: 85–180.

Wild, R. 1973. Die Triasfauna der Tessiner Kalkalpen. XXIII. *Tanystrophaeus longobardicus* (Bassani) (Neue Ergebnisse). *Schweiz. Pal. Abh.* 95: 1–162.

6 The ichnogenus *Atreipus* and its significance for Triassic biostratigraphy

PAUL E. OLSEN AND
DONALD BAIRD

Introduction

Reptile footprint faunules from the early Mesozoic Newark Supergroup of eastern North America (Fig. 6.1) have been studied since the the early 1800s. Most are from the Hartford and Deerfield basins of the Connecticut Valley; thanks to many works, particularly those of E. Hitchcock (1836, 1843, 1847, 1858, 1865) and Lull (1904, 1915, 1953), the Connecticut Valley faunules are relatively well known (although now desperately in need of revision). Long thought to be of Late Triassic age, all of the Connecticut Valley tracks come from strata above the oldest extrusive basalt flows in the Hartford and Deerfield basins and are now thought to be Early Jurassic in age. Similar Early Jurassic faunules have been more recently identified in the Newark, Culpeper, and Fundy basins (Olsen, McCune, and Thomson 1982; Olsen and Baird 1982) as well as in the Glen Canyon Group of the southwestern United States and the upper Stormberg Group of southern Africa (Olsen and Galton 1977, 1984).

Older, undoubtably Late Triassic footprint assemblages of the Newark have received scant attention, although neither localities nor specimens are rare. Tracks from the preextrusive intervals of the Newark Basin were first reported by Eyerman in 1886, but it was not realized that these faunules were distinctly different from those of the Connecticut Valley until Baird's revisions of the assemblages in 1954 and 1957. Within the past thirty years many new footprint faunules have been found in Newark horizons of Late Triassic age, and, consequently, the composition of the older faunules has become much clearer. The purpose of this chapter is to describe an entirely new ichnogenus that is one of the most common and distinctive elements of these older assemblages, widespread in eastern North America and present as well in the European Middle Keuper.

We begin by describing the morphology and systematics of the new form and follow with a discussion of its biostratigraphic significance. Detailed locality and stratigraphic information for the material discussed in the systematic section is given in the biostratigraphic section. Finally, we attempt to reconstruct the osteology of the ichnites and try to find a likely trackmaker among known osteological taxa.

Materials and methods

All of the specimens described in this chapter are footprints, either natural casts or natural molds, not actual osseous remains of animals. We use the traditional latinized binominals for the track taxa, but the taxonomy is one parallel to that of the biological Linnaean hierarchy, not part of it. These ichnotaxa do not correspond to osteologically based fossil taxa either materially or in concept. The names are used as handles and for the classification of tracks of different morphology.

With the exception of the European form, *Atreipus metzneri*, all of the material described here comes from seven localities in the Passaic and Lockatong formations of the Newark Basin, two localities in the Gettysburg Shale in the Gettysburg Basin, one locality in the Cow Branch Formation of the Dan River Group, and one in the Wolfville Formation of the Fundy Basin (Fig. 6.2). Details of the localities are given in the section on geological occurrences and associated fossils (Appendix 2).

We have analyzed these footprints by preparing line drawings of all the reasonably well-preserved material and producing composites of these drawings. We define "reasonably well-preserved material" as tracks that show impressions of pads and that are not strongly distorted; thus, they are as close to the trackmakers' morphology as possible. We have purposely stayed away from the analysis of tracks that do not meet this criterion. The line draw-ings were made by tracing photographs of the tracks themselves or of inked latex or polysulfide peels of the tracks. The outlines were drawn along the sur-faces of maximum change in curvature of the track. This method produces very accurate, reproducible renderings. Composites of the best material from successive tracks of single individuals (where pos-

sible) or populations were produced by visual inter-polation of the superimposed track outlines.

Osteological reconstructions were developed from the composite outlines by placing joint artic-ulations at the center of the pads. This conforms to the pedal structure of living cursorial dinosaurs (birds) (Heilmann 1926; Peabody 1948; Baird 1954, 1957) as well as to the pedal structure of crocodiles (Chapter 20) and of the largest living terrestrial rep-tile, the Komodo dragon (Padian and Olsen 1984).

Systematics

Ichnofamily Atreipodidae nov.

Diagnosis. Habitually quadrupedal ichnites; pes tulip-shaped with digit three longest; manus small, digitigrade, tridactyl or tetradactyl.

Ichnogenus Atreipus nov.
(Fig. 6.3A–C)

Type ichnospecies. Atreipus milfordensis (Bock), 1952.

Included ichnospecies. The type species; *Atrei-pus sulcatus* (Baird 1957); *A. acadianus* nov.; *A. metzneri* (Heller 1952).

Range of Ichnogenus. Late Carnian to ?middle Norian (Late Triassic) strata of the Newark Super-group of eastern North America.

Diagnosis. As for monotypic ichnofamily. Small (9–14 cm), tulip-shaped pes impression with metatarsal–phalangeal pads of digits II and IV oval to circular and often impressed. Relative proportions very similar to those of *Grallator*. Distal phalangeal pads often indistinct from more proximal pads. Hal-lux not impressed even in very deep tracks. Manus digit III longest followed in length by II, IV, and I (the last impressing only in *A. acadianus*).

Etymology. Named in honor of Atreus Wan-ner, who uncovered many fine footprints, including *Atreipus*, in Goldsboro, York County, Pennsylvania (Wanner 1889).

Discussion

The pedes of all the ichnospecies of *Atreipus* would fit quite comfortably within the Ichnofamily Gral-latoridae, as they did in the analyses of Baird (1957), were it not for the functionally tridactyl manus impressions (Figs. 6.4–6.10). Such a manus has never been described from a Newark grallatorid walking trackway, despite what must be many thousands of Newark tracks collected over the last 150 years. [The supposed "manus prints of *Anchisauripus*" reported by Willard (1940) are apparently pes imprints of a smaller individual.] The manus, in fact, is more like what might be expected in a chirotheriid. The com-bination of this manus with a *Grallator*-like pes is so unusual that it cannot reasonably be included in any known ichnofamily, so we erect a new ichnofamily for this distinctive ichnotaxon.

Figure 6.1. The Newark Supergroup of Eastern North America (from Olsen, McCune, and Thomson 1982). *Atreipus* has been found in basins marked 18, 14, 13, and 2. Key to basins as follows: 1, Wadesboro Subbasin of Deep River Basin; 2, Sanford Subbasin of Deep River Basin; 3, Durham Subbasin of Deep River Basin; 4, Davie County Basin; 5, Dan River–Danville Basin; 6, Scottsburg Basin; 7, Briery Creek Basin and subsidiary basin to south; 8, Farmville Basin; 9, Richmond Basin; 10, exposed part of Taylorsville Basin; 11, Scottsville Basin; 12, Culpeper Basin; 13, Gettysburg Basin, 14, Newark Basin; 15, Pomperaug Basin; 16, Hartford Basin; 17, Deerfield Basin; 18, Fundy Basin; 19, Chedabucto Basin (= Orpheus Graben).

Figure 6.2. Localities for *Atreipus* in the Newark, Gettysburg, Dan River and Fundy basins. **A**, Newark Basin. Main lacustrine body is Lockatong Formation (shown in gray except as marked); Jurassic sedimentary rocks (mostly lacustrine) are shown by hatched marking. Igneous rocks ajacent to the Jurassic sedimentary rocks are extrusive flows; all others are intrusive plutons; all igneous rocks are shown in black. Localities are: 1, Perkasie Member, Passaic Formation, Milford, New Jersey; 2, Graters Member, Passaic Formation, near Frenchtown, New Jersey; 3, Members E–F, Passaic Formation, near Frenchtown, New Jersey; 4, Passaic Formation, Lyndhurst–Rutherford area, New Jersey; 5, lower Lockatong Formation, Gwynedd, Pennsylvania; 6, Weehawken Member, Lockatong Formation, Arcola, Pennsylvania. Abbreviations as follows: J, Early Jurassic extrusive flows and sediments in isolated synclines; NO, New Oxford Formation = Stockton Formation where Lockatong is absent; P, Passaic Formation; Pk, Perkasie Member, Passaic Formation; S, Stockton Formation. **B**, Gettysburg Basin. Principally gray lacustrine rocks are shown in gray; igneous rocks are shown in black. Large gray area is New Oxford Formation, which has a very large fluvial component. Localities are: 1, Gettysburg Shale, Trostle Quarry near York Springs, Pennsylvania; 2, Gettysburg Shale, near Goldsboro, Pennsylvania. Abbreviations as follows: G, Gettysburg Shale; H, Heidlersburg Member of Gettysburg Shale; J, Early Jurassic extrusive flows and sediments. **C**, Dan River Group in Dan River and Danville basins. Gray areas indicate outcrops of lacustrine Cow Branch Formation. 1 indicates position of Solite Quarry locality for *Atreipus*, Leaksville Junction, Virginia and North Carolina. **D**, Onshore and offshore Fundy Basin of New Brunswick and Nova Scotia, Canada, and the Gulf of Maine. Black areas indicate outcrops of extrusive basalts; gray areas show the offshore extent of the same basalts. Hatched lines show area of offshore extent of inferred Early Jurassic sediments (Scots Bay and McCoy Brook formations). Abbreviations are: JM, onshore outcrops of Early Jurassic McCoy Brook Formation; pTr, pre-Triassic rocks present offshore; Tr, offshore Triassic rocks; TrWB, onshore Triassic Wolfville and Blomidon formations. 1 shows the position of the Fundy Basin *Atreipus* near Paddy Island, Kings County, Nova Scotia. Maps adapted from Klein (1962), Keppie et al. (1979), Ballard and Uchupi (1980), Swift and Lyall (1968), Tagg and Uchupi (1966), Robbins (1982), Thayer (1970), Meyertons (1963), Olsen (1980a, 1984a), Faille (1973), and Nutter (1978).

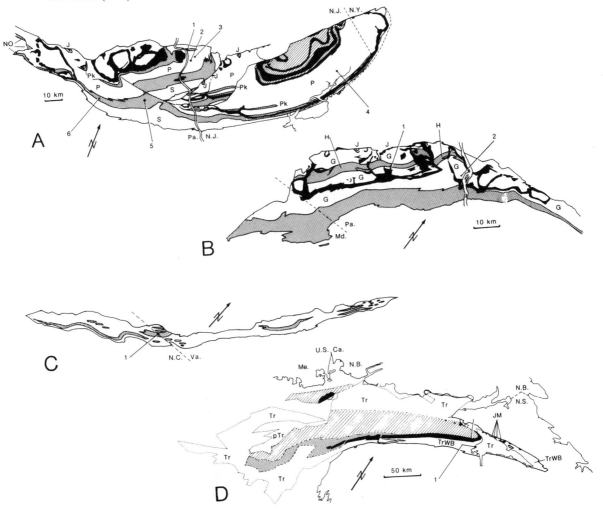

Atreipus milfordensis (Bock) nov.
comb. (Fig. 6.3A)

Gigandipus (?Anchisauripus) milfordensis.
Bock (1952, p. 403, Plate 43, Fig. 3) (in legend, for
"No. 15210" read S488): Unit O, Perkasie Member,
Lower Passaic Formation, New Jersey, early Norian
of Late Triassic.

(?)Anchisauripus gwyneddensis. Bock (1952,
pp. 406–7, Plate 44): lower Lockatong Formation,
Pennsylvania, late Carnian of the Late Triassic.

Type. LC. S488, a left pes impression slightly
distorted by the paratype of *Brachychirotherium parvum* (Fig. 6.4A).

Previous hypodigm. The type and two other
imprints on the same slab; MCZ 135, 229; AMNH
1981 and 1984, all collected by John Eyerman,
1885–7. Type of *A. gwyneddensis*, ANSP 15222
(missing).

Distribution. Late Carnian and early Norian
of the Late Triassic in the Newark, Gettysburg, and
Dan River basins of the Newark Supergroup.

Locality of type. Clark Brothers quarry, 1.5
km north of the Delaware River Bridge at Milford,
Hunterdon County, New Jersey.

Referred material. Additional material from
the abandoned Clark Quarry: a slab with several
quadrupedal trackways (PU 18581), manus–pes sets
(PU 23486) and miscellaneous tracks (PU 23649)
collected by D. Baird and Robert M. Salkin in 1963;
manus–pes sets from the base of member F of the
Passaic Formation (PU 23640), from Nishisakawick
creek 4.7 km east of Frenchtown, New Jersey, collected by P. Olsen in 1984; isolated pes impression
from the lower division (G) of the Graters Member
(YPM 7554) from Little Nishasakawick Creek,
Frenchtown, New Jersey, collected by P. Olsen,
1971; two overlapping trackways from the lower Pas-
saic Formation (YPM 9960) near Rutherford, New
Jersey, collected by Larry Black in 1969; quadrupedal trackways from the Gettysburg Shale of the
Gettysburg Basin from near Goldsboro, York
County, Pennsylvania, collected by Atreus Wanner
(Wanner 1889) in 1888 and accessioned by the
United States National Museum, but later lost
(Baird 1957); trackways from the Trostle Quarry
near Bermudian Creek, Bermudian Springs, Penn-
sylvania, Adams County, Pennsylvania, collected by
Elmer R. Haile, Jr., 1937 (Stose and Jonas 1939)
including CM 12081 and 12087; isolated manus–pes
set (YPM 9961) from the long trackway from lower
Lockatong in cut of Schuykill Expressway at Arcola,
Pennsylvania, collected by P. Olsen and Cynthia
Banach in 1983; latex peel (YPM 9962) of large slab
of poor trackways from the upper member of the

Figure 6.3. Examples of the three recognized species of *Atreipus* from the Newark Supergroup: **A**, Latex cast
of right manus–pes set of *Atreipus milfordensis* from the middle Gettysburg Shale, Trostle Quarry, Adams
County, Pennsylvania (CM 12081). **B**, Natural cast of right pes of *Atreipus sulcatus*, the type specimen from
the lower Passaic Formation of Milford, New Jersey (MCZ 215). **C**, Natural cast of right manus–pes set of
Atreipus acadianus, from the upper Wolfville Formation of Paddy Island, Kings County, Nova Scotia (PU
23635). Scales in cm.

Cow Branch Formation of the Dan River Group, Solite Corporation Quarry, Leaksville Junction, Virginia and North Carolina.

Diagnosis. Atreipus in which digits II and IV of the pes are nearly equal in their forward projection, with the bases of the claws of II and IV lying almost opposite the crease between the first and second phalangeal pads of digit III. Digit I of manus not impressed.

Discussion

An examination of all the specimens in the previous hypodigm and new material from the type locality, in addition to the referred material from the other localities and horizons, allows Baird's (1957) description of this ichnospecies to be enlarged (Figs. 6.4, 6.5, 6.7, 6.8). Most significant is the realization that this ichnospecies represents a habitually quadrupedal form, a condition never seen before in tracks otherwise similar to members of the Ichnofamily Grallatoridae. The block containing the holotype is unfortunately broken in such a way that it is impossible to tell if there ever was an associated manus impression. The same is true of most of the hypodigm with the exception of MCZ 135 in which a partial manus is present (Fig. 6.4B). The trackways of the smaller individual from the type locality (PU 18581, Fig. 6.4j–n) have clear manus impressions, as do most trackways from other localities. One crucial trackway, of which YPM 9962 is representative, shows that *Atreipus milfordensis* trackways are not consistently quadrupedal (Fig. 6.6o).

Three digits are always present in the manus, and digits II and III of some specimens (such as YPM 9960) have impressions of small claws (Figs. 6.4–6.6). The manual digits represented can be understood only with reference to *Atreipus acadianus*, where they prove to be digits I–IV (see below). The pads of the manus are poorly differentiated even in the clearest of impressions; however, there are often creases at about the midpoints of digits II and III. On many manus impressions, there is also a trans-

Figure 6.4. Type material of *Atreipus milfordensis* (**A–E**) and *A. sulcatus* (**F–I**) and new material of *A. milfordensis* from the Smith–Clark Quarry (**J–N**). **A–M** all drawn as right pes impressions. **A**, *Atreipus milfordensis*, type specimen (LC488), a left pes impression slightly distorted by the paratype of *Brachychirotherium parvum* (dotted outline); **B**, MCZ 135, a right pes and partial manus impression; **C**, MCZ 229, a left pes impression; **D**, paratype of *A. milfordensis* (LC 488), a left pes impression; **E**, plesiotype of *A. milfordensis* (AMNH 1984), a left pes impression; **F**, type of *Atreipus sulcatus* (MCZ 215), a left pes impression; **G**, paratype *A. sulcatus* (AMNH 1982), a left manus and pes impression, pes overlapped by another conspecific pes impression (dotted outline); **H**, paratype *A. sulcatus* (AMNH 1983), a left pes impression; **I**, paratype *A. sulcatus* (MCZ 216), a left pes impression; **J–N**, *Atreipus milfordensis* (PU 18581): **J–K**, left manus and pes impressions; **L–M**, right manus and pes impressions; **N**, a trackway composed of successive left (above) and right (below) manus and pes impression sets. Scale is 4 cm.

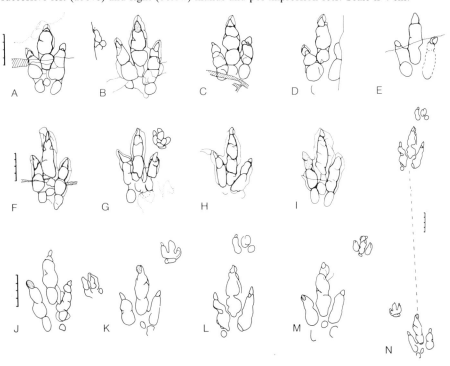

versely elongate pad at the posterior end of digit III, which is usually broader than the adjacent digits. Other creases appear, but these are variable in sequential manus impressions of the same individual and were probably due to interaction of the animal with the substrate. The available trackways show that the manus impressions are usually placed anterior and lateral to the pes impressions, but this pattern is far from consistent [as the Arcola, Pennsylvania trackway shows (Fig. 6.6)].

In the pes, the flexed and hastate claws described by Bock (1952) and Baird (1957) prove discernible only on the deeply impressed type material from Milford, and as a diagnostic feature they are unreliable. As Baird (1957) noted, each proximal pad of the pes is bounded distally by a shallow sulcus rather than a crease, and the more distal pads are more or less confluent. This differs from the situation in the Grallatoridae, in which all of the pads are distinct and separated by creases except those on the distal portion of digit IV. Also, as Baird (1957) noted, in *Atreipus* the metatarsal–phalangeal pads of both digits II and III are often impressed, whereas in grallatorids only that of digit IV is usually present. These last two features are synapomorphies of all three ichnospecies of *Atreipus*.

Composites of the *Atreipus milfordensis* specimens from the major localities are shown in Figure 6.8. There is as much variation in the pedal impressions among specimens of the hypodigm as among any of the specimens from the other localities or their composites. Baird (1957) commented on this variation, noting that in one specimen (MCZ 135), digit IV is relatively shorter than the others and as a consequence digit II projects more anteriorly than IV. In addition, there is also considerable variation in digit divarication. In the smallest specimens from the type locality (PU 18581), the digits do not appear to be as robust as in the hypodigm, but this is almost certainly because their impressions are relatively shallow. Here again, digit II projects further than IV in some specimens, but not others, *even in successive tracks of the same individual* (Fig. 6.5B and 6.6). The Lyndhurst–Rutherford material shows little variation within the sample, and, despite the individual tracks' greater size, there are no obvious proportional differences between the composite and that of the Milford material (Fig. 6.8). The pads on the pedes, however, are even less well differentiated than in the type material. Pedal impressions from Arcola, Pennsylvania are not as clearly impressed as those from the rest of the Newark Basin material of

Figure 6.5. *Atreipus milfordensis* from the lower middle Passaic Formation of the Lyndhurst–Rutherford area, Bergen County, New Jersey (**A, C–E**) and the middle Gettysburg Shale of the Trostle Quarry, Adams County, Pennsylvania (**B, F, G**). **C–G** drawn as if all were right manus pes sets. **A,** Trackway (YPM 9960) of successive right and left manus–pes impression sets (drawn reversed). **B,** Trackway (CM 12081) of successive left and right manus–pes impression sets (drawn reversed). **C,** Left manus–pes impression set of trackway in (**A**). **D,** Right manus–pes impression set of trackway in (**A**). **E,** Isolated left manus–pes impression set on same slab as (**A**). **F,** left manus–pes impression set from trackway in (**B**). **G,** right manus–pes impression set from trackway in (**B**).

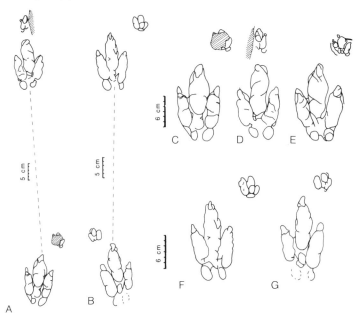

Atreipus. As might be expected in sloppy tracks, the amount of variation between successive tracks is large. The composite (Fig. 6.8), however, still clearly resembles *A. milfordensis*.

Bock's (1952) *Anchisauripus gwyneddensis* appears to be closely comparable to *Atreipus milfordensis*, judging from Bock's plate, and seems especially close to the Trostle Quarry forms described below (Baird, 1957). Unfortunately, careful searches by each of us and by the museum staff of the Academy of Natural Sciences of Philadelphia failed to locate the type (ANSP 15222). In addition, Bock's plate of the specimen and the negative from which it was printed are cropped too tightly to show whether or not there is an associated manus impression. Our assignment of *A. gwyneddensis* to *Atreipus milfordensis* is thus based solely on our interpretation of Bock's plate of the pes and thus must be

Figure 6.6. *Atreipus milfordensis* from the lower Lockatong Formation of Arcola, Pennsylvania. **A–N**, Successive pes and manus–pes impressions all drawn as if right impressions, from trackway shown in **O**. A is a right pes impression, **B** is a left pes impression, etc. **O**, Trackway showing position of tracks (A–N above). P in **O** shows trackway of *Rhynchosauroides hyperbates*. **J** is YPM 9961.

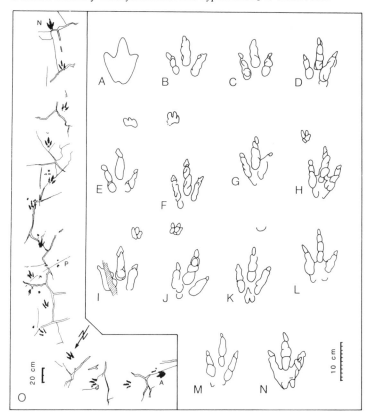

Figure 6.7. Cow Branch *Atreipus milfordensis*. Tracing of a photographed but not collected block from Leaksville Junction, North Carolina–Virginia. Peel of specimen is YPM 9962.

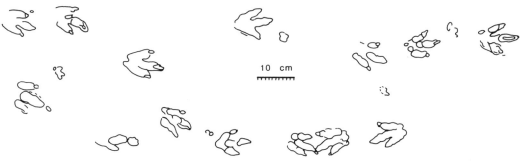

regarded as tentative. Fortunately, *Atreipus milfordensis* has page priority over *A. gwyneddensis*.

The composite (Fig. 6.8D) of the Gettysburg Basin (Trostle Quarry) specimens (Fig. 6.5) is closely comparable to the type of *Atreipus milfordensis* and especially to the Lyndhurst–Rutherford material, but here again there are surprising differences in proportions between the left and right pedes of the two successive tracks with respect to the apparent relative projection of digits II and IV. Wanner's (1889) undesignated material from near Goldsboro, York County is unfortunately lost, but what can be seen from his figures (especially Plates 9 and 10) is indistinguishable from *A. milfordensis*.

Available material from the Cow Branch Formation (Fig. 6.7) is rather poor and difficult to interpret. The tracks lack the distinguishing features of *A. sulcatus* or *A. acadianus*, and comparison with the composite of all the Cow Branch material suggests that it too belongs in *Atreipus milfordensis*.

Atreipus sulcatus (Baird)
(Fig. 6.3B)

Grallator sulcatus. Baird (1957, pp. 453–61, Fig. 1, Plate 1); full synonymy, and a list of incorrect assignments are given in Baird (1957).

Type. MCZ 215, a left pes impression.

Previous hypodigm. MCZ 215 to 228 inclusive, the type and thirty-six other pes impressions. AMNH 1982 and 1983, six imprints. LC S487, parts of three overlapping pes imprints. All collected by John Eyerman, 1885–7.

Horizon. Early Norian of the Late Triassic, Newark Supergroup, Newark Basin, Passaic Formation, Perkasie Member, Unit O.

Locality. Quarry of the Messrs. Clark about one-half mile east of the Smith-Clark Quarry, near Milford, Hunterdon County, New Jersey (from Eyerman, 1889, p. 32). Subsequent collecting indicates that the type horizon occurs a little below that of *A. milfordensis* in the same quarry.

Referred Material. Topotypes from Clark Quarry, PU 20743 A-C and a large slab at Upsala College, collected by James Sorensen, 1969; PU 23642, two pedes collected by Olsen and Baird, 1984; from Perkasie Member, Unit O, Passaic Formation.

Diagnosis. Atreipus distinguished from other ichnospecies by having the impressions of the bases of pes digits II and III closely united, while IV remains separated from III by a deep sulcus that extends back to the metatarsal–phalangeal pad of III.

Discussion

The description of Baird (1957) requires little modification except to add the presence of a manus indistinguishable from that of *A. milfordensis*. A clear manus impression is present on the paratype of *A. sulcatus* (AMNH 1982, Figs. 6.4g and 6.9). Other slabs from the type locality have manual impressions, but none are worth figuring.

New material from the Clark Quarry is evidently from the type horizon; it resembles Eyerman's specimens in all observable features, but includes many associated manus impressions. The quality of most of this material is poor and adds little to our knowledge of pedal morphology of the ichnospecies except to show that it is more similar to *A. milfordensis* than was originally believed. The fact that the *A. sulcatus* impressions were made on softer mud may account for some of the apparent differences. However, on the basis of the available material, the two ichnospecies are morphologically

Figure 6.8. Composites of *Atreipus milfordensis* from each major locality. Tracks arranged in order of increasing size: **A**, PU 18581, from the Smith–Clark Quarry, Milford, New Jersey. **B**, Hypodigm of *Atreipus milfordensis* from the Smith–Clark quarry, Milford, New Jersey [modified from Baird (1957)]. **C**, Lockatong material from Arcola, Pennsylvania. **D**, CM 12081, from the Trostle Quarry, near Bermudian Springs, Adams County, Pennsylvania. **E**, YPM 9960, from the Passaic Formation, Lyndhurst–Rutherford area, New Jersey. Scale is 2 cm.

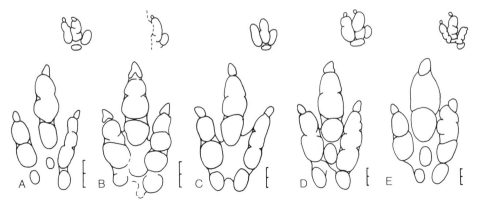

distinct, and *A. sulcatus* cannot be proved to fall within the range of variation of *A. milfordensis*. We, therefore, retain them as separate ichnospecies, while acknowledging the possibility that a larger sample size might show them to be parts of a morphological continuum.

Atreipus acadianus nov. (Fig. 6.3C)

Type. PU 21713B, a slab bearing five manus–pes sets; collected by D. Baird and P. Olsen, 1973.

Hypodigm. PU 21808, 23511, 23512 A-C, 23606, 23607, topotype slabs collected by Princeton parties, 1975, 1976, 1982; PU 23632–23637, collected by P. Olsen, J. Smoot, and M. Anders, 1984.

Locality. Cliffs and beach just south-southeast of Paddy Island, Kings County, Nova Scotia.

Age and Horizon. Late Carnian or early Norian of Late Triassic, Newark Supergroup, Fundy Group, Wolfville Formation, about 8.7 m below the contact with the overlying Blomidon Formation (Baird and Olsen 1983).

Etymology. From Acadia, early name for Nova Scotia.

Diagnosis. Atreipus in which digit IV of pes is very small with the center of its most distal phalangeal pad lying opposite the center of first phalangeal pad on digit III, and the manus is consistently tetradactyl.

Discussion

The combination of a small, functionally tridactyl manus and a tridactyl pes of *Grallator*-like form with poorly differentiated distal phalangeal pads marks all of the collected dinosaurian tracks from the upper Wolfville Formation at Paddy Island, Nova Scotia. All of these specimens have a pes with a proportionally long digit III, a robust digit II, an unusually slender digit IV, and less prominent metatarsal–phalangeal pads than other ichnospecies of *Atreipus* (Figs. 6.9 and 6.10). The center of the metatarsal–

phalangeal pad of digit III lies nearly opposite the most anterior part of the first phalangeal pad of digit IV. The center of the most distal pad on digit II lies about opposite the center of the first phalangeal pad on digit III. The center of the first phalangeal pad on digit III is opposite the center of the most distal (pad 4) of digit IV. Finally, digit II projects consistently farther anterior than IV. The last two points reflect how relatively delicate digit IV is. The general proportional resemblance to *A. sulcatus* (Figs. 6.9, 6.10) is especially obvious in the relative length of pedal digit III and the relative projection of digit II. In this new ichnospecies, however, digits II and III remain well separated as in *A. milfordensis*, but digit IV appears closely appressed to III. As in other ichnospecies of *Atreipus*, there is considerable variation, but the Nova Scotian tracks are consistent in the characters cited.

The manus differs from those of other ichnospecies of *Atreipus* in consistently having an impression of an additional digit medial to the three present in the other ichnospecies. This digit has only one pad. We recognize it as digit I and thereby infer that digits I–IV impress in *A. acadianus*. Only digits II, III, and IV impress in the other ichnospecies, and a definite impression of digit V has not been seen in any *Atreipus* specimens.

Atreipus metzneri (Heller)

Coelurosaurichnus metzneri. Heller (1952, pp. 135–7; includes his second and third ichnite type), Tafel 9, Figs. 2 and 3.

Type. Specimen in the Geologisches Institut of Erlangen (Heller 1952, Tafel 9, Fig. 2).

Hypodigm. Pes impressions on the type slab of *Chirotherium wondrai* in the Geologisches Institut von Erlangen and in the private collection of Dr. R. Metzner, in Markt Erlbach (Heller 1952, Tafel 9, Figs. 1 and 3). From the same locality and horizon as the type.

Figure 6.9. Composites of, **A**, *Atreipus milfordensis*; **B**, *A. sulcatus*; **C**, *A. acadianus*; and **D**, *A. metzneri*. Scale is 2 cm.

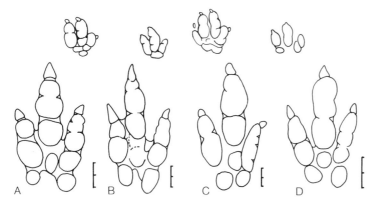

Horizon. Carnian of the Late Triassic, Ansbacher Sandstone of the Germanic Middle Keuper.

Locality. Altselingsbach, Franconia, Federal Republic of Germany.

Discussion

A small footprint faunule from the Ansbacher Sandstone of the Middle Keuper of Germany described by Heller in 1952 contains an almost certain example of *Atreipus* (see also Chapter 15). The assemblage, according to Heller (1952), consists of three kinds of ichnites. The first type is a manus–pes set that he designated *Chirotherium wondrai*. The second type consists of a number of tridactyl pes impressions that Heller named *Coelurosaurichnus metzneri*. The third type consists of a small, three-toed track to which Heller declined to apply a new name. The figured pes impressions [Heller's second type (Heller 1959, Plate 9, Figs. 2 and 3) (composite in Fig. 6.9 in this chapter) suggest a tulip-shaped track reminiscent of the pes of *Atreipus*. Heller's Plate 9, Figure 3 shows one of these pes impressions directly behind his third type of track (Heller's "*eine weitere Reptilfaehrtes*"), which bears an uncanny resemblance to the manus of *Atreipus* (Chapter 15). No described ichnite pes resembles Heller's third type of track, and we conclude that it is in fact the manus of his second track type, *Coelurosaurichnus metzneri*. Thus, the latter taxon is, pending actual examination of the material, almost certainly referable to *Atreipus*. The figured material bears a definite resemblance to *A. milfor-*

densis (Fig. 6.9), but we feel it is prudent to retain Heller's specific epithet at least until a more detailed comparison can be made.

Comparison of Atreipus with other dinosaurian ichnites

The combination of a manus impression with a tridactyl pes of the *Grallator* type is almost unknown in the ichnological literature. The Early Jurassic Connecticut Valley forms *Ammopus, Corvipes, Plectropterna, Xiphopeza, Arachnichnus, Orthodactylus, Sauropus barrattii,* and *Tarsodactylus caudatus* (see Lull 1953) show various combinations of tridactyl to tetradactyl pes impressions with tridactyl to pentadactyl manus impressions. All but the last two ichnotaxa were made in very muddy substrate. The tracks show little constancy of digit number or footprint form even in their types and, therefore, should be regarded as indeterminate. *Sauropus barrattii* and *Tarsodactylus caudatus*, however, are figured by Lull with a *Grallator*-like pes in association with a five-toed manus, and they, therefore, need to be considered in relation to *Atreipus*.

Sauropus barrattii as figured by Lull (1953) is comparable to *Atreipus* in having a pes virtually indistinguishable from that of *Grallator* (*Anchisauripus*). It is combined with a manus that in this case bears five toes. Lull (1953) placed *Sauropus* in the ichnofamily Anomoepodidae because of the manus and because of the supposed sitting posture in which "heel" imprints occur. Restudy of the original spec-

Figure 6.10. Holotype (**A**) and hypodigm (**B–H**) of *A. acadianus*. **A–H** all drawn as right manus and pes. **A–E**, PU 21713B; **F**, PU 23636; **G–I**, PU 23635.

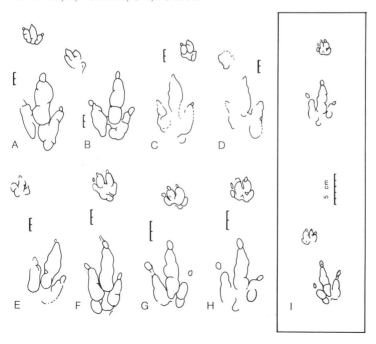

imens, however, reveals that *Sauropus barrattii* is an unnatural composite of two ichnotaxa that has perpetuated factual and nomenclatural misinterpretations for nearly 120 years. The *Sauropus* of Lull's rendering has been often reprinted and used as the basis of additional inferences about trackmakers, posture, and behavior. It has also formed the basis of an ichnofamily, the Sauropodidae (Haubold 1969). In Appendix 1, we review the confused history of *Sauropus*, declare it indeterminate, and conclude that it requires no additional attention.

Lull's (1904, 1915, 1953) figures of *Tarsodactylus caudatus* Hitchcock (1858) indicate a walking trackway of small, somewhat *Grallator*-like pedal impressions associated with five-toed, outwardly rotated manual imprints. On reexamination, the type (AC 42/5), from the Turners Falls Sandstone of Turners Falls, Massachusetts, appears to be a rare quadrupedally walking example of *Anomoepus*. (The manus impressions of *Anomoepus* are usually impressed only in sitting position.) We believe that the resemblance between *Anomoepus* and *Atreipus* may indicate phylogenetic relationship. as we discuss in the section on osteological correlations.

As we know of no other forms that resemble *Atreipus* in having a three- or four-toed manus associated with a *Grallator*-like pes, additional ichnological comparisons seem unwarranted.

Biostratigraphy of Atreipus

The isolated basin sections of the Newark Supergroup can be broadly divided into the lower and upper Newark Supergroup (Froelich and Olsen 1984). The upper Newark Supergroup is characterized by extrusive basalt formations at its base and is Early Jurassic in age, while the lower Newark Supergroup completely lacks extrusive igneous rocks and is Late Triassic or older in age (Cornet 1977a; Cornet and Olsen 1985; Froelich and Olsen 1984).

Atreipus is widely distributed in the lower Newark Supergroup. Details of the stratigraphy of each of the basins in which *Atreipus* occurs, along with details of the stratigraphy of each locality and lists of associated biota are given in Appendix 2. Only the outlines of the distribution will be given here.

Newark Basin

Together, the Stockton, Lockatong, and Passaic formations make up the lower Newark Supergroup in the Newark Basin. There are seven younger formations making up the upper Newark Supergroup (Fig. 6.11). Thus far, although *Atreipus* has been found only in the Lockatong and Passaic formations, the range of the ichnogenus comprises more than 30 percent of the cumulative thickness of the basin section. Within the Lockatong and Passaic formations,

definite *Atreipus* specimens have been found at five horizons spread through about 1,700 m from the Weehawken Member of the Lockatong Formation to the Mettlars Brook Member of the Passaic Formation (Fig. 6.11). Another locality that produces pes impressions of the *Atreipus* type, but as yet has yielded no manus impressions, is from about 700 m higher (Ukrainian Member of the Passaic Formation), bringing the total range of *Atreipus* in the basin to 2,400 m. This is the greatest stratigraphic thickness over which *Atreipus* is known in the Newark Supergroup.

Gettysburg Basin

The Gettysburg Basin section is similar to that of the Newark Basin. There, *Atreipus* has been found at two horizons about 2,400 m apart, both within the lower Gettysburg Formation (Appendix 2), which is laterally equivalent and connected to the Passaic Formation of the Newark Basin. The Gettysburg Formation rests below the Aspers Basalt, the basal portion of the upper Newark within the basin.

Dan River Basin

Unlike the Newark, Gettysburg, and Fundy basins, the Dan River Basin contains no extrusive basalt flows and consists entirely of lower Newark Supergroup. Only one locality has produced *Atreipus* in the basin, and it is within the upper member of the Cow Branch Formation. The Cow Branch Formation is lithologically very similar and apparently the same age as the Lockatong Formation of the Newark Basin.

Fundy Basin

Atreipus is known from one locality in this basin as well. In this case, the horizon is within the uppermost Wolfville Formation about 360 m below the North Mountain basalt, the basal formation of the upper Newark Supergroup in the Fundy Basin. This horizon seems to correlate with the Lockatong or lower Passaic of the Newark Basin.

Stratigraphic relationship of Atreipus assemblages

Atreipus and its associated ichnites make up a consistent assemblage, usually dominated by *Atreipus*, small *Grallator* spp. and *Rhynchosauroides* spp., but also including *Apatopus* and *Brachychirotherium* spp. Only in the Newark and Fundy basins can the stratigraphic relationship between *Atreipus* faunules (which are the oldest footprint assemblages in those basins) and subsequent ichnite faunules be demonstrated.

In the Newark Basin, the stratigraphically highest occurrence of the *Atreipus* assemblage appears to be a trackway of *Apatopus* (PU 21235) in the Passaic Formation at a level about 10 m below the Orange Mountain Basalt at Llewellyn Park, West Orange, New Jersey. In the uppermost meter of the same formation, however, there appears a distinctly different ichnofauna that heralds a change to the classic, dinosaur-dominated ichnofaunas of the Connecticut Valley. This intermediate fauna was recovered by Chris Laskowich during excavation for the Montclair State College parking lot in Little Falls, Passaic County, New Jersey. As a carryover from the earlier, typically Triassic ichnofaunas, it includes *Rhynchosauroides brunswickii* (PU 22005, 22006, 22204, 22214); this genus is rare in the higher (postextrusive) beds of the Newark Basin and is unknown in the Connecticut Valley. On the other hand, the small crocodyloid ichnogenus *Batrachopus* makes its earliest known appearance at Little Falls, New Jersey in the form of abundant and well-preserved trackways of *Batrachopus* cf. *B. deweyi* (PU 21902, 21933, 22001, 22004–22007, 22205, 22213–22214), the common species in the Connecticut Valley. Theropod-type tracks include the large *Grallator* (*Anchisauripus*) *minusculus* (PU 21900, 21901), the

Figure 6.11. Stratigraphy of the Newark Basin showing the positions of the major footprint faunules discussed in the text and the known ranges of *Atreipus* and the Connecticut Valley–type faunules. Black horizontal bars and lines indicate gray and black clusters of Van Houten cycles, with the most obvious spacing being the 100 m (400,000 year) compound cycle. The gross aspects of the time scale are derived from the DNAG scale (Palmer 1983), but the fine calibration is based on the inferred durations of Van Houten cycles from Olsen (1984a, b). Members from Olsen (1984a).

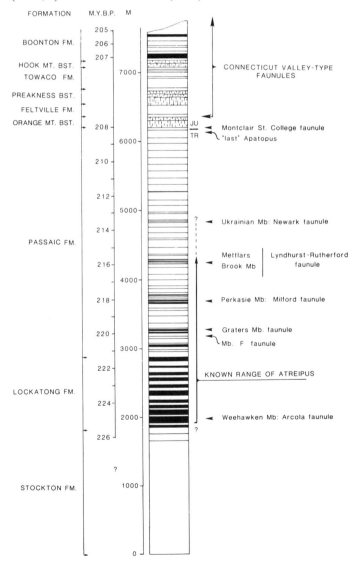

medium-sized *G. (A.) sillimani* (PU 21900, 21903, 21904, 21933, 22003, 22202, 22210, 22313, 22366), and *Grallator* sp. (PU 21903, 22201, 22313, 22366). (Duplicated numbers indicate slabs bearing several species.) The Little Falls faunule is decidedly different from any older Newark Basin assemblage and is the only preextrusive assemblage from the Newark Supergroup that bears a definite Connecticut Valley aspect.

The sedimentary sequence – Feltville, To-waco, and Boonton formations – that overlies and is interbedded with the Newark Basin extrusives of the upper Newark Supergroup of the Newark Basin has yielded many footprint faunules at different horizons. These faunules are very similar and as a whole are indistinguishable from the Connecticut Valley assemblage described by Hitchcock (1858, 1865), Lull (1904, 1915, 1953), and Olsen (1980a). The characteristic ichnogenus *Anomoepus* makes its first appearance in the Feltville Formation (PU 18565), and in this and higher formations is accompanied by *Grallator* (*Eubrontes*) spp., *G. (Anchisauripus)* spp., *G. (Grallator)* spp., and *Batrachopus* spp. The Towaco Formation has also yielded examples of cf. *Rhynchosauroides* sp. and *Ameghinichnus* sp. Citations of specimens and a complete faunal list are given in Olsen (1980b,c).

In the Fundy Basin of Nova Scotia, we have discovered several footprint faunules that are younger than the *Atreipus* assemblage (Baird 1984). A small faunule comprising *Grallator* (*Grallator*) spp., *G. (Anchisauripus)* spp., and *Rhynchosauroides* sp. occurs in the Blomidon Formation (which overlies the Wolfville Formation) on the north shore of St. Marys Bay, Digby County, Nova Scotia, about 250 m below the North Mountain Basalt. This horizon, therefore, falls roughly 110 m above the level of the *Atreipus* faunule. In the Fundy Basin, the upper Newark consists of the North Mountain Basalt and the overlying Scots Bay Formation and McCoy Brook formations. In the type Scots Bay Formation, theropod-type tracks occur on the top surface of an eolian brown sandstone that overlies the fish-bearing lake beds: *Grallator* (*Eubrontes*) *giganteus*, *Grallator* (*Anchisauripus*) *sillimani*, and *G. (A.)* cf. *tuberosus*. These have been photographed but not collected. On the north shore of the Minas Basin at Five Islands Provincial Park (Colchester County), and at Blue Sac and McKay Head (Cumberland County), the lower McCoy Brook Formation has yielded tracks of *Anomoepus* spp., *Grallator* (*Eubrontes*) spp., *G. (Anchisauripus)* spp., *Batrachopus* spp., and *Otozoum moodii*. Specific localities and specimen citations for these localities are given in Olsen (1981). The Scots Bay and McCoy Brook formations are lateral equivalents, and their footprint faunules are indistiguishable from

those of the Connecticut Valley and those of the post-Passaic sediments of the Newark Basin.

To find footprint faunules older and different from those containing *Atreipus*, we must go to the Sanford Subbasin of the Deep River Basin in North Carolina, where the Pekin Formation of the Chatham Group has produced a small but significant ichnofauna. This Pekin assemblage can be placed in stratigraphic perspective, vis-à-vis the *Atreipus* faunules of the Dan River Group, Gettysburg Basin, Newark Basin, and Fundy Group, by correlative means independent of footprints.

Only one locality in the Pekin Formation has produced a reasonable ichnofaunule: the Pomona Pipe Products clay pit near Gulf, North Carolina. So far the identifiable forms include relatively short-toed, tridactyl, presumably dinosaurian footprints; very large- to medium-sized bipedal tracks resembling *Brachychirotherium* sp.; and *Apatopus* sp. Apart from *Apatopus* there are no forms in common with the *Atreipus* and younger assemblages. Associated skeletal remains include the dicynodont *Placerias*, the phytosaur *Rutiodon*, the aetosaur *Typothorax*, and large carnosaur-like teeth that presumably represent a rauisuchid pseudosuchian (Baird and Patterson 1967). This assemblage lies roughly 300 m below the base of the Cumnock Formation and is Middle Carnian on the basis of associated mega- and microfossil florules (Hope and Patterson, 1969; Cornet 1977a). The overlying Cumnock Formation has produced a microflorule that correlates with the basal New Oxford Formation of the Gettysburg Basin and the lower member of the Cow Branch Formation of the Dan River Group (Cornet 1977a; Robbins 1982; Cornet and Olsen 1985). The middle New Oxford Formation contains mega- and microfloras correlating with the Lockatong Formation and the upper member of the Cow Branch Formation (Cornet 1977a, b; Robbins 1982). These floral correlations are completely supported by correlation by vertebrate skeletal evidence (Olsen et al. 1982). The Lockatong, in turn, correlates with the middle Wolfville of the Fundy Group on the basis of skeletal remains (Olsen 1981; Olsen et al. 1982; Baird and Olsen 1983). These correlations show that the Pekin faunule is significantly older than those containing *Atreipus*.

The palynology and skeletal remains of the *Atreipus*-bearing beds in the Newark and Gettysburg basins allow the younger ichnofaunules to be correlated. Microflorules from the Heidlersburg Member of the Gettysburg Shale of the Gettysburg Basin (which overlies the *Atreipus* faunules) are Norian in age and correlate with the Perkasie Member of the Newark Basin (Cornet 1977a). Pollen and spore assemblages from near the Mettlars Brook Member in the Newark Basin indicate a middle Norian age, and those from the Ukrainian Member indicate an early late Norian (early Rhaetian of older works). Likewise, in the Fundy Basin, the Wolfville *Atreipus* faunule lies some 360 m below a latest Norian or earliest Jurassic florule (Olsen 1981) and, therefore, is of late Carnian or Norian age.

To summarize for the entire Newark Supergroup, the oldest *Atreipus* occurs in the Lockatong Formation and upper member of the Cow Branch Formation of late Carnian age and the youngest occurs near the Mettlars Brook Member of the Passaic Formation of middle Norian age or possibly in the overlying Ukrainian Member, embracing a stratigraphic thickness of about 2,400 m. At an approximate sedimentation rate of 400,000 years per 100 m, calibrated both by varves in the characteristic and permeating lacustrine rhythmic sedimentary sequences called Van Houten cycles (a detailed description of which is given in Appendix 2) and by published radiometric scales (Olsen 1984a,b), this

works out to roughly 9,600,000 years (or possibly 12,400,000 years if the Ukrainian Member is included) for the known temporal range of *Atreipus*. This is roughly three-quarters of the estimated duration of the Late Triassic.

Atreipus (Coelurosaurichnus) metnerzi comes from the Ansbacher Sandstone of the German Middle Keuper. This unit lies between the Schilfsandstein and the Blasensandstein, both thought to be of late Carnian age (Laemmlen 1956; Gall, Durand, and Muller 1977: but see Chapters 24 and 25). We can, therefore, conclude that both in Europe and in eastern North America, the known range of *Atreipus*, is late Carnian to at least middle Norian (Fig. 6.12).

It is worth noting that the manus impressions of *Atreipus* were overlooked for nearly a century: they were either not recognized or thought to be tracks of another animal. It is possible, therefore, that a number of *Grallator*-like tracks in the litera-

Figure 6.12. Correlation of Sanford, Dan River, Gettysburg, Newark, Hartford, Fundy basins of the Newark Supergroup with the European standard stages, the radiometric scale, and ranges of the Pekin and Connecticut Valley faunules, *Atreipus*, and *Apatopus*. Position of the German *Atreipus metzneri* shown by "A" in the standard stages column. Abbreviations for formations of the Newark basins as follows: AB, Aspers Basalt; B, Boonton Formation; BL, Blomidon Formation; C, Cumnock Formation; CB1, lower member of Cow Branch Formation; CB2, upper member of Cow Branch Formation; G, Gettysburg Shale; HB, Hartford Basin extrusive zone (Talcott Basalt, Shuttle Meadow Formation, Holyoke Basalt, East Berlin Formation, Hampden Basalt); L, Lockatong Formation; LE, Lower Economy beds of Wolfville Formation; MH, McCoy Brook and Scots Bay Formations; NA, New Haven Arkose; NB, North Mountain Basalt; NE, Newark Basin extrusive zone (Orange Mountain Basalt, Feltville Formation, Preakness Basalt, Towaco Formation, Hook Mountain Basalt); NO, New Oxford Formation; P, Pekin Formation; PH, Pine Hall Formation; PH–ST, Pine Hall plus Stoneville Formations; PL, Portland Formation; PS, Passaic Formation; S, Sanford Formation; SF, Stockton Formation; ST, Stoneville Formation; W, Wolfville Formation.

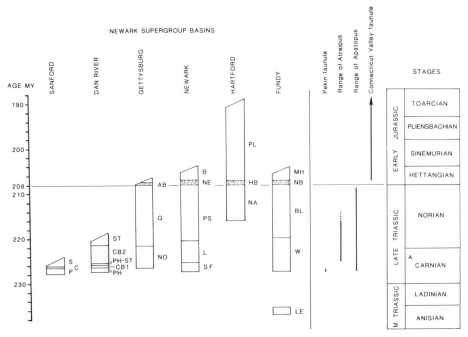

ture may be *Atreipus* with manus impressions that have gone unnoticed. All such Late Triassic material should accordingly be reexamined.

Osteological correlations

Attempts to assign a footprint to a trackmaker should consist of a search for derived and primitive features, just as the placement of a biological taxon within a phylogenetic classification does. These features of the footprint should reflect the osteology of the maker. Only two classes of trackmaker characters are important: first, those that are primitive (plesiomorphic) for a group and that serve *only* to bar membership in the groups possessing derived states of that character and, second, those that are shared with and derived for a group (synapomorphic) and hence uniquely identify membership in a group. Thus, merely showing that an osteological reconstruction from a footprint is identical to the feet of small theropod dinosaurs does not mean that the track might not have been made by a primitive bird, because many of the typical "avian" characters are primitive for theropod dinosaurs *including* birds (Gauthier 1984; Gauthier and Padian, 1985). However, such an osteological reconstruction does indicate that the trackmaker was a dinosaur (including birds) because such structure was attained *uniquely* by that group. Thus, assignment of an ichnotaxon to a particular biological group demands some a priori knowledge of the phylogenetic relationships of the groups under consideration. It is not enough to show that a track and a known skeleton are similar.

The foregoing consideration becomes a special problem with *Atreipus* because its combination of features is unknown in osteological taxa. As we show below, it apparently represents a new kind of dinosaur, at this stage known only by its feet.

According to the rules of osteological reconstruction worked out by Heilmann (1926), Peabody (1948), and Baird (1957), joints between bones in the pes are reconstructed in the center of the pads as they are in cursorial birds (Heilmann 1926; Peabody, 1948; Baird 1957) and in crocodiles (Chapter 20) and large lizards (Padian and Olsen 1984). In many ways, the pes imprints of *Atreipus* and *Grallator* spp. could be attributed to small theropod dinosaurs. It comes as a great surprise, therefore, that *Atreipus* proves to have a manus completely incompatible with those of any known theropod. The *Atreipus* manus imprint shows three to four short digits with only tiny claw imprints, if there are claw imprints at all. All known theropods (except post-*Archaeopteryx* birds) have large trenchant claws on the manus. In addition, the manus of theropods is thought to be a grasping hand, not a walking hand; but *Atreipus* has a theropod-like pes in combination

with a short-clawed, functionally tridactyl manus that was habitually used in locomotion.

The osteological reconstruction of the pes of the North American species of *Atreipus* poses no problems (Fig. 6.13). Obviously, the pedal skeleton was functionally tridactyl, a unique feature of dinosaurs (and their immediate relative *Lagosuchus*). The dinosaurs are divided into ornithischians and saurischians. The proportions of the pes correspond to those of theropods that have a comparatively long digit III; ornithischian pedes tend to have a relatively shorter digit III. It is not known, however, whether theropod-like proportions are primitive for dinosaurs in general or derived for theropods. Therefore, we cannot conclude that the pedal structure excludes ornithischians.

The manus of *Atreipus* is clearly highly derived with respect to the primitive condition in archosaurs (Peabody 1948; Krebs 1976). In the primitive archosaur manus, digits I, II, and III are more robust than IV and V, but all five toes are functional. Less derived archosaur footprints, such as *Chirotherium*, show a five-toed imprint (Peabody 1948). Digit III (or more rarely IV) tends to be the longest, and digit I or V the shortest. The manus of *Atreipus* retains some of this primitive pattern, but is clearly not functionally pentadactyl, and there is a clear shift of emphasis to digits II, III, and IV rather than I, II, and III. In order of increasing length, the digits are I, IV, II, III, with no impression of V. The weakly differentiated pads and very short toes suggest strong digitigrady. The tiny claws of the manus suggest the presence of small, pointed unguals. The manus of saurischians was not primitively tridactyl, although digits I, II, and III were already dominant (as in ornithosuchians and crocodiles). In later theropods, however, there is a continued reduction of the number of digits to three (and two in tyrannosaurs). All

Figure 6.13. Osteological reconstructions of *Atreipus*: **A**, *Atreipus milfordensis*; **B**, *Atreipus sulcatus*; **C**, *Atreipus acadianus*. Scale is 2 cm.

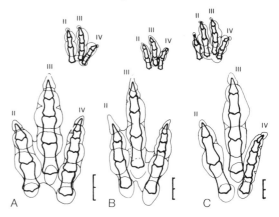

saurischians retain large claws on the dominant manual digits, and all but some carnosaurs retain a large manus, however.

Ornithischians also had a tendency toward functional tridactyly and reduction of trenchant claws in the manus; however, this tendency took at least two different directions. *Heterodontosaurus*, from the Early Jurassic of southern Africa, has a manus that is very similar proportionally to those of some prosauropods and theropods (Bakker and Galton 1974), with large trenchant claws on the non-reduced digits I, II, and III. According to Bakker and Galton (1974), this resemblance is a synapomorphy uniting ornithischians and saurischians, and hence such a manus is primitive for dinosaurs in general. This hypothesis has been severely criticized by a number of authors (Thulborn 1975; Walker 1977), although Gauthier (1984) and other workers accept it.

In contrast, the manus of a number of ornithischians, such as *Leptoceratops, Tenontosaurus,* and *Thescelosaurus* (Brown and Schlaikjer 1942; Galton 1974a) are similar in proportions to the manus in nondinosaurian archosaurs, such as *Euparkeria* (except for the dinosaurian reduction in phalangeal formula of digit IV and V). This resemblance could be interpreted as a shared primitive condition between ornithischians and their nondinosaurian ancestors. A manus with elongated digits I, II, and III with trenchant claws would thus be a synapomorphy of saurischians, and the proportions and massive claws of *Heterodontosaurus* would be convergent. *Atreipus* would thus share the condition of small nontrenchant unguals with predinosaurian archosaurs and ornithischians primitively.

On the other hand, if the prosauropod–heterodontosaur–like manus is primitive for dinosaurs, the manus of *Atreipus* would be derived for

Figure 6.14. Comparison of dorsal views of manus and pedes of various archosaur taxa: *Crocodylus* (adapted from Romer 1956); the prosauropod *Anchisaurus* (adapted from Galton 1976); the heterodontosaurid ornithischian *Heterodontosaurus tuckii* (adapted from Santa Luca 1980); reconstruction of *Atreipus milfordensis* from Figure 6.13, with the addition of hypothetical digits I and IV of the manus (the former based on *A. acadianus*) and digits I and V of the pes; the fabrosaurid ornithischian *Lesothosaurus* (loosely adapted from Thulborn 1972) with the manus reconstruction from illustrations of the actual material restored to the pattern seen in the ichnogenus *Anomoepus* in Figure 6.15; the ornithopod ornithischian *Thescelosaurus* (adapted from Galton 1974a; Romer 1956); and the ornithopod ornithischian *Hypsilophodon foxii* (adapted from Galton 1974b).

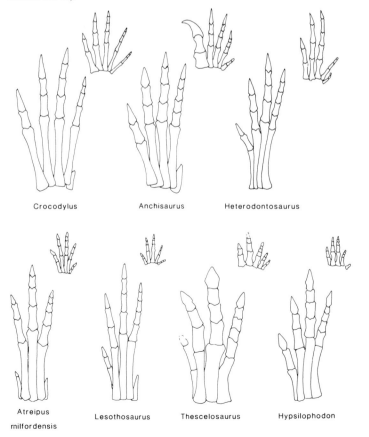

Crocodylus   Anchisaurus   Heterodontosaurus

Atreipus
milfordensis   Lesothosaurus   Thescelosaurus   Hypsilophodon

either group unless it is not a dinosaur. Reduced unguals could be a synapomorphy uniting nonheterodontosaurid ornithischians, including *Atreipus*.

The configuration of the manus in the other early ornithischians adds some weight to the first hypothesis. Unfortunately, the earliest (Carnian) forms, including *Pisanosaurus* (Bonaparte 1976) and new forms from the Newark Supergroup of Nova Scotia and North Carolina (Galton 1983) and the Dockum Group of Texas (S. Chatterjee pers. comm.), are fragmentary, without a well-preserved manus or pes. The next oldest ornithischians, including *Scutellosaurus* (from the Early Jurassic of the Glen Canyon Group of Arizona) and *Lesothosaurus* (from the Early Jurassic Upper Stormberg Group of southern Africa), are represented by much more complete material, although neither has a complete manus. Thulborn (1972) reconstructed the manus of the latter genus to fit the pattern seen in the Cretaceous ornithischian *Hypsilophodon* (Galton 1974b). The manus of *Scutellosaurus* is even less complete than that of *Lesothosaurus*, although Colbert (1981) reconstructed it after the pattern seen in *Lesothosaurus* and *Hypsilophodon*. *Scutellosaurus* provides no additional information except that it does have small pointed unguals.

A different reconstruction of *Lesothosaurus* and *Scutellosaurus* is possible if a different phylogenetic model is used. The very reduced digit V of

Hypsilophodon is surely a derived feature because a more complete digit V is seen in ceratopsians, ankylosaurs, stegosaurs, *Camptosaurus, Tenontosaurus, Thescelosaurus*, and other ornithischians and is almost certainly primitive for the group (e.g., *Heterodontosaurus*). It seems more prudent to reconstruct the manus of *Lesothosaurus* and *Scutellosaurus* according to the primitive rather than a derived pattern. If the manus of *Lesothosaurus* is restored after these latter forms, particularly *Thescelosaurus* (see Galton 1974a), it resembles the reconstructed manus of the the common Early Jurassic ichnite *Anomoepus* (Figs. 6.14 and 6.15). The manus of *Anomoepus* also closely resembles that of the chirotheres (including *Brachychirotherium*) and crocodiles (including *Batrachopus* and *Otozoum*), except for the presence of the dinosaurian specialization of a reduced number of pads on digits IV and V. In addition, the reconstructed pes of *Anomoepus* bears a very strong resemblance to the pedes of all the well-preserved early ornithischians.

Nothing in the manus of the early ornithischians suggests the pronounced tridactyl digitigrady of *Atreipus*. The Early Jurassic *Anomoepus* makes a much better trackway for Early Jurassic gracile ornithischians than does the Late Triassic *Atreipus*. Nonetheless, the differences between the manus of *Atreipus* and *Anomoepus* is basically one of degree, not of kind, with the basic difference being that digitigrady is more prounounced in *Atreipus*.

Digitigrady of the manus is well developed in some later Mesozoic ornithischians, such as iguanodontids and hadrosaurs, but digit I is very reduced in these forms. The manus of *Atreipus* could, however, be regarded as an early expression of that trend. Interestingly, digit I of the pes, which does not impress in *Atreipus*, is lost in these same later Mesozoic groups.

In summary, *Atreipus* is a dinosaur (or very close to being a dinosaur) because of the birdlike, tridactyl pes. However, according to which hypothesis of dinosaur monophyly is followed (Fig. 6.16), *Atreipus* could be either an ornithischian with a more derived condition than *Heterodontosaurus* or a dinosaur less derived than saurischians. In the latter case, it could be either an ornithischian or a very plesiomorphic dinosaurian belonging neither to ornithischians nor saurischians. In any case, the manus of *Atreipus* finds a close dinosaurian counterpart in the ornithischians. However, there are no really close skeletal correlatives now known, and our best guess is that *Atreipus* may have been made by a very early ornithischian with marked quadrupedal, cursorial adaptations, in which a somewhat hadrosaur-like manus was combined with a coelurosaur-like pes that had a short digit I. We have thus reconstructed the maker of *Atreipus* as a gracile ornithischian dinosaur (Fig. 6.17).

Figure 6.15. *Anomoepus*: **A**, composite manus–pes set of *Anomoepus crassus*, left side of standard sitting trackway [based principally on RU main display slab (figured in Olsen 1980a)]; **B**, osteological reconstruction of **A**.

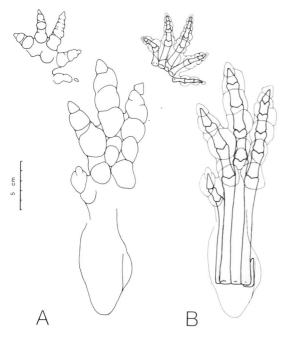

A   B

If this reconstruction is accepted, *Atreipus* and *Anomoepus* reflect the presence of two temporally distinct ornithischian taxa in Newarkian time. *Atreipus*, the Triassic form, has certain derived features, while *Anomoepus*, the Jurassic form, has no apparently derived features (within ornithischians). Both ichnogenera are characteristic of their times, whatever the zoological affinities of their makers.

Appendix 1: the problem of Sauropus barrattii

In 1837, Edward Hitchcock gave the name *Sauroidichnites barrattii* to a natural cast of a manus on a paving stone from Middletown, Connecticut, presumably from the Portland Formation. This specimen (AC 20/4) is illustrated by Olsen and Padian (Chapter 20). On the same surface are one fairly clear small pes impression and two small, poor tracks of the *Grallator* type, as well as at least one *Batrachopus* trackway. There is no obvious relation between the relatively large manus on the slab and the relatively small pedes. In 1841, Hitchcock redescribed and refigured the same manus. He renamed it *Sauropus barrattii* in 1845, referred it to *Anomoepus* in 1848, and renamed it *Chimaera barrattii* in 1858. His son, Charles H. Hitchcock, for good measure, renamed it *Chimaerichnus barrattii* in 1871.

Meanwhile, in 1843 Edward Hitchcock had given the name *Ornithichnites lyellii* to an isolated medium-sized track of what we would call *Grallator* from the Turners Falls Sandstone of Turners Falls, Massachusetts (AC 31/85). In succession, he renamed the same track *Fulicopus lyellianus* (1845), then *Aethyopus lyellianus* (1848), and

finally he referred the species to *Amblonyx* in 1858. In 1858, E. Hitchcock also referred trackways of a grallatorid on a large slab (AC 1/1) from the Turners Falls Sandstone at Gill, Massachusetts to *Amblonyx lyellianus*. Both specimens were designated as types of *Amblonyx lyellianus* by E. Hitchcock in 1865.

Edward Hitchcock (1858) referred one other trackway on AC 1/1 to the genus *Anomoepus*, giving it and another specimen (AC 1/7) the new specific name *A. major*. Hitchcock thought that this trackway was made by an animal sitting down, in the standard sitting pose of *Anomoepus*, because he thought he could see two parallel manus–pes sets and two "heel" impressions posterior to the pedal impressions, as well as a more posterior central "ischial" impression. Let us now trace how this interpretation evolved.

In Plate VIII of Hitchcock (1858), a line drawing, the left pes is drawn entirely with dashes except for two small circular depressions. Both manus impressions were drawn with solid lines and clearly resemble Hitchcock's 1837 rendering of the manus of *Sauropus barrattii*. On the excellent lithograph in the same volume (E. Hitchcock 1858, Plate XXXVIII) only two irregular impressions are shown where Hitchcock had drawn the manus imprints, and only two small depressions show in the line drawing for the left pes.

In 1904, Lull traced the left half of Hitchcock's line drawing of AC 1/1, but turned the dashed lines into solid lines, labeling the illustration *Fulicopus lyellianus*. He also provided a reconstruction showing an ornithopod-like dinosaur in sitting position and listed the type specimens of *F. lyellianus* as AC 1/1 and 31/85 (only the latter, of course, is the true type). At the same time, Lull correctly noted

Figure 6.16. Alternate cladograms of the relationships of *Atreipus* to other dinosaur groups and crocodiles, mostly based on characters of the manus. Characters as follows: 1, "archosaurian" nonpedal characters (from Thulborn 1975; Walker 1977); 2, reduction in number of phalanges in digit IV of the manus (may be synapomorphy of crocodiles plus dinosaurs and may be related to character 4); 3, tridactyl dinosaurian pes; 4, reduction of number of phalanges in manual digits IV and V; 5, "ornithischian" nonpedal characters (from Thulborn 1975; Walker 1977); 6, enlargement of digits I, II, and III of manus; 7, reduction in length of digits I, II, and III of manus; and 8, great reduction of digit I of pes.

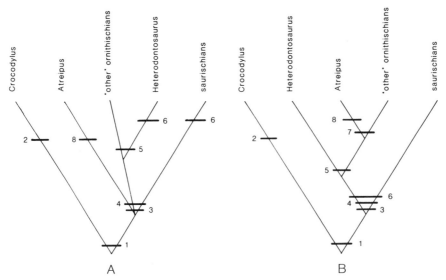

A   B

that the left pes of the specimen referred to *Amblonyx lyellianus* on AC 1/1 and the left pes of the trackway of *Anomoepus major* on the same slab both had a unique deformity or injury and, therefore, must have been made by the same individual. In 1915, Lull synonymized *Sauropus barrattii* with *Anomeopus major* and *Amblonyx lyellianus*, the first having priority, and listed the types as AC 1/1 and AC 1/7 with AC 31/85 being dropped from mention. Lull (1953) retraced all of Hitchcock's (1858) line drawing of *Anomoepus major*, again filling in the dotted lines, and repeated his previous opinion (Lull 1915) on the synonymy of *Sauropus* and the supposed type specimens; he also provided a new reconstruction of the trackmaker.

This convoluted situation requires rectification. First, only the Middletown slab (AC 31/81) can be the holotype of *S. barrattii*, because it was the only specimen figured in 1836 when the genus was named. In regard to this specimen, C. H. Hitchcock (1858, p. 55) stated, "This is the true type of the genus and species Chimaera [*Sauropus*] . . . ." Thus, the type of *Sauropus barrattii* is an isolated manus (as it can hardly belong with any of the several grallatorid tracks on the same slab), and, because it is so

worn as to be indeterminate, we regard the genus and species as *nomina vana*.

Second, the often-figured specimen AC 1/1, which is both a syntype of *Anomoepus major* and incorrectly listed as a type of *Amblonyx lyellianus*, shows no indication whatsoever of manus impressions except for a single irregular mark in the position of the left manus in Hitchcock's line drawing of *A. major*. There are several similar impressions over the surface in other locations. What Hitchcock (1858) figured as the left "heel" of *A. major* proves to be a complete left pes, and the supposed left pes belonging to the same "heel" consists, as shown in Hitchcock's lithograph, of two small impressions that are like dozens of others dotting the slab. The "ischial" impression is very faint and seems to bear no connection to what is clearly a perfectly normal walking trackway, apparently made by the same individual that made the adjacent "*Amblonyx*" trackway. These are clearly two excellent trackways of the *Grallator (Anchisauripus) tuberosus* type and have nothing to do with *Anomoepus*.

Third, the other syntype of *Anomoepus major* (AC 1/7) is a poor but typical *Anomoepus* sitting trackway with a genuine "ischial" impression. With the removal of AC 1/1, AC 1/7 becomes the lectotype of *A. major* by elimination.

Fourth, all the other specimens cited as examples of *Sauropus* (E. Hitchcock, 1858) prove to be either very sloppy and indeterminate tracks (13/2, 13/14) or good specimens of *Anomoepus* (37/9). Therefore, we can conclude that the concept of *Sauropus barrattii* as envisioned by Lull (1904, 1915, 1953) and shown in his figures has no basis in the morphology of any actual footprints. *Sauropus* and its type species, *S. barrattii*, are therefore *nomina vana* based on indeterminate material.

Appendix 2: geological occurrences and associated fossils
Newark Basin

Material referred to *Atreipus* has been found at five localities in the Passaic Formation and two localities in the Lockatong Formation. These localities represent six separate main horizons within the Newark Basin spanning roughly 2,400 m of the Newark Basin section. We begin by discussing the general stratigraphic framework of the Lockatong and Passaic and follow this with descriptions of each Newark Basin locality, its stratigraphic position, and the associated faunal and floral remains.

Stratigraphy of the Lockatong and Passaic formations

As currently understood (Van Houten 1969; Olsen 1980a,b) the Newark Basin is divided into ten formations (Fig. 6.2A) that are in ascending order: the Stockton Formation, Lockatong Formation, Passaic Formation and its lateral equivalent the Hammer Creek Conglomerate, Orange Mountain Basalt, Feltville Formation, Preakness Basalt, Towaco Formation, Hook Mountain Basalt, and Boonton Formation. Only the Lockatong and Passaic formations need be considered here.

The Lockatong and Passaic formations together make up a natural lithologic facies package united by a common theme of repetitive and permeating transgressive–

Figure 6.17. Reconstruction of *Atreipus*: **A**, ventral view of right restored manus–pes set of *A. milfordensis* with hypothetical metapodials and digits I and V in foreshortened perspective (digit I based on *A. acadianus* and digit V is hypothetical); **B**, lateral view of left manus and pes; **C**, hypothetical flesh reconstruction of *A. milfordensis* based on a gracile relatively unspecialized ornithischian and proportioned after the Arcola, Pennsylvania trackways of *A. milfordensis* (Figure 6.5).

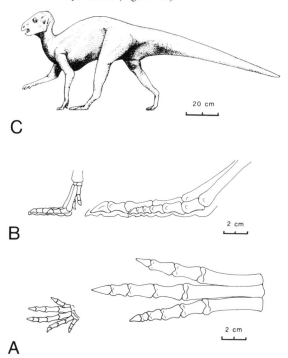

regressive lacustrine cycles (Van Houten 1964, 1969, 1980; Olsen 1980a,b, 1984a,b). The fundamental sedimentary cycle consists of:

1. A lower thin (<1 m), shallow water, transgressive gray siltstone or fine sandstone (division 1) overlain by
2. A 0.5–2-m deeper water (high stand), gray to black, often laminated siltstone (division 2), which grades up into
3. A shallow water, regressive, thicker (1–4 m) gray or red siltstone and sandstone (division 3).

These are called Van Houten cycles (Olsen 1985) after their discoverer, and have a mean thickness of 4–6 m throughout much of the Newark Basin. Very similar cycles characterize most of the finer-grained sediments of the Newark Supergroup.

These transgressive–regressive cycles make up larger compound cycles that are mainly expressed as ordered sequences of Van Houten cycles. They differ principally in the degree to which their division 2 approaches an organic-rich, microlaminated siltstone – the deepest water lithology. Two orders of compound cycles are obvious: a cycle 25 m thick and one 100 m thick. The 25-m cycles consist of four to five Van Houten cycles in which division 2 is better developed in the lower cycles. There is a strong tendency for the upper Van Houten cycles in these sequences to be mostly red. The 100-m cycles consist, in turn, of about four of the 25-m cycles. Division 2 of the contained Van Houten cycles is best developed in the lower 25-m cycles, and the upper 25-m cycles tend to be red. The main difference between the Lockatong and Passaic formations is that most of the 100-m cycles of the Passaic tend to be red, and these red intervals have less well-developed Van Houten cycles. The boundary between the Lockatong and Passaic is operationally defined as where red beds become dominant over gray.

Van Houten cycles and the compound cycles they make up apparently can be traced over very large areas. In the Passaic, the 100-m cycles comprise relatively easily mapped units because of the alternation of thick gray and red units. McLaughlin (1941, 1946, 1948) has traced all of the 100-m cycles of the lower 1,000 m of the Passaic Formation throughout the Hunterdon Plateau Fault Block and has designated the gray portions of the 100-m cycles as individual informal members of the Passaic. These members have since been recognized in the adjacent fault blocks and provide a relatively precise stratigraphy for much of the Newark Basin. A similar framework of members has been proposed for the Lockatong (Olsen 1984a,b).

Some Van Houten cycles have varved beds in division 2. Varve counts from these beds can be used to calibrate the duration of the cycles. Using this method Van Houten (1964, 1969, 1980) and Olsen (1984a,b) estimated the duration of Van Houten cycles at about 21,000–23,000 years and the duration of the compound cycles at 100,000 years for the 25-m cycles and 400,000 years for the 100-m cycles. Fourier analysis of long sections in the Lockatong, comparison of the number of cycles in the Lockatong plus Passaic Formation, and published radiometric scales confirm these estimates of the cycles' duration (Olsen 1984a, b).

Van Houten cycles have a paleontological as well as lithological expression. Fossil fish and crustaceans generally occur in division 2 and reptile footprints in divisions 1 and 3. In those Van Houten cycles with an especially poorly developed division 2, however, reptile footprints can occur in the deepest water portions of the unit. At most localities, *Atreipus* occurs in division 1 or the lower parts of division 3 of Van Houten cycles. These facts allow the bulk of Lockatong and Passaic footprint faunules to be placed in stratigraphic and temporal position with unusual precision.

Atreipus localities and associated biota in the Newark Basin of Newark, New Jersey

Within the city of Newark, a footprint faunule in which *Atreipus* is questionably present was collected in 1971 by Neal K. Resch. It occurs in the Ukrainian Member of Olsen (1984a), which is the Easton Avenue member and Second Precinct Station member of Cornet (1977a) of the Passaic Formation, about 700 m above the Mettlars Brook Member (Fig. 6.11) and 1,000 m below the Orange Mountain Basalt. This poorly preserved faunule, part of which was mentioned by Olsen (1980b, p. 30), is dominated by small grallatorids of uncertain affinites (PU 21516) but also contains single footprints comparable to *Grallator (Anchisauripus) sillimani* (Hitchcock) (PU 21518) and a *Grallator* of the "*A. tuberosus* group" (PU 21516). Pes imprints of the *Atreipus* type (PU 21517) are present, but in each of the three cases the slab terminates immediately anterior to the pes, so that the presence or absence of a manus cannot be determined. Associated quadrupeds include the ubiquitous *Rhynchosauroides brunswickii* (PU 21520), several sizes of *Brachychirotherium* cf. *B. eyermani* (Baird) (PU 23645–23647), and obscure tracks of *Apatopus* sp. (PU 23643–23644).

The associated biota includes conchostracans and a palynoflorule dominated by *Corollina meyeriana*, the oldest such florule in the Newark Basin. This florule can be dated as probably late middle Norian (late Norian of Cornet and Traverse (1975, pp. 27–8), who recognized the Rhaetian as a separate stage). The lateral equivalent of this member at Ukrainian Village, Mercer County, New Jersey has produced well-preserved *Semionotus* sp. and an identical florule. In the same area, red siltstones within 100 m below this member have produced a single, small *Brachychirotherium* sp. (PU 23423).

Lyndhurst–Rutherford area, New Jersey

A diverse collection of footprints was made at a lower Passaic locality during the late 1960s by Mr. Larry Black. Unfortunately, the exact locality was not recorded at the time of collection, and the area has changed so much in the following twenty years that the original site cannot be located. Apparently, the locality was on the east side of the ridge that underlies the towns of Lyndhurst and Rutherford, Bergen County, New Jersey, which places it roughly 1,400 m above the local base of the Lockatong [if the thickness of the Palisades Sill (300 m) is subtracted] and 1,700 m below the Orange Mountain Basalt. The local lithology consists of beds of platy red siltstone and fine sandstone separated by more massive red siltstones and sandstones. At about this level, in the New Brunswick area, 48 km to the southwest, is the Mettlars Brook Member of the Passaic Formation (Olsen 1984a) (Fig. 6.11).

Associated ichnites include *Grallator (Grallator)*

spp., *Grallator (Anchisauripus)* spp., *Coelurosaurichnus* sp. (*sensu stricto*, characterized by a long digit IV), *Brachychirotherium parvum*, *B. eyermani*, *Apatopus* sp., and *Rhynchosauroides brunswickii*.

Perkasie Member, Milford Area, New Jersey

The Perkasie Member (Drake, McLaughlin, and Davis 1961) of the Passaic Formation consists of two sequential 25-m compound cycles, each consisting of two well-developed Van Houten cycles and one red and purple, weakly developed Van Houten cycle. McLaughlin named the lower 25-m cycle N and the upper one O. All of the *Atreipus* from the Milford area comes from division 3 of the second gray and black Van Houten cycle of unit O of the Perkasie Member (Fig. 6.11), and all of those specimens apparently come from the Clark Quarry, not the two supposedly different localities mentioned by Baird (1957).

The Perkasie Member occurs in the highest exposed beds of the Passaic along the Delaware River. Once thought to be the youngest strata in the Newark Basin (Fig. 6.2A), it is now clear that this interval actually lies in the lower Passaic Formation, roughly 2,800 m below the top of the formation. All the overlying strata have been eroded in the Milford area. Correlation of the Milford area with the Sourland Mountain Fault Block and the Newark Basin area (Olsen 1984a) places the Perkasie roughly 800 m below the Mettlars Brook member.

The gray footprint bearing horizons in unit O of the Perkasie that produce the type material of *A. milfordensis* and *A. sulcatus* have also produced the types of *Brachychirotherium parvum* (C. H. Hitchcock 1889), *B. eyermani* (Baird 1957), *Apatopus lineatus* (Bock 1952) and *Rhynchosauroides hyperbates* (Baird 1957), as well as examples of *Grallator (Anchisauripus) parallelus* (E. Hitchcock), *Rhynchosauroides* cf. *R. brunswickii* (Ryan and Willard 1947) (Baird 1957), *Coelurosaurichnus* sp. (same ichnospecies as the Mettlars Brook Member) (PU 23651), and an uncertain small tridactyl form (Baird 1957). Gray sandstones of undetermined position within unit N have produced an important megafossil plant assemblage (Newberry 1888; Bock 1969) including *Glyptolepis platysperma* and *G. keuperiana* (Cornet 1977a), *G. delawarensis* (Bock 1969) (which may be equivalent to one of the preceding species), *Pagiophyllum* spp., and (?) *Cheirolepis munsteri*, *Clathropteris* sp., and *Equisetites* spp. In addition, conchostracans have been found in division 2 of the upper cycle of unit O.

The purplish and red footprint-bearing units overlying the Perkasie have produced, in addition to *A. milfordensis*, the type of *Chirotherium lulli* Bock 1952, cf. *Coelurosaurichnus* sp. [still *sensu stricto* but a different ichnospecies than the two previous occurrences (Baird 1954)], and numerous tracks of a small *Grallator* (PU 19910) associated with *C. lulli*. Gray and red claystones form division 2 of a Van Houten cycle about 50 m above the base of unit O of the Perkasie in the same area (Fig. 6.11). These contain *Semionotus* sp., the conchostracans *Cyzicus* sp. and cf. *Ellipsograpta*, and nondarwinulid ostracodes. Outcrops of these same two units along Route 519 are Cornet's (1977a) localities M–4 and M–3. These, along with a gray unit about 100 m above the Perkasie (Cornet's locality PF–3) contain a pollen and spore assemblage belonging to Cornet's (1977a) and Cornet and Olsen's (1985) lower Passaic–Hiedlersburg Palynoflora. The age of this assemblage is Early Norian (Cornet 1977a; Cornet and Olsen 1985).

Graters Member, Frenchtown, New Jersey

Footprints are common in the red and gray platy siltstones that comprise McLaughlin's (1941, 1946, 1948) Graters Member along Nishisackawick and Little Nishisackawick Creek, east of Frenchtown, New Jersey. This set of two 25-m cycles lies about 400 m below the Perkasie Member. The single *Atreipus* pes from this locality was collected from gray siltstones underlying division 2 of the upper cycle of member H of the Graters Member in Little Nishisakawick Creek adjacent to the boro boundary of Frenchtown, Hunterdon County, New Jersey. Associated ichnotaxa include *Gwyneddichnium minor* and *Rhynchosauroides brunswickii*. Cornet (1977a) has described a palynomorph florule from the underlying cycle of H exposed along Nishisackawick Creek. This assemblage belongs in the transitional zone between the New Oxford–Lockatong Palynoflora and the lower Passaic–Heildersburg Palynoflora and thus appears to be latest Carnian or earliest Norian (Cornet 1977a; Cornet and Olsen 1985).

A probable occurrence of *Atreipus* is mentioned by Lyman (1902) from Fisher's Quarry on the south side of Lodel Creek about 1.6 km northwest of Graterford (Grater's Ford), Pennsylvania. The quarry exposes beds that appear to be just above the Graters Member of the Passaic Formation. Lyman describes several dinosaur footprints with small manus impressions; unfortunately, their present whereabouts are uncertain.

Member F, Frenchtown, New Jersey

Atreipus occurs in abundance in red and purple siltstones about 1.5 m below McLaughlin's (1941, 1948) member F. This horizon is about 75 m below the Graters Member (Fig. 6.11) and is part of a pair of 25-m cycles, which, like the Perkasie and Graters Members, have been traced over a large part of the Newark Basin. In the area north of Frenchtown, along Nishisackawick Creek, Member E becomes wholly red and difficult to separate from the rest of the Passaic. Member F, however, still retains a thin gray-green and purple unit, above which are flaggy red and purple siltstones with abundant reptile footprints. In addition to *Atreipus*, this horizon has produced *Brachychirotherium eyermani* (PU 23639, 23641).

Gwynedd (North Wales), Pennsylvania

The slab containing the lost type of Bock's (1952) *Anchisauripus gwyneddensis* was collected in rubble excavated from the deep railroad cut at Gwynedd, Montgomery County, Pennsylvania (Fig. 6.2A). The precise bed that was the source of this specimen is unknown. Footprints occur at a number of levels within the cut, a detailed section of which is given in Olsen (1984a). The exposed interval represents the lower, but not lowest Lockatong [Gwynedd I and II members of Olsen (1984a, b)], which occurs about 1,000 m below members E and F (Fig. 6.11).

Fossils are abundant in the forty or so Van Houten cycles exposed in the Gwynedd cut, and the types of a large number of Newark vertebrate taxa come from this site. These include the types of *Anchisauripus gwyneddensis*, *Gwyneddichnium elongatum* Bock 1952 (= cf. *Ryncho-*

sauroides sp.), and *G. minor* Bock 1952 (a minute chirotheriid), as well as the osseous taxa *Rhabdopelix longispinis* Cope (1869–70) [now lost specimens possibly consisting of a composite of the tanystropheid *Tanytrachelos* Olsen (1979) and possibly the flying lizard *Icarosaurus* Colbert (1966)]; *Gwynneddosaurus erici* Bock (1945) (another composite, this of a coelacanth and possibly *Tanytrachelos*); *Lysorocephalus eurei* Huene and Bock (1954) [thought to be an amphibian but really a skull roof of *Turseodus* (Baird 1965)] *Gwynneddichtis major, G. gwynneddensis*, and *G. minor* Bock (1959) (the last three may all belong in *Turseodus*); *Cionichthys (Redfieldius) obrai* (Bock 1959), (?) *Semionotus howelli* Bock (1959), *Diplurus (Osteopleurus) newarki* (Bryant 1934), *Diplurus (Pariostegus = Rhabdolepis) gwynneddensis*, and *D. striata* Bock (1959); and *Carinacanthus jepseni* Bryant (1934). Phytosaur teeth, *Synorichthys* sp., conchostracans, ostracodes, and coprolites have also been found. Bock (1946) has described a supposed phyllocarid crustacean from this locality, but the specimen has proved to be the anal plate and associated scales of a fish.

Scraps of plants are not uncommon, but tend to be poorly preserved. An unfortunately large number of taxa have been founded on mostly indeterminate remains from Gwynedd by Bock (1969), including the types of *Brachyphyllum conites, Thujatostrobus triassicus, Gloeotrichata formosa, Stolophorites lineatus, Cycadenia elongata, Cycadospadix gwynneddensis, Zamiostrobus minor, Z. minor, Z. rhomboides, Carpolithus carposerratus, C. amygdalus, Albertia gwynneddensis, Araucarites cylindroides*, and *Pagiophyllum crassifolium*. Most of these are conifer shoots, cone scales, or invertebrate ichnofossils. One of Bock's (1961) forms, *Diplopororundus rugosus* from Gwynedd, is definitely the arthropodan ichnite *Scoyenia*; namely, Wanner's (1889) *Ramulus rugosus*, was based on *Scoyenia* from his *Atreipus* locality. Bryant (1934) cites "*Podozamites*" and Bock (1952) also lists *Pterophyllum powelli* (? = *Zamites powelli*) without locality, and "*Neocalamites* from Gwynedd."

Pollen and spores also occur within portions of Van Houten cycles at Gwynedd. The palynomorph assemblage is dominated by *Patinosporites densus* and nonstriate bisaccates and belongs to Cornet's (1977a,b) New Oxford–Lockatong palynoflora (Cornet and Olsen, 1985). According to Cornet (1977a) and Olsen and Cornet (in press), this assemblage is of late Carnian age.

Arcola, Pennsylvania

A long series of road cuts (now covered) for the Schuylkill Expressway near the towns of Arcola and Oaks, Montgomery County, Pennsylvania exposed roughly 200 m of lower Lockatong Formation [Weehawken and Hoboken Members of Olsen (1984a, b)]. About 100 m of this section were repeated by faults within the outcrops. At the northernmost cut, an enormous (900 m²) area of a single footprint-bearing bed was exposed during construction. This unit was the source for the *Atreipus* trackway shown in Figure 6.6. It lies roughly 100 or 200 m below the Gwynedd horizon and 1,100 m below members E–F of the Passaic Formation.

The exposed sections showed a long sequence of unusually fossiliferous Van Houten cycles. Footprints occur at several intervals in these sections. The main footprint-bearing layer occurs in red siltstones of a division 1 of a Van Houten cycle near the top of the section, with only a green division 2. Very abundant *Rhynchosauroides* cf. *R. brunswickii* and rare, poor dinosaur tracks (?*Atreipus*?) occur in the lower parts of this division. At the main track locality, about 7 cm below the top of division 1, is a very laterally persistent parting surface, traceable over the entire exposure, which is so unusual that it deserves special description. The bed forming the track-bearing surface is about 1 to 2 cm thick. It is finely laminated at the base and ripple laminated at the top; the ripple troughs are filled with fine siltstone so that the upper surface shows no ripples. There are no internal parting planes, but there are small calcareous nodules within the oscillation–ripple bedded portion of the bed. This unit is broken by narrow but deep (+ 30 cm) mud cracks propagated down through at least the overlying 5 cm of division 1. The track-bearing surface is covered by irregular patches of very fine, short (1–4 mm) wavy lines, which we suggest could be the impression of a filamentous algal scum. Large (4–20 cm) cylindrical siltstone tubes puncture the surface at irregular intervals. Because these tubes branch downward in the underlying units, we assume that they represent small trees. Detailed impressions of conifer fronds (cf. *Pagiophyllum simpsoni*) drape around a number of these tubes where they intersect the main footprint surface, proving that the trees lived at the time the footprints were made. Conifer shoots occur sporadically over the rest of the surface. We infer that the trees puncturing the footprint surface produced the foliage present on the surface, although this cannot be proved.

The most common ichnite on the main surface appears to be *Rhynchosauroides hyperbates* Baird (1957), originally known from the Perkasie Member of the Passaic Formation. Many of the individual manus–pes sets have scaly plantar surfaces completely preserved in exquisite detail. Sinuous trackways of this form crisscross the surface, and some trackways come full circle. Because all the material is almost precisely the same size, it is possible that all of these trackways were made by the same individual wandering (stalking?) back and forth. Some of the trackways have belly impressions where the trackmaker seems to have rested; others show a transition from walking to swimming, strongly suggesting that the footprint surface was under a few centimeters of water at the time of impression.

Only two trackways belonging to other ichnotaxa crossed the exposed part of the surface. One faint trackway appears to be referable to *Brachychirotherium eyermani* (YPM 9963), known elsewhere from the Perkasie Member and Stockton Formation of the Newark Basin and the Gettysburg Formation of the Gettysburg Basin. The other trackway was, of course, the *Atreipus milfordensis* of Figure 6.6. The most western tracks of this series are much deeper and sloppier than the more eastern ones, so that the track surface was not of uniform competence. No other dinosaurian tracks could be found on this surface.

No fossils were found in the green claystone of division 2 of this cycle, but small, poor *Rhynchosauroides* occur sporadically in division 3. Other cycles exposed in these cuts contained abundant fragmentary to complete fish [*Diplurus (Osteopleurus) newarki, Diplurus (Pariostegus)* sp., *Turseodus, Synorichthys*, and *Semionotus* spp.], as well as conchostracans, ostracodes, phytosaur teeth and bones, and the burrow *Scoyenia*. Nearby, the old Reading Railroad cut between Arcola and Oaks exposes the same cycles with the same fossils, including abundant *Rhynchosauroides* cf. *brunswickii* from division 1 of the same cycle that produced the *Atreipus* and other ichnites at the road cut. This exposure is Bock's (1959) "Yerkes" locality.

The context of the main track-bearing surface within the cyclic section suggests that the reptiles walked in very shallow water upon a drowned soil that still supported living trees. This surface was already draped with several centimeters of lacustrine mud deposited by a recent transgression of what was shortly to become a perennial lake.

Gettysburg Basin

Newark strata of the Gettysburg Basin of Pennsylvania and Maryland are divided into three formations, which are (from the bottom up): the Gettysburg Formation, made up of mainly red clastics; the New Oxford Formation, consisting of gray and red sandstones and siltstones; and the Hammer Creek Conglomerate, composed principally of red conglomerate (Glaeser, 1963). Near the top of the Gettysburg Formation are the Aspers Basalt and thin, overlying, red and gray clastics, and in the upper middle of the Gettysburg is the gray, black, and red Heidlersburg Member (Stose 1932; Stose and Jonas 1939). The Gettysburg Formation is the fine lateral equivalent of the Hammer Creek Conglomerate, and the latter is the coarse equivalent of the Passaic Formation; together they make up a single cohesive lithosome. Likewise, the New Oxford Formation is laterally continuous with the Stockton and Lockatong formations of the Newark Basin, and these make up another major lithosome.

Atreipus occurs at two localities in the Gettysburg Basin. Both are in the Gettysburg Formation between the Heidlersburg Member and the New Oxford Formation. Wanner's (1889) material was collected from a small quarry opened in a hillside about 1.6 km (1 mi) south of Goldsboro, York County, Pennsylvania. This locality is a calculated 4,120 m above the base of the New Oxford Formation and 2,850 m below the Heidlersburg Member of the Gettysburg Formation. Based solely on Wanner's illustrations, associated fossils include ?*Apatopus* sp., the same small short-toed form that occurs at Milford (Baird 1957), ?*Rhynchosauroides* sp., and the arthropodan burrow *Scoyenia* (Wanner's *Ramulus rugosus*).

The Trostle Quarry ichnofauna was collected by Elmer R. Haile, Jr. in 1937 from the quarry located on Burmudian Creek, 4.8 km (3 mi) east of York Springs near Burmudian Springs (Pondtown), Adams County, Pennsylvania. The horizon is roughly 6,000 m above the base of the New Oxford Formation, within 450 m of the base of the Heidlersburg Member of the Gettysburg shale, and about 3,900 m below the base of the Jurassic Aspers Basalt. Other ichnotaxa from this locality include *Brachychirotherium eyermani*, an odd pentadactyl stubby-toed form suggestive of a dicynodont foot (Stose and Jonas 1939, Plate 22 lower right; Baird 1957), *Rhynchosauroides brunswickii*, and *Scoyenia*.

Dan River Basin

Four formations comprise the Dan River Group in the contiguous Dan River and Danville basins of Virginia and North Carolina: the Pine Hall and Dry Fork formations, of gray and buff sandstone and red siltstone; the Cow Branch Formation, made up of gray and black siltstones and sandstones; and the Stoneville Formation of red siltstone and conglomerate (Meyertons 1963; Thayer 1970). The Pine Hall and Stoneville formations are the basal units and complexly interfinger with the other three formations in all directions.

The Cow Branch Formation *Atreipus* was found in the Solite Company Quarry in Leakville Junction, Virginia–North Carolina. The Solite Quarry exposes more than 300 m of the upper member of the Cow Branch Formation (Olsen et al. 1978). The footprint-bearing block was not found in place, but does come from the currently active part of the quarry, which limits its origin to the middle 150 m of the section measured by Olsen et al. (1978) and Olsen (1979, 1984a). Like the Lockatong, the Cow Branch Formation consists of transgressive–regressive Van Houten cycles (Olsen 1984a), but at this point the exact cycle from which the *Atreipus* originated would be very difficult to determine.

A large and varied faunal and floral assemblage has been found in the Cow Branch Formation in the Solite Quarry. Ichnites from this locality include *Atreipus* sp. and *Grallator (Grallator)* spp. (Olsen et al. 1978). Vertebrate skeletal taxa found in association include *Rutiodon* sp., the type of *Tanytrachelos ahynis*, *Turseodus* spp., *Cionichthys* sp., *Synorichthys* sp., *Semionotus* sp., *Diplurus (Pariostegus)* sp., and a new fish, possibly a pholidophorid. Conchostracans, ostracodes, a large new crustacean, and diverse insects are abundant. Nineteen nominal megafloral taxa and a small palynoflora have also been recovered (Robbins 1982). The diverse faunule and megaflorule have been reviewed in detail by Olsen et al. (1978).

Fundy Basin

Klein (1962) and Keppie et al. (1979) divide the Nova Scotian portion of the Fundy Group into five formations, which are from the bottom up: the Wolfville Formation (0–400 m), of coarse red and brown clastics; the Blomidon Formation (10–350 m), of red siltstone and sandstone; the North Mountain Basalt (250 m); and the Scots Bay Formation (0–20 m) of gray and white limestone chert and brown sandstone, which is laterally equivalent to the McCoy Brook Formation (+ 300 m) of red clastics.

We have recovered slabs of footprints from just below the contact between the Wolfville and Blomidon formations along the cliffs and on the beach just southeast of Paddy Island, Kings County, Nova Scotia (Fig. 6.10). The footprint-bearing unit appears to be a single bed of brown sandstone overlying a thin red siltstone in the uppermost Wolfville 8.5 m below what is locally the basal, horizontally bedded sandstones of the Blomidon Formation. The footprint level is 18.6 m below the lowest red mudstone of the Blomidon in the same section. This works out to about 360 m below the North Mountain Basalt. Abundant *Rhynchosauroides* occur on the same slabs as the *Atreipus*, and two trackways of *Brachychirotherium parvum* have also been found. In addition, a procolophonid skull with partial skeleton (apparently *Hypsognathus* sp.) have been found about 6.8 m below the track-bearing horizon. This faunule will be the subject of a more detailed future paper.

Abbreviations of cited repositories

AC, Pratt Museum of Geology, Amherst College, Amherst, Massachusetts

AMNH, American Museum of Natural History, New York, New York

ANSP, Academy of Natural Sciences, Philadelphia, Pennsylvania

CM, Carnegie Museum of Natural History, Pittsburgh, Pennsylvania

LC, Geology Department of Lafayette College, Easton, Pennsylvania

MCZ, Museum of Comparative Zoology of Harvard University, Cambridge, Massachusetts

PU, Museum of Natural History, Princeton University, Princeton, New Jersey, collection now housed at Yale University

RU, Rutgers University, Geological Museum, New Brunswick, New Jersey

YPM, Peabody Museum of Natural History of Yale University, New Haven, Connecticut

Acknowledgments

We thank the following for their invaluable assistance in the field at various times over the past twenty years: Mark Anders, Cynthia J. Banach, Richard Boardman, Alton Brown, George R. Frost, Eldon George, Robert Grantham, C. H. Gover, Donald Hoff, John R. Horner, Chris Laskowitz, Patrick Leiggi, James Leonard, Anthony Lessa, E. F. X. Lyden, Harold Mendryk, O. R. Patterson III, Gustav Pauli, Neal K. Resch, Robert Salkin, Robert F. Salvia, Joseph Smoot, Steven Steltz, Gilbert Stucker, William F. Take, and Richard S. Upright. We gratefully thank Lawrence Black for access to his personal collection and thank the staff of the cited repositories for access to their collections. Bruce Cornet, Franklyn B. Van Houten, Warren Manspeizer, and Alfred Traverse have been most generous in sharing with us considerable unpublished information. Finally, research for this work by P.O. was supported by National Science Foundation Grants to K.S. Thomson (Nos. BMS 75–17096m, BMS 74–07759, GS 28823X, DEB 77–08412, and DEB 79–21746) from 1975 to 1983, and by a fellowship from the Miller Institute for Basic Research in Science at the University of California at Berkeley during 1983 and 1984. P.O. also thanks Kevin Padian and the staff of the Paleontology Department at the University of California at Berkeley for hospitality and support during 1983–4. Support for research for D. B. was supplied by the William Berryman Scott Fund of Princeton University.

References

Baird, D. 1954. *Chirotherium lulli*, a pseudosuchian reptile from New Jersey. *Mus. Comp. Zool. (Harvard Univ.), Bull.* 111: 163–92.

1957. Triassic reptile footprint faunules from Milford, New Jersey. *Mus. Comp. Zool. (Harvard Univ.), Bull.* 117: 449–520.

1965. Paleozoic lepospondyl amphibians. *Am. Zool.* 5: 287–94.

1984. Lower Jurassic dinosaur footprints in Nova Scotia. *Ichnol. Newsl.* 14: 2.

Baird, D., and P. E. Olsen. 1983. Late Triassic herpetofauna from the Wolfville Fm. of the Minas Basin (Fundy Basin) Nova Scotia, Can. *Geol. Soc. Am., Abst. Prog.* 15(3): 122.

Baird, D., and O. F. Patterson, III. 1967. Dicycnodont-archosaur fauna in the Pekin Formation (Upper Triassic) of North Carolina. *Geol. Soc. Am., Spec. Paper.* 115: 11.

Bakker, R. T., and P. M. Galton. 1974. Dinosaur monophyly and a new class of vertebrates. *Nature (London)* 248: 168–72.

Ballard, R. D., and E. Uchupi. 1980. Triassic rift structure in the Gulf of Maine. *Am. Assoc. Petrol. Geol.* 59(7): 1041–72.

Bock, W. 1945. A new small reptile from the Triassic of Pennsylvania. *Notulae Naturae* 154: 1–8.

1946. New crustaceans from the Lockatong of the Newark Series. *Notulae Naturae* 183: 16 pp.

1952. Triassic reptilian tracks and trends of locomotive evolution. *J. Paleontol.* 26: 395–433.

1959. New eastern American Triassic fishes and Triassic correlations. *Geol. Cent. Res. Ser. (North Wales, Pa.).* 1: 1–189.

1961. New fresh water algae of the Eastern American Triassic. *Proc. Penn. Acad. Sci.* 35: 77–81.

1969. The American Triassic flora and global correlations. *Geol. Cent. Res. Ser.* 3–4: 1–340.

Bonaparte, J. F. 1976. *Pisanosaurus mertii* Casamiquela and the origin of the Ornithischia. *Jr. Paleontol.* 50: 808–20.

Brown, B., and E. M. Schlaikjer. 1942. The skeleton of *Leptoceratops* with the description of a new species. *Am. Mus. Novitates* 1169: 1–15.

Bryant, W. L. 1934. New fishes from the Triassic of Pennsylvania. *Am. Phil. Soc. Proc.* 73: 319–26.

Colbert, E. H. 1966. A gliding reptile from the Triassic of New Jersey. *Am. Mus. Novit.* 2246: 1–23.

1981. A primitive ornithischian dinosaur from the Kayenta Formation of Arizona. *Mus. N. Ariz. Press., Bull. Ser.* 53: 1–61.

Cope, E. D. 1869–1870. Synopsis of the extinct Batrachia, Reptilia and Aves of North America. *Trans. Am. Phil. Soc.* 14: 1–252.

Cornet, B. 1977a. The palynostratigraphy and age of the Newark Supergroup, Ph.D. thesis, Department of Geosciences, University of Pennsylvania.

1977b. Preliminary investigation of two Late Triassic conifers from York County, Pennsylvania. *In* Romans, R. C. (ed.), *Geobotany*. (New York: Plenum), pp. 165–72.

Cornet B., and A. Traverse. 1975. Palynological contribution to the chronology and stratigraphy of the Hartford Basin in Connecticut and Massachusetts. *Geosci. Man.* 11: 1–33.

Cornet, B., A. Traverse, and N. G. McDonald. 1973. Fossil spores, pollen, fishes from Connecticut indicate Early Jurassic age for part of the Newark Group. *Science* 182: 1243–7.

Cornet, B., and P. E. Olsen. 1985. A summary of the biostratigraphy of the Newark Supergroup of eastern North America, with comments on early Mesozoic provinciality. III Congreso Latin-Amer. Paleontología. Mexico. Simposio sobre floras del Triasico tardio, su fitogeografía y paleoecología. Memoria, pp. 67–81.

Drake, A. A., D. B. McLaughlin, and R. E. Davis. 1961. Geology of the Frenchtown Quadrangle, New Jersey - Pennsylvania. *U. S. Geol. Surv. Quad.*, Map: GQ 133.

Eyerman, J. 1886. Footprints on the Triassic sandstone (Jura-Trias) of New Jersey. *Am. J. Sci.* 131: 72.

1889. Fossil footprints from the Jura(?)–Trias of New Jersey. *Acad. Nat. Sci. Phil., Proc.* 1889: 32–3.

Faille, R. T. 1973. Tectonic development of the Triassic Newark-Gettysburg Basin in Pennsylvania. *Geol. Soc. Amer. Bull.* 84: 725–40.

Froelich, A. J., and P. E. Olsen. 1984. Newark Supergroup, a revision of the Newark Group in eastern North America. *U.S. Geol. Surv. Bull.* 1537-A: A55–8.

Gall, J.-C., M. Durand, and E. Muller. 1977. Le Trias de part d'autre du Rhin. Correlations entre les marges et le centre du bassin germanique. *Bur. Rech. Geol. Geophys. Min., 2nd ser. sec. IV*, no. 3: 193–204.

Galton, P. M. 1974a. Notes on *Thescelosaurus*, a conservative ornithopod dinosaur from the upper Cretaceous of North America, with comments on ornithopod classification. *J. Paleontol.* 48: 1048–67.

1974b. The ornithischian dinosaur *Hypsilophodon* from the Isle of Wight. *Brit. Mus. (Nat. Hist.), Geol. Bull.* 25: 1–152.

1976. Prosauropod dinosaurs (Reptilia - Saurischia) of North America. *Postilla* 169: 1–98.

1983. The oldest ornithischian dinosaurs in North America from the Late Triassic of Nova Scotia, North Carolina and Pennsylvania. *Geol. Soc. Amer., Abst. Prog.* 15(3): 122.

Gauthier, J. A. 1984. A Cladistic Analysis of the Higher Systematic Categories of the Diapsida. Ph.D. thesis, Department of Paleontology, University of California, Berkeley.

Gauthier, J. A., and K. Padian. 1985. Phylogenetic, functional, and aerodynamic analyses of the origin of birds and their flight. *In* Hecht, M. K., J. H. Ostrom, G. Viohl, and P. Wellnhofer (eds.), *The Beginnings of Birds. Proceedings of the International Archaeopteryx Conference, Eichstatt, 1984*, pp. 185–97.

Glaeser, P. 1963. Lithostratigraphic nomenclature of the Newark–Gettysburg Basin. *Pa. Acad. Sci., Proc.* 37: 179–88.

Haubold, H. 1969. Ichnia Amphibiorum et Reptiliorum fossilium. *Handbuch der Paleoherpetologie* (Stuttgart: Gustav Fischer).

Heilmann, G. 1926. *The Origin of Birds* (Appleton, New York).

Heller, F. 1952. Reptilfaehrten-Funde aus dem Ansbacher Sandstein des Mittleren Keupers von Franken. *Geol. Blaett. NO-Bayern.* 2: 129–41.

Hitchcock, C. H. 1871. [Account and complete list of the Ichnozoa of the Connecticut Valley.] Walling and Gray's Official Topographic Atlas of Massachusetts (Boston: Walling and Gray), pp. xx–xxi.

Hitchcock, C. H. 1889. Recent progress in ichnology. *Proc. Boston Soc. Nat. Hist.* 24: 117–27.

Hitchcock, E. 1836. Ornithichnology. Description of the footmarks of birds (*Ornithichnites*) on New Red Sandstone in Massachusetts. *Am. J. Sci.* 29: 307–40.

1837. Fossil footsteps in sandstone and graywacke. *Amer. J. Sci.* 32(1): 174–6.

1841. *Final Report on the Geology of Massachusetts*, Pt. III. *Amherst and Northampton*, Amherst, pp. 301–714.

1843. Description of five new species of fossil footmarks, from the red sandstone of the valley of the Connecticut River. *Assoc. Am. Geol. Natural. Trans.* 1843: 254–64.

1845. An attempt to name, classify, and describe the animals that made the fossil footmarks of New England. *Proc. 6th Mtg., Am. Assoc. Geol. Naturalists, New Haven, Conn.*, pp. 23–65.

1847. Description of two new species of fossil footmarks found in Massachusetts and Connecticut, or of the animals that made them. *Am. Jour. Sci.* 4(2): 46–57.

1848. An attempt to discriminate and describe the animals that made the fossil footmarks of the United States, and especially New England. *Mem. Am. Acad. Arts Sci.* 3(2):129–256.

1858. *Ichnology of New England. A Report on the Sandstone of the Connecticut Valley, Especially Its Fossil Footmarks* (Boston: William White).

1865. *Supplement to the Ichnology of New England* (Boston: Wright and Potter).

Hope, R. C., and O. F. Patterson, III. 1969. Triassic flora from the Deep River Basin, North Carolina. *N. C. Dept. Conserv. Dev., Spec. Publ.* 2: 1–22.

Huene, F. von, and W. Bock. 1954. A small amphibian skull from the Upper Triassic of Pennsylvania. *Wagner Free Inst. Sci., Bull.* 29: 27–34.

Keppie, D., D. J. Gregory, A. K. Chatterjee, N. A Lyttle, and G. K. Muecke. 1979. *Geologic Map of Nova Scotia* (Nova Scotia Dept. Mines Energy, Halifax).

Klein, G. deV. 1962. Triassic sedimentation, Maritime provinces of Canada. *Geol. Soc. Am., Bull.* 73: 1127–46.

Krebs, B. 1976. *Pseudosuchia. Handbuch der Paläoherpetology*. Pt. 13, *Thecodontia* (Stuttgart: Gustav Fischer), Verlag. pp. 40–98.

Laemmlen, M. 1956. Keuper. *Lex Stratigr., Intern., I, Eur.* 5(2): 1–335.

Lull, R. S. 1904. Fossil footmarks of the Jura–Trias of North America. *Boston Soc. Nat. Hist.* 5: 461–557.

1915. Triassic Life of the Connecticut Valley. *Conn. State Geol. Nat. Hist. Surv., Bull.* 24: 1–285.

1953. Triassic Life of the Connecticut Valley. *Conn. State Geol. Nat. Hist. Surv., Bull.* 81: 1–331.

Lyman, B. S. 1902. Lodel and Skippack Creek. *Acad. Nat. Sci., Proc.* 53(1901): 604–7.

McLaughlin, D. B. 1941. The Revere Well and Triassic stratigraphy. *Pa. Acad. Sci. Proc.* 17: 104–10.

— 1946. The Triassic rocks of the Hunterdon Plateau, New Jersey. *Penn. Acad. Sci. Proc.* 20: 89–98.

— 1948. Continuity of strata in the Newark Series. *Mich. Acad. Sci. Paps.* 32(1946): 295–303.

Meyertons, C. T. 1963. Triassic formations of the Danville Basin. *Va. Div. Min. Res. Rept. Inv.* 6: 1–65.

Newberry, J. S. 1888. Fossil fishes and fossil plants of the Triassic rocks of New Jersey and the Connecticut Valley. *U.S. Geol. Surv., Mono.* 14: 1–152.

Nutter, L. J. 1978. Hydrogeology of the Triassic rocks of Maryland. *Md. Geol. Surv., Rept. Inv.* 26: 1–37.

Olsen, P. E. 1979. A new aquatic eosuchian from the Newark Supergroup (Late Triassic–Early Jurassic) of North Carolina and Virginia. *Postilla.* 176: 1–14.

— 1980a. A comparison of the vertebrate assemblages from the Newark and Hartford Basins (early Mesozoic, Newark Supergroup) of eastern North America. *In* Jacobs, L. L. (ed.), *Aspects of Vertebrate History: Essays in Honor of Edwin Harris Colbert* (Flagstaff, Arizona: Museum of Northern Arizona Press), pp. 35–53.

— 1980b. Triassic and Jurassic Formations of the Newark Basin. Field Studies in New Jersey Geology and Guide to Field Trips. *52nd Ann. Mtg. New York State Geol. Assoc., Newark Coll. Arts Sci.,* Rutgers Univ., Newark, pp. 2–39.

— 1980c. Fossil great lakes of the Newark Supergroup in New Jersey. Field Studies in New Jersey Geology and Guide to Field Trips, *52nd Ann. Mtg. New York State Geol. Assoc. Newark Coll. Arts Sci., Rutgers Univ., Newark,* pp. 352–98.

— 1981. Comment on "Eolian dune field of Late Triassic age, Fundy Basin, Nova Scotia." *Geology* 9: 557–61.

— 1984a. Comparative Paleolimnology of the Newark Supergroup: A Study of Ecosystem Evolution. Ph.D. thesis, Biology Department, Yale University.

— 1984b. Periodicity of lake-level cycles in the Late Triassic Lockatong Formation of the Newark Basin (Newark Supergroup, New Jersey and Pennsylvania). Milankovitch and Climate. NATO Symposium (Dordrecht: D. Reidel Publishing).

— 1985. Distribution of organic-rich lacustrine rocks in the early Mesozoic Newark Supergroup rocks. *U.S. Geol. Surv. Circ.* 946: 61–4.

Olsen, P. E., and D. Baird. 1982. Early Jurassic vertebrate assemblages from the McCoy Brook Fm. of the Fundy Group (Newark Supergroup), Nova Scotia, Canada. *Geol. Soc. Am., Abst. Progr.* 14(1–2): 70.

Olsen, P. E., and P. M. Galton. 1977. Triassic–Jurassic tetrapod extinctions: are they real? *Science* 197: 983–6.

— 1984. A review of the reptile and amphibian assemblages from the Stormberg of southern Africa, with special emphasis on the footprints and the age of the Stormberg. *Palaeontol. Afr. (Haughton Memorial Volume)* 25: 87–110.

Olsen, P. E., A. R. McCune, and K. S. Thomson. 1982. Correlation of the early Mesozoic Newark Supergroup by Vertebrates, principally fishes. *Am. J. Sci.* 282: 1–44.

Olsen, P. E., C. L. Remington, B. Cornet, and K. S. Thomson. 1978. Cyclic change in Late Triassic lacustrine communities. *Science* 201: 729–33.

Padian, K., and P. E. Olsen. 1984. Footprints of the Komodo Dragon and the trackways of fossil reptiles. *Copeia* 1984: 662–71.

Palmer, A. R. 1983. The Decade of North American Geology 1983 Time Scale. *Geology* 11: 503–4.

Peabody, F. E. 1948. Reptile and amphibian trackways from the Lower Triassic Moenkopi Formation of Arizona and Utah. *Univ. Calif. Dept. Geol. Sci. Bull.* 27: 295–468.

Robbins, E. I. 1982. "Fossil Lake Danville": The Paleoecology of a Late Triassic Ecosystem on the North Carolina–Virginia Border. Ph.D. thesis, Department of Geosciences, Pennsylvania State University.

Romer, A. S., 1956. *Osteology of the Reptiles* (Chicago: University of Chicago Press).

Ryan, J. D., and Willard, B. 1947. Triassic footprints from Bucks County, Pennsylvania. *Penn. Acad. Sci., Proc.* 21: 91–3.

Santa Luca, A. P. 1980. The postcranial skeleton of *Heterodontosaurus tucki* (Reptilia, Ornithischia) from the Stormberg of South Africa. *Ann. S. Afr. Mus.* 79: 159–211.

Stose, G. W. 1932. Geology and mineral resources of Adams County, Pennsylvania. *Penn. Geol. Surv., 4th Ser., Bull.* C1: 1–153.

Stose, G. W., and A. I. Jonas. 1939. Geology and Mineral resources of York County, Pennsylvania. *Penn. Geol. Surv., 4th. Ser., Bull.* C67: 1–199.

Tagg, A. R., and E. Uchupi. 1966. Distribution and geologic structure of Triassic rocks in the Bay of Fundy and the northern part of the Gulf of Maine. *U.S. Geol. Soc., Prof. Pap.* 550B: B95–8.

Thayer, P. A. 1970. Stratigraphy and geology of Dan River Triassic Basin, North Carolina. *Southeast. Geol.* 12: 1–31.

Thulborn, R. A. 1972. The post-cranial skeleton of the Triassic ornithischian dinosaur *Fabrosaurus australis. Palaeontology* 15: 20–60.

— 1975. Dinosaur polyphyly and the classification of archosaurs and birds. *Aust. J. Zool.* 23: 249–70.

Van Houten, F. B. 1964. Cyclic lacustrine sedimentation, Upper Triassic Lockatong Formation, central New Jersey and adjacent Pennsylvania. *Geol. Surv. Kansas, Bull.* 169: 497–531.

— 1969. Late Triassic Newark Group, north central New Jersey, and adjacent Pennsylvania and New York. *In* Subitzki, S. S. (ed.), *Geology of Selected Areas in New Jersey and Eastern Pennsylvania* (Rutgers University Press, New Brunswick, New Jersey), pp. 314–47.

— 1980. Late Triassic part of the Newark Supergroup, Delaware River Section, West-Central New Jersey. Field Studies in New Jersey Geology and

Guide to Field Trips, *52nd Ann. Mtg. New York State Geol. Assoc., Newark Coll. Arts Sci. Rutgers Univ. Newark*, pp. 264–76.

Walker, A. 1977. Evolution of the pelvis in birds and dinosaurs. *In* Andrews, S., R. Miles, and A. Walker (eds.), *Problems in Vertebrate Evolution.* (London: Academic Press), pp. 319–57.

Wanner, A. 1889. The discovery of fossil tracks, algae, etc., in the Triassic of York County, Pennsylvania, *Penn. Geol. Surv. Ann. Rept. (1887)*, pp. 21–35.

Willard, B. 1940. Manus impressions of *Anchisauripus* from Pennsylvania. *Proc. Penn. Acad. Sci.* 14:37–9.

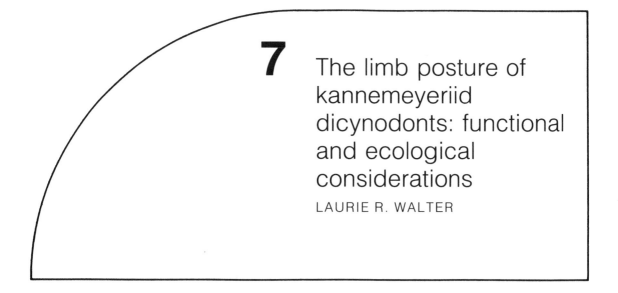

7

The limb posture of kannemeyeriid dicynodonts: functional and ecological considerations

LAURIE R. WALTER

Introduction

The dicynodonts were among the most abundant of Permian tetrapods and were also important, although less numerous, in Triassic terrestrial communities. The unique, double-condylar structure of their jaw joint and replacement of teeth with a horny "beak" may have allowed dicynodonts to feed on plants that earlier herbivores could not process (Crompton and Hotton 1967).

A change in the dicynodont fauna occurred at or near the Permo-Triassic boundary. Typical Permian dicynodonts (the genus *Dicynodon* and its allies) disappeared, to be replaced in the earliest Triassic by the amphibious dicynodont *Lystrosaurus* and in the Early through Late Triassic by dicynodonts of the Family Kannemeyeriidae (Anderson and Anderson 1970).

In addition to being the last of the dicynodonts, kannemeyeriids were the giants of the group (body size 1–3 m). They were the largest herbivores in many Triassic ecosystems long after small- to medium-sized herbivore niches previously occupied by dicynodonts has been taken over by cynodonts, bauriamorphs, and rhynchosaurs (Anderson and Anderson 1970; Gow 1978).

Some authors (Charig 1972, 1980; Bonaparte 1982) have attributed the decline of dicynodonts and other therapsids in the Triassic to competitive displacement by or predation pressure from archosaurs. They asserted that archosaurs had locomotor superiority over therapsids because archosaurs had attained an upright hindlimb posture and were thus able to run quickly with little metabolic expenditure. Therapsids were, on the other hand, conservative in limb stance and locomotion. Such differential locomotor advancement could have been a potent factor in competitive interaction between the groups.

I will attempt to answer the question of whether a particular family of therapsids, the kan-

nemeyeriids, were inferior in locomotor ability, as demonstrated by hindlimb posture, to contemporary archosaurs and will discuss the larger question of whether kannemeyeriids were driven to extinction by these archosaurs. In order to place the locomotor adaptations of kannemeyeriids within the proper context, I will begin with a brief review of the evolution of locomotion in early tetrapods, pelycosaurs, and therapsids and then discuss locomotor posture in kannemeyeriids and compare it with posture of archosaurs.

The evolution of locomotion
Locomotion in early tetrapods

Romer (1922) discussed the locomotion of the first tetrapods. Their limb stance was sprawling, and the propodials extended laterally from the body in a horizontal plane. The pectoral girdle was massive; the platelike scapulocoracoids and strutlike clavicles accommodated the transverse stresses generated by the sprawling limb posture (Jenkins 1971a; Sues 1986). The ischium and pubis were also massive and the ilium small; the only expansion of the iliac blade was a posterior flange to accommodate caudal musculature. The glenoid and the acetabulum were broad, laterally facing, shallow cavities.

Edwards (1977) demonstrated in work on salamanders that the three modes of progression in tetrapod locomotion are girdle rotation, limb retraction, and limb rotation. Girdle rotation requires a sprawling limb posture and considerable lateral undulation of the axial skeleton; it is a "primitive" mode of progression for tetrapods and relies on a rhipidistian grade of muscular and neural control. Active limb retraction is most effective in a tetrapod with a nonundulating vertebral column and well-developed limbs and is a more "advanced" component of locomotion than girdle rotation (Zug 1971). Propodial rotation with a fixed position of the epipodials, which Edwards compared to the action of a wheel and axle, functions best in animals with a straight vertebral column and propodials held perpendicular to the sagittal plane.

In primitive tetrapods, only a small proportion of the limb stride length was due to active retraction. Most of the stride was the result of rotation of the propodials about their longitudinal axes and girdle rotation by lateral undulation of the axial skeleton (Zug 1971; Edwards 1977).

Locomotion in pelycosaurs and therapsids

Despite many modifications of cranial morphology and dentition, the pelycosaurs were conservative in limb structure. Their limb girdles remained massive, with a broad, shallow, laterally facing glenoid and acetabulum. The propodials developed a longer shaft region but retained other primitive features to the extent that Romer (1956) used the humerus and femur of *Dimetrodon* as examples of the primitive morphology of these bones.

In the therapsids, a number of changes in limb osteology took place. The scapula narrowed, and the glenoid became a semilunar concavity at the posterior edge of the scapulocoracoid. The precoracoid contributed less to glenoid formation in therapsids and eventually was excluded completely. These changes in glenoid morphology and position allowed the elbow to be "swung in," bringing the humerus

closer to a sagittal plane, though it remained horizontal (Fig. 7.1)

Several changes in pelvic girdle osteology and myology occurred in therapsids. The primitive posterior flange of the ilium was retained (Fig. 7.2), but because of the reduction in tail length that took place in therapsids, it no longer functioned primarily as an attachment for caudal ligaments. The posterior portion of the ilium may instead have been the origin of an important hindlimb retractor muscle, the iliofibularis. The mammalian homologue of the iliofibularis is the biceps femoris, originating on the ischial tuberosity. Romer (1922) felt that the iliofibularis origin could have migrated to the ischial tuberosity at any point in therapsid history; I have restored the iliofibularis origin on the posterior ilium in kannemeyeriids (see Fig. 7.11), as Romer did for *Cynognathus*, while acknowledging that an origin on the ischium is also plausible.

An innovative aspect of pelvic morphology in therapsids was the development of a large anterior expansion of the iliac blade (Fig. 7.2). The iliofemoralis, originating over most of the lateral surface of the ilium (see Fig. 7.11), inserted on the greater trochanter of the femur, which was also large in advanced therapsids (Figs. 7.3, 7.4).

The mammalian homologue of the iliofemoralis is the gluteal complex of muscles. The gluteal muscles and the biceps femoris (the iliofemoralis and iliofibularis of therapsids) are two of the major fem-

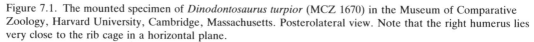

Figure 7.1. The mounted specimen of *Dinodontosaurus turpior* (MCZ 1670) in the Museum of Comparative Zoology, Harvard University, Cambridge, Massachusetts. Posterolateral view. Note that the right humerus lies very close to the rib cage in a horizontal plane.

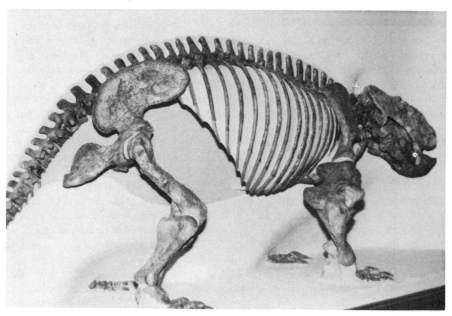

oral retractors in mammals. The remaining "hamstring" retractor muscles of mammals, the semimembranosus and semitendinosus, are represented in therapsids by the flexor tibialis internus and flexor tibialis externus, running from the ischial tuberosity to the proximal tibia (Figs. 7.5, 7.6). All of these muscles were well developed in therapsids, the iliofemoralis becoming particularly large in such advanced groups as kannemeyeriids and cynodonts.

Concomitant changes in femoral morphology in therapsids included medial inflection of the head, which brought the femur to a semiupright position, more directly beneath the body (Fig. 7.3). In the classic mammalian stance, the limbs are vertical and limb excursion is in a parasagittal plane [e.g., Charig (1972); see, however, Jenkins (1970, 1971b) for a demonstration that many mammals do not employ this "mammalian" limb posture]. Thus, the ther-

Figure 7.2. Right ilium, ischium, femur, and proximal tibia of the mounted specimen of *Dinodontosaurus turpior*. Lateral view.

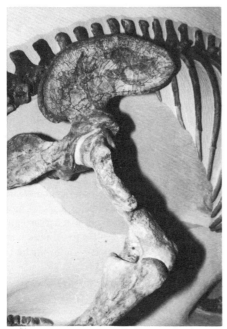
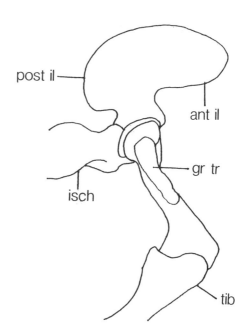

Figure 7.3. Right ilium and proximal femur of the mounted specimen of *Dinodontosaurus turpior*. Anterior view. Note that abduction of the hindlimb would cause dislocation of the hip joint.

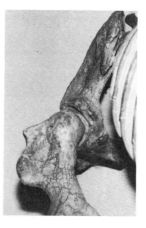
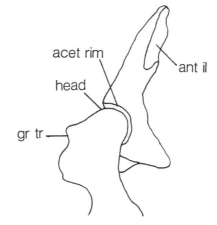

apsids had made numerous advances toward the mammalian condition both in pelvic girdle morphology and in hindlimb stance.

Locomotion in Permian dicynodonts

The postcranial skeleton and locomotion of *Robertia broomiana*, a Middle Permian dicynodont, were analyzed by King (1981b), who demonstrated that the posture of both fore- and hindlimbs was sprawling. In *Dicynodon trigonocephalus*, a Late Permian dicynodont, the forelimb remained sprawling; the hindlimb, by contrast, began and ended its stride in a semiupright position, while adopting a sprawling posture during much of the retractive phase (King 1981a).

The differences in posture between the fore- and hindlimbs in *Dicynodon trigonocephalus* suggested to King that the function of the two sets of limbs might also have differed. The forelimbs generated little or no locomotor force; only the latissimus dorsi was properly placed to generate any significant power in humeral retraction, but its action tended to cause dislocation of the glenohumeral joint, so it was probably little used (King 1981a, p. 285). Muscles for postural maintenance, on the other hand, were robust, and King felt that the forelimb functioned primarily for support of the anterior half of the body. A supportive function of the forelimb may have been particularly necessary because the dicynodont head was unusually large and heavy.

The hindlimb of *Dicynodon trigonocephalus* supplied most of the propulsive force. The iliofemoralis and other muscles retracted the hindlimb and contributed a powerful forward thrust. Kemp (1980a) documented a similar functional dichotomy in the limbs of *Procynosuchus*, an Early Triassic cynodont.

Locomotor thrust generated in the hindlimb is transmitted to the anterior half of the body by the axial skeleton. In early tetrapods, lateral undulation of the vertebral column contributed much of the stride length. More advanced tetrapods rely primarily on limb retraction to propel the body and tend to have a less flexible axial skeleton (Zug 1971; Edwards 1977). Hoffstetter and Gasc (1969) demonstrated in work on snakes that zygapophyses set oblique to the horizontal plane preclude significant lateral flexion of the axial skeleton; in *Dicynodon trigonocephalus*, such flexion appears to have been possible only at the dorsosacral junction, and locomotor force generated by the hindlimbs was transmitted anteriorly relatively undiminished (King 1981a).

Locomotion in kannemeyeriid dicynodonts

The Triassic kannemeyeriids retained a horizontal forelimb posture, although with the humerus in a near-sagittal plane (see Fig. 7.1). The pectoral appendage of kannemeyeriids was little modified from its condition in Permian dicynodonts; exclusion of the precoracoid from participation in the glenoid

Figure 7.4. Right femur of the mounted *Stahleckeria potens*, showing muscle origins and insertions. **a,** Anterior view, traced from a photograph. **b,** Posterior view, traced from a photograph.

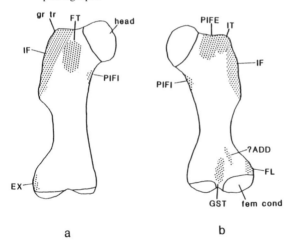

Figure 7.5 Right ischiopubis of the mounted *Stahleckeria potens*, showing muscle origins and insertions. **a,** Ventrolateral view, traced from a photograph. **b,** Dorsomedial view, reverse of ventrolateral view.

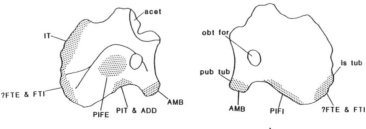

articulation was the only major derived character of this region of the skeleton in kannemeyeriids.

More noticeable changes took place in the kannemeyeriid pelvic girdle and hindlimb. Extreme medial inflection of the femoral head and the development of a robust greater trochanter constrained the hindlimb to a semiupright posture, with abduction limited to approximately 15° from the vertical in most genera (see Fig. 7.3).

The postural dichotomy is echoed by the reconstructed musculature of kannemeyeriids. The most powerful forelimb muscles were the pectoralis and the deltoid, inserting, respectively, on the ventral and dorsal sides of the enormous deltopectoral crest (Figs. 7.7, 7.8a, 7.9). Edwards (1977) showed that rotation of the propodials in salamanders occurs because retractor muscles insert on a *crista ventralis*, a ventral process of the propodial. The deltopectoral crest of therapsids was a *crista ventralis* homologue; its large size in kannemeyeriids was an adaptation to increase the ability of the pectoralis and the deltoid to rotate the humerus about its own longitudinal axis.

Muscles for humeral protraction and retraction, on the other hand, were poorly developed in kannemeyeriids (Figs. 7.8–7.10). Only the clavicular deltoid originated far enough anteriorly to cause more protraction than abduction or adduction. The latissimus dorsi, a humeral retractor, may have been of moderate size, but it is unlikely to have been as large as the pectoralis or the deltoid (compare the sizes of their insertions in Figure 7.9; because the latissimus dorsi originated primarily on dorsal fascia, the sizes of their origins cannot be compared). Humeral rotation, rather than retraction, is likely to have contributed most of the forelimb stride length in kannemeyeriids. A similar reliance of the forelimb stride upon humeral rotation was demonstrated by Jenkins (1970) in the echidna (*Tachyglossus aculeatus*).

Figure 7.6. Right tibia and fibula of the mounted *Stahleckeria potens*, showing muscle origins and insertions. **a.** Anterior view, traced from a photograph. **b.** Posterior view, reverse of anterior view.

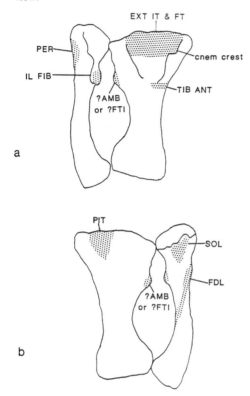

Figure 7.7. Left forelimb of the mounted specimen of *Stahleckeria potens* in the Institut und Museum für Geologie and Paläontologie der Universität Tübingen, Tübingen, Federal Republic of Germany. Anterior view.

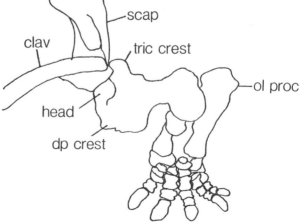

Some of the same muscles that contributed longitudinal humeral rotation were also important for postural maintenance in kannemeyeriids. The pectoralis, aided by the coracobrachialis, functioned as an adductor, preventing collapse of the thorax between the forelimbs. Support of the anterior half of the body may have been a major function of the forelimbs.

Kannemeyeriid hindlimb musculature was developed in a different fashion (see Figs. 7.4–7.6, 7.11). The large size of the iliac blade, particularly anteriorly, and of the greater trochanter indicate that the iliofemoralis was robust. The posterior corner of the ilium and the ischial tuberosity, origins of the other hindlimb retractor muscles, were also large; the available evidence indicates that active retraction was the major component of the hindlimb stride.

The dichotomies in posture and musculature between the two sets of limbs in kannemeyeriids may have been indicative of a dichotomy in function. If this idea is correct, the forelimbs would have been more important for support of the body, especially the head, while the hindlimbs would have generated

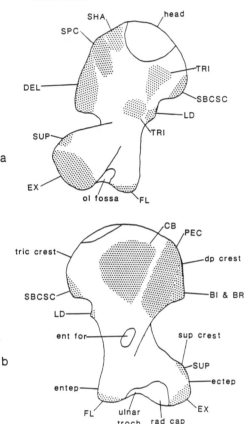

Figure 7.9. Left humerus of the mounted *Stahleckeria potens*, showing muscle origins and insertions. **a,** Dorsal view, traced from a photograph. **b,** Ventral view, reverse of dorsal view.

Figure 7.8. Right scapula and coracoids of the mounted *Stahleckeria potens*, showing muscle origins and insertions. **a,** Lateral view, traced from Huene (1935–42, Plate 7, Figs. 1a and 4). **b.** Medial view, reverse of lateral view.

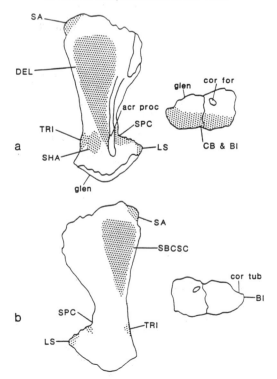

Figure 7.10. Right ulna of the mounted *Stahleckeria potens*, showing muscle origins and insertions. **a,** Anterior view, traced from Huene, (1935–42, Plate 7, fig. 8a). **b,** Posterior view, reverse of anterior view.

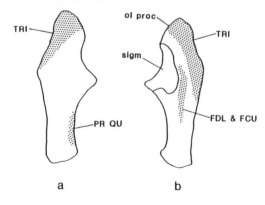

most of the locomotor force. A dichotomy of a similar type, with the hindlimbs producing the primary thrust, is demonstrated by the tendency of lizards to lift the forelimbs off the substrate in order to run bipedally at high speeds (Snyder 1962).

An additional feature of kannemeyeriids that supports this analysis of dichotomous limb function is the postulated rigidity of the axial skeleton, which allowed transmission of posteriorly generated locomotor forces to the anterior half of the body. The zygapophyses in an articulated vertebral column of *Dinodontosaurus turpior* were set at angles ranging from 30° to 90° from the horizontal. Huene (1935–42, pp. 35–7) noted two classes of vertebrae in *Stahleckeria potens*, those with *getrennte* (separate) and those with *unten verwachsene* (ventrally contiguous) zygapophyses. These had zygapophysial angles from the horizontal of 40° to 60° and of 60° to 80°, respectively. Such obliquely set zygapophyses conferred stability on the axial skeleton but precluded side-to-side undulation [as demonstrated by Hoffstetter and Gasc (1969) in snakes].

N. Hotton, III (pers. comm.) suggests that this picture of the extent of possible intervertebral flexibility may be oversimplified and that the zygapophyses of successive vertebrae "hug" one another in dicynodonts and other tetrapods for the prevention of torsion rather than of lateral flexion. If his ideas are correct, the axial skeleton of kannemeyeriids was more flexible that I have assumed. The analysis of limb function here presented, that of a possible dichotomy between the primary functions of fore- and hindlimbs mirroring their different postures and reconstructed musculatures, does not, however, depend upon absolute axial rigidity.

A similar postural dichotomy between an upright hindlimb and a sprawling forelimb was outlined by Kemp (1980b) in *Luangwa*, a Late Triassic cynodont, and by Hatcher, Marsh, and Lull (1907) and Lull (1933) in ceratopsian archosaurs. Studies of the limb musculature and locomotor cycles of *Luangwa*

and of ceratopsians would help to establish whether differences in locomotor function between fore- and hindlimbs analogous to those that I have discussed in kannemeyeriids also existed in these unrelated reptilian groups.

Discussion

The kannemeyeriids had a hindlimb of nearly upright posture. They had lost the ability to abduct the hindlimb more than approximately 15° from the vertical because of the medial inflection of the femoral head and the large size of the greater trochanter (see Fig. 7.3).

Several groups of tetrapods, including mammals and many archosaurs, have attained a fully erect hindlimb stance, allowing limb excursion in a parasagittal plane. Charig (1972, 1980) and Bonaparte (1982) cited the upright hindlimb posture of Mesozoic archosaurs as a factor giving them competitive superiority over the therapsids, whose limb morphology and posture were more conservative.

The most advanced archosaurs did not, however, appear until the Jurassic and Cretaceous. In the Late Triassic, most archosaurs either had a sprawling hindlimb posture (phytosaurs), alternated between sprawling and semierect (crocodilians), or had attained an obligatory semierect posture no more upright than that of kannemeyeriids (pseudosuchians) (Schaeffer 1941; Charig 1972; Brinkman 1980). Only the prosauropods, the coelurosaurs, and the rauisuchid and ornithosuchid thecodonts had attained an upright hindlimb stance at this early stage of archosaurian history.

The prosauropods, early experiments in large archosaurian herbivory, were a rare element of some Late Triassic faunas (e.g., Huene 1926) but they did not occur in the same deposits as kannemeyeriids (P.E. Olsen pers. comm.). Fully grown kannemeyeriids were too large to be preyed upon effectively by most carnivores, including the diminutive coelurosaurs; the rauisuchid and ornithosuchid thecodonts, however, had a large enough body size to have been a predatory threat to kannemeyeriids. Predation pressure from these groups may have contributed to kannemeyeriid extinction.

Conclusions

The kannemeyeriids demonstrated a rather ungainly, "push-up" stance, characterized by a sprawling forelimb and a nearly upright hindlimb posture. The musculature of the two sets of limbs also differed: in the forelimb, muscles for humeral rotation were the most robust, while muscles for active retraction of the hindlimb were well developed.

This postural and muscular dichotomy between the two sets of limbs in kannemeyeriids may

Figure 7.11. Right ilium of the mounted *Stahleckeria potens*, lateral view, traced from a photograph, showing muscle origins and insertions.

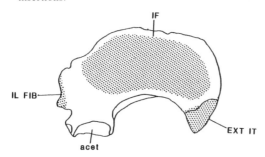

have been echoed by a difference in their primary locomotor functions, the forelimb functioning mainly in support of the body while the hindlimb generated most of the locomotor thrust. Kemp (1980a) and King (1981a) have suggested that similar functional dichotomies occurred in other therapsids.

Charig's (1972, 1980) and Bonaparte's (1982) claims that competition from or predation by archosaurs with advanced locomotion drove the therapsids to extinction applied to kannemeyeriids only in the case of predation by rauisuchid and ornithosuchid thecodonts. Other late Triassic carnivorous archosaurs either had a more primitive hindlimb stance than kannemeyeriids or were too small to have been effective predators upon them. Large, herbivorous archosaurs did not occur in the same communities as kannemeyeriids; it is probable that archosaurs replaced kannemeyeriids only in the sense of filling the same roles at a later time.

Abbreviations

acet, acetabular facet
acet rim, acetabular rim
acr proc, acromion process
ADD, femoral adductors
AMB, ambiens
ant il, anterior flange of ilium
BI, biceps
BR, brachialis
CB, coracobrachialis
clav, clavicle
cnem crest, cnemial crest
cor for, coracoid foramen
cor tub, coracoid tubercle
DEL, deltoid
dp crest, deltopectoral crest
ectep, ectepicondyle
entep, entepicondyle
ent for, entepicondylar foramen
EX, extensor muscles
EXT IT, extensor iliotibialis
FCU, flexor carpi ulnaris
FDL, flexor digitorum longus
fem cond, femoral condyles
FL, flexor muscles
FT, femorotibialis
FTE, flexor tibialis externus
FTI, flexor tibialis internus
glen, glenoid facet
gr tr, greater trochanter
GST, gastrocnemius
head, humeral/femoral head
IF, iliofemoralis
IL FIB, iliofibularis
isch, ischium
is tub, ischial tuberosity
IT, ischiotrochantericus
LD, latissimus dorsi
LS, levator scapulae
obt for, obturator foramen
ol fossa, olecranon fossa

ol proc, olecranon process
PEC, pectoralis
PER, peroneus longus and brevis
PIFE, puboischiofemoralis externus
PIFI, puboischiofemoralis internus
PIT, puboischiotibialis
post il, posterior flange of ilium
PR QU, pronator quadratus
pub tub, pubic tubercle
SA, serratus anterior
SBCSC, subcoracoscapularis
scap, scapula
SHA, scapulohumeralis anterior
sigm, sigmoid notch
SOL, soleus
SPC, supracoracoideus
SUP, supinator
sup crest, supinator crest
tib, tibia
TIB ANT, tibialis anterior
TRI, triceps
tric crest, triceps crest

Acknowledgments

The work reported here is a portion of a dissertation submitted to fulfill in part the requirements for the Ph.D. degree in Yale University. Many people helped with access to specimens and discussion; of these I have space only to acknowledge F.A. Jenkins, Jr., and C.R. Schaff (Harvard) and F.W. Westphal (Tübingen) for allowing me to take the photographs reproduced here. Comments by K. Padian and three anonymous reviewers on an earlier version of this manuscript were most helpful. Financial support from Grant 190, Joseph Henry Fund, National Academy of Sciences, is also gratefully acknowledged.

References

Anderson, H. M., and J. M. Anderson. 1970. A preliminary review of the biostratigraphy of the uppermost Permian, Triassic, and lowermost Jurassic of Gondwanaland. *Palaeontol. Afr.* (Suppl.) 13:1–22 and charts.

Bonaparte, J. F. 1982. Faunal replacement in the Triassic of South America. *J. Vert. Paleontol.* 2:362–71.

Brinkman, D. E. 1980. The hindlimb step cycle of *Caiman sclerops* and the mechanics of the crocodile tarsus and metatarsus. *Can. J. Zool.* 58:2187–200.

Charig, A. J. 1972. The evolution of the archosaur pelvis and hindlimb: an explanation in functional terms. *In* Joysey, K. A., and T. S. Kemp (eds.), *Studies in Vertebrate Evolution* (Edinburgh: Oliver and Boyd), pp. 121–55.

1980. Differentiation of lineages among Mesozoic tetrapods. *Mem. Soc. Géol. Fr., N.S.* 59:207–10.

Crompton, A. W., and N. Hotton, III. 1967. Functional morphology of the masticatory apparatus of two dicynodonts (Reptilia, Therapsida). *Postilla.* 109:1–51.

Edwards, J. L. 1977. The evolution of terrestrial locomotion. *In* Hecht, M. K., P. C. Goody, B. M. Hecht (eds.), *Major Patterns in Vertebrate Evolution*. NATO Advanced Study Institute, Ser. A, Life Science 14 (New York: Plenum Press), pp. 553–77.

Gow, C. E. 1978. The advent of herbivory in certain reptilian lineages during the Triassic. *Palaeontol. Afr.* 21:133–41.

Hatcher, J. B., O. C. Marsh, and R. S. Lull. 1907. The Ceratopsia. *U.S. Geol. Surv. Mono.* 49:1–300.

Hoffstetter, R., and J.-P. Gasc. 1969. Vertebrae and ribs of modern reptiles. *In* Gans, C. (ed.), *Biology of the Reptilia*, Vol. 1 (New York: Academic), pp. 201–310.

Huene, F. F. von. 1926. Vollständige Osteologie eines Plateosauriden aus dem schwäbischen Keuper. *Geol. Pal. Abhandl., Jena, N.S.* 15:1–43.

——— 1935–42. *Die fossilen Reptilien des südamerikanischen Gondwanalandes an der Zeitenwende* (Tübingen: Franz F. Heine).

Jenkins, F. A., Jr. 1970. Limb movements in a monotreme (*Tachyglossus aculeatus*): a cineradiographic analysis. *Science* 168:1473–5.

——— 1971a. The postcranial skeleton of African cynodonts. *Bull. Peabody Mus. Nat. Hist.* 36:1–216.

——— 1971b. Limb posture and locomotion in the Virginia opossum (*Didelphis marsupialis*) and in other noncursorial mammals. *J. Zool.* 165:303–15.

Kemp, T. S. 1980a. The primitive cynodont *Procynosuchus*: structure, function and evolution of the postcranial skeleton. *Phil. Trans. Roy. Soc. London B* 288:217–58.

——— 1980b. Aspects of the structure and functional anat-

omy of the Middle Triassic cynodont *Luangwa*. *J. Zool.* 191:193–239.

King, G. M. 1981a. The functional anatomy of a Permian dicynodont. *Phil Trans. Roy. Soc. London B* 291:243–322.

——— 1981b. The postcranial skeleton of *Robertia broomiana*, an early dicynodont (Reptilia, Therapsida) from the South African Karoo. *Ann. South Afr. Mus.* 84:203–31.

Lull, R. S. 1933. A revision of the Ceratopsia or horned dinosaurs. *Mem. Peabody Mus. Nat. Hist.* 3:1–175.

Romer, A. S. 1922. The locomotor apparatus of certain primitive and mammal-like reptiles. *Bull. Amer. Mus. Nat. Hist.* 46:517–606.

——— 1956. *Osteology of the Reptiles* (Chicago: The University of Chicago Press).

Schaeffer, B. 1941. The morphological and functional evolution of the tarsus in amphibians and reptiles. *Bull. Am. Mus. Nat. Hist.* 78:395–472.

Snyder, R. C. 1962. Adaptations for bipedal locomotion in lizards. *Am. Zool.* 2:191–203.

Sues, H.-D. 1986. Locomotion and body form in early therapsids (Dinocephalia, Gorgonopsia and Therocephalia). *In* Hotton, N., P. D. Mclean, J. J. Roth, and F. C. Roth (eds.), *Ecology and Biology of Mammal-like Reptiles* (Washington, DC: Smithsonian Press).

Zug, G. R. 1971. Buoyancy, locomotion, morphology of the pelvic girdle and hindlimb, and systematics of cryptodiran turtles. *Misc. Publs. Mus. Zool., Univ. Mich.* 142:1–98.

8 A new family of mammals from the lower part of the French Rhaetic

D. SIGOGNEAU-RUSSELL,
R.M. FRANK, AND J. HEMMERLÉ

The French Rhaetic locality of Saint-Nicolas-de-Port (hereafter referred to as SNP) has furnished so far about 500 mammalian teeth that constitute a very diversified fauna. Theria and non-Theria alike are represented, both with new genera and species, of which very few have as yet been described. Sigogneau-Russell (1983) described two multi-tuberculate teeth (SNP 78 Wouters and SNP Marignac 2) that she considered in the theropsid lineage. A third tooth of this type has since been found (SNP 335 W), as well as four teeth of somewhat different morphology that we regard as representing a related animal. Divergent opinions of their systematic attribution have been expressed to us by some specialists; therefore, we decided to analyze these teeth more thoroughly from morphological and functional points of view and to submit one of them (SNP 335 W) to ultramicroscopic examination. We have compared it in this way not only to teeth of contemporary mammals, but also to dermal denticles of Rhaetic selachians from Württemberg, to which a distant resemblance in general morphology and in wrinkling of the enamel was noted. Even so, the consistency of the structural organization of the three mammalian teeth contrasts with the variability of the aggregates found in these selachian denticles.

Morphology

Detailed descriptions of specimens SNP 78 W and SNP Ma 2 [Fig. 8.1 (1,2)] have already been published (Sigogneau-Russell 1983) using Hahn's (1973) terminology, and will not be repeated here. Their crowns (Fig. 8.2a) are characterized by three unequal rows (designated A, B, and C) of low, rounded cusps, separated from each other by valleys, as well as by wrinkling of the enamel, which [as in selachians (Reif 1973)] may have acted as a reinforcement. The new tooth SNP 335 W also has a subcircular shape. The crown is composed essentially of two rows of cusps; however, a shallow valley isolates a part of the tooth that is comparable in position to row C of the two other teeth [see Fig. 8.1(3)]. Moreover, on this tooth the wrinkling of the enamel is only perceptible at the base of the main cusps. Finally, at the "front" of the two rows, the morphology is simpler than in SNP 78 W and SNP Ma 2: There is only a single accessory cusp instead of two or three, although a faint transverse sulcus incises the anterior flank of the large cusp 1, row B. Otherwise, the disposition of the cusps and sulci is strikingly similar to those of the two teeth previously described.

The roots are not preserved in any of these teeth. However, SNP 78 W shows a definite indication of division of its root into at least two, and more probably three, elements (Fig. 8.2b). Unfortunately, we do not know the disposition of these roots, which would help to orient the teeth.

Wear

Of the three teeth considered here, SNP 78 W is the most heavily worn (Fig. 8.3a). Maintaining the orientation proposed in the original description (Sigogneau-Russell 1983), wear first affects the posterior part of the tooth. In lateral view, the profile of the crown shows a "peak" at the level of the main cusps 2A and 1B, with a short anterior part and a longer posterior part (see Fig. 8.2c). On the posterior part, the cusps of the so-called B row are abraded perpendicular to the surface of the crown, but those of the A and C rows show abrasion surfaces that slope toward the valleys separating them from row B (more so on A than on C). On the anterior part, the cusps are again abraded perpendicularly to the crown.

In addition to abrasion surfaces, there are distinct facets. The main facet is formed in the posterior end of the two valleys, mostly in the valley between A and B, which is excavated as a basin. Moreover, the anterior flank of cusp 1A has a secondary concavity. Finally, a facet appears to truncate the external flanks of 2A and 3A (see below and Fig. 8.4).

SNP 335 W (see Fig. 8.3c) shows the same type of abrasion, but it is much less advanced. Also, excavation of the posterior border of the A–B valley and an incipient triangular wear facet on the anterior side of 1A are similarly developed. On SNP Ma 2 (see Fig. 8.3b), the cusps are barely blunted; but again there is the beginning of an excavation at the rear of the A–B valley.

Scanning election microscopic (SEM) examination of the abrasion surfaces of the valleys did not reveal any clear striations, in contrast to the pattern of wear on the central valley of a haramiyid tooth.

Figure 8.1. *Theroteinus nikolai,* gen. nov., sp. nov. Upper? molars, occlusal view. **1,** SNP 78 W, type specimen. **2,** SNP Ma 2. **3,** SNP 335 W. Stereophotos by D. Serrette, U.A. 12.

At the most, some pitting of the enamel can be detected. Thus, the jaws that carried these teeth probably did not make longitudinal movements. Rather, chewing must have involved orthal movement, with a transverse component in the "grinding cycle" of mastication, as indicated by the presence of leading and trailing edges (Costa and Greaves 1981) on row A of SNP 78 W [see Fig. 8.1(1)]. In any case, the similarity of occlusal facets and abrasion surfaces presented by these three teeth suggests a consistency in pattern of jaw motion, and hence a consistency of upper and lower molar relationship, which is essentially a mammalian or premammalian characteristic, because it supposes a diphyodont mode of dental replacement.

These observations, however, do not help us to determine the orientation of these isolated teeth, so we looked for interdentinal contact facets. No convincing facets could be detected on SNP 78 W or SNP 335 W; but on SNP Ma 2 [see Figs. 8.1(2),

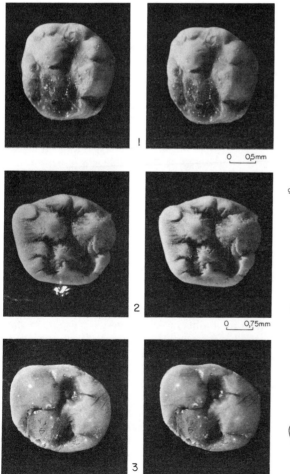

Figure 8.2. **a,** Schematic morphological plan of the molars (uppers?) of *Theroteinus nikolai* gen. nov. and sp. nov. 1a–c = rows of cusps; g = situation of the radicular grooves. **b,** Radicular view of SNP 78 W, type specimen of *Theroteinus nikolai,* gen. nov., sp. nov. **c,** Lateral view of SNP 78 W, showing the partition of the crown into an anterior and a posterior part ("front" at the left). **d,** DP6/ of *Paulchoffatia delgadoi,* from Guimarota, Portugal. (After Hahn 1969.) **e,** *Thomasia* sp., SNP 65, showing rows A and B. **f,** Illustration showing the planes in which the teeth were ground.

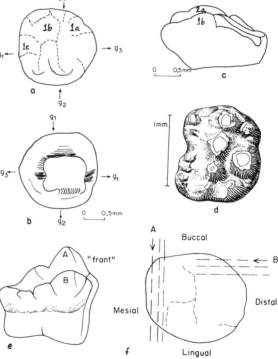

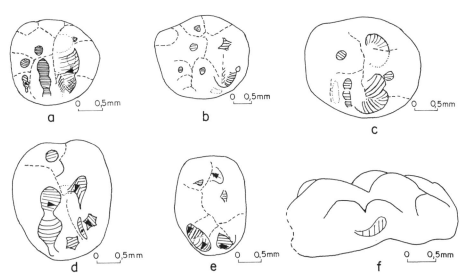

Figure 8.3. Wear surfaces on the teeth attributed to *Theroteinus nikolai* gen nov., sp. nov. The arrow indicates the direction of slope of the wear surfaces. **a,** SNP 78 W; **b,** SNP Ma 2; **c,** SNP 335 W; **d,** SNP 487 W; **e,** SNP 309 W; all in occlusal view, with the "front" at the top; **f,** SNP Ma 2, anterior view. Dotted lines encircle uncertain or faint wear surfaces.

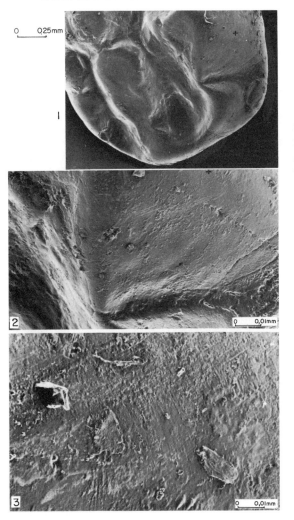

Figure 8.4. *Theroteinus nikolai*, gen. nov., sp. nov., SNP 78 W, type specimen; upper? molar, occlusal view, posterior part. **2** and **3** are successive enlargements of areas marked (+) on **1** and **2**, respectively. Photographs by C. Weber-Chancogne, U.A. 12.

8.3f], a small and little emphasized but indisputable facet flattens the base of the anterior side of the anteromedial cusp, confirming the orientation initially proposed for these teeth (Sigogneau-Russell 1983).

We did not examine this facet with the SEM because SNP Ma 2 is particularly fragile. SEM observation of the other two teeth confirmed the absence of contact facets at the front and rear; however, we noted a network of striations on the external flank of the A rows [see Figs. 8.4, 8.5(1,2)].These striations seem to be the same as those discernible by SEM on the contact facet situated at the front of a haramiyid tooth [SNP 330 W, Fig. 8.5(3,4)], where they create a network. On the latter tooth, their absence outside this facet eliminates the hypothesis that they are artifacts produced during or after fossilization. Do the striations on the external flank of SNP 78 W and SNP 335 W testify to a contact with the adjacent tooth? This does not seem likely because they are situated so high on the side of the crown. An interdental contact would occur at the point of maximum width of the crown (in this case, at the base). As suggested above, these striations may be made by occlusion with teeth in the opposing quadrant of the dentition, or by hard particles in the food.

Ultrastructure

For this SEM study of SNP 335 W, the preparation technique described earlier by Sigogneau-Russell, Frank, and Hemmerlé (1984) was used. The tooth was embedded in Epon 812 (E.F. Fullam, Inc., Schenectady, New York). Using a grinding machine consisting of a rotating circular disk, a superficial buccolingual section was cut in the coronal part of the tooth (see A in Fig. 8.2f). The prepared surface was polished with a fine diamond powder and slightly etched for 30 sec with 20 percent citric acid solution. After coating with Au–Pd in a Hummer Junior evaporator (Siemens, Karlsruhe), the tooth surface was examined with a Jeol (Tokyo) JSM 35C scanning

Figure 8.5. **1** and **2**, *Theroteinus nikolai*, gen. nov., sp. nov., SNP 335 W; upper? molar, occlusal view. **3** and **4**, *Thomasia* sp., SNP 330 W, anterior view showing a contact facet. **2** and **4** are enlargements of areas marked (+) on **1** and **3**, respectively. Photographs by C. Weber-Chancogne, U.A. 12.

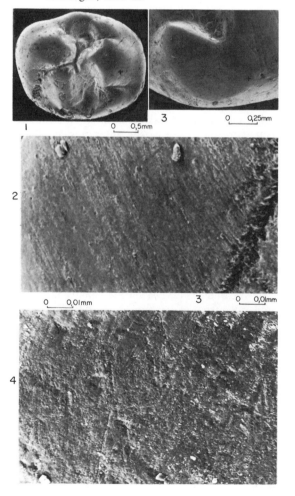

electron microscope (SEM) under 10–15 kV. A new polished surface was then prepared by grinding off a small portion of the tooth about 200 μm thick.

Following the same process of etching and coating, the new surface was studied in the SEM. This procedure was pursued for half the thickness of the tooth. The other half was processed in the same way from one superficial side inward in a plane perpendicular to the first sequence (see B in Fig. 8.2f). This sectioning procedure enabled the whole thickness of the tooth to be successively studied in different planes: mesiodistal, buccolingual, and oblique.

In order to answer various suggestions as to the identification of these teeth, we also prepared and studied by SEM a tooth of an actinopterygian from the same locality (SNP) and a selachian dermal denticle from the Rhaetian of Wurttemberg.

The SEM study of the selachian denticle showed that the "coronal" part consisted of typical dentine, which in cross section showed dentinal tubules with open lumina [Fig. 8.6(1)]. This dentine layer was covered by a densely calcified layer devoid of incremental lines and prismatic structure and had the structural characteristics of aprismatic enamel [Fig. 8.6(1–3)]. This "enamel" consisted of elongate inorganic crystals oriented in all directions with no preferential orientation [Fig. 8.6(2,3)]. These observations, together with the heavy mineralization, do not agree with Reif's (1973, p. 249) observations of dermal denticles; that is, low mineralization and crystallites perpendicular to the surface of the "enamel" (see also Schmidt and Keil 1971, pp. 202–11). Our results also differ from observations of "enameloid" tissues, where individual crystals are hardly identifiable in transmission or scanning electron microscopy. We found that individual crystals appeared in transverse sections with rounded or flattened outlines of about 90–155 Å diameter [Fig. 8.6(3)], and reached more than 1 μm in length [Fig. 8.6(2)]. Preprismatic or prismatic patterns were not observed in this "enamel," and no dentinal tubules were present.

A differentiated "enamel" layer also was observed in the teeth of an actinopterygian fish [Fig. 8.7(1)]. This layer is also aprismatic and lacks tubules and incremental lines [Fig. 8.7(1,2)]. These observations are contrary to what has been described in the teeth of piranhas (Shellis and Berkovitz 1976); however, the latter are carnivorous fish whose teeth parallel those of sharks, which may account for the difference.

Returning to the theropsid teeth under consideration, SNP 335 W has a well-differentiated layer of enamel covering the coronal dentine [Fig. 8.8(1–3)]; the latter contains numerous dentinal tubules that were sectioned in transverse or longitudinal

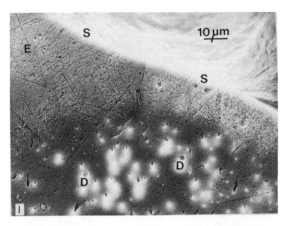

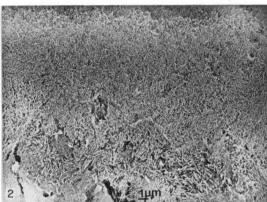

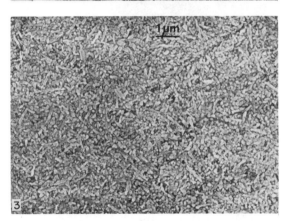

Figure 8.6. Rhaetic selachian denticle from Württemberg. **1,** Enamel (E) and dentine (D). Note the aprismatic character of the enamel, and the presence of cross-sectioned dentinal tubules in the dentine. S, enamel surface. **2,** Presence of unoriented apatite crystals in the enamel. The preprismatic pattern is absent. **3,** Presence of densely packed apatite crystals in aprismatic enamel. Photographs from U.E.R. Dentaire, Strasbourg.

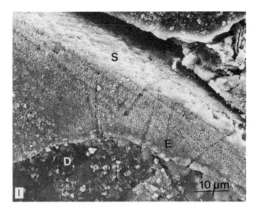

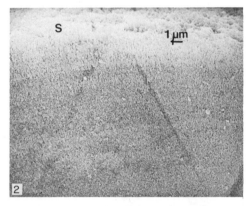

Figure 8.7. A palatal actinopterygian tooth, showing aprismatic character of the enamel. S, enamel surface. Photographs from U.E.R. Dentaire, Strasbourg.

planes. Numerous lateral branches radiate from the main lumen of the tubule [Fig. 8.8(3,4)]. Tubules were also observed in enamel, where they were easily identified as empty cross-sectioned lumina [Fig. 8.8(2)]. Sometimes the lumen of the tubules in dentine and enamel was filled with a calcitic deposit. The tubule marked T in Figure 8.8(3) can be followed from the outer dentine into the inner enamel.

In transverse [Figs. 8.9(1,2), 8.10(2), 8.11(1)], as well as in oblique [Fig. 8.10(1)] and longitudinal, sections, typical preprismatic enamel as described earlier by Frank, Sigogneau-Russell, and Voegel (1984) and Sigogneau-Russell et al. (1984) was observed. Even at low magnifications, the delineations of directly apposed "prisms" without interprismatic enamel were visible [Figs. 8.9(1), 8.11(1)].

In longitudinal enamel sections [Fig. 8.11 (2,3)], a repetitive pattern is present in the orientation of the apatite crystals: the *c* axes of the crystals are oriented at divergent angles like the barbs of a feather. These "prisms," about 4–5 μm wide, were directly in contact with each other and not separated by interprismatic substance.

The enamel tubules are oriented generally at right angles to the enamel dentine junction and extend to the enamel surface [Fig. 8.11(3)], and therefore parallel the longitudinal axes of the enamel prisms. In oblique [see Fig. 8.10(1)] and transverse [Fig. 8.10(2)] sections, the tubules clearly appear to be located in the center of the "prisms" with apatite crystals radiating like "sunbeams" all around the tubular lumina. However, in longitudinal sections

[Fig. 8.11(3)], it was apparent that, in addition to tubules in the center of prisms, some tubules were located between enamel "prisms" and ran exactly parallel to them.

Finally, in the superficial part of the enamel layer of SNP 335 W, wavy lines of apparently lower electron density were observed running parallel to the enamel surface [see Fig. 8.9(1)]. At higher magnification [see Fig. 8.9(2)], these lines were less rich in organic material, which suggests that they are incremental lines.

Figure 8.8. SNP 335 W. **1,** Enamel and dentine with numerous cross-sectioned dentinal tubules. **2,** Detail of cross-sectioned dentinal tubules in the enamel. **3,** Enamel and dentine with longitudinal tubules. A dentinal tubule with a calcified lumen can be followed across the dentino-enamel junction in the enamel. **4,** Cross-sectioned dentine showing dentinal tubules with lateral branches. D, dentine; S, enamel surface; E, enamel; T, tubule. Photographs from U.E.R. Dentaire, Strasbourg.

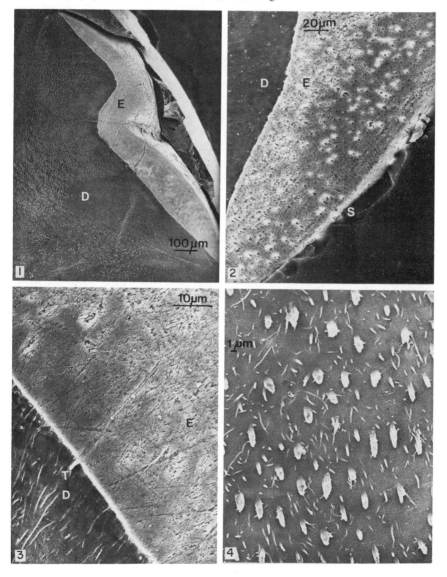

Conclusion

The structure of the enamel of SNP 335 W definitely seems to exclude any attribution to selachians and to lower vertebrates in general. It is undeniable, of course, that true enamel is also known in crossopterygians and dipneustes (Smith 1979) and that, as Poole (1971) reported, the transformation of enameloid into enamel must have occurred in the evolution of fish. It also should be remembered that the presence of both types of "covering," enamel and enameloid, has been encountered in the same organism at different stages of development (Kerr 1959).

Moreover, a preprismatic or prismatic enamel has been recognized in several groups outside the theropsid line, including placodonts (Schmidt and Keil 1971), an agamid lizard (Cooper and Poole 1982), even crossopterygians (M. Smith pers. comm.), cyprinid actinopterygians (Sauvage 1972), and finally in crocodiles (Dauphin 1984). However, as Schmidt and Keil (1971) indicated, it is not certain whether these different kinds of enamel are really homologous to that of mammals.

The similarity between the images furnished by SNP 335 W and the structure that we have detected in teeth of Haramiyidae and *Kuehneotherium* (Frank et al. 1984; Sigogneau-Russell et al. 1984) from the same locality is so great that it is very difficult to avoid the conclusion that these three groups were at the same evolutionary stage in dental ultrastructure. This conclusion is further confirmed because SNP 78 W is probably multirooted, that is to say, the teeth considered here would represent a

Figure 8.9. SNP 335 W. **1,** Enamel and dentine, showing a preprismatic pattern clearly visible on the right in cross-sections. Several wavy incremental lines are seen in the enamel surface. **2,** Slightly higher enlargement of the surface enamel shown in **1.** Incremental lines are visible. Note the different orientation of groups of apatite crystals, suggesting a preprismatic pattern. The surface of the enamel is at the top. D, dentine; E, enamel; S, enamel surface. Photographs from U.E.R. Dentaire, Strasbourg.

Figure 8.10. SNP 335 W. **1,** Oblique section of inner enamel. Groups of apatite crystals appear to be centered on cross-sectioned tubules in a fanlike disposition. Dentine is at the bottom. **2,** Cross-sectioned enamel, showing "sunbeam" configuration of apatite crystals around cross-sectioned tubules. T, tubules. Photographs from U.E.R. Dentaire, Strasbourg.

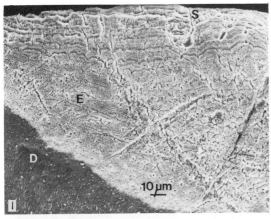

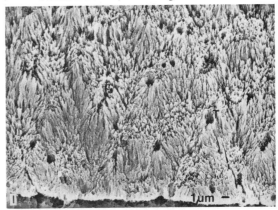

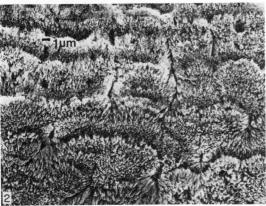

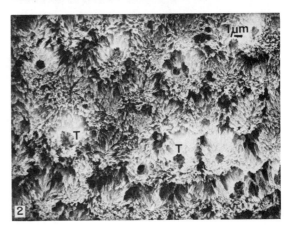

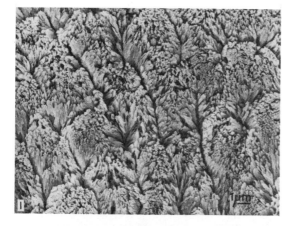

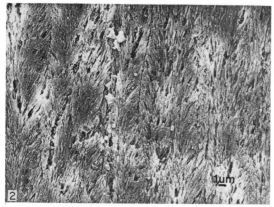

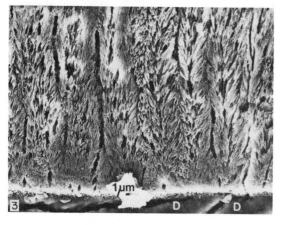

Figure 8.11. SNP 335 W. **1,** Cross-sectioned preprismatic pattern of tooth enamel. The outlines of the prisms are clearly visible. **2,** Same enamel in longitudinal section with preprismatic pattern. In each prism a repetitive orientation of the *c* axes of the apatite crystals with a pinnate disposition is clearly shown. **3,** Same enamel in longitudinal section with typical preprismatic pattern. Longitudinal tubules can be seen between prisms as well as in the center of prisms. D, dentine. Photographs from U.E.R. Dentaire, Strasbourg.

member of the theropsid line, and one close to the mammalian threshold.

However, an important characteristic of these teeth is the presence of tubules in the enamel. Tubules have not been recorded either in *Kuehneotherium* or, even more significantly, in the Haramiyidae, which definitely eliminates any basis for reference of the teeth described here to a member of the latter family. This conclusion is reinforced by observations of the occlusal facets; in haramiyids, tooth wear patterns indicate longitudinal masticatory movements (P.M. Butler, cited in Krause 1982), whereas in the animals represented by the three teeth described here, orthal movement seems to have been more likely. The only possibility left is that the three teeth are milk teeth of a haramiyid. However, if this is the case, it would be difficult to account for their low abundance in the sample (1 tooth for 100 haramiyid teeth), their notably larger size, their dissimilar morphology, their different pattern of wear, and, finally, the abundance of tubules in the enamel. Nothing permits the assertion that deciduous teeth have enamel richer in tubules than do permanent teeth.

A phylogenetic relationship with multituberculates could be envisioned; the taxon represented by the three teeth might be from one of the true precursors of the Multituberculata, rather than from a haramiyid. The morphological resemblance between our scheme (see Fig. 8.2a) and the DP6/ of *Paulchoffattia delgadoi* (see Fig. 8.2d) figured by Hahn (1969) is assuredly very suggestive. Also, tubules are numerous in the enamel of multituberculates (Fosse, Risnes, and Holmbakken 1973), even if they do not extend, as in these three teeth, parallel to the prisms but cross them after a short distance and then taper. Finally, as is well known in multituberculates, the wear pattern of the teeth also indicates longitudinal jaw motion from back to front (Krause 1982), but such wear does not necessarily affect all elements of the dentition. However, the probability of this relationship is lessened by the fact that tubules are "very scarce" in Late Jurassic plagiaulacid multituberculates (Fosse and Kielan-Jaworowska 1985). This would require the hypothesis that after having been abundant in the enamel of Rhaetic forms, tubules became rare in teeth of Jurassic multituberculates before again increasing in number in Cretaceous forms.

Comparison with the Tritylodontidae appears to indicate an even greater phylogenetic separation. The transverse widening of the cheek teeth in this family, the crescentic shape of the cusps, the pattern of wear, and the different enamel ultrastructure (Grine, Gow, and Kitching 1974) render such a relationship as unlikely as the grouping of tritylodonts and gomphodonts as reproposed by Hopson and Kitching (1972), which they do not substantiate by unique synapomorphies.

The possibility remains that the animal represented by our teeth is related to gomphodont cynodonts, an idea already proposed (Sigogneau-Russell 1983) and supported by the abundance of tubules in the enamel of diademodonts (Osborn and Hillman 1979). However, typical gomphodont postcanines are characterized by a transverse widening, emphasized by a similarly oriented crest (even if it is depressed in the middle), and do not have a mesiodistal valley. In any case, the apparent presence of several roots supporting the teeth described here and the structure of their enamel [which is aprismatic in *Diademodon*, according to Grine, Vrba, and Cruickshank (1979)] would oblige us to put the teeth in a different systematic unit.

Thus, the three teeth described here appear to represent a new family of mammals, or of "things very much like them" (Simpson 1984), whose affinities assuredly remain very uncertain, but whose discovery accentuates several points. First, the study of dental ultrastructure contributes to research on evolutionary processes as well as to systematic analyses. Second, these teeth add evidence supporting the presumption of repetitive evolution of preprismatic enamel. Third, these few specimens add evidence about the fermentation of the therapsid world at the Triassic–Jurassic boundary, the competition that must have taken place, and how very little we know about it.

Systematic paleontology

Class Mammalia
Order incertae sedis
Family Theroteinidae nov.
Type. Theroteinus gen. nov. and only genus.
Distribution. Lower part of the Early Rhaetic.
Diagnosis. As for the species.
Theroteinus gen. nov.
Type species. Theroteinus nikolai sp. nov.
Diagnosis. As for the genus.
Etymology. Τείνειν, Greek to tend toward; θήρ, wild beast; a term universally adopted to designate mammals, in the absence of a Greek substantive for this class.
Theroteinus nikolai sp. nov.
Type specimen. SNP 78 W, upper? molar, crown complete but roots very incomplete.
Attributed material. SNP Ma 2, 335 W (upper molars?), SNP 487 W, SNP 309 W, SNP 497 W (lower molars?) (Fig. 8.12; see also Fig. 8.3d,e).
Locality. The quarry of Saint-Nicolas-de-Port, situated on the commune of Rosières-aux-Salines, Department of Meurthe-et-Moselle, France.
Diagnosis. Cheek teeth with cusps aligned in three rows separated by a valley; jaw movements essentially orthal; preprismatic enamel.
Etymology. Nikolai, from the name of the quarry that has furnished so many mammalian teeth.

Measurements

SNP 78 W	1.88 mm	1.78 mm
SNP Ma 2	2.65 mm	2.34 mm
SNP 335 W	2.45 mm	2.10 mm
SNP 487 W	2.50 mm	2.10 mm
SNP 309 W	1.35 mm	2.14 mm
SNP 497 W	1.60 mm	1.65 mm

Figure 8.12. *Theroteinus nikolai*, gen. nov., sp. nov. Lower? molars, occlusal view. **1,** SNP 487 W. **2,** SNP 309 W. **3,** SNP 497 W. Stereophotos by D. Serrette, U.A. 12.

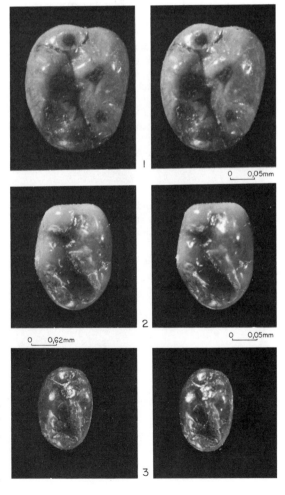

Acknowledgments

The authors are first of all greatly indebted to the owners, G. Wouters and C. Marignac, for long-term loan of the teeth studied here. Dr. W.-E. Reif kindly authorized me to section a selachian denticle from the collections of the Institut und Museum für Geologie und Palaeontologie, Universitaet Tübingen. Mrs. Weber-Chancogne patiently and ably contributed to the ultramicroscopic study. Miss Vrain and Mr. Serrette helped with the illustrations, and Mrs. Poncy with the typing. W.A. Clemens, K. Padian, and N. Simmons reviewed and helped with the text.

References

Cooper, J. S., and D. F. G. Poole. 1982. The dentition and dental tissues of the agamid lizard *Uromastix*. *J. Zool, London* 169: 85–100.

Costa, R. L., Jr., and W. S. Greaves. 1981. Experimentally produced tooth wear facets and the direction of jaw motion. *J. Paleontol*. 55: 635–8.

Dauphin, Y. 1984. Mise en evidence d'une structure prismatique dans l'émail d'un Reptile archosaurien actuel: *Alligator mississipiensis* (Daudin). *Compt. Rend. Acad. Sci. Paris*, 298(II), 20: 911–14.

Fosse, G., and Z. Kielan-Jaworowska. 1985. The microstructure of tooth enamel of multituberculate mammals. *Palaeontology*. 28(3): 435–9.

Fosse, G., S. Risnes, and N. Holmbakken. 1973. Prisms and tubules in Multituberculate enamel. *Calcif. Tiss. Res*. 11: 113–50.

Frank, R. M., D. Sigogneau-Russell, and J. C. Voegel. 1984. Tooth ultrastructure of Late Triassic Haramiyidae. *J. Dent. Res*. 63(5): 661–4.

Grine, F. E., C. E. Gow, and J. W. Kitching. 1974. Enamel structure in the cynodonts *Pachygenelus* and *Tritylodon*. *Elect. Microsc. Soc. S. Afr., Proc*. 9: 99–100.

Grine, F. E., E. S. Vrba, and E. R. S. Cruickshank. 1979. Enamel prisms and diphyodonty: linked apomorphies of Mammalia. *S. Afr. J. Sci*. 75: 114–20.

Hahn, G. 1969. Beitrage zur fauna der Grube Guimar-ota. III. Die Multituberculata. *Paleontographica, Abt. A*. 133(1–3):1–100.

1973. Neue Zähne von Harimiyiden aus der deutscher ober-Trias und ihre Beziehungen zu den Multituberculaten. *Palaeontographica* A142(1–3):1–15.

Hopson, J. A., and J. W. Kitching. 1972. A revised classification of Cynodonts (Reptilia; Therapsida). *Paleontol. Afr*. 14: 71–85.

Kerr, T. 1959. Development and structure of some Actinopterygian and Urodele teeth. *Proc. Zool. Soc. London* 133: 401–22.

Krause, D. W. 1982. Jaw movement, dental function and diet in the Paleocene multituberculate *Ptilodus*. *Paleobiology*. 8(3): 265–81.

Osborn, J. W., and J. Hillmann. 1979. Enamel structure in some Therapsids and Mesozoic mammals. *Calcif. Tis. Intern*. 29: 47–61.

Poole, D. F. G. 1971. An Introduction to the phylogeny of calcified tissues, *In* Dahlberg, A. A. (ed.), *Dental Morphology and Evolution* (Chicago: University of Chicago Press, pp. 65–79.

Reif, W. E. 1973. Morphologie und Ultrastruktur der Hai-"Schmelzes." *Zool. Scrip*. 2: 231–50.

Sauvage, C. 1972. Ultrastructure des dents de *Carassius auratus* (Cyprinidae). Thesis Université L. Pasteur (Strasbourg 4), UER Odontologie.

Schmidt, W. J., and A. Keil. 1971. *Polarizing Microscopy of Dental Tissues* (New York: Pergamon).

Shellis, R. P., and B. K. Berkovitz. 1976. Observations on the dental anatomy of piranhas (Characidae) with special reference to tooth structure. *J. Zool. London* 180: 69: 69–84.

Sigogneau-Russell, D. 1983. Nouveaux taxons de Mammiféres rhétiens. *Acta Paleontol. Polon*. 28(1–2): 233–49.

Sigogneau-Russell, D., R. M. Frank, and J. Hemmerlé. 1984. Enamel and dentine ultrastructure in the early Jurassic therian *Kuehneotherium*. *Zool. J. Linn. Soc*. 82: 207–15.

Simpson, G. G. 1984. Preface, *Zool. J. Linn. Soc*. 82(1/2): 3–5.

Smith, M. M. 1979. SEM of the enamel layer in oral teeth of fossil and extant crossopterygian and dipnoan fishes. *Scan. Elect. Microsc*. 11: 483–90.

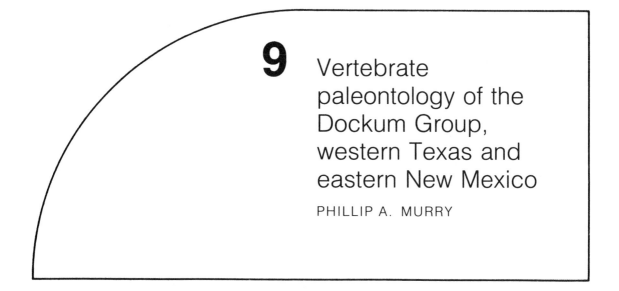

9 Vertebrate paleontology of the Dockum Group, western Texas and eastern New Mexico

PHILLIP A. MURRY

Introduction

The Upper Triassic–age Dockum Group of western Texas outcrops as an almost continuous north-trending band from an area north of San Angelo to the Canadian River Valley northwest of Amarillo. Its surface exposures are broken only by the Ogallala cap along the Llano Estacado of eastern New Mexico and the Texas Panhandle. In eastern New Mexico, a tremendous area is encompassed by the Dockum beds, especially in Quay and Guadalupe counties, with small bands of exposures extending almost to Taos to the north and nearly to Carlsbad to the south (Fig. 9.1).

The Dockum has been studied in some detail by geologists and paleontologists, but many basic questions remain unanswered. This chapter outlines previous geological and paleontological research in the Dockum Group and some of the current work that is being conducted in these beds. Sedimentological, lithostratigraphic, biostratigraphic, and paleoecological studies are beginning to reveal the complex history of the Dockum depositional basin, although the interpretations of various workers have often conflicted on specific details. Many fossil vertebrates have also been described from the Dockum, but studies of the older material and recent recovery of many more vertebrate specimens have resulted in a reevaluation of the taxonomic relationships on many of the vertebrate taxa.

In this chapter, the study of the Dockum, has been approached in several ways. A geological overview discusses current views on Dockum sedimentology and lithostratigraphy. The majority of the chapter constitutes a review of taxonomic questions and distributional patterns for the various species of fossil vertebrates. Finally, there are short summaries of proposed Dockum biostratigraphy and paleoecology.

Geological background

The character of the Dockum depositional systems is related to Pangean tectonism. The initiation of doming along the Ouachita Foldbelt during the Upper Triassic reactivated deposition in the relict Paleozoic basins, such as the Midland and Palo Duro, and resulted in the formation of a large la-

Figure 9.1. Geologic map of the Triassic in western Texas, eastern New Mexico, western Oklahoma, and southern Colorado.

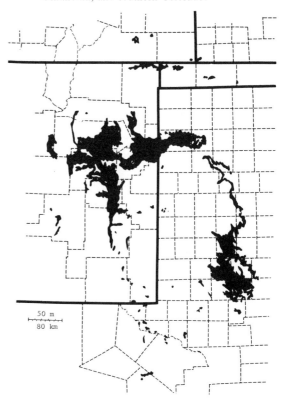

50 m
80 km

custrine basin with internal drainage. To the north, sediments were derived from the remnants of the Paleozoic Amarillo Uplift, Bravo Dome, and Matador Arch. In the west, the Central Basin Platform provided a source for Dockum sedimentation. To the south and east, the rejuvenated Ouachita Uplift also provided a source of sediments for the Dockum basins (Walper, 1980; Granata 1981).

The history of Dockum sedimentation is inextricably linked to the reactivation of tectonism and the formation of this large structural basin. In the southeastern portion of the basin, the oldest Dockum units, deeply buried beneath younger Triassic and Cretaceous sediments, represent prodelta and delta front/delta plain facies of a lobate deltaic system. Other parts of the Midland Basin formed minor lacustrine deposits, and within the northwestern portion of the basin, reworked Permian sediments were redeposited as a small apron of braided alluvial and sheet wash sediments. This latter unit has been designated the Lower Sandstone Unit of the Santa Rosa Formation (Gorman and Roebeck 1946; Granata 1981). The fauna and flora of this early phase of Dockum sedimentation are unknown.

The next phase of sedimentation, designated the "major early progradational sequence" by Granata (1981), represents nearly the entire history of the Dockum as exposed in outcrops and, therefore, is most important when reviewing Dockum paleontology. Along the eastern portion of the Dockum Basin, exposed in outcrop, and within the west central basin there is a thin (less than 20 m), uniform mudstone that probably represents lacustrine and prodelta deposits. These mudstones thicken toward the Midland Basin and exhibit a coarsening upward sequence, indicating delta progradation. The east central portions of the Dockum exposed in outcrop reveal a history of lacustrine transgression and regression, probably reflecting variable rainfall, with concomitant changes in the fluviodeltaic systems feeding them. Meandering streams supplied sediment for feeding the high-constructive lobate delta systems within the east central portion of the basin. In the southern and northern portions of the Dockum Basin, braided streams and fan deltas were the dominant depositional systems. During times of aridity, many of the ephemeral lakes dried, and the base level within the larger lakes was lowered. During these periods, fan deltas evolved along the lake margins, and evaporites, calcretes, silcretes, and soils developed (McGowen, Granata, and Seni 1979).

Within the northwestern part of the Dockum Basin, a widely distributed coarse-grained sandstone meanderbelt outcrops widely over DeBaca, Guadalupe, San Miguel, and Quay counties, New Mexico. This unit has been designated the Middle Sandstone Member of the Santa Rosa Sandstone (McGowen and Garner 1970). Granata (1981) suggests that deposition of this sandstone was disrupted along the Matador Arch, which may have been the boundary for the extensive lacustrine sedimentation during early Dockum times. However, the Middle Sandstone Member of the Santa Rosa Sandstone is overlain by deltaic and lacustrine deposits that may indicate a basinwide expansion of the lacustrine environment during the uppermost part of the lower Dockum.

Along the eastern boundary of the Dockum Basin in Texas, the upper Dockum has not been preserved. In eastern New Mexico, this portion of the Dockum is preserved and is overlain by Jurassic dune deposits. When the margin of the Gulf of Mexico lowered, it is assumed that the Dockum sedimentation shifted eastward, and eventual erosional truncation of much of the upper Dockum in Texas probably took place during Upper Jurassic times. Subsurface studies by Granata (1981) indicate an easterly prograding fluviodeltaic system within the northeastern and north central Dockum Basin during the Upper Triassic. The eastern shift of Dockum deposition may be indicated by the depositional sequence within the Redonda Formation in the Tucumcari Basin of New Mexico. The Redonda shows cyclic sedimentation of lacustrine mudstones and winnowed, marginal lacustrine sandstones. These features, along with the presence of evaporites in the Redonda and the Jurassic dune-formed Exeter sandstone, have been suggested as evidence of increased aridity and thus decreased alluvial input during upper Dockum times (Granata 1981).

The Dockum Group of the northwestern Texas Panhandle was divided into two formations by Gould (1907), and his stratigraphic nomenclature has been used by most later workers. The Tecovas Formation, consisting primarily of mudstones, is overlain by the Trujillo Formation, composed primarily of sandstones and conglomerates. Although this nomenclature may be used in the Canadian River area, the deposition of sandstones upon mudstones is related to deposition within fluviodeltaic systems within the Dockum Basin. These systems are difficult to correlate and, in fact, represent largely diachronous sedimentological events. However, it appears that there has been some recent progress in Dockum lithostratigraphy in the central and southern portions of the basin. For example, Chatterjee (Chapter 10) has correlated the stratigraphic units near Post, Garza County, Texas with the type section along the Canadian River. He has demoted the Tecovas and Trujillo to member status and proposed the addition of a younger member, the Cooper, to the Dockum stratigraphic sequence. Until recently, a detailed correlation of the Dockum within the southern portion of the basin has also been largely unsuccessful. However, Gene Greenwood, an independent consulting geologist from Midland, has recognized a vertical sequence of four distinct conglomeratic units of extensive geographic distribution in that area. Although the relationship of the Otis Chalk strata to these conglomerates has not yet been determined, Greenwood and S. W. Bishop, of Exxon Corporation, discovered a new Dockum vertebrate locality in the Iatan–East Howard Oil Field, which lies immediately below Greenwood's conglomerate 3. The uppermost unit, conglomerate 4, is also exposed in this area (G. Greenwood pers. comm.). The Iatan–East Howard field is situated immediately north of the Otis Chalk locality. These discoveries may aid in interpreting the complex stratigraphy of the Dockum basin.

Although significant progress has been made in local correlation of the Dockum, attempts at detailed basinwide correlation have been largely unsuccessful. Therefore, use of the names "Trujillo" and "Tecovas" has been avoided in this chapter, although "upper" and "lower" Dockum have been used as defined by Granata (1981).

Dockum paleontology

Some of the first descriptions of Upper Triassic fossils from west Texas and eastern New Mexico were by E. D. Cope in the early 1890s. Some of the type specimens described by Cope, largely based on fragmentary remains, are in the University of Texas collections and the American Museum of Natural History. The substantial Dockum collection of the University of Michigan Museum of Paleontology was made by E. C. Case and his collectors, especially William Buettner, primarily during the 1920s and 1930s. During the late 1930s, the Works Progress Administration, under the supervision of vertebrate paleontologists from the University of Texas and West Texas State University, assembled large collections that formed the nucleus of the substantial Upper Triassic collections at these institutions. The University of Texas collections have been studied by a number of workers, including J. T. Gregory, H. J. Sawin, J. A. Wilson, and, more recently, P. Parks and R. L. Elder.

The collection of Triassic fossils by Texas Tech University was initiated in the early 1950s by F. E. Green, and large collections have recently been made by Sankar Chatterjee from localities in the Texas Dockum. Bobb Schaeffer collected fish for the American Museum of Natural History at localities near Otis Chalk in 1953 and 1954 (Schaeffer 1967).

B. H. Slaughter of Southern Methodist University began collecting and wet sieving for Dockum fossils during the early 1970s, and P. A. Murry has continued collecting Triassic vertebrate fossils in western Texas and eastern New Mexico; these fossils are housed in the Shuler Museum collection.

Eastern New Mexico has tremendous potential for aiding in the understanding of the southwestern Upper Triassic because the exposures are spectacular and the stratigraphy seemingly simpler than in the Dockum of Texas. However, few collections have been described from that region. Collections were made in 1916 and 1919 by M. G. Mehl of the University of Missouri and by J. W. Stovall of the University of Oklahoma during the 1930s. Collections for the University of California Museum of paleontology and Yale University have been made by J. T. Gregory within east central New Mexico from the late 1940s to the present.

Institutions that house Triassic collections mentioned in this chapter include the following:

American Museum of Natural History (AMNH)

British Museum of Natural History (BMNH)

Cleveland Museum of Natural History (CMNH)

Dallas Museum of Natural History (DMNH)

Field Museum of Natural History (FMNH)

Museum of Comparative Zoology, Harvard University (MCZ)

Museum of Northern Arizona (MNA)

Stovall Museum of the University of Oklahoma (OU)

Princeton University Geology Collections (PU)

Shuler Museum of Paleontology, Southern Methodist University (SMUSMP)

The Museum at Texas Tech University (TTU)

The University of California Museum of Paleontology (UCMP)

The University of Michigan Museum of Paleontology (UMMP)

The University of Missouri Paleontology Collections (UMo)

The United States National Museum (USNM)

The University of Texas/ Texas Memorial Museum (UT/ TMM)

University of Wisconsin (UW)

West Texas State University (WTSU)

Yale Peabody Museum (YPM)

Numerous species of sharks, fish, amphibians and reptiles have been recovered from the Dockum Group, but much is yet to be learned about the taxonomy, ecology, and biostratigraphy of the Texas–New Mexico Upper Triassic vertebrates. The following section reviews the known Dockum vertebrates and some of the problems encountered in studies of the Dockum fauna.

Systematic paleontology
Class Selachii
Cohort Hybodontoidei
Order Hybodontiformes
Family Lonchidiidae
Genus *Lissodus* Brough 1935
Lissodus humblei (Murry 1981)

The teeth of *L. humblei* are minute, low-crowned with a well-developed labial buttress and longitudinal rows of large foramina along the labial and lingual faces of the root. The only specimens (at SMUSMP) (Fig. 9.2A–E) of *L. humblei* from the Dockum Group are from the Kalgary locality in Crosby County, Texas (Murry 1981). Other specimens of *Lissodus* from the North American Triassic were recovered from the Chinle Formation near St. Johns, Arizona and are probably referable to *L. humblei* (Jacobs and Murry 1980; Tannenbaum 1983).

Figure 9.2. *Lonchidion humblei* specimens from the Kalgary local fauna, lower Dockum Group. **A,** Tooth, SMUSMP 67951, 1, Labial view; 2, lingual view; 3, occlusal view. Scale = 0.034 cm. **B,** Thin section of tooth (SMUSMP 67954). Scale = 0.025 cm. **C,** Dorsal fin spine (SMUSMP 67469). Scale = 0.1 cm. **D,** Dorsal fin spine (SMUSMP 67955). Scale = 0.1 cm. **E1,2,** Cephalic spine (SMUSMP 67955). Scale = 0.1 cm. **F-***Xenacanthus moorei.* 1, Denticle (SMUSMP 67984). Scale = 0.083 cm. 2, Tooth (SMUSMP 67328). Scale = 0.059 cm. Both specimens from Kalgary local fauna.

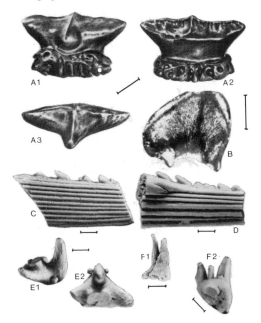

Lissodus africanus is a freshwater hybodont from the Lower Triassic of South Africa. According to Duffin (1981), *Lonchidion* and *Lissodus* are synonymous, with *Lissodus* Brough 1935 having priority over *Lonchidion* Estes 1964. Other, as yet undescribed specimens (BMNH P47166–69) of *Lissodus* are from the Muschelkalk of Germany. Several specimens (BMNH P29503–6) from the Upper Keuper of England referred to *"Palaeobates" keuperinus* have a tooth morphology similar to that seen in *Lissodus*, although other specimens referred to the latter taxon are very different. The other described species of *Lissodus* are from the Cretaceous of Great Britain (Patterson 1966), Belgium (Herman 1977), Alberta (Fox 1977), and Wyoming (Estes 1964), Texas (Thurmond 1971), New Jersey (Cappetta and Case 1975), New Mexico (Armstrong-Zeigler 1978), Colorado (Carpenter 1979), and Montana (Carpenter 1979).

A number of elasmobranch dorsal fin spines were recovered from the Kalgary locality by collectors from SMU (Murry 1982) and Texas Tech (Green 1954). Schaeffer (1967) also reported a large fin spine recovered from this area. These fin spines are oval in cross section, with a row of alternating denticles along the dorsal side and well-developed longitudinal striations on the lateral surface (Fig. 9.2C–D). This morphology is similar to that of *Lissodus africanus* (Brough 1935), as well as of other hybodontiforms such as *Hybodus, Acrodus,* and a number of Paleozoic taxa (including *Tristychius*) (Maisey 1978). Unornamented hybodontiform cephalic spines (SMUSMP) similar to those of *Lissodus* were recovered from Kalgary, although such spines are characteristic of many hybodontiform sharks besides *Lissodus*.

Genus *Lissodus*?
A small tooth (SMUSMP 69650), somewhat similar in morphology to *Lissodus humblei*, was recovered from Quarry 1 near Otis Chalk, Howard County, Texas. However, this tooth possesses both a labial and lingual buttress, and the anteroposterior extremities of the tooth are curved labially (Fig. 9.3B). This tooth may belong to a new species. Unfortunately, the root was not preserved in the specimen, and, therefore, the precise taxonomic position of the tooth must remain unresolved until further specimens are obtained.

Cohort Neoselachii
Order Orectolobiformes?
A neoselachian shark tooth (SMUMP 69649) was recovered from the Rotten Hill locality along Sierrita de la Cruz Creek, Potter County, Texas (Fig. 9.3A).

The referral of this tooth to the Neoselachii is based primarily on root morphology. The base of the root is bisected by a prominent medial groove [the holaulacorhize condition as defined by Casier (1947)], and a "rhinobatoid" root vascularization is present by which the central portion of the root is pierced by a large foramen. The high tooth crown is labiolingually compressed, and the morphology of SMUSMP 69649 is similar to that of symphyseal teeth found in other neoselachians.

Among Triassic–Liassic selachians, no known genus exhibits tooth characteristics similar to those of the Dockum specimen. Presumed Upper Triassic–Liassic neoselachian taxa based on teeth include *Hueneichthys costatus* and *Reifia minuta* from Germany, *Vallisia coppi* from southwest England, and *Palaeospinax* from Great Britain and Germany. Neither *Reifia* or *Paleospinax* possesses a medial groove (Maisey 1977; Duffin 1980); this characteristic distinguishes them from the Rotten Hill tooth. *Vallisia coppi* exhibits a holaulacorhize root morphology, but teeth of this taxon, as well as those of *Reifia*, *Hueneichthys*, and *Palaeospinax*, develop lateral cusplets (Maisey 1977; Reif 1977; Duffin 1980, 1982).

The Rotten Hill specimen shows a surprising

Figure 9.3. **A**, Neoselachian tooth (SMUSMP 69649) in oblique occlusal (A1), lateral (A2), and basal (A3) views. Rotten Hill local fauna. Oblique occlusal view, lateral view, basal view. Scale = 0.063 cm (A1), 0.077 (A2), and 0.054 cm (A3). **B**,*Lissodus* sp.? tooth (SMUSMP 69650) in occlusal (B1) and labial (B2) views. Otis Chalk local fauna. Occlusal view labial view. Scale = 0.04 cm (B1) and 0.033 (B2).

similarity to *Cretorectolobus olsoni*, a Cretaceous (Campanian) orectolobid from Wyoming (G. R. Case 1978), which may indicate a close relationship of the Dockum specimen to the orectolobid sharks.

Order Xenacanthodii
Family Xenacanthidae
Genus *Xenacanthus* (Beyrich 1848)
Xenacanthus moorei (Woodward 1889)

Teeth and dermal denticles (SMUSMP, AMNH, BMNH) of *X. moorei* were recovered from the lower Dockum Group near Kalgary in Crosby County, from Sunday Canyon in Randall County, and from the Rotten Hill locality in Potter County.

The teeth of *Xenacanthus moorei* have two large, conical, carinated cusps and a smaller central cusp attached to a roughly triangular base (see Fig. 9.2F2). Xenacanth dermal denticles were recovered from Kalgary that were generally less than one mm in length, with one to three slender, slightly curved prongs fused to a bulbous, irregularly shaped base (see Fig. 9.2F1). Because xenacanth teeth are the most numerous single element of the Kalgary fauna, it is surprising that no cephalic spines of xenacanth sharks have been recovered from the Dockum. Their absence suggests that these xenacanths may have lost these peculiar spines during some late period of their evolutionary history.

Numerous teeth of *Xenacanthus moorei* were reported from the Petrified Forest Member of the Chinle Formation near St. Johns, Arizona (Jacobs and Murry 1980; Tannenbaum 1983) and from Petrified Forest National Park (Jacobs 1980). Other specimens of *Xenacanthus moorei* are from the Keuper of Somersetshire, England (Woodward 1889) and the Gipskeuper near Gaildorf, North Wurttemberg, West Germany (Seilacher 1943). Also, the morphology of the teeth of *Xenacanthus indicus* from the Maleri Formation in the Godavari River Valley of Andhra Pradesh, India (Jain 1980) is identical to those of *Xenacanthus moorei* and the former should be considered a junior synonym.

Class Osteichthyes
Subclass Actinopterygii
Infraclass Chondrostei
Order Palaeonisciformes
Suborder Palaeoniscoidei
Family Palaeoniscidae
Genus Turseodus? Leidy 1857

These small- to medium-sized fusiform fishes may prove to be useful biostratigraphically because they seem to be restricted to deposits of late Carnian age (Fig. 9.4A). *Turseodus acutus*, *T. gyrolepis*, *T. minor*, *T. major*, and *T. gwyneddensis* were recovered from the Lockatong Formation, Pennsylvania (Schaeffer 1952; Bock 1959). Other Newark *Turseodus* specimens have been found in the upper

Figure 9.4. **A**, *Turseodus* sp. (after Schaeffer 1952, 1967). Scale = 20 mm. **B**, *Ceratodus* sp. (after Colbert 1972). Scale = 50 mm. **C**, *Cionichthys* sp. (after Schaeffer 1967). Scale = 10 mm. **D**, *Lasalichthys* sp. (after Schaeffer 1967). Scale = 10 mm. **E**, *Hemicalypterus weiri* (after Schaeffer 1967). Scale = 10 mm. **F**, Ichthyolith morphotype A. **G**, Ichthyolith morphotype D. **H**, *Hemicalypterus* premaxilla **I**, Ichthyolith morphotype E. Scales in F-I equal 0.5 mm. **J**, *Chinlea sorenseni* (skull after Schaeffer 1967), hypothetical postcranial skeleton based on *Diplurus* after Schaeffer (1952). Scale = 50 mm.

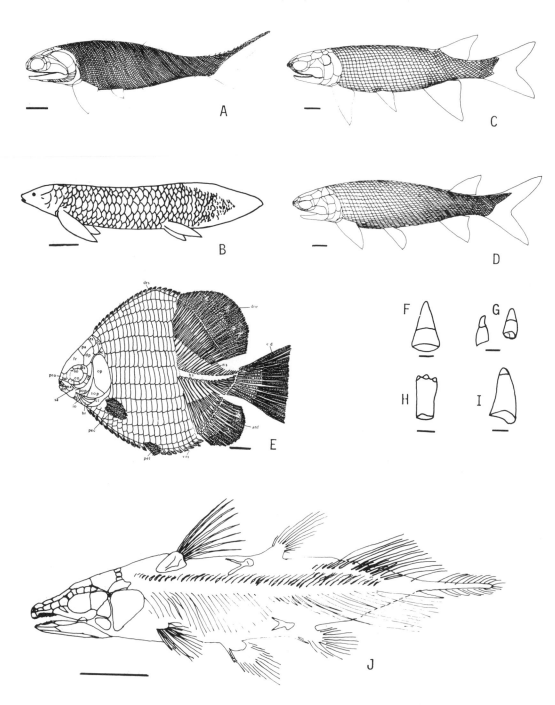

Stockton, lower Lockatong, Cow Branch, and New Oxford formations; in the "Durham Basin Fish Bed"; and possibly in the Cumnock Formation (Schaeffer and McDonald 1978; Olsen, McCune, and Thomson 1982). Specimens of *Turseodus dolorensis*, which is very similar in morphology to *T. minor*, are from the upper portion of the Chinle Formation near Bedrock, Colorado (Schaeffer 1967). Annular (ring-shaped) centra referred to *Turseodus* and *Turseodus*-like scales were also recovered from the Chinle Formation near St. Johns, Arizona (Jacobs and Murry 1980; Tannenbaum 1983) and from Petrified Forest National Park (Jacobs 1980).

The presence of *Turseodus* in the Dockum is suggested by rhomboidal to crescentic, ridged scales in the SMU collection from the Kalgary (Crosby County) and Rotten Hill (Potter County) localities as well as from the Schaeffer fish quarry and Quarry 3 near Otis Chalk, Howard County, Texas. In the Kalgary and Rotten Hill collections there are scales with the morphology seen in *Turseodus gyrolepis*, including both the large, rhomboidal, many-ridged types and the slender crescentic scales characteristic of that species (Bock 1959). However, there are also rhomboidal *Turseodus*-like scales with less than five ridges in both the Kalgary and Rotten Hill sites that may indicate the presence of more than one species of fish with *Turseodus*-like scale morphology at those localities. These smaller scales and those from Otis Chalk are very similar in morphology and size to *T. minor* and *T. dolorensis*. *Turseodus*-like scales (SMUSMP) have also been recovered from the "Shark Tooth Hill" locality near San Jon, New Mexico.

Order Redfieldiiformes
Family Redfieldiidae
Genus *Cionichthys*
Cionichthys greenei Schaeffer 1967
The only Dockum specimens (AMNH) of *Cionichthys greenei*, a small fusiform redfieldiid fish, are from the Schaeffer fish quarry near Otis Chalk, Howard County, Texas (Fig. 9.4C). A closely related species, *Cionichthys dunklei*, was recovered from the upper portion of the Chinle in Utah and Colorado (Schaeffer 1967). *Cionichthys* specimens similar in morphology to *C. greenei* were collected from the lower Barren Beds at Manakin, Virginia. Other Newark *Cionichthys* have been collected from the Pekin, Lockatong, Cumnock, and lower Cow Branch formations (Olsen, McCune, and Thomson 1982).

Cionichthys greenei was distinguished from *C. dunklei* by the dimensions of the dermosphenotic, adnasal, and antorbital bones and by the denticulated posterior borders of the opercular and scales (Schaeffer 1967). However, *Cionichthys dunklei* and

C. greenei specimens in the AMNH both have denticulated scales, although the scales appear to be relatively few in number and restricted to a small region on the anterolateral portion of the body. The other scales are very difficult to distinguish when isolated, because they are simple, rhomboidal, and chagrenate with few diagnostic characters. However, the absence of posterior denticulated scales at most Dockum localities may be of some significance because *Cionichthys* seems to have a rather limited stratigraphic distribution.

Genus *Lasalichthys* Schaeffer 1967
Lasalichthys/Synorichthys specimens (Fig. 9.4D) have been recovered from UT *Trilophosaurus* Quarry 1 and the Schaeffer fish quarry near Otis Chalk, Howard County, Texas (Schaeffer 1967). In addition, rhomboidal, punctate scales with prominent peg and socket articulations have been recovered by SMU field crews from Kalgary, Otis Chalk, and Rotten Hill, and robust pectoral fin rays with numerous fringelike fulcra were found at the Kalgary locality. These features are characteristic of the *Lasalichthys / Synorichthys / Rushlandia*–type fishes (Bock 1959; Schaeffer 1967), although a number of colobodontids also possess robust fin fulcra.

It is extremely difficult to distinguish *Synorichthys* from *Lasalichthys* because the only difference between the genera is the absence of a postrostral bone in *Synorichthys*. Considering the extreme reduction of this element in *Lasalichthys* and the similar morphology of *Synorichthys* and *Lasalichthys*, these genera are very closely related and may yet prove to be synonymous.

Other *Synorichthys* and *Lasalichthys* specimens have been recovered from the upper Chinle Formation in Utah and Colorado (Schaeffer 1967). In the Newark Supergroup, specimens of *Synorichthys/Lasalichthys* have been found in the Lockatong, lower Passaic, Cumnock, Pekin, Cow Branch, and the New Oxford formations (Olsen et al. 1982). The only other redfieldiid known that is of similar morphology to *Lasalichthys* is *Mauritanichthys rugosus*, from the Upper Triassic of Morocco (Martin 1982a).

Other redfieldiid specimens

Denticulated rostral and antorbital bones (SMUSMP) referrable to redfieldiid fishes have been recovered from the Rotten Hill locality in Potter County and the Kalgary locality in Crosby County, Texas. Isolated redfieldiid elements (SMUSMP) were recovered from the "Shark Tooth Hill" locality near San Jon, New Mexico and have been reported from the Chinle Formation near St. Johns (Tannenbaum 1983) and in Petrified Forest National Park, Arizona (Jacobs 1980).

Order Perleidiformes

Family Colobodontidae

Colobodontid toothplates (Fig. 9.5A) in the SMU collections from the lower Dockum Group were recovered from the Kalgary, Rotten Hill, and Otis Chalk (Quarry 1) localities. There are two tooth morphologies present within the colobodontids. Sharply pointed teeth are present on the premaxillary and on the anterior and labial margins of the maxillary and dentalosplenial. On the inner edge of the maxillary and dentalosplenial toothplates, as well as on the prearticulars, ectopterygoids and pterygoids, low, rounded, blunt crushing teeth are found in many genera (Stensiö 1921; Schaeffer 1955). Both tooth morphologies are present in the Dockum, and there are also several morphotypes, which probably indicates the presence of several taxa. Similar toothplates were recovered from the Chinle Formation at the Placerias Quarry (Jacobs and Murry 1980) and

Figure 9.5. **A**, Colobodontid toothplates. 1, SMUSMP 67488; 2, SMUSMP 67969, Scale = 0.077 cm. **B**, *Ceratodus dorotheae* toothplate, UMMP 7324 (holotype). Scale = 0.5 cm. **C**, Holotype of *Colognathus obscurus* (UMMP 7506) in occlusal (1) and lateral view (2). Scale = 0.435 cm. **D**, Sphenodontid partial left maxilla, SMUSMP 67983. Scale = 0.083 cm. **E**, Juvenile sphenodontid? partial dentary, SMUSMP 67959. Scale = 0.083 cm. **F**, Kuehneosaurid? mandible fragment, SMUSMP 67960. Scale = 0.05 cm. **G**, *Trilophosaurus buettneri* holotype, UMMP 2338. Scale = 0.25 cm. **H**, *Eudimorphodon?* sp.; 1, partial mandible, SMUSMP 69125; 2, partial maxilla, SMUSMP 69124. Scale = 0.1 cm.

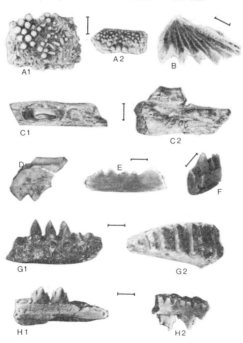

a large colobodontid toothplate was found in the Blue Hills northeast of St. Johns, Arizona (R. A. Long pers. comm.). In New Mexico, colobodontid toothplates (SMUSMP) have been recovered from the Red Peak and Apache Canyon localities in Quay County.

Hutchinson (1973) noted that although the colobodontids were rather restricted temporally (Scythian–Norian/Rhaetic?), they were a very diverse and widely distributed group with many phyletic lines. Most of the toothplates found at Rotten Hill, Kalgary, and Otis Chalk are similar to those seen in *Mendocinia*, *Perleidus*, and *Meidiichthys* in which the teeth are strongly heterodont but unornamented. Other colobodontids have been found in South America, Europe, Africa, and Australia (Stensiö 1921, 1932; Zittel 1932; Schaeffer 1955; Lehman 1966; Martin 1982a,b).

Order Semionotiformes

Suborder Semionotodiei

Family Semionotidae

Genus *Hemicalypterus?* Schaeffer 1967

A small premaxilla (SMUSMP; Fig. 9.4H) from the lower Dockum Group at the Rotten Hill locality is similar in morphology to the unique premaxilla of *Hemicalypterus weiri*, although three styliform teeth are present on the premaxilla rather than the usual four or five (Schaeffer 1967, p. 320 and Plate 25, Fig. 4). Well-preserved specimens of this deep-bodied, unusual semionotid (AMNH, USNM) are from the upper Chinle Formation in Utah (see Fig. 9.4H). Another premaxilla in the MNA collection from the St. Johns Quarry, Chinle Formation is identical in morphology to the Rotten Hill specimen and is therefore probably referable to *H. weiri*.

Genus *Semionotus* Agassiz 1832

The only material positively identified as *Semionotus* from the Dockum Group consists of isolated skull elements (frontals and a preoperculum in the YPM) from the "Trujillo" in Sunday Canyon, Randall County, Texas (Olsen 1984). A few large, rhomboidal semionotid-like scales and teeth (SMUSMP) have also been recovered from the Otis Chalk localities, although these are not positively referred to *Semionotus*. The taxonomy of the semionotids in the Chinle and Newark, as well as that in the Triassic–Jurassic of Europe, is confusing, especially because the morphology changes drastically during the ontogeny of taxa such as *Semionotus* (Schaeffer 1967; A. R. McCune pers. comm.). Therefore, the classification of the semionotids, especially the genus *Semionotus*, will change as many of the current taxa are synonymized.

Dockum ichthyoliths

Many sedimentary units devoid of most fossils contain the skeletal debris of fishes, which seems to be very resistant to chemical dissolution. These skeletal and tooth morphotypes, termed ichthyoliths, have been used to correlate Cenozoic and Paleozoic strata. Comparisons of fish teeth within the Dockum using only a few descriptive characters and some dimensional data (such as the length of the enameloid cap, and width and length of the tooth) indicate that several ichthyolith morphotypes were present in the Dockum Triassic.

Ichthyolith morphotype A, found only at the Otis Chalk Quarry 1, is characterized by its relatively large size and by the very long enameloid cap, which extends over one-half the total length of the tooth (see Fig. 9.4F). Similar teeth are found in semionotids and in the Early Triassic palaeoniscid *Birgeria* (Stensiö 1921).

Ichthyolith morphotype D is from the lower Dockum Group at the Kalgary and Rotten Hill localities. Similar teeth (MNA) have been recovered from the Chinle Formation in the Placerias Quarry near St. Johns, Arizona. These teeth are unique in their external morphology in that the enameloid tip is flattened and there is a very bulbous lingual surface (see Fig. 9.4G). Although incisiform fish teeth from the Triassic have been attributed to the genus *Sargodon* (Zittel 1932), *Sargodon* teeth are constricted proximally as opposed to the bulbous Dockum teeth.

Ichthyolith morphotype E is represented by numerous teeth (SMUSMP) from the Kalgary and Rotten Hill localities. This morphotype includes typical palaeoniscoid teeth that are acicular with a short enameloid cap (see Fig. 9.4I). Redfieldiid marginal teeth, such as those found in *Cionichthys*, and palaeoniscid teeth, such as those seen in *Turseodus*, are of this morphotype.

Subclass Sarcopterygii
Infraclass Crossopterygii
Order Coelacanthini
Family Coelacanthidae
Genus *Chinlea* Schaeffer 1967
Chinlea sorenseni Schaeffer 1967

Chinlea sorenseni was based on a nearly complete specimen (AMNH 5652) from the upper Chinle Formation of Utah. University of Michigan coelacanth specimens from the lower Dockum Group include a cleithrum fragment collected at Walkers Tank (near Kalgary) and a quadrate and partial pterygoid, also from Crosby County. These latter specimens were identified as *Macropoma* by Warthin (1928). A skull (YPM 3928) was found in Gold Canyon in Randall County, Texas (see Fig. 9.4J), and other *Chinlea* specimens have been recovered from the Chinle Formation near Bedrock, Utah; at Ghost Ranch, New Mexico (Schaeffer 1967); and near St. Johns, Arizona (Tannenbaum 1983). Two coelacanth quadrates have been recovered from the upper

unit of the Petrified Forest Member, Chinle Formation in Petrified Forest National Park, Arizona (R. A. Long pers. comm.) and isolated coelacanth scales have been recovered from the Dockum group in Sunday Canyon, Randall County, Texas (Olsen 1984).

Chinlea and *Diplurus*, a Newark taxon, are quite similar in a number of key characters, such as the shape of the basisphenoid and the presence of long pleural ribs; they appear to be more similar to each other than to the remaining coelacanths (Schaeffer 1967). *Diplurus longicaudatus* has been recovered in the Newark Supergroup from the Lockatong, Shuttle Meadow, and East Berlin formations and the Culpeper Group; *Diplurus newarki* has been found in the Lockatong, Pekin, Cumnock, and Cow Branch formations (Bock 1959; Schaeffer 1967; Olsen et al. 1982).

Order Ceratodontida
Family Ceratodontidae
Genus *Ceratodus*
Ceratodus dorotheae Case 1921

The holotype of *Ceratodus dorotheae* (see Fig. 9.5B) is from a locality near the headwaters of Holmes Creek, probably within the area of the Shuler Museum's Kalgary locality. The metatype and Case's other specimens are from the Holmes Creek/Walker's Tank localities in Crosby County (Case 1921). *Ceratodus* teeth (UC, SMUSMP) were also found in the Redonda Formation of Apache Canyon, Quay County, New Mexico.

These specimens, as well as all of the southwestern Triassic ceratodontid toothplates, probably represent a single taxon. The characters that Warthin (1928) used to separate *Ceratodus crosbiensis* toothplates (also from Walker's Tank) from those of *C. dorotheae* (i.e., horizontal length, distal sloping, separation, and intersection of ridges) actually represent ontogenetic variations in which the "*C. crosbiensis*" toothplate represents a middle stage in the life history of the lungfish. However, although there is a relative decrease in ridge height versus width from juvenile to adult forms, *Ceratodus dorotheae* never forms the broad, very low-crowned crushing plates seen in many ceratodontids. Other Dockum lungfish specimens have been recovered from Trilophosaurus Quarry 1 near Otis Chalk in Howard County (SMUSMP), from the Kalgary locality (Green 1954), and from the Sierrita (Cerita) de la Cruz, Potter County (UMMP). These specimens are quite similar in morphology to toothplates recovered from the Chinle Formation at the Placerias Quarry (MNA) near St. Johns (Jacobs and Murry 1980; Tannenbaum 1983) and at Adamana, Arizona (AMNH). A large number of *Ceratodus* toothplates have been found in both the lower and upper units of the Petrified Forest Member, Chinle Formation in Petrified Forest National Park (R. A. Long pers. comm.).

The only ceratodontid body specimen recovered from the North American Triassic is a nearly complete skeleton from the Monitor Butte Member, Chinle Formation near Fort Wingate, New Mexico. This specimen is currently being studied by Russell Dubiel. The other North American Triassic ceratodontids, represented only by isolated toothplates and scales from the Southwest, are quite similar in morphology to a number of taxa from Europe, Asia, and Africa. These toothplates are characterized by relatively small size, sharp cutting edges, and generally five or more ridges. Toothplates assigned to *Ceratodus arganensis* from the Upper Triassic of Morocco (Martin 1979) are almost exactly like those of *C. dorotheae*. Martin (1982b) placed *Ceratodus arganensis* and *C. dorotheae* in *Arganodus*, for which he created the Family Arganodontidae. However, *Arganodus atlantis* Martin 1979 has a greater anterointerior angle than *Ceratodus* and is therefore distinct from both *C. arganensis* and *C. dorotheae*. Several other *Ceratodus* species, known only from isolated toothplates, are of the "*Ceratodi excisi*" morphotype with sharp, high cutting ridges. This includes *Ceratodus kaupi* from the upper Muschelkalk and Lettenkohle of Germany and the upper Keuper of England, *C. capensis* from the Stormberg Beds of South Africa, and *C. parvus* from the upper Keuper of West Germany and the Rhaetic of England (Woodward 1891; Peyer 1968).

Class Amphibia
Subclass Labyrinthodontia
Order Temnosponydyli
Suborder Stereospondyli
Family Metoposauridae
Genus *Metoposaurus* Lydekker 1890
Metoposaurus fraasi Lucas 1904

Numerous taxa of metoposaurid amphibians have been described over a wide geographic range. However, many of these taxa have proved to be synonymous, as illustrated in studies by Branson (1905), Case (1946), Romer (1947), Colbert and Imbrie (1956), Roy-Chowdhury (1965), and Gregory (1980). Gregory (1980) divided the Metoposauridae into two genera, *Anaschisma* and *Metoposaurus*, primarily on the basis of the presence or absence of an otic notch.

There is some confusion as to whether *Anaschisma* actually possessed an otic notch, although Gregory (1980), working from photographs in Branson (1905), believed that *Anaschisma* lacked both otic notch and tabular horns. The relatively posterior placement of the orbits, the separation between the orbital and the supratemporal and postorbital canals, and the probable absence of the otic notch in *An-*

aschisma may very well be characters for generic separation from *Metoposaurus*. The only *Anaschisma* specimen that has been found in the Dockum is a partial skull (YPM 4201) (Fig. 9.6D) from the Redonda Formation (upper Dockum) of Apache Canyon, southeast of Tucumcari, Quay County, New Mexico (Gregory 1980). Roy-Chowdhury (1965) includes *Metoposaurus jonesi*, *Buettneria perfecta*, *Buettneria bakeri*, and *Buettneria howardensis* as synonyms of *Metoposaurus fraasi jonesi*. *Metoposaurus jonesi* is based on an associated interclavicle and clavicles (UMMP 3814) from the lower Dockum Group along Sand Creek in Crosby County, Texas (Case 1920a). Colbert and Imbrie (1956) found that these elements matched those in the numerous specimens described as *Buettneria perfecta* by Case (1922), also from Sand Creek. The type of "*B. perfecta*" (UMMP 7475) (Fig. 9.6B) is a skull, and the paratypes include UMMP mandibular and postcranial material. The skull of *Buettneria howardensis* (Fig. 9.6C) from the Trilophosaurus Quarry near Otis Chalk, Texas, is almost exactly like that of "*Buettneria perfecta*." The features Sawin (1944) used to separate the two (broader skull with larger palatal vacuities in "*B. howardensis*") seem insignificant in view of the large intraspecific variation seen in the metoposaurids. However, Sawin notes that in the pattern of sensory

Figure 9.6. Metoposaurid amphibians: diagrammatic dorsal and palatal views of left half of skull. **A**, *Metoposaurus bakeri* (after Case 1931). **B**, *Metoposaurus perfecta* (after Branson and Mehl 1929). **C**, *Metoposaurus howardensis* (after Sawin 1944). **D**, *Anaschisma* sp. (after Gregory 1980). Scale = 10 cm.

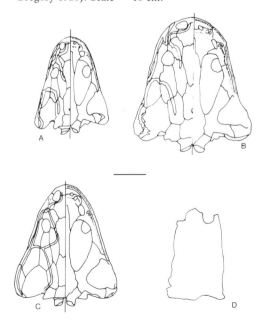

(or slime) canals, *Buettneria howardensis* is more like *Anaschisma* than like other species of "*Buettneria*," although *Buettneria bakeri* also has an incomplete sensory canal system. *Buettneria bakeri* was described on the basis of numerous skulls and skeletons from approximately 38 km north of Snyder in Scurry County, Texas (Case 1932a). The elongation of the palatial vacuities and the narrowness and small size of the skull (Fig. 9.6A) noted by Case (1931) are again ontogenetic and probably not taxonomically significant. Until the significance of these features is better understood, it is probably best to include all of the above in the species *Metoposaurus fraasi*.

Metoposaurus material (SMUSMP, UMMP, WTSU, AMNH, UT/TMM, TTU) has been found at many other localities in the Dockum Group, including Walker's Tank and Home Creek in Crosby County, the Palo Duro Canyon area in Randall County, the Rotten Hill locality along Sierrita de la Cruz Creek in Potter County, and the Miller Site near Post in Garza County.

Metoposaurus fraasi has been collected from the Chinle Formation from many sites in the southwestern Triassic. Type specimens (UMo, USNM) of the various now synonymized taxa were found in the Chinle Formation in Arizona (Colbert and Imbrie 1956). Numerous specimens (USNM, MCZ) of a "*Buettneria perfecta*" morphology were found in the Chinle Formation south of Lamy, New Mexico (Romer 1947). Other New Mexico *Metoposaurus* material has been found along Revuelto Creek, Quay County and Bull Canyon in Guadalupe County (Gregory 1972). Metoposaurid material (SMUSMP) was also recovered from Apache Canyon and above the Redonda Ledge at Red Peak in Quay County, New Mexico. A very interesting find is a skull (PU) from the Wolfville Formation in Nova Scotia with a "*Buettneria bakeri*" morphology (Baird and Olsen 1983). This shows a relationship between the Nova Scotia Triassic faunas and the Dockum faunas. *Metoposaurus maleriensis*, from the Maleri Formation at Achlapur, India, is almost exactly like *Metoposaurus fraasi* in its morphology (Roy-Chowdhury 1965).

Other *Metoposaurus* species distinct from *M. fraasi* include *Metoposaurus durus* from the upper Stockton and lower Lockatong Formations of the Newark Basin in Pennsylvania (Olsen 1980), *Metoposaurus ouazzouri* from Argana, Morocco (Dutuit 1976a), and *Metoposaurus diagnosticus* from the Keuper Formation of West Germany and the Lettenkohle of East Germany (Colbert and Imbrie 1956; Roy-Chowdhury 1965).

A detailed review of the ontogeny of the metoposaurids is needed, and the use of certain morphological features, especially in the development of the otic notch and the sensory canals, should be tested in further studies of this group.

Order Temnospondyli?
Suborder Stereospondyli
Family Latiscopidae
Genus *Latiscopus* Wilson 1948
Latiscopus disjunctus Wilson 1948
Latiscopus disjunctus is based on a very small (34 mm long) skull and jaws (UT) from Otis Chalk Quarry 1 in Howard County, Texas. A skull (TTU) has also been recovered from the Miller locality in Garza County (S. Chatterjee, pers. comm.). The skull of *Latiscopus* is high, narrow, and triangular with relatively large, laterally placed orbits, and it lacks the pitted cranial sculpture characteristic of the metoposaurs (Wilson 1948). Cosgriff and Zawiskie (1979) consider it to be Temnospodyli *incertae sedis* with certain resemblances to *Almasaurus habbazi*. The arrangement of the bones in the skull roof of the Texas Tech University specimen suggests possible relationships to the Metoposauridae.

Class Reptilia
Order Cotylosauria
Suborder Procolophonoidea?
Family Incertae sedis
Genus *Colognathus* Case 1928
Colognathus obscurus (Case 1928)
Colognathus obscurus was described on the basis of partial jaw fragments (see Fig. 9.5C) from Crosby County, Texas (Case 1928). Other UMMP specimens of *Colognathus* were discovered in approximately the same fossil several miles away in Crosby County. A single jaw, referred to *Colognathus obscurus*, was found in the Kalgary Quarry by the SMU crews, and Green (1954) also mentioned specimens found by an amateur collector near Kalgary. A lower jaw (UCMP cast) was recovered by an amateur collector approximately five miles southeast of Crosbyton in Crosby County, Texas and Case (1932b) referred several other specimens from Palo Duro Canyon. *Colognathus* has been considered a fish (Case 1928) and a procolophonoid (Baird and Take 1959), and similarities have even been noted with diadectic amphibians (Case 1922). I tentatively consider *Colognathus* a procolophonoid on the basis of its tooth morphology and occlusal wear pattern. The procolophonoids are unique in that most have teeth differentiated into incisiform and molariform types. The molariform cusps become worn through tooth to tooth occlusion. The crown becomes progressively flattened, and a fossette appears on the center of the tooth (Gow 1977). This tooth morphology and occlusal pattern is present in *Procolophon* and *Myocephalus* from South Africa (Broom 1936) and in the Russian procolophonids *Coelodontognathus*, *Tichvinskia*, *Orenburgia*, *Burtensia*, *Kapes*, and *Macrophon* (Ivachnenko 1973a,b, 1975, 1979). Most of the European, Chinese, and Amer-

ican procolophonoids, such as *Koiloskiosaurus*, *Sclerosaurus*, *Leptopleuron*, *Hypsognathus*, and *Paoteodon*, also exhibit extreme heterodonty and tooth to tooth occlusal wear as seen in *Colognathus* (Huene 1912, 1920; Colbert 1946; Chow and Sun 1960). However, *Colognathus* is unique among the procolphonoids in that the teeth are aligned antero-posteriorly rather than transversely. The only pro-colophonoid known with teeth aligned parallel to the jaw axis is an unnamed species from the Wolfville Formation of Hants County, Nova Scotia (Baird and Take 1959). Also, a UT *Colognathus* specimen from the Dockum has a long, swordfish-like edentulous beak (D. Baird pers. comm.), a feature unknown in any other procolophonoid. Therefore, if *Colognathus* is a procolophonoid, it is of such unique morphology that it probably should be placed in a separate family. Another distinct possibility is that *Colognathus* belongs to another reptile (or fish?) group that is yet to be defined.

Subclass Lepidosauria
Order Rhynchocephalia
Family Sphenodontidae

A left maxillary fragment (SMUSMP 67983)(see Fig. 9.5D) referable to the Sphenodon-tidae was recovered from the Rotten Hill locality. In New Mexico, sphenodontid jaw fragments (SMUSMP) were recovered from Apache Canyon in Quay County. In addition, other maxillary and mandibular fragments recovered from the Rotten Hill and Kalgary localities are tentatively identified as belonging to juvenile sphenodontids.

It is apparent that several lineages of rhyn-chocephalians evolved during the Mesozoic. Several genera, such as *Anisodontosaurus*, *Santaisaurus*, *Elachistosuchus*, and *Monjurosuchus*, possessed a simple tooth morphology (Koh 1940; Welles 1947; Young 1948; Huene 1956). Most other Triassic taxa were probably short-snouted forms in which roughly pyramidal teeth were developed. These include *Brachyrhinodon*, *Glevosaurus*, *Palacrodon*, *Planocephalosaurus*, *Sigmala*, and *Polysphenodon*, as well as specimens from Zimbabwe and specimens from the Dockum of Texas and the Chinle of Arizona (Huene 1910; Robinson 1973; Gow and Raath 1977; Fraser 1982; Fraser and Walkden 1983). The tooth morphology of *Opisthias* from the Morrison Formation and the extant *Sphenodon* is also similar to those of the aforementioned genera.

The Dockum jaw fragments referred to the Sphenodontidae are quite similar in morphology to many of the aforementioned taxa and may be compared with the genera *Opisthias* and *Sphenodon*. The teeth on the Rotten Hill maxillary fragment (SMUSMP 67983) and the Apache Canyon jaw fragments are acrodont, broad-based cones that increase in size posteriorly. Although the teeth of *Opisthias*, *Sphenodon*, and the Dockum specimens are of similar form in lateral view, they are somewhat different in occlusal view. The adult dentitions of *Opisthias* and *Sphenodon* have roughly quadrangular bases with sharp oblique shearing edges on the posterior edge of the maxillary dentition. In *Opisthias*, the posterior surface of the teeth is more convex than in *Sphenodon*, and the teeth are more closely spaced (Simpson 1926; Throckmorton, Hopson, and Park 1981). In the Rotten Hill maxillary fragment (SMUSMP 67699), the preserved teeth are closely spaced, and the occlusal surfaces are more rounded than in either *Sphenodon* or *Opisthias*. The sharp shearing edges present in the aforementioned taxa are not evident on the Rotten Hill specimen, and the apex of the tooth is placed lingually on the Dockum teeth instead of centrally on the tooth row as in *Sphenodon* and *Opisthias*. Specimens SMUSMP 67959 (see Fig. 9.5E) and 67490 from the Kalgary locality exhibit characters common to juvenile sphenodontids. These have acrodont dentition and a well-developed Meckelian canal, and in both the splenial is absent. In addition, SMUSMP 67959 has a well-developed coronoid process.

Family Rhynchosauridae
Subfamily Rhynchosaurinae

Elder's (1978) study of the material from the UT *Trilophosaurus* quarries at Otis Chalk, Howard County revealed the presence of partial premaxillae from Quarry 3a, as well as two humeri and two femora from Quarry 1 that were referrable to rhyn-chosaurs. Elder (1978) stated that the humeri show considerably more torsion than in *Stenaulorhynchus* from the Manda Formation of Tanzania, and are less massive than in *Paradapedon* or *Hyperodapedon* from the Elgin Formation (Lower Keuper) of Scotland. Also, although the femora resemble those of *Stenaulorhynchus* and *Paradapedon*, they are less robust and straighter, with poorer development of the adductor ridge (Elder 1978).

An undescribed rhynchosaur from the Wolfville Formation in Nova Scotia (Carnian) is also similar to *Hyperodapedon* (Baird 1963). Considering the similarities of these faunas, a comparison between these fossils would be interesting. A slender femur missing its distal extremity and the distal end of a humerus (UCMP) were recovered from the Placerias Quarry near St. Johns, Arizona; these are identical to the UT specimens, and the femur is quite similar to that of a Middle Triassic species from Madagascar, *Isalarhynchus genovetae* Buffetaut 1983 (R. A. Long pers. comm.). Also, the Upper Triassic of Morocco has yielded a rhynchosaur, *Acrodenta irerhi* (Dutuit 1976b) that should be compared to the Dockum/Nova Scotia material in view of the similarity of the faunas.

Lepidosauria incertae sedis

Two dentary fragments and a number of isolated teeth from the Kalgary locality are of lepidosaurian affinity. One slender, delicate jaw fragment (SMUSMP 67960)(see Fig. 9.5F) contains conical, closely set, homodont, subpleurodont teeth similar to those of the kuehneosaurids *Kuehneosaurus* and *Kuehneosuchus* from the English fissure fills (Robinson 1962), *Icarosaurus* from the Lockatong Formation at North Bergen, New Jersey (Colbert 1970), and jaw fragments (PU 17173) recovered from the Chinle Formation near Winslow, Arizona.

Teeth from a partial dentary (SMUSMP 67980) from the Kalgary locality show greater similarity in dental morphology to those of the modern lizard infraorders than to those of the kuehneosaurs. The dental implantation is pleurodont, and the teeth are tricuspid, with a large central and two smaller accessory cusps. This differs considerably from the isodont, subpleurodont dentition seen in the Eolacertilia.

A number of isolated teeth (SMUSMP) from the Kalgary locality consist of labiolingually compressed, heterodont, denticulated teeth. This tooth morphology is identical to that from a lepidosaurian jaw fragment (MNA P13196) from the Placerias Quarry near St. Johns, Arizona. This morphology is also similar to that seen in *Fulengia youngi* from the Upper Triassic of China (Carroll and Galton 1977), although the lepidosaurian affinity of *Fulengia* is questionable. Other isolated teeth from the Dockum and Arizona Chinle are of the same general appearance as MNA P13196 but possess heavy serrations upon each of the denticles. This suggests the presence of a second, closely related lepidosaurian taxon.

Although subpleurodont/pleurodont tooth implantation is characteristic of most of the early "eolacertilians," subpleurodont implantation is a primitive character for lepidosaurs and lizard-like tooth implantation is also found in the eosuchian (or lepidosauromorph) genera *Daedalosaurus* and *Coelurosauravus* (Carroll 1978). Therefore, until better specimens are obtained and more is known of lepidosaurian morphology and taxonomy, it is not known as to which nonsphenodontoid lepidosaurian groups these specimens may be referred.

Subclass Euryapsida
Order Trilophosauria
Family Trilophosauridae

All specimens of *Trilophosaurus* found in the Dockum Group have been referred to a single species, *T. buettneri* (see Figs. 9.5G, 9.13A). However, a review of southwestern trilophosaurids indicated that more than one species may have been present. Although several vertebrae recovered from the lower unit of the Petrified Forest Member in Petrified Forest National Park are virtually identical to those of *Trilophosaurus* from Otis Chalk Quarry 1 in Texas (R. A. Long pers. comm.), within the Placerias Quarry at St. Johns, Arizona, a number of small trilophosaurid jaw fragments and teeth have been recovered that indicate a taxon with strongly cingulated teeth and possibly with penultimate teeth similar to the British Upper Triassic taxon *Variodens inopinatus* (Robinson 1957). Interestingly, teeth in the Yale Peabody Museum (YPM 4225, 4227, 4230) from the upper Redonda Formation, North Apache Canyon Quarry 2, Quay County, New Mexico, are almost identical to the Chinle posterior teeth but are much larger.

A comparison of tooth width (Gregory 1945; Parks 1969) and femoral length (Elder 1978) from Quarry 1 at Otis Chalk indicates the presence of much larger trilophosaurids than those at Quarry 2. Although previous studies concluded that Quarry 2 specimens were juveniles of Quarry 1 specimens, studies of specimens from other localities suggest that more than one trilophosaurid species may be present. The size of the teeth in Quarry 2 are similar to those found on the type of *Trilophosaurus buettneri* from Walker's Tank and the other Crosby County specimens from Kalgary. Of the 71 specimens measured from Crosby County, none were of a size comparable to the mean of the Quarry 1 specimens measured by Gregory (1945) or Parks (1969). The Crosby County specimens were larger than those from the St. Johns Quarry and only 6 percent possessed cingula, a much lower figure than that found for the St. Johns trilophosaurids. However, in both tooth morphology and size, the Crosby County specimens are almost identical to those from Quarry 2 at Otis Chalk. It is tempting to conclude that the Crosby County and Potter County specimens, referable to the type of *Trilophosaurus buettneri*, were of a trilophosaurid that was slightly larger than the Chinle taxon but smaller than "*Trilophosaurus buettneri*" from Quarry 1. However, because very large individuals referred to *Trilophosaurus buettneri* by Gregory (1945) are found at Otis Chalk, there must have been juveniles somewhere in close proximity. DeMar and Bolt (1981) reported juvenile "*Trilophosaurus buettneri*" from the Otis Chalk localities that possessed well-developed cingula. The cingulated St. Johns specimens may be distinguished from the adult Otis Chalk material because no edentulous beak is developed and the teeth are oriented slightly diagonal to the jaw in the Chinle specimens. Also, the presence of isolated *Variodens*-like penultimate teeth, probably referable to the Chinle taxon and large teeth of similar morphology in the Redonda Formation at Apache Canyon, also indicate the probable presence of several trilophosaurids in the southwestern Triassic.

The only other trilophosaurids from North America are two undescribed taxa from the Wolfville Formation at Burncoat, Nova Scotia. One of these is a short-skulled genus related to *Trilophosaurus*, and the other taxon is probably related to *Variodens* (Carroll et al. 1972). *Variodens* and *Tricuspisaurus* are from the Upper Triassic fissure fills of England. *Variodens* is similar to the Chinle specimens in that it lacks the edentulous beak and possesses peculiar penultimate teeth possibly present in the Chinle and Apache Canyon trilophosaurids. *Tricuspisaurus* possesses an edentulous beak, but the labiolingual and anteroposterior measurements of the posterior teeth are relatively greater than in other trilophosaurids (Robinson 1957).

Subclass Archosauria
Order Thecodontia
Suborder Proterosuchia

Elder (1978) reported two proterosuchid mandibular fragments (TMM 31185-94, 31025-267) from Otis Chalk, Howard County, Texas. These specimens exhibited an ankylothecodont dentition with ovate, conical teeth. Another specimen (TMM 31025–269) was also tentatively referred to the Proterosuchia based on its ankylothecodont tooth implantation, although it appears that this mandible terminated in a slender point rather than the deep triangular morphology of TMM 31185-94 and TMM 31025-267 (Elder 1978, p. 81). If these specimens are proterosuchians, they would extend the stratigraphic range of the group from the Early Triassic into the Carnian.

Suborder Aetosauria
Family Stagonolepididae

Until recently, the taxonomy of the Triassic aetosaurs in the American Southwest was very confused. Two genera of aetosaurs were recognized, *Desmatosuchus* and *Typothorax*, and all collected elements of stagonolepidids were either classified with these genera or were misidentified as phytosaurs. Long and Ballew (1985), in their study of aetosaur dermal armor, have solved many of these taxonomic problems. They recognize four genera of aetosaurs in the southwestern Triassic, three of which occur within the Dockum. These include *Desmatosuchus* and the newly proposed genera *Paratypothorax* and *Calyptosuchus*. *Typothorax coccinarum* Cope 1875, a large aetosaurid with a flat, discoid carapace in which only the lateral cervical plates bear spines, seems to be restricted to the upper Petrified Forest Member (Norian?) of the Chinle Formation in eastern Arizona and north-central New Mexico. *Typothorax coccinarum* never occurs with the other aetosaurs and has never been positively identified from the Dockum.

Genus *Paratypothorax* Long and Ballew 1985
Paratypothorax ornatus Long and Ballew 1985 has a discoid carapace with straplike paramedian scutes similar to *Typothorax*. However, the paramedian scutes of *P. ornatus* are ornamented with prominent grooves and ridges that radiate from posteromedially placed bosses. These large, posteriorly hooked eminences have sharp anterodorsal margins and concave posterior margins. According to Long and Ballew (1985), *Paratypothorax* paramedian scutes from the Dockum in Crosby County include specimens from the east bank of the White River near the old Spur–Crosbyton mail road (UMMP 9600) and from Home Creek (UMMP 8858, 8859). *Paratypothorax ornatus* scutes from the Stubensandstein of Heslach, Germany have been generally attributed to the phytosaur, *Nicrosaurus kapffi*. The American material seems to be allied to the German genus. The only other identified *Paratypothorax* remains from North America consist of isolated scutes (UMMP V82238/126839, 7043/126881) from the lower unit, Petrified Forest Member, Chinle Formation (late Carnian) in Petrified Forest National Park, Arizona (Long and Ballew 1985).

Genus *Desmatosuchus* Case 1920
Desmatosuchus haplocerus (Cope 1892)
Desmatosuchus haplocerus was a massive narrow-bodied aetosaur that possessed a pair of enormous dorsolaterally projecting shoulder spines on its pectoral dermal armor (see Fig. 9.12A). Posterior to these pectoral "horns," spines and knobs that decreased in size posteriorly were developed on the presacral lateral scutes. The cervical paramedian scutes of *D. haplocerus* are elongate and rectangular, with truncated anterolateral margins and a raised boss or bar developed on the anterior margins. The postcervical paramedian scutes are not as elongate as the cervical plates, and the anterior bar is not developed. Instead, a thin lamina of bone projects anteriorly beneath the preceding plate. Scute ornamentation in *D. haploceras* typically consists of randomly placed pits and grooves showing little if any radiate pattern. *Desmatosuchus haplocerus* is widely distributed in the Dockum. The type specimen (ANSP 14688), which Cope (1892) designated *Episcoposaurus haplocerus*, consisted of isolated bones and dermal scutes collected approximately three miles north of Dockum, Dickens County, Texas. Possible topotypes are in the UT collections (Wilson 1950). A well-preserved carapace and skeleton (UMMP 7476), recovered near the old Spur–Crosbyton mail road in Crosby County, was designated the holotype of *Desmatosuchus spurensis* (Case 1920b, 1922). This specimen is considered to represent a junior synonym of *Desmatosuchus haplocerus* by Long and Ballew (1985), and bones of a second *D. haplocerus* specimen (UMMP 7504) were

found associated with UMMP 7476. Other Crosby County specimens of *D. haplocerus* include shoulder spines (UCMP, UT) from Home (Holmes) Creek and scutes (UCMP, UT) recovered in the vicinity of Cedar Mountain. *Desmatosuchus haplocerus* specimens have also been recovered (WTSU, UMMP) from the Rotten Hill Locality in Potter County, from Palo Duro Canyon in Randall County, from Borden County, and from the vicinity of Otis Chalk (UMMP) in Howard County (Long and Ballew 1985). The supposed Bissett Conglomerate specimens are probably Cretaceous dinosaurs. Texas Tech University material of *Desmatosuchus* has been recovered from the Miller locality in Garza County (Chapter 10), and a number of other *D. haplocerus* specimens were recovered by the Dallas Museum of Natural History from the Miller locality. In the Dockum of eastern New Mexico, *Desmatosuchus* has been recovered from the Revuelto Creek area in Quay County (Chapter 11). In Arizona, *D. haplocerus* has been recovered from the lower unit of the Petrified Forest Member of the Chinle Formation in Petrified Forest National Park, and dermal armor (MNA, UCMP) has been reported from the Placerias Quarry, Downs Quarry, Big Hollow Wash, and Blue Hills near St. Johns in Apache County (Camp and Welles 1965; Jacobs and Murry 1980; Long and Ballew 1985). *Desmatosuchus* shoulder spines have also been recovered near Winslow, Navajo County and at Ward's Bonebed near Tanner's Crossing in Coconino County, Arizona (Long and Ballew 1985).

Genus *Calyptosuchus* Long and Ballew 1985
Calyptosuchus wellesi Long and Ballew 1985
Calyptosuchus wellesi has a narrow carapace in which the cervical armor is segmented in accordance to the underlying vertebrae. The cervical paramedian scutes are wide and thin, with a slight ridge present, and they possess radiate ornamentation. Posteriorly, large bosses are present near the midline on the posterior margin of the dorsal paramedian scutes from which a series of well-defined grooves and pits radiate. The cervical lateral scutes are strongly angular, with faint radial ornamentation and dorsal and lateroventral flanges. There are posterolaterally directed horns in the cervical region, but no lateral horns are present posterior to those on the neck. The dorsal lateral scutes possess an anterior bar with an anteroposteriorly directed ridge from which a raylike ornamentation emanates. The lateral caudal scutes are strongly arched, with an anterior bar and strong bosses, but the radially incised ornamentation is indistinct in both the lateral and paramedian caudals (Long and Ballew 1985).

Calyptosuchus wellesi is found at a number of localities in the Dockum. The holotype (UMMP 13950) of *C. wellesi*, consisting of armor, vertebrae, and a crushed pelvis, was recovered northeast of

Rotten Hill, near the breaks of the Cerita (Sierrita) de la Cruz Creek, probably in easternmost Oldham County, Texas. A partial skeleton with associated armor (UMMP 7470) was found along Home Creek, and a partial paramedian scute (UCMP 102347) was found in the vicinity of Cedar Mountain, Crosby County, Texas (Long and Ballew 1985). Scutes of similar morphology (DMNH 1160-17) were recovered from the Miller locality in Garza County, Texas, although these were previously assigned to the phytosaur *Nicrosaurus* (Murry 1982, p. 241). Paramedian scutes (YPM 3695, 3696) previously assigned to *Phytosaurus* by Gregory (1962a) are also referable to *Calyptosuchus wellesi*. Scutes of *C. wellesi* have been recovered from twenty sites in the lower unit of the Petrified Forest Member (late Carnian) of the Chinle Formation in Petrified Forest National Park, Arizona. Numerous specimens (MNA, UCMP) have also been found in Apache County, Arizona at the Placerias and Downs quarries near St. Johns, at Big Hollow Wash, in the Blue Hills near St. Johns, and at a locality seven miles north of the Nazlini Trading Post. In Coconino County, numerous broken scutes (UCMP) of *C. wellesi* were found at Ward's Bonebed near Cameron (Long and Ballew 1985).

Genus *Typothorax?*
Typothorax? meadei Sawin 1947
Typothorax meadei is known from excellent specimens (UT) from Otis Chalk quarries 3 and 3A, Howard County, Texas (see Fig. 9.12B). Long and Ballew (1985) believe that these specimens represent a form very different from the genotype of *Typothorax* (*T. coccinarum*) and that they are a different genus. The length of these specimens is less than 3 m, and they possessed quite gracile limbs. The carapace is discoidal, the dorsal paramedian plates possess prominent pyramidal bosses, and the ornamentation is radiate. Lateral spines are present, but there is no development of prominent shoulder spines as seen on *Desmatosuchus*.

Suborder Rauisuchia
Remains of at least two genera of rauisuchians seem to have been recovered from the Dockum Group of Texas. Walker (1969) believed that some of the ilia figured by Case (1922, Plate 12 A,B,F, Figs. 27D,E, 1943) were referable to poposaurs. One of these (UMMP 7266) is from the head of Holmes Creek within the Kalgary area (see Fig. 9.13D). Elder (1978) also described a number of elements from the Texas Dockum Group that she believed to be referable to the Poposauridae. These include an almost complete pelvis (UT 31172-21) from Site 2 in Crosby County. Other specimens from Otis Chalk, Howard County, include an ischium and vertebrae from Quarry 1. According to R. A. Long (pers. comm.), all of the above specimens may be

referable to an undescribed species of rauisuchian. The same taxon is relatively abundant in the lower unit of the Petrified Forest Member of the Chinle Formation in Arizona. Specimens (UCMP, MNA) have been recovered from the Placerias and Downs quarries and the Blue Hills near St. Johns, in the Petrified Forest National Park, and in the vicinity of Cameron, Arizona. Excavations at the Miller locality in Garza County by Texas Tech University field parties have recovered excellent material from a second rauisuchian genus. This material, as well as the braincase recovered along the Spur–Crosbyton mail road in Crosby County and possibly some of the Otis Chalk specimens, may be referable to Chatterjee's (1985) *Postosuchus*. Parrish and Carpenter (Chapter 11) have also recovered remains of a poposaurid from the Revuelto Creek Local Fauna, Quay County, New Mexico.

The taxonomic relationships of the rauisuchians and poposaurs is uncertain. The two groups are certainly closely related to each other, but they have been variously interpreted as also related to phytosaurs, ornithischians, pseudosuchians, and carnosaurs. Chatterjee (1985) considers *Postosuchus* related to *Albertosaurus*-like carnosaurs on the basis of the structure of the pubis and the peg and socket articulation of the calcaneum. However, most workers believe that the rauisuchian tarsal structure, including that of *Postosuchus*, is of crocodiloid morphology and, therefore, is more closely related to crocodiles than to saurischians.

As Walker (1969) suggested, form genera based on teeth, such as *Cladeiodon*, *Zatomus*, and *Teratosaurus*, may actually be referable to the poposaurs or rauisuchians. Large, laterally compressed serrated teeth (SMUSMP) similar to these tooth taxa have been recovered from the Kalgary locality and may possibly be referred to the rauisuchids or poposaurs. However, considering the similarities exhibited in the above archosaurian groups, the referral of most Triassic taxa to any higher taxonomic group will be subject to much revision. Also, species previously referred to some tooth form genera may belong to other taxonomic categories.

Suborder Phytosauria
Family Phytosauridae

Since the early nineteenth century, numerous taxa of phytosaurs have been described from North America, Europe, Asia, and Africa. Many classifications have been proposed as new attempts have been made to understand the phylogenetic relationships of the group. The significance of the morphological characters within the Phytosauria is not understood, although there is certainly much sexual and ontogenetic variation within that suborder (Camp 1930; Colbert 1947; Langston 1949; Gregory 1962a) that has greatly contributed to the taxonomic

confusion. An accurate revision of the phytosaurs is needed, especially because they have been extensively used for Triassic biostratigraphy, but this will require a detailed study of material from many collections throughout the world.

Several genera of phytosaurs have been recovered from the Dockum. These include *Paleorhinus*, *Angistorhinus* (including "*Brachysuchus*"?), *Rutiodon*, and possibly *Nicrosaurus*.

Genus *Paleorhinus* Williston 1904

Species of *Paleorhinus* recognized by Gregory (1962a) and Westphal (1976) included *Paleorhinus bransoni*, *Paleorhinus* (*Promystriosuchus*) *ehlersi*, *Paleorhinus parvus*, and *Paleorhinus scurriensis*. In addition, a species from the Upper Triassic of the Atlas Mountains in Morocco, *Paleorhinus magnoculus*, is very similar in morphology to *P. ehlersi* (Dutuit 1977b). A partial skull referred to *Paleorhinus* has been found in the Chinle Formation near St. Johns, Arizona (K. Padian and R. A. Long pers. comm.).

Paleorhinus scurriensis Langston 1949 is based on the posterior portion of a skull (TTU 539) from the Camp Springs Conglomerate (lower Dockum Group) approximately 4 km northeast of Camp Springs in Scurry County, Texas. *Paleorhinus scurriensis* shows many characteristics typical of *Paleorhinus*: anterior position of external nares, orbits directed upward, low ratio of prenarial to postnarial length, low ratio of skull height to skull width (0.33), and relatively homodont dentition (Fig. 9.7B). *Paleorhinus scurriensis* does show a couple of characters that do not follow the diagnosis of Gregory (1962a), Westphal (1976), or Chatterjee (1978). All of these authors state that the posttemporal arcade is at the level of the skull roof in *Paleorhinus*, and there is no posterior hooklike process on the squamosal. In Langston's (1949) reconstruction of *P. scurriensis*, there is a noticeable (although slight) depression of the posttemporal arcade, and the squamosal is extended posteriorly into a hooklike process. Langston noted trends from *Paleorhinus scurriensis* to *P. bransoni* that he believed represented ontogenetic stages as suggested by Colbert (1947) for "*Machaeroprosopus*." These included an increase in robustness, narrowing in the spacing of the teeth, relative decrease in the size of the palatine foramen, and obliteration of the pineal foramen in the adult. *Paleorhinus scurriensis* showed evidence of a pineal foramen, a condition previously reported only in *Mesorhinus* and juvenile specimens of *Rutiodon* (*Machaeroprosopus*) *lithodendrorum* (Camp 1930). Considering the apparent ontogenetic variability in the presence or absence of a pineal foramen, the validity of using this as a diagnostic character (as in Chatterjee 1978) is questionable. However, *Paleorhinus scurriensis* shows several derived characters, including an increased length of the palatine vacuity and the posterior extension of the squamosal into a hooklike process not seen in other *Paleorhinus* species. The Borden County *Paleorhinus* (UT 31213-16) is a very large specimen that Gregory (1962a) believed may represent a new species. There were approximately thirty-nine teeth in the skull in this specimen, which agrees well with the number in *Paleorhinus bransoni*, *P. parvus*, and *Paleorhinus scurriensis*; also, the ratio of prenarial to postnarial length (1.01) is similar to that seen in *Paleorhinus bransoni* (1.09). The Borden County specimen differs from the type of "*Promystriosuchus*" *ehlersi* (UMMP 7487) in which there are thirty-one premaxillary and sixteen maxillary teeth, and the Borden County specimen possesses considerably fewer teeth than that seen in the four Otis Chalk specimens (UT 31100) assigned to *P. ehlersi* by Gregory (1962a). The Otis Chalk specimens, although having a dental count more similar to that of "*Promystriosuchus*" and a more posterior placement of the external nares, are well within the range of the ratio of prenarial to postnarial length (1.08–1.22) of the Borden County specimen (1.01), *Paleorhinus bran-*

soni (1.09), and the type of *Promystriosuchus ehlersi* (1.43). Until the ontogenetic and sexual variability among the taxa of phytosaurs is known, I have tentatively placed the Otis Chalk specimens within the genus *Paleorhinus*, but as a taxon distinct from *Paleorhinus ehlersi*. The differences between *P. ehlersi* and most taxa of *Paleorhinus* include a heterodont dentition, a long interpterygoid vacuity, possibly a slightly more elevated posttemporal arcade (at approximately the level of the skull roof), and a rugose palatine ridge (Fig. 9.7A).

Colbert and Gregory (1957), Gregory (1962a), and Gregory and Westphal (1969) have used the genera of phytosaurs to correlate the terrestrial faunas of the western United States. The genus *Paleorhinus*, supposedly more primitive in morphology, was considered to be of earlier age than the more derived *Nicrosaurus* and *Rutiodon*. The type "*Promystriosuchus*" was found at the head of Holmes Creek in Crosby County in an area where fossils referable to the "derived" genera have also been found. Considering that "*Promystriosuchus*" has been placed within the genus *Paleorhinus* by Gregory (1962a) and Westphal (1976), it becomes apparent that either the distribution of the phytosaurian taxa cannot be explained simply in temporal terms or the taxonomic identifications are so confused that the correct temporal sequence is not apparent.

Genus *Angistorhinus* Mehl 1913

Angistorhinus has been recovered from the Popo Agie Formation of Wyoming (Mehl 1913, 1915, 1928; Eaton 1965), from the Dockum Group of west Texas (Case 1929, Stovall and Wharton 1936), and from the Upper Triassic (T5) of Argana, Morocco (Dutuit 1977a). *Angistorhinus* is the only phytosaurian genus with the combination of a posttemporal arcade at the level of the skull roof and external nares between the antorbital fenestrae. Other characters of this taxon include orbits that are directed outward as much as upward, a large infratemporal fenestra at the level of the orbit, a small interpterygoid vacuity, a moderate ratio of prenarial to postnarial length (1.4–1.9), a rounded, short posterior squamosal process, and a heterodont dentition (Case 1922; Gregory 1962a; Eaton 1965; Westphal 1976; Chatterjee 1978).

"*Brachysuchus*" *megalodon* is based on a skull (UMMP 10336) from a red clay–pellet conglomerate near Otis Chalk, Howard County, Texas (Case 1929). Lower jaws (UMMP 10336A) collected near the skull probably belonged to the same individual (Case 1930). Another skull (UMMP 14366) and a lower jaw were found in the "Blue Hills" along Sierrita (Cerita) de la Cruz Creek in Potter County (Case and White 1934). Gregory (1962a) placed "*Brachysuchus*" within the genus *Phytosaurus*, probably on

Figure 9.7. *Paleorhinus* skulls of Dockum specimens. Skull of *Paleorhinus ehlersi* in superior view (**A1**), lateral view (**A2**), posterior view (**A3**), inferior view (**A4**), after Case (1922). Skull of *Paleorhinus scurriensis* in superior view (**B1**), posterior view (**B2**), lateral view (**B3**), after Langston (1949). In phytosaur illustrations, scale = 10 cm.

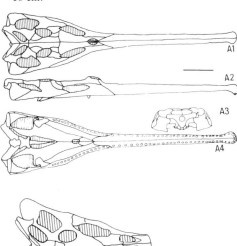

the basis of the seemingly depressed posttemporal arcade. Later, Gregory (cited in Eaton 1965) reported that the posterior portion of the skull had been crushed. In reality, the posterior temporal arches were at the level of the skull roof and showed the posterior widening that is characteristic of *Angistorhinus*. The ratio of prenarial to postnarial length in the large skull (1,243 mm) of *Angistorhinus megalodon* is approximately 1.7. Contrary to the "weakly heterodont " condition noted for *Angistorhinus* by previous authors (Gregory 1962a; Chatterjee 1978), the teeth of *Angistorhinus megalodon* are as strongly differentiated as those in *Rutiodon* and *Nicrosaurus*. The rostrum on the type of *A. megalodon* is quite massive, and the ratio of length to width on the specimen figured by Case (1929) at the rostral midpoint is approximately 6.0. The mandible is equally massive, and both the dentary and premaxilla are greatly expanded distally. However, the prenarial crest in *A. megalodon* is better developed than in other figured *Angistorhinus* specimens (Fig. 9.8A). Although there are certain differences between *A. megalodon* and other *Angistorhinus* species, until the morphology of the posttemporal arcade is understood, the referral of "*Brachysuchus*" as a synonym of *Angistorhinus* is tentatively followed.

The type of *Angistorhinus alticephalus* Stovall and Wharton 1936 was recovered approximately 40 km southeast of Big Spring in Howard County, Texas, very near the type locality of *Angistorhinus megalodon*. The skull of *A. alticephalus* (Fig. 9.8B) is slightly smaller (1,220 mm) than the type of *A. megalodon* (1,243 mm), and the ratio of prenarial to postnarial length (2.0) is larger than that seen in *A. megalodon* (1.7). From the drawing in Stovall and Wharton (1936), the ratio of length to width at the rostral midpoint is approximately 7.7, considerably different from that of *A. megalodon* (6.0). This reflects the relative slenderness of the rostrum in *Angistorhinus alticephalus*. Gregory (1962a) tentatively counted thirty-six upper teeth in the type of *A. alticephalus*, and Stovall and Wharton (1936) showed widely spaced teeth in that specimen. The presence of a slender rostrum with relatively few, widely spaced teeth has been suggested as a juvenile character by Colbert (1947) and Langston (1949). The type of *Angistorhinus alticephalus* may have been a juvenile, because two other UT skulls assigned to this species from Otis Chalk have a larger number of teeth (Gregory 1962a). However, certain differences have been noted between *A. megalodon* and the type of *A. alticephalus* that may or may not be distinguishing features. These include the obstruction in lateral view of the quadrate by the quadratojugal in *A. alticephalus*; the long, splintlike septomaxillaries in *A. megalodon* versus the short, broad septomaxillaries in *A. alticephalus*; the high ratio of quadrate height to skull width of *A. alticephalus* (0.71–0.78) versus that of *A. megalodon* (approximately 0.50); and the rhomboid shape of the lateral temporal fenestra versus the somewhat oval shape in *Angistorhinus megalodon* (Case 1929, 1930; Gregory 1962a; Eaton 1965). However, considering that the specimen of *A. megalodon* was crushed dorsoposteriorly, the effect on the size and shape of the lower temporal opening may have been considerable. Also, the validity of using the morphology of the septomaxillaries in phytosaur taxonomy is questionable because they seem to be affected during otogeny by the increased robustness of the rostrum.

Genus *Rutiodon* Emmons 1856

Case (1922) and Case and White (1934) referred a number of skulls from the Dockum to the genus *Leptosuchus*. Gregory (1962a) synonymized *Leptosuchus* with *Rutiodon*, and his diagnosis for the latter genus is identical to the condition seen in "*Leptosuchus*": posterior portion of the external nares between the antorbital fenestra, posterior border of the supratemporal fenestra depressed, rela-

Figure 9.8. Skulls of *Angistorhinus* specimens. **A**, Skull of *Angistorhinus? megalodon* in superior (**A1**) and lateral view (**A2**), after Huene (1956). **B**, Skull of *Angistorhinus alticephalus* in superior (**B1**) and lateral view (B2), after Stovall and Wharton (1936).

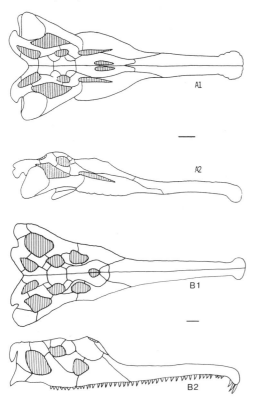

tively large posttemporal fenestra, long posterior processes present on squamosals, orbits directed obliquely upward and outward, tall quadrate, heterodont dentition, and narrow interpterygoid vacuities.

Rutiodon crosbiensis and *Rutiodon imperfecta* are based on skulls found in Crosby County, Texas, probably from the Swenson Ranch (Case 1922). The skull (UMMP 7522) of *R. crosbiensis* (Fig. 9.9B) is 875 mm long with a ratio of prenarial to postnarial length of 1.25. The restored skull of *R. imperfecta* (UMMP 7523) is much larger (1,120 mm) than that of *R. crosbiensis*, and the ratio of prenarial to postnarial length is approximately 1.58. Gregory (1962a) reported that this ratio varies inversely with skull length in *Rutiodon*, from approximately 1.7 in skulls 700 mm long to 1.1 in 1,400 mm skulls. This is not the condition seen in *Rutiodon crosbiensis* and *R. imperfecta*, where the longer skull also has the proportionally long snout. However, the skull of *Rutiodon imperfecta* was fragmentary, and a portion of the snout that was missing had been restored on the specimen, which may account for the inconsistent

Figure 9.9. Skulls of Dockum *Rutiodon* specimens. Skull of *Rutiodon* (*Leptosuchus*) *studeri* in lateral view (**A1**), posterior view (**A2**), and superior view (**A3**), after Case and White (1934). Skull of *Rutiodon* (*Leptosuchus*) *crosbiensis* in inferior view (**B1**), lateral view (**B2**), and superior view (**B3**), after Case (1922).

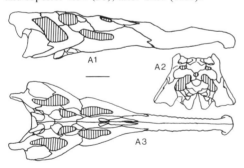

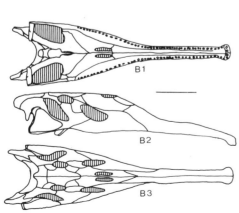

ratio. The only features by which Case separated the two taxa were a posterior rostral crest seen on *R. imperfecta* and a large pit present on the upper surface of its snout that extends into the symphysis of the premaxillary bones. Considering the ontogenetic changes and sexual dimorphism seen in phytosaurians, as reported in Camp (1930) and Colbert (1947), these features are probably insignificant and cannot be used to differentiate the two taxa. Both authors postulated that the females in *Rutiodon* ("*Machaeroprosopus*") are those in which the external nares are raised as a crater-like structure (as in *R. imperfecta*), whereas in males the external nares were not conspicuously elevated (as in *R. crosbiensis*).

The type locality of *Rutiodon* ("*Leptosuchus*") *studeri* is along Sierrita de la Cruz Creek in Potter County, Texas. Case and White (1934) distinguished this species by the convex contour of the lower edge of the rostrum, the prominent palatal ridges, extremely large septomaxillaries, and the long prenarial crest on the rostrum (Fig. 9.9A). The total length of the skull is 900 mm, and the ratio of prenarial to postnarial length is 1.71. The character of the septomaxillary, the elevation of the inner edge of the pterygoid ("palatal ridges"), and the length of the prenarial crest are not diagnostic in the Phytosauria as Gregory indicated. Colbert (1947) believed that the development of the rostral crest is primarily ontogenetic, although juvenile females apparently lacked a crest. Concerning the upward curvature of the rostrum, one can see similar variability in *Rutiodon carolinensis* illustrated by Colbert (1947). The evidence suggests that the Dockum species referred to "*Leptosuchus*" are syntypic, with the "diagnostic characters" presented for each representing intraspecific variability.

I also find no characters to distinguish "*Phytosaurus*" *doughtyi* from the aforementioned Dockum *Rutiodon* specimens (see Fig. 9.11A). This species was based on a skull fragment (AMNH 4919) collected by George D. Doughty of Post, Texas. A comparison of the superior view of "*Phytosaurus*" *doughtyi* (Case 1920b, Fig. 5) with that of *Rutiodon studeri* (Case and White 1934, Fig. 2) shows similar development in the elongate posterior process of the squamosal; the presence of a "tongue-and-groove" articulation of quadratojugal and jugal; the jugal extending posteriorly beneath the quadratojugal nearly to the quadrate; and the rhomboidal shape of the lateral temporal fenestrae.

Cope (1881) described *Belodon buceros* from the Chinle Formation of northwestern New Mexico, probably in the area of Gallina Creek in Rio Arriba County. The type specimen (AMNH 2318) of "*Belodon*" *buceros* has the same characteristics seen in *Rutiodon*: posttemporal arcade depressed, external nares posterior in position and between antorbital fenestrae, orbits directed obliquely more outward than upward, a well-developed posterior process of the squamosal, and a heterodont dentition (Cope 1881, 1887; Huene 1913). The skull of *Rutiodon buceros* is very similar in morphology and size (815 mm) to the type of *Rutiodon crosbiensis* from the Dockum of Texas. However, the prenarial to postnarial length was probably longer in *Rutiodon buceros* (approximately 1.55), and the rostral crest was much better developed. This skull is intermediate in dimensions between *R. crosbiensis* and *R. studeri*, although because the anterior portion of the rostrum is missing in the type of *R. buceros*, these measurements are approximate. Because there is also much intraspecific variation in the ratio of prenarial to postnarial length and in the development of the rostral crest in *Rutiodon* (Colbert 1947; Gregory 1962a,b), there seems to be little to separate *R. buceros* from the Dockum *Rutiodon* specimens. If they are conspecific, then *Rutiodon buceros* Cope 1881 has priority over *Rutiodon crosbiensis* Case 1922 and the other "*Leptosuchus*" specimens from Texas.

I can see even less difference between the type specimen (FMNH) of "*Machaeroprosopus*" *andersoni* Mehl 1922 and *Rutiodon crosbiensis*. The skull in *Rutiodon andersoni* is 865 mm in length, and although the ratio of prenarial versus postnarial length (1.45) is greater than that calculated for *R. crosbiensis* (1.25), no diagnostic features separate the two (Fig. 9.10B). In fact, the illustration of Mehl (1922, Figs. 1–3) for *Rutiodon andersoni* are practically indistinguishable from those of Case (1922, Fig. 25), except that the skull is more elevated at the external nares and above the orbits in *R. andersoni* and the antorbital fenestra are more oval in the illustration of *Rutiodon crosbiensis*. Gregory (1972) indicated that *R. andersoni* was probably recovered from the Bull Canyon area in Guadalupe County, New Mexico. As noted above, both *R. andersoni* and *R. crosbiensis* may very well be junior synonyms of *Rutiodon buceros*. Camp (1930) noted a very close similarity between *R. andersoni* and "*Machaeroprosopus*" *tenuis* from the Chinle Formation of Arizona.

I find the skull described by Stovall and Savage (1939) as "*Machaeroprosopus* sp." and attributed to *R. validus* by Gregory (1972) to be of the same morphology as the other Texas/New Mexico *Rutiodon* specimens. This skull (OU 1250) was from the Sloan Canyon Formation of Sloan Canyon Creek, Union County, New Mexico (Fig. 9.10A). The rostral swelling on OU 1250 is not as well developed as those seen in *Rutiodon buceros* or *Rutiodon* "*studeri*," but the nares seem more elevated than in *Rutiodon* "*crosbiensis*" or *R.* "*imperfecta*." It is somewhat larger than the above specimens (1,170 mm), with

a ratio of prenarial to postnarial length of 1.63. There are forty-three to forty-five teeth in each upper jaw (Stovall and Savage 1939). However, such differences are within the range of variability expected from a single taxon of *Rutiodon*, and there is very little to separate this from the "*Rutiodon buceros*" type of morphology. Stovall and Savage found that this specimen is also very close morphologically to "*Machaeroprosopus tenuis*." The scutes described by Stovall and Savage in the Union County specimen are as those in other *Rutiodon* specimens: elliptical to subtriangular in outline, with pitted sculpture and a large central ridge.

Gregory (1972) indicated that the *Rutiodon* skull (YPM) collected at "Shark Tooth Hill" South of San Jon in Quay County is the most specialized in the structure of the temporal region. Parrish and Carpenter (Chapter 11) refer material collected along Revuelto Creek, Quay County, to *R. gregorii*. Other *Rutiodon* material has been found in Apache Canyon, Quay County, and possible *Rutiodon* specimens were found at the Lamy location in Santa Fe County.

Figure 9.10. Skulls of Dockum *Rutiodon* specimens. Skull of *Rutiodon* (*Machaeroprosopus*) sp. in superior (**A1**) and lateral views (**A2**), after Stovall and Savage (1939). Skull of *Rutiodon* (*Machaeroprosopus*) *andersoni* in lateral (**B1**) and inferior views (**B2**), after Mehl (1922).

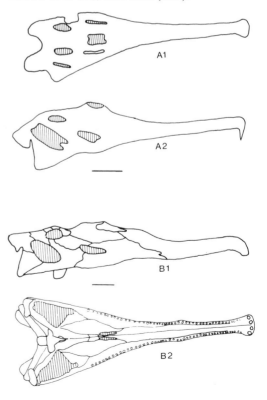

Genus *Nicrosaurus* Meyer 1860

Nicrosaurus Meyer 1860 is very similar to *Belodon/Rutiodon* in many features, but the former is distinguishable by the broader and heavier rostrum, the strong prenarial crest, the larger prenarial–postnarial ratio (approximately 1.2), and the relatively deep, short, broadly rounded posterior squamosal process (Fig. 9.11B). Although isolated *Nicrosaurus* scutes (DMNH) were reported from the Miller locality in Garza County, Texas (Murry 1982), many of these are probably referable to aetosaurs. However, Chatterjee (Chapter 10) recognizes the presence of *Nicrosaurus* specimens (TTU) in the Cooper Member of the Dockum at the Miller locality. This is the only locality in Texas from which *Nicrosaurus*-type phytosaurs have been positively recognized.

Order Pterosauria?

A mandibular fragment and partial maxilla (see Fig. 9.5H), possibly referable to the Pterosauria, were recovered from the Kalgary locality (SMUSMP). Very few possible Triassic pterosaurians have been recovered from North America, and these are all very fragmentary. Isolated teeth (Jacobs and Murry 1980) and phalanges (K. Padian pers. comm.) possibly referable to pterosaurs have been found in strata of comparable age near St. Johns, Arizona. The Chinle isolated teeth are similar to the genus *Eudimorphodon* (Norian, Italy) and may be-

Figure 9.11. Partial skull of *Rutiodon doughtyi* in superior view (**A**), after Case (1920b). Skull of *Nicrosaurus gregorii* in lateral view (**B**), after Westphal (1976). Ang, angular; D, dentary; F, frontal; J, jugal; L, lacrimal; Mx, maxilla; N, nasal; PF, postfrontal; PM, premaxilla; PO, postorbital; PR, prefrontal; QJ, quadratojugal; SP, splenial; SQ, squamosal; SUR, surangular.

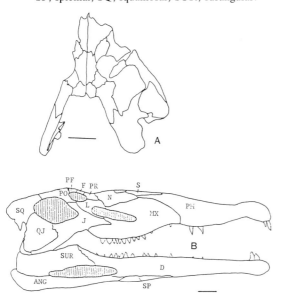

long to the same species as the Dockum specimens. Although these elements are morphologically similar to those of pterosaurs, other taxa such as therapsids are similar in general form, and misidentification is possible.

The Dockum specimens possibly referable to *Eudimorphodon* Zambelli 1973 consist of two jaw fragments and a number of isolated teeth. A mandibular fragment (SMUSMP 69125) has preserved two five-cusped teeth of thecodont implantation. There are two small accessory cusps anterior and two posterior to the larger centrally placed primary cusp on each tooth. A comparison of this specimen to *Eudimorphodon ranzii* (Wild 1978) indicates that this fragment is probably from the middle of the dentary.

A maxillary fragment (SMUSMP 69124) from Kalgary contains the major portion of two thecodont teeth and a fragment of a third. The most complete tooth in this partial maxilla is four-cusped, with one primary cusp and one anterior and two posterior accessory cusps. The morphology of the teeth is most similar to that of the two large fang teeth in the middle of the maxilla of *Eudimorphodon ranzii*. However, the basal portions of the Dockum teeth are relatively shorter than those of the Italian specimens, which may indicate at least specific distinction from *Eudimorphodon ranzii*. Measurements of the Dockum specimen also suggest that this taxon may have been slightly larger than the European species, but with a relatively shallower dentary. However, more complete material from the southwestern Triassic must be recovered before the identification and specific taxonomic relationships of this Triassic material can be made with confidence.

Order Saurischia

The only saurischian remains positively identified from the Dockum Group are those of a procompsognathid from Revuelto Creek, Quay County, New Mexico (Chapter 11). However, a relatively common genus of procompsognathid, *Coelophysis*, has been collected from other southwestern Triassic localities. The type specimen of *C. bauri* Cope 1889 is believed to have been collected from the Petrified Forest Member of the Chinle Formation at Ghost Ranch, New Mexico (Williston and Case 1913; Huene 1915; Colbert 1964; and Chapter 5). Although several fragmentary specimens recovered from the Dockum have been referred to *Coelophysis*, no specimens have yet been found that can positively be referred to this taxon. Case (1927, 1932b) described a number of teeth, a braincase, an ilium, and a series of vertebrae from a few miles north of Cedar Mountain in Crosby County, Texas that he believed were referable to *Coelophysis*. Although some of the teeth are of similar size and morphology to *Coelophysis bauri*, a wide variety of archosaurs also possessed laterally compressed, serrated teeth of similar morphology. The vertebrae and braincase described by Case belong to a different taxon later designated as *Spinosuchus caseanus* by Huene (1932) and are discussed below (see also Chapter 5). The ilium that Case collected is not referable to *Coelophysis*, but appears to be most similar to that of the herrerasaurid saurischians (R. A. Long pers. comm.). In Howard County, specimens referred to *Coleophysis* were reported from the Trilophosaurus quarries (Gregory 1945; Elder 1978), although these have not been studied in detail and their taxonomic assignment has not been confirmed. Laterally compressed, serrated teeth (SMUSMP) have also been found at the Kalgary Locality in Crosby County, Texas and at the Apache Canyon, Red Peak, and "Shark Tooth Hill" localities in Quay County, New Mexico. The Wolfville Formation, which has yielded a number of taxa identical to or very similar to those of the Dockum–Chinle, has also produced teeth very similar to those of *Coelophysis* (Carroll et al. 1972). *Coelophysis* (*Podokesaurus*) *holyokensis* from the Liassic portion of the Newark Supergroup may be closely related to *C. bauri* (Colbert 1964), but they have no diagnostic characters below the level of Theropoda (Chapter 5). Other Newark material (USNM) questionably referred to *Coelophysis* was recovered near Ashland, Virginia. A partial skeleton referred to *Coelophysis* (UCMP) has been recovered from the upper unit of the Pe-

Figure 9.12. **A**, *Desmatosuchus haplocerus*, after Case (1922). **B**, *Typothorax? meadei*, after Sawin (1947). **C**, *Spinosuchus caseanus* vertebrae, after Case (1927). Scale on **A** and **B** equals 0.5 m; scale on **C** equals 0.25 m.

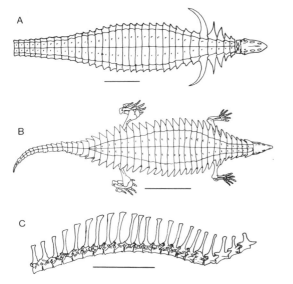

trified Forest Member of the Chinle Formation at Petrified Forest National Park (Chapter 5), and tarsi and other limb elements of theropods have been recovered from the Placerias Quarry near St. Johns, Arizona (R. A. Long pers. comm.).

Order Saurischia?
Family Incertae sedis
Genus *Spinosuchus* Huene 1932
Spinosuchus caseanus Huene 1932
Spinosuchus caseanus (Fig. 9.12C) is based on five cervical, fourteen dorsal, and three sacral vertebrae (UMMP 7507) that were earlier tentatively referred to the genus *Coelophysis* (Case 1927; Huene 1932). These vertebrae were collected from the Dockum Group north of Cedar Mountain in Crosby County, Texas. This species is readily separated from other Triassic archosaurs by the very long neural spines. However, its dinosaurian affinities are questionable (Chapter 5). A braincase recovered south of the old Spur–Crosbyton mail road by Case and tentatively referred by him to *Spinosuchus* is more probably rauisuchian or poposaurian (Chapter 5); Chatterjee (1985) refers it to *Postosuchus*.

Order Ornithischia
Suborder Ornithopoda
Family Fabrosauridae
Chatterjee (Chapter 10) refers material from the Miller locality, Garza County, to the fabrosaurid ornithischians. Chatterjee regards the locality as Norian, although Dockum pollen suggests a late Carnian age (Dunay and Fisher 1979). If the locality is Carnian, then these are among the oldest ornithischian specimens in the world and are of major importance. Although bulbous, waisted teeth (SMUSMP) similar in morphology to those of fabrosaurids have been recovered from the Kalgary locality, the Miller locality specimens represent the only positively identified Triassic ornithischian material from the Dockum.

Subclass Synapsida
Order Therapsida
Infraorder Cynodontia
Family Tritheledontidae
Genus *Pachygenelus* Watson 1913
Pachygenelus milleri Chatterjee 1983
A fragmentary right dentary (TTU P 9020) of an ictidosaur was recovered from the Miller locality near Post in Garza County, Texas (Fig. 9.13B). Chatterjee (1983) noted the similarity of this taxon to *Pachygenelus monus* from South Africa and *Chaliminia musteloides* from Argentina. The Dockum species is small and differs from *P. monus* in the development of an accessory posterior cusp on the lower postcanines. Tannenbaum (1983) reported

possible cynodont teeth similar to those of *Pachygenelus* from the lower Petrified Forest Member (late Carnian) of the Chinle Formation near St. Johns, Arizona.

Dockum ichnogenera

Baird (1964) described reptile footprints (CMNH) from the Sloan Canyon Formation (Dockum Group) of Peacock Canyon, Union County, New Mexico.

A large footprint (Fig. 9.13C3) tentatively identified as *Anomoepus* is now believed to be a chirotheriid (D. Baird pers. comm.) footprint that is believed to have been formed by pseudosuchian archosaurs (Haubold 1971). Other Sloan Canyon footprints were referred to cf. *Anchisauripus hitchcocki* (footprints of a small coelurosaurian dinosaur; see Fig. 9.13C2), and to what Baird (1964, p. 120) believed to be a new species of *Rhynchosauroides* [probably representing a small lepidosaurian (Fig. 9.13C1)].

Figure 9.13. **A**, *Trilophosaurus "buettneri"* skull in superior (**A1**) and lateral view (**A2**), after Parks (1969). Scale = 5 cm. **B**, *Pachygenelus* skull (**B1**) and *P. milleri* mandible (**B2**), after Chatterjee (1983). Scale = 5 mm. **C**, Dockum ichnogenera. *Rhynchosauroides* (**C1**), cf. *Anchisauripus hitchcocki* (**C2**), and chirotheriid (**C3**) footprints; after Baird (1964). Scale = 5 cm. **D**, Poposaurid? UMMP 7266 ilium (**D1**) and sacrum (**D2**); after Case (1922). Scale = 5 cm.

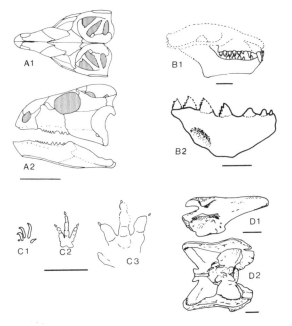

Biostratigraphy of the Dockum Group

Dunay and Fisher (1979) collected samples for palynological correlation in the central and northern portions of the Texas Dockum. Two of the localities collected in these studies, in Crosby County and Potter County, were within the same limited geographic area as the Kalgary and Rotten Hill localities. Based on the ranges of the palynomorphs *Vallasporites ignacii*, *Enzonalasporites vigens*, *Patinasporites densus*, *Brodispora striata*, and *Infernopollenites sulcatus*, a Carnian age for all of the Dockum localities studied was indicated.

However, the distribution of vertebrates within the central and northern portion of the Dockum Basin in Crosby, Randall, and Potter Counties may suggest a slightly different age for these surficial units. For example, Long and Ballew (1985) noted the presence of phytosaurs ("*Rutiodon* Group A" and "*Nicrosaurus*"), stagonolepidids (*Paratypothorax*), and rauisuchians in the central and northern Dockum Basin that are similar if not identical to those within the Norian lower Stubensandstein at Heslach, Germany. It is apparent that the data from palynology and vertebrate paleontology conflict, and therefore, more detailed stratigraphic studies need to be conducted in the Dockum to resolve these problems. Concerning Otis Chalk, the presence of well-preserved specimens of *Cionichthys* (middle–upper Carnian) and *Lasalichthys* (middle Carnian–middle Norian?) suggest a middle to upper Carnian age for that portion of the Dockum Basin. However, no productive palynological samples have been recovered from the southern portion of the Dockum Basin to compare with the vertebrate assemblages.

The localities in Crosby, Randall, and Potter counties show a great similarity to those of the lower Petrified Forest Member (late Carnian) of the Chinle Formation and also are quite similar to several of the Newark formations, such as the Cow Branch, New Oxford, and especially the Lockatong Formation. Detailed stratigraphic and paleontological studies by Paul Olsen indicate a late Carnian age for these Newark formations.

It is highly unlikely that all the Dockum beds are precisely contemporaneous with each other. The Carnian stage encompassed a period of some 4.5 to 6 MY with the Carnian–Norian boundary placed anywhere from 205 to 225 MYA (Anderson and Anderson 1970; Anderson and Cruickshank 1978). This indicates a possible error in Dockum correlation of over 3 MY, a substantial time in which many evolutionary and ecological changes could occur, and also indicates the poor chronostratigraphy for the Triassic. Until substantial lithostratigraphic, biostratigraphic, and chronostratigraphic dates are available, it can be stated that at least some of the Dockum is Carnian in age, with some units in the central and northern Dockum basin possibly referable to the Norian.

The Dockum vertebrate community

An ecological community is defined as a group of organisms living together within a definite locality (Olson 1980). Because paleontologists can only determine which taxa have been preserved together, recognition of fossil communities is rather difficult and depends largely on taxonomic analogy with modern communities or recurrent death assemblages. The presence of both terrestrial and aquatic vertebrates within each Dockum locality studied suggests a mixture of elements from at least two living communities. Therefore, any conclusions stated

Figure 9.14. Trophic relationships of the Rotten Hill and Kalgary fossil vertebrate community, lower Dockum Group of Texas.

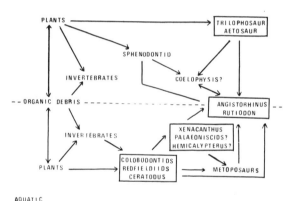

ROTTEN HILL

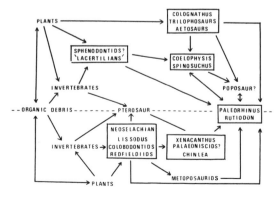

KALGARY

herein are tentative and should be tested by future studies of functional anatomy and paleoecology.

Diagrams (Figs. 9.14, 9.15) showing theoretical trophic relationships of the Dockum community imply the existence of five general categories, which include aquatic omnivores, aquatic carnivores, terrestrial herbivores, terrestrial carnivores, and aerial carnivores. Within the categories of carnivores and aquatic omnivores, I include taxa that may have been insectivorous or malacophagous.

The distribution and composition of faunas in the Texas Triassic reflect the structural, stratigraphic, and sedimentological development of the Dockum paleoenvironment. To obtain a better understanding of paleoecological relationships of localities within the Dockum, screen sieving techniques were used for this study at sites in the southern, central, and northern portions of the basin. By this method, many of the smaller members of the Dockum fauna were recovered. The pie diagrams (Fig. 9.16) showing the relative numbers of faunal remains found obviously do not reflect actual populations. In this study, a single fish scale or a phytosaur skull would each count as a single unit. Considering that many animals, especially fish and sharks, possess many more potentially countable elements than other taxa causes considerable sampling bias, especially toward the aquatic members of the faunas. This is reflected in the pie diagrams, in which aquatic vertebrates constitute the major portion of total elements recovered. As a measure of differences between faunas, the use of relative numbers of elements is invaluable. However, the counting of elements for a detailed population study at a single site would be inaccurate and could only vaguely approximate natural conditions. This must be considered when analyzing the faunal composition of the Dockum.

Figure 9.15. Trophic relationships of the Otis Chalk fossil vertebrate community, lower Dockum Group of Texas.

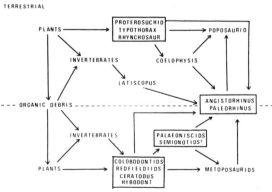

OTIS CHALK

All of the Dockum localities contain both aquatic and terrestrial vertebrates. Otis Chalk is superficially similar to Kalgary and Rotten Hill in composition; all have a variety of sharks and palaeoniscid, redfieldiid, semionotid and colobodontid fish, and lungfish. Also, metoposaurid amphibians, aetosaurs, phytosaurs, and a variety of other small and large carnivorous thecodonts are present in all of the faunas. However, the terrestrial fauna at Otis Chalk is unique in the presence of a possible proterosuchid, a rhynchosaur, and the amphibian *Latiscopus* (although *L. disjunctus* is found at the Miller locality), and the specific faunal composition is markedly different from that of Kalgary and Rotten Hill. No xenacanth sharks and only a single hybodontoid tooth are known from Otis Chalk (Figs. 9.15, 9.16C). This differs from Kalgary, in which shark elements total approximately twenty percent of the total fauna recovered (Figs. 9.14, 9.16B). Rotten Hill has some xenacanth sharks, although they represent only three percent of the total elements (Figs. 9.14, 9.16A). Therefore, even though the three localities are similar on a familial level, there is very little to relate Otis Chalk to the Kalgary and Rotten Hill localities at a species level. However, such differences may be temporal as well as paleoecological.

Considering the large faunal differences between the Dockum localities, future studies of Triassic biostratigraphy should be concerned with the ecological differences of faunas as well as their relative ages. Using wet sieving techniques in future studies will certainly aid in better understanding Triassic biostratigraphy and paleoecology, but it is not a panacea, especially for taxonomic problems.

Figure 9.16. Comparative abundance of vertebrate elements in the Rotten Hill (**A**), Kalgary (**B**), and Otis Chalk (**C**) fossil localities.

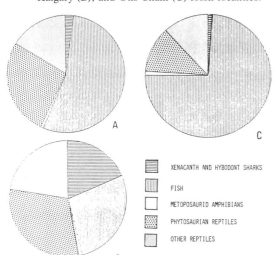

XENACANTH AND HYBODONT SHARKS

FISH

METOPOSAURID AMPHIBIANS

PHYTOSAURIAN REPTILES

OTHER REPTILES

Furthermore, the use of specific taxonomic groups (such as phytosaurs) for biostratigraphic correlation should be avoided when the systematic relationships and range zones of those taxa are not established. Triassic biostratigraphic and paleoecological problems will only be resolved by a concerted effort of many scientists, and must involve many scientific disciplines besides vertebrate paleontology.

Acknowledgments

I would like to thank Sankar Chatterjee, Robert Long, Paul Olsen, Kevin Padian, and Michael Parrish for their thorough reviews of this chapter. Many others have provided valuable data concerning Triassic vertebrate taxonomy and distribution, and hopefully I have cited their contributions in this chapter. I have been ably assisted in field and laboratory by many people, including Robert Acree, Bryce Bell, Arthur Endress, Roy Frosch, Henry Huggins, Kirk Kennedy, Karen Murry, Kent Newman, Al Santa Luca, Stefan Stamoulis, and Richard Wolfe. Karen Murry and Norman MacLeod assisted in preparation of the illustrations, and the manuscript was typed by Lisa Mitchell and Connie Murry. I would especially like to thank Bob Slaughter for his support of these projects and to all of the landowners for access to their land. Funding for this project was provided by National Science Foundation Grant EAR-82184872 and Tarleton State University Organized Research Grant 15-613.

References

Anderson, H. M., and J. M. Anderson. 1970. A preliminary review of the uppermost Permian, Triassic, and lowermost Jurassic of Gondwanaland. *Palaeontol. Afr.* 13:1–22.

Anderson, J. M., and A. R. I. Cruickshank. 1978. The biostratigraphy of the Permian and the Triassic; Part 5, A review of the classification and distribution of Permo-Triassic tetrapods. *Palaeontol. Afr.* 21:15–44.

Armstrong-Ziegler, J. G. 1978. An aniliid snake and associated vertebrates from the Campanian of New Mexico. *J. Paleontol.* 52(2):480–3.

Baird, D. 1963. Rhynchosaurs in the Late Triassic of Nova Scotia. *Geol. Soc. A. Spec. Pap.* 73:107.

1964. Dockum (Late Triassic) Reptile Footprints from New Mexico. *J. Paleontol.* 38(1):118–25.

Baird, D., and P. E. Olsen. 1983. Late Triassic herpetofauna from the Wolfville Formation of the Minas Basin (Fundy Basin), Nova Scotia, Canada. *Geol. Soc. Am.* Abst. Prog. 15:122.

Baird, D., and W. F. Take. 1959. Triassic reptiles from Nova Scotia. *Bull. Geol. Soc. Am.* 70:1565–6.

Bock, W. 1959. New Eastern American Triassic Fishes and Triassic Correlations. *Geol. Cent. Res. Ser.* 1:1–184.

Branson, E. B. 1905. Structure and Relationships of American Labyrinthodontidae. *J. Geol.* 13(7):568–610.

Branson, E. B., and M. G. Mehl. 1929. Triassic amphibians from the Rocky Mountain region. *Univ. Missouri Studies* 4:155–255.

Broom, R. 1936. The South African Procolophonia. *Ann. Transvaal Mus.* 18(4):387–91.

Brough, J. 1935. On the structure and relationships of the hybodont sharks. *Mem. Proc. Manchester Lit. Phil. Soc.* 69:35–47.

Buffetaut, E. 1983. *Isalarhynchus genovoetae* n. g., n. sp., (Reptilia:Rhynchocephalia), un nouveau rhynchosaure du Trias de Madagascar. *N. Jb. Geol. Palaeont. Mh.* 8:465–80.

Camp, C. L. 1930. A study of the phytosaurs, with description of new material from North America. *Mem. Univ. Calif.* 10:1–174.

Camp, C. L., and S. P. Welles. 1965. Triassic Dicynodont Reptiles. *Mem. Univ. Calif.* 13:255–348.

Cappetta, H., and G. R. Case, 1975. Contribution a l'étude des selaciens du groupe Monmouth (Campanien-Maestrichtien) du New Jersey. *Palaeontographica (A)* 151(1–3):1–46.

Carpenter, K. 1979. Vertebrate Fauna of the Laramie Formation (Maestrichtian)Weld County, Colorado. *Contrib. Geol. Univ. Wyo.* 17:37–49.

Carroll, R. L. 1978. Permo-Triassic "lizards" from the Karoo System; Part 2, A gliding reptile from the Upper Permian of Madagascar. *Palaeontol. Afr.* 1:143–59.

Carroll, R. L., and P. M. Galton. 1977. "Modern" lizard from the Upper Triassic of China. *Nature (London)* 266(5599):252–5.

Carroll, R. L., E. S. Belt, D. I. Dineley, et al. 1972. Vertebrate paleontology of eastern Canada. *24th, Canada, 1972, Int. Geol. Congr., Guidebk. Field Excursion, Montreal.* pp. 1–113.

Case, E. C. 1920a. On a very perfect thoracic shield of a large labyrinthodont in the geological collections of the University of Michigan. *Occ. Pap. Univ. Mich. Mus. Zool.* 82:1–3.

1920b. Preliminary description of a new suborder of Phytosaurian reptiles, with a description of a new species of *Phytosaurus. J. Geol.* 28(6):524–35.

1921. A new species of *Ceratodus* from the Upper Triassic of Western Texas. *Occ. Pap. Mus. Zool. Univ. Mich.* 101:1–2.

1922. New reptiles and stegocephalians from the Upper Triassic of Western Texas. *Carnegie Inst. Wash. Pub.* 321:1–84.

1927. The vertebral column of *Coelophysis* Cope. *Contr. Mus. Pal. Univ. Mich.* 2(10):209–22.

1928. Indications of a *Cotylosaur* and of a new form of fish from the Triassic beds of Texas, with remarks on the Shinarump Conglomerate. *Univ. Mich. Contr. Mus. Paleontol.* 3 (1):1–14.

1929. Description of the skull of a new form of phytosaur. *Mich. St. Univ. Pal. Mem.* 2:1–56.

1930. On the lower jaw of *Brachysuchus megalodon. Contr. Mus. Paleontol. Univ. Mich.* 3(8):155–61.

1931. Description of a new species of *Buettneria*, with a discussion of the brain case. *Contr. Mus. Paleontol. Univ. Mich.* 3(11):187–206.

1932a. A collection of stegocephalians from Scurry County, Texas. *Univ. Mich. Cntr. Mus. Paleontol.* 4 (1):1–56.

1932b. On the caudal region of *Coelophysis* sp. and on some new or little known forms from the Upper Triassic of Western Texas. *Cntr. Mus. Paleontol. Univ. Mich.* 4(3):81–91.

1943. A new form of phytosaur pelvis. *Am. J. Sci.* 241(3):201–3.

1946. A census of the determinable genera of the *Stegocephalia. Am. Phil. Soc. Trans.* 35(4):325–420.

Case, E. C., and T. E. White, 1934. Two new specimens of phytosaurs from the Upper Triassic of Western Texas. *Contr. Mus. Paleontol. Univ. of Mich.* 4(9):133–42.

Case, G. R. 1978. A new selachian fauna from the Judith River Formation (Campanian) of Montana. *Palaeontographica, Abt. A*, 160:176–205.

Casier, E. 1947. Constitution et évolution de la Racine Dentaire des Euselachii. I. Note Préliminaire. *Bull. Mus. Roy. Hist. Nat. Belg.*, 23(13):1–15.

Chatterjee, S. 1978. A primitive Parasuchid (Phytosaur) reptile from the Upper Maleri Formation of India. *Paleontology* 21(1):83–127.

1983. An Ictidosaur fossil from North America. *Science*, 220:1151–3.

1985. *Postosuchus*, a new thecodontian reptile from the Triassic of Texas and the origin of tyrannosaurs. *Phil. Trans. Roy. Soc. London B* 309:395–460.

Chow, M., and A. I. Sun. 1960. A new procolophonid from north-western Shansi. *Vert. Palasiatica* 4:11–13.

Colbert, E. H. 1946. *Hypsognathus*, a Triassic reptile from New Jersey. *Bull. Am. Mus. Nat. Hist.* 86:225–74.

1947. Studies of the phytosaurs *Machaeroprosopus* and *Rutiodon. Bull. Am. Mus. Nat. Hist.* 88 (2):53–96.

1964. The Triassic dinosaur genera *Podokesaurus* and *Coelophysis. Am. Mus. Novit.* 2168: 1–12.

1970. The Triassic gliding lizard *Icarosaurus. Bull. Am. Mus. Nat. Hist.* 143(2):89–142.

1972. Vertebrates from the Chinle Formation. *Mus. North. Ariz. Bull.* 47:1–11.

Colbert, E. H., and J. T. Gregory. 1957. Correlations of continental Triassic sediments by vertebrate fossils. *In* Reeside et al., Correlation of the Triassic formations of North America exclusive of Canada. *Bull. Geol. Soc. Am.* 68:1456–67.

Colbert, E. H., and J. Imbrie. 1956. Triassic metoposaurid amphibians. *Bull. Am. Mus. Nat. Hist.* 110(6):399–452.

Cope, E. D. 1881. *Belodon* in New Mexico. *Am. Nat.* 15:922–3.

1887. A contribution to the history of the Vertebrata of the Trias of North America. *Proc. Am. Phil. Soc.* 24:209–28.

1892. A contribution to the vertebrate paleontology of Texas. *Proc. A. Phil. Soc.* 30:123–31.

Cosgriff, T. W., and J. M. Zawiskie. 1979. *Pneumatostega potamia*, new genus new species of the Rhyteosteidae from the *Lystrosaurus* Zone and a review of the Rhytidosteodiea. *Palaeontol. Afr.* 22:1–28.

Demar, R., and J. R. Bolt. 1981. Dentitional organization and function in a Triassic reptile. *J. Paleontol.* 55(5):967–84.

Duffin, C. J. 1980. A new euselachian shark from the Upper Triassic of Germany. *N. Jahrb. Geol. Palaeont. Monatshefte* 1980(1):1–16.

———. 1981. Comments on the selachian genus *Doratodus* Schmid (1861) (Upper Triassic, Germany). *Neues Jahrb. Geol. Palaeont. Monatshefte* 1981(5):289.

———. 1982. Teeth of a new selachian from the Upper Triassic of England. *N. Jahrb. Geol. Palaeont. Monatshefte* 1982(3):156–66.

Dunay, R. E., and M. J. Fisher. 1979. Palynology of the Dockum Group (Upper Triassic), Texas, U.S.A. *Rev. Palaeobot. Palyn.* 28:61–92.

Dutuit, J. M. 1976a. Introduction to the paleontological study of the Moroccan continental Trias: Description of the first stegocephalians collected in the Argana Corridor (Western Atlas). *Mem. Mus. Nat. Hist. Nat. Ser. C Sci. Terre (Paris)* 36:1–253.

———. 1976b. Il est probable que les Rhychocéphales sont représentés dans la fauna du Trias Marocain. *Compt. Rend. Acad. Sci. Paris, Ser. D.* 283:483–6.

———. 1977a. Description du crane de *Angistorhinus talainti* n. sp., un nouveau phytosaure de Trias Atlantique Marocain. *Bull. Mus. Nat. Hist. Sci. Terre* 66:277–37.

———. 1977b. *Paleorhinus magnoculus*, Phytosaure du Trias Supérieur de l'Atlas Marocain. Provence, *Univ. Ann. Geol. Mediter.* 4(3):225–68.

Eaton, T. H. 1965. A new Wyoming phytosaur. *Univ. Kan. Paleontol. Contrib.* 2:1–6.

Elder, R. L. 1978. Paleontology and paleoecology of the Dockum Group, Upper Triassic, Howard County, Texas. M.S.Thesis, University of Texas at Austin.

Estes, R. 1964. Fossil vertebrates from the late Cretaceous Lance Formation, Eastern Wyoming. *Univ. Calif. Pub. Geol. Sci.* 49:1–180.

Fox, R. C. 1977. A primitive therian mammal from the Upper Cretaceous of Canada. *Can. J. Earth Sci.* 9:1474–94.

Fraser, N. C. 1982. A new rhynchocephalian (*Planocephalosaurus robinsonae*, new genus new species) from the British Upper Trias. *Palaeontology* 25(4):709–26.

Fraser, N. C., and G. M. Walkden. 1983. The ecology of a Late Triassic reptile assemblage from Gloucestershire, England (U.K.). *Palaeogeogr. Palaeoclimatol. Palaeoecol.* 42(3/4):341–66.

Gorman, J. M., and R. C. Roebeck. 1946. Geology and asphalt deposits of north-central Guadalupe County, New Mexico. 1946: U.S. Geol. Surv. Oil. and Gas Inves. Prelim. Map No. 44.

Gould, C. N. 1907. The geology and water resources of the western portion of the panhandle of Texas. *U.S. Geol. Surv. Water Supply Pap.* 191:1–70.

Gow, C. E. 1977. Tooth function and succession in the Triassic reptile *Procolophon trigoniceps. Palaeontology* 20(3):695–704.

Gow, C. E., and M. A. Raath. 1977. Fossil vertebrate studies in Rhodesia: sphenodontid remains from the Upper Triassic of Rhodesia. *Palaentol. Afr.* 20:121–2.

Granata, G. E. 1981. Regional sedimentation of the Late Triassic Dockum Group, West Texas and Eastern New Mexico. MA thesis, University of Texas, Austin.

Green, F. E. 1954. The Triassic deposits of northwestern Texas. Ph.D. thesis, Texas Tech University, Lubbock, Texas.

Gregory, J. T. 1945. Osteology and relationships of *Trilophosaurus. Univ. Tex. Publ.* 4401:273–359.

———. 1962a. The genera of phytosaurs. *Am. J. Sci.* 260:652–90.

———. 1962b. The relationships of the American phytosaur *Rutiodon. Am. Mus. Novit.* 2095:1–22.

———. 1972. Vertebrate faunas of the Dockum Group, Triassic Eastern New Mexico and West Texas. *N. Mex. Geol. Sci. Guidebk., Field Conf.*, 23:120–3.

———. 1980. The otic notch of metoposaurid labyrinthodonts. *In* Jacobs, L. L. (ed.), *Aspects of Vertebrate History* (Flagstaff, Arizona; Museum of Northern Arizona Press), pp. 125–36.

Gregory, J. T., and F. Westphal. 1969. Remarks on the phytosaur genera of the European Trias. *J. Paleontol.* 43:1296–8.

Haubold, H. 1971. Ichnia Amphiborum et Reptiliorum fossilum. *In* Kuhn, O. (ed.), *Handbuch der Palaoherpetologie* (Stuttgart, Gustav Fischer): pp. 1–124.

Herman, J. 1977. Les selaciens des terrains néocretacés et paléocènes de Belgique et des contrées limitrophes. *Mem. Serv. Geol. Belg.* 15:1–450.

Huene, F. von. 1910. Über einen echten Rhynchocephalen aus der Trias von Elgin. *Brachyrhinodon taylori. N. Jahrb. Min. Geol. Pal.* 1910:29–62.

———. 1912. Die cotylosaurier der Trias. *Palaeontographica.* 59:69–102.

———. 1913. A new Phytosaur from the Palisades near New York. *Bull. Am. Mus. Nat. Hist.* 32(15):275–82.

———. 1915. On Reptiles of the New Mexican Trias in the Cope Collection. *Bull. Am. Mus. Nat. Hist.* 34:485–507.

———. 1920. Ein *Telerpeton* mit gut erhaltenem Schadel. *Centralbl. Min. Geol. Palaontol.* 1920(11–12):189–92.

———. 1932. Die fossile Reptil-Ordnung Saurischia, ihre Entwicklung und Geschiehte. *Monog. Geol. Pal.* 1(4):1–362.

———. 1956. *Palaontologie und Phylogenie der Niederen Tetrapoden* (Stuttgart: Gustav Fischer), pp. 1–716.

Hutchinson, P. 1973. A revision of redfieldiiform and perleidiform fishes from the Triassic of Bekker's Kraal (South Africa) and Brookvale (New South Wales). *Bull. Brit. Mus. Nat. Hist. Geol.* 22(3):233–354.

Ivachnenko, M. F. 1973a. Structure of the skull of the early Triassic procolophonid *Tichvinskia vjatkensis. Paleontol. J.* 4:74–83.

———. 1973b. New Cotylosaurs from the Ural Mountains. *Paleontol. J.* 2:131–4.

———. 1975. Early Triassic procolophonid genera of Cisuralia. *Paleontol. J.* 1:86–91.

———. 1979. The Permian and Triassic procolophonids from the Russian Platform. *Akad. Nauk. SSSR, Paleontol. Inst. Tr.* 164:1–80.

Jacobs, L. L. 1980. Additions to the Triassic vertebrate fauna of Petrified Forest National Park, Arizona.

J. Ariz.-Nev. Acad. Sci. 1980:247.

Jacobs, L. L., and P. A. Murry, 1980. The vertebrate community of the Triassic Chinle Formation near St. Johns, Arizona, In Jacobs, L. L. (ed.), Aspects of Vertebrate History (Flagstaff, Arizona: Museum of Northern Arizona Press), pp. 55–71.

Jain, S. L. 1980. Freshwater xenacanthid (= Pleuracanth) shark fossils from the Upper Triassic Maleri Formation, India. J. Geol. Sci. India 21:39–47.

Koh, T. P. 1940. Santaisaurus yuani n. gen. et spec. nov., ein neues Reptil aus der Unteren Trias von China. Bull. Geol. Soc. China 20:73–92.

Langston, W. 1949. A new species of Paleorhinus from the Triassic of Texas. Am. J. Sci. 247(5):324–41.

Lehman, J. P. 1966. Actinoptergyii, In Piveteau, J. (ed.), Traité de Paléontologie Vol. 4 (Paris: Masson), pp. 1–242.

Long, R. A., and K. L. Ballew. 1985. Aetosaur dermal armor from the Late Triassic of southwestern North America, with special reference to material from the Chinle Formation of Petrified Forest National Park. Mus. N. Ariz. Bull. 54:45–68.

Maisey, J. G. 1977. The fossil selachian fishes Palaeospinax Eagerton, 1872 and Nemacanthus Agassiz, 1837. Zool. J. Linn. Soc. 60:259–73.

1978. Growth and form of finspines in hybodont sharks. Palaeontology 21(3):657–66.

Martin, M. 1979. Arganodus atlantis et Ceratodus arganensis, deux nouveaus Dipneustes du Trias supérieur continental marocain. Comp. Rend. Acad. Sci. Paris, Ser. D 289:89–92.

1982a. Les Actinopterygiens (Perleidiformes et Redfieldiiformes) du Trias supérieur continental du couloir d'Argana (Atlas occidental, Maroc.) N. Jahrb. Geol. Palaont. Abh. 162(3):352–72.

1982b. Les Dipneustes et Actinistiens du Trias supérieur continetal marocain. Stuttgarter Beitr. Naturl. (B) 69:1–29.

McGowen, J. H., and L. E. Garner, 1970. Physiographic features and stratification types of coarse-grained pointbars: modern and ancient examples. Sedimentology 14(1/2):77–111.

McGowen, J. H., G. E. Granata, and S. J. Seni. 1979. Depositional framework of the lower Dockum Group (Triassic), Texas Panhandle. Bur. Econ. Geol. Univ. Texas, Austin, Rept. Invest. No. 97.

Mehl, M. G. 1913. Angiostorhinus, a new genus of Phytosauria from the Trias of Wyoming. J. Geol. 21:186–91.

1915. The Phytosauria of the Trias. J. Geol. 23:129–65.

1922. A new phytosaur from the Trias of Arizona. J. Geol. 30:144–57.

1928. The Phytosauria of the Wyoming Triassic. Dennison Univ. Bull. 23:141–72.

Murry, P. A. 1981. A new species of freshwater hybodont from the Dockum Group (Triassic) of Texas. J. Paleontol. 55(3):603–7.

1982. Biostratigraphy and Paleoecology of the Dockum Group, Triassic, of Texas. Ph.D. thesis, Southern Methodist University, Dallas, Texas.

Olsen, P. E. 1980. A comparison of the vertebrate assemblages from the Newark and Hartford Basins (Early Mesozoic, Newark Supergroup) of Eastern North America. In Jacobs, L. L. (ed.), Aspects of Vertebrate History (Flagstaff, Arizona: Museum of Northern Arizona Press) pp. 35–53.

1984. Comparative Paleolimnology of the Newark Supergroup: A study of ecosystem evolution. Ph.D. dissertation. Department of Biology, Yale University.

Olsen, P. E., A. McCune, and K. S. Thomson. 1982. Correlation of the early Mesozoic Newark Supergroup by vertebrates, principally fishes. Am. J. Sci. 282:1–44.

Olson, E. C. 1980. Permian lake faunas; a study in community evolution. J. Paleontol. Soc. India 20:146–63.

Parks, P. 1969. Cranial anatomy and mastication of the Triassic reptile Trilophosaurus. M. A. Thesis. Department of Geology, University of Texas, Austin.

Patterson, C. 1966. British Wealden sharks. Bull. Br. Mus. Nat. Hist. Geol. 11:283–350.

Peyer, B. 1968. Comparative Odontology (University of Chicago Press, Chicago) pp. 1–347.

Reif, W. -E. 1977. Tooth enameloid as a taxonomic criterion: 1. A new euselachian shark from the Rhaetic–Liassic boundary. N. Jahrb. Palaeont. Monatshefte 1977:565–76.

Robinson, P. L. 1957. An unusual sauropsid dentition. J. Linn. Soc. (Zool.) 43:282–93.

1962. Gliding lizards from the Upper Keuper of Great Britain. Proc. Geol. Soc. Lond. 1601:137–46.

1973. A problematical reptile from the British Upper Trias. J. Geol. Soc. Lond. 129(5):457–79.

Romer, A. S. 1947. Review of the Labyrinthodontia. Bull. Mus. Comp. Zool. Harvard 99(1):1–368.

Roy-Chowdhury, T. 1965. A new metoposaurid amphibian from the Upper Triassic Maleri formation of central India. Phil. Trans. Roy. Soc. London, Ser. B 250:1–52.

Sawin, H. J. 1944. Amphibians from the Dockum Triassic of Howard County, Texas. Univ. of Tex. Publ. 4401:361–99.

1947. The pseudosuchian reptile Typothorax meadei. J. Paleontol. 21:201–38.

Schaeffer, B. 1952. The paleoniscoid fish Turseodus from the Upper Triassic Newark Group. Am. Mus. Novit. 1581:1–24.

1955. Mendocinia, a subholostean fish from the Triassic of Argentina. Am. Mus. Novit. 1737:1–23.

1967. Late Triassic fishes from the western United States. Bull. Am. Mus. Nat. Hist. 135(6):287–342.

Schaeffer, B., and N. G. McDonald. 1978. Redfieldiid fishes from the Triassic–Liassic Newark Supergroup of Eastern North America. Bull. Am. Mus. Nat. Hist. 159(4):131–73.

Seilacher, A. 1943. Elasmobranchier – Reste aus dem oberen Muschelkalk und dem Keuper Württembergs. N. Jahrb. Miner. Geol. Palaeont. (B) 1943:256–92.

Simpson, G. G. 1926. American terrestrial rhynchocephalia. Am. J. Sci. Ser. 5 12:12–16.

Stensiö, E. A. 1921. Triassic Fishes from Spitzbergen

(Vienna: Adolf Holzhausen), pp. 1–307.

1932. Triassic fishes from East Greenland. *Meddel. Grønland.* 83(3):1–305.

Stovall, J. W., and D. E. Savage. 1939. A phytosaur in Union County, New Mexico. *J. Geol.* 47:759–66.

Stovall, J. W., and J. B. Wharton. 1936. A new species of phytosaur from Big Spring, Texas. *J. Geol.* 44(2):183–92.

Tannenbaum, F. 1983. The microvertebrate fauna of the Placerias and Downs Quarries, Chinle Formation (Upper Triassic) near St. Johns, Arizona. M.S. thesis, Department of Paleontology, University of California, Berkeley, pp. 1–111.

Throckmorton, G. S., J. A. Hopson, and P. Parks, 1981. A redescription of *Toxolophosaurus cloudi* Olson, a lower Cretaceous herbivorous sphenodontid reptile. *J. Paleontol.* 55(3):586–97.

Thurmond, J. T. 1971. Cartilaginous fishes of the Trinity Group and related rocks (Lower Cretaceous) of North Central Texas. *Southeast. Geol.* 13:217–18.

Walker, A. D. 1969. The reptile fauna of the Lower Keuper sandstone. *Geol. Mag.* 106:470–6.

Walper, S. L. 1980. Tectonic evolution of the Gulf of Mexico. *In* Pilger, R. H., Jr. (ed.), *The Origin of the Gulf of Mexico and the Early Opening of the Central North Atlantic; Proceedings of a Symposium* (Baton Rouge: Louisiana State University Press), pp. 87–98.

Warthin, A. S. 1928. Fossil fishes from the Triassic of Texas. *Contrib. Mus. Paleont. Univ. Mich.* 3(2):15–18.

Welles, S. P. 1947. Vertebrates from the Upper Moenkopi Formation of Northern Arizona. *Bull. Dept. of Geol. Sci. Univ. Calif.* 27(7):241–94.

Westphal, F. 1976. Phytosauria. *Handb. Palaeoherpetol. Stuttgart* 13:99–120.

Wild, R. 1978. Die Flugsaurier (Reptilia, Pterosauria) aus der Oberen Trias von Cene Bei Bergamo, Italien. *Est. Boll. Soc. Paleontol. Ital.* 17(2):176–256.

Williston, S. W., and E. C. Case, 1913. Description of the vertebrate-bearing beds of north-central New Mexico. *Publ. Carnegie Inst. Wash.* 181:1–6.

Wilson, J. A. 1948. A small amphibian from the Triassic of Howard County, Texas. *J. Paleontol.* 22(3):359–61.

1950. Cope's types of fossil reptiles in the collection of the Bureau of Economic Geology, the University of Texas. *J. Paleont.* 24:113–5.

Woodward, A. S. 1889. Paleoichthyological notes 2. On *Diplodus moorei* sp. nov., from the Keuper of Somersetshire. *Ann. Mag. Nat. Hist.* 6(3)297–302.

1891. *Catalogue of the Fossil Fishes of the British Museum: Acanthodii, Holocephali, Ichthyodorulites, Ostracodermi, Dipnoi, and Teleostomi. (Crossopterygii and Chondrostean Actinopterygii.)* (London: Taylor and Francis), pp. 1–567.

Young, C. C. 1948. A review of Lepidosauria from China. *Am. J. Sci.* 246:711–19.

Zittel, K. A. von. 1932. *In* Woodward, A. S. (ed.), *Textbook of Paleontology*, Vol. 2, 2nd Eng. ed., rev. (London: MacMillan), pp. 1–464.

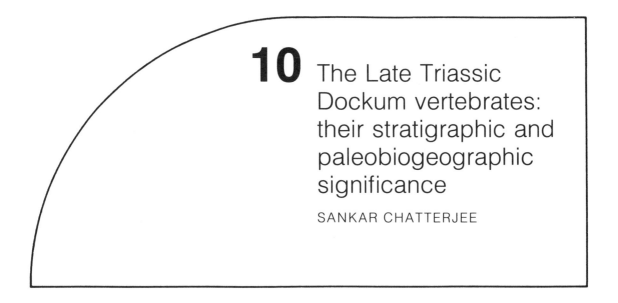

10 The Late Triassic Dockum vertebrates: their stratigraphic and paleobiogeographic significance

SANKAR CHATTERJEE

Introduction

The Late Triassic Dockum deposits of eastern New Mexico and western Texas are continental red bed sequences, composed of 70–700 m of terrigenous clastics that accumulated in a mosaic of fluvial, lacustrine, and floodplain environments (Fig. 10.1) (McGowen, Granata, and Seni 1979). They lie unconformably on the Late Permian Quartermaster Group and are capped by Cretaceous, Tertiary, or Quaternary sediments.

Although the lithology and the vertebrate fossils of the Dockum rocks have been known almost for a century, their stratigraphic nomenclature is in a state of confusion, and their paleontological importance is not yet fully appreciated. It is generally believed that the Dockum fauna of Texas is somewhat primitive, corresponding to the "lower fauna" of the Chinle Formation of Arizona, and its age has been correlated with that of the Carnian of the standard European marine sequence. The correlative "upper fauna" of Chinle is thought to be absent in the Texas Dockum (Gregory 1957; Colbert and Gregory 1957; Dunay 1972; Murray 1982; Olsen, McCune, and Thomson 1982).

Figure 10.1. Outcrops of Late Triassic deposits in the American Southwest.

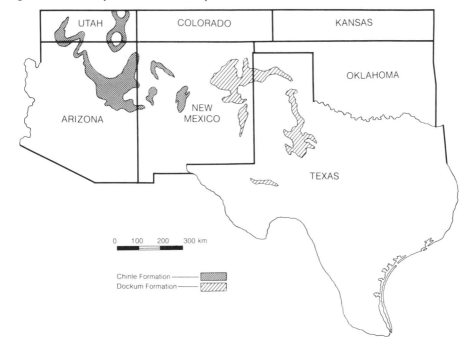

Against this interpretation is the recently discovered "upper fauna" found in the Dockum mudstone, near Post, Garza County, Texas (Chatterjee 1983, 1984, 1985). A quarry here has produced a varied tetrapod fauna, including new representatives of taxa hitherto unknown from the Triassic of North America. The fauna includes parasuchids (= phytosaurs), metoposaurs, coelurosaurs, aetosaurs, pterosaurs, protorosaurs, squamates, poposaurs, ictidosaurs, and fabrosaurs. The parasuchids, metoposaurs, and aetosaurs are highly advanced over the previously known forms from Texas. The ictidosaurs and fabrosaurs are similar to those of the Elliot and Clarens formations of the Upper Stormberg Series (Late Triassic–Early Jurassic) of South Africa, whereas poposaurs are closely related to *Teratosaurus*, known from the Stubensandstein of Germany. This new find clearly indicates that the age of the Texas Dockum is not restricted to the Carnian as is usually believed, but may extend to the Early Norian.

An attempt is made here to clear up the nomenclatorial confusion of the Dockum units and to evaluate their vertebrate fauna for zonation and correlation of Late Triassic continental formations. The recognition of Gondwana elements, such as fabrosaurs and ictidosaurs, in the Dockum fauna may have significant paleobiogeographic implication for determining the extent of the connection between Laurasia and Gondwana during the Late Triassic.

Lithostratigraphic problem
Previous work (1890–1983)
In 1890, W. F. Cummins, a geologist for the Texas Geological Survey, recognized a distinctive Triassic red bed sequence of clay, sandstone, and conglomerate in the vicinity of Dockum (Dickens County, Texas), unconformably overlying the Permian rocks. He named these Triassic red beds the Dockum Formation. In 1892, his colleague, N. F. Drake, conducted the first comprehensive study of the Dockum deposits in Texas and New Mexico. He subdivided the Dockum into three rock units: the lower shale unit, the middle sandstone and conglomerate unit, and the upper shale unit. Drake measured the sections along different areas of the Dockum Basin and found that the thickness and character of his subdivisions varied indiscriminately from locality to locality. He reported that his youngest unit was absent in the Canadian River area of the Texas Panhandle.

Gould (1907) studied the Dockum stratigraphy in the Texas Panhandle area in connection with a water resources survey. He raised the status of the Dockum to a group, and recognized two formations within it. The lower unit, the Tecovas Formation (type locality: Tecovas Creek, Potter County), consists primarily of red and variegated shales and siltstones. The upper unit, the Trujillo Formation (type locality: Trujillo Creek, Oldham County) is dominated by massive sandstones and conglomerates. In Palo Duro Canyon of the Texas Panhandle, the entire sequence is well exposed; this is not so in other areas. Baker (1915) concluded later that Drake's two lower units were synonymous with Gould's Tecovas and Trujillo formations. Later Adams (1929), Green (1954), and Kiatta (1960) extended the term "Chinle" Formation of Arizona to designate the upper shaly unit of Drake.

Although this subdivision of the Dockum is locally useful, it has encouraged haphazard naming of strata when these formational names have been extended to correlate the rock units of central and southern basins of the Dockum. The various shale units have been referred to Tecovas or Chinle, and sandstones and conglomerates given names such as Trujillo, Santa Rosa, Quito, Taylor, Dickens, Dripping Springs, Barstrow, or Camp Springs (Hoots 1925; Case 1928; Adams 1929; Jones 1953; Green 1954).

In 1957, Reeside et al. published a comprehensive review of the Triassic formations of North America. Their Chart 8a shows possible extensions of the name "Dockum" across Texas, New Mexico, Oklahoma, and Colorado. This report included some of the valuable remarks by other authors and should be seriously considered in conjunction with the problems of the Dockum stratigraphy. These are as follows:

1. J. T. Gregory (in Reeside et al. 1957, p. 1477): Although the Dockum is "generally accorded group status, its subdivisions, except possibly the Santa Rosa Sandstone, are all local. Terminology would be simplified, if the Dockum were called a formation, and its subdivisions were termed members."

2. E. D. McKee (in Reeside et al. 1957, p. 1476): "Triassic strata of eastern New Mexico, like those of central western Texas, are included in the Dockum Group and are commonly subdivided into formations, the uppermost of which is called the 'Chinle.' The extension of this name from the west and its restrictions to the uppermost unit of the Triassic in this area is unfortunate, for here the name connotes a meaning different from that in the type locality."

3. E. H. Colbert (quoted by McKee in Reeside et al. 1957, p. 1476): The Dockum Group of eastern New Mexico is "almost the exact equivalent of Chinle" in Arizona.

Surprisingly, in spite of these helpful remarks, Reeside et al. (1957, Chart 8a) treated the Dockum as a group, incorporated the name "Santa Rosa

Sandstone" to designate the middle unit of Drake in west central Texas, and retained the name "Chinle" as the upper unit of the Dockum (see Table 10.1).

In 1972, Dunay commented on the nomenclatorial problem of the Dockum and suggested that its rank should be reduced to a formation, with all the subdivisions functioning as local members.

Kelley (1972) recognized three formations within the Dockum of eastern New Mexico: Santa Rosa, Chinle, and Redonda. He divided the Chinle into three members: Lower shale, Cuervo Sandstone, and Upper shale.

Recently, the Bureau of Economic Geology and graduate students at the University of Texas at Austin have done excellent work on the depositional framework of the Dockum sediments (McGowen et al. 1979, 1983). These authors carefully avoided the nomenclatorial problems of the Dockum, and continued to treat it as a group. However, their use of stratigraphic nomenclatures is not entirely satisfactory (see McGowen et al. 1983, Fig. 8).

Persistent problems of the Dockum nomenclature and some proposed solutions

Improper usage of the name "Chinle" in the Dockum stratigraphy

McKee (in Reeside et al. 1957) deplored the different usage of the term "Chinle" in Texas and eastern New Mexico, where it is used to designate the upper shale unit of Drake. This seems an undesirable extension of the term "Chinle," because the Dockum as a whole appears to be equivalent to the Chinle Formation of Arizona. The term "Chinle" should be abandoned in reference to the upper unit of the Dockum; a new name should be coined for it.

Age of the Santa Rosa Sandstone

In the Tucumcari Basin of eastern New Mexico, the Santa Rosa Sandstone lies unconformably over the Permian sediments, and its age and proper stratigraphic position have been debated for a long time. Reeside et al. (1957) extended the name "Santa Rosa Sandstone" in the Midland Basin of Texas to designate the middle unit of the Dockum. This correlation is highly misleading and unwarranted, because in Texas, the middle unit occurs fairly high stratigraphically above the Permian rocks, and there is a distinctive shale unit between them. Recently, Dr. Spencer Lucas has discovered a beautiful skull of a benthosuchid (= *Eocyclotosaurus*) from the basal bed of the Santa Rosa Sandstone in the Tucumcari Basin. An identical benthosuchid has been long known from the Holbrook Member of the Moenkopi Formation of Arizona (Welles 1967), suggesting a Middle Triassic age (early Anisian) for the Santa Rosa Sandstone (S. Lucas and M. Morales pers. comm.). This implies that there is a great hiatus between the Santa Rosa Sandstone (early Anisian) and the basal Dockum (early Carnian). The Santa Rosa should be regarded as a distinctive formation in the Tucumcari Basin and should not be grouped with the Dockum. It is not correlative with the middle member of the Dockum in Texas, and its designation as such should be discontinued.

The rank and nomenclature of the Dockum

Cummins (1890) originally named the Dockum a formation. Later Gould (1907) raised its rank to a group, but his formational units are local and cannot be traced extensively from one basin to another. Because there are no marker beds and index fossils

Table 10.1. *Correlation of Late Triassic formations in the American Southwest*

ARIZONA			NEW MEXICO			TEXAS	
Middle-northern part	Northeastern part	Northwestern part	Central-northern part	Northeastern part	Southeastern part	Central-western part	Panhandle
CHINLE FORMATION — Owl Rock m.	CHINLE FORMATION — Owl Rock m.	CHINLE FORMATION — Owl Rock m.	CHINLE FORMATION — Correo SS. m.	DOCKUM GROUP — "Chinle" fm. — Redonda m.	DOCKUM GROUP — "Chinle" fm.	DOCKUM GROUP — "Chinle" fm.	DOCKUM GROUP — Trujillo fm.
Petrified Forest m.	Petrified Forest m.	Petrified Forest m.	Unnamed shale m.	Unnamed red shale m.			Tecovas fm.
Unnamed m.	Unnamed m.	Unnamed m.	Poleo ss. lentil	Santa Rosa sandstone	Santa Rosa sandstone	Santa Rosa sandstone	
	Shinarump m.		Salitral shale tongue		Pierce Canyon red beds (?Permian)	Tecovas fm.	
Shinarump m.		Shinarump m.	Agua Zarca sandstone m.			Camp Springs conglom.	

Source: Adapted from Reeside et al. (1957).

and because the subunits are lenticular, precise correlation becomes difficult.

I concur with Gregory (in Reeside et al. 1957) and Dunay (1972) that the Dockum should be regarded as a formation, because this is the primary mappable unit; its subdivisions should be demoted to members. This is logical because the correlative Chinle in Arizona has always been regarded as a formation, and its subunits as members.

There are already a number of lithostratigraphic names for the members of the Dockum. To preserve the stability of the nomenclature, Gould's Tecovas and Trujillo should be retained for the lower and middle units of Drake. However, Drake's upper shale unit never received any formal name other than the "Chinle," which now should be suppressed. A new name for the upper member of the Dockum will be proposed here.

Drake noticed that his upper unit is fairly well developed in the Borden and Garza counties, but it is absent farther north. It thickens again in the Tucumcari Basin. This is corroborated by the discovery of a new "upper fauna" in the upper shale unit, 8 km southeast of Post, Garza County, on the slope of a ridge. This is considered the type area for the upper shale member because no stratotype has ever been designated for this unit. This site is 2 km due south of Cooper Creek on the Millers' Ranch within the Getty oil lease (Cooper Creek quadrangle; latitude 33°03′N, longitude 101°19′E) (Fig. 10.2). The upper rock unit is accordingly named the Cooper Member after Cooper Creek. Here the highly crossbedded Trujillo Sandstone is visible at the base of the Cooper Member. It is extensively developed in Boren's Ranch, Justiceburg, where the estimated thickness is 60 m. Similar thick exposures of the

Trujillo Sandstone have been traced to Silver Falls, Crosby County. The thickness of the Cooper Member in its type area is about 16 m. It is capped by Quaternary sediments.

In summary, then, the Dockum of west Texas and eastern New Mexico is hereafter called a formation. It is subdivided in Texas into three members: Tecovas, Trujillo, and Cooper, in ascending order. A tentative correlation of the different Dockum members of Texas is shown in Fig. 10.3. It will be shown that these lithostratigraphic units correspond to vertebrate faunal zones.

Biostratigraphic problems
Stratigraphy of the Late Triassic epoch

The type section of the Triassic is in Germany, where it is divisible into Early, Middle, and Late phases, known as the Buntsandstein, Muschelkalk, and Keuper, respectively. The Buntsandstein and Keuper are continental, whereas the intervening Muschelkalk is marine.

Before analyzing the fauna of the Dockum, it is important to define as precisely as possible the stratigraphy of the Keuper of Germany. Unfortunately, this is not a straightforward matter. Ammonites, which are one of the best biostratigraphic tools available, are almost absent from the Keuper section. The continental Keuper fauna cannot be directly correlated to the ammonite zones of the standard marine sequence. It is much better to use a set of continental vertebrate zones for the Late Triassic rather than the imprecisely correlated marine stages. Tozer (1967) proposed "a standard for Triassic time" in which better represented North American ammonite zones were used to characterize

Figure 10.2. **a**, Outcrops of the Dockum Formation in West Texas. **b**, Type section of the new Cooper Member (named after the Cooper Creek) of the Dockum Formation, near Post, Garza County, West Texas. The Cooper Member is well-exposed in Garza County, overlying the Trujillo Member.

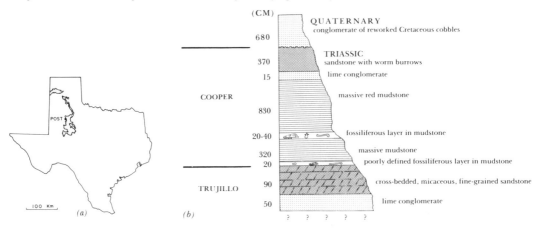

the standard stages more precisely. There have been lengthy arguments about the validity of the uppermost Triassic deposit, the Rhaetian stage, because it is restricted to one ammonite zone. Some workers (Hallam 1981) prefer to drop this stage and include its ammonite zone with the Norian. In this discussion, the correlation of Late Triassic deposits of Germany with the Alpine sequence is largely adapted from Harland et al. (1982).

Biozones of the Dockum by parasuchids (= phytosaur) genera

The parasuchid nomenclature in this discussion is after Chatterjee (1978), except that the European genus *Francosuchus* is considered here a junior synonym of *Parasuchus* (= *Paleorhinus*), as suggested by Gregory (1962, 1969). The Dockum Formation contains metoposaurs, aetosaurs, parasuchids, and recently found poposaurs very similar to those of the German Keuper fauna, indicating a Late Triassic age. Of all these tetrapods, parasuchids permit more detailed correlation and paleontological subdivisions of the Late Triassic because of two factors: (1) They are restricted temporally over a short geologic time span, and (2) they have a fairly wide geographic distribution.

In the type Triassic section of Germany, the primitive parasuchid genus *Parasuchus* is restricted to the Rotewand (= Blasensandstein), while *Nicrosaurus* and *Mystriosuchus* are restricted to the Stubensandstein. *Rutiodon* has a long range (Gregory 1969; Westphal 1976) (see Table 10.2). In North America, four parasuchid genera are known from Late Triassic deposits. These are *Parasuchus*, *Angistorhinus*, *Rutiodon*, and *Nicrosaurus*. Gregory (1957, 1972) recognized several faunal zones within the Dockum Formation on the occurrence of primitive and advanced parasuchid genera. Chatterjee (1978) presented a worldwide correlation of the Late Triassic continental deposits using Gregory's scheme. The apparent clarity of this scheme is marred by the overlap of primitive (*Parasuchus* and *Angistorhinus*) and advanced (*Rutiodon*) genera in Crosby, Potter, and Howard counties, and their assumed temporal separation has been questioned (Elder 1978; Murry 1982; Shelton 1984).

Elder pointed out another anomalous assemblage in Howard County, where *Nicrosaurus* (= *Brachysuchus megalodon*) is found along with *Parasuchus*. *Nicrosaurus* previously identified from Howard County is now considered *Angistorhinus* (Eaton 1965; Murry 1982), and, as a result, there is no conflict as these two genera occur together. However in Crosby and Potter counties, *Angistorhinus* or *Parasuchus* occurs with *Rutiodon*.

The presence of *Rutiodon* throughout the Dockum and Chinle sections is disturbing, and its validity for zonation is questionable. However, different species of *Rutiodon* may show temporal distribution. In Germany, *Rutiodon* has a long range and is of little stratigraphic value. Contrary to this, there is a clear stratigraphic separation between *Parasuchus* and *Nicrosaurus*. *Parasuchus* is restricted to the Carnian, and, therefore, there is no problem in

Figure 10.3 Left: map showing the Dockum outcrops in West Texas and the locations of sections measured; 1, Otis Chalk, Howard County; 2, Miller's and Boren's Ranch, Garza County; 3, Kalgary locality, Crosby County; 4, Palo Duro Canyon, Randall County. Right: a tentative correlation of the Dockum members in west Texas; thickness of the members in each locality compiled from several measured sections.

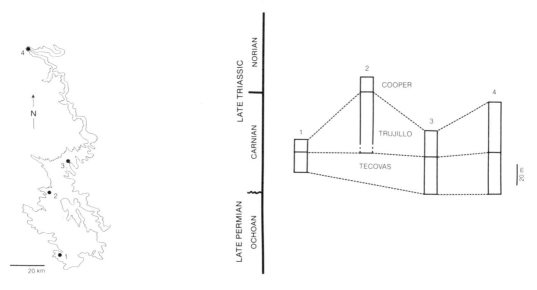

delineating the lower faunal zone (Gregory's basal zone) of the Late Triassic deposits in North America. The main problem is how to demarcate the upper faunal zone. On this point, I think that *Nicrosaurus* is a better index fossil than *Rutiodon*, because there is no overlap in time between *Nicrosaurus* and *Parasuchus*.

In the Chinle Formation of Arizona and the Dockum Formation of eastern New Mexico, *Nicrosaurus* is a very common form, but the primitive genera *Parasuchus* and *Angistorhinus* have never been found there. In the Dockum of Texas and the Popo Agie Member of the Chugwater Formation of Wyoming, the reverse is the case. Here *Parasuchus* and *Angistorhinus* are common elements, but no definite *Nicrosaurus* has been reported. The same is true for the Maleri Fauna of India and the Argana fauna of Morocco, where *Parasuchus* has been found, but not *Nicrosaurus*.

The so-called *Nicrosaurus* from the Texas Dockum is based on indeterminate fragments. Cope (1893) described several isolated teeth as *Palaeoctonus orthodon, Palaeoctonus dumblinus* from Palo Duro Canyon, and a few fragments as *Belodon superciliosus* from Dickens County. Similar compressed and isolated teeth, described by Cope, are known in *Angistorhinus, Rutiodon, Nicrosaurus*, and poposaurs, and they have little diagnostic value. Thus, all Cope's taxa are *nomina dubia*. In parasuchid taxonomy, unless the skull is found, it is very difficult to ascertain its relationship. Gregory (1962, p. 679) provisionally listed all Cope's taxa as *Nicrosaurus* "without regard to their probable synonymy or biological validity." Recently, Murry (1982) con-

sidered that these teeth belong to *Angistorhinus*. Thus, *Nicrosaurus* in a strict sense has never been reported from the Texas Dockum.

The apparent absence of *Nicrosaurus* in the Texas Dockum may be an artifact of biased collections. Previous collections for the past eighty years were made essentially from the lower Dockum (Tecovas Formation), which is correlative with the *Parasuchus* zone. *Nicrosaurus* occurs at a higher level (Stubensandstein equivalent).

Recently, a distinctive *Nicrosaurus* zone fauna has been discovered from the upper Dockum (Cooper Member), near Post, Garza County, Texas. In addition to *Nicrosaurus*, the fauna includes metoposaurs, aetosaurs, protorosaurs, podokesaurs, poposaurs, pterosaurs, ictidosaurs, fabrosaurs, and squamates (Fig. 10.4). The presence of *Nicrosaurus* suggests that the Cooper Member is correlative with the Petrified Member of the Chinle Formation of Arizona and the Stubensandstein of Germany. However, some of the vertebrates at Post quarry are more similar to vertebrates from higher stratigraphic levels elsewhere. For example, the new poposaur *Postosuchus* is very similar to *Teratosaurus*, from the Stubensandstein of Germany (Chatterjee 1985). The Dockum ictidosaur *Pachygenelus* is congeneric with that of the Elliot Formation (Lias) of the Stormberg Group, South Africa (Chatterjee 1983). Recently, an identical ictidosaur has been discovered from the McCoy Brook Formation (Lias) of the Newark Supergroup of Nova Scotia (C. Schaff pers. comm.). The new fabrosaur *Technosaurus* shows strong allies to *Lesothosaurus* of the Elliot Formation of South Africa and to *Scutellosaurus* of the Kayenta For-

Table 10.2. *Parasuchid zonation in Germany*

EPOCH	AGE (MARINE SUCCESSION)	Ma	GERMANY			RANGE OF PARASUCHIDAE
LIAS (J₁)	HETTANGIAN					
		213				
	RHAETIAN	219	RHÄTKEUPER			
LATE TRIASSIC (Tr₃)	NORIAN		KNOLLENMERGEL	STEINMERGEL	U P P E R	*Nicrosaurus Mystriosuchus Rutiodon*
			STUBENSANDSTEIN			
		225	ROTEWAND			*Parasuchus*
	CARNIAN		SCHILFSANDSTEIN		K E U P E R	
			GIPSKEUPER			
		231	LETTENKEUPER			
MIDDLE TRIASSIC (Tr₂)	LADINIAN		MUSCHELKALK			

mation (Lias) of Arizona (Chatterjee 1984). The aetosaur seems to be a highly specialized and advanced species of *Desmatosuchus*. In the skull, the premaxilla and the anterior part of the dentary are completely edentulous, and the lower temporal opening is partially closed. This is the only known aetosaur in which the premaxilla lacks teeth. The trend is very similar to early ornithischians, in which the anterior part of the jaw shows gradual reduction and loss of teeth. From the construction of the pelvis and femur, it appears that the Post aetosaur could attain erect posture (B. J. Small pers. comm.). Two advanced aetosaurs, *Paratypothorax* and *Typothorax*, have also been recognized in the Post Quarry. These genera occur in the Petrified Forest Member of the Chinle Formation and the Stubensandstein of Germany (R. A. Long pers. comm.).

The significance of a solitary skull of *Metoposaurus* from this quarry is yet to be determined. The skull is very small, about 8 cm long. The otic notch is very shallow, contrary to the lower Dockum form, and the pineal foramen is indistinct. It is not clear at this stage whether this specimen represents a juvenile stage of the known *Metoposaurus* or a distinct species.

Other interesting finds from this quarry include a juvenile specimen referable to *Coelophysis*;

this genus is known from the Chinle Formation and the Newark Supergroup. A pterosaur, very similar to *Eudimorphodon* from the Norian of Italy, has been found here. The Dockum specimens exhibit partial jaws with multicuspid teeth, a string of vertebrae, humerus, wing bones, femur, tarsals, and phalanges. The protorosaur is identical to *Malerisaurus* from the Late Triassic of India (Chatterjee, 1986). A small labyrinthodont skull (34 mm long), similar to *Latiscopus* (Wilson 1948) has also been recovered from this quarry. *Latiscopus* is so far restricted to the Dockum of Texas, and its affinity and stratigraphic range are unknown. The squamates, represented by several skeletons, are currently under study, and their taxonomic positions are yet to be determined.

It thus appears that a simpler and more meaningful picture emerges if we use *Parasuchus* and *Nicrosaurus* for paleontological subdivisions of the Late Triassic deposits of North America. The *Parasuchus* zone corresponds to the lower Dockum (Tecovas Member), and the *Nicrosaurus* zone to the upper Dockum (Cooper Member). The intervening Trujillo Member lacks fossils and seems to be a barren interzone.

Gregory (1957, 1972) considered the Redonda fauna of eastern New Mexico the uppermost faunal

Figure 10.4. Reconstruction of a Late Triassic Dockum landscape showing some of the Cooper tetrapods in the Post Quarry. 1, the new poposaur, *Postosuchus* (Chatterjee 1985); 2, the new ornithischian dinosaur, *Technosaurus* (Chatterjee, 1984a); 3, the protorosaur, *Malerisaurus* (Chatterjee, 1986); 4, the new squamate, unnamed; 5, the ictidosaur, *Pachygenelus*; 6, the labyrinthodont, *Latiscopus*; 7, the pterosaur, *Eudimorphodon*; 8, the parasuchid, *Nicrosaurus*; 9, the metoposaur, *Metoposaurus*; 10, the aetosaur, *Desmatosuchus*.

sequence on the basis of an advanced species of *Rutiodon* that shows the complete concealment of its upper temporal opening in dorsal view. Although the Redonda Member occurs stratigraphically in the uppermost Dockum section, the faunal composition suggests that its age is no younger than that of the Cooper Member. Because the Redonda fauna is poorly known, no precise age could be assigned to it other than the upper part of the Late Triassic. Biostratigraphically, the Dockum fauna of eastern New Mexico is equivalent to the Cooper fauna of Texas; both zones are characterized by *Nicrosaurus* (Table 10.3).

Age of the Dockum Formation

The Dockum Formation has never been dated with accuracy. The virtual absence of invertebrate remains, coupled with the presence of a long hiatus both above and below this unit, have presented considerable difficulties in determining its age. The Dockum flora is poorly known; so far nine genera of plant fossils, including remains of three ferns, five gymnosperms, and one possible cycad have been recorded, indicating a Late Triassic age (Ash 1972).

Traditionally, the vertebrates, especially the tetrapods, provide the best means of determining the age of the Dockum. It may be useful at the outset to present a list of the Dockum tetrapods known to date with their stratigraphic ranges, as a basis for discussion (Table 10.4). Although the aetosaurs, me-

Table 10.3. *Tentative correlation of the Dockum members*

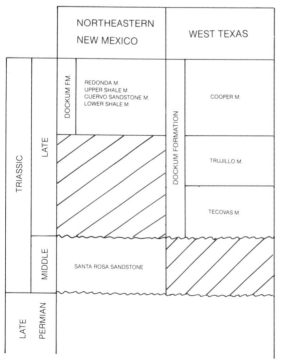

toposaurs, and poposaurs all indicate a Late Triassic age for the Dockum, the parasuchids so far offer the finest resolution.

In Germany, *Parasuchus* occurs in the Rotewand (= Blasensandstein). The lower limit may extend down to the Schilfsandstein, as a doubtful record of a parasuchid jaw is known from this horizon (see Gregory 1969). The advanced genera, such as *Mystriosuchus* and *Nicrosaurus*, are common higher up in the Stubensandstein, whereas *Rutiodon* continues to be Rhatkeuper (Westphal 1976). Thus, the Dockum is homotaxially equivalent to the German Keuper, ranging from the Schilfsandstein to the Stubensandstein.

In Germany, the Knollenmergel overlies the Stubensandstein, and it contains two distinctive tetrapods: plateosaurs and tritylodonts. So far these two animals have not been found in the Dockum. However, there is indirect evidence that the Dockum may be younger than the Stubensandstein. The fabrosaurs and ictidosaurs found in the Dockum are unknown in the German Keuper, so a direct correlation cannot be made. However, they have been found in the Early Jurassic (Olsen and Galton 1977) deposits of the Stormberg and the Kayenta along with plateosaurs and tritylodonts. An identical ictidosaur has been found recently from the Early Jurassic McCoy Brook Formation of the Newark Supergroup (C. R. Schaff pers. comm.). Thus, the presence of ictidosaurs and fabrosaurs in the Dockum indicate that its upper limit can extend at least to the Knollenmergel. This statement has to be regarded as provisional, and only the future discovery of plateosaurs and tritylodonts from the Dockum can help in arriving at a more definite conclusion. The tentative age of the Dockum is probably late Carnian to early Norian of the standard marine sequence, that is, equivalent to the Schilfsandstein and the Steinmergel of the German continental Trias.

Dunay (1972) and Dunay and Fisher (1974) studied the palynomorphs from Crosby and Potter counties of the lower Dockum (i.e., Tecovas Member), and inferred a middle to late Carnian age. This is consistent with the range of *Parasuchus*, which is restricted to the lower Dockum. It would be interesting to check independently the age of the upper Dockum (Cooper Member) on the basis of palynomorphs, because no floral data are yet available for it.

Correlation of the Dockum Formation

The parasuchid zones, defined for the Keuper and the Dockum, provide the chief biostratigraphic foundation for correlation of the Late Triassic continental deposits. In addition to North American and Germanic records, the parasuchids have been found in Morocco (Dutuit 1977), India (Chatterjee 1978), China (Young 1951; Cheng 1980), and Thailand (Buffetaut and Ingavat 1982). A tentative correlation of the parasuchid-bearing sediments is given in Table 10.5.

Chronostratigraphic problem

Fossils constitute one of the best and most widely used tools for tracing and correlating beds and thus determining the relative age of one to another. Thus, fossils are extremely important both to biostratigraphy and chronostratigraphy. Isotopic dating methods provide another independent parameter of chronostratigraphy. Unfortunately, radiometric dates are very sparse for the continental Triassic deposits. So far, the Ladinian–Anisian boundary has been accepted as the only tie point of 238 MY (Harland et al. 1982). Due to lack of volcanic material, no radiometric dates are available for the Dockum or the Keuper, and the absolute time scale for these horizons has been extrapolated.

Magnetostratigraphy has become an additional tool for regional and global correlation, particularly from the Late Jurassic (Oxfordian) to the present. This is the seafloor spreading time scale that exhibits a fairly continuous record of geomagnetic reversals (Larson and Pitman 1972). Prior to the Late Jurassic, the magnetic polarity time scale is fragmentary, and the correlation is not very accurate.

McElhinny and Burek (1971) suggested a geomagnetic polarity time scale for the Mesozoic in which they assigned the Keuper–Chinle zone as the Middle Triassic. They recognized a long period of predominantly normal polarity from Late Triassic to Late Jurassic. This is in conflict with the results of other workers. For example, Reeve and Helsley (1972) recognized at least eight separate polarity intervals for the upper part of the Dockum (including the Redonda Member) in eastern New Mexico. They pointed out that the Middle Triassic age of the Dockum suggested by McElhinny and Burek is in-

Table 10.4. *Stratigraphic ranges of the Dockum tetrapods.*

| TETRAPODS | MIDDLE TRIASSIC | LATE TRIASSIC | | | LOWER JURASSIC |
		CARNIAN	NORIAN	RHAETIAN	
CLASS AMPHIBIA					
Family Metoposauridae					
Genus *Metoposaurus*					
Family Unknown					
Genus *Latiscopus*					
CLASS REPTILIA					
Family Procolophonidae					
Genus *Colognathus*					
Family Rhynchosauridae					
Family Trilophosauridae					
Genus *Trilophosaurus*					
Family Parasuchidae					
Genus *Parasuchus*					
Genus *Angistorhinus*					
Genus *Rutiodon*					
Genus *Nicrosaurus*					
Family Stagonolepididae					
Genus *Typothorax*					
Genus *Desmatosuchus*					
Family Poposauridae					
Genus *Postosuchus*					
Family Podokesauridae					
Genus *Coelophysis*					
Genus *Spinosaurus*					
Family Fabrosauridae					
Genus *Technosaurus*					
Family Eudimorphodontidae					
Genus *Eudimorphodon*					
Family Tritheledontidae					
Genus *Pachygenelus*					
Family Protorosauridae					
Genus *Malerisaurus*					

consistent with available paleontological evidence. They could not correlate the Dockum with the Newark Supergroup on the basis of magnetic polarity. Thus, accurate stratigraphic control is needed in order to use this magnetic polarity sequence. It is a promising tool, but it needs careful sampling. Unfortunately, the paleomagnetic data from the Late Triassic deposits are too sparse to make a correlation meaningful on a regional and global scale.

Paleobiogeographic significance of the Dockum vertebrates
Laurasia–Gondwana connection

Large terrestrial tetrapods are more reliable indicators of past land connections than plants, invertebrates, or fish, because of their general inability to cross major sea barriers. It is generally believed that Pangea remained an integral geographic unit throughout the Triassic; the North Atlantic began opening about 200 MY separating North American and African plates (Heirtzler 1973; van Houten 1977). If this is so, a strong faunal correlation between North America and Africa is expected in pre-Jurassic time. So far no such correlation has been documented before. Case (1922), while searching for Permo-Triassic vertebrates in west Texas, devoted much of his energies to seeking faunas correlative with the Karroo beds of South Africa. He spent twenty years collecting fossils from this area. He claimed he never succeeded in recovering any pre-Keuper assemblages similar to those of South Africa.

From a cursory view, one is tempted to conclude that during Permo-Triassic time there were some barriers that prevented faunal migrations between North America and Africa. A careful study, however, reveals that gaps in the Permo-Triassic records between these two continents are too many to arrive at such a conclusion. Anderson and Cruickshank (1978) recognized five phases of evolutionary development in the Permo-Triassic tetrapods of the world. These phases are roughly equivalent to some of the stratigraphic chronofaunas or "empires." These are as follows:

1. Edaphosaurid empire (Late Permian)
2. Tapinocephalid empire (Middle Permian)
3. Endothiodontid empire (Upper Permian)
4. Lystrosaurid empire (Late Triassic)
5. Kannemeyeriid– Diademodontid empire (Middle Triassic)

Out of these, only the Edaphosaurid empire is restricted to the United States; the other four empires are found in South Africa, and their correlative forms are unknown from North America. There is no record of Middle Permian to Early Triassic continental sediments in North America. Contrary to this, a well-documented sequence of vertebrate faunal zones ranging in age from Middle Permian to Middle Triassic is preserved in the Karroo Basin of South Africa. Thus, the poor faunal correlation between North America and Africa during Permo-Triassic time is due to the lack of contemporary biotas between these two areas. It is only during the

Table 10.5. *Provenance of the Parasuchidae*

EPOCH	ALPINE FACIES	BIOZONE	EUROPE		NORTH AMERICA						ASIA			AFRICA
			GERMANY		ATLANTIC COAST	WEST TEXAS	EAST NEW MEXICO	ARIZONA	WYOMING	INDIA	CHINA	THAILAND	MOROCCO
LIASSIC	HETTANGIAN										LUFENG SERIES		
	RHAETIAN												
LATE TRIASSIC	NORIAN	NICROSAURUS ZONE	STEINMERGEL	KNOLLENMERGEL	NEWARK SUPERGROUP								
				STUBENSANDSTEIN		(COOPER M.) DOCKUM FM.	DOCKUM FM.	CHINLE FM.			UPPER KELAMAYI FM.	HUAI HIN LAT FM.	
	CARNIAN	PARASUCHUS ZONE		ROTEWAND		(TECOVAS M.)			POPO AGIE M. (CHUGWATER FM.)	MALERI FM. TIKI FM.			ARGANA FM
				SCHILFSANDSTEIN									

Late Triassic that we first encounter correlative beds between these two continents allowing us to compare their degree of faunal similarities.

Dockum vertebrates, such as *Metoposaurus, Parasuchus*, and *Angistorhinus*, are congeneric with those of the Argana fauna of Morocco (Dutuit 1977). Similarly, two Dockum members, *Coelophysis* and *Technosaurus*, are matched by similar South African genera, such as *Syntarsus* and *Lesothosaurus, Pachygenelus* is common to both regions. The extended distribution of the Dockum fauna provides the earliest paleontological evidence for a land connection between North America and Africa during the Late Triassic.

The Dockum fauna also shows a striking resemblance to certain tetrapods from the Maleri Formation of India. Four tetrapod genera (*Parasuchus, Metoposaurus, Typothorax*, and *Malerisaurus*) are identical in both areas, and rhynchosaurs and podokesaurs show close similarities. Possibly, the route of faunal migration between India and North America was via Morocco (Chatterjee, 1984b).

Surprisingly, the degree of faunal similarities between the Dockum and the contemporary Ischigualasto and Santa Maria formations of South America is very poor. So far aetosaurs and rhynchosaurs are the two common elements between these two continents during the Late Trias. It is generally assumed that North and South America were in touch with each other until the Early Tertiary (Romer 1973). However, the early tectonic history of Central America and the whole Carribean region is very complicated. The nature of the connection, if any, between North and South America during the Triassic is not clear at this stage. Probably, there were some barriers or filters that prevented tetrapod migrations back and forth between these two continents during the Triassic.

Although continental drift theory has, triumphed in recent years, many details are uncertain. There is a long-standing debate concerning whether there were two unconnected supercontinents, Laurasia and Gondwana, or one supercontinent Pangea during the beginning of the Mesozoic. The close similarities of the Dockum tetrapods with those of Africa and India clearly attest to a Laurasia–Gondwana connection during the Late Triassic and support the concept of Pangea.

Acknowledgments

I thank Spencer G. Lucas, Michael A. Morales, Charles R. Schaff, Robert A. Long, A. W. Crompton, and Bryan J. Small for providing me valuable information on the newly discovered Triassic–Jurassic tetrapods; Joseph T. Gregory, James E. Barrick, David D. Proctor, Kevin Padian, Michael Parrish, and Phillip A. Murry for helpful discussions and critical appraisals of the manuscript; Michael W. Nickell for illustrations; and Shirley Burgeson for typing the manuscript. This research was supported by a series of grants from the National Geographic Society and Texas Tech University.

References

Adams, J. E. 1929. Triassic of West Texas. *Bull. Am. Assoc. Petrol. Geol.* 12(8):1045–55.

Anderson, J. M., and A. R. I. Cruickshank. 1978. The biostratigraphy of the Permian and the Triassic. Part 5. A Review of the classification and distribution of Permo-Triassic tetrapods. *Paleontol. Afr.* 21:15–44.

Ash, S. R. 1972. Upper Triassic Dockum flora of eastern New Mexico and Texas. *In* Kelley, V. C., and F. D. Trauger (eds.), *Guidebook of East-Central New Mexico, 23rd Field Conference* (Albuquerque: New Mexico Geological Society), pp. 124–8.

Baker, C. L. 1915. Geology and underground waters of the northern Llano Estacado. *Univ. Texas Bull.* No. 57, pp. 1–225.

Buffetaut, E., and R. Ingavat. 1982. Phytosaur remains (Reptilia, Thecodontia) from the Upper Triassic of north-eastern Thailand. *Geobios* 15:7–17.

Case, E. C. 1922. New reptiles and stegocephalians from the Upper Triassic of western Texas. *Carnegie Inst. Wash. Publ.* No. 321, pp. 1–84.

1928. Indications of a cotylosaur and of a new form of fish from the Triassic beds of Texas, with remarks on the Shinarump conglomerate. *Univ. Mich. Contr. Mus. Paleontol.* 3(1):1–14.

Chatterjee, S. 1978. A primitive parasuchid (phytosaur) reptile from the Upper Triassic Maleri Formation of India. *Palaeontology* 21:83–127.

1983. An ictidosaur fossil from North America. *Science* 220:1151–3.

1984a. A new ornithischian dinosaur from the Triassic of North America. *Naturwiss.* 71:630–1.

1984b. The drift of India: a conflict in plate tectonics. *Mem. Soc. Géol. France.*

1985. *Postosuchus*, a new thecodontian reptile from the Triassic of Texas and the origin of tyrannosaurs. *Phil. Trans. Roy. Soc. London B* 309:395–460.

1986. *Malerisaurus langstoni*, a new diapsid reptile from the Triassic of Texas. *J. Vert. Paleontol.* 6(4).

Cheng, Z. 1980. Permo-Triassic continental deposits and vertebrate faunas of China. *In* Creswell, M. M. and P. Vella (eds.), *Gondwana Five* (Rotterdam: A. A. Balkema), pp. 65–70.

Colbert, E. H., and J. T. Gregory. 1957. Correlation of continental Triassic sediments by vertebrate fossils. *Bull. Geol. Soc. Am.* 68:1456–67.

Cope, E. D. 1893. A preliminary report on the vertebrate paleontology of the Llano Estacado. *Texas Geol. Surv. Ann. Rept. No. 4*, pp. 11–87.

Cummins, W. F. 1980. The Permian of Texas and its overlying beds. *Texas Geol. Surv. 1st Ann. Rept.*, pp. 183–97.

Drake, N. F. 1892. Stratigraphy of the Triassic of West

Texas. *Geol. Surv. Texas 3rd Ann. Rept.* pp. 227–47.

Dunay, R. E. 1972. The palynology of the Triassic Dockum Group of Texas, and its application to stratigraphic problems of the Dockum Group, Ph.D. dissertation. Pennsylvania State University, pp. 1–382.

Dunay, R. E., and M. J. Fisher. 1974. Late Triassic palynofloras of North America and their European correlations. *Rev. Palaeobot. Palynol.* 17:179–86.

Dutuit, J. M. 1977. Description du crane de *Angistorhinus talainti* n. sp., un nouveau phytosaure de Trias Atlantique Marocain. *Bull. Mus. Nat. Hist. Sci. Terre (Paris)* 66:277–337.

Eaton, T. H. 1965. A new Wyoming phytosaur. *Univ. Kansas Paleontol. Contr.* 2:1–6.

Elder, R. L. 1978. Paleontology and paleoecology of the Dockum Group, Upper Triassic, Howard County, Texas, M.S. thesis, University of Texas at Austin, pp. 1–205.

Gould, C. N. 1907. The geology and water resources of the western portion of the Panhandle of Texas. *U. S. Geol. Surv. Water Supply Pap.* 191:1–70.

Green, F. E. 1954. The Triassic deposits of northwestern Texas, Ph.D. dissertation, Texas Tech University, pp. 1–196.

Gregory, J. T. 1957. Significance of fossil vertebrates for correlation of Late Triassic continental deposits of North America. *Int. Geol. Cong. 20th Sess. Mexico. Sec. II*, pp. 7–25.

1962. The genera of phytosaurs. *Am. J. Sci.* 260(9):652–90.

1969. Evolution and interkontinentale Bezeihunger der Phytosauria (Reptilia). *Palaeontol. Z.* 43:37–51.

1972. Vertebrate faunas of the Dockum Group, Triassic, eastern New Mexico and West Texas. *In* Kelley, V. C., and F. D. Trauger (eds.), *Guidebook of East-Central New Mexico, 23rd Field Conference* (Albuquerque: New Mexico Geological Society), pp. 120–30.

Hallam, A. 1981. The end-Triassic bivalve extinction event. *Palaeogeog. Palaeoclimatol. Palaeoecol.* 35:1–44.

Harland, W. B., A. V. Cox, P. G. Llewllyn, C. A. G. Pickton, A. G. Smith, and R. Walters. 1982. *A Geologic Time Scale* (Cambridge: Cambridge University Press).

Hoots, H. W. 1925. Geology of a part of western Texas and southeastern New Mexico, with special reference to salt and potash. *Bull. U. S. Geol. Surv.* 780:33–126.

Jones, T. S. 1953. *Stratigraphy of the Permian Basin of West Texas.* (Midland, Texas: West Texas Geological Society.), pp. 1–63.

Kelley, V. C. 1972. Triassic rocks of the Santa Rosa Country. *In* Kelley, V. C., and F. D. Trauger (eds.), *Guidebook of East-Central New Mexico, 23rd Field Conference* (Albuquerque: New Mexico Geological Society), pp. 84–90.

Kiatta, H. W. 1960. A provenance study of the Triassic deposits of northwestern Texas. M.A. thesis, Texas Tech University, pp. 1–63.

Larson, R. L., and W. C. Pitman III. 1972. Worldwide correlation of Mesozoic magnetic anomalies, and its implications. *Bull. Geol. Soc. Am.* 83(12):3645–62.

McElhinny, M. W., and Burek, P. J. 1971. Mesozoic paleomagnatic stratigraphy. *Nature (London)* 232:98–102.

McGowen, J. H., G. E. Granata, and S. J. Seni. 1979. Depositional framework of the lower Dockum Group (Triassic), Texas Panhandle. *Texas Bur. Econ. Geol. Rept. Inv.* 97:1–60.

McGowen, J. H., G. E. Granata, and S. J. Seni. 1983. Depositional setting of the Triassic Dockum Group, Texas Panhandle and eastern New Mexico. *In* Reynolds, M. W., and E. D. Dolly (eds.), *Mesozoic Biostratigraphy of the West-Central United States* (Denver: Society of Economic Paleontologists and Mineralogists), pp. 13–38.

Murry, P. 1982. Biostratigraphy and paleoecology of the Dockum Group (Triassic) of Texas. Ph.D. dissertation. Southern Methodist University, pp. 1–459.

Olsen, P. E., and P. M. Galton. 1977. Triassic–Jurassic tetrapod extinctions: are they real? *Science* 197:983–6.

Olsen, P. E., A. R. McCune, and K. S. Thomson. 1982. Correlation of the early Mesozoic Newark Supergroup by vertebrates principally fishes. *Am. J. Sci.* 282: 1–44.

Reeside, J. B., P. C. Applin, E. H. Colbert, J. T. Gregory, H. D. Hadley, B. Kummel, P. J. Lewis, J. D. Love, M. Maldonadokoerdel, E. D. McKee, D. B. McLaughlin, S. W. Muller, J. A. Reinemund, J. Rodgers, J. Sanders, N. J. Silberling, and K. Waage. 1957. Correlation of the Triassic formations of North America exclusive of Canada. *Bull. Geol. Soc. Am.* 68:1451–514.

Reeve, S. C., and C. E. Helsley. 1972. Magnetic reversal sequence in the upper portion of the Chinle Formation, Montoya, New Mexico. *Bull. Geol. Soc. Am.* 83:3795–812.

Romer, A. S. 1973. Vertebrates and continental connections: an introduction. *In* Tarling, D. H., and S. K. Runcorn (eds.), *Implications of Continental Drift to Earth Sciences*, Vol. 1 (London: Academic Press), pp. 345–9.

Shelton, S. Y. 1984. Parasuchid reptiles from the Triassic Dockum Group of West Texas. M.A. Thesis, Texas Tech University, pp. 1–101.

Tozer, E. T. 1967. A standard for Triassic time. *Bull. Can. Geol. Surv.* 156:1–103.

Van Houten, F. B. 1977. Triassic–Liassic deposits of Morocco and eastern North America: comparison. *Bull. Am. Assoc. Petrol. Geol.* 61:79–99.

Welles, S. P. 1967. Arizona's giant amphibian. *Pacific Discovery* 20(4):10–15.

Westphal, V. E. 1976. Phytosauria. *In* Kuhn, O. (ed.), *Handbuch der Palaoherpetrologie* (Stuttgart: Gustav Fischer), pp. 99–120.

Wilson, J. A. 1948. A small amphibian from the Triassic Howard County, Texas. *J. Paleont.* 22(3):359–61.

Young, C. C. 1951. The Lufeng saurischian fauna of China. *Palaeontol. Sin. C* 13:19–96.

11 A new vertebrate fauna from the Dockum Formation (Late Triassic) of eastern New Mexico

J. MICHAEL PARRISH AND
KENNETH CARPENTER

Introduction

Fossil vertebrates have been known from the Dockum Formation of New Mexico for many years (e.g., Wright and Fitch 1971; Gregory 1972 and references therein), but little recent work has been devoted to prospecting for vertebrates in this region. A preliminary reconnaissance along Revuelto Creek, Quay County, by the University of Colorado Museum (UCM) in 1981 resulted in the discovery of a major concentration of vertebrate material where the Yale Peabody Museum (YPM) collected a large phytosaur skull (Gregory 1972). The new material was collected from one of the numerous irregular channels capping the escarpment east of Revuelto Creek and is here referred to as the Revuelto Creek Local Fauna (UCM Loc. 82021) (Fig. 11.1).

Figure 11.1. Locality for Revuelto Creek Local Fauna, Quay County, New Mexico. Dashed lines separate lower reddish orange unfossiliferous mudstone from upper purple bone-bearing unit. Quarry in foreground.

The fossils occur in a purple weathering mudstone containing cross-stratified, reworked lenses of caliche as BB- to pea-sized carbonate nodules. One such lens was found to truncate the rear of the main quarry, while another lens truncated the snout of a nearby phytosaur skull. The unweathered mudstone of the main quarry is a gray and pink mottle, with red streaks of iron stain. Numerous gray mudstone intraclasts are also present.

The vertebrate concentration consisted mostly of isolated bones of several taxa of large vertebrates. In addition to the macrofauna, a diverse and abundant microfauna was found by washing the matrix surrounding the prepared elements. This material will be described in conjunction with J. R. Bolt. We have already tentatively identified numerous paleoniscid scales, an amphibian atlas, numerous reptilian jaw fragments containing teeth, lacertiloid vertebrae, and an ictidosaur(?) jaw fragment.

Systematic paleontology
Class Osteichthyes
Subclass Actinopterygii
Order Paleonisciformes
Fam., gen. et sp. indet.
Material. UCM 52079, partial body composed of articulated scales; numerous uncataloged scales.
Discussion. A partial skeleton lacking the skull and tail fins was discovered during the course of quarrying. The body form is preserved as slightly disarticulated scales. Without the skull and fins, it is not possible to assign this specimen to any of the known genera of paleonisciform fish. Numerous isolated scales were also found scattered throughout the quarry.

Class Reptilia
Subclass archosauria
Order Thecodontia
Suborder Parasuchia
Family Phytosauridae
Rutiodon gregorii (Fig. 11.2)
Referred specimens. YPM 3293, complete skull; UCM 48441, posterior skull roof.
Discussion. Phytosaurs are represented from the Revuelto Creek locality by a complete skull (YPM 3293) collected by J. T. Gregory in 1947, and by the posterior part of a skull (UCM 48441) that was found about fifty feet southwest of the main quarry. The second skull lacks the face anterior to the orbits, which was lost during downcutting of a channel into older sediments. Later this channel was filled by cross-stratified carbonate nodules in a mudstone matrix. The ventral part of the skull is poorly preserved; the braincase is almost completely eroded away. Both squamosals are present, and they are of the broad, hooked type characteristic of the strati-

graphically older taxa of the genus *Rutiodon* (Camp 1930; R. A. Long pers. comm.). This squamosal morphology and the shape of the temporal fenestrae allow referral of the specimen to *Rutiodon gregorii*. It should be noted that members of this taxon comprise some of the largest known phytosaurs. The largest currently known is apparently the *R. gregorii* skull (UCMP A272/27200) at the University of California Museum of Paleontology.

Rutiodon sp.
Material. One caudal vertebra, many teeth, juvenile to adult.
Discussion. the caudal vertebra is large and well ossified, indicating that it is from an adult. The teeth, on the other hand, range in size from 2 to 33 mm. All are apparently shed teeth because roots are not present.

Suborder Aetosauria
Family Stagonolepididae
Desmatosuchus sp. (Fig. 11.3)
Material. UCM 47225. Left lateral pectoral plate, two uncataloged paramedial plates.
Discussion. Comparisons with the University of Michigan specimen (Case 1922) indicate that the lateral plate is probably a third or fourth cervical. It is unusual among aetosaur plates in that the spine is sharply curved posterolaterally, whereas those of the type of *D. haploceros* are only slightly recurved (Case 1922). However, it otherwise conforms closely to the morphology of those of the type specimen.

Two paramedial plates are present. One is almost complete, although broken somewhat anteriorly and posteriorly and eroded dorsally. The other consists of the central part of a plate, but the sculpturing, flat profile, and anteroposterior breadth indicate that it is also a *Desmatosuchus* paramedial. Similarities in plate length among the two paramedials indicate that they could belong to a single specimen.

Suborder Rauisuchia
Family Poposauridae
Poposaurus sp. (Fig. 11.4)
Material. UCM 52080. Small right ilium, probably immature; left pubic foot.
Discussion. On the basis of compatibility in size and proximity of preservation, these two elements are referred to the same individual. Both are somewhat different than in other *Poposaurus* specimens. The Poposauridae are united largely on the basis of their distinctive pelves, with the following characters:
1. A long, low iliac blade with prominent anterior and posterior projections
2. A prominent flange located on the lateral side of the iliac blade, just above the acetabulum (Colbert 1961; Galton 1977)

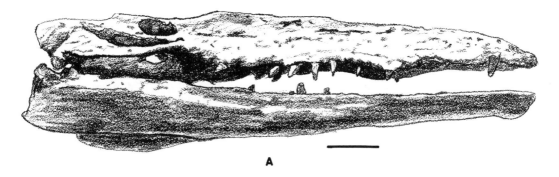

A

Figure 11.2. *Rutiodon gregorii*, scale = 10 cm. **A**, Skull collected by Yale Peabody Museum (YPM 3293). Lateral view. Scale = 10 cm. **B**, Posterior skull roof (UCM 48441). Scale = 1 cm. Abbreviations: Stf = supratemporal fenestra; ltf = lateral temporal fenestra; o = orbit.

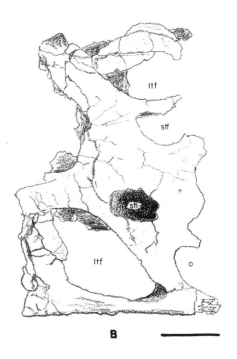

B

Figure 11.3. Lateral cervical plate of *Desmatosuchus* (UCM 47225). Lateral view. Scale = 1 cm.

Figure 11.4. *Poposaurus* sp. Right ilium (UCM 52080). Lateral view. Scale = 1 cm.

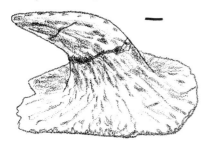

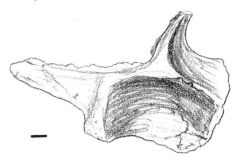

3. A prominent, ventrally projecting "foot" on the distal end of the pubis

In the genus *Poposaurus*, the supraacetabular flange originates above the anteroposterior midpoint of the acetabulum, veers anteriorly to run across the top of the iliac blade, and terminates posterior to the anterior tip of the blade. The ventral margin of the acetabulum is excavated slightly, forming a perforate acetabulum in some specimens (e.g., FMNH UR–357, TMM 31025–257). The ischiadic peduncle of the ilium forms a distinctive, ventrally projecting knob. Variations in these features are seen in the other poposaurids (R. A. Long pers. comm.; Chatterjee 1985).

The Revuelto Creek ilium has several characteristics in common with those of *Poposaurus*. The flange of the iliac blade curves anterodorsally, the ventral margin of the acetabulum is excavated slightly, and the ischiadic peduncle of the ilium forms a knob. The most distinctive characteristic of the new specimen is the anteroposteriorly foreshortened blade relative to those of other taxa. Although the blade is broken dorsally, its anterior and posterior extents appear to be present. Furthermore, the preserved portion of the dorsal margin of the iliac blade indicates that the ilium was much taller in relation to iliac length than in the other poposaurids. The fact that this ilium is smaller than those of other known poposaurids suggests that the specimen may be juvenile, and the differences observed in blade shape may be allometric.

The pubic fragment consists of the distal "foot" and the most distal part of the pubic rod. The proximal part, comprising the distal extent of the pubic rod, is concave medially, whereas the foot itself is directed ventrally and slightly laterally. Poposaurid pubic feet are rarely well preserved. The foot of the left pubis of the holotype of *Poposaurus gracilis* is directed ventrally without posterior prolongation, whereas that of an undescribed poposaurid (UMMP 7244) projects posteroventrally. The foot of the new specimen projects ventrally, again suggesting affinities with the genus *Poposaurus*.

The lack of other small poposaurid ilia, coupled with the paucity of described poposaurid specimens, makes determination of specific affinities of this specimen impossible. Long (pers. comm.) and Chatterjee (1985) have worked on other poposaurid material, and their studies may shed light on the affinities of the Revuelto Creek specimen.

Order Saurischia
Suborder Theropoda
Family Procompsognathidae
    gen. et sp. indet. (Figs. 11.5b,d, 11.6a–d, 11.7b,c,e,h,j)
Material. UCM 47221. Partial skeleton, consisting of right pubis, scapula, left tibia, four dorsal centra, one dorsal neural arch, four ribs, one metatarsal, one chevron, and one tooth.

Discussion. The single tooth is complete including its root. This indicates that the tooth was not shed, but slipped from its alveolus when decay loosened the ligaments. The tooth has eighteen serrations per 5 mm on the anterior edge and twenty serrations per 5 mm on the posterior edge.

In Figures 11.5 and 11.7, the scapula and tibia of the Revuelto Creek specimen are illustrated with the corresponding elements of *Halticosaurus liliensterni* (Huene 1934), the primitive theropod most closely matching UCM 47221 in size. Both specimens display the primitive theropod pattern for the morphology of the figured elements, as described below.

The scapula has a straight posterior border, with the anterior border emarginated between proximal and distal extremities. The bone curves sharply posteriorly at its ventral extent for the glenoid expansion. It is broken anteroventrally, but is otherwise complete. The blade is straight in anterior view, but markedly recurved medially toward the glenoid. No trace of the coracoid is present, suggesting that it was not fused to the scapula.

Figure 11.5. Procompsognathid limb girdle elements. Scale = 1 cm. **A**, *Halticosaurus liliensteri*, holotype. Right scapulocoracoid, lateral view. **B**, Procompsognathid, indet. (UCM 47221). Right scapula, lateral view. **C**, *Halticosaurus liliensterni*, holotype. Right pubis, lateral view. **D**, Procompsognathid, indet. (UCM 47221). Right pubis, lateral view.

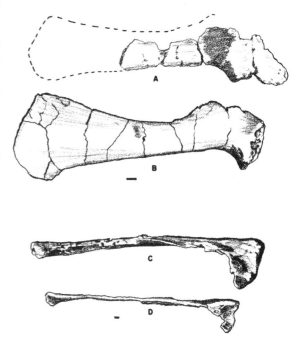

The right pubis is complete except for the medial part of the pubic plate. It is straight in lateral view and expanded only slightly at its distal end; no pubic foot is present. Instead, a knoblike swelling is observed, with a vertically aligned groove on its distal face. The proximal surface consists of two flat surfaces, comprising the iliac and ischiadic peduncles, that meet at a 120° angle. The obturator foramen is located just distal to the junction of the two peduncles. The pubic rod is thickened laterally, but narrows markedly medially to form a plate that is eroded at its medial extent.

A complete left tibia is preserved. This bone is of the primitive theropod pattern. The proximal end is broadened anteroposteriorly, with the anterior side projecting 2 cm higher than the posterior side. The central part of the proximal face is occupied by a bowllike depression into which the cartilagenous meniscus of the knee joint presumably fit. Laterally, there is a groove for articulation with the fibula at both the proximal and distal ends of the bone. The anterior face of the proximal part of the bone is occupied by a prominent cnemial crest. The center of the proximal part of the lateral face is occupied by a proximodistally oriented fibular trochanter about 6 cm in length. A posterior groove is seen at the proximal end of the bone, presumably for passage of the femoral head of the gastrocnemius onto the crus. The shaft of the tibia is somewhat recurved medially. Its distal end is occupied by a

steplike articular facet for the astragalus. The anterior part of the facet consists of a nearly horizontal shelf, angled slightly proximolaterally. The posterior part of the facet is a vertically oriented shelf projecting 4 cm below the level of the horizontal part.

Four ribs are present, one cervical and three thoracic. The cervical rib is almost complete proximally, although the distal part is absent. One complete thoracic rib is present, apparently undistorted. This rib is rather straight and elongate distal to a ventral curvature near its head, suggesting a deep-chested animal, and its size matches well with those of the other elements referred to this taxon. The tubercular and capitular facets are subequal in size, which suggests that the rib is from the anterior part of the thoracic series. Another thoracic rib is present, but its proximal and distal ends are absent, and thus more specific identification is impossible. The fourth rib is apparently a posterior thoracic. The capitular facet is smaller than the tubercular facet, and the strongly curving body of the rib is narrower

Figure 11.7. Procompsognathidae, tibiae. **A,D,F,G,I**, *Halticosaurus liliensterni*, holotype right tibiae (reversed for comparison). **B,C,E,H,J**, Procompsognathid indet. (UCM 47221), left tibia. Scale = 1 cm. **A,B**, Medial view. **C,D**, Distal view. **E,F**, Proximal view. **G,H**, Lateral view. **I,J**, Anterior view.

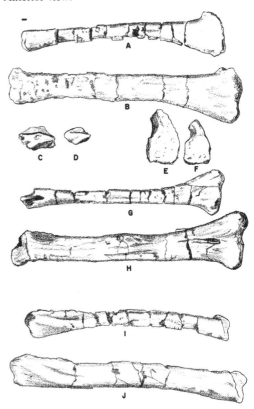

Figure 11.6. Indeterminate procompsognathid, UCM 47221. Lateral view of four dorsal centra, **(A)** with neural arch and spine. Scale = 1 cm.

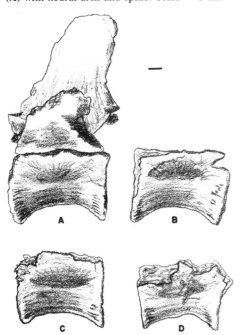

than in the other preserved ribs. It is also broken distally.

A single, complete gastralium is present. Although it lacks defined articular surfaces, it has an anteroventral, mediolaterally aligned groove.

There are four dorsal centra. Although disarticulated, their similar size and morphology indicate that all are referable to one individual. Each centrum is pinched medially at midbody, and twin depressions are present on the ventral sides of the body adjacent to the floor of the neural canal. One disarticulated neural arch and spine is present and seems to fit one of the centra, although it is somewhat crushed laterally. The location of tubercular facets near the base of the transverse processes indicates that the arch is an anterior dorsal. The transverse processes project dorsolaterally, although their

position may be due to crushing. Well-developed hyposphenes and hypantra are seen on the one preserved neural arch.

One hemal chevron is present, very well preserved. Its large size suggests an anterior location in the tail.

This specimen is referred to the Procompsognathidae on the basis of its close similarity to the type of *Halticosaurus liliensterni* (Huene 1934). Pubic, tibial, and scapular morphologies are virtually identical between the two specimens, although the New Mexican specimen is somewhat larger and more robust (Table 11.1). There is not sufficient material preserved of the New Mexican specimen to permit generic or specific identification. Although it conforms closely to the morphology of *H. liliensterni*, the morphology observed in both represents a prim-

Table 11.1. *Measurements (mm) of specimens discussed in text*

Procompsognathids[a]	UCM 47221	Type of Halticosaurus liliensterni	Coelophysis longicollis
Pubis			
Proximodistal L	497	415	226
Proximal H	120	73.7	56
Distal H	41.8	31.0	12
Tibia			
.L	469	415	
Max. Prox. W	112.2	91.0	
Prox. W perp. to max.	69.0	53.3	
Max. distal W	78.0	51.0	
Distal W perp. to max.	54.3	41.2	
Min. shaft diameter	43.0	27.6	
W perp. to min.	55.2	35.0	
Scapula			**Petrified Forest Scapula**
Max. L	314	240 (est.)	
Max. W	110.6	58.0	78.0 (min.)
Min. W	37.7	31.3	34.0
Transverse W	16.0	12.0	18.6
Centra			
A L	68.4		
Max. ant. H	51.2		
Max. ant. W	64.7		
B L	74.2		
Max. ant. H	52.1		
Max. ant. W	58.1		
C L	66.0		
Max. ant. H	50.0		
Max. ant. W	58.5		

[a]H, height; L, length; W, width; perp., perpendicular; Max., maximum; Min., minimum; ant., anterior; post., posterior; prox., proximal.

itive theropod pattern. In the absence of more complete material, it seems imprudent to attempt a generic or specific classification at this time, although pubic, tibial, and scapular morphology suggests that it is closer to *Halticosaurus* than to the other well-known primitive theropods *Coelophysis* (Chapter 5) and *Syntarsus* (Raath 1977). These taxa are assigned to the Procompsognathidae following Ostrom (1981). However, as noted by Padian (Chapter 5), the characters uniting the Triassic theropods are apparently primitive for all Theropoda. Many of the types are poorly preserved, poorly diagnosed, or both. Thus, the familial assignment should be regarded as provisional, pending further study of the Triassic theropods.

Suborder Theropoda
Fam, gen., et sp. indet.
Material. Two uncataloged caudal vertebrae smaller than those of the above specimen.
Discussion. The two caudal vertebrae are much smaller in diameter and more laterally compressed than the centra of UCM 47221. Their short lengths relative to their diameters, coupled with the presence of prominent neural spines, indicates that they were anterior caudals. The similarities in size, morphology, and preservation suggests that they belong to the same individual. Similarities in diameter, length, and neural spine size suggest that the two vertebrae were located near one another, and were possibly adjacent.

Class Reptilia indet. (Fig. 11.8)

Material. Edentulous right premaxilla, UCM 52081
Discussion. One of the more enigmatic specimens from the Revuelto Creek locality is a complete left premaxilla of an edentulous reptile. The specimen is short, high, and gently rounded anteriorly. It is strongly excavated posteriorly by the anterior margin of the naris. A groove located along the dorsal arm of the posterior end of the bone presumably represents the site of sutural contact with the maxilla. A depression is seen on the posteroventral part of the external side of the bone, terminating as a shallow groove at its posterior margin. Medially a strong, horizontally aligned ridge is continuous with the ventral margin of the naris. Below this ridge the bone is sharply excavated, forming a depression where a horny beak may have inserted; numerous nutrient foramina also occur. The medial edge of the bone is hollowed dorsal to the ridge, especially just anterior to the naris, where a dishlike depression is present.

This bone cannot be readily matched with the premaxillae of any of the known taxa lacking premaxillary teeth (e.g., Chelonia, Aetosauria, Trilo-

phosauridae, Rhynchosauria, Dicynodontia, or *Lotosaurus*). Thus, it must be referred to Reptilia *incertae sedis* for the present.

Comparison with other Dockum and Chinle faunas

The fauna represented from the Revuelto Creek locality is interesting in that it contains a few taxa unknown elsewhere in the Late Triassic of the southwest, in addition to typical Dockum and Chinle taxa. Comparisons with other notable Dockum and Chinle faunas are made in Table 11.2.

The presence of *Rutiodon gregorii* indicates affinities with the faunas of the lower part of the vertebrate-bearing Chinle in Arizona and the upper part of the Dockum Formation as evidenced by the fauna from Crosby County, Texas. Reexamination of the Chinle and Dockum phytosaurs (R. A. Long pers. comm.) has suggested that the phytosaurs with posteriorly positioned nares from the Chinle and Dockum Formations comprise two groups without biostratigraphic overlap. One group, consisting of forms from the upper Dockum Formation of Texas and the lower Chinle Formation of Arizona, consists of forms with broad squamosals. The second group, known from the upper part of the Chinle Formation, consists of taxa with rodlike squamosals. Pending a formal presentation of this classification, these two groups are, respectively, assigned to the informal taxonomic groupings "type A and B" within the genus *Rutiodon*, with *Rutiodon gregorii* referable to the stratigraphically lower "type A."

The presence of *Rutiodon* and the absence of any of the North American phytosaur taxa with anteriorly located nares (e.g., *Paleorhinus, Angistorhinus*) presumably places the age of the Revuelto Creek locality as younger than that of the lower Dockum Formation of Texas and the Popo Agie Formation of Wyoming, both of which are characterized by phytosaurs of the latter type. However, the discovery of a *Paleorhinus* snout in the lower Chinle Formation of the Downs quarry near St. John's, Arizona (MNA PL 2698) (W. Downs pers. comm.) indicates that the range of these phytosaurs

Figure 11.8. Reptilia indet. (UCM 52081). Edentulous right premaxilla. Scale = 1 cm. **A**, Lateral view. **B**, Medial view.

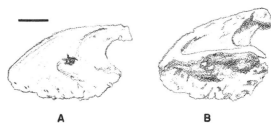

A   **B**

overlapped with that of "type A" *Rutiodon*. Phytosaur taxonomy worldwide is currently in a confused state, and a wholesale revision of the Suborder Parasuchia is badly needed.

The aetosaur *Desmatosuchus* is also a characteristic element of each of these faunas. While *Desmatosuchus* is also known from the lower part of the Dockum in Howard County, Texas, it is unknown in the upper part of the Chinle Formation.

The poposaurid *Poposaurus* is known from the lower Dockum Formation of Howard County, Texas, from the Popo Agie Formation of Wyoming, and from the lower Chinle Formation in Arizona. The Poposauridae have an extensive stratigraphic range, as they are known from the ?Middle Triassic continental facies of the Moenkopi Formation of Arizona, and also from facies of Late Triassic age, including the upper part of the Dockum Formation of Texas, the Stubensandstein of Germany, the Lower Keuper of Germany and the "Lower Keuper" Sand-

Table 11.2 *Late Triassic faunas of the southwestern United States*

Arizona	New Mexico	Texas	Wyoming
Chinle F., NW AZ **Upper Fauna**	**Chinle Fm. Ghost Ranch**		
Aetosauria	Aetosauria		
Typothorax	*Typothorax*		
Phytosauria	Phytosauria		
Rutiodon B	*Rutiodon B*		
Saurischia	Saurischia		
Coelophysis	*Coelophysis*		
Thecodontia indet.	Rauisuchia indet.		
Lower fauna	**Dockum Fm., Rev. Cr. Loc.**	**Dockum Fm. Crosby Co.**	
Aetosauria	Aetosauria	Aetosauria	
Desmatosuchus	*Desmatosuchus*	*Desmatosuchus*	
Calyptosuchus	Rauisuchia	*Calyptosuchus*	
Rauisuchia	*Poposaurus*	Phytosauria	
Poposaurus	Phytosauria	*Rutiodon A*	
undescr. rauisuchian	*Rutiodon A*	Poposauridae indet.	
Phytosauria	Saurischia		
Rutiodon A	large		
Paleorhinus	procompsognathid		
Dicynodontia			
Placerias			
Saurischia			
large			
procompsognathid			
Trilophosauridae			
		Dockum Fm., Howard, **Post counties**	**Popo Agie Fm.**
		Aetosauria	
		Desmatosuchus	
		"*Typothorax meadei*"	
		Phytosauria	Phytosauria
		Paleorhinus	*Paleorhinus*
		Angistorhinus	Rauisuchia
			Dolichobrachium
		Poposauridae	Poposauridae
		Postosuchus	*Poposaurus*
		Trilophosauridae	Rauisuchidae
		Trilophosaurus	*Heptasuchus*
		Rhynchosauria indet.	Dicynodontia
			Eubrachiosaurus

stone of Warwickshire, and possibly the upper Chinle of New Mexico (Meyer 1861; Walker 1969; Galton 1977; R. A. Long pers. comm.).

The procompsognathid is not of great utility for regional biostratigraphy because members of this family are known throughout the vertebrate-bearing Chinle and well into the Liassic formations worldwide. Specimens that have been referred to *Halticosaurus*, the genus displaying the greatest similarities to the new specimen, are known from the Knollenmergel of Thuringen and Halberstadt (Huene 1921, 1934), from the Stubensandstein of Pfaffenhofen (Huene 1932), and from the "Rhaetian" of Normandy (Larsonneur and de Lapparent 1966). Thus, the known range for the genus in Europe is Norian through the uppermost Triassic. However, the generic synonymy of the various *Halticosaurus* specimens has not been convincingly demonstrated.

The best known theropod from the southwest is *Coelophysis bauri*, described by Cope (1889) on the basis of fragmentary material and known from numerous referred specimens from New Mexico (Colbert 1964). The specimens described by Colbert are considerably smaller and more gracile than the specimen considered here. *Coelophysis "longicollis"* was redescribed and illustrated for the first time by Huene (1906, 1915) on the basis of good material from the upper Chinle Formation of New Mexico. For a discussion of the type material of *Coelophysis* see Chapter 5. Although some similarities are seen between the Revuelto Creek procompsognathid and *C. "longicollis"* [as illustrated by Huene (1915)] in the pubis and dorsal vertebrae, significant differences are observed as well. Even discounting breakage, the pubic rods of *C. "longicollis"* (Huene 1915, Fig. 39) and *C. bauri* are much more strongly curved ventrally than is that of the UCM specimen. Also, the tuberosity on the distal pubis is much more pronounced in the latter specimen than in the former. Finally, although not necessarily of taxonomic significance, the pubis of the type of *C. "longicollis"* is roughly half the size of that of UCM 42771 (see Table 10.1). Thus the specimen described here is apparently the first record of a large theropod from the Carnian of North America. An uncataloged theropod scapula from the lower part of the Petrified Forest Member of the Chinle Formation in the Petrified Forest National Park compares favorably with the specimen from New Mexico in size (see Table 10.1), although poor preservation of the Arizona specimen prevents giving it a generic assignment. It is interesting that the two records of large theropods from the Upper Triassic of the southwest United States occur in biostratigraphically correlative localities.

The presence of the stratigraphically restricted phytosaur *Rutiodon* "type A" and the aetosaur *Des-* *matosuchus* places the age of the Revuelto Creek locality as older than that of the upper Chinle Formation, from which only the phytosaur *Rutiodon* "type B" and the aetosaur *Typothorax* are known (Parrish and Long 1983). The presence of *Rutiodon* and the absence of the phytosaurs *Paleorhinus* and *Angistorhinus* suggest that the fauna is younger than those of the Popo Agie Formation and of the lower Dockum Formation of Howard County, Texas. Thus, the Revuelto Creek fauna is most closely correlative biostratigraphically with the Dockum Formation of Crosby County, Texas and with the lower Chinle Formation of northeastern Arizona. Thus, palynological (Dunay and Fisher 1979) and vertebrate paleontological (Chapter 9) evidence suggests an upper Carnian age for the Revuelto Creek local fauna.

Acknowledgments

We would like to thank Dr. H. Jaeger of the Humboldt Museum für Naturkunde in East Berlin for providing casts of the *Halticosaurus liliensterni* specimen, and Dr. Rupert Wild of the Staatliches Museum für Naturkunde in Stuttgart for providing photographs of *Halticosaurus longotarsus*. J. T. Gregory and Glenn Storrs kindly provided information about the Yale collections from Revuelto Creek. Discussions with R. T. Bakker, R. A. Long, and K. Padian contributed significantly to this chapter, although any errors are clearly our own. Thanks are also due to S. Chatterjee, P. Murry, and K. Padian for their careful reviews of the manuscript.

References

Camp, C. L. 1930. A study of the phytosaurs, with description of new material from western North America. *Mem. Univ. Calif.* 10:1–174.

Case, E. C. 1922. New reptiles and stegocephalians from the Upper Triassic of western Texas. *Carnegie Inst. Wash. Pub.* No. *321*, pp. 1–84.

Chatterjee, S. 1985. *Postosuchus*, a new thecodontian reptile from the Triassic of Texas, and the origin of Tyrannosaurs. *Phil. Trans. Roy. Soc. London B* 309:395–460.

Colbert, E. H. 1961. The Triassic Reptile *Poposaurus*. *Fieldiana: Geol.* 14:59–78.

1964. The Triassic dinosaur genera *Podokesaurus* and *Coelophysis*. *Am. Mus. Novit.* 2168:1–12.

Cope, E. D. 1889. On a new genus of Triassic Dinosauria. *Am. Nat.* 23:626.

Dunay, R. E., and M. J. Fisher. 1979. Palynology of the Dockum Group (Upper Triassic), Texas, U.S.A. *Rev. Paleobot. Palynol.* 28:61–92.

Galton, P. M. 1977. On *Staurikosaurus pricei*, an early saurischian dinosaur from the Triassic of Brazil, with notes on the Herrarasauridae and Poposauridae. *Paläont. Z.* 51(3–4):234–45.

Gregory, J. T. 1972. Vertebrate faunas of the Dockum Group, Triassic, eastern New Mexico and west Texas. *New Mex. Geol. Soc. 23rd Guidebook*, pp. 120–3.

Huene, F. von. 1906. Die dinosaurier der Europaischen Triasformen. *Geol. Pal. Abh. Suppl. Bd.*, p. 1.

1915. On reptiles of the New Mexican Trias in the Cope collection. *Am. Mus. Nat. Hist. Bull.* 34:500–17.

1921. Coelurosaurier-Reste aus dem obersten Keuper von Halberstadt. *Centralbl. Min. Geol. Pal.* 1921:315–20.

1932. Die fossil Reptilien Ordnung Saurischien, ihre entwicklung und Geschichte. *Verl. Gebr. Borntragen Berlin*, pp. 1–364.

1934. Ein neuer Coelurosaurier in der thuringischen Trias. *Palaeontol. Z.* 16:145–70.

Larsonneur, C., and A. F. de Lapparent. 1966. Un dinosaurien carnivore, *Halticosaurus* dans le Rhetien d'Ariel (Manche). *Bull. Linn. Soc. Normandie* 7:108–17.

Meyer, H. von. 1861. Reptilien aus dem Stubensandstein des Oberen Keupers. *Palaeontographica* 7: 253–351.

Ostrom, J. H. 1981. *Procompsognathus* – theropod or thecodont? *Palaeontograph. Abt. A* 175:179–95.

Parrish, J. M., and R. A. Long. 1983. Vertebrate paleoecology of the Late Triassic Chinle Formation, Petrified Forest and vicinity, Arizona (Abstract). *Geol. Soc. Am. Abst. Progr.* 15(5):285.

Raath, M. A. 1977. The anatomy of the Triassic theropod *Syntarsus rhodesiensis* (Saurischia: Podokesauridae) and a consideration of its biology. Ph.D. thesis, Rhodes University.

Walker, A. J. 1969. The reptile fauna of the "Lower Keuper" sandstone. *Geol. Mag.* 106:470–6.

Wright, J. C., and W. I. Finch. 1971. An annotated bibliography of flora and fauna from the Dockum Group of Triassic Age in eastern New Mexico and Western Texas. USGS Open File Report. 26 pp.

12 Vertebrate biostratigraphy of the Late Triassic Chinle Formation, Petrified Forest National Park, Arizona: preliminary results

R. A. LONG AND KEVIN PADIAN

Introduction

Within the past several years, knowledge of the terrestrial vertebrate fossil record during the interval generally recognized as the beginning of the "Age of Dinosaurs" has undergone radical change. According to the traditional view, during the Late Triassic, faunas of primitive archosaurs ["thecodontians," here including primarily phytosaurs, aetosaurs, rauisuchids, poposaurids, ornithosuchids, and several other "pseudosuchian" families *sensu* Krebs (1976)] coexisted with a variety of early dinosaur faunas until, by the end of the Triassic, the "thecodontians" had completely given way to the dinosaurs (Romer 1966; Colbert 1972, 1981; Charig 1979). At this point, the Early Jurassic, the terrestrial record seemed to disappear, to be found again only in the great dinosaurian faunas of the Morrison Formation and the Tendaguru beds (both Upper Jurassic). The dearth of Early Jurassic vertebrate-bearing horizons has often been lamented by paleontologists, because of the importance of unwitnessed faunal changes that took place prior to the Late Jurassic.

Two recent advances have shifted the focus of research on the beginning of the Age of Dinosaurs (Padian and Clemens 1985). The first was the realization that many horizons, notably the Newark Supergroup of eastern North America, were not entirely Triassic in age but were largely Early Jurassic, including the extensive footprint-bearing beds of the Connecticut Valley (Olsen and Galton 1977). Vertebrate, palynological, and geophysical evidence converged on the conclusion that the faunas of many terrestrial Newark beds were simply not represented in the European Jurassic marine type sections. The second advance was that, in many beds legitimately considered of Triassic age in North America, dinosaurs were present, but low in abundance and diversity and small in size; these faunas were fully dominated by "thecodontians" and other strictly Triassic reptilian groups. Faunas dominated by larger dinosaurs and crocodiles did not appear until the Early Jurassic, when all traces of "thecodontians" vanished from the record.

In this chapter, we report preliminary results from three years of paleontological field work in the Petrified Forest National Park, Arizona (PFNP). The goal of our preliminary research was to begin to test the understanding of the second major advance – the taxonomic composition of Late Triassic faunas and the sequence of change that occurred in them – in the most extensive fossiliferous geologic section in the world from the Late Triassic. The revised correlation of Triassic–Jurassic terrestrial boundary sediments was proposed in detail by Olsen and Galton (1977) for the Newark Supergroup and was hypothetically extended to horizons in South Africa (Olsen and Galton 1984), Europe, and the southwestern United States. Testing of this revised correlation in all these areas is now required. Because of (1) the remarkably thick, continuous exposures of the Upper Triassic geologic section in the region of the Petrified Forest National Park, and (2) the extensive but only recently curated vertebrate collections made fifty years ago from that region, the Petrified Forest National Park is an ideal study site for these purposes. Correlation of outlying horizons in the region can best be based on the known sequence at PFNP (work in progress by Dr. S. R. Ash). The principal value of collections from the southwestern United States bearing on this general question is the wealth of osseous remains, including complete skeletons, from the beds of the Chinle Formation and the Dockum Group. Although the Newark Supergroup contains abundant footprint faunas (Olsen 1980), as well as skeletons of many tiny reptiles (Olsen 1978) and fishes (Olsen 1980; Olsen, McCune, and Thomson 1982), there is a paucity of skeletal material from the large vertebrates that made the footprints on which much of the correlation is based. Estimates of taxonomic identification and of diversity are, therefore, largely limited to families of footprints, which may not reflect the extent of lower level taxonomic diversity obtainable from skeletal material. In this chapter, we will review the geologic setting of the Petrified Forest Member and the history of paleontologic work in these and stratigraphically equivalent beds. We will then summarize results of taxonomic analysis of earlier collections in the Museum of Paleontology, University of California (UCMP), relocation of earlier sites, and taxonomic and biostratigraphic results of our recent fieldwork in these horizons.

Paleogeographic and environmental setting

Exposures of Upper Triassic terrestrial deposits in the southwestern United States include the Chinle Formation of Arizona, New Mexico, Utah, and Colorado (see map, Fig. 12.1); the laterally equivalent Dockum Group of Texas; and the geographically restricted Popo Agie Formation of Wyoming. Of these, the Dockum is a partial rock equivalent of the Chinle, but the Chinle is geographically more extensive, it has many more superposed time-transgressive exposures, and it has a greater known paleontologic diversity, although this is not evident in the published literature. The Chinle has four members commonly recognized in Northern Arizona: the Shinarump, the Monitor Butte, the Petrified Forest, and the Owl Rock (Repenning, Cooley, and Akers 1969), in ascending order. Of these, the Petrified Forest Member is much greater in thickness than the sum of the other three and is by far the most fossiliferous, calibrated both in absolute terms and in abundance per unit exposure. The most complete section of this member is in the Petrified Forest National Park itself, which comprises the thickest set of continuous exposures in the Triassic of the southwest (see map, Fig. 12.2). Other quarries, ranging from the immediate vicinity of the Park to north central New Mexico, can be tied biostratigraphically to Petrified Forest deposits (Camp 1930; Camp and Welles 1956; Colbert 1972, 1974;

Figure 12.1. Map of Chinle Formation exposures. After Stewart et al. (1972).

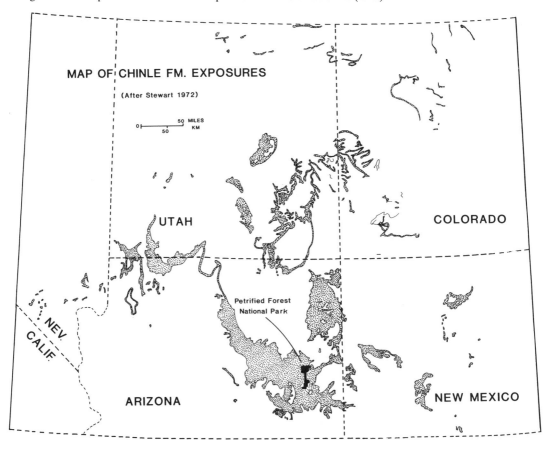

Long and Ballew 1985). Yet, despite the completeness of the geologic section and the abundant fossil riches of the park, no systematic analysis of the vertebrate paleontology of the park has been done since Camp's preliminary work of the 1920s and 1930s, which was based on collections only half-completed at that time. They are only now being studied for taxonomic diversity and stratigraphic occurrence of vertebrate fossils.

The Petrified Forest Member of the Chinle Formation is a series of variegated bentonitic shales and siltstones, subdivided by a series of distinct sandstone units (Billingsley 1985). The white, gritty Sonsela Sandstone separates the upper and lower units of the Petrified Forest Member. Within the lower unit, the Newspaper Rock and Rainbow Forest Sandstones are local features, and four lithologically similar local sandstones (numbered First through Fourth), apparently representing cyclic sedimenta-

Figure 12.2. Map of Petrified Forest National Park, Arizona, with some geographical names of fossiliferous areas noted in the text.

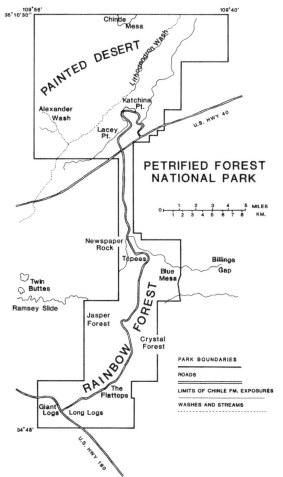

tion, occur above the Sonsela in the upper unit. Another local sandstone, the Black Forest Sandstone, is correlated with the Fourth Sandstone (Billingsley 1985). The sandstones generally range from 1 to 5 m in thickness; the thicknesses of the interposed shales and siltstones in the Petrified Forest section are under study by Ash. Preliminary petrographic analysis (Elzea 1983) suggests no distinguishing petrographic characteristics for any individual sandstones, probably because they derive from similar source beds and were deposited under similar regimes.

Upper and lower units do not always cooccur throughout the Park, but a combination of lithologic, vertebrate, and paleobotanic evidence capable of testing the established lithostratigraphic framework has produced encouraging preliminary results. Geographically, the PFNP actually consists of two distinct areas. The north end (containing the Painted Desert) is separated from the south end (containing the Rainbow Forest) by a minimum of 6 km of Tertiary alluvium, although the actual distance to the main mappable geologic sections between north and south is nearly 15 km (Fig. 12.2). Upper and lower units are not equally represented in both regions. The north end is almost entirely made up of upper unit beds, with some lower unit beds in the west of the Painted Desert remaining to be refined stratigraphically. The south end contains both lower and upper units, including all seven commonly recognized sandstone units. Billingsley's (1985) geologic map of the Park identifies all units mentioned above and is the primary reference against which hypotheses of biostratigraphic correlation can be tested. Faunas and floras distinctive of recognized upper and lower unit beds are now known from over 150 localities within the park, the result of reconnaissance in 1981–3 plus relocation of localities worked by Camp and others. Up to this point, the several lines of evidence have converged on the same stratigraphic correlations in nearly all cases, with some uncertainty due to insufficient exploration.

The depositional setting of the Petrified Forest Member is generally recognized as low relief floodplain cut by braided streams in a hot, humid, perhaps seasonally arid environment frequently punctuated by volcanic activity (Stewart, Poole, and Wilson 1972; Ash 1972; Cifelli, Breed, and Billingsley 1979). Freshwater gastropods and unionid pelecypods are common in the park, according to our preliminary work (*contra* Bryant 1965), and many kinds of insect burrows are known from the petrified logs (Walker 1938, Caster 1944). Over thirty genera of plant megafossils are known from the park in a variety of associations (review in Ash 1980); these remains are complemented by a diverse pollen and spore flora (130 species) concentrated mainly in the lower unit

(Peabody and Kremp 1964; Gottesfeld 1965, 1972a, b, 1975). Paleoecological analysis of the plant composition, including studies of pollen, leaves, and their cuticles, suggest several environments of deposition, including a floodplain swamp dominated by ferns and bennettitaleans, an *Araucarioxylon* forest, and perhaps an upland gymnospermous community hypothesized from a monosaccate–disaccate pollen record likely derived from conifers, cordaitaleans, and caytonialeans (Gottesfeld 1972a,b; but see Ash 1972). The patchy distribution of plant megafossils throughout the park is being integrated into the stratigraphic sequence by Ash. Ultimately, studies of vertebrate fossils will be combined with the plant data to arrive at a complete picture of ecological and evolutionary change in the plants and animals during Petrified Forest time (late Carnian–early Norian). Preliminary evidence bearing on the question of faunal change through time is discussed below.

Fossil record

Petrified wood from the Chinle Formation was first discovered in the 1850s by United States Army expeditions in the regions of the Canyon de Chelly and the Petrified Forest; since then, many workers have studied the fossil flora of the region (review in Ash 1972). Of the fifty species known from the Chinle flora, over thirty have been recorded from the Petrified Forest, including material based on wood, leaves, pollen, and reproductive structures. Daugherty (1941) is primarily responsible for compiling and revising diversity of these plants, but new discoveries continue to be made, and new occurrences continue to be integrated into the stratigraphic section by Ash.

In contrast, comparable work on the vertebrate fossils has tended to lag, mainly because of neglect of older collections and the absence of any complete systematic study of Chinle horizons. E. D. Cope, D. Baldwin, M. G. Mehl, and Barnum Brown were among the earliest workers to collect vertebrate fossils systematically from the Chinle Formation, but it was not until the 1920s and early 1930s that C. L. Camp of the University of California Museum of Paleontology (UCMP), spurred by discoveries made by Annie Alexander and Louise Kellogg of Berkeley, began an intensive reconnaissance and collection of fossil vertebrates from the Chinle Formation. He concentrated his attention primarily on three areas: the Petrified Forest, the region of St. Johns, Arizona, and the Chama Basin of New Mexico. Though the last two regions yielded many impressive remains, including complete skeletons, it was within the Petrified Forest that Camp realized the beginnings of a biostratigraphic system of correlation, based on large vertebrates, that still appears valid today. In his classic study of the phytosaurs (Camp

1930), based on only half the material that he eventually collected and prepared, Camp recognized seven species from the Chinle Formation, of which three (*Rutiodon adamanensis, R. lithodendrorum,* and *R. tenuis*) were found at distinct stratigraphic levels within the Petrified Forest. The constant differences in morphological structure seen in large samples from various stratigraphic levels convinced Camp of evolution "from species to species within a single genus," in which variation due to size, age, and sex could be determined. Diagnostic characters of these species were mainly recognized in the temporal region, of which the conformation of the squamosal bones was of principal importance. In his monograph, Camp planned further work on the "thecodontians" *Typothorax, Desmatosuchus,* and "*Stagonolepis,*" which are now recognized as aetosaurs although Camp included them with the phytosaurs. In the 1930s, Camp carried out extensive excavations in the St. Johns area, the Round Rock area of the Navajo lands, and the Canjilon region of New Mexico. Unfortunately, his interests soon turned to other pursuits, and the Chinle work was never completed. Camp and Welles (1956) eventually published a study of the dicynodont *Placerias* from St. Johns, previously considered a cotylosaur; but no other major publications on the Chinle issued from the work of Camp and his UCMP crews in Arizona.

Other paleontologists aided materially in understanding the taxonomic relationships of Chinle reptiles and amphibians. Case (1920–43) worked on phytosaurs, aetosaurs, and metoposaurs from the Dockum Group of Texas; Sawin followed Case with monographs on the metoposaurs (1945) and aetosaurs (1947). Colbert studied a range of Chinle taxa collected by American Museum of Natural History field crews, including phytosaurs (Colbert 1947a), metoposaurs (Colbert and Imbrie 1956), and other archosaurs (Colbert 1952, 1961) and dinosaurs (Colbert 1947b). J. T. Gregory (1953) described aetosaurs, phytosaurs (1962a,b, 1969, 1972), and metoposaurs (1980) from principal collections in North America and Europe. Most recently, Jacobs and Murry (1980) and Tannenbaum (1983) have analyzed the microvertebrate faunas of the Placerias and Downs quarries near St. Johns. These and other studies (e.g., Camp et al. 1947; Colbert 1950) have expanded faunal lists for the Petrified Forest Member and have contributed to the paleoecological understanding of faunal associations in the Chinle.

Preliminary results

In 1981, UCMP carried out a preliminary reconnaissance of vertebrate-producing localities previously worked by UCMP field parties in the Triassic and Jurassic strata of northeastern Arizona. In the region of the PFNP and St. Johns, Arizona, ten localities were recovered, including the three phytosaur type localities, twenty new localities were discovered, four taxa previously unknown in the park were recovered, and a large, nearly complete phytosaur skull was excavated (Ballew 1986). In 1982, reconnaissance of the Petrified Forest Member continued, concentrating on deposits within the Park. Because earlier occurrences and sites had been tied into the lithostratigraphic framework, each identifiable discovery of fossil bone could be immediately assessed in the field for biostratigraphic value. Only new taxa and new stratigraphic and geographic occurrences of known taxa were retained. Even with this preliminary culling, approximately 1,200 fragmentary specimens, including six new taxa and sixteen not previously found in the Park, were recovered from eighty-six new localities. Noteworthy were the remains of three reasonably complete skeletons collected from the upper unit in the north end of the park, including a small crocodylomorph archosaur (currently under study by J. M. Parrish), a metoposaurid amphibian (currently under study by J. R. Bolt), and the small theropod *Coelophysis*, the first dinosaur skeleton recovered in the park (Chapter 5). Additionally important were many discoveries of fresh water invertebrates, including forty currently known localities yielding a variety of unionid clams and four occurrences of "pockets" containing dozens of gastropods similar to those described by Yen and Reeside (1946) and Yen (1951). During the field season of 1983, approximately one-hundred new localities were found in the upper unit, mostly representing new occurrences of taxa previously thought rare in the park.

The preliminary results of these initial investigations can be summarized as follows.

Diversity and abundance of fossil vertebrates

Camp correctly recognized five fossil vertebrate taxa from the Petrified Forest, including the three previously mentioned species of the phytosaur *Rutiodon* (his *Machaeroprosopus*), the giant amphibian *Metoposaurus*, and the aetosaur *Typothorax*. Revision of Camp's collections and identification of material recently discovered indicates the presence of an additional thirty fossil vertebrate taxa within the boundaries of the PFNP. These consist of four aetosaurs [*Typothorax, Desmatosuchus, Calyptosuchus*, and *Paratypothorax*, the latter two genera described on the basis of armor (Long and Ballew 1985)]; a new rauisuchid; a poposaurid; several new archosaurs of indeterminate affiliation; a crocodylomorph; the small theropod dinosaur *Coelophysis* (Chapter 5); the peculiar archosauromorph *Trilophosaurus* (Gregory 1945); an

indeterminate procolophonid reptile; the dicynodont *Placerias*; and a new metoposaurid amphibian; as well as lungfish teeth, coelacanth quadrates, and teeth and scales of paleoniscid (*Turseodus*), redfieldiid, and semionotid fishes (Jacobs 1981). In addition, there are extensive remains of phytosaurs and metoposaurs. Isolated skeletal elements not pertaining to the above taxa may represent taxa known from equivalent horizons elsewhere. Camp's primary interest in the Petrified Forest was to collect reasonably complete skeletons, mainly in connection with his monographic work on the phytosaurs. In returning to his localities, and in discovering new ones, the abundance of material and the taxonomic associations that recur indicate a richness in the patterns of diversity and abundance of fossil vertebrates that has previously been overlooked.

Faunal associations within the Petrified Forest Member

Gottesfeld (1972) recognized several plant associations within the Petrified Forest that appear to reflect different environments of deposition. Preliminary work on the faunal associations suggests a similar division of fossil vertebrates that may provide baseline data for correlation with changes in plant composition through the Petrified Forest section (Parrish and Long 1983).

1. In the lower unit, phytosaurs and metoposaurs are commonly found together in what appears to be a mainly aquatic association, as in Camp's (1930) "Crocodile Hill" locality on Blue Mesa in the PFNP. Such associations are found outside the park, for example, on Ward's Terrace near Cameron, Arizona, where extensive remains of phytosaurs and metoposaurs are commonly preserved together.

2. Also in the lower unit, associations dominated by aetosaurs, with various admixtures of phytosaurs or other archosaurs (but seldom with both) suggest a less fully aquatic regime, perhaps reflecting a complex of environmental or taphonomic circumstances. The Downs Quarry, adjacent to Camp's Placerias Quarry near St. Johns, Arizona, is a particularly rich locality outside the park that preserves a diverse association of reptiles, amphibians, and fishes.

3. The upper unit contains many indications of a different aquatic environment. Although metoposaurs are rare, and phytosaurs and vertebrate remains of all kinds are generally less abundant than in the lower unit, phytosaurs are the most common vertebrate fossils apart from the aetosaur *Typothorax*, and over one hundred pelecypod localities and four gastropod localities were discovered in the upper unit in 1982. A rich fauna of archosaurs, dinosaurs, fishes, and amphibians was found in the quarry worked in 1982 and 1983 in the north end

near Lacey Point, in association with gastropods and pelecypods. A somewhat different association characterizes the famous Ghost Ranch *Coelophysis* Quarry in New Mexico (Colbert 1948), generally considered laterally equivalent to the upper unit of the Petrified Forest Member (Colbert 1974; O'Sullivan 1974). Further collecting and data analysis are required to determine the statistical significance of these associations and their validity as faunal units, prior to testing their value in stratigraphic correlation.

Biostratigraphic utility of fossil vertebrates in the Chinle Formation

Camp (1930) undertook to show the value of fossil vertebrates in establishing biostratigraphic correlations within the Chinle Formation, despite the claims of Branson and Mehl (1928) that vertebrates were unreliable, even locally. In his monograph, Camp not only recognized the stratigraphic utility of phytosaurs, but made observations on occurrences, absences, and restrictions of various other vertebrates within the park and its surrounding area, as well as in the divisions recognized by Camp as the upper and lower Chinle.

Preliminary evidence from the 1981–3 field seasons and curation of Camp's UCMP collections suggests that aetosaurs may be at least as useful in biostratigraphic correlation as Camp believed the phytosaurs to be (Long and Ballew 1985). Two aetosaurian genera, *Desmatosuchus* (six localities) and *Calyptosuchus* (twenty-five localities), appear to be wholly restricted to the lower unit. *Paratypothorax* (three localities) is rare and so far occurs in beds close to the Sonsela Sandstone (which separates the upper and lower units), both above and below it. *Typothorax* has so far been found only in the upper unit (thirty localities), but in considerable abundance. Metoposaurs are common in the lower unit, but are very rare in the upper unit; a new, unusual metoposaurian appears in the upper unit, but not in abundance (except at one quarry in the north end of the park). The large dicynodont *Placerias* occurs rarely in the lower unit and never in the upper unit, as does an undescribed rauisuchian. Probable remains of dinosaurs are poorly known from the lower unit within the boundaries of the park, although they are present (if rare) in lower unit beds of the Placerias Quarry, near St. Johns.

These data impart support to the hypothesis that the upper and lower units of the Petrified Forest Member are characterized by distinct nonoverlapping faunas through most of their stratigraphic and geographic extents. Further taxonomic, geographic, and stratigraphic analyses are required to test the consistency of these biostratigraphic units and to examine their utility in stratigraphic correlation with horizons outside the vicinity of the PFNP. Correlations with faunas from other Upper Triassic formations may then be proposed.

Conclusions

Exposures of the Upper Triassic Chinle Formation in the Petrified Forest National Park provide the best opportunity for examining the succession of faunal change in the southwest United States in the Late Triassic (late Carnian). Because of its extensive vertical and horizontal exposures, and its paleontological richness, it is an ideal reference section for studies of evolutionary and paleoenvironmental change in the region during this time; other, outlying sections rich in fossil remains but of uncertain stratigraphic position may be tied into the PFNP section through biostratigraphic correlation.

Stratigraphic correlation within the PFNP is based on a series of extensive sandstone beds; one, the Sonsela Sandstone, approximately corresponds to a break in faunal zones that characterize the lower and upper units, respectively. Paleobotanic evidence and preliminary analysis of deposition suggest some environmental change corresponding to these divisions, in addition to evolutionary change recognized in several vertebrate taxa. The biostratigraphic units have been reinforced by analysis of approximately 2,000 fossil vertebrate specimens in the UCMP collections that can be placed in the stratigraphic section. These patterns suggest that the taxonomic changes are real and, therefore, that vertebrate faunas from outlying beds may potentially be correlated with those from the PFNP. Ultimately, correlations with faunas from other formations may be proposed; at present, the PFNP sequence appears to fit best with the late Carnian sequence of the Newark Supergroup recognized by Olsen and Galton (1977), and dated by fossil pollen.

In the southwestern United States, the Upper Triassic is mainly represented by the Dockum and Chinle formations. Above the Chinle in the Four Corners region lie the Moenave and Wingate formations, locally extensive sandstone units that have produced the earliest known crocodiles and turtles in the Southwest (Chapter 23; Padian et al. 1981). Above these lies the Kayenta Formation, which appears to have a fully "Jurassic" complement of dinosaurs (Welles 1984), crocodiles, turtles, tritylodontids (Chapter 22), pterosaurs (Padian 1984), and mammals [Jenkins, Crompton, and Downs 1983; see Clark and Fastovsky (Chapter 23) for a general review of the Kayenta]. Olsen and Galton (1977) tentatively correlated the Kayenta Formation with Newark Supergroup deposits of Early Jurassic age; the Moenave is not as well known, but what little biostratigraphic evidence we have suggests correlation with basal Jurassic- or uppermost Triassic-age deposits in the Newark Supergroup (Chapter 20). So far, no "thecodontians" have been found above the Chinle beds, and they have not been diagnostically identified in Jurassic horizons anywhere else in the world. Conversely, though all the members of the Early Jurassic faunas have representatives in Late or Latest Triassic deposits (late Carnian–Norian), they are generally not recognized as abundant or diverse faunal elements until the Early Jurassic, at which time the "Triassic" archosaurs and many other lower vertebrate lineages disappear. In fact, the vertebrate paleontological "boundary" between the Triassic and Jurassic is largely based on negative evidence: the loss of the "thecodontian" groups of the Late Triassic. The taxa that dominate Early Jurassic terrestrial faunas are nearly all present, at least at the family level, in the Upper Triassic, and very few diagnostic appearances of taxa of any sort unequivocally mark this boundary in terrestrial faunas (Chapters 24 and 25). The Petrified Forest deposits preserve some of the most diverse and abundant faunas of Late Triassic age – perhaps the beginning of the transition marking the start of the "Age of Dinosaurs" in North America. They also, as our preliminary results show, provide the key to vertebrate biostratigraphic correlation of many Upper Triassic vertebrate-bearing horizons in the southwestern United States.

Acknowledgments

Many individuals contributed in many ways in the development of this chapter. R. A. Long was principally responsible for background research, supervision of the fieldwork, the bulk of the collecting effort, and primary identification and curation. Without his knowledge of Chinle faunas and history the program would not have been possible. Kevin Padian was principally responsible for grants, contracts, permits, administration, report writing, and (with Long) fieldwork and research objectives of the project. J. M. Parrish also helped in research design and was a prime mover of the paleoenvironmental and biostratigraphic aspects, as well as a primary participant in fieldwork. Karen Ballew provided invaluable help with fieldwork, organization, identification, and curation of collected materials, and drafted the two figures in this report. J. R. Bolt donated much field time and his expertise in geology and paleontology during the field seasons. We are most grateful also to G. R. Billingsley, who allowed free use of his unpublished lithologic map of the Petrified Forest National Park, which provided a baseline framework for stratigraphic correlations and tests of paleovertebrate distributions, and S. R. Ash, who was extremely generous with his time and knowledge of the geology and paleobotany of the Petrified Forest, and without whose help many stratigraphic mysteries would still remain. Field help and advice from W. R. Downs, J. Elzea, S. M. Gatesy, M. Morales, P. E. Olsen, R. Stout, and J. Woodcock are gratefully acknowledged. We owe a special debt of thanks to S. P. Welles, whose efforts, time, fieldwork, and vast knowledge and experience of southwestern Mesozoic formations provided the impetus for much of this work. Drs. J. T. Gregory, J. M. Parrish, and H.-D. Sues kindly provided helpful comments on the manuscript. Finally, we must credit the inspiration provided by the work of Charles L. Camp: As so much of Western philosophy can be said to be a footnote to Plato, so stands the relationship of any modern biostratigraphic work in the Chinle Formation to the pioneering efforts of Camp. For greatly facilitating our research efforts and for invaluable help, friendship, and hospitality, we thank former Superintendent Roger F. Rector and Mrs. Betty Rector, and Chief Ranger Chris Andress and Mrs. Paula Andress, as well as the entire staff of the Petrified Forest National Park. This work was made possible by funds from the Museum of Paleontology, University of California, and Grants 13577-G2 from the Petroleum Research Fund of the American Chemical Society to K. P.

References

Ash, S. R. 1972. Late Triassic plants from the Chinle Formation in northeastern Arizona. *Palaeontology* 15: 598–618.

———— 1980. Upper Triassic fossil floras of North America. *In* Dilcher, D. L., and T. N. Taylor (eds.), *Biostratigraphy of Fossil Plants* (Stroudsburg, Pennsylvania: Dowden, Hutchinson & Ross).

Ballew, K. L. 1986. A phylogenetic analysis of the Phytosauria (Reptilia: Archosauria) of the Late Triassic of the western United States. M.A. thesis, Dept. of Paleontology, University of California, Berkeley.

Billingsley, G. R. 1985. General stratigraphy of the Petrified Forest National Park, Arizona. *Bull. Mus. N. Ariz.* 54: 3–8.

Branson, E. B., and M. G. Mehl. 1928. Triassic vertebrate fossils from Wyoming. *Science* 67: 325–6.

Bryant, D. L. 1965. Paleontological reconnaissance in the Petrified Forest National Park during August, 1964. *In* Bryant, D. L., and J. E. Roadifer (eds.), *Preliminary Investigations of the Microenvironments of the Chinle Formation, Petrified Forest National Park, Arizona*, II; Interim Res. Rept. No. 7, Geochronology Laboratories, University of Arizona, Tucson, pp. 1–2.

Camp, C. L. 1930. A study of the phytosaurs, with descriptions of new material from western North America. *Mem. Univ. Calif.* 10: 1–174.

Camp, C. L., E. H. Colbert, S. P. Welles, and E. D. McKee. 1947. A guide to the continental Triassic of Northern Arizona. *Plateau* 20: 1–9.

Camp, C. L., and S. P. Welles. 1956. Triassic dicynodont reptiles. I. The North American genus *Placerias Mem. Univ. Calif.* 13: 255–348.

Case, E. C. 1920. Preliminary description of a new suborder of phytosaurian reptiles, with description of a new species of *Phytosaurus. J. Geol.* 28: 524–35.

———— 1921. On an endocranial cast from a reptile, *Desmatosuchus spurensis*, from the Upper Triassic of West Texas. *J. Comp. Neurol.* 33: 133–47.

1922. New reptiles and stegocephalians from the Upper Triassic of West Texas. *Publ. Carnegie Inst. Wash.* No. 321, pp. 1–84.

1927a. The vertebral column of *Coelophysis* Cope. *Contrib. Mus. Geol. Univ. Mich.* 2: 209–22.

1927b. A complete phytosaur pelvis from the Triassic beds of West Texas. *Contrib. Mus. Geol. Univ. Mich.* 2: 227–9.

1929. Description of the skull of a new form of phytosaur, with notes on the characters of described North American phytosaurs. *Mem. Mus. Pal. Univ. Mich.* 2: 1–56.

1930. On the lower jaw of *Brachysuchus megalodon*. *Contrib. Mus. Pal. Univ. Mich.* 3: 155–61.

1931. Description of a new species of *Buettneria*, with a discussion of the braincase. *Contrib. Mus. Pal. Univ. Mich.* 3: 187–206.

1932a. A collection of stegocephalians from Scurry County, Texas. *Contrib. Mus. Pal. Univ. Mich.* 4: 1–56.

1932b. A perfectly preserved segment of the carapace of a phytosaur, with associated vertebrae. *Contrib. Mus. Pal. Univ. Mich.* 4: 57–80.

1932c. On the caudal region of *Coelophysis* sp. and on some new or little known forms from the Upper Triassic of West Texas. *Contrib. Mus. Pal. Univ. Mich.* 4: 81–91.

1934. Description of a skull of *Kannemeyeria erithrea* Haughton. *Contrib. Mus. Pal. Univ. Mich.* 4: 115–27.

Charig, A. J. 1979. *A New Look at the Dinosaurs* (Heinemann, London).

Cifelli, R. L., G. R. Billingsley, and W. J. Breed. 1979. *The Paleontologic Resources of the Petrified Forest*. A report to the Petrified Forest Museum Associates, pp. 1–9, with two appendices.

Colbert, E. H. 1947a. Studies of the phytosaurs *Machaeroprosopus* and *Rutiodon*. *Bull. Am. Mus. Nat. Hist.* 88: 53–96.

1947b. Little dinosaurs of Ghost Ranch. *Nat. Hist.* 56: 392–9.

1948. Triassic life in the southwestern United States. *Trans. N.Y. Acad. Sci.* 10: 229–35.

1950. Mesozoic vertebrate faunas and formations of northern New Mexico. *Guidebook, 4th Field Conf., Soc. Vert. Paleontol.*, pp. 56–73.

1952. A pseudosuchian reptile from Arizona. *Bull. Am. Mus. Nat. Hist.* 99: 565–92.

1961. The Triassic reptile *Poposaurus*. *Fieldiana Geol.* 14: 59–78.

1972. Vertebrates from the Chinle Formation. *Mus. N. Ariz. Bull.* 47: 1–12.

1974. Mesozoic vertebrates of Northern Arizona. *In* Karlstrom, T. N. V., et al. (eds.), *Geology of Northern Arizona, Part 1*. Geol. Soc. Amer., Rocky Mtn. Sect., pp. 208–19.

1981. A primitive ornithischian dinosaur from the Kayenta Formation of Arizona. *Mus. N. Ariz. Bull.* 53: 1–61.

Colbert, E. H., and J. Imbrie. 1956. Triassic metoposaurid amphibians. *Bull. Am. Mus. Nat. Hist.* 110: 399–452.

Daugherty, L. H. 1941. The Upper Triassic flora of Arizona. *Carnegie Inst. Wash. Publ.* No. 526, pp. 1–108.

Elzea, J. 1983. A petrographic and stratigraphic analysis of the Petrified Forest Member (Triassic Chinle Formation) sandstones, Petrified Forest National Park, Arizona. B.A. honors thesis, Department of Paleontology. University of California, Berkeley.

Gottesfeld, A. S. 1965. M.S. thesis, University of Arizona.

1972a. Paleoecology of the lower part of the Chinle Formation in the Petrified Forest. *Mus. N. Ariz. Bull.* 47: 59–74.

1972b. Palynology of the Chinle Formation. *Mus. N. Ariz. Bull. Suppl.*, pp. 13–18.

1975. Upper Triassic palynology of the Southwestern United States. Ph.D. thesis, Department of Paleontology, University of California, Berkeley.

Gregory, J. T. 1945. Osteology and relationships of *Trilophosaurus*. *Univ. Texas. Publ. No. 4401*, pp. 273–359.

1953. *Typothorax* and *Desmatosuchus*. *Postilla* 16: 1–26.

1962a. The genera of phytosaurs. *Am. J. Sci.* 260: 652–90.

1962b. The relationships of the American phytosaur *Rutiodon*. *Am. Mus. Novit.* 2095: 1–22.

1969. Evolution und interkontinental Beziehungen der Phytosauria (Reptilia). *Palaeontol. Z.* 43(1/2): 37–51.

1972. Vertebrate faunas of the Dockum Group, Triassic, eastern New Mexico and western Texas. *N. Mex. Geol. Soc. Guidebk., Field Conf.*, 23: 120–3.

1980. The otic notch of metoposaurid labyrinthodonts. *In* Jacobs, L. L. (ed.), Aspects of Vertebrate History. (Flagstaff, Arizona: Museum of Northern Arizona Press), pp. 125–36.

Jacobs, L. L. 1981. Additions to the Triassic vertebrate fauna of Petrified Forest National Park, Arizona. *J. Ariz. Nev. Acad. Sci.*, p. 247.

Jacobs, L. L., and P. A. Murry. 1980. The vertebrate community of the Triassic Chinle Formation near St. Johns, Arizona. *In* Jacobs, L. L. (ed.), *Aspects of Vertebrate History* (Flagstaff, Arizona: Museum of Northern Arizona Press), pp. 55–72.

Jenkins, F. A., Jr., A. W. Crompton, and W. R. Downs. 1983. Mesozoic mammals from Arizona: new evidence of mammalian evolution. *Science.* 222: 1233–5.

Krebs, B. 1976. Pseudosuchia. *In* O. Kuhn (ed.), *Handbuch der Palaeoherpetologie*, Part 13: *Thecodontia*. (Stuttgart: Gustav Fischer), pp. 40–98.

Long, R. A., and K. L. Ballew. 1985. Aetosaur dermal armor from the Late Triassic of southwestern North America, with special reference to material from the Chinle Formation of Petrified Forest National Park. *Bull. Mus. N. Ariz.* 54: 35–68.

Olsen, P. E. 1978. A new aquatic eosuchian from the Newark Supergroup (Late Triassic–Early Jurassic) of North Carolina and Virginia. *Postilla (Yale Peabody Mus.)* 176: 1–14.

1980. A comparison of vertebrate assemblages from the Newark and Hartford Basins (Early Mesozoic, Newark Supergroup) of Eastern North America. *In* Jacobs, L. L. (ed.), *Aspects of Vertebrate History* (Flagstaff, Arizona: Museum of Northern Arizona Press), pp. 35–53.

Olsen, P. E., and P. M. Galton. 1977. Triassic–Jurassic tetrapod extinctions: are they real? *Science* 197: 983–6.

1984. A review of the reptilian and amphibian assemblages from the Stormberg Group of South Africa, with special emphasis on the footprints and the age of the Stormberg. *Palaeontol. Afr.* 25: 87–110.

Olsen, P. E., A. R. McCune, and K. S. Thomson. 1982. Correlation of the early Mesozoic Newark Supergroup (Eastern North America) by vertebrates, principally fishes. *Am. J. Sci.* 282: 1–44.

O'Sullivan, R. B. 1974. The Upper Triassic Chinle Formation in north-central New Mexico. *N. Mex. Geol. Soc. Guidebk. Field Conf.*, 25: 171–4.

Padian, K. 1984. Pterosaur remains from the Kayenta Formation (?Early Jurassic) of Arizona. *Palaeontology.* 27(2): 407–13.

Padian, K., J. M. Clark, D. E. Foster, and C. Hotton. 1981. Preliminary biostratigraphic-sedimentologic exploration of the Kayenta Formation of Arizona. *National Geogr. Res. Rept.*, in press.

Padian, K., and W. A. Clemens. 1985. Terrestrial vertebrate diversity: episodes and insights. *In* Valentine, J. W. (ed.), *Factors in Phanerozoic Diversity* (Princeton, New Jersey: Princeton University Press), pp. 41–96.

Parrish, J. M., and R. A. Long. 1983. Vertebrate paleoecology of the Late Triassic Chinle Formation, Petrified Forest and vicinity, Arizona. *Abstr. Prog., Geol. Soc. Am., Rocky Mt. Cordilleran Sects.*, p. 285.

Peabody, D. M., and G. O. W. Kremp. 1964. Preliminary studies of the polynology of the Chinle Formation, Petrified Forest. *Interim Res. Rept. No. 3, Geochronological Lab., Univ.* Ariz., pp. 11–20.

Repenning, C. A., M. E. Cooley, and J. P. Akers. 1969. Stratigraphy of the Chinle and Moenkopi Formations, Navajo and Hopi reservations, Arizona, New Mexico, and Utah. *U.S. Geol. Soc. Prof. Paper* No. 521-B, pp. 1–34.

Romer, A. S. 1966. *Vertebrate Paleontology*, 3rd ed. (Chicago: University of Chicago Press).

Sawin, A. J. 1945. *Amphibians from the Dockum Triassic of Howard County, Texas.* Univ. Texas Publ. No. 4401, pp. 361–99.

1947. The pseudosuchian reptile *Typothorax meadei. J. Paleontol.* 21: 201–38.

Stewart, J. H., F. G. Poole, and R. F. Wilson. 1972. Stratigraphy and origin of the Chinle Formation and related Upper Triassic strata in the Colorado Plateau region. *U.S. Geol. Soc. Prof. Pap.* 690: 1–335.

Tannenbaum, F. A. 1983. The microvertebrate fauna of the Placerias and Downs Quarries, Chinle Formation (Upper Triassic), near St. Johns, Arizona. M.S. thesis, Department of Paleontology, University of California, Berkeley.

Walker, M. V. 1938. Evidence of Triassic insects in the Petrified Forest National Monument, Arizona. *U.S. Nat. Mus. Proc.* 85: 137–41.

Welles, S. P. 1984. *Dilophosaurus wetherilli* (Dinosauria, Theropoda): osteology and comparisons. *Palaeontographica* 185A: 85–180.

Yen, T.-C. 1951. Some Triassic fresh-water gastropods from Northern Arizona. *Am. J. Sci.* 249: 671–5.

Yen, T.-C., and J. B. Reeside, Jr. 1946. Triassic fresh-water gastropods from southern Utah. *Am. J. Sci.* 244: 49–51.

III Taxa and trends across the Triassic–Jurassic boundary

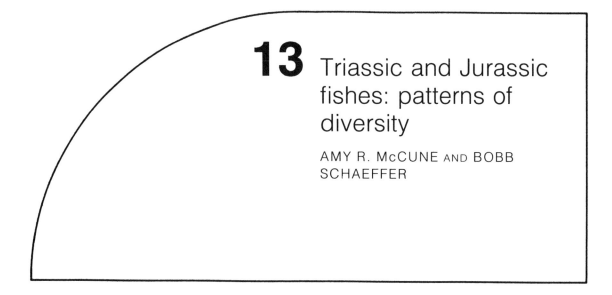

13 Triassic and Jurassic fishes: patterns of diversity

AMY R. McCUNE AND BOBB SCHAEFFER

Introduction

The Triassic and Jurassic periods are of particular interest in the history of fishes because it is during this interval that the dominant groups of living fishes, teleosts and modern sharks, first appeared and diversified. Although the overall patterns of diversification of fishes through the Phanerozoic have been well documented (Schaeffer 1973a; Thomson 1977; Sepkoski 1981), the Triassic and Jurassic have not been examined in detail. In the most comprehensive study to date, Thomson (1977) summarized the Phanerozoic diversity of both marine and nonmarine fishes, using the basic classificatory scheme and stratigraphic range data from Romer (1966) and Harland (1967). Although Thomson recognized the drawbacks of analyzing diversity within taxa that are not necessarily monophyletic (e.g., Holostei and Chondrostei, both of which have long been known to be grades), the relationships of fishes were not sufficiently known at that time to do otherwise. In the past fifteen years, however, there has been substantial progress in our understanding of the relationships of fishes, and it is now possible to begin to reexamine diversity curves based largely on monophyletic taxa.

In this chapter, we describe the patterns of fish diversity throughout the Triassic and Jurassic periods. We have not only tried to update the basic taxonomic and stratigraphic data from that which was compiled by Romer (1966), but we have attempted, so far as possible, to group taxa in currently recognized monophyletic groups. Even with the many remaining uncertainties about the relationships of some fishes, interesting patterns of diversity are beginning to emerge.

The prevailing picture of diversification of actinopterygian (ray-finned) fishes during the Mesozoic is familiar to most students of vertebrate evolution: the Paleozoic chondrostean radiation gave way to the holosteans during the Triassic; holosteans dominated in the Jurassic, but were effectively replaced in turn by teleosts during the Cretaceous. By focusing on the Triassic and Jurassic, we are taking a more detailed look at this critical period of actinopterygian evolution when chondrosteans decline, holosteans radiate, and teleosts originate. We will attempt to show here that when genera are grouped into monophyletic taxa, the pattern of diversification of the actinopterygian fishes is not one of the successive replacement of chondrosteans by holosteans by teleosts. Rather, the pattern is better described by the replacement of the paleopterygians (approximately equivalent to "Chondrostei" in traditional classifications) by the Teleostei.

Methods

We compiled range data for Triassic and Jurassic genera from: (1) Romer's (1966) *Vertebrate Paleontology*; (2) computer searches of three bibliographies – *Georef* (1980–4), *Geoarchive* (1974–84), and *ZooRecord* (1978–81) – all from the Dialog Information Service; and (3) a summary of the literature of Mesozoic fishes kept by Schaeffer until about 1975. We edited the resulting data set on the basis of our knowledge of the taxa concerned. It has become increasingly evident that many descriptions and identifications in the literature are suspect, particularly in groups that have not been revised recently, and many groups included in the data set require a critical reexamination of adequately preserved specimens in the light of current systematic practice. In addition, some taxa have probably been missed inadvertently, and some ranges are surely incorrect. Despite these difficulties, we feel that our final data set is reasonably accurate and complete. We include the complete listing of genera (Appendix Table 13.A) so that others may make corrections and improvements to both the systematic and stratigraphic components of the data for future analyses.

The second step of our analysis was to group these genera into monophyletic taxa (Table 13.1). Despite recent advances in our understanding of the relationships of fishes from the Mesozoic (e.g., Patterson 1973, 1977, 1981, 1982; Patterson and Rosen 1977; Maisey 1982a; Olsen 1984; Schaeffer and Patterson 1984; McCune 1986), not all genera can be easily assigned to monophyletic groups at a relatively low level of universality. Some taxa, known to be monophyletic, can only be placed *incertae sedis* in higher categories or related to other taxa at higher levels of universality [e.g., *Hulettia* in Halecostomi inc. sed., *Todiltia* in Teleostei inc. sed. (Schaeffer and Patterson 1984)]. A number of geographically widespread, long-ranging polyphyletic or paraphyletic taxa (e.g., pholidophorids, leptolepids) are sorely in need of revision. These taxa generally present problems concerning the number of genera they include, but they can usually be assigned to a monophyletic group at the levels of universality that concern us here.

The two notable nonmonophyletic groups we have had to include are the Halecostomi inc. sed. (halecostomes include semionotids, amiids, caturids, macrosemiids, teleosts, etc.) and the "Paleopterygii." In the Halecostomi inc. sed. are taxa once included in the traditional grade "Holostei" that cannot yet be assigned to a monophyletic group within the Halecostomi. We use the term "Paleopterygii" to refer informally to the large array of paleonisciformes and other lower actinopterygians whose relationships remain undetermined. We use the term "Paleopterygii" rather than "Chondrostei" because this latter term has been restricted to living sturgeons, paddlefishes, and their immediate fossil relatives (Patterson 1982).

Results

Data for stratigraphic ranges of individual genera (Appendix) are summarized in Table 13.2. The diversities within higher taxa are represented graphically in Figure 13.1. From these data, we offer the following generalizations about the history of sharks and the higher bony fishes during the Triassic and Jurassic:

1. Hybodont sharks have a relatively low, fairly constant diversity throughout the Triassic and Jurassic. *Hybodus*, is, however, a form genus, perhaps along with other genera in this monophyletic group.

2. Neoselachians are first known from the Early Triassic. By the Early Jurassic, numerous subgroups of modern neoselachians (hexanchoids, orectoloboids, squatinoids, and batoids) were present (Maisey 1982a).

3. The chimaeriforms or ratfishes may be as old as the Pennsylvanian, and various problematic subgroups referred to this group existed during the Late Paleozoic. They are not definitely known from the Triassic, but are well represented in the Jurassic.

4. A large paraphyletic group of relatively generalized actinopterygians, which we refer to here as "Paleopterygians," is represented in the Triassic and Jurassic by various paleonisciforms and by a variety of more advanced but paraphyletic groups, the "subholosteans" of earlier workers. The relationships of these "paleopterygians" have not yet been resolved, although they have been discussed by a variety of authors (e.g., Patterson 1973; Schaeffer 1973b; Schaeffer and Mangus 1976; Lauder and Liem 1983; Gardiner 1984; Olsen 1984; Schaeffer and Patterson 1984). The Acipenseriformes, which first appear in the Jurassic, include living sturgeons and paddlefishes.

5. The fossil record for the monophyletic neopterygians extends back to the Late Permian (*Acentrophorus*). One subgroup, the halecomorphs,

Table 13.1 *Classification of higher taxa*

Chondrichthyes	Osteichthyes
Elasmobranchii	Actinopterygii
Ctenacanthoidea	"Paleopterygii"; see text
Hybodontoidea	Neopterygii
Neoselachii	Halecostomi
Holocephali	Semionotiformes
Squalorajiformes	Pycnodontiformes
Myriacanthoidei	Halecomorphi
Chimaeroidei	Teleostei
	Sarcopterygii
	Actinistia
	Dipnoi

Table 13.2. *Summary of the generic diversity of fishes in the Triassic and Jurassic*[a]

	Triassic			Jurassic		
	E	M	L	E	M	L
Ctenacanthoidea	1	1	1	—	—	—
Hybodontoidea	3	9	12	5	4	4
Neoselachii	1	1	5	7	13	21
Holocephali	—	—	2	9	6	5
"Paleopterygii"	53	53	39	16	7	3
Halecostomi inc. sed.	11	10	13	4	5	9
Semionotiformes	—	—	2	4	3	10
Pycnodontiformes	—	—	4	3	6	8
Halecomorphi	1	4	4	5	5	8
Teleostei	—	3	9	15	21	50
Actinistia	11	2	5	5	3	7
Dipnoi	1	1	4	1	1	1
Totals	82	83	99	73	74	126
Duration of epochs	5	12	18	32	18	19
No. of genera/MY	16.4	6.9	5.5	2.3	4.1	6.6
Sediment volume[b] (km³)	6.0	8.0	15.5	13.5	16	18.3
No. of genera/km³ sediment	13.7	10.4	6.4	5.4	4.6	6.9

[a]*E*arly, *M*iddle, and *L*ate.
[b]Data from Raup (1972).

which includes *Amia*, had diversified considerably by the Middle Triassic. The systematics of two other extinct but characteristic Mesozoic neopterygian groups, the semionotiformes, restricted to macro-semiids, lepisosteids, and semionotids by Olsen (1984) and Olsen and McCune (in prep.), and the pycnodontiformes (R. Nursall pers. comm.) are currently being investigated.

    6. The earliest representatives of the largest and most diversified neopterygian group, the monophyletic Teleostei (Patterson and Rosen 1977; Patterson 1981; Rosen 1982), are the Middle or Late Triassic pholidophorids. Some authors report a Middle Triassic origin for these teleosts but provide no original citation (Romer 1966; Gardiner 1967; Patterson 1973; Lauder and Liem 1983), while others report a Late Triassic origin (Patterson 1982). Our attempt to find the original literature supporting a Middle Triassic origin was only partly successful. Stolley (1920) reports the occurrence of *Pholidophorous* from Gandersheim in the German Muschelkalk, and Tanner (1925) gives an account of *Pholidophorous* from the Lower Triassic of Utah. Neither Stolley nor Tanner, however, provided a figure, description, or reference to a specimen sufficient to evaluate his identification. Even if one dis-

Figure 13.1 Diversity of fishes in the Triassic and Jurassic (Early, Middle, and Late). The width of each band at the middle of a division reflects the number of genera known from that division.

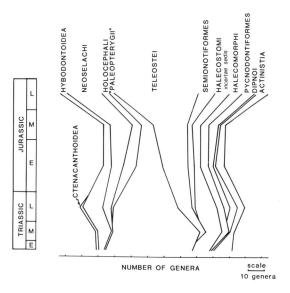

counts the Middle Triassic records of *Pholidophorous*, there are records of other pholidophorids of probably mid-Triassic [i.e., *Archaeomaene* and *Prohalecites* (Romer 1966)].

During the Jurassic, numerous other early groups of teleosts appeared, but most of these disappeared from the record by the end of the Cretaceous. The osteoglossomorphs (represented in the Jurassic by the lycopterids) and the Clupeocephala (represented in the Jurassic by several species of "*Leptolepis*") are extant today.

7. The actinistians, or coelacanths, were perhaps more diversified during the Triassic than at any other interval in their long history. The dipnoans were represented in the Triassic and Jurassic by the tooth form genus *Ceratodus* and related genera.

Discussion

Overall generic diversity of fishes appears to be relatively constant through the Triassic and Jurassic periods. In absolute numbers, counts of generic diversity fluctuate between seventy-three and eighty-three in most epochs of these two periods (see Table 13.2). Two exceptions are a moderate peak in diversity in the Late Triassic (99 genera) and a larger peak in the Late Jurassic (126 genera; see Table 13.2). The Late Jurassic peak is easily accounted for by the marked increase in the diversity of teleosts and neoselachians, trends which continue into the Recent. The reason for the Late Triassic peak is not so readily apparent. The diversity of the "Paleopterygii" appears to decrease from the Middle to Late Triassic; however, this decrease is more than compensated for by slight increases in diversity of all other groups but the Halecostomi inc. sed. (see Table 13.2). This Late Triassic peak may actually be an artifact. The apparent pattern in overall diversity changes markedly when the variable length of the six divisions of the Triassic and Jurassic is considered. If the diversity data are converted to number of genera per MY for each division, the diversity decreases steadily from the Early Triassic to the Early Jurassic and then increases in the Middle and Late Jurassic (see Table 13.2).

The most obvious additional bias in interpreting the peaks of diversity in the Late Triassic and the Late Jurassic is the inflating effect of *Lagerstaetten*, the excellent preservation of large assemblages such as the Alpine Late Triassic and the Solenhofen Limestone of the Late Jurassic. *Lagerstaetten* may explain the slight increases from the Middle to Late Triassic and the Middle to Late Jurassic for most groups; but against this background, the marked decrease of "Paleopterygii" in the Late Triassic and the increase of teleosts in the Late Jurassic are still pronounced.

Other possible biases in the data include collecting effort, monographic effects, taxonomic interest, and volume of sediment available for sampling [see Raup (1972) for a general discussion of these biases in the fossil record]. It is difficult to evaluate the specific effects of these various biases here except for sediment volume. The estimated volume of marine sediments (Raup 1972) is low for the Early and Middle Triassic and higher, but fairly constant, in the Late Triassic and throughout the Jurassic (see Table 13.2). The greatest volume of sediment exists for the Late Jurassic and, in fact, may be sufficient to account for the Late Jurassic peak in diversity (see number of genera per km^3 of sediment in Table 13.2). However, the relevant data concerning sediment volume are the amounts sampled, not those available for sampling, and these data are lacking. Considering all the possible biases, one should probably be very cautious about interpreting changes in absolute diversity in any given group.

It is perhaps more appropriate to use these data to examine the proportional representation of different groups of fishes through these two periods, and here the patterns are striking. The most interesting pattern apparent in Figure 13.1 is the inverse correlation in diversity of "paleopterygians" and teleosts. Paleopterygian diversity decreases markedly from the Late Triassic through the Jurassic, a trend that is more the result of a decreasing rate of origination than an increasing rate of extinction (Table 13.3). At the same time, the rate of origination of teleosts increases, resulting in progressively increasing diversity of teleosts from the Middle Triassic through the Jurassic. Thus, contrary to the classic, textbook picture of the succession of "chondrosteans," "holosteans," and teleosts (e.g., Futuyma 1979), teleosts appear to be replacing "paleopterygians" (traditional "chondrosteans"). The Neopterygii, excluding teleosts (traditional "holosteans"), persist at moderate levels of diversity throughout this interval of Late Triassic–Early Jurassic turnover.

This new pattern has emerged because our understanding of the relationships of fishes has changed. The earliest occurrence for a group is as much dependent on an opinion about the character(s) that defines the group in question as it is on simply discovering a fossil. For example, by defining the teleosts by characters that are shared by pholidophorids (Patterson 1977, 1982), the first occurrence of teleosts may be pushed back to the Middle Triassic. Thus, teleosts appear to be more diverse earlier in their history, and "holosteans," which no longer include pholidophorids, are less diverse in the same period. Our picture of patterns of diversity may continue to change as the relationships of fishes are better resolved, especially when the relationships of

Table 13.3. *Summary of the number of first and last occurrences of genera for each division of the Triassic and Jurassic*[a]

	First occurrences						Last occurrences					
	Triassic			Jurassic			Triassic			Jurassic		
	E	M	L	E	M	L	E	M	L	E	M	L
Ctenacanthoidea	—	0	0	0	0	0	0	0	1	0	0	—
Hybodontoidea	—	6	3	1	0	0	0	0	8	1	0	—
Neoselachi	—	1	4	7	7	12	0	0	4	2	4	—
Holocephali	—	0	2	7	3	1	0	0	0	6	2	—
"Paleopterygii"	—	31	11	10	4	1	31	24	33	13	5	—
Halecostomi inc. sed.	—	8	8	3	2	6	9	5	12	1	2	—
Semionotiformes	—	0	2	2	1	7	0	0	0	2	0	—
Pycnodontiformes	—	0	4	1	3	2	0	0	2	0	0	—
Halecomorphi	—	3	1	1	0	3	0	2	0	0	1	—
Teleostei	—	3	6	11	13	38	0	0	5	7	9	—
Actinistia	—	1	4	2	0	4	10	1	2	2	0	—
Dipnoi	—	0	3	0	0	0	0	0	3	0	0	—

[a]Early, Middle, Late.

the "paleopterygians," the largest nonmonophyletic group considered here, are understood.

Summary

In contrast to the traditional picture of teleosts replacing holosteans, which had earlier replaced chondrosteans, it now appears that teleosts replaced the "paleopterygians" ("chondrosteans" in traditional classifications), while nonteleostean Neopterygii ("holosteans" in traditional classifications) persisted at moderate but relatively constant diversity. At the end of the Triassic, "paleopterygians" in particular decreased in diversity due to decreasing origination rates, but, even so, the overall diversity of fishes in the Early Jurassic was about the same as the diversity of fishes throughout most of the Triassic and Jurassic. The exceptions to this relatively invariant diversity are apparent peaks in the Late Triassic and the Late Jurassic. The former may be an artifact of the variable duration of the divisions of these periods, but the latter appears to be caused by the radiation of the teleosts. Both peaks may be emphasized by *Lagerstaetten*. In any event, all these patterns must be interpreted cautiously, owing to the probable influences of biases in the fossil record and the probability that our understanding of the relationships of nonneopterygian actinopterygians will change.

Acknowledgments

We thank Dr. John Maisey for his help in compiling the data on elasmobranchs. Drs. Paul Olsen, Kevin Padian, and David Winkler read the manuscript and suggested numerous improvements. Bruce Young and Michael Bonda gave invaluable assistance in compiling the data. This research was supported in part by Hatch Project 183–7421 to A.R.M.

Note: Appendix Table 13.A appears on pp. 176–9; References begin on p. 180.

Appendix Table 13.A. *Stratigraphic ranges of genera of Triassic and Jurassic fishes*

	ET	MT	LT	EJ	MJ	LJ	Reference
CTENACANTHOIDEA							
Xenacanthus	X	X	X				Murry, this volume; Johnson 1980
HYBODONTOIDEA							
Acrodonchus							
Acrodus		X	X	X	X	X	
Acronemus		?X	X				Rieppel 1982
Asteracanthus		?X	X	X	X	X	Barthel 1978
Bdellodus		X	X				
Carinacanthus		X	X				
Doratodus		X	X				
Hybodus	?X	X	X	X	X	X	Maisey 1976
Lissodus (=Lonchidion)	?X	X	X	X	X	X	Murry 1981; Maisey 1982a
Palaeobates		X	X				
Polyacrodus	X	X	X				
Scoliorhiza			X				
Steinbachodus			X				
NEOSELACHII							
Hueneichthys	X	X					
Nemacanthus	X	X	?X	X	X		Reif 1977
Palaeospinax			?X	X	X		
Vallisia			X	X		X	Duffin 1982
GALEOMORPHA							
Agaleus				X			Duffin & Ward 1983
Annea					X		Thies 1983
Corysodon						X	
Crossorhinops						X	
Heterodontus				X	X	X	Capetta 1977
Orectoloboides					X	X	
"Orectolobus"				X	X		
Palaeobrachaelurus				X	X	X	Thies 1983
Palaeocarcharias						X	
"Palaeoscyllium"						X	
Paracestracion					?X	X	Maisey 1982b
Phorcynus						X	
Pristiurus						X	
Reifia			X				Duffin 1980
Sphenodus (=Orthacodus)				X	X	X	
Synechodus				?X	?X	X	
SQUALOMORPHA							
Eonotidanus				?X	X		Pfeil 1983
"Notidanus"					X	X	
Notorynchus					X	X	Capetta 1975
Protospinax							
Pseudodalatias			X				Sykes 1971
Pseudorhina					X	X	
Squatina					X	X	Barthel 1978
BATOMORPHA							
Asterodermus						X	
Belemnobatis						X	
Breviacanthus				X	X	X	Maisey 1976
Spathobatis (=Aellopos)				X	X	X	Desroches 1972
HOLOCEPHALI							
SQUALORAJIFORMES							
Squaloraja				X			Reif 1980
MYRIACANTHOIDEI							
Acanthorhina				X			
Agkistracanthus			X	X			Duffin & Furrer 1981
Alethodontus			X	X	X	X	Duffin 1983
Chimaeropsis				X	X		
Metopacanthus				X	X		
Myriacanthus				X	X		
Recurvacanthus			X	X			Duffin 1981
CHIMAEROIDEI							
Brachymylus					X		
Elasmodectes						X	
Ganodus					X	X	
Ischyodus					X	X	
Pachymylus					X		
OSTEICHTHYES							
ACTINOPTERYGII							
"PALEOPTERYGII"							
Acrolepis	X	X					
Acrorhabdus	X	X	X				
Aegicephalichthys		X					

Left table

Genus	ET	MT	LT	EJ	MJ	LJ	Reference
Aethodontus		X					
Anatoia	X						
Andalusias		X	X				Sykes & Simon 1979
Aneurolepis		X	X				
Apateolepis	X						
Arctosomus	X	(X)					
Atherstonia	X	X					
Atopocephala	X	X					
Australosomus		X	X				
Beaconia		X					
Belichthys		X					
Besania		X					
Birgeria	X	X	X				
Bobasatrania	X	X	X				Selezneva 1982
Boreichthys*	X	X					
Boreosomus	X	X					
Brookvalia		X					
Broometta	X						
Browneichthys				X			
Caminchaia	X						
Caruichthys	X						
Centrolepis		X		X	X		
Cephaloxenus			X				
Challaia	X	X		X			
Chondrosteus				X	X		Su 1974
Chrotichthys	X						Schaeffer 1967
Chungkingichthyes			X		X		
Cionichthys			X				
Cleithrolepis	X	X	X				
Coccolepis	X	X	X	X	X	X	Schaeffer & Patterson 1984
Colobodus	X	X		X			
Cosmolepis (=Oxygnathus)		X	?X				
Crenilepis	X						
Daedalichthys	X						
Dicellopyge	X						
Dictyopleurichthys		X					
Dictyopyge			X				
Dimorpholepis*	X	X					
Dipteronotus	X	X					
Dollopterus	X	X					
Echentaia	X						
Ecrinesomus	X						
Engycolobodus		X					
Errolichthys	X	(X)					
Eurynothus	X						
Evenkia	X						
Geitonichthys		X					
Gigantopterus			X				
Guaymayenia	X						
Gyrolepidoides	X						

Right table

Genus	ET	MT	LT	EJ	MJ	LJ	Reference
Gyrolepis	X	X	X				
Gyrosteus		X	X	X			Schaeffer 1967
Habroichthys			X				
Helichthys	X						
Helmolepis	X						
Hydropessum	X						
Ischnolepis	X						
Lambeichthys			X				
Lasalichthys			X				
Leighiscus		X	X				
Leptogenichthys		X					Martin 1980
Luganoia		X	X				
Macroaethes	X	X					
Manlietta	X						
Mauritanichthys	X	X	X				
Meidichthys		X					
Mendocinichthys		X					
Meridensia	X	X	X				
Mesembroniscus		X					
Molybdichthys		X					
Myriolepis	X	X	X				
Neochallaia	X	X					
Ohmdenia					X		
Palaeoniscionotus					X		
Palaeoniscus	X	X		X			
Pasamhaya	X						
Peipiaosteus						X	Liu & Zhou 1965
Peltopleurus		X	X				
Perleidus	X	X	X				
Phlyctaenichthys		X					
Pholidopleurus		X					
Placopleurus		X	X	X			
Platysiagum		X	X	X			
Platysomus	X	X	X				
Plesiococcolepis				X			Wang 1977b
Pristisomus	X	X					
Procheirichthys		X					
Pseudobeaconia		X					
Pteronisculus	X						
Pteroniscus			X	X	(X)		
Ptycholepis	X	X	X	X		X	
Pygopterus	X	X	X	X			
Redfieldius				X			
Sakamenichthys	X						
Saurichthys	?X	X	?X	X			
Saurorhynchus	?X	X		X			
Scanilepis	X	X	X				
Schizurichthys		X					
Sinkiangichthys*	X	X	X				Martinson 1973
Stichopterus					X		

Appendix Table 13.A. *(cont.)*

Taxon	ET	MT	LT	EJ	MJ	LJ	Reference
Synorichthys	X		X				Schaeffer 1967
Tanaocrossus	X		X				Schaeffer 1967
Thoracopterus		X					
Tripelta		X					
Turseodus			X				
Yuchoulepis					X		Su 1974
Zeuchthiscus	X						
HALECOSTOMI incertae sedis							
Acentrophorus	X		X				
Alleiolepis	X	X	X				
Angolaichthys	X	X					
Archaeolepidotus	X	X					
Asialepidotus		X					
Broughia	X						
Callopterus						X	
Corunegenys							
Dandya			X				
Dapedium			X	X	X		
Enigmatichthys		X		X			
Eoeuganthus		X	X				
Eugnathides						X	
Hemicalypterus			X				Schaeffer 1967
Heterostrophus					X	X	
Hulettia					X		Schaeffer & Patterson 1984
Ionoscopus						X	
Jacobulus	X						
Oligopleurus						X	
Orthurus			X				
Ospia	X						
Paracentrophorus	X						
Paradapedium				X			Jain 1973
Parasemionotus	X						
Prionopleurus				X	X	X	
Pristiosomus				X			
Promecosomina		X					
Sargodon		X	X				
Serrolepis		X	X				
Sinoeugnathus		X	X				
Sinosemionotus		X					
Songanella						X	
Stensiönotus	X						
Tetragonolepis				X	X		
Tungusichthys	X			X			
Urocles						X	
Watsonulus	X						
Woodthorpea			X				

Taxon	ET	MT	LT	EJ	MJ	LJ	Reference
SEMIONOTIFORMES							
Austrolepidotes						X	Bocchino 1974
Enchelyolepis						X	
Eusemius						X	
Histionotus					X	X	
Ischypterus							McCune 1982
Lepidotes			X	X	X	X	
Macrosemius				X	X	X	
Neosemionotus						X	Bocchino 1973
Notagogus						X	
Propterus						X	
Semionotus			X	X	X	X	Olsen et al. 1982; Olsen & McCune MS
"S. elegans group"							
PYCNODONTIFORMES							
Athrodon						X	
Brembodus			X			X	Tintori 1980
Coelodus					X	X	
Eomesodon				X	X	X	
Gibbodon			X			X	Tintori 1980
Gyrodus				X	X	X	
Macromesodon					X	X	
Mesturus				X	X	X	
Proscinetes					X	X	
HALECOMORPHI							
Amiopsis			X	X	X	X	
Caturus			X	X	X	X	
Eosemionotus		X	X	X	X	X	
Furo	X	X	X	X	X	X	
Heterolepidotus		X	X	X	X	X	
Ikechaoamia					X	X	
Liodesmus					X	X	
Ophiopsis		?X		(X)	X		
Osteorachis			X	X	X		
Sinamia						X	
TELEOSTEI							
Aetheolepis						X	
Allothrissops					X	X	
Anaethalion					X	X	
Ankylophorus						X	
Aphnelepis				X			

	ET	MT	LT	EJ	MJ	LJ	Reference
Archaeomaene		?X	?X		X		
Ascalobos						X	
Aspidorhynchus					X	X	
Asthenocormus					X	X	
Australopleuropholis						X	
Baleiichthyes					X	X	Su 1980
Belonostomus					X	X	
Calamopleurus						X	
Catervariolus						X	
Ceramurus						X	
Chongichthys						X	Arratia 1982
Clupavus						X	
Ctenolepis						X	
Enchelyolepis						X	
Eoprotelops						X	
Eurycormus						X	
Eurystichthys					X		
Euthynotus				X	X		
Flugopterus				X			
Fuchunkiangia			X				Zhang & Zhou 1974
Galkinia							
Hengnania				X			Wang 1977a
Huashia							
Hungkiichthys			X	X	?X	X	
Hypsocormus				X	X	X	
Ichthyokentema					?X		
Leedsichthys					X		
Leptolepides						X	Nybelin 1974
"Leptolepis"			X	X	X	X	
Ligulella						X	
Luisichthys						X	
Lycoptera						X	
Madariscus					X	X	
Majokia						X	
Mesoclupea						X	
Neolycoptera					X	X	Schaeffer & Patterson 1984
Occithrissops				X			Schaeffer 1972
Oreochima				X			
Orthocormus					X	X	
Pachycormus				X	X		
Pachythrissops						X	
Paraclupea			X				Zambelli 1975
Parapholidophorus			X				
Parapleuropholis						X	
Pholidaphoristion						X	Zambelli 1978
Pholidoctenus			X				
Pholidolepis				X			Nybelin 1966
Pholidophoroides				X			
Pholidophoropsis			X	X			Nybelin 1966
Pholidophorus	?X	?X	X	X	X	X	

	ET	MT	LT	EJ	MJ	LJ	Reference
Pholidorhynchodon			X				Zambelli 1980
Pleuropholis						X	
Prohalecites		X	X				Nybelin 1974
Proleptolepis			X				
Protoclupea					X	X	Arratia Fuentes et al. 1975
Pterothrissus					X	X	
Sauropsis				X	X	X	
Saurostomus				X			
Sinolycoptera						X	Gaudant 1966
Tharsis					X	X	
Thrissops					X	X	
Toditia			X				Schaeffer & Patterson 1984
Tongxinichthys						X	Fengchen 1980
Varasichthys						X	Arratia 1981
Vidalemia						X	

SARCOPTERYGII

ACTINISTIA

	ET	MT	LT	EJ	MJ	LJ	Reference
Axelia	X						
Bunoderma						X	
Chinlea			X			X	Schaeffer 1967
Coccoderma						X	
Coelacanthus	?X						
Diplurus			X	X		X	
Holophagus			X	X	X	X	Jain 1974
Indocoelacanthus					(X)	X	
Laugia	X						
Libys						X	
Lualabaea						X	Saint-Seine 1955
Moenkopia	X						
Mylacanthus	X						
Piveteauia	X						
Rhipis	X						
Sassenia	X						
Scleracanthus	X						Schaeffer & Patterson 1984
Sinocoelacanthus		(X)	(X)				Schaeffer 1972
Ticinepomis		X	X				Rieppel 1980
Undina	X		X	X	X	X	
Whiteia	X						
Wimania	X						

DIPNOI

	ET	MT	LT	EJ	MJ	LJ	Reference
Arganodus				X	X	X	Martin 1979
Ceratodus	X	X	X	X	X	X	
Ptychoceratodus		X	X	X	X	X	
Sagenodus	?X		?X				

Notes: For discussion of higher taxa, see text. ET = Early Triassic, MT = Middle Triassic, LT = Late Triassic, EJ = Early Jurassic, MJ = Middle Jurassic, LJ = Late Jurassic. Asterisk (*) denotes taxa known from a single locality in either the Triassic or Jurassic, but the division is unknown. (×) indicates an inferred occurrence based on there being occurrences of that genus both earlier and later. References for the first occurrences of taxa described since Romer's (1966) compilation are given in the table. References are not given for taxa included in Romer.

References

Arratia, G. F. 1981. *Varasichthyes ariasi* n. gen. et sp. from the Upper Jurassic of Chile (Pisces, Teleostei, Varasichthyidae n. fam.). *Palaeontogr. Abt. A Palaeozool. Stratigr.* 175:4–6.

——— 1982. *Chongichthys dentata*, new genus and species from the Late Jurassic of Chile (Pisces, Teleostei; Chongichyidae new family). *J. Vert. Paleontol.* 2:133–149.

Arratia Fuentes, G., A. Chang Garrido, and G. Chong Dias. 1975. Sobre un pez fosil de Jurasico de Chile y sus posibles relaciones con clupeidos sudamericanos vivientes. *Rev. Geol. Chile* 2:10–21.

Barthel, K. W. 1978. *Solnhofen: ein Blick in die Erdgeschichte* (Thus, Switzerland: Ott).

Bocchino R., A. 1973. Semionotidae (Pisces, Holostei, Semionotiformes) of the Lagarcito Formation (Upper Jurassic), San Luis, Argentina. *Ameghiniana* 10:254–68.

Bocchino R., A. 1974. *Austrolepidotes cuyanus* gen. et sp. nov. and other fossil fish from the Lagarcito Formation (Upper Jurassic), San Luis, Argentina. *Ameghiniana* 11:237–48.

Capetta, H. 1975. Selaciens et Holocephale du Gargasien de la region de Gargas (Vaucluse). *Geol. Mediterr.* 2:115–34.

——— 1977. Selaciens nouveaux de l'Albien Superieur de Wissant (Pas-de-Calais). *Geobios* 10:967–72.

Desroches, A. 1972. Catalogue des poissons fossiles du gisement de Cerin (Ain); conserves dans les collections du Museum d'Histoire Naturelle de Lyon (Premiere Partie). *Mus. Hist. Nat. Lyon, Nouv. Arch.* 9:13–108.

Duffin, C. J. 1980. A new euselachian shark from the Upper Triassic of Germany. *N. Jahrb. Geol. Palaeontol. Monatsh.* 1:1–16.

——— 1981. The fin spine of a new holocephalan from the Lower Jurassic of Lyme Regis, Dorset, England. *Geobios* 14:469–75.

——— 1982. Teeth of a new selachian from the Upper Triassic of England. *N. Jahrb. Geol. Palaeontol. Monatsh.* 3:156–66.

——— 1983. Holocephalans in the state museum for natural science in Stuttgart 2. A myriacanthid tooth plate from the Hettangian Lower Lias of northern Bavaria West Germany. *Stuttg. Beitr. Naturkd. Ser. B (Geol. Palaeontol.)* 98:1–7.

Duffin, C. J., and H. Furrer. 1981. Myriacanthid holocephalan remains from the Rhaetian (Upper Triassic) and Hettangian (Lower Triassic) of Graubuenden (Switzerland). *Eclogae Geol. Helv.* 74(3):803–29.

Duffin, C. J., and D. J. Ward. 1983. Teeth of a new neoselachian shark from the British Lower Jurassic. *Palaeontology* 26(4):839–44.

Fengchen, M. A. 1980. A new genus of Lycopteridae from Ningxia, China. *Vert. Palasiat.* 18(4):286–95.

Futuyma, D. J. 1979. *Evolutionary Biology* (Sunderland, Massachusetts: Sinauer Associates).

Gardiner, B. 1967. Subclasses Chondrostei and Holostei. *In* Harland, W. B., et al. (eds.), *The Fossil Re-*

cord (London: Geological Society), pp. 644–54.

——— 1984. The relationships of the palaeoniscid fishes, a review based on new specimens of *Mimia* and *Moythomasia* from the Upper Devonian of Western Australia. *Bull. Brit. Mus. (Nat. Hist.) Geol.* 37:173–432.

Gaudant, J. 1966. Les Actinopterygiens du Mésozoique continental d'Asie central et orientale et le problème de l'origine des Téléostéens. *Bull. Soc. Geol. Fran.* 8:107–13.

Harland, W. B. 1967. *The Fossil Record* (London: Geological Society of London).

Jain, S. L. 1973. New specimens of Lower Jurassic holostean fishes from India. *Palaeontology* 16:149–77.

——— 1974. *Indocoelacanthus robustus* n. gen. n. sp. (Coelacanthidae, Lower Jurassic), the first fossil coelacanth from India. *J. Paleontol.* 48:49–62.

Johnson, G. D. 1980. Xenacanthodii (Chondrichthyes) from the Tecovas Formation (Late Triassic) of West Texas. *J. Paleontol.* 54(5):923–32.

Lauder, G. V., Jr. and K. F. Liem. 1983. The evolution and interrelationships of the actinopterygian fishes. *Bull. Mus. Comp. Zool.* 150(3):95–197.

Liu, H.-T., and J.-J. Zhou. 1965. A new sturgeon from the Upper Jurassic of Liaonning, North China. *Vert. Palasiat.* 9:237–47.

Maisey, J. G. 1976. The Middle Jurassic selachian fish *Breviacanthus* n.g. *N. Jahrb. Geol. Palaeontol. Monatsh.* 7:432–8.

——— 1982a. The anatomy and interrelationships of Mesozoic hybodont sharks. *Am. Mus. Novit.* 2724:1–48.

——— 1982b. Fossil hornshark fin spines (Elasmobranchii, Heterodontidae) with notes on a new species (*Heterodontus tuberculatus*). *N. Jahrb. Geol. Palaeontol. Abh. B* 164(3):393–413.

Martin, M. 1979. *Arganodus atlantis* et *Ceratodus arganensis*, deux nouveaux dipneuster due Trias Superieur continental marocain. *Compt. Rend. Hebd. Seances Acad. Sci. Ser. D Sci. Nat.* 289(2):89–92.

——— 1980. *Mauritanichthys rugosus* n. gen. et n. sp., Redfieldiidae (Actinopterygi, Chondrostei) du Trias Superieur continental marocain. *Geobios* 13(3):437–40.

Martinson, G. G. 1973. O stratigrafii yurskikh i melovykh otlozheniy Mongolii. *Akad. Nauk SSSR, Izv., Ser. Geol.* 12:89–95.

McCune, A. R. 1982. Early Jurassic Semionotidae (Pisces) from the Newark Supergroup: systematics and evolution of a fossil species flock. Ph.D. dissertation. Yale University.

——— 1986. A revision of *Semionotus* (Pisces: Semionoridae) from the Triassic and Jurassic of Europe. *Palaeontology* 29: in press.

Murry, P. A. 1981. A new species of freshwater hybodont from the Dockum Group (Triassic) of Texas. *J. Paleontol.* 55(3):603–7.

Nybelin, O. 1966. On certain Triassic and Liassic representatives of the family Pholidophoridae s. str. *Bull. Brit. Mus. (Nat. Hist.) Geol.* 11:353–432.

1974. A revision of the leptolepid fishes. *Acta R. Soc. Sci. Litt. Gothoburg (Zoologica)* 9:1–202.

Olsen, P. E. 1984. The skull and pectoral girdle of the parasemionotid fish *Watsonulus eugnathoides* from the Early Triassic Sakamena Group of Madagascar, with comments on the relationships of the holostean fishes. *J. Vert. Paleontol.* 4(3):481–99.

Olsen, P. E., A. R. McCune, and K. S. Thomson. 1982. Correlation of the Newark Supergroup by vertebrates, especially fishes. *Am. J. Sci.* 282:1–44.

Patterson, C. 1973. Interrelationships of holosteans. *In* Greenwood, P. H., R. S. Miles, and C. Patterson (eds.), *Interrelationships of Fishes* (London: Academic Press), pp. 233–305.

1977. The contribution of paleontology to teleostean phylogeny. *In* Hecht, M. K., P. C. Goody, and B. M. Hecht (eds.), *Major Patterns in Vertebrate Evolution* (New York: Plenum), pp. 579–643.

1981. Agassiz, Darwin, Huxley, and the fossil record of teleost fishes. *Bull. Brit. Mus. (Nat. Hist.) Geol.* 35(3):213–24.

1982. Morphology and interrelationships of primitive actinopterygian fishes. *Am. Zool.* 22:241–59.

Patterson, C., and D. Rosen. 1977. Review of ichthyodectiform and other Mesozoic fishes and the theory and practice of classifying fossils. *Bull. Am. Mus. Nat. Hist.* 158(2):81–172.

Pfeil, F. H. 1983. Zahnmorphologische Untersuchungen an rezenten und fossilen Heinen der Ordnungen Chlamydoselachiformes und Echinrhiniformes. *Palaeoichthyologica* 1:1–355.

Raup, D. M. 1972. Taxonomic diversity during the Phanerozoic. *Science* 177:1065–71.

Reif, W.-E. 1977. Tooth enameloid as a taxonomic criterion: 1. A new euselachian shark from the Rhaetic–Liassic boundary. *N. Jahrb. Geol. Palaeontol. Monatsh.* 9:565–76.

Reif, W. 1980. Tooth enameloid as a taxonomic criterion, 3: A new primitive shark family from the lower Keuper. *N. Jahrb. Geol. Palaeontol. Abh. B* 160(1):61–72.

Rieppel, O. 1980. A new coelacanth from the Middle Triassic of Monte San Giorgio, Switzerland. *Eclogae Geol. Helv.* 73(3):921–39.

1982. A new genus of shark from the Middle Triassic of Monte San Giorgio, Switzerland. *Palaeontology* 25(2):399–412.

Romer, A. S. 1966. *Vertebrate Paleontology* (Chicago: University of Chicago Press).

Rosen, D. E. 1982. Teleostean interrelationships, morphological function, and evolutionary inference. *Am. Zool.* 22(2):261–74.

Saint-Seine, P. De. 1955. Poissons fossiles de l'étage de Stanleyville (Congo Belge). Première parti. La faune des argilites et schistes bitumineaux. *Ann. Mus. Congo Belge, Sér. in 8, Sci. Géol.* 14:1–126.

Schaeffer, B. 1967. Late Triassic fishes from the western United States. *Bull. Am. Mus. Nat. Hist.* 135(6):289–342.

1972. A Jurassic fish from Antarctica. *Am. Mus. Novit.* 2495:1–17.

1973a. Fishes and the Permian-Triassic boundary. *Canad. Soc. Petrol. Geol. Mem.* 2:493–7.

1973b. Interrelationships of chondrosteans. *In* Greenwood, P. H., R. S. Miles, and C. Patterson (eds.), *Interrelationships of Fishes* (London: Academic Press), pp. 207–26.

Schaeffer, B., and M. Mangus. 1976. An Early Jurassic fish assemblage from British Columbia. *Bull. Am. Mus. Nat. Hist.* 156(5):517–63.

Schaeffer, B., and C. Patterson. 1984. Jurassic fishes from the western United States with comments on Jurassic fish distribution. *Am. Mus. Novit.* 2796:1–86.

Selezneva, A. A. 1982. Triassic fish finds in the Franz Josef Land archipelago. *Paleontol. J.* 16:131–4.

Sepkoski, J. J., Jr. 1981. A factor analytic description of the Phanerozoic marine fossil record. *Paleobiology* 7(1):36–53.

Stolley, E. 1920. Beiträge zur Kenntnis der Ganoiden des deutschen Muschelkalks. *Paleontographica* 63:25–86.

Su, D. 1980. A new species of *Perleidus* from Anhui. *Vert. Palasiat.* 19(2):107–12.

Su, T. T. 1974. New Jurassic ptycholepid fishes from Szechuan, southwest China. *Vert. Palasiat.* 12:1–20.

Sykes, J. H. 1971. A new dulatiid fish from the Rhaetic bone bed at Barnstone, Nottinghamshire. *Mercian Geol.* 4:13–22.

Sykes, J. H., and O. J. Simon. 1979. A new colobodont fish from the Trias of Spain. *Mercian Geol.* 1:65–74.

Tanner, V. M. 1925. A study of Utah fossil fishes with description of a new genus and species. *Utah Acad. Sci. Arts Lett.* 15(6):81–5.

Thies, D. 1983. Jurazeitliche Neoselachier aus Deutschland und S. England. *Cour. Forsch. Senckenberg* 58:1–116.

Thomson, K. S. 1977. The pattern of diversification among fishes. *In* Hallam, A. (ed.), *Patterns of Evolution* (Amsterdam: Elsevier), pp. 377–404.

Tintori, A. 1980. Two new pycnodonts (Pisces, Actinopterygii) from the Upper Triassic of Lombardy (N. Italy). *Riv. Ital. Paleontol. Stratigr.* 86(4):795–824.

Wang, N. 1977a. A new pholidophorid fish from Hengnan, Hunan. *Vert. Palasiat.* 15:177–83.

1977b. Jurassic fishes from Lingling-Hengyang, Huan and their stratigraphical significance. *Vert. Palasiat.* 15:233–43.

Zambelli, R. 1975. Note sui Pholidophoriformes: I. Parapholidophorus nybelini gen. n. sp. n. *Rend. Ist. Lomb. Sc. e Lett. (Scienze B), Milano* 109:3–49.

1978. Note sui Pholidophoriformes. II. *Pholidocténus serianus* gen. n. sp. n. *Accad. Nazion. Dei XL. Estr. Rend. Ser. V* III: 101–24.

1980. Note sui Pholidophoriformes. IV. Contributo: *Pholidorhynchodon malzanni* gen. nov. sp. nov. *Riv. Mus. Sc. Nat. BG* 2:129–59.

Zhang, M., and J. Zhou. 1974. Fish fossils from the latter part of the Mesozoic of Zhejiang Province (brief report). *Vert. Palasiat* 12:183–6.

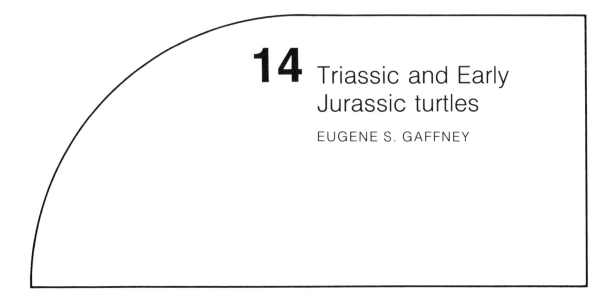

14 Triassic and Early Jurassic turtles

EUGENE S. GAFFNEY

Introduction

The history of turtles extends back 200 MY into the Late Triassic, and specimens of that age have been known for more than 100 years (Meyer 1863; Cope 1884). The first discovered specimens, however, consisted of shell fragments and steinkerns that did not reveal a great deal about the animals themselves. Even when the first skull of a Triassic turtle was discovered (Jaekel 1918), strong lateral crushing of that specimen led to erroneous interpretations of its morphology. It was not until 1961 that Parsons and Williams revealed some information about a well-preserved turtle skull found in the 1930s in southern Germany. The first accurate description of that specimen did not appear until 1983 (Gaffney and Meeker 1983). Detailed descriptions of shells (Gaffney 1985) of well-preserved Triassic turtles are also becoming available for the first time. In addition to these new efforts to describe and study the older material, the past few years have coincidentally seen the discovery in the field of new records of Triassic and Early Jurassic turtles that greatly extend the geographic range and provide interesting new taxa that will allow new insight into early chelonian evolution. Much of the new material is under study at present, and this report is intended to summarize briefly the current status of early Mesozoic cheloniology. Table 14.1 summarizes the known pre–Late Jurassic turtle record.

The German Proganochelys

The best-preserved specimens of any pre–Late Jurassic turtle is the German genus *Proganochelys* Baur (= *Triassochelys* = *Stegochelys* Jaekel). There are seven specimens known to date, among them three skull–shell associations. The best specimen, SMNS 16980 (Figs. 14.1, 14.2), is a nearly complete skeleton with a well-preserved skull (Gaffney 1983;

Gaffney and Meeker 1983). This specimen, along with two other partial skeletons (Gaffney 1985) was found in 1932 in Trossingen, Federal Republic of (West) Germany, in a large quarry developed for the collection of *Plateosaurus* specimens.

In addition to the Trossingen–Aixheim area in southern West Germany, a skull, shell, and cervical series were collected in Halberstadt, near the Harz Mountains of the German Democratic Republic (East Germany), about 500 km north of Trossingen. The Halberstadt material was described by Jaekel in 1918 as *Triassochelys*, but a new study of the material (Gaffney 1985) synonymized it with *Proganochelys*. The horizons of the German *Proganochelys* specimens vary from the Knollenmergel to the Stubensandstein, but it is thought that all are mid-Keuper, Late Triassic (Norian) in age.

The skull of *Proganochelys* shows a number of generalized amniote features not found in other turtles. It retains the supratemporal bone, lacrimal bone, lacrimal duct, movable basipterygoid articulation, and has a middle ear lacking the expansion and bony walls found in all other turtles. All remaining turtles form a monophyletic group, the Casichelydia (Gaffney 1975, 1984), with the following derived features: supratemporal bone absent, lacrimal bone and duct absent, basipterygoid articulation fused, interpterygoid vacuity closed, and middle ear with lateral and ventral wall.

The postcranial features of *Proganochelys* are also interesting. It retains two generalized amniote features absent in other turtles: The cleithrum is present and the clavicle (= epiplastron) has a dorsal process. The shell of *Proganochelys* is typically chelonian but has a greater number of bones and scales than in other turtles [see Gaffney (1985) for figures and detailed description].

Table 14.1. *Triassic and Early Jurassic turtles*

Taxon	Consists of	Locality	Horizon	Remarks
Proganochelys quenstedti	Seven shells, three partial skeletons, and three skulls	Trossingen and other West German localities; Halberstadt, East Germany	Stubensandstein and Knollenmergel; Keuper, Late Triassic	The only Triassic turtle known from nearly complete skeletons; sister group to all other turtles
Proterochersis robusta	At least one dozen shells	Various localities in southern West Germany	Stubensandstein; Keuper, Late Triassic	The oldest pleurodire, contemporary with *Proganochelys*
"Proganochelys" ruchae DeBroin 1984	Shell fragments	Vicinity of Chum Phae-Lom Sak highway, northeast Thailand	Huai Hin Lat Formation, Late Triassic	A possible proganochelyid roughly contemporary with the German turtles, the oldest turtle from Asia
unnamed	Skull	Clocolan district, Orange Free State, South Africa	Elliot Formation (Red Beds), Late Triassic–Early Jurassic?	Definite proganochelyid, the oldest turtle from Africa
unnamed	At least four skulls, two shells, and various postcranial elements	Various localities in northeastern Arizona, USA	Kayenta Formation, Early Jurassic	The oldest turtle from North America, a cryptodire

The German Proterochersis

The first Triassic turtle described was *Chelytherium obscurum* (Meyer 1863). The type consists only of fragments, but these fragments appear to be identical to *Proterochersis robustum* (Fraas 1913). *Chelytherium* appears to fall under the fifty-year rule for "forgotten names," therefore *Proterochersis* is the proper name for this form. There are about a dozen specimens of *Proterochersis* (Fig. 14.3) known at present; most of these are partial shells. The pelvis is sutured to the carapace and plastron, and only this element has been found in addition to the shell. *Proterochersis* is under study at present by the author.

Proterochersis occurs in the Stubensandstein of southern Germany, in the Murr Valley, and the vicinity of Stuttgart. It is therefore mid-Keuper and Late Triassic (Norian) in age. Some of the *Progan-*

ochelys specimens also come from the Stubensandstein of this region, and both *Proganochelys* and *Proterochersis* appear in the fossil record at the same time.

Because *Proterochersis* is known only from the shell and pelvis, definitive hypotheses about its relationships are not practical until something is known about its skull. However, the sutured shell in *Proterochersis* is a derived character for the Pleurodira, the side-necked turtles (Gaffney 1975). The Pleurodira and the Cryptodira are the two monophyletic groups making up the Casichelydia (Gaffney 1975, 1983). The identification of *Proterochersis* as a pleurodire has interesting consequences. In terms of tree construction and evolutionary scenarios, the contemporaneity of *Proganochelys* (the sister taxon to all turtles) and *Proterochersis* (a member of a group within the Casichelydia) suggests that by the Late

Triassic the Casichelydia had already split into the Cryptodira and the Pleurodira. The eventual discovery of a Triassic cryptodire could be predicted from these ideas.

The Thailand Triassic turtle

In 1982, DeBroin et al. announced the discovery of turtles in Late Triassic rocks from Thailand. The described material consists of shell fragments that are diagnostic enough to be identified as chelonian, but further identification is speculative.

Turtle fragments have been found at a number of localities in the vicinity of the Chulabhorn Dam and Khon Kaen in northeastern Thailand (see Ingavat and Janvier 1981, for map). The turtle fragments occur in the Huai Hin Lat Formation, which

seems to be well documented as Norian in age, based on plants and ostracods (see DeBroin et al. 1982, for references), as well as a vertebrate fauna of capitosaurids and phytosaurs.

DeBroin (1984) described an epiplastron from the Thai material that has a dorsal process very similar to that seen on the epiplastron of *Proganochelys* (Gaffney 1985). This feature, although at present unique to *Proganochelys* among turtles, is a retained primitive feature and does not really allow identification of the Thai turtle as *Proganochelys* or a near relative. Nonetheless, the discovery of a bona fide Triassic turtle outside Germany more than one hundred years after the first discovery of a Triassic turtle is very exciting and hopeful for future discoveries.

Figure 14.1. The skull of *Proganochelys quenstedti* (SMNS 16980), Late Triassic of Trossingen, FRG. Abbreviations: bo, basioccipital; bs, basisphenoid; ex, exoccipital; fr, frontal; ju, jugular; la, lacrimal; mx, maxilla; na, nasal; op, opisthotic; pa, parietal; pal, palatine; pfr, prefrontal; pmx, premaxilla; po, postorbital; pr, prootic; pt, pterygoid; qj, quadratojugal; qu, quadrate; soc, supraoccipital; sq, squamosal; st, supratemporal; vo, vomer; XII, foramen nervi hypoglossi. From Gaffney and Meeker (1983).

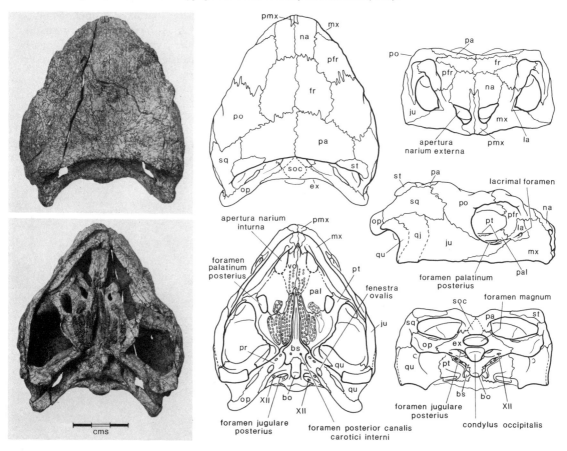

Figure 14.2. The carapace of *Proganochelys quenstedti* (SMNS 16980), Late Triassic, Trossingen, FRG. Maximum carapace length, 56 cm.

The South African Proganochelyid

In 1980, James Kitching announced the discovery of a turtle skull from the early Mesozoic of South Africa (*Society of Vertebrate Paleontology News Bulletin*, No. 120, Oct. 1980, p. 37). Work on this specimen is still underway by Kitching, but he has kindly sent me photographs and a cast.

The skull is from the Elliot Formation (Red Beds) in the Clocolan district, Orange Free State. I am not yet aware of its age, as the Elliot Formation contains both Triassic and Jurassic sediments.

On the basis of Kitching's photographs and casts it can be seen that his specimen has a basioccipital tubercle, identified as a *Proganochelys* autapomorphy by Gaffney and Meeker (1983). The South African skull does differ from the German *Proganochelys*, but the two taxa may be tentatively hypothesized as sister taxa forming their own monophyletic group.

The Kayenta turtle

The Kayenta Formation of northern Arizona has recently yielded numerous turtle specimens, including two shells, partial skeletons, and at least four skulls. This material is still in the preliminary stages of study, but it appears that there is only one taxon represented and that it is a casichelydian, probably the sister taxon to the cryptodires. Although nearly all of the specimens are crushed, the bone preser-

Figure 14.3. Shell of *Proterochersis robusta* (SMNS 17561), Late Triassic, Murrhardt, FRG. Maximum carapace length, 33 cm.

vation is good, and dependable reconstructions are possible.

The specimens have been found at various localities in northeastern Arizona that have recently yielded an interesting and diverse tetrapod fauna, including dinosaurs, crocodilians, therapsids, lizards, amphibians, pterosaurs, and mammals (see Jenkins, Crompton, and Downs 1983; Padian 1984, for references). The horizon yielding the turtles is the Kayenta Formation, traditionally considered to be Late Triassic in age, but recent work presents evidence favoring an Early Jurassic date (Olsen and Galton 1977).

The Kayenta turtle has palatal teeth, similar to those in *Proganochelys*, and an open interpterygoid vacuity. However, the basipterygoid articulation is fused, a synapomorphy for the Casichelydia. Although specimens are still being prepared, there does seem to be one cryptodiran synapomorphy in the basicranium. It is likely that the Kayenta turtle will provide significant information about character patterns in Casichelydia.

Conclusions

The oldest known turtles are *Proganochelys* and *Proterochersis* from Germany and indeterminant shell fragments from Thailand, all of Norian (Keuper) age. *Proganochelys* is the best known Triassic turtle, being represented by three skulls and a number of skeletons. Apparently younger turtles are known from the Elliot Formation of South Africa and the Kayenta Formation of Arizona, both records represented by skull material currently under study. Phylogenetic studies completed to date argue that *Proganochelys* is the sister group to all other turtles while its contemporary, *Proterochersis*, is a pleurodire, suggesting that the major groups of turtles diversified before the Late Triassic.

Acknowledgments

The opportunity to study *Proganochelys* is due to the encouragement of Dr. Rupert Wild and Professor Bernhard Ziegler, Staatliches Museum für Naturkunde in Stuttgart (SNMS) and Dr. Karl-Heinz Fischer and Dr. Hermann Jaeger, Museum für Naturkunde, Berlin. I am very grateful to Dr. Farish Jenkins and his associates at the Museum of Comparative Zoology, Harvard University; Dr. Kevin Padian, Museum of Paleontology, Berkeley; and to the Museum of Northern Arizona, Flagstaff, for the opportunity to study the Kayenta turtles. Dr. France DeBroin and Dr. Philippe Janvier provided access to the Thailand specimens housed in the Museum National d'Histoire Naturelle, Paris. Dr. James Kitching, Bernard Price Institute, Johannesburg, sent information on the South African turtle. I am very grateful to all these people for their assistance.

The work summarized in this paper was supported by NSF Grants DEB 8002885 and BSR 8314816.

References

Cope, E. D. 1884. The Vertebrata of the Tertiary formations of the west. Book I. *Rept. U.S. Geol. Surv. Terr., F. V. Hayden, 1884*, 3:1–1009.

DeBroin, F. 1984. *Proganochelys ruchae* n. sp., chelonien du Trias supérieur de Thailande. *In* DeBroin, F., and Jimenez-Fuentes (eds.), *Studia Palaeocheloniologica* (Salamanca: Studia Geologica Salamanticensia).

DeBroin, F., R. Ingavat, P. Janvier, and N. Sattayarak. 1982. Triassic turtle remains from northeastern Thailand. *Vert. Paleontal.* 2(1):41–6.

Fraas, E. 1913. *Proterochersis*, eine pleurodire Schildkröte aus dem Keuper. *Jahresh. Ver. Vaterland. Naturk. Württemberg*, 80:13–30.

Gaffney, E. S. 1975. A phylogeny and classification of the higher categories of turtles. *Bull. Am. Mus. Nat. Hist.* 155(5):387–436.

    1983. The basicranial articulation of the Triassic turtle, *Proganochelys*. *In* Rhodin, A. G. J., and K. Mayata (eds.), *Advances in Herpetology and Evolutionary Biology, essays in honor of Ernest E. Williams* (Cambridge, Massachusetts: Museum of Comparative Zoology Special Publication, pp. 190–4.

    1984. Historical analysis of theories of chelonian relationships. *Syst. Zool.* 33(3): 283–301.

    1985. The shell morphology of the Triassic turtle, *Proganochelys*. *N. Jahrb. Geol. Paläont. Abh.* 170:1–26.

Gaffney, E. S., and L. J. Meeker, 1983. Skull morphology of the oldest turtles: a preliminary description of *Proganochelys quenstedti*. *J. Vert. Paleontal.* 3(1):25–8.

Ingavat, R., and P. Janvier. 1981. *Cyclotosaurus* cf. *posthumus* Fraas (Capitosauridae, Stereospondily) from the Huai Hin Lat Formation (Upper Triassic), northeastern Thailand, with a note on capitosaurid biogeography. *Geobios* 6(14):711–25.

Jaekel, O. 1918. Die Wirbeltierfunde aus dem Keuper von Halberstadt. Serie II. Testudinata. Teil 1. *Stegochelys dux* n.g., n. sp. *Paläont. Z.* 2:88–214.

Jenkins, F., A. W. Crompton, and W. R. Downs. 1983. Mesozoic mammals from Arizona; new evidence on mammalian evolution. *Science* 222:1233–5.

Meyer, H. von. 1863. Letter on various fossils. *N. Jahrb. Min. Geol. Paleontal.* 1863:444–50.

Olsen, P. E., and P. M. Galton. 1977. Triassic–Jurassic tetrapod extinctions: are they real? *Science* 197:983–6.

Padian, K. 1984. Pterosaur remains from the Kayenta Formation (?Early Jurassic) of Arizona. *Palaeontology* 27(2):407–13.

Parsons, T. S., and E. E. Williams. 1961. Two Jurassic turtle skulls: a morphological study. *Bull. Mus. Comp. Zool.* 125:43–107.

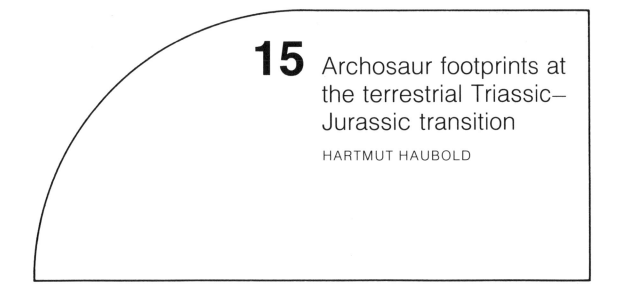

15 Archosaur footprints at the terrestrial Triassic–Jurassic transition

HARTMUT HAUBOLD

Introduction

The footprints of reptiles, especially dinosaurs, are common indicators of the continuous existence of terrestrial life in the Late Triassic and Early Jurassic. Footprints of "Upper Triassic" (now Lower Jurassic) age of eastern North America were first discovered in 1802 and described by Edward Hitchcock (1836). The fossil footprint faunas of the Hartford Basin in Connecticut and Massachusetts became well known through the publications of Hitchcock (1836, 1845, 1858, 1865), whose work was continued by R. S. Lull (1904, 1915, 1953). Additional records of the Newark footprints in related deposits from Nova Scotia to Virginia were described by Baird (1954, 1957), Olsen (1980), and Olsen and Baird (1982 and Chapter 6).

This chapter is a brief review of the footprint faunas across the Triassic–Jurassic boundary, and some of their associated problems. The nomenclatural status of the faunas examined here is inconsistent, as are the taxonomic methods employed to describe them. Nevertheless, clearly similar forms occur in Europe, North America, and southern Africa, and from these we can recognize generic congruences and extensive, uniform evolutionary trends. The faunas can be attributed to three main sections: Carnian–Norian, Rhaetian, and Early Jurassic (Hettangian–Toarcian), which will be characterized sequentially. In addition to many transitory forms, tridactylous grallatorids (theropod dinosaur tracks), and crocodiloid footprints (e.g., *Batracho-pus*), a gradual reduction in the representation of various "thecodont" ichnogenera is important to recognize. In the formations of Rhaetian age, new types of tridactyls (e.g., *Eubrontes*) occur, but the faunas generally seem to be reduced to a few tridactyls and crocodiles and the last "thecodonts."

During the Early Jurassic, a nearly worldwide increase in different dinosaurian footprint types is recorded.

The accuracy of the stratigraphic correlation of footprint faunas was increased by the palynomorph studies of Cornet and Traverse (1975). The range of the Newark Supergroup formations is now known to be from Carnian to Toarcian in age. Olsen (1983) suggested a succession of three footprint zones, corresponding roughly with the three palynomorph zones of Cornet and Traverse. The three footprint zones are (1) the *Brachychirotherium–Atreipus [Anchisauripus] milfordensis* zone, (2) the *Batrachopus–Grallator [Anchisauripus]* zone, and (3) the *Anomoepus–Grallator [Eubrontes]* zone. These zones correspond to the European intervals of Carnian–Norian (Upper Triassic), Rhaetian (Uppermost Triassic), and Hettangian–Toarcian (Lower Jurassic). The Passaic (= Lower Brunswick), Lockatong, and Stockton Formations of New Jersey and Pennsylvania represent the Carnian (Fig. 15.1).

In western Europe, Upper Triassic as well as Lower Jurassic occurrences are known. Carnian–Rhaetian–age footprint faunas come from the German Keuper (Ruehle von Lilienstern 1938; Beurlen 1950; Heller 1952; Haubold 1971b); from related formations in England and Wales (Sollas 1879; Tucker and Burchette 1977); from the Rhaetian of the Isle of Schonen (Boelau 1952); from the Carnian–Norian in the Toscana region of Italy (Huene 1941); and from the Norian–Rhaetian of Languedoc, Sanary, and Toulon in France (Ellenberger, Ellenberger, and Ginsburg 1970). Lower Jurassic footprint faunas occur in the Hettangian of France (Thaler 1962; Lapparent and Montenat 1967), and in the Lotharingian of Aveyron (Ellenberger and Fuchs 1965).

The first publications of equivalent faunas of the South African Stormberg Series in Lesotho (P. Ellenberger 1955; Ellenberger and Ellenberger 1958) were followed by the first parts of monographic descriptions (P. Ellenberger 1970, 1972, 1974). F. Ellenberger et al. (1970) dated the sediments from the Molteno Formation to the Drakensberg Lavas as Upper Triassic to Lower Jurassic (Pliensbachian), and suggested a correlation to the Western European footprint faunas of France. Following the first stratigraphic determination of the Toscana footprints as Late Triassic in age (Huene 1940), F. Ellenberger et al. (1970) demonstrated one of the first stratigraphic uses of dinosaur footprints during the Triassic–Jurassic transition (see also Demathieu 1977).

Many other more isolated or not so well-known occurrences of the same age from Morocco, Iran, China, Australia, South America, and North America (Utah, Arizona, and Colorado) are mentioned in Haubold (1984). They will not be discussed here in detail, but all have about the same faunal content and reflect the same tendencies. A general overview of the stratigraphic positions of the footprint-bearing horizons is summarized in Fig. 15.1.

Taxonomic problems

The nomenclature of reptilian footprints generally follows the principles of binominalism and parataxonomy, and is intended to reflect relationships with osteological classifications (Peabody 1955;

Haubold 1971a, 1984). The first scientific description of vertebrate footprints, *Chirotherium barthii* Kaup 1835, from the German Lower Triassic, followed this procedure. Binominalism for the Newark footprints was first introduced by Hitchcock (1845), and their taxonomy and nomenclature were consolidated over the following decades, first by Hitchcock and later by Lull (1904, 1915, 1953). However, at present, the taxonomic position of many forms is not clear and needs much refinement (Baird 1957; Olsen 1983). Olsen and Baird (Chapter 6) and Olsen and Padian (Chapter 20) provide current contributions with particular revisions, and their definition of *Atreipus* and their revision of *Batrachopus* are accepted here. In this chapter, I am not attempting additional revision or stabilization of the nomenclature of Triassic–Jurassic footprint taxa, desirable though that may be. I will try only to indicate the complexity of the problems. Detailed interpretation of many dinosaur footprints is also far from clear. This problem is too large to discuss here, but some aspects have recently been discussed elsewhere (Haubold 1984): Of some 450 published genera of fossil amphibian and reptilian footprints, only 310 of them were suggested as valid, and of these, about 160 genera are known from Triassic and Lower Jurassic horizons.

In Europe, the description of Triassic–Jurassic footprints has a longer tradition, but no monographic study of the Keuper–Liassic material exists (Haubold 1971a,b). However, there are clear similarities with the Newark Supergroup faunas, al-

Figure 15.1. Relative position of footprint-bearing horizons in Upper Triassic and Lower Jurassic formations. Abbreviations: Fm. = Formation; Gr. = Group; L. = Lower; Sst. = Sandstone; U. = Upper. Adapted from Haubold (1984).

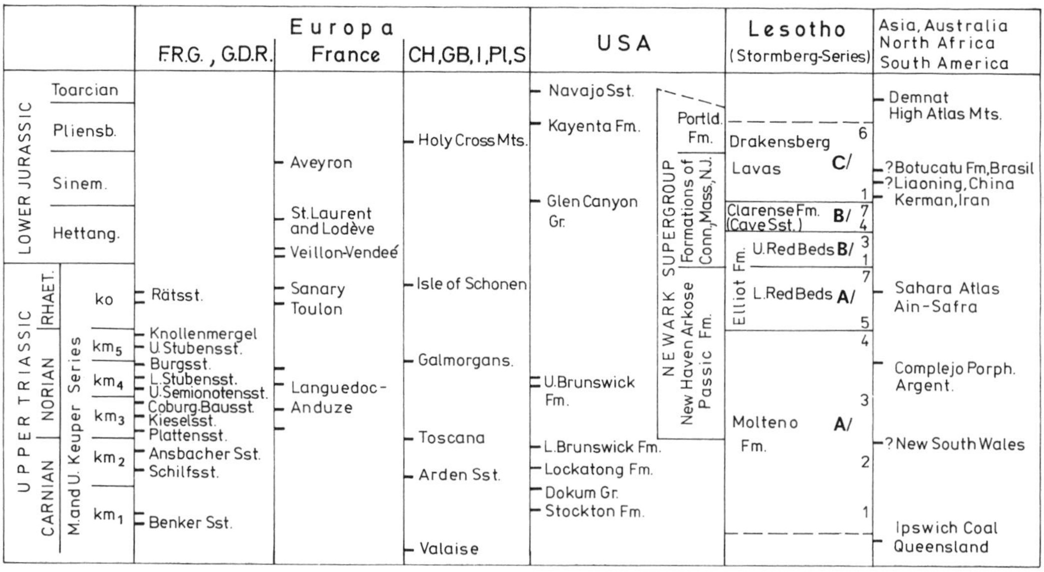

though only compilations and some more recent studies show the connections. Some of the most important genera that occur in Upper Triassic and Lower Jurassic sediments of both North America and Europe are *Anchisauripus, Grallator, Eubrontes, Brachychirotherium, Chirotherium, Batrachopus, Apatopus/Thecodontichnus, Coelurosaurichnus,* and *Atreipus [Anchisauripus] milfordensis* resp. *Grallator sulcatus/Coelurosaurichnus metzneri.*

In contrast, the footprints of Lesotho nominally have practically no relationship to those of the Newark or the European Keuper and Liassic formations. P. Ellenberger (1970, 1972, 1974) described the material without closely comparing it to previously known taxa. As a result, his classifications show only minor agreement with European and North American faunas of the same age. One reason for the nomenclatural separation of the Stormberg footprints might be the partly unclear status of some of the Newark ichnotaxa resulting from differentiations and definitions in the ichnological literature through Lull (1953). This is demonstrated, for example, by the problem of *Batrachopus*, discussed in the introduction to Chapter 20. Other reasons could be the paleogeographic and evolutionary separation of Lesotho, as P. Ellenberger argued (1970, p. 353). However, his completely independent description of about 75 new genera and 150 new species, without any consideration for earlier published material and priorities from other parts of the world, must be considered problematic. Furthermore, Ellenberger leaned considerably toward the practice of "splitting" when he differentiated many of his taxa (Figs. 15.3, 15.9); yet as late as 1974 only the footprints from the Molteno Formation through the basal Upper Red Beds (now called the Elliot Formation; see Fig. 15.1) were published and adequately defined. All the remaining names of Jurassic age noted by P. Ellenberger (1970, 1972) are still undefined. These names are placed in quotation marks in this chapter. Insofar as the Lesotho nomenclature is restricted to the Stormberg footprints, it may be provisionally accepted. However, without major revision, any transfer of this nomenclature to the European Triassic–Jurassic faunas, as P. Ellenberger (1970) did for Anduze in France, and as Demathieu and Weidmann (1982) did for Valais in Switzerland, seems unjustifiable. The following examples suggest some possible relationships or synonymies of some tridactylous Lesotho footprints with genera from the Newark Supergroup:

> **Grallator**: *Pseudotrisauropus, Tritotrisauropus, Deuterosauropus, Masitisisauropus,* and *Trisaurodactylus*
>
> **Anchisauropus**: *Prototrisauropodiscus* and *Trisauropodiscus*

> **Eubrontes**: *Qemetrisauropus, Neotrisauropus,* and *Plastisauropus*
>
> **Anomoepus**: *Moyenisauropus*

Setting aside the problems of their separate nomenclature, the knowledge of the Lesotho footprints, carefully studied and published in detail by P. Ellenberger, are of great importance for the Triassic–Jurassic transition, and were it not for the activities of P. Ellenberger, the material would still not be known for comparison and general analyses. Because a comprehensive revision is unlikely in the near future, we have to seek a compromise that is appropriate to the older taxonomic priorities as well as to the great work of P. Ellenberger.

Zonation of footprints

The succession of footprint associations suggested by Olsen (1980, 1983) for the Newark Supergroup is generally applicable to formations of similar age in western Europe and southern Africa. The following three levels, corresponding to zones I–III of Olsen (1983), contain characteristic elements of archosaurian footprints, especially those of dinosaurs.

I: Carnian–Norian (Figs. 15.2, 15.3)

North America

Lockatong, Stockton, Passaic (= Lower Brunswick), and Wolfville formations of the Newark Supergroup:

Anchisauripus sp., *Atreipus [Anchisauripus] milfordensis – Grallator sulcatus* (n.g. of Olsen and Baird, this volume), *Coelurosaurichnus* sp., *Chirotherium lulli, Brachychirotherium parvum, B. eyermanni, Apatopus lineatus,* and *Gwyneddichnium minor*

Europe

Middle Keuper formations, km1–km4:

Anchisauripus sp., *Coelurosaurichnus* sp., *Chirotherium* sp., *C. wondrai, Brachychirotherium thuringiacum, Parachirotherium postchirotherioides, Gwyneddichnium* sp., and *Thecodontichnus* spp. Ellenberger's taxa are mentioned as *Anatrisauropus, Deuterotrisauropus, Mafatrisauropus, Prototrisauropus, Deuterosauropodopus, Parasauropodopus,* and *Pseudotetrasauropus*

Lesotho (South Africa)

Molteno Formation, Zones A/1–4:

Anatrisauropus, Bosiutrisauropus, Deuterotrisauropus, Mafatrisauropus, Paratrisauropus, Prototrisauropus, Qemetrisauropus, Seakatrisauropus, Sauropodopus, Deuterosauropodopus, Pentasauropus, Pseudotetrasauropus, and *Tetrasauropus*

II: Rhaetian (Fig. 15.4)

North America

Uppermost Passaic Formation:

Anchisauripus sillimani, A. minusculus, Batrachopus sp., and *Eubrontes giganteus*

Europe

Rhaetic Sandstone and Höganäs Series:

Coelurosaurichnus spp. and *Eubrontes* sp.

Lesotho

Lower Red Beds (= Lower Elliot Formation), Zones A/5–7:

Trisauropodiscus spp. and *Tritotrisauropus medius*

III: Lower Jurassic, Hettangian–Toarcian (Figs. 15.5–15.9)

North America

Upper formations of the Newark Supergroup, Hartford, Newark, and Fundy basins:

Anchisauripus sillimani, A. exsertus, A. hitchcocki, A. minor, A. parallelus, Grallator cursorius, G. cuneatus, G. gracilis, G. tenuis, Eubrontes giganteus, E. approximatus, E. divaricatus, E. platypus, Apatichnus, Gigandipus, Hyphepus, Otozoum, Sauropus, Anomoepus isodactylus, A. scambus, A. cuneatus, A. curvatus, A. minimus, A. crassus, A. intermedius, Stenonyx, Selenichnus, Platypterna, Argoides, Trihamus, Batrachopus deweyi, B. dispar,

Figure 15.2. Carnian–Norian footprint types of Europe and North America. **A**, *Chirotherium lulli*; **B**, *C. wondrai*; **C**, *Brachychirotherium parvum*; **D**, *B. eyermani*; **E**, *Parachirotherium postchirotherioides*; **F**, *Gwyneddichnium*; **G** and **H**, *Anchisauripus*; **I**, *Coelurosaurichnus toscanus*; **K**, *Atreipus [Anchisauripus] milfordensis*; **L**, *Atreipus [Coelurosaurichnus] metzneri*; **M**, *Apatopus lineatus*. Redrawn from Baird 1954 (**A**), Heller 1952 (**B**), Baird 1957 (**C**, **D**, and **M**), Kuhn 1958 (**E**), Olsen 1980 (**F**, **K**, and **L**), Tucker and Burchette 1977 (**G** and **H**), and Huene 1941 (**I**).

B. parvulus, Cheirotheroides, Shepardia, and *Tarsodactylus*

Formations of the Glen Canyon Group, Kayenta Formation, and Navajo Sandstone:

Batrachopus, Dilophosauropus, Hopiichnus, Kayentapus, and *Navahopus*

Europe
Infralias, Hettangian:

*Grallator variabilis, G. oloensis, G. maximus, Eu-*brontes, Saltopoides, Talmontopus, Anatopus, Batrachopus,* and *Dahutherium*

Lesotho
Upper Red Beds (= Upper Elliot Formation) and Cave Sandstone (= Clarens Formation), Zones B/1–5:

Neotrisauropus, Plastisauropus, Aetonychopus, Masitisisauropus levicauda, M. palmipes, M. angustus, M. exiguus, Masitisisauropezus, Masitisisauropodiscus,

Figure 15.3. Tridactyl footprints of bipeds and pentadactyl footprints of quadrupeds, from the Molteno Formation of Lesotho. The tridactylous prints show the extreme taxonomic splitting at the generic level. **A**, *Qemetrisauropus*; **B**, *Anatrisauropus*; **C**, *Prototrisauropus*; **D**, *Paratrisauropus*; **E**, *Seakatrisauropus*; **F**, *Bosiutrisauropus*; **G**, *Deuterotrisauropus*; **H**, *Mafatrisauropus*; **I**, *Pseudotrisauropus*; **K**, *Pseudotetrasauropus*; **L**, *Tetrasauropus*; **M**, *Paratetrasauropus*; **N**, *Deuterosauropodopus*; **O**, *Sauropodopus*. After P. Ellenberger (1972).

Moyenisauropus natator, M. natatilis, M. vermivorus, M. dodai, M. minor, M. longicauda, Suchopus, Trisaurodactylus, "*Katosauropus*," "*Neotripodiscus*," "*Otouphepus magnificus*," "*O. minor*," *O. palustris*," "*O. declivis*," "*Platysauropus*," "*Platytrisauropus*," "*Cridotrisauropus*," "*Gyrotrisauropus*," "*Megatrisauropus*," "*Kleitotrisauropus*," "*Kainotrisauropus*," "*Grallator damanei*," "*G. molapoi*," "*Paragrallator*" "*Kainomoyenisauropus*," and "*Qomoqomosauropus*"

General tendencies of the Triassic–Jurassic transition

The footprint assemblages demonstrate special trends in all lines of archosaurs (Fig. 15.10). The Carnian–Norian section (I) contains the last chirotherians (usually presumed to be the tracks of quadrupedal "thecodontians"), associated with a number of smaller tridactylous footprints and the

Figure 15.4. Rhaetian footprints of Europe, North America, and Southern Africa. **A**, *Anchisauripus sillimani*; **B**, *A. minusculus*; **C**, *Eubrontes giganteus*; **D**, *Coelurosaurichnus* spp.; **E**, *Trisauropodiscus*. A–C after Lull (1953); **D**, after Kuhn (1958); **E** after P. Ellenberger (1974). Scale = 5 cm.

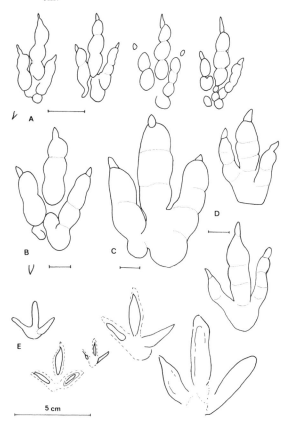

presumed footprints of phytosaurs (*Apatopus*). Most of these elements are descendants of the widely distributed first radiation of archosaurs, including the first dinosaurian lineages of the earlier Triassic (Demathieu and Haubold 1978). All of the very different lineages of chirotherians start in the Lower Triassic and are in many regards above the "thecodontian" level (Haubold 1983). The first tridactylous bipeds are recorded from the Anisian–Ladinian. The faunas of the Lower Stormberg in Lesotho have a somewhat different aspect: both small and large tridactylous bipeds are very common. The occurrence here of different kinds of quadrupedal, sauropod-like footprints (the Pentasauropodia, Melanorosauria, and Sauropoda of Ellenberger) is also remarkable. Chirotherians and crocodylomorph types are absolutely missing in the Carnian–Norian equivalents of Lesotho.

In the Rhaetic section (II), all faunas from footprint-bearing horizons are both less abundant and less diverse, and mainly consist of small tridactylous bipeds (grallatorids) of coelurosaurian origin. *Apatopus*, the presumed phytosaur trackway, persists from the former section. In the Passaic Formation of the Newark Supergroup and the Rhaetian of Europe, larger tridactyls (*Eubrontes*) appear, but they are missing in the lower Red Beds of Lesotho.

The Lower Jurassic section (III) is marked by an enormous increase of some tridactylous footprints, especially in the Newark Supergroup. Some common grallatorids are also found in the European Infralias. The presumed crocodylomorph trackway *Batrachopus* was also continuous and transitional. The record of *Batrachopus* (*sensu stricto*) starts in the uppermost Rhaetian horizon. Important relatives in the Lower Jurassic of North America and southern Africa are *Anomoepus* and *Moyenisauropus*, which are possibly the trackways of ornithischians. They are also good indicators for the correlation of basal Jurassic footprint beds. Larger tridactylous animals increased in abundance later in the uppermost Red Beds of South Africa and in the Cave Sandstone. A similar increase of protomammalian–mammalian footprints has also been documented for the Lower Jurassic of Lesotho, but none of these taxa has been defined. The earliest true quadrupedal sauropod footprints are first observed in the higher Liassic horizons of Morocco [High Atlas Mountains (S. Ishigaki pers. comm.)].

Conclusions

Tridactylous grallatorids and (to a lesser degree) *Batrachopus* are transitional archosaur footprints of the Triassic–Jurassic formations. The first group appears in the Middle Triassic, and in the Rhaetian formations these two types are the only footprints known. Many other ichnites show a gradual change in the intercontinental record. A general reduction in the abundance and diversity of all faunas is documented at the end of the Norian. After the relatively short interval of the Rhaetian, an increase in abundance and diversity occurs in the basal Jurassic.

The correlation between footprints and skeletal remains is not the main intention of this chapter. However, the amount of morphological and stratigraphic information now available suggests that some comments and overview of this problem should

be added. Unfortunately, in the Upper Triassic and Lower Jurassic the fossil records of both kinds of information have been influenced by quite different criteria, depending on the paleogeographic and sedimentologic conditions of their environments of preservation. It is possible that chronologic discontinuities exist between the records of footprints and the skeletal remains of their originators. These problems are demonstrated by the example of the earlier Triassic (Demathieu and Haubold 1974). At present, I feel unable to analyze, in detail and with good arguments, the reasons for the terrestrial faunal

Figure 15.5. Smaller dinosaurian footprints of the upper Newark Supergroup, Lower Jurassic. **A**, *Grallator cursorius*; **B**, *G. tenuis*; **C**, *G. gracilis*; **D**, *G. cuneatus*; **E**, *G. formosus*; **F**, *Anchisauripus hitchcocki*; **G**, *A. parallelus*; **H**, *A. exsertus*; **I**, *Anomoepus intermedius*; **K**, *A. isodactylus*; **L**, *A. curvatus*; **M**, *A. crassus*; **N**, *Hyphepus fieldi*; **O**, *Apatichnus minor*; **P**, *A, circumagens*. Redrawn from Lull (1953).

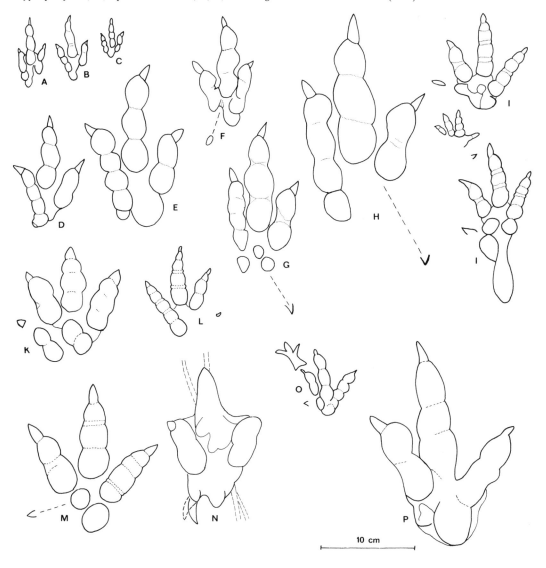

10 cm

changes from the Triassic to the Jurassic. The intercontinental reduction of all Rhaetian footprint faunas virtually to only a few tridactylous forms seems significant, and I would suggest this as a problem for broader future discussion before reaching any additional conclusions.

Osteological correlation

Every student of vertebrate footprints must also admit to interpreting the identity of the trackmakers, if the evidence to relate tracks to osteological groups of about the same age is poor. In contrast to a lot of other ichnofossils, the parataxonomy of vertebrate footprints often seems discernible to the level of species and genera, and, therefore, can be ordered with some justification into osteological categories. Ichno- or morphofamilies only have formal value, however, and are more for convenience.

At the present state of knowledge, a general relationship among higher taxonomic categories of tracks and trackmakers is available. Any greater separation of the parasystems, as sometimes suggested, seems unjustifiable. On the other hand, a direct connection at the species level between skeletal remains and footprints is not practicable. The level of taxonomic significance shows a shift or disproportion: Footprint species approximate osteological genera. In other words, all genera of tetrapods, fossil and living, show clear "species"-level differences in footprints and trackway patterns. (Although modern

Figure 15.6. Larger dinosaurian footprints of the upper Newark Supergroup, Lower Jurassic. **A**, *Eubrontes approximatus*; **B**, *E. divaricatus*; **C**, *E. tuberatus*; **D**, *E. platypus*; **E**, *Tarsodactylus caudatus*; **F**, *Platypterna deanii*; **G**, *Sauropus barrattii* (but see Chapter 6, on this taxon and also *Tarsodactylus*, represented in **E**); **H**, *Otozoum moodii*; **I**, *O. minor*; **K**, *Gigandipus caudatus*. Redrawn from Lull (1953). Scale = 10 cm, except **E** and **F** where scale = 5 cm.

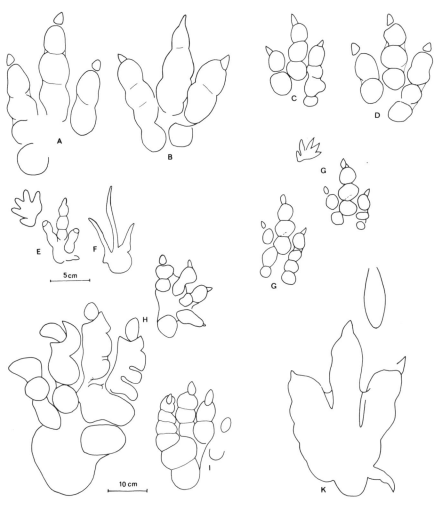

hunters can separate in more detail the tracks of living animals of a local fauna, this cannot be transferred to the field of paleontology because of its large scale – in both vertical and horizontal dimensions. In recent faunas, the experience with trackways started with a knowledge of the animals.)

The optimal, exceptional case in the fossil record might be to use diagnostic evidence to correlate footprints with the osteological genus, as done with *Iguanodon*. However, the interpretation of the archosaurian footprints of the Upper Triassic and Lower Jurassic formations is not so clear in detail. With the beginning of the Triassic, the archosaurs underwent a large-scale adaptive radiation; even now their osteological systems are not clear in all respects, because many phyletic lines are osteologically not well documented. Moreover, similar foot

structures and their respective footprints have often been formed by parallel evolution, a situation that complicates interpretation. For example, tridactylism and bipedalism are common trends that evolved independently in many lines, and consequently the methods of differentiating tridactylous dinosaur footprints are sometimes open to question.

With these restrictions in mind, the ichnogenera mentioned above are listed with their presumed trackmakers.

Figure 15.8. Hettangian footprints (Infralias) of Veillon (Vendée), France. **A**, *Grallator oloensis*; **B**, *G. variabilis*; **C**, *G. maximus*; **D**, *Anatopus palmatus*; **E**, *Saltopoides igalensis*; **F**, *Talmontopus tersi*; **G**, *Eubrontes veillonensis*; **H**, *Batrachopus gilberti*; **I**, *Dahutherium* (manus and pes not in natural position). Redrawn from Lapparent and Montenat (1967). Scale = 5 cm, except **A**, where scale = 2 cm.

Figure 15.7. Presumed crocodylomorph and some supposed coelurosaurian footprints of the upper Newark Supergroup, Lower Jurassic. **A**, *Batrachopus dispar*; **B**, **C**, **D**, **H**, *B. deweyi*; **E**, *Shepardia palmipes*; **F**, *Stenonyx lateralis*; **G**, *Plesiornis pilulatus*; **I**, *Argoides minimus*; **K**, *A. macrodactylus*. Redrawn from Lull (1953).

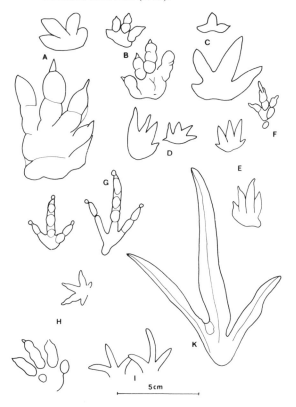

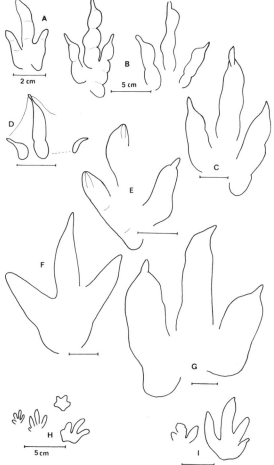

"Thecodontia"

Pseudosuchians and ornithosuchians (undifferentiated): *Chirotherium lulli* Bock 1952, *C. wondrai* Heller 1952, *Gwyneddichnium* Bock 1952
Ornithosuchids: *Parachirotherium* Kuhn 1958
Aetosaurids and rauisuchids: *Brachychirotherium* Beurlen 1950
Parasuchians: *Apatopus* Baird 1957, *Thecodontichnus* Huene 1941

Crocodylia

Protosuchians: *Batrachopus* Hitchcock 1845 (including *Cheirotheroides* and *Shepardia*), *Otozoum* Hitchcock 1847, *Dahutherium* Montenat 1970, *Deuterosauropodopus, Pentasauropus, Pseudotetrasauropus*, and *Sauropodopus* Ellenberger 1972, and *Suchopus* Ellenberger 1974

Ornithischia

Ornithopods and fabrosaurs: *Anomoepus* Hitchcock 1848, *Apatichnus* Hitchcock 1858, *Anatopus* Lap-

parent and Montenat 1967, *Atreipus* Olsen and Baird (Chapter 6), *Moyenisauropus* P. Ellenberger 1974, *Paratrisauropus* and *Psilotrisauropus* P. Ellenberger 1972, and *"Kainomoyenisauropus"* of Ellenberger

Prosauropoda

Anchisaurids, plateosaurids, and melanorosaurids: *Aetonychopus* P. Ellenberger 1974, *Tetrasauropus* P. Ellenberger 1972, and *Navahopus* Baird 1980.

Theropoda

Coelurids, podokesaurids, segisaurids, and megalosaurids: *Argoides* and *Eubrontes* Hitchcock 1845, *Gigandipus* Hitchcock 1855, *Grallator, Hyphepus*, and *Selenichnus* Hitchcock 1858, *Trihamus* Hitchcock 1865, *Anchisauripus* and *Stenonyx* Lull 1904, *Coelurosaurichnus* Huene 1941, *Saltopoides* and *Talmontopus* Lapparent and Montenat 1967, *Dilophosauripus, Hopiichnus,* and *Kayentapus* Welles 1971, *Anatrisauropus, Bosiutrisauropus, Deuterotrisauropus, Mafatrisauropus, Neotrisauropus, Plastisauropus,*

Figure 15.9. Dinosaur footprints of the Lower Jurassic upper Red Beds and Cave Sandstone of Lesotho. **A**, *Plastisauropus ingens*; **B**, *Neotrisauropus deambulator*; **C**, *Aetonychopus digitigradus* and *A. rapidus*; **D**, *Masitisisauropus angustus*; **E**, *M. exiguus*; **F**, *M. palmipes*; **G**, *M. minimus*; **H**, *Moyenisauropus dodai*; **I**, *M. natator*; **K**, *M. vermivorus* (**H** and **K** with seated manus–pes set and different impressions of the standing pes); **L**, *"Platysauropus"*; **M**, *"Qomoqomosauropus"*; **N**, *"Kainotrisauropus"*; **O**, *"Megatrisauropus"*; **P**, *"Kleitotrisauropus"*; **Q**, *"Gyrotrisauropus."* Redrawn from P. Ellenberger (1972, 1974).

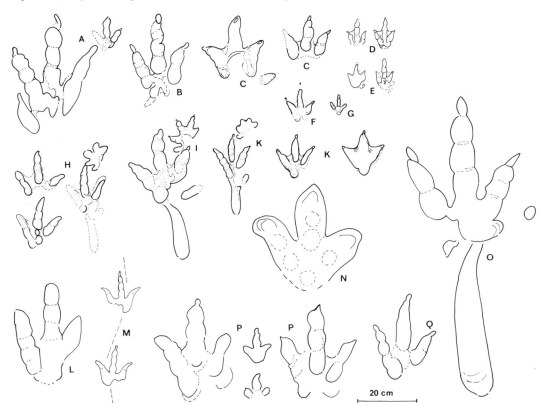

Qemetrisauropus, Seakatrisauropus, Trisauropodiscus, and *Tritotrisauropus* P. Ellenberger 1972, *Masitisisauropezus, Masitisisauropus, Masitisisauropodiscus,* and *Trisaurodactylus* P. Ellenberger 1974, and without definition "*Cridotrisauropus,*" "*Gyrotrisauropus,*" "*Kainotrisauropus,*" "*Kleitotri-* *sauropus,*" "*Megatrisauropus,*" "*Neotripodiscus,*" "*Paragrallator,*" "*Platysauropus,*" "*Platytrisauropus,*" and "*Qomoqomosauropus*"

Finally, the great difference in the respective number of footprint taxa presumed of prosauropod

Figure 15.10. Distribution of characteristic archosaur footprint types in the Triassic and Lower Jurassic. Data from material of Europe and North America. Note: The Scythian–Ladinian types of the different chirotherian ichnotaxa and the first Anisian–Ladinian tridactylous ichnites have no representative skeletal record; for example, Scythian chirotherians are mostly above the *Euparkeria* (skeletal) level, and the reconstructed pedes of the ichnogenus *Rotodactylus* resembles those of the Middle Triassic lagosuchians (skeletal). Adapted from Haubold (1984).

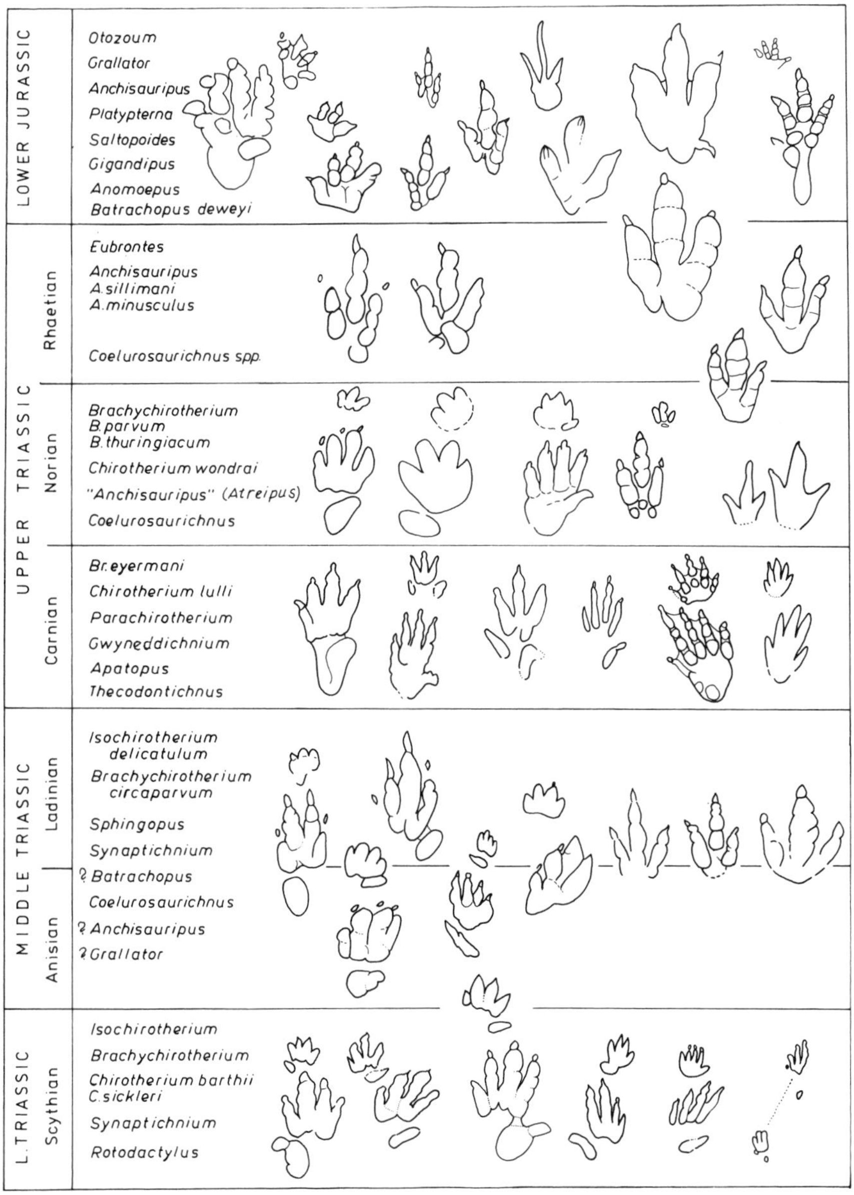

and of theropod origin should be noted, because this relation contradicts the relative taxonomic proportions of the skeletal record.

Note added in manuscript. According to Olsen and Baird (Chapter 6), the ichnotaxa *"Anchisauripus" milfordensis, "Grallator" sulcatus,* and *Coelurosaurichnus metzgeri* mentioned in this text and shown in Figures 15.2K,L and 15.10 are synonymized with the new ichnogenus *Atreipus,* and *Sauropus barrattii* (Fig. 15.6G) should be considered a *nomen vanum.*

Acknowledgments

Dr. Donald Baird and another reviewer were very helpful in sharing information and correcting the English of the original manuscript. I would like to express my best thanks to the editor, Dr. Kevin Padian, for preparing the final version of the manuscript, including discussions of some modifications related to recent progress concerning *Atreipus* and *Batrachopus.*

References

Baird, D. 1954. *Chirotherium lulli,* a pseudosuchian reptile from New Jersey. *Bull. Mus. Compar. Zool.* 111:165–92.

1957. Triassic reptile footprint faunules from Milford, New Jersey. *Bull. Mus. Compar. Zool.* 117:449–520.

Beurlen, K. 1950. Neue Fährtenfunde aus der fränkischen Trias. *N. Jahrb. Geol. Paleontol., Monatshefte* 308–20.

Bölau, E. 1952. Neue Fossilfunde aus dem Rät Schonens und ihre palaeogeographisch-ökologische Auswertung. *Geol. Fören. Stockholm Förhandl.* 74:44–50.

Cornet, B., and A. Traverse. 1975. Palynological contributions to the chronology and stratigraphy of the Hartford Basin in Connecticut and Massachusetts. *Geosci. Man* 11:1–33.

Demathieu, G. 1977. La palichnologie des vertébrés. Dévelopment récent et role dans la stratigraphie du Trias. *Bull. B.R.G.M. (2 Ser.), Ser. IV, No. 3,* pp. 269–78.

Demathieu, G., and H. Haubold. 1974. Evolution und Lebensgemeinschaft terrestrischer Tetrapoden nach ihren Fährten in der Trias. *Freiberger Forsch.* C 298:51–72 [reprinted in *Benchmark Pap. Geol.* 76:178–99].

1978. Du probleme de l'origine des dinosauriens d'après les donnés de l'ichnologie du Trias. *Geobios* 11:409–12.

Demathieu, G., and M. Weidmann. 1982. Les empreintes de pas de reptiles dans le Trias du Vieux Emosson (Finhaut, Valais, Suisse). *Eclogae geol. Helv.* 75:721–57.

Ellenberger, F., and P. Ellenberger. 1958. Principaux types de pistes de vertébrés dans les couches du Stormberg au Basutoland (Afrique du Sud). Note préliminaire. *Compt. Rend. Geol. Soc. France* 1958:65–7.

Ellenberger, F., and Y. Fuchs. 1965. Sur la présence de pistes de vertébrés dans le Lotharingien marin de la région de Severac-le-Chateau (Aveyron). Compt. Rend. Séanc. Soc. Geol. France, Fasc. 2:39–40.

Ellenberger, F., and P. and L. Ginsburg. 1970. Les dinosaures du Trias et du Lias en France et en Afrique du Sud, d'aprés les pistes qu' ils ont laissées. *Bull. Soc. Geol. France, 7 Ser.,* 12:151–9.

Ellenberger, P. 1955. Note préliminaire sur les pistes et les restes osseux de vertébrés du Basutoland (Afrique du Sud). *Comp. Rend. Seanc. Acad. Sci. Paris* 240:889–91.

1970. Les niveaux paléontologiques de première apparition des Mammifères primordiaux en Afrique du Sud et leur ichnologie. *Proc. Papers, 2nd Gondwana Symp., C.S.I.R. Pretoria,* pp. 343–70.

1972. Contribution à la classification des pistes de vertébrés du Trias: Les types du Stormberg d'Afrique du Sud (I). *Palaeovertebrata, Mem. Extraord.,* 1972:1–117.

1974. Contribution à la classification des pistes de vertébrés du Trias: Les types du Stormberg d'Afrique du Sud (II). *Palaeovertebrata, Mem. Extraord.,* 1974:1–141.

Haubold, H. 1971a. Ichnia Amphibiorum et Reptiliorum fossilium. *In* Kuhn, O. (ed.), *Handbook of Palaeoherpetology,* Vol. 18 (Stuttgart: Fischer).

1971b. Die Tetrapodenfährten des Buntsandsteins in der Deutschen Demokratischen Republik und in Westdeutschland und ihre Äquivalente in der gesamten Trias. *Palaeontol. Abh.* A4(3):397–548.

1983. Archosaur evidence in the Buntsandstein. *Acta Palaeontol. Polon.* 28(1/2):123–32.

1984. *Saurierfährten.* (Wittenberg: Ziemsen), pp. 1–231.

Heller, F. 1952. Reptilfährtenfunde aus dem Ansbacher Sandstein des Mittleren Keupers von Franken. *Geol. Bl. No-Bayern* 2:129–41.

Hitchcock, E. 1836. Ornithichnology. Description of the footmarks of birds on the New Red Sandstone in Massachusetts. *Am. J. Sci.* 29:307–40.

1845. An attempt to name, classify and describe the animals that made the fossil footmarks of New England. *Proc. Assoc. Am. Geol. Naturalists (6th Meet.),* pp. 23–5.

1858. *Ichnology of New England* (Boston: William White), pp. 1–220.

1865. *Supplement to the Ichnology of New England* (Boston: Wright and Potter), pp. 1–96.

Huene, F. von. 1940. Saurierfährten aus dem Verrucano des Monte Pisano. *Centrbl. Min. B,* pp. 349–52.

1941. Die Tetrapoden-Fährten im toskanischen Verrucano und ihre Bedeutung. *N. Jahrb. Min. Geol. Paleontol.* B 86:1–34.

Kuhn, O. 1958. *Die Fährten der vorzeitlichen Amphibien und Reptilien* (Bamberg: Meisenbach).

Lapparent, A. F. de, and C. Montenat. 1967. Les empreintes de pas de reptiles de l'Infralias du Veillon

(Vendeé). *Mem. Soc. Geol. France, N. S., 107*:1–44.

Lull, R. S. 1904. Fossil footprints of the Jura-Trias of North America. *Mem. Boston Soc. Nat. Hist.* 5:461–557.

1915. Triassic life of the Connecticut Valley. *Bull. State Geol. Nat. Hist. Surv. Conn.* 24:1–285.

1953. Triassic life of the Connecticut Valley, revised. *Bull. State Geol. Nat. Hist. Surv.* 81:1–331.

Olsen, P. E. 1980. A comparison of the vertebrate assemblages from the Newark and Hartford Basins (Early Mesozoic, Newark Supergroup) of Eastern North America. *In* Jacobs, L. L. (ed.), *Aspects of Vertebrate History* (Flagstaff, Arizona: Museum of Northern Arizona), pp. 35–53.

1983. Relationship between biostratigraphic subdivisions and igneous activity in the Newark Supergroup. *Geol. Soc. Am., Abst. Progr.* 15:93.

Olsen, P. E., and D. Baird. 1982. Early Jurassic vertebrate assemblages from the McCoy Brook Formation of the Fundy Group (Newark Supergroup, Nova Scotia, Ca.). *Geol. Soc. Am., Abst. Progr.* 14(1/2):70.

Olsen, P. E., and P. M. Galton. 1977. Triassic–Jurassic tetrapod-extinctions: are they real? *Science* 197:983–6.

Peabody, F. E. 1955. Taxonomy and the footprints of tetrapods. *J. Paleontol.* 29:915–18.

Rühle v. Lilienstern, H. 1938. Fährten aus dem Blasensandstein (km4) des Mittleren Keupers von Südthüringen. *N. Jahrb. Geol. Paläontol.* B80:63–71.

Sollas, W. 1879. On some three-toed footprints from the Triassic conglomerate of South Wales. *Quart. J. geol. Soc.* 35:511–16.

Thaler, L. 1962. Empreintes de pas de dinosaures dans les dolomies du Lias inférieur de Causses (Note prelim.). *C. R. Somm. Seanc. Soc. geol. France*, 62:190–92.

Tucker, M. E., and T. P. Burchette. 1977. Triassic dinosaur footprints from South-Wales: their context and preservation. *Palaeogeogr. Palaeoecol. Palaeoclimatol.* 22(3):195–208.

16 Herbivorous adaptations of Late Triassic and Early Jurassic dinosaurs

PETER M. GALTON

Introduction

The fossil record of terrestrial herbivorous dinosaurs shows several changes in passing from the Late Triassic to the Middle Jurassic (Fig. 16.1). The first records of prosauropods and ornithischians are from the Carnian. From the Norian, prosauropods are the dominant herbivores, whereas ornithopods are rare. The prosauropods continue into the Early Jurassic [until Pliensbachian or Toarcian (Olsen and Galton 1977, 1984)], ornithopods become more common, and the first sauropods and thyreophorans (*sensu* Gauthier 1984; Norman 1984: stegosaurs, ankylosaurs and allies) appear (Fig. 16.1). Middle Jurassic terrestrial faunas are rare, but there is no evidence for any prosauropods; the first true ankylosaurs and stegosaurs appear; ornithopods are rare; and sauropods are more common. The excellent terrestrial faunas of the Late Jurassic show that ankylosaurs are still rare, whereas sauropods, ornithopods, and stegosaurs are more diverse and common. The herbivorous adaptations of these dinosaur groups are discussed below following the usual systematic sequence.

Prosauropoda

Prosauropods may represent a paraphyletic grade of early sauropodomorphs because the taxa involved were usually not large and did not have columnar limbs as in the true sauropods. Prosauropods are the most common dinosaurs in the Norian (Late Triassic) and are also important in the Early Jurassic (Fig. 16.1) (Olsen and Galton 1977, 1984). Cooper (1981) suggests that prosauropods were scavenger-predators with occasional cannibalism. However, his arguments are based on controversial ideas, incomplete data, and inappropriate comparisons with the adaptations of herbivorous and carnivorous mammals. In many cases, he shows that prosauropods had reptilian feeding systems, but he does not provide any information on their diet. Galton (1984, 1985a) discusses the evidence for the diet of prosauropods and concludes that all were herbivorous (Table 16.1). The herbivorous adaptations of the prosauropod families Anchisauridae (includes Plateosauridae, Galton 1971; Cooper 1981), Yunnanosauridae, and Melanorosauridae (Fig. 16.1) are discussed below.

Anchisauridae (Table 16.1, includes Plateosauridae of most authors)

Skull form

The skull is lightly built; the size of the coronoid eminence of the lower jaw varies from low to medium height, and the jaw articulation is set below the line of the tooth row (Fig. 16.2A–H). In herbivorous mammals, the jaw articulation is dorsal to the tooth row. Cooper (1981) discounts the ventral articulation of prosauropods as an herbivorous adaptation because he considers that the primary function of an offset articulation is to allow for a more effective lateral movement to the jaw. Transverse movements of the jaws were not possible in anchisaurids, but two functions proposed for the dorsally offset articulation of mammals that do not involve transverse movements also apply to the ventrally offset articulation of anchisaurids. First, offsetting increases the angle between the lever arm of the bite force and the plane of the teeth, which is important in dealing with resistant plant material (Crompton and Hiiemae 1969). Second, it allows a more even distribution of the biting force by insuring that the tooth rows are almost parallel at occlusion, so that contact is made along the complete length of the tooth row by a "nutcracker"-like action (Colbert 1951, p. 89). The amount of ventral offset is small in *Anchisaurus* and *Thecodontosaurus* (Fig. 16.2A,B) and quite large in *Plateosaurus*, in which the skull is more heavily built (Fig. 16.2G). Aetosaurian thecodonts (see Fig. 16.5J; Walker 1961; Charig et al. 1976) and ornithischian dinosaurs (see Fig. 16.5A–I; Galton 1973) are two other groups of herbivorous reptiles with a ventral jaw articulation. In carnivorous thecodonts (see Charig et al. 1976) and theropod dinosaurs (Steel 1970), the jaw artic-

ulation is in line with the tooth row to give a "scissor"-like action (see Colbert 1951, p. 89).

Two cranial features indicate that in *Plateosaurus* (Fig. 16.2G; Galton 1984, 1985c; Paul 1984) cheeks were present that, as was also the case in most ornithischians (Galton 1973), were probably analogous to those of mammals in retaining food in the mouth after it was bitten off: first, the presence of a diagonally inclined ridge lateral to the dentary teeth that were slightly inset and, second, the presence of only a few large lateral nerve foramina on

Figure 16.1. Stratigraphic ranges for families of Triassic and Jurassic herbivorous dinosaurs. Ranges in the Early Jurassic are based on the youngest estimate for the beds concerned (mostly from Olsen and Galton 1977, 1984), whereas the reverse is the case for the Middle Jurassic so that the faunal transition may have been more gradual than shown. Ages of subdivisions of periods in MY from Weishampel and Weishampel (1983). European marine stages in third column: B, Bathonian; Ba, Bajocian; C, Callovian; Ca, Carnian; H, Hettangian; K, Kimmeridgian; N, Norian; O, Oxfordian; P, Purbeckian; Pl, Pliensbachian; Po, Portlandian; R, Rhaetic (uppermost part of Norian); S, Sinemurian; T, Toarcian. Families: Anch., Anchisauridae (Figs. 16.2A–H, 16.3A,J,K, 16.4A,B; includes Plateosauridae). Blik., Blikanasauridae (Galton and Van Heerden 1985) for *Blikanasaurus*, a partial hindlimb. Brach., Brachiosauridae (Fig. 16.4J), earliest records teeth and postcrania of *Bothriospondylus* (Thevénin 1907). Camar., Camarasauridae (Figs. 16.2L, 16.4I). Cetio., Cetiosauridae (Figs. 16.2J,K), earliest record teeth and postcrania of *Amygdalon* (Fig. 16.4F,G). Diplod., Diplodocidae, earliest record postcrania of *Cetiosauriscus* (Woodward 1905). Fabr., Fabrosauridae (Figs. 16.3L,M, 16.4K, 16.5A, 16.6A), earliest record possibly mid-Carnian, teeth from Morocco (Figs. 16.3L,M, Galton 1985a,b) and Texas (Chatterjee in press), most recent record possibly Late Cretaceous teeth from Hell Creek Formation of Montana (Galton in prep.). Heter., Heterodontosauridae (Figs. 16.5B,C, 16.6K–T). Hypsil., Hypsilophodontidae (Figs. 16.5D, 16.6V), earliest record possibly mid-Carnian, ?*Pisanosaurus* (Fig. 16.6K,L) if not a heterodontosaurid, teeth from Nova Scotia and Pennsylvania (Galton 1983a; Galton and Olsen in prep.). Iguan., Iguanodontidae, earliest record femur of *Callovosaurus* (Galton 1980a). Melan., Melanorosauridae, all postcrania: earliest record *Melanorosaurus* Haughton 1924; latest definite record from Rhaetic of England (Huene 1907–8, pp. 112–17; Galton 1985d) but family may be represented in lower Lufeng Beds (Early Jurassic) of China (Galton 1985d). Nodos., Nodosauridae (Figure 16.5; Galton 1983c; Ankylosauria). Ohmd., *Ohmdenosaurus* based on a tibia and astragalus (Wild 1978; Galton and Wild in prep.). Scel., Scelidosauridae (Figs. 16.5E, 16.6U). Scut., *Scutellosaurus* Colbert (1981) based on jaws with teeth (Fig. 16.6F,G) plus skeleton, possibly more recent forms *Trimucrodon* (Fig. 16.6H,I) and *Echinodon* (Fig. 16.6B–E) (Galton in prep.). Steg., Stegosauridae (Fig. 16.5F–H, Stegosauria), earliest record postcrania of *Lexovisaurus ? vetustus* (Galton and Powell 1983). Thyreoph, Thyreophora. Vulc., Barapasauridae Halstead and Halstead 1981 (=Vulcanodontidae Cooper 1984), teeth and postcrania of *Barapasaurus* (Jain et al. 1975, 1979), *Vulcanodon* postcrania (Cooper 1984). Yunn., Yunnanosauridae (Figs. 16.2I, 16.3B–I, 16.4C,D).

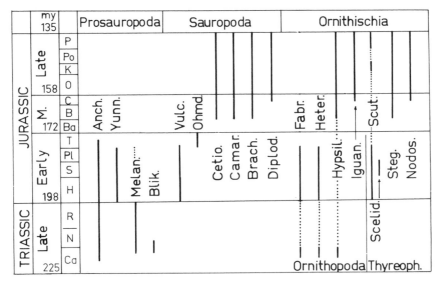

the lateral surface of the jaws rather than a series of more numerous small foramina in forms without cheeks (Paul 1984). The attachment area for cheeks probably extended along the dentary ridge as far as tooth position 9, and Paul (1984) reconstructs the corner of the mouth level with tooth positions 12 to 13. Cheeks were probably also present in *Anchisaurus* (Fig. 16.2A; YPM 1883), *Massospondylus* (Fig. 16.2E; Crompton pers. comm.), *Sellosaurus*, and *Thecodontosaurus* (Galton 1985e), so they were probably present in most anchisaurids.

Tooth form

The spatulate crowns of the maxillary and dentary teeth of anchisaurids are laterally compressed and expanded anteroposteriorly, so the maximum width of the crown is greater than that of the root (Fig. 16.3A,J, 16.4A,B), and the teeth are symmetrical and straight in anterior or posterior views (Fig. 16.3K). This crown shape is similar to those of the herbivorous lizard *Iguana iguana* (Fig. 16.3L). In herbivorous lizards of the families Iguanidae and Agamidae, the shearing edge of the crown is expanded anteroposteriorly so that, with a reduction of the space between adjacent teeth in the tooth row, there is a more nearly continuous cutting edge than in insectivorous lizards (Fig. 16.3L,M; Ray 1965; Montanucci 1968; Throckmorton 1976).

The maxillary and dentary teeth have prominent denticles (high notches or *Spitzkerbung* of Huene 1926) set at an angle of about 45° to the anterior and posterior cutting edges (Figs. 16.3A,J, 16.4A,B) as in the herbivorous lizard *Iguana iguana* (Fig. 16.4L).

The cutting edges of the crowns are obliquely inclined with respect to the long axis of the maxilla and dentary, so the anterior and posterior edges are slightly more medial and lateral, respectively (Fig. 16.3A) (for ventral views of maxillary tooth rows of *Lufengosaurus* and *Plateosaurus* see Young 1951; Galton 1984). The teeth of *Iguana iguana* (Fig. 16.4L,M) have the same *en echelon* arrangement. Throckmorton (1976, p. 387) notes that, because of the oblique orientation of the teeth in *Iguana*,

the anterior end of the perforation made by one tooth lies medial to the posterior end of the one preceding it. If the food item is thin, then the perforations of the upper and lower dentition will overlap, freeing the piece of food item in the mouth. If the food item is thicker, the perforations, although not overlapping, still allow the food item to be torn by a quick movement of the head.

He also notes that the amount of head movement increases with the difficulty of cropping the food.

Gastric mill

The lack of wear facets on the teeth of anchisaurids suggests that there was no tooth-to-tooth occlusion. Throckmorton (1976) shows that *Iguana iguana* uses its teeth, which also lack wear facets, to bite off a piece of a plant and that no further mechanical breakdown of the food occurs once it is in the oral cavity. Presumably, the situation was similar in anchisaurids. However, a gastric mill may have provided a supplementary mechanical breakdown of the

Table 16.1. *Some contrasting cranial characters of carnivorous reptiles and herbivorous lizards, anchisaurid prosauropods, and ornithischian dinosaurs*

Character	Carnivorous reptiles[a]	Herbivorous reptiles[b]
Jaw articulation	In line with tooth rows	Ventral to tooth rows[c] (Figs. 16.2, 16.5)
Spacing of teeth	Prominent gaps (Fig. 16.4)	Small or no gaps (Figs. 16.3A, 16.4B,L, 16.6A–E,K–T,V)
Orientation of crowns	Along middle of jaw	*en echelon* (Fig. 16.3A, 16.4L,M, 16.6D,E,P,Q,V)
Shape of cheek teeth crowns	Continous taper from root (Figs. 16.4O,P)	Widens before tapering (Figs. 16.3A,J,L, 16.4A,B,K,L, 16.6A–J,M–Q,U,V)
Form of serrations	Fine and perpendicular to edge (Fig. 16.4O,P)	Coarse, at 45° to edge and toward apex (Figs. 16.3A,J,L, 16.4A,B,K,L, 16.6A,C–J,M–O,U)

[a]Theropods (see Steel 1970), carnivorous thecodonts (see Charig et al. 1976), and varanid lizards (see Auffenberg 1981; Burden 1928; Mertens 1942).
[b]Iguanid lizards (see Montanucci 1968; Ray 1965; Throckmorton 1976), prosauropods, ornithischians.
[c]Not present in herbivorous lizards, present in herbivorous aetosaurian thecodonts (Fig. 16.5J).

plant material once it was ingested. A well-preserved gastric mill consisting of a concentrated mass of small stones has been found associated with the stomach contents of several specimens of *Massospondylus* (Bond 1955; Raath 1974; Chapter 17). Cooper (1981, p. 824) notes that the main function of the gastroliths "was certainly the trituration of food,"

and this is true regardless of diet. The use of stomach stones for the breakdown of ingested mice has been observed (and filmed) at Yale University in 1968 using cinefluroscopy in the crocodile *Caiman sclerops* [mentioned by Bakker (1971a) and Chapter 17, fuller account by Crompton and Hiiëmae in Darby and Ojakangas (1980, p. 553)]. Modern birds, many

Figure 16.2. Left lateral views of skulls of sauropodomorph dinosaurs. **A–I**, Prosauropoda, **A–H**, anchisaurids from the Late Triassic (**B,C,F,G**) and Early Jurassic (**A,D,E,H**). **A**, *Anchisaurus* from Portland Beds (Pliensbachian or Toarcian) of the Connecticut Valley (YPM 1883), modified from Galton (1976). **B**, *Thecodontosaurus* from Rhaetic of Wales, United Kingdom, after Kermack (1984). **C**, *Mussaurus* from El Tranquilo Formation (Norian) of Argentina, an extremely juvenile individual, after Bonaparte and Vince (1979). **D**, ?juvenile *Massospondylus* from Clarens Formation (Sinemurian or Pliensbachian) of South Africa, after Cooper (1981). **E**, *Massospondylus* from Kayenta Formation (Sinemurian or Pliensbachian) of Arizona, United States, after Attridge, Crompton, and Jenkins (1985). **F**, *Coloradia* from Los Colorados Formation (Norian) of Argentina, after Bonaparte (1978). **G**, *Plateosaurus* from Knollenmergel (Norian) of West Germany, after Galton (1985c). **H**, *Lufengosaurus* from lower Lufeng Beds (Sinemurian or Pliensbachian) of Lufeng, China, adapted from Young (1941). **I**, yunnanosaurid *Yunnanosaurus* from lower Lufeng Beds of Lufeng, China, adapted from Young (1942). **J–L**, Sauropoda from Middle Jurassic. **J,K**, cetiosaurids from lower Shaximiao Formation (Bathonian–Callovian) of Sichuan, China. **J**, *Shunosaurus*, after Zhang, Yang, and Peng (1984). **K**, *Datousaurus*, after Dong and Tang (1984). **L**, Camarasaurid *Omeiosaurus* from early Shaximiao Formation of Sichuan, China, after He et al. (1984). Scale lines represent 1 cm (**C**), 5 cm (**A,B,D–J**) and 10 cm.

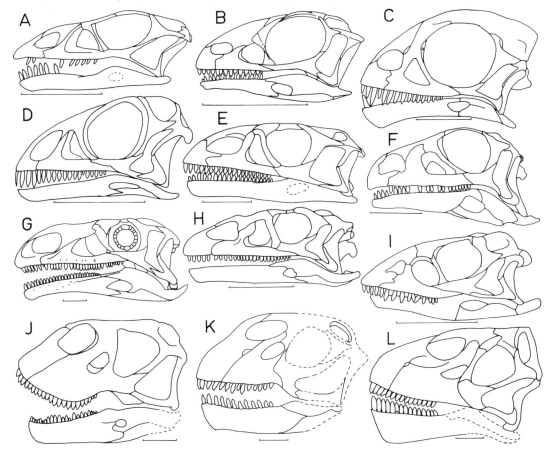

of which are herbivorous, lack teeth but use stones in the gizzard for the further mechanical breakdown of the food (Welty 1975, p. 92).

Adaptations for high browsing

All saurischians have long necks, unlike most other terrestrial reptiles. Bakker (1978, p. 662) suggests that

> from the late Carboniferous until the early Late Triassic, all the big herbivores were short-necked, short-limbed, low-browsers (diadectids, edaphosaurs, dinocephalians, pareiasaurs, dicynodonts, rhynchosaurs, gomphodonts and aetosaurs). But in the mid-Late Triassic, this trophic role was taken over by prosauropod dinosaurs – long-necked forms with long hindlimbs, powerful pelvis and tail. Prosauropods could probably feed tripodally – supporting their weight on the hindlimbs and stout tail, much as modern varanid lizards do when threatened.

The long neck extended the vertical feeding range so that vegetation at higher levels could be reached as in giraffes. The proportionally small skull would have reduced its leverage at the end of the long neck, especially when the neck was horizontal.

Yunnanosauridae

Many of the herbivorous adaptations of anchisaurids occur in *Yunnanosaurus huangi*. For example, the teeth are spatulate (Figs. 16.3B–I, 16.4C–E); the crowns are obliquely inclined with respect to the long axis of the maxilla and dentary (Young 1951); the jaw articulation is offset ventrally (Fig. 16.2I); and the skull is small with a proportionally long neck (Young 1942, 1951). However, the tooth crowns (Figs. 16.3B–I, 16.4C–E) differ from those of anchisaurids (Figs. 16.3A,J, 16.4A,B) and resemble those of some sauropods (Galton 1985a). The edges of the flat and obliquely inclined wear surface on the maxillary tooth of *Yunnanosaurus* (Figs. 16.3C,D, 16.4D,E) was self-sharpened by wear against the corresponding surface on the opposing dentary tooth as also occurred in the sauropod *Brachiosaurus* (Janensch 1935–6). Consequently, a series of relatively coarse, 45° inclined marginal denticles, as seen in anchisaurids, was no longer necessary for cutting plant material, and vestiges of this system are retained on only a few teeth (Young 1942, 1951; Galton 1985a). The tooth crowns of *Yunnanosaurus* are characterized by an almost complete lack of marginal denticles, apices that are slightly medially directed, a partly concave medial surface, and wear surfaces that were formed by tooth-to-food and tooth-to-tooth contact (Figs. 16.3B,C, 16.4D). However, the rest of the anatomy of *Yunnanosaurus* (Young 1942, 1951) is comparable to that of anchisaurids rather than sauropods so this genus is retained in the Prosauropoda in its own family (Galton 1985a).

Melanorosauridae

The supposed teeth of melanorosaurids described to date were shed by carnivorous theropods feeding on the carcass; no skull bones have been described (see Galton 1985a). However, the elongate neck in *Riojasaurus* (Bonaparte 1971), the only adequately known melanorosaurid, indicates that the skull was probably proportionally small as in other prosauropods. This large (up to 10 m) quadrupedal animal with sauropod-like proportions formed about 40 percent of the total number of individual vertebrate specimens collected from the Los Colorados Formation (Upper Triassic) of Argentina (Bonaparte 1982). It is difficult to envision it as anything other than a herbivore, and Bonaparte (pers. comm.) reports that the teeth in a newly discovered skull of *Riojasaurus* are typically prosauropod.

Sauropoda

It has long been assumed that sauropods were swamp- or lake-dwelling forms, but Bakker (1971a,b) argues that they were terrestrial herbivores, and Coombs (1975, p. 1), who provides a review of the evidence for both interpretations, concludes that "where firm morphologic interpretations are possible, they usually point to terrestrial behavior." The earliest records are from the Early Jurassic (Fig. 16.1) of Africa (*Vulcanodon*, Cooper

Figure 16.3. Teeth of prosauropod (**A–K**) and fabrosaurid dinosaurs (**L,M**). **A**, Anchisaurid *Plateosaurus* from middle Keuper (Late Triassic) of Frick, Switzerland, middle left dentary teeth, MSF 2, in external view. **B–I** Yunnanosaurid *Yunnanosaurus* from lower Lufeng Beds (Early Jurassic) of Yunnan, China, isolated teeth. **B–D**, FMNH 2051 in **B**, anterior, **C**, internal, and **D**, posterior views. **E,F**, FMNH 2340c in **E**, anterior, and **F**, internal views. **G–I**, FMNH 2056 in **G**, anterior, **H**, internal, and **I**, external views. **J–M**, Isolated teeth from early Carnian of Morocco: **J,K**, *Azandohsaurus* MNHN ALM 508 in **J**, external, **K**, anterior views. **L,M** Fabrosaurid MNHN ALM 509 in **L**, external, **M**, anterior views. Scale lines represent 2 mm.

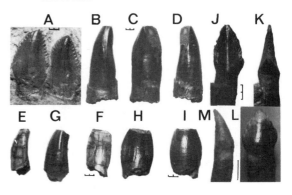

1984), West Germany (*Ohmdenosaurus*, Wild 1978), India (*Barapasaurus*, Jain et al. 1975, 1979), and China (*Zizhongosaurus*, Dong, Zhou, and Zhang 1983). The material includes a few teeth (*Barapasaurus*) plus postcrania that show a mosaic of pro-sauropod and sauropod characters. In addition, there are undescribed sauropods from the Early Jurassic of China (*Chinshakiangosaurus* from Yunnan (Yeh 1975; Chapter 21; J. McIntosh pers. comm.) and Tibet (*Damalosaurus* Zhao 1983) plus a sauropod-like vertebra from the Late Triassic of Argentina (Chapter 19). Although sauropods were

geographically widespread during the Early Jurassic, the record is so poor that it is necessary to refer to the better Middle Jurassic record (Figs. 16.2J–L, 16.4F–I), and even the rich Late Jurassic, to obtain a more complete idea of the herbivorous adaptations of sauropods.

Skull form

The premaxilla and maxilla that support the teeth are proportionally much deeper in sauropods than they are in prosauropods (Fig. 16.2), and, as a result, the nasal opening is more dorsal in position as is the

Figure 16.4. **A,B**, teeth of anchisaurid prosauropods from the Late Triassic: A, *Azendohsaurus laaroussi* Dutuit from the Argana Formation (mid-Carnian) of the Atlas Mountains, Morocco, cotype tooth MNHN MTD XVI 2, after Dutuit (1972). **B**, *Thecodontosaurus* from the Rhaetic near Bristol, England, two dentary teeth with detail of edge, after Huene (1907–8). **C–E**, Tooth of yunnanosaurid *Yunnanosaurus* from lower Lufeng Beds (Early Jurassic) of Yunnan, China, FMNH 2051 (see Fig. 16.3B–D) in C, external; **D**, internal; **E**, posterior views. **F–J**, Teeth of sauropods from the Middle (**F–I**) and Late Jurassic (**J**). F,G, Isolated teeth of cetiosaurid *Amygdalodon* from Bajocian of Argentina; F, in external and anterior views and G, in external view, after Cabrera (1947). **H**, Internal view of tooth of *Datousaurus* from early Shaximiao Formation (Bathonian–Callovian) of Sichuan, China, after Dong and Tang (1984). **I**, Internal view of maxillary tooth of camarasaurid *Omeiosaurus* from early Shaximiao Formation, after He et al. (1984). **J**, Left dentary tooth in medial view of brachiosaurid *Brachiosaurus* from Late Jurassic of Tanzania, after Janensch (1935–6). **K**, Fabrosaurid ornithopod *Lesothosaurus* from upper Elliot Formation (Early Jurassic) of Lesotho, isolated right dentary tooth in medial view, after Thulborn (1970a). **L,M**, Recent herbivorous lizard *Iguana iguana*. L, Right dentary teeth in medial view, after Montanucci (1968). **M**, Dentary teeth in occlusal view, after Throckmorton (1976). **N,O**, Recent carnivorous lizard *Varanus komodensis*. N, Dentary, after Mertens (1942). **O**, Tooth, after Burden (1928) with detail of posterior edge, after Auffenberg (1981). **P**, Isolated tooth of carnosaurian theropod *Megalosaurus* from Middle Jurassic of England, after Owen (1857). Scale lines represent 10 mm (**C–J,N,P**) and 1 mm.

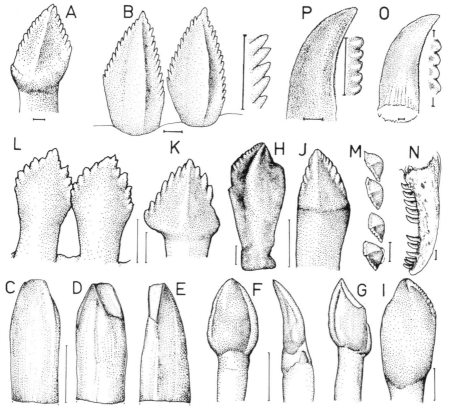

reduced antorbital opening. The coronoid eminence is of medium height in most sauropods (Fig. 16.2J–L), except in *Camarasaurus* in which it is high (see Gilmore 1925). The jaw articulation is approximately in line with the tooth rows in Middle Jurassic sauropods (Fig. 16.2J–L) and in Late Jurassic diplodocids in which the teeth are restricted to the anterior end of the jaws (Janensch 1935–6; Berman and McIntosh 1978). However, in Late Jurassic camarasaurids (Gilmore 1925) and brachiosaurids (Janensch 1935–6), the jaw articulation is set ventrally as in most prosauropods (Fig. 16.2C–I).

Tooth form

The crowns of *Barapasaurus* are spoon-shaped (Jain et al. 1975, p. 223) as are those of the better represented Middle Jurassic sauropods (Fig. 16.4F–I). The crowns are proportionally much broader than those of *Yunnanosaurus* (Figs. 16.3C,F,H,I, 16.4C,D), and the earliest sauropods known with teeth of a comparable form are brachiosaurids from the Late Jurassic (Janensch 1935–6; Lapparent 1943). In Late Jurassic diplodocids, the crowns are unexpanded and the teeth are slender and pencil-like (Holland 1924; Janensch 1935–6). The crowns of *Barapasaurus* have coarse denticles on both edges (Jain et al. 1975), and coarse denticles are present on some teeth of *Omeiosaurus* and *Datousaurus* (Fig. 16.4H,I). Even an occasional tooth of *Brachiosaurus* (Fig. 16.4J) from the Late Jurassic has a few coarse, 45° oriented denticles. However, most sauropod teeth completely lack denticles, and, as in *Yunnanosaurus*, this is probably correlated with the development of an irregular series of wear facets formed between teeth that curve slightly medially.

Tooth wear

An irregular series of surfaces formed by tooth-to-tooth wear are well documented for teeth of camarasaurids (White 1958; Carey and Madsen 1972), brachiosaurids (Janensch 1935–6) and diplodocids (Holland 1924; Janensch 1935–6) from the Late Jurassic. The only Middle Jurassic teeth that possibly show such facets are from the Bajocian cetiosaurid *Amygdalodon* (Fig. 16.4G) (Casamiquela 1963). However, the teeth in the well-preserved sauropod skulls (and isolated teeth) from the Middle Jurassic of China (Fig. 16.2J–L) are not illustrated in enough detail to determine if such facets are present, and the teeth of *Barapasaurus* are still to be illustrated. Consequently, there is insufficient information available to determine the incidence of tooth-to-tooth wear in early sauropods and when it first occurred.

Gastric mill

The occurrence of a concentrated mass of small stones among the ribs is documented by Brown (1941) for one sauropod skeleton, and Bakker (1971a) concludes that this find represents a gastric mill. Janensch (1929) describes several gastroliths that were found associated with sauropods from the Late Jurassic of Tanzania. Darby and Ojakanges (1980), who discuss the literature on gastroliths, note Cannon's (1906) report that rounded and highly polished pebbles were found among the bones of a skeleton of *Atlantosaurus* (now *Apatosaurus*), and no comparable material of similar size, form, surface markings, or composition occurred elsewhere in the *Atlantosaurus* clays. Consequently, sauropods probably resembled prosauropods in the use of a gastric mill for supplementary mechanical breakdown of ingested plant material.

Adaptations for high browsing

Vulcanodon (Cooper 1984), *Barapasaurus* (Jain et al. 1975, 1980), cetiosaurids (Woodward 1905), and camarasaurids (Gilmore 1925) were basically quadrupedal with long necks like prosauropods but, because the forelimbs were proportionally longer, the shoulder region was relatively higher. In brachiosaurids, the forelimbs were even more elongate, raising the shoulders above the hips, and the neck was exceptionally long so *Brachiosaurus* could feed at 42 ft (13 m) (see Janensch 1950). Bakker (1971b) points out that the skeletons of the diplodocids *Apatosaurus* and *Diplodocus* (Romer 1956) show basically bipedal specializations: The very long and heavy tail served as a counterbalance and a third leg as in kangaroos, the tall neural spines over the hips improved the transmission of the body weight to the hindlimbs, and the shorter back and forelimbs brought the center of gravity closer to the hips. Consequently, diplodocids probably often fed in a tripodal pose, as did prosauropods, and *Apatosaurus* and *Diplodocus* could browse at 50–60 ft (15.2–18.3 m) among conifers and long-trunked cycads (Bakker 1971b).

Ornithopoda

The earliest records of ornithopods are rare finds from the Carnian (Fig. 16.1): isolated teeth (Galton 1983a) and jaws with teeth from North America (Chatterjee in press) and North Africa (Fig. 16.3L,M) plus jaws with teeth (Fig. 16.6K,L) from the partial skeleton of *Pisanosaurus* (Casamiquela 1967; Bonaparte 1976). However, ornithopods are practically unknown in the Norian and are not common until the Early Jurassic, when good material of fabrosaurids and heterodontosaurids is known from southern Africa (Figs. 16.5A–C, 16.6A,M–T,W) (Chapter 17) plus fragmentary material from China (Simmons 1965; Young 1982). The earliest undoubted hypsilophodontid is *Yandusaurus* (Figs. 16.5D, 16.6V) from the Middle Jurassic (Bathonian–Callovian) of China (He and Cai 1984). However, on the basis of the material described to date, the nonfabrosaurid *Pisanosaurus* (Fig. 16.6K,L; Casamiquela 1967; Bonaparte 1976) could equally well be a hypsilophodontid (Galton 1972; Colbert 1981) as a heterodontosaurid (Charig and Crompton 1974; Bonaparte 1976). The earliest undoubted iguanodontid is *Camptosaurus* from the Late Jurassic of United States and England [Lower Kimmeridgian (Galton and Powell 1980)], but a femur (*Callovosaurus* Galton 1980a) from the Callovian of England may be referable to this family.

Fabrosauridae

Skull form

The skull is lightly built in *Lesothosaurus* (Fig. 16.5A; Chapter 17); the jaw articulation is slightly ventral to the tooth rows; the teeth are marginal; and the coronoid process is low. The mandibular rami are linked together at the symphysis by a single predentary bone that acted as a cropping device against the opposing teeth of the premaxillae (Figs. 16.5A, 16.6A). The teeth leave no room for a horny beak on the premaxilla of *Lesothosaurus* (Fig.

Figure 16.5. **A–I**, Ornithischian skulls in left lateral view from the Early (**A–C,E**), Middle (**D,J,M**), and Late Jurassic (**K,L**). **A**, Fabrosaurid *Lesothosaurus* from upper Elliot Formation of Lesotho, southern Africa, after Thulborn (1970a). **B,C**, Heterodontosaurids from the upper Elliot Formation. **B**, *Heterodontosaurus* from South Africa, after Charig and Crompton (1974). **C**, *Abrictosaurus* from Lesotho, after Thulborn (1974). **D**, hypsilophodontid *Yandusaurus* from early Shaximiao Formation (Bathonian–Callovian) of Sichuan, China, after He and Cai (1984). **E**, Scelidosaurid thyreophoran *Scelidosaurus* from Sinemurian of England, modified from Owen (1861). **F–H**, stegosaurids. **F**, *Huayangosaurus* from early Shaximiao Formation of Zigong, Sichuan, China, after Zhou (1983). **G**, *Tuojiangosaurus* from Shangshaximiao Formation, Sichuan, China, after Dong, Zhou and Zhang (1983). **H**, *Stegosaurus* from Morrison Formation of western United States, after Gilmore (1914). **I**, Ankylosaur, ?nodosaurid, incomplete mandible of *Sarcolestes* from lower Oxford Clay (Callovian) of England, after Galton (1983c). **J**, Aetosaurian thecodont reptile *Stagonolepis* from Upper Triassic of Scotland, after Walker (1961). **K–M**, kinematic diagrams of ornithopod jaw mechanics, arrows indicate sites of movement, after Weishampel (1984). **K**, Lateral view of fabrosaurid *Lesothosaurus* with strictly orthal closing of lower jaws. **L**, Lateral and transverse views of heterodontosaurid *Heterodontosaurus* with medial rotation of lower jaws during transverse power stroke. **M**, Lateral and transverse views of hypsilophodontid *Hypsilophodon* with lateral movement of maxillae during transverse power stroke. Scale lines represent 5 cm (**A,E,I,J**) and 10 cm.

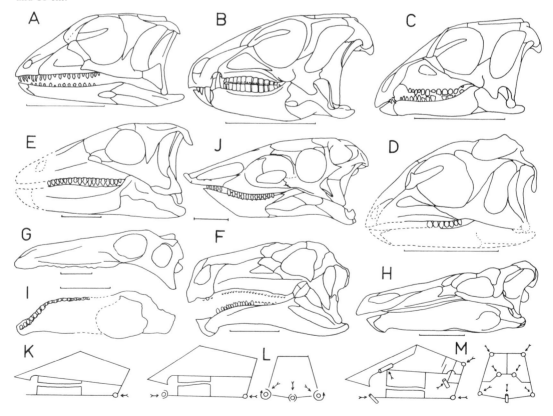

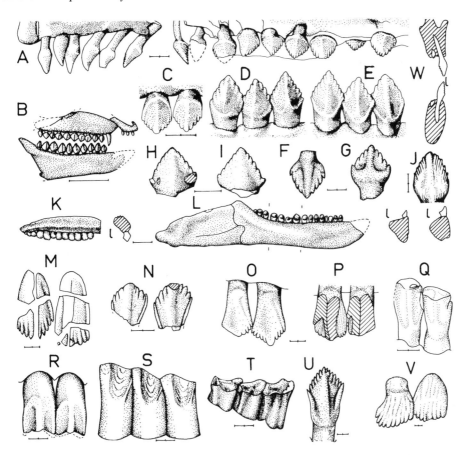

Figure 16.6. Teeth of ornithischian dinosaurs from the Late Triassic (**K,L**), Early (**A,F,G,M–U,W**), Middle (**J,V**), and Late Jurassic (**B–E**). **A**, Fabrosaurid ornithopod *Lesothosaurus* from upper Elliot Formation of Lesotho, teeth of left premaxilla and maxilla in lateral view, after Thulborn (1970b, right maxilla reversed). **B–E**, ?Scutellosaurid thyreophoran *Echinodon* from Lulworth Beds of England, after Galton (1978). **B**, Reconstruction of tooth-bearing bones. **C**, Middle maxillary teeth in lateral view. **D,E**, Anterior dentary teeth in **D**, lateral and **E**, medial views. **F,G**, Scutellosaurid thyreophoran *Scutellosaurus* from Kayenta Formation of western United States, after Colbert (1981). **F**, Maxillary tooth in medial view. **G**, Dentary tooth in medial view. **H,I**, ?Scutellosaurid *Trimucrodon* from late Kimmeridgian of Portugal, after Thulborn (1973), cheek tooth in **H**, lateral and **I**, medial views. **J**, Stegosaurid *Huangosaurus* from the lower Shaximiao Formation (Bathonian–Callovian) of Zigong, Sichuan, China, dentary tooth in lateral view; after Dong, Tang, and Zhou (1982). **K,L**, ?Heterodontosaurid or ?hypsilophodontid, *Pisanosaurus* from Ischigualasto Formation of Argentina, after Bonaparte (1976). **K**, Incomplete right maxilla in lateral view with transverse section. **L**, Right mandible in lateral view with transverse sections as indicated. **M–T**, Heterodontosaurids from upper Elliot Formation (= Red Beds) of Lesotho (M,N) and South Africa. **M,N**, *Abrictosaurus* Hopson (1975). **M**, Left maxillary teeth 5 and 6, in lateral view, after Thulborn (1970c). **O,P**, *Lanasaurus*, after Gow (1975), left maxillary teeth 7 and 8 in **O**, lateral and **P**, medial views. **Q**, *Lycorhinus*, left dentary teeth 3 and 4 in lateral view, after Hopson (1980). **R–T**, *Heterodontosaurus*. **R**, Left maxillary teeth 7 and 8 in lateral view (right reversed), after Charig and Crompton (1974). **S**, Right dentary teeth in medial view, after Crompton and Charig (1962). **T**, Right dentary teeth 7 to 9 in lateral view, after Hopson (1980). **U**, Scelidosaurid *Scelidosaurus* from Liassic of England, maxillary tooth in lateral view, after Owen (1861). **V**, Hypsilophodontid *Yandusaurus* from lower Shaximiao Formation (Bathonian–Callovian) of Sichuan, China, right maxillary teeth in lateral view, after He (1979). **W**, Diagrammatic transverse section through right maxilla and dentary of fabrosaurid *Lesothosaurus*, after Thulborn (1970b) but with dentary tooth drawn in reverse, so laterally rather than medially directed to agree with descriptions of isolated dentary teeth (Thulborn 1970b) and their form in *Fabrosaurus* (Galton 1978). l, lateral. Scale lines represent 10 mm (**B,K,L**), 1 mm (**H,I**), and 2 mm.

16.5A) as reconstructed by Thulborn (1970a), but Crompton and Attridge (Chapter 17) consider that a small horny beak was present, and the small predentary definitely bore a horny beak as in other ornithopods. However, actual remains of a horny beak are known in only a few exceptionally well-preserved hadrosaurid ornithopods (Morris 1970; but for function see Norman 1980). A horny beak has several advantages over teeth as a cropping device because it always forms a continuous edge that can be self-sharpening, it can be molded into cutting edges and/or flat crushing areas, and it is rapidly replaced as it is worn.

On the basis of the marginal position of the teeth in *Lesothosaurus* (Figs. 16.5A, 16.6A), Galton (1973) presumed the absence of cheeks in fabrosaurids. However, Paul (1984) points out that a maxilla of *Lesothosaurus* figured by Thulborn (1970a) has a shallow horizontal ridge with a limited number of large nerve openings; therefore, small cheeks were probably present in this genus.

Tooth form

The tooth crowns of fabrosaurids are anteroposteriorly expanded and triangular to rhomboidal in outline (Figs. 16.3L, 16.4K, 16.6A), with four to seven coarse marginal denticles on either side of the apex. Both sides of the crown are uniformally enameled, the central part is thickened to form a central vertical ridge, and the base of the crown is thickened to form a transverse cingulum that is connected to the most basal denticles by anterior and posterior ridges. The cinguli are at slightly different levels so the teeth are asymmetrical, being slightly curved in anterior or posterior view (Fig. 16.3M), with the maxillary teeth pointing slightly medially and the dentary teeth pointing slightly laterally (Fig. 16.6W) as in other ornithopods.

Tooth wear and jaw action

The isolated teeth of *Lesothosaurus* with double wear facets formed by tooth-to-tooth wear (cf. Fig. 16.6P) described by Thulborn (1970b) are not from the two individuals that provided the skull bones that Thulborn (1970a) described (see Galton 1973; Hopson 1980; Weishampel 1984). However, the presence of double wear facets on in situ teeth of *Lesothosaurus* is confirmed by Crompton and Attridge (Chapter 17). Their presence indicates that the upper and lower tooth positions were staggered so each tooth wore against two opposing teeth and jaw action was strictly vertical. On the basis of a kinematic analysis, Weishampel (1984, p. 74) concludes that the jaw mechanism of fabrosaurids (Fig. 16.5A,K) resembles

a two-link chain formed by the [rigid] skull (*sensu stricto*) and the rigid mandible pairs. The chewing cycle consists

of adduction of the lower jaws (i.e., synchronous vertical rotation around both quadrate–mandibular joints, each of which has one degree of freedom), bilateral occlusion of the teeth, vertical or near vertical tooth–tooth shearing motion, and lastly abduction of the lower jaws. Through isognathous occlusion, the teeth of the dentary sheared past those of the maxilla along a highly inclined wear surface. Such a mechanism most closely resembles those of lepidosaurs (e.g., agamids and iguanids) – [and] is best viewed as one involving little more than simple shearing.

Heterodontosauridae
Skull form

In comparison with *Lesothosaurus* (Fig. 16.5A), the skulls of heterodontosaurids (Fig. 16.5B,C) are more heavily built, the jaw articulation is ventrally offset, and the coronoid process is much larger. The maxillary and dentary teeth are inset with a large lateral space that is roofed by the overhanging maxilla and floored by the massive dentary (Figs. 16.5B,C, 16.6K,L). The mouth was probably small, and the space was delimited laterally by well-developed cheeks that prevented the loss of food from the sides of the jaws during chewing (Galton 1973). The elongate, narrow tongue was probably quite large, as shown by the long rodlike first ceratobranchial of several different ornithopods (hypsilophodontids, iguanodontids, hadrosaurids). It presumably removed food as it accumulated in the cheek space.

Weishampel (1984) notes that the broad, fabrosaurid-style symphysial joint became a well-developed spheroidal joint, with three degrees of rotational freedom, and that it was surrounded by the concave posterior surfaces of the predentary.

Secondary specializations, i.e. strengthening of the cranial support to the mandible (prequadratic bracing of the quadrate, strengthening of the postorbital bar, broadening of the ventral quadrate head), evolved in order for the skull to insure occlusal relations and resistance during the [transverse] power stroke. (Weishampel 1984, p. 76; see "Tooth wear and jaw action," below)

Tooth form

The premaxillary teeth leave room for only a small horny beak in *Heterodontosaurus* (Fig. 16.5B), but there is more room in *Abrictosaurus* (Fig. 16.5C). Wear on the inside of the premaxillary teeth shows that the horny beak on the predentary bit inside the premaxillary teeth that, along with the smaller premaxillary beak, acted as a cropping device (Thulborn, 1970c, 1974).

Canines are present on the premaxilla and dentary in presumed males (Fig. 16.5B) (Thulborn 1970c; Hopson 1975), which probably used them for intraspecific fighting and as defense weapons as in peccaries and tragulids (Thulborn 1974; Molnar 1977). These canines were absent in presumed females (Fig. 16.5C).

In *Pisanosaurus* (Fig. 16.6K,L; Casamiquela 1967; Bonaparte 1976), from the Carnian of Argentina (Fig. 16.1), the teeth are close together and asymmetrical, the uppers curving slightly medially and the lowers slightly laterally. The crowns are not laterally compressed, are well worn, and show very few details, and the occlusal surfaces are almost continuous. This genus is a nonfabrosaurid, but, because the cranial features described could be those of either a primitive heterodontosaurid or a primitive hypsilophodontid, more cranial material is needed to determine its systematic position.

The teeth of the sympatric genera *Abrictosaurus* (Figs. 16.5C, 16.6M,N), *Lanasaurus* (Fig. 16.6O,P), *Lycorhinus* (Fig. 16.6Q), and *Heterodontosaurus* (Figs. 16.5B, 16.6R–T), from the Lower Jurassic of Lesotho and South Africa, show differing degrees of specialization for an herbivorous diet. The cheek teeth of *Abrictosaurus* (Fig. 16.6M,N; Thulborn 1970c, 1974) are well separated, and the symmetrical crowns are distinctly set off from the unswollen roots by a distinct cingulum. The teeth of *Lanasaurus* (Fig. 16.6O,P; Gow 1975) are similar, but the crowns are closer together so some overlap occurs. In *Lycorhinus* (Fig. 16.6Q; Hopson 1975, 1980), the teeth are still symmetrical, and both sides have a convex dorsoventral profile. However, the roots are wide with a slight constriction, adjacent to the cingulum, that is represented by a dorsoventrally deep swelling, and the crowns are proportionally deeper relative to their width. In *Heterodontosaurus* (Fig. 16.6R–T; Crompton and Charig 1962; Charig and Crompton 1974; Hopson 1975, 1980; Chapter 17), the roots are broad and squarish rather than circular in cross section. Because there is no constriction or cingular swelling, they merge into the asymmetrical crowns. The dorsoventral crown profile is straight externally and convex internally for dentary teeth, and the reverse is found for maxillary teeth. The convex surface is enameled and the straight surface is not, whereas in *Abrictosaurus* and *Lanasaurus* both sides are enameled.

The trend within this family is toward the acquisition of an uninterrupted battery of high-crowned grinding teeth. The increase in the height and useful lifetime of the crown is correlated with a slowing in the rate of replacement of cheek teeth that ceased altogether in mature individuals of *Heterodontosaurus* (Hopson 1975, 1980).

Tooth wear and jaw action
The presence of double, obliquely inclined wear facets that meet along a transverse edge in *Lanasaurus* (Fig. 16.6P) shows that, as in *Lesothosaurus*, the jaw action was strictly orthal.

Thulborn (1978) argues that jaw movements were fore-and-aft in the other heterodontosaurids because the wear surfaces are usually confluent from one tooth to the next, and the dentition appears to have worn down as a unit. Hopson (1980) analyzes the less obvious double facets on teeth of *Lycorhinus* (Fig. 16.6Q) and on the posterior teeth of *Heterodontosaurus* (Fig. 16.6T) and shows that the jaw action was orthal. He concludes that, in all heterodontosaurids, a grinding action was accomplished by an oblique dorsal and medial shifting of the lower dentition while in occlusal contact with the upper dentition, and that the differences in the pattern of wear and of the angle of the occlusal surface depends on how worn down the teeth are.

By means of a kinematic analysis, Weishampel (1984) shows that heterodontosaurids retained the primitive isognathous, rigid skull condition but mobilized the lower jaws to rotate medially to form a three-link kinematic chain (Fig. 16.5L) to facilitate a maxillary–dentary transverse power stroke. Weishampel (1984, p. 76) notes that

the chewing cycle consists of mandibular adduction, bilateral occlusion, power stroke through medial rotation of the mandibles during continued adduction, and lastly mandibular abduction to their original position at the beginning of the cycle. Rotation provided for reduced inclination of the occlusal surface.

He also notes that the transverse power stroke, resulting from the medial excursion of the tightly packed dentary teeth against those of the maxilla, indicates a shift from the fabrosaurid retention-cut function (*sensu* Osborn and Lumsden 1978) with shearing to one of tearing and cutting, perhaps initiated by compression, of heterodontosaurids.

Hypsilophodontidae
Skull form
The jaw articulation is ventrally offset; the coronoid process is large; the cheek teeth are inset; and the outline of the skull is similar to that of heterodontosaurids (Fig. 16.5D; Galton 1974, 1983b). However, Weishampel (1984) points out several important differences from fabrosaurids that relate to the mobilization of two joints within the skull. Mobilization of the quadrate–squamosal joint, to give a streptostylic quadrate, involved reduction of the prequadratic process of the squamosal and of the quadratojugal–quadrate joint. Mobilization of the maxilla–premaxilla joint, to give a maxilla that could rotate laterally, involved development of a rostral maxillary process that fitted into the posterior part of the premaxilla, and a loosening of the dorsal contact between the maxilla and premaxilla, maxilla and nasal, and lacrimal and prefrontal. "Secondary modifications include elongation and depression of the basipterygoid processes, broadening of the ventral head of the quadrate, and emargination of the tooth rows" (Weishampel 1984, p. 77).

Tooth form

Apart from *Hypsilophodon* (Galton 1974) and possibly *Yandusaurus* (Fig. 16.5D), all hypsilophodontids lack premaxillary teeth that were functionally replaced by a large horny beak. The crowns of the cheek teeth are laterally compressed, leaf-shaped, close together, and arranged *en echelon*. One crown surface is convex dorsoventrally and thickly enameled, with a well-developed series of vertical ridges, and it forms the leading edge of the tooth once an obliquely orientated wear surface is present. This is the lateral surface for maxillary teeth, which curve slightly medially and have ventromedially facing wear surfaces, and the medial surface for dentary teeth, which curve slightly laterally and have dorsolaterally facing wear surfaces. The ridges formed a series of serrations on the leading edge that increased its efficiency as a cutting edge and for holding the food while it was being cut. The opposite surface of the crown is dorsoventrally concave, usually lacks vertical ridges, and is less thickly enameled.

Tooth wear and jaw action

Double wear facets are known in hypsilophodontids (Galton 1973), therefore jaw action was vertical. Weishampel (1984) shows that hypsilophodontids modified the isognathous, rigid skull condition of fabrosaurids to allow mobility of the upper jaws to rotate laterally to produce a transverse power stroke analogous to that of heterodontosaurids. Hypsilophodontids have a four-link mechanism (Fig. 16.5M):

Paired mobile maxillopalatal segments (1 degree of rotational freedom at each premaxillary–maxillary joint) rigidly joined to corresponding streptostylic quadrates (2 degrees of rotational freedom at each quadrate-squamosal joint), enabling them to utilize a transverse power stroke through lateral motion of the maxilla . . . between adducted dentaries. Hence, the chewing cycle begins with mandibular adduction and bilateral occlusion, with continued adduction the power stroke is accomplished through the lateral rotation of the maxillae and adjoining bones. During jaw opening, the mandibles are abducted and the upper jaw complex returns to its original position. The effect upon food items and tooth wear in hypsilophodontids is much the same as that in heterodontosaurids: food is compressed by contact between maxillary and dentary teeth, followed by tearing and cutting during excursion of maxillary over dentary teeth, especially in those regions where enamel contacts enamel. [The iguanodontid] jaw systems are very little modified from their hypsilophodontid ancestors. (Weishampel 1984, pp. 76–7)

Suborder Thyreophora

Gauthier (1984) refers all the armored ornithischians to the Thyreophora (Fig. 16.1) in which, in addition to well-known ankylosaurs and stego-

saurs (Fig. 16.5F–I), he includes a couple of genera from the Early Jurassic: *Scutellosaurus* from Arizona, and *Scelidosaurus* from England (Figs. 16.5, 16.6F,G,U).

Scutellosaurus

The small skull of *Scutellosaurus* was presumably lightly built and is represented by jaws; the coronoid eminence is low (Colbert 1981). The teeth resemble those of fabrosaurids (Fig. 16.6F,G) and are not inset, but, because there are only a few large lateral nerve foramina, small cheeks were probably present. A couple of teeth show slight signs of wear on the denticles along their edges (Colbert 1981) that probably resulted from tooth-to-food wear. Two "fabrosaurids" from the Late Jurassic of Europe are probably related to *Scutellosaurus*: *Trimucrodon* (Fig. 16.6H,I; Thulborn 1973) from the Kimmeridgian of Portugal and *Echinodon* (Figures 16.6B–G; Galton 1978) from the Portlandian of England and the United States [Morrison Formation, Fruita, Colorado (Galton in prep.)].

The two cinguli are at the same level in the teeth of *Trimucrodon* and *Echinodon* that are symmetrical and straight in anterior or posterior views (Fig. 16.6C,E,H,I), like the teeth of anchisaurid prosauropods (Fig. 16.3K). In *Echinodon* there are three premaxillary teeth, a caniniform first maxillary tooth, and a posteriorly deep dentary with a high coronoid process; cheeks were probably present (Fig. 16.6B). Small dermal scutes from the same horizon and locality, originally described as "granicones" of the "lizard" *Nuthetes* [now regarded as a megalosaurian theropod (Steel 1970)] by Owen (1879), are probably referable to *Echinodon*.

Scelidosauridae

Scelidosaurus (Figs. 16.5E, 16.6U) (Owen 1861) was long referred to the Stegosauria (Romer 1956), but more recently it has been referred to the Ankylosauria (Romer 1968; Galton 1975), Ornithopoda (Thulborn 1974, 1977), suborder *incertae sedis* (Charig 1979), and to the Thyreophora as a sister group of ankylosaurs and stegosaurs (Gauthier 1984; Norman 1984). The only other scelidosaurid is a poorly prepared piece of jaw with teeth from the Lower Jurassic of Portugal that was described as *Lusitanosaurus* by Lapparent and Zbyszewski (1957). The cheek teeth of *Scelidosaurus* are well inset, the coronoid eminence is deep, and the jaw articulation is ventrally offset (Fig. 16.5E). The tooth crowns (Fig. 16.6U) have prominent anterior, central, and posterior ridges, resembling slightly those of *Echinodon* (Fig. 16.6C–E) (Galton 1978), and each slightly overlaps the one behind it. Charig (1979, p. 128) gives a photograph of the fully prepared skull. The left dentary is shown in lateral view (although articulated with right quadrate), and double wear facets that meet along a transverse edge are visible on some of the teeth. The wear facets are steeply inclined because they occupy most of the tall crown. The ventral process of the squamosal has an extensive suture with the dorsal process of the quadratojugal so that the quadrate was firmly fixed. Consequently, although the symphysis is not preserved, jaw action was probably as described above for *Lesothosaurus*.

Stegosauria

The skull of *Huayangosaurus* (Fig. 16.5F; Dong, Tang, and Zhou 1982; Zhou 1984) from the Middle Jurassic (Bathonian–Callovian) of China is similar to that of *Scelidosaurus* (Fig. 16.5E); in *Tuojiangosaurus* (Fig. 16.5G; Dong, Zhou, and Zhang 1983) and *Stegosaurus* (Fig. 16.5H; Gilmore 1914), it is proportionally lower. There are three premaxillary teeth in *Huayangosaurus* (Zhou 1984), but none in the other stegosaurs, and the cheek teeth are simple with only a slight central ridge (Fig. 16.6J; Dong et al. 1982; Zhou 1984). The teeth of *Tuojiangosaurus* and *Stegosaurus* have a squarer outline, the vertical ridging is more prominent, and there are prominent cinguli (see Galton 1980b). Some of the teeth of *Stegosaurus* have obliquely inclined wear facets, the symphysial surface of the dentary is extensive and rugose, and in adult individuals the quadrate is fused to the adjacent bones (Gilmore 1914). Consequently, jaw action in stegosaurs was as described above for *Lesothosaurus*.

Ankylosauria

The only ankylosaurian cranial material described to date from the Jurassic is the incomplete lower jaw of *Sarcolestes* (Fig. 16.5I; Galton 1983c), the earliest record of the group (Fig. 16.1), and the maxilla of *Priodontognathus* from the Oxfordian of England (Galton 1980b, 1983c). The lower jaw is deep with a very small sutural surface for a predentary, the teeth are inset, the coronoid eminence is low, and the jaw articulation is ventrally offset (Fig. 16.5I). The teeth are slightly recurved with only a slight central ridge, the denticles continue onto the crown as short curved ridges, and the exposed medial surface bears a prominent bulbous cingulum that is denticulate in *Priodontognathus*, whereas in *Sarcolestes* there is no cingulum. Only replacement teeth are preserved, but obliquely inclined double wear surfaces are present on the teeth of some Cretaceous ankylosaurs (Russell 1940). The symphysis is extensive and rugose in *Sarcolestes* (Galton 1983c) and Cretaceous ankylosaurs in which the maxilla is solidly fused to the surrounding bones and the overlying dermal scutes (Coombs 1978). Consequently, jaw action was as in *Lesothosaurus*.

Discussion
Proportional representation

The earliest records of anchisaurid, fabrosaurid and heterodontosaurid–hypsilophodontid dinosaurs are approximately contemporaneous at about the mid-Carnian (Fig. 16.1), and each group is rare. However, in the lower Elliot Formation (late Carnian or early Norian) of South Africa, the large (up to 6 m) anchisaurid *Euskelosaurus* is common (Van Heerden 1979; Kitching and Raath 1984), the large (up to 6 m) melanorosaurid *Melanorosaurus* (Haughton 1924; Galton 1985d) and the large (about 5 m) Blikana prosauropod (Charig et al. 1965; Galton and Van Heerden 1985) are rare (two and one individuals, respectively), and there are no records of ornithischians despite extensive collecting recently (Kitching and Raath 1984). The numerical predominance of anchisaurid specimens in the fossil record continues for the rest of the Norian and into the Early Jurassic. From the data given by Benton (1983a), *Plateosaurus* (with junior synonym *Gresslyosaurus*) represents at least 75 percent of the individual animals from the Knollenmergel (Upper Triassic) of Germany, and *Lufengosaurus* (together with *Yunnanosaurus*) represents 82 percent of the individual animals from the upper Lower Lufeng Series (Lower Jurassic) of China. In each case, the anchisaurids are by far the largest animals and, although calculating the percentages of individuals in the original population and the percentage biomass that they represented is fraught with uncertainties, anchisaurids certainly constituted a large percentage of the original biomass.

An exception to the predominance of anchisaurids in well-represented fossil assemblages from the Norian is the Los Colorados Formation of Argentina in which the large (up to 10 m) melanorosaurid prosauropod *Riojasaurus* constitutes about 40 percent of the individual animals collected (Bonaparte 1982). The smaller anchisaurid *Coloradia* (Fig. 16.2F) is rare (Bonaparte 1978), and ornithischian dinosaurs are unknown (Bonaparte 1982).

Most of the records of sauropods from the Early Jurassic are based on one or two specimens, and, apart from the record from the upper Lower Lufeng Series (Chapter 21), none are sympatric with either prosauropods or ornithischians. However, the very large (up to 25 m) *Barapasaurus* from India is extremely abundant, forming extensive bone beds at two localities about 36 km apart (Jain et al. 1975, 1979; Yadagiri, Prasad, and Satsangi 1980), and it was certainly the dominant, very large terrestrial herbivore, a niche that the sauropods occupied exclusively in the Middle and Late Jurassic.

The absence of fabrosaurids and heterodontosaurids in the Norian fossil record may represent an ecological difference, with the ornithischians occupying a more arid and/or upland habitat. The sediments represented by the Elliot (= Red Beds) and Clarens (= Cave Sandstone) formations of southern Africa were deposited under progressively increasingly arid conditions (Haughton 1924; Charig et al. 1965; Cooper 1981). The anchisaurid *Massospondylus* (up to 4 m) occurs throughout the middle and upper Elliot Formation, fabrosaurids occur from the top of the middle and through the upper Elliot Formation, and heterodontosaurids occur at the top of the upper Elliot (Kitching and Raath 1984), with anchisaurids representing 21 percent and ornithischians 7 percent of the individual animals (Benton 1983a). All three families occur at the base of the Clarens Formation, with anchisaurids representing 40 percent and ornithischians 22 percent of the individual animals. Most fabrosaurids and heterodontosaurids represent animals with a body length of about 1 m, but at least one fabrosaurid equaled a medium-sized *Massospondylus* (Chapter 17). Hypsilophodontids and iguanodontids are rare in the Middle Jurassic, but are locally abundant in the Late Jurassic of the United States (*Dryosaurus*, *Camptosaurus*) and Tanzania (*Dryosaurus*).

Thyreophorans are represented by several different individuals from the Early Jurassic of the United States (*Scutellosaurus*) and England (*Scelidosaurus*). Stegosaurs are fairly common in the Middle Jurassic of China (*Huayangosaurus*) and England (*Lexovisaurus*) but ankylosaurs are rare (*Sarcolestes*); stegosaurs are common in the Late Jurassic, whereas ankylosaurs are rare (Galton 1983c). This is the reverse of the situation in the Cretaceous (Galton 1980c).

Possible reasons for success of different groups

The predominance of prosauropods in the Norian record of terrestrial herbivores is especially surprising when the poor herbivorous adaptations of the large prosauropods (up to 10 m) are compared to the better adaptations for dealing with resistant plant material by the much smaller herbivores that occur in the Carnian and Norian, such as chelonians (Wild 1974; Gaffney and Meeker 1983); procolophonids (Colbert 1946; Gow 1978); trilophosaurids (Gregory 1945); rhynchosaurs (Benton 1983b); aetosaurids (Figure 16.5J; Walker 1961); heterodontosaurids (Fig. 16.6K,L); mammal-like reptiles (Gow 1978; Kemp 1982), such as dicynodonts (Cruickshank 1978), diademodontids, and tritylodontids; and the very small mammals (Kemp 1982). Herbivorous adaptations in these groups include horny cropping beaks on the anterior parts of the premaxilla and dentary for cropping food (tortoises, trilophosaurids, aetosaurids, heterodontosaurids, dicynodonts); a trough-shaped, more posterior

horny triturating surface on the maxilla and dentary (chelonians and dicynodonts); leaf-shaped teeth with tooth-to-tooth wear (aetosaurs); transversely expanded and interdigitating cheek teeth with tooth-to-tooth wear (procolophonids and trilophosaurids); multiple rows of ankylothecodont teeth, with additional teeth added to the end of the series during growth to form dental batteries with a precision-shear bite (rhynchosaurs); and cheek teeth with crushing or grinding plus a loss of alternate tooth replacement so as not to interfere with the efficiency of the crushing–grinding–shearing mechanism (heterodontosaurids, diademodontids, tritylodontids, and mammals). Compared to the skulls of most of these herbivores, those of prosauropods are lightly constructed, with noninterdigitating, overlapping, or abutting sutures that tended to separate prior to preservation (see Galton 1984, 1985c), and there is proportionally much less room for the adductor muscles (see Chapter 17). Tooth-to-tooth wear facets are only present in *Yunnanosaurus* (Figs. 16.3B–D, 16.4D,E). However, as shown for *Massospondylus* (Bond 1955; Raath 1974; Cooper 1981), anchisaurids probably had a gastric mill that, along with chemical and/or bacterial action, may have compensated for the relative lack of herbivorous specializations of the masticatory apparatus.

Most of the nonarchosaurian herbivores discussed above were "sprawlers," with the propodials (humerus, femur) held horizontally while running, whereas archosaurs are either semierect (most thecodontians including aetosaurs), with the propodials held at 45° to the vertical, or fully erect (rauisuchid thecodontians, dinosaurs), with the propodials held vertically to bring the limbs under the body (Charig 1972, 1984; Bonaparte 1984). Although the hindlimb was held more upright, the femur was probably at an angle of about 20° to the parasagittal plane in anchisaurids (Cooper 1981; 30°–45° according to Van Heerden 1979) and probably also in *Yunnanosaurus* (see Young 1942, 1951). This angle was probably closer to 10° in melanorosaurids, in which the femur has a more medially set head and is straighter in anterior or posterior view (Bonaparte 1971; Galton 1985d). However, even in mammals, a fully erect pose is only present in cursorial and graviportal mammals, and the femur is at an angle of 20°–50° to the parasagittal plane in other mammals (Jenkins 1971). The improved gait and the development of a gastric mill may be sufficient to explain the success of prosauropod dinosaurs even though there were few major floral changes toward the end of the Triassic (Chapter 2). It is also possible that, as discussed by Crompton and Attridge (Chapter 17), competition from other terrestrial herbivores may not have been a factor in the initial radiation of prosauropods. However, it is difficult to understand why prosauropods rather than ornithischians dominate the Norian record, especially as *Pisanosaurus* (Fig. 16.5K,L) of the mid-Carnian was already specialized to deal with more resistant plant material and probably had a fully erect hindlimb (Casamiquela 1967; Bonaparte 1976).

By the Middle Jurassic, sauropods had replaced prosauropods as the large high browsers, and even some of the Early Jurassic sauropods were extremely large [e.g., *Barapasaurus* up to 25 m (Jain et al. 1979)]. Sauropods were fully erect with graviportal limbs in which the femur has a medially set head and a straight shaft in anterior or posterior view. Locomotory improvements are probably sufficient to explain the replacement of prosauropods by sauropods, but, extrapolating from the situation in Middle Jurassic forms, improvements in the masticatory apparatus were probably also involved.

The masticatory apparatus of fabrosaurids is similar to that of prosauropods in having a loosely built skull with a proportionally small area for the jaw adductor muscles (Fig. 16.5A). However, cranial improvements for processing slightly more resistant plant material include the small horny premaxillary and predentary beaks and the curvature of opposing cheek teeth toward each other (Fig. 16.6W) for a more efficient shearing action with tooth-to-tooth wear in at least one case (see Chapter 17). Heterodontosaurids (Figs. 16.5B,C, 16.6M–T) could process even more resistant plant material because of the more solid construction of the skull, the proportionally larger area for jaw adductor muscles, the presence of a regular series of tooth-to-tooth wear surfaces that were confluent across adjacent teeth, and the transverse power stroke produced by medial rotation of the lower jaws while the teeth were in occlusion (Fig. 16.5L).

The hypsilophodontids (up to 4 m, *Dryosaurus*) and the iguanodontids (up to 6 m, *Camptosaurus*) of the later Jurassic expanded the size range of ornithopods feeding on more resistant plant material. However, the transverse power stroke while the teeth were in occlusion was produced by a lateral movement of the maxillae (Fig. 16.5M). Their success may have resulted from the development of a series of vertical ridges on the opposing thicker enameled convex surface of the cheek teeth crowns and the retention of a normal reptilian pattern of tooth replacement. This ensured that the teeth were replaced as they were worn down. Locomotory improvements were probably not involved because both heterodontosaurids and hypsilophodontids were cursorial with fully erect hindlimbs.

Two thyreophoran ornithischians with dermal armor and fully erect limbs occur in the Early Jurassic; they range in size from 1.4 m for *Scutellosaurus*, which was probably less bipedal than the fabrosaurids (Colbert 1981), to the fully quadrupedal *Scelidosaurus* at about 4 m (Charig 1979). The skull of *Scelidosaurus* (Fig. 16.5E) was more solidly built than those of prosauropods (Fig. 16.2A–I) and fabrosaurids (Fig. 16.5A), and tooth-to-tooth wear was present. However, the area for the jaw-closing muscles is proportionally small, and jaw action was strictly vertical with no transverse power stroke. The situation was similar in the Stegosauria, the common armored quadrupeds in the later Jurassic, and the Ankylosauria that were rare then but common in the Cretaceous. All were adapted to eat fairly soft plant food and presumably took over part of the niche that was previously occupied by the prosauropods. However, it is not readily apparent why ankylosaurs largely replaced stegosaurs across the Jurassic–Cretaceous boundary.

Acknowledgments

I thank the following people for all their assistance while I studied the specimens cited from their respective institutions (with abbreviations used): J. Bolt, Field Museum of Natural History, Chicago, for S.V.D. Catholic University of Peking collection (FMNH); J. M. Dutuit, Museum National d'Histoire Naturelle, Paris (MNHN); W. Walchli, Museum Saurierkommission Frick, Townwerke Keller A. G. Frick, Switzerland (MSF); J. H. Ostrom and M. A. Turner, Peabody Museum of Natural History, Yale University, New Haven (YPM); and all the other people at numerous institutions who have assisted me since 1967 in my research on the cranial material of herbivorous dinosaurs that (although not specifically cited in this chapter) has been utilized in the writing of this chapter. I thank J. S. McIntosh, Wesleyan University, Middletown, Connecticut for information on sauropod dinosaurs; J. Bonaparte, Museo "B. Rivadavia," Buenos Aires, Argentina for information on *Riojasaurus*; F. A. Jenkins, Jr., Harvard University, Cambridge for a drawing used as basis for Fig. 16.2E; and M. Benton, Queen's University, Belfast, Northern Ireland; K. Padian, University of California, Berkeley; and D. Weishampel, Johns Hopkins University, Baltimore for their extensive comments on this chapter. Figures 16.4 and 16.6 were drawn by Barbara Whitman, and the photographs were printed by Joseph Sousza, both of the University of Bridgeport, the manuscript was typed by Natalie Susan Galton-Rawls and Mary Beth Brubaker, and this research was partly supported by N.S.F. Research Grants DEB–8101969 and BSR–8500342.

References

Attridge, J., A. W. Crompton, and F. A. Jenkins. 1985. The southern African Liassic prosauropod dinosaur *Massospondylus* in North America. *J. Vert. Paleontol.* 5:128–32.

Auffenberg, W. 1981. *The Behavioral Ecology of the Komodo Monitor* (Gainesville, Florida: University Presses of Florida).

Bakker, R. T. 1971a. Ecology of brontosaurs. *Nature*

(London) 229:172–4.

1971b. Brontosaurs. *McGraw-Hill. Yearbk. Sci. Technol.* 1971:179–181.

1978. Dinosaur feeding behavior and the origin of flowering plants. *Nature (London)* 274:661–3.

Benton, M. J. 1983a. Dinosaur success in the Triassic: a noncompetitive ecological model. *Quart. Rev. Biol.* 58:29–55.

1983b. The Triassic reptile *Hyperodapedon* from Elgin: functional morphology and relationships. *Phil. Trans. Roy. Soc. London B* 302:605–720.

Berman, D. S., and J. S. McIntosh. 1978. Skull and relationships of the Upper Jurassic sauropod *Apatosaurus* (Reptilia, Saurischia). *Bull. Carnegie Mus.Nat. Hist.* 8:1–35.

Bonaparte, J. F. 1971. Los Tetrapodos del sector Superior de la Formación Los Colorados, La Rioja, Argentina (Triásico Superior), I Parte. *Opera Lilloana* 22:1–183.

1976. *Pisanosaurus mertii* Casamiquela and the origin of the Ornithischia. *J. Paleontol.* 50:808–20.

1978. *Coloradia brevis* n. g. et n. sp. (Saurischia Prosauropoda), dinosaurio Plateosauridae de la Formación Los Colorados, Triásico Superior de La Rioja, Argentina. *Ameghiniana* 15:327–32.

1982. Faunal replacement in the Triassic of South America. *J. Vert. Paleontol.* 2:362–71.

1984. Locomotion in rauisuchid thecodonts. *J. Vert. Paleontol.* 3:210–18.

Bonaparte, J. F., and M. Vince. 1979. El hallazgo del primer nido de Dinosaurios Triásicos (Saurischia, Prosauropoda), Triásico superior de Patagonia, Argentina. *Ameghiniana* 16:173–82.

Bond, G. 1955. A note on dinosaur remains from the Forest Sandstone (Upper Karroo). *Arnoldia (Rhodesia)* 2(20):795–800.

Brown, B. 1941. The last dinosaurs. *Nat. Hist.* 48:290–5.

Burden, W. D. 1928. Results of the Douglas Burden Expedition to the Island of Komodo. V. – Observations on the habits and distribution of *Varanus komodoensis* Ouwens. *Am. Mus. Novit.* 316:1–10.

Cabrera, A. 1947. Un sauropodo nuevo del Jurássic de Patagonia. *Not. Mus. La Plata* 12:1–17.

Cannon, G. L. 1906. Sauropodan gastroliths. *Science N.* 24:116.

Carey, M. A., and J. H. Madsen, Jr. 1972. Some observations on the growth, function and differentiation of sauropod teeth from the Cleveland–Lloyd Quarry. *Proc. Utah Acad. Sci.* 49(1):41–3.

Casamiquela, R. M. 1963. Consideraciones acerca de *Amygdalodon* Cabrera (Sauropoda, Cetiosauridae) del Jurásico Medio de la Patagonia. *Ameghiniana* 3:79–95.

1967. Un nuevo dinosaurio ornitisquio triásico, (*Pisanosaurus mertii*; Ornithopoda) de la formación Ischigualasto, Argentina. *Ameghiniana* 5:47–64.

Charig, A. J. 1972. The evolution of the archosaur pelvis and hind-limb; an explanation in functional terms. *In* Joysey, K. A., and T. S. Kemp (eds.), *Studies in Vertebrate Evolution* (Oliver and Boyd, Edinburgh), pp. 121–55.

1979. *A New Look at the Dinosaurs* (New York: Mayflower), pp. 1–160.

1984. Competition between therapsids and archosaurs during the Triassic period: a review and synthesis of current theories. *Symp. Zool. Soc. London* 52:597–628.

Charig, A. J., and A. W. Crompton. 1974. The alleged synonymy of *Lycorhinus* and *Heterodontosaurus*. *Ann. S. Afr. Mus.* 64:167–89.

Charig, A. J., J. Attridge, and A. W. Crompton. 1965. On the origin of the sauropods and the classification of the Saurischia. *Proc. Linn. Soc. London* 176:197–221.

Charig, A. J., B. Krebs, H.-D. Sues, and F. Westphal. 1976. Thecodontia. *Hdb. Paläoherpetol.* 13:1–137.

Chatterjee, S. in press. A new ornithischian dinosaur from the Triassic of North America. *Geolog. Rundschau.*

Colbert, E. H. 1946. *Hypsognathus*, a Triassic reptile from New Jersey. *Bull. Am. Mus. Nat. His.* 86:225–74.

1951. *The Dinosaur Book: The ruling reptiles and their relatives* (New York: McGraw-Hill), pp. 1–156.

1981. A primitive ornithischian dinosaur from the Kayenta Formation of Arizona. *Bull. Ser. Mus. N. Ariz. Press.* 53:1–61.

Coombs, W. P., Jr. 1975. Sauropod habits and habitats. *Palaeogeogr., Palaeoclimatol., Palaeoecol.* 17:1–33.

1978. The families of the ornithischian dinosaur order Ankylosauria. *Palaeontology* 21:143–70.

Cooper, M. R. 1981. The prosauropod dinosaur *Massospondylus carinatus* Owen from Zimbabwe: its biology, mode of life and phylogenetic significance. *Occ. Pap. Natn. Mus. Monum. Zimbabwe Ser. B. Nat. Sci.* 6(10):689–840.

1984. A reassessment of *Vulcanodon karibaensis* Raath (Dinosauria: Saurischia) and the origin of the Sauropoda *Palaeontol. Afr.* 25:203–31.

Crompton, A. W., and A. J. Charig. 1962. A new ornithischian from the Upper Triassic of South Africa. *Nature (London)* 196:1074–7.

Crompton, A. W., and K. Hiiëmae. 1969. How mammalian molar teeth work. *Discovery, New Haven* 5:23–34.

Cruickshank, A. R. I. 1978. Feeding adaptations in Triassic dicynodonts. *Palaeont. Afr.* 21:121–32.

Darby, D. G., and R. W. Ojakangas. 1980. Gastroliths from an Upper Cretaceous plesiosaur. *J. Paleontol.* 54:548–56.

Dong, Z., and Z. Tang. 1984. Note on a new mid-Jurassic sauropod (*Datousaurus bashanensis* gen. et sp. nov.) from Sichuan Basin, China. *Vert. Palasiatica.* 22:69–75.

Dong, Z., Z. Tang, and S. Zhou. 1982. Note on the new mid-Jurassic stegosaur from Sichuan Basin, China. *Vert. Palasiatica.* 20:83–87.

Dong, Z., S. Zhou, and Y. Zhang. 1983. The dinosaurian remains from Sichuan Basin, China. *Paleontol. Sinica.* 162(n.s. c,23):1–145.

Dutuit, J. M. 1972. Découverte d'un Dinosaure ornithischian dans le Trias supérieur de l'Atlas occidental marocain. *Compt. Rend. Acad. Sci. Paris (D)* 275:2841–4.

Gaffney, E. S., and L. J. Meeker. 1983. Skull morphology of the oldest turtles: a preliminary description of *Proganochelys quenstedti. J. Vert. Paleontol.* 3:25–8.

Galton, P. M. 1971. The prosauropod dinosaur *Ammosaurus*, the crocodile *Protosuchus*, and their bearing on the age of the Navajo Sandstone of northeastern Arizona. *J. Paleontol.* 45:281–95.

   1972. Classification and evolution of ornithopod dinosaurs. *Nature (London)* 239:464–6.

   1973. The cheeks of ornithischian dinosaurs. *Lethaia* 6:67–89.

   1974. The ornithischian dinosaur *Hypsilophodon* from the Wealden of the Isle of Wight. *Bull. Brit. Mus. (Nat. Hist.)* 25:1–152c.

   1975. English hypsilophodontid dinosaurs (Reptilia : Ornithischia). *Palaeontology* 18:741–52.

   1976. Prosauropod dinosaurs (Reptilia : Saurischia) of North America. *Postilla* 169:1–98.

   1978. Fabrosauridae, the basal family of ornithischian dinosaurs (Reptilia : Ornithopoda). *Paläontol. Z.* 52:138–59.

   1980a. European Jurassic ornithopod dinosaurs of the families Hypsilophodontidae and Iguanodontidae. *N. Jahrb. Geol. Paläontol. Abh.* 160:73–95.

   1980b. *Priodontognathus phillipsii* (Seeley), an ankylosaurian dinosaur from the Upper Jurassic (or possibly Lower Cretaceous) of England. *N. Jahrb. Geol. Paläontol. Mh.* 1980:477–89.

   1980c. *Craterosaurus pottonensis* Seeley, a stegosaurian dinosaur from the Lower Cretaceous of England, with a review of Cretaceous stegosaurs. *N. Jahrb. Geol. Paläontol. Abh.* 161:28–46.

   1983a. The oldest ornithischian dinosaurs from North America from the Late Triassic of Nova Scotia, N. C., and PA. *Geol. Soc. Am. Abst. Prog. March* 1983:122.

   1983b. The cranial anatomy of *Dryosaurus*, a hypsilophodontid dinosaur from the Upper Jurassic of North America and East Africa, with a review of hypsilophodontids from the Upper Jurassic of North America. *Geol. Palaeontol.* 17:207–43.

   1983c. Armored dinosaurs (Ornithischia : Ankylosauria) from the Middle and Upper Jurassic of Europe. *Palaeontographica* A182:1–25.

   1984. Cranial anatomy of the prosauropod dinosaur *Plateosaurus* from the Knollenmergel (Middle Keuper, Upper Triassic) of Germany. I. Two complete skulls from Trossingen/Württ. with comments on the diet. *Geol. Palaeontol.* 18:139–71.

   1985a. Diet of prosauropod dinosaurs from the late Triassic and early Jurassic. *Lethaia.* 18:105–23.

   1985b. An early prosauropod dinosaur from the Upper Triassic of Nordwürttemberg, West Germany. *Stuttgarter Beitr. Naturk. B,* 106:1–26.

   1985c. Cranial anatomy of the prosauropod *Plateosaurus* from the Knollenmergel (Middle Keuper, Upper Triassic) of Germany. II. All the cranial material and details of soft-part anatomy. *Geol. Palaeontol.* 19:119–59.

   1985d. Notes on the Melanorosauridae, a family of large prosauropod dinosaurs (Saurischia: Sauropodomorpha). *Geobios.* 18:671–6.

   1985e. Cranial anatomy of the prosauropod dinosaur *Sellosaurus gracilis* from the Middle Stubenstein (Upper Triassic) of Nordwürttemberg, West Germany. *Stuttgarter Beitr. Naturk B.* 118:1–39.

Galton, P. M., and H. P. Powell. 1980. The ornithischian dinosaur *Camptosaurus prestwichii* from the Upper Jurassic of England. *Palaeontology.* 23:411–43.

   1983. Stegosaurian dinosaurs from the Bathonian (Middle Jurassic) of England, the earliest record of the family Stegosauridae. *Geobios* 16:219–29.

Galton, P. M. and J. Van Heerden. 1985. Partial hindlimb of *Blikanasaurus cromptoni* n. gen. and n. sp., representing a new family of prosauropod dinosaurs from the Upper Triassic of South Africa. *Geobios.* 18:509–16.

Gauthier, J. A. 1984. A cladistic analysis of the higher systematic categories of the Diapsida. Ph.D. thesis, University of California, Berkeley, Department of Paleontology.

Gilmore, C. W. 1914. Osteology of the armored Dinosauria in the United States National Museum, with special reference to the genus *Stegosaurus*. *Bull. U.S. Nat. Mus.* 89:1–136.

   1925. A nearly complete articulated skeleton of *Camarasaurus*, a saurischian dinosaur from the Dinosaur National Monument, Utah. *Mem. Carnegie Mus.* 10:347–84.

Gow, C. E. 1975. A new heterodontosaurid from the Red Beds of South Africa showing clear evidence of tooth replacement. *Zool. J. Linn. Soc. London* 57:335–9.

   1978. The advent of herbivory in certain reptilian lineages during the Triassic. *Palaeontol. Afr.* 21:133–41.

Gregory, J. T. 1945. *Osteology and Relationships of Trilophosaurus*. University of Texas Publ. No. 4401, pp. 273–359.

Halstead, L. B., and J. Halstead, 1981. *Dinosaurs* (Poole: Blanford Press).

Haughton, S. H. 1924. The fauna and stratigraphy of the Stormberg fauna. *Ann. S. Afr. Mus.* 12:323–497.

He, X. 1979. A new discovered ornithopod dinosaur, *Yandusaurus* from Zigong, Sichuan. *Contrib. Int. Exch. Geol. Strat. Paleontol.* 2:116–23.

He, X., and K. Cai. 1984. *The Middle Jurassic Dinosaurian Fauna from Dashanpu, Zigong, Sichuan*, Vol. I, The ornithopod dinosaurs (Sichuan: Science and Technical Publishing).

He, X., K. Li, K. Cai, and Y. Gao. 1984. *Omeisaurus tianfuensis* – a new species of *Omeisaurus* from Dashanpu, Zigong, Sichuan. *J. Chengdu College Geol. Suppl.* 2:13–32.

Holland, W. J. 1924. The skull of *Diplodocus. Mem. Carnegie Mus.* 9:379–403.

Hopson, J. A. 1975. On the generic separation of the ornithischian *Lychorinus* and *Heterodontosaurus* from the Stormberg Series (Upper Triassic) of South Africa. *S. Afr. J. Sci.* 71:302–5.

   1980. Tooth function and replacement in early Mesozoic ornithischian dinosaurs: implications for aestivation. *Lethaia* 13:93–105.

Huene, F. von. 1907–8. Die Dinosaurier der europäischen Triasformation mit Berucksichtigung der aussereuropäischen Vorkomnisse. *Geol. Paläontol. Abh. Suppl.* 1:1–419.

1926. Vollständige Osteologie eines Plateosaurian aus dem schwäbischen Trias. *Geol. Paläontol. Abh. N. F.* 15(2):139–79.

Jain, S. L., T. S. Kutty, T. Roy-Chowdhury, and S. Chatterjee. 1975. The sauropod dinosaur from the Lower Jurassic Kota Formation of India. *Proc. Roy. Soc. London A* 188:221–8.

1980. Some characteristics of *Barapasaurus tagorei*, a sauropod dinosaur from the Lower Jurassic of Deccan, India. *Intern. Gondwana Symp. IV, 1977, Calcutta*, pp. 204–16.

Janensch, W. 1929. Magensteine bei Sauropoden der Tendaguru-Schichten. *Palaeontographica Suppl. 7* 1(2):135–43.

1935–6. Die Schädel der Sauropen *Brachiosaurus, Barasaurus* and *Dicraeosaurus* aus den Tendaguru-Schichten Deutsch-Ostafrikas. *Palaeontolographica Suppl. 7* 1(2):149–297.

1950. Die Skelettrekonstruktion von *Brachiosaurus brancai*. *Palaeontographica Suppl. 7* 1(3):95–103.

Jenkins, F. A. 1971. Limb posture and locomotion in the Virginia opossum (*Didelphis marsupialis*) and in other non-cursorial mammals. *J. Zool. London* 165:303–15.

Kemp, T. S. 1982. *Mammal-Like Reptiles and the Origin of Mammals* (New York: Academic Press).

Kermack, D. 1984. New prosauropod material from South Wales. *Zool. J. Linn. Soc.* 82:101–17.

Kitching, J. W., and M. A. Raath. 1984. Fossils from the Elliot and Clarens Formations (Karoo Sequence) of the Northeastern Cape, Orange Free State and Lesotho, and a suggested biozonation based on tetrapods. *Palaeontol. Afr.* 25:111–25.

Lapparent, A. F. de. 1943. Les dinosauriens Jurassiques de Damparis (Jura). *Mém. Soc. Géol. France (n.s.)* 47:1–20.

Lapparent, A. F. de, and G. Zbyszewski. 1957. Les dinosauriens du Portugal. *Mém. Serv. Geol. Portugal (n.s.)* 2:1–64.

Mertens, R. 1942. Die Familie der Warane (Varanidae). *Abh. Senck. Naturf. Ges.* 462:1–116.

Molnar, R. E. 1977. Analogies in the evolution of combat and display structures in ornithopods and ungulates. *Evol. Theory.* 3:165–90.

Montanucci, R. R. 1968. Comparative dentition in four iguanid lizards. *Herpetologica.* 24:305–15.

Morris, W. J. 1970. Hadrosaurian dinosaur bills – morphology and function. *Contr. Sci. Los Angeles Co. Mus.* 193:1–14.

Norman, D. B. 1980. On the ornithischian dinosaur *Iguanodon bernissartensis* of Bernissart (Belgium). *Mem. Inst. Roy. Soc. Nat. Belg.* 178:1–105.

1984. A systematic reappraisal of the reptile order Ornithischia. *In* Reif, W. E., and F. Westphal (eds.), *Third Symposium on Mesozoic Terrestrial Ecosystems*, Short Papers (Tubingen: ATTEMPTO), pp. 157–62.

Olsen, P. E., and P. M. Galton. 1977. Triassic–Jurassic tetrapod extinctions: are they real? *Science* 197:983–86.

1984. A review of the reptile and amphibian assemblages from the Stormberg Group of southern Africa with special emphasis on the footprints and the age of the Stormberg. *Palaeontol. Afr.* 25:87–110.

Osborn, J. W., and A. G. S. Lumsden. 1978. An alternative to "thegosis" and a re-examination of the ways in which mammalian molars work. *N. Jahrb. Geol. Paläont. Abh.* 156:371–92.

Owen, R. 1857. Monograph on the fossil Reptilia of the Wealden and Purbeck formations. Part III. *Megalosaurus bucklandi. Palaeont. Soc. Monogr.* 9:1–26.

1861. Monograph of the British fossil Reptilia from the Oolitic formations. Part first, containing *Scelidosaurus harrisonii* and *Pliosaurus grandis. Palaeont. Soc. Monog.* 8:1–16.

1879. Monograph of the fossil Reptilia of the Wealden and Purbeck formations. Suppl. 9. Crocodilia. *Palaeont. Soc. Monogr.* 33:1–19.

Paul, G. S. 1984. The segnosaurian dinosaurs: relics of the prosauropod-ornithopod transition? *J. Vert. Paleontol.* 4:507–15.

Raath, M. A. 1974. Fossil vertebrate studies in Rhodesia: further evidence of gastroliths in prosauropod dinosaurs. *Arnoldia (Rhodesia)* 7(5):1–7.

Ray, C. E. 1965. Variation in the number of marginal tooth positions in three species of iguanid lizards. *Breviora* 236:1–15.

Romer, A. S. 1956. *Osteology of the Reptiles* (Chicago: University of Chicago Press), p. 1–772.

1968. *Notes and Comments on Vertebrate Paleontology.* (Chicago: University of Chicago Press), pp. 1–304.

Russell, L. S. 1940. *Edmontonia rugosidens* (Gilmore) an armoured dinosaur from the Belly River Series of Alberta. *Univ. Toronto Stud., Geol. Ser.* 43:1–28.

Simmons, D. J. 1965. The non-therapsid reptiles of the Lufeng Basin, Yunnan, China. *Fieldiana, Geol.* 15(1):1–93.

Steel, R. 1970. Saurischia. *Handb. Paläoherpetol.* 14:1–88.

Thevénin, A. 1907. Paléontologie de Madagascar. IV. Dinosauriens. *Ann. Paleontol. (Paris)* 2:121–36.

Throckmorton, G. S. 1976. Oral food processing in two herbivorous lizards, *Iguana iguana* (Iguanidae) and *Uromastix aegyptius* (Agamidae). *J. Morphol.* 148:363–90.

Thulborn, R. A. 1970a. The skull of *Fabrosaurus australis*, a Triassic ornithischian dinosaur. *Palaeontology.* 13:414–32.

1970b. Tooth wear and jaw action in the Triassic ornithischian dinosaur *Fabrosaurus. J. Zool. London* 164:165–179.

1970c. The systematic position of the Triassic ornithischian dinosaur *Lycorhinus angustidans. Zool. J. Linn. Soc. London* 49:235–45.

1973. Teeth of ornithischian dinosaurs from the Upper Jurassic of Portugal. *Mem. Serv. Geol. Port. N.S.* 22:91–134.

1974. A new heterodontosaurid dinosaur (Reptilia: Ornithischia) from the Upper Triassic Red Beds of Lesotho. *Zool. J. Linn. Soc. London* 55:151–75.

1977. Relationships of the Lower Jurassic dinosaur *Scelidosaurus harrisonii*. *J. Paleontol.* 51:725–39.

1978. Aestivation among ornithopod dinosaurs of the African Trias. *Lethaia* 11:185–98.

Van Heerden, J. 1979. The morphology and taxonomy of *Euskelosaurus* (Reptilia : Saurischia; Late Triassic) from South Africa. *Navors. Nasion. Mus. Bloenfontein.* 4:21–84.

Walker, A. D. 1961. Triassic reptiles from the Elgin area: *Stagonolepis, Dasygnathus* and their allies. *Phil. Trans. Roy. Soc. London B* 244:103–204.

Weishampel, D. B. 1984. Evolution of jaw mechanisms in ornithopod dinosaurs. *Adv. Anat. Embryol. Cell Biol.* 87:1–110.

Weishampel, D. B., and J. B. Weishampel. 1983. Annotated localities of ornithopod dinosaurs: implications to Mesozoic paleobiogeography. *Mosasaur* 1:43–87.

Welty, J. C. 1975. *The Life of Birds*, 2nd ed. (London: Saunders), pp. 1–623.

White, T. C. 1958. The braincase of *Camarasaurus lentus* (Marsh). *J. Paleontol.* 32:477–94.

Wild, R. 1974. Lebensbilder württembergischer Triassaurier. *Stuttgarter. Beitr. Naturk. C,* 1:20–7.

1978. Ein Sauropoden – Rest (Reptilia, Saurischia) aus dem Posidonienschiefer (Lias, Toarcium) von Holzmaden. *Stuttgarter Beitr. Naturk. B,* 41:1–15.

Woodward, A. S. 1905. On parts of the skeleton of *Cetiosaurus leedsi*, a sauropodous dinosaur from the Oxford Clay of Peterborough. *Proc. Zool. Soc. London* 1905:232–43.

Yadagiri, P., K. N. Prasad, and P. P. Satangi. 1980. The sauropod dinosaur from Kota Formation of Pranhita-Godavari Valley, India. *Intern. Gondwana Symp. IV, 1977, Calcutta*, pp. 199–203.

Yeh, H. K. 1975. *Mesozoic Redbeds of Yunnan* (Academica Sinica, Beijing).

Young, C.-C. 1941. A complete osteology of *Lufengosaurus huenei* Young (gen. et sp. nov.) from Lufeng Yunnan, China. *Palaeont. Sinica (c)* 7:1–53.

1942. *Yunnanosaurus huangi* (gen. et sp. nov.), a new Prosauropoda from the Red Beds at Lufeng, Yunnan. *Bull. Geol. Soc. China* 22:63–104.

1951. The Lufeng saurischian fauna in China. *Palaeont. Sinica* 134:1–96.

1982. *Selected Works of Yang Zhungjian* (Beijing: Academia Sinica), pp. 1–219.

Zhang, Y., D. Yang, and G. Peng. 1984. New materials of *Shunosaurus* from Middle Jurassic of Dashanpu, Zigong, Sichuan. *J. Chengdu College Geol. Suppl.* 2:1–12.

Zhao, X. 1983. Phylogeny and evolutionary stages of Dinosauria. *Acta Palaeontol.* 28:295–306.

Zhou, S. 1983. A nearly complete skeleton of stegosaur from Middle Jurassic of Dashanpu, Zigong, Sichuan. *J. Chengdu College Geol. Suppl.* 1:15–26.

1984. *The Middle Jurassic Dinosaurian Fauna from Dashanpu, Zigong, Sichuan*, Vol. 2, Stegosaurs (Sichuan: Scientific and Technical Publishing), pp. 1–52.

17 Masticatory apparatus of the larger herbivores during Late Triassic and Early Jurassic times

A. W. CROMPTON AND J. ATTRIDGE

Introduction

Major changes in the terrestrial vertebrate faunas toward the end of the Triassic and beginning of the Jurassic have been documented by Cox (1967), Crompton (1968), Ostrom (1969), Colbert (1971, 1980) Bakker (1977), Olsen and Galton (1977), Benton (1979, 1983), Charig (1980), Bonaparte (1982), and Cooper (1982). These papers give an account of the transition from a terrestrial fauna dominated by mammal-like reptiles, rhynchosaurs, and pseudosuchians to one dominated by theropod, sauropodomorph, and ornithischian dinosaurs and crocodiles. All authors, with the exception of Benton (1983), suggest that superior structural and/or physiological features gave these archosaurs a competitive edge over other archosaurs and nonarchosaurs. The adaptive advantages of these features are often correlated with changing climatic conditions or changes in the composition and distribution of floras. Benton (1983) refers to this type of change as "differential survival (competitive) replacement."

Charig (1984) has critically reviewed Benton's (1983) and other reasons proposed to account for the dominance of archosaurs by Late Triassic times. He stressed his earlier view (Charig 1980) that a probable cause for the dramatic change was the competitive advantage of the "improved" stance and gait of archosaurs. Charig has suggested the following course of events during Carnian and Norian times. Initially (early Carnian), the carnivorous archosaurs (Rauisuchidae, a group already well established in the earlier Anisian) replaced the dominant therapsid carnivores (cynodonts). Carnivorous dinosaurs arose during the Carnian, and these forms, with the Rauisuchidae, were [according to Charig (1980)] so successful that, as a result of "overkill," they caused the virtual extinction of Norian times of the traversodont cynodonts, anomodonts, and rhyncho-saurs. This left less animal food for the carnivores, but plenty of plant food was available. Competition between carnivorous archosaurs was intense, with the result that some groups became extinct, some became "better" carnivores, and others became herbivorous (prosauropod dinosaurs and aetosaurs). The prosauropods with their improved gait could better avoid predation by carnivorous thecodontians than the rhynchosaurs, anomodonts, and traversodonts. Charig (1980) states, "The prosauropods at once became the dominant element in the herbivore fauna, apparently causing the extinction of the dicynodonts." According to Charig (1980, 1984), the transition from a fauna dominated by carnivorous therapsids to one dominated by carnivorous archosaurs seems to have been gradual, whereas the change from a fauna in which therapsids and rhynchosaurs formed the dominant herbivores to one in which the prosauropods were the dominant herbivores appears to have been fairly sudden, with very little overlap between the two faunas. Benton (1983) agrees with the view that some groups in the Late Triassic became extinct as a result of direct competition, but he does not agree that this was the reason for the early (Norian) dinosaur radiation. In his view, the extinction of the rhynchosaurs and herbivorous therapsids (anomodonts and traversodont cynodonts) was not related to direct competition with archosaurs, and, in fact, dinosaurs only radiated after the extinction of rhynchosaurs and herbivorous therapsids. He refers to this as "opportunistic replacement."

In this chapter, we wish to contrast the structure and function of the masticatory apparatus of therapsids and rhynchosaurs, on one hand, with those of prosauropods and ornithischian dinosaurs, on the other.

An important aspect of this debate is the relative age of specific continental deposits near and at the Triassic–Jurassic boundary. The relevant literature is reviewed in Olsen and Galton (1977), Colbert (1980, 1981), Cooper (1982), Benton (1983), Hopson (1984), Olsen and Galton (1984), Attridge, Crompton, and Jenkins (1985), N. Shubin, A. W. Crompton, P. E. Olsen, and W. W. Amaral (in prep.). Some workers (e.g., Colbert, 1980, 1981, for a review) placed the Elliot Beds and Clarens Sandstone of southern Africa, the upper Lower Lufeng Beds of southern China, the fissure deposits of Wales, the Glen Canyon Group, and Zone 3 of the Newark Basin of North America in the Late Triassic. By allocating these deposits to the Liassic, Olsen and Galton (1977, 1984) have increased the time range of some groups, for example, prosauropods. We have accepted the ages and stratigraphic correlations of continental deposits proposed by Olsen and Galton (1977, 1984).

Brief review of terrestrial herbivores of the Late Triassic and Early Jurassic

During Carnian and early Norian times (Fig. 17.1), the main terrestrial herbivores were large anomodonts, traversodont cynodonts, and rhynchosaurs (Olsen and Galton 1977; Charig 1984). Some of the early aetosaurs showed carnivorous tendencies, while others appear to have been the first archosaurs to have adopted an herbivorous diet (Casamiquela 1960, 1961; Bonaparte, 1971, 1978). Prosauropod dinosaurs appeared for the first time during mid-Carnian times (Dutuit 1972; Galton 1985a), but only during Norian times did they become the dominant herbivores (Olsen and Galton 1984; Galton 1984 and 1985a). In some locations, such as the Chinle Formation of Arizona and the Los Colorados Formation of South America (Bonaparte 1982), aetosaurs formed a small but significant part of the herbivore fauna, together with the prosauropods. There are very few locations where prosauropods occur together with traversodont cynodonts and anomodonts. One of these is the lower Elliot Formation, where possible anomodont footprints (Olsen and Galton 1984), a traversodont cynodont *Scalenodontoides* (Crompton and Ellenberger 1957), plateosaurid (*Euskelosaurus*), and melanorosaurid (*Melanorosaurus*) prosauropods, as well as prosauropod trackways occur together (see also Kitching and Raath 1984). Hopson (1984) has described *Scalenodontoides* from the lower Wolfville Formation of Nova Scotia, where it occurs together with plateosaurid prosauropods and possibly a small ornithischian. Other survivors of an earlier (Carnian) radiation of anomodonts are *Jachaleria* (Bonaparte 1978) from the basal Los Colorados Formation and *Placerias* (Camp and Welles 1956; Cox 1965) from the Chinle Formation of North

America. The earliest tritylodontids (Bonaparte 1971) appear in the uppermost strata of the Los Colorados Formation. Tritylodontids are considered by Charig (1980, 1984) to be the direct replacement of traversodont cynodonts. Although tritylodontids may have arisen from traversodonts (Crompton and Ellenberger 1957; Crompton 1972a; Hopson 1984; Sues 1985), there are some important differences between these two groups (Kemp 1982), and tritylodontids are not known in deposits generally considered to be Norian, that is, basal Elliot Formation, Middle Keuper of Europe, Dockum and Chinle of North America, and most of the lower part of the Los Colorados Formation.

The prosauropods continued to be the dominant herbivores in the Liassic of southern Africa and China (Young 1951; Charig, Attridge, and Crompton 1961; Cooper 1985) and possibly in North America (Galton 1976; Attridge, Crompton, and Jenkins 1985) and Europe (Kermack 1984) as well. The tritylodontids formed a significant part of the total fauna of large terrestrial herbivores during the Liassic (Hopson 1964; Sues 1983, 1984; Clark and Hopson 1985). Undoubted ornithischians are found in several Liassic deposits (Crompton and Charig 1962; Thulborn 1970, 1972, 1974; Charig and Crompton, 1974; Gow 1975; Hopson 1975; Santa Luca 1980, 1984; Colbert 1981; Charig 1982). *Pisanosaurus* (Bonaparte, 1976) from the midsection of the Ischigualasto Formation (Carnian) is very poorly preserved,

Figure 17.1. Occurrence of the larger terrestrial herbivores during Late Triassic and Early Jurassic times. Groups that developed powerful masticatory mechanisms are shaded.

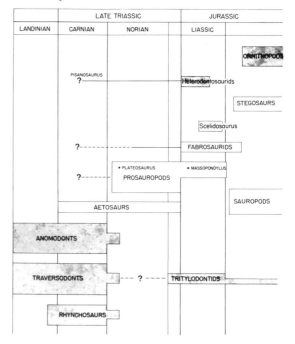

but it is generally accepted as a heterodontosaurid ornithischian on the basis of the dentition (upper teeth directed ventromedially, continuous wear surface, with both upper and lower teeth set in medially from the external surface). Chatterjee (pers. comm.) is describing what appears to be a fabrosaurid from the Dockum (Norian), and Dutuit (1972) and Galton (1985a) have claimed the presence of fabrosaurid dinosaurs in the mid-Carnian of Morocco. If the identification of these forms as ornithischians is correct, it means that the earliest ornithischians were contemporaneous with the earliest known saurischians from South America (mid-Carnian). However, it must be stressed that this conclusion is based upon both poorly preserved and limited material. Current evidence suggests that ornithischians did not form a significant element in the terrestrial fauna until the Liassic. The relative ages of the fossiliferous continental deposits of the Late Triassic and Early Jurassic are still subject to debate, but current evidence suggests that there was very little recorded overlap within geographic regions between the anomodonts, traversodonts, and rhynchosaurs, on one hand, and prosauropods, on the other. This is contrary to Charig's (1984) summary diagram that shows an overlap between prosauropods (which are shown to have arisen in mid-Norian times) and rhynchosaurs and dicynodonts.

Masticatory apparatus of anomodonts, traversodonts, and rhynchosaurs

Common functional characteristics of anomodonts, traversodonts, and rhynchosaurs were a very large volume of jaw musculature relative to the biting or chewing area, and clear adaptive features for dealing with tough food.

The mechanism of the jaw action in anomodonts (Fig. 17.2) has been demonstrated in an Early Triassic form, *Lystrosaurus* (Crompton and Hotton 1967). The front of the snout and lower jaw were massive, and the symphysis fully fused. The fronts of the upper and lower jaws were covered with horny beaks, and, when the jaws were closed, the lower horny beak fitted snugly inside the upper beak. Upper and lower beaks provided sharp cutting edges for the slicing of plant material. Surfaces suitable for crushing or grinding of vegetation were not present. The adductor mass was large relative to skull size; this was in part due to the development of a lateral external adductor mass that originated on the expanded external surface of the zygoma and suspensorium. This suggests that these forms must have been capable of slicing or cropping hard food. Muscular cheeks do not appear to have been present in anomodonts, and this suggests that food was not manipulated or progressively broken down in the oral cavity as it is, for example, in modern mammals.

Food was probably cut and swallowed much as it is in modern herbivorous lizards (Throckmorton 1976; Smith 1984). Both articulating surfaces of the jaw joint were convex in lateral view, and this permitted not only extensive posterior movements of the lower jaw during closing, but also a rocking of the lower joint around a point slightly behind the lower beak. Carnian anomodonts (Romer and Price 1944; Cox 1965; Bonaparte 1971) retained basically the same mechanism for cutting food as earlier anomodonts, and the volume of adductor muscles relative to the cutting area remained high.

Some of the Carnian traversodont cynodonts, such as *Exaeretodon*, *Proexaeretodon* and *Ischignathus* (Bonaparte 1963a,b), and *Scalenodontoides* (Crompton and Ellenberger 1957; Hopson, 1984), had large and massive skulls, and the body lengths of some of these traversodonts reached up to six feet. They had fully differentiated dentitions (Fig. 17.3A) with large upper and lower molariform teeth. These were characterized by several shearing surfaces. During occlusion, the shearing surfaces of the lower molariforms were dragged posteriorly across the matching surfaces of the upper teeth. The lower jaw symphysis was massive, and occlusion must have been bilateral (Crompton 1972a; Crompton and Hylander 1986). The enamel layer was very thin, and the teeth show extensive wear. To compensate for this, the molariform teeth were lost progressively from the front of the dentition, and new teeth were

Figure 17.2. Masticatory apparatus of the early Triassic anomodont *Lystrosaurus* (after Crompton and Hotton 1967). The stippled area indicates the position of the beak.

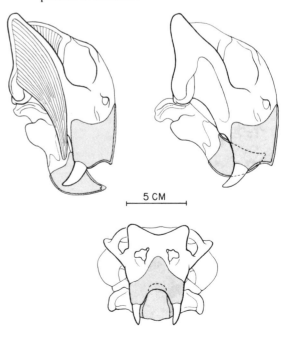

5 CM

added at the back of the tooth row (Hopson 1971; Crompton 1972b). The teeth are inset from the outer surface of the face, suggesting the presence of muscular cheeks. It is likely that the structure of the tongue and soft palate of advanced cynodonts was similar to that of mammals. If, in addition, they possessed muscular cheeks, they must have been capable of manipulating and repositioning food within the oral cavity. Wear of the "molariform" teeth suggests that, as in mammals, food was repeatedly repositioned between the teeth and progressively broken down in a masticatory sequence. In contrast to mammals in which occlusion is unilateral and the

power stroke is in a dorsomedial direction, occlusion in traversodonts was bilateral and the power stroke was directed posteriorly. The adductor muscles of traversodonts were massive (Barghusen 1968; Crompton 1972b), and the ratio of muscle volume to the occlusal surface of the "molariform" teeth was large. In contrast to anomodonts, therefore, traversodonts were probably capable of triturating food effectively within the oral cavity, but like anomodonts they could deal with tough vegetation.

Rhynchosaurs (Sill 1971; Chatterjee 1980; Benton 1983) were characterized by dental batteries with short ankylothecodont teeth that seemed to have broken down food by simultaneously crushing and grinding it (Fig. 17.3B). The volume of jaw muscles relative to the dental batteries was large, and, in common with anomodonts and traversodonts, they appear to have been capable of breaking down tough plant food.

It is not known exactly what these three different types of herbivores ate, although several plant groups have been suggested (Bonaparte 1982; Cruickshank 1978; Weishampel 1984b).

Figure 17.3. **A**, Masticatory apparatus of the late Carnian traversodont *Exaeretodon* (after Chatterjee 1982). **B**, Masticatory apparatus of the rhynchosaur *Scaphonyx* (after Sill 1971). **C**, Skull of the Norian prosauropod *Plateosaurus* (after Romer 1966).

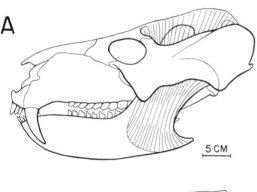

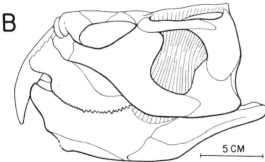

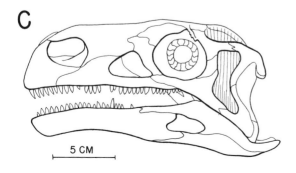

Masticatory apparatus of prosauropods

The skulls of the Norian prosauropods such as *Plateosaurus* (Galton 1984) stand in sharp contrast to those of the Carnian herbivores discussed above. Although some prosauropods had fairly large skulls (Fig. 17.3C), reaching up to 45 cm in length, they were lightly built, and many of the cranial elements met in noninterdigitating, overlapping, or abutting sutures. In contrast to the larger Carnian herbivores, the skull maintained an open basipterygoid joint. In addition, as in all archosaurs, the laterosphenoid and frontal appear to have been separated by a thin layer of cartilage. The skull of *Plateosaurus* is long and deep, with the upper dentition occupying more than one-half of the skull length. The upper tooth row is longer than the lower, and the teeth were laterally compressed. The teeth are similar in structure from front to back. When the jaws were closed, the anterior teeth lay immediately behind the corresponding uppers, and the remainder of the lower teeth lay internal to the uppers. However, as there are no signs of wear on the inner surface of the uppers or outer surface of the lowers (Galton 1984, 1985a), the teeth did not occlude. The volume of adductor muscle relative to skull length is considerably less than in anomodonts, traversodonts, and rhynchosaurs. In sharp contrast to these forms, therefore, *Plateosaurus* and related forms did not require either a powerful bite or occlusion between upper and lower teeth to deal with the food they were eating.

The melanorosaurids (Charig, Attridge, and Crompton 1965; Bonaparte 1971; Van Heerden

1979; Olsen and Galton 1984; Galton 1985b) were quadrupedal prosauropods. Until recently, no skull had been found in association with an undoubted melanorosaurid skeleton. On the basis of isolated teeth, it was suggested (Charig et al. 1965) that they may have been carnivorous forms. Bonaparte (pers. comm.) has recently found a melanorosaurid skull, and the dentition is similar to that of the prosauropods rather than to that of carnivorous dinosaurs

Figure 17.4. Lateral view of the skull of *Massospondylus harriesi*.

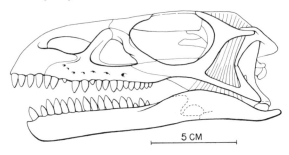

Figure 17.5. Upper and lower teeth of *Massospondylus* to illustrate the differentiation along the tooth row.

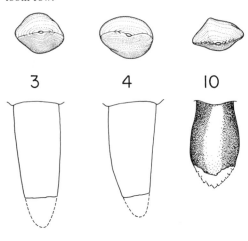

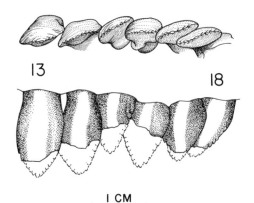

or thecodontians. This suggests [as has been claimed by Hopson (1984)] that the melanorosaurids were also herbivores and that the carnivorous teeth found in close association with melanorosaurids and large plateosaurids were more likely to have been the teeth of carnivorous rauisuchids that preyed upon the melanorosaurids.

In the Liassic, the prosauropods continued to be the dominant terrestrial herbivores. However, in contrast to the Norian, the ornithischian dinosaurs formed a large part of the total fauna of terrestrial herbivores. The best known prosauropod of the Liassic is *Massospondylus* (Cooper 1981; Attridge et al. 1985). The skull structure (Fig. 17.4) is basically the same as that of *Plateosaurus*, but there are some important differences. The skull is not as deep, and the preorbital region is shorter relative to skull length. There is a fairly marked differentiation of the teeth along the tooth row (Fig. 17.5). The anterior teeth are circular in cross section at the alveolar border. They have fine serrations on their anterior and posterior edges; they are tall and stout. The posterior teeth are similar to those of *Plateosaurus*, but are shorter and laterally compressed; the anterior edge of one tooth overlaps the posterior surface of the tooth in front, and the serrations on the anterior and posterior edges are larger than those of the anterior teeth. We observed no wear that would have resulted from tooth-to-tooth contact.

A peculiar feature of the two *Massospondylus* skulls available to the authors is that the lower jaw is shorter than the upper (Figs. 17.4, 17.6), so that the upper teeth extend anteriorly beyond the lower teeth. Initially this was thought to be the result of crushing, but careful measurement of individual skull elements confirmed the difference in length of upper and lower jaws. Cooper (1981) figured in lateral view a small skull that he referred to *Massospondylus*. The lower jaw of the specimen is the same length as that of the upper, but the skull proportions are quite unlike those of the *Massospondylus* skull

Figure 17.6. Anterior region of the skull and lower jaw of *Massospondylus* sp. from the Kayenta formation. RP, resorption pit; B, beak.

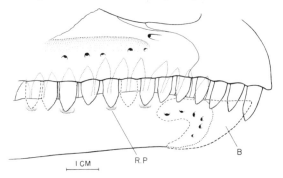

illustrated in Figure 17.4. The latter is based upon a skull in the South African Museum (K 1314). Cooper also figured this skull as it is preserved, and also refers it to the genus *Massospondylus*. Consequently, the small skull figured by Cooper may be a juvenile; on the other hand, it may not be referable to *Massospondylus*. In the North American *Massospondylus* (Attridge et al. 1985; Fig. 17.6), there is additional evidence for the different lengths of the upper and lower jaws. Small depressions are present on the outer surface of the mandible. These appear to have been formed by older teeth on the tooth row pressing against the gum surface. This resulted in resorption of bone immediately below the area of pressure. In this specimen, some of the teeth were preserved fitting into these pockets, thereby indicating the relative position of the lower jaw. Similar pockets associated with upper postcanine teeth have also been observed in *Thrinaxodon*, a Lower Triassic cynodont (Crompton 1972a). If the lower jaw of *Massospondylus* is positioned by these pits, at least three upper teeth lie in front of the tip of the lower jaw. Despite the "overhang" of the anterior teeth, a small vertical wear facet is present on the lingual surface of the first upper premaxillary teeth. This facet could not have been caused by a lower tooth. The outer surface of the tip of the lower jaws is rugose, and numerous vascular canals are present. The mandibular symphysis is extremely slender. It is suggested that the tips of the mandibles supported a horny beak and that this structure was responsible for the small facet on the lingual surface of the upper anterior tooth. A lower horny beak and long conical upper teeth may have provided a more effective cropping device than that present in *Plateosaurus*, where spatulate teeth were present in both the premaxilla and lower jaw.

Because the anterior teeth of *Massospondylus* are tall and robust, Cooper (1981) suggested that this genus and other prosauropods were carnivores. In an earlier paper (Attridge et al. 1985), we disputed this view and pointed out the similarities of the dentition of prosauropods to those of herbivorous lizards (*Iguana*) and fabrosaurid ornithischians, the latter generally being accepted as herbivores. Galton (1984, 1985a) has critically reviewed Cooper's (1981) arguments and given numerous reasons why prosauropods must have been herbivores. Part of the reason for the debate on the diet preference of prosauropods is that, compared with the earlier traversodonts, dicynodonts and rhynchosaurs and later ornithopod dinosaurs, the masticatory apparatus of prosauropods is poorly adapted for breaking down plant food. The teeth do not possess matching shearing surfaces, the ratio of adductor muscle to tooth surface is low, and the skull is lightly and loosely built.

Although the sauropods formed the largest terrestrial herbivores ever to have lived (Colbert 1961; Charig 1979), they too, in common with the prosauropods, did not develop a masticatory apparatus designed to crop finely or break down food within the oral region, although randomly organized wear facets indicate dental occlusion. In contrast to traversodonts and mammals. sauropod cheek teeth did not possess complex matching shearing surfaces that trap and shear food and could not, therefore, effectively and efficiently break down plant material (Crompton 1972, 1974).

Masticatory apparatus of ornithischians

The common ornithischians of the Upper Elliot Beds of southern Africa are *Lesothosaurus* (*Fabrosaurus*) (Thulborn 1970, 1971, 1972; Galton 1978). This form is similar to the Kayenta *Scutellosaurus* (Colbert 1981). Most of the remains of *Lesothosaurus* are smaller than those of *Massospondylus*, but at least one undescribed skeleton of what appears to be a fabrosaurid in the South African Museum reaches the dimensions of a medium-sized *Massospondylus*. *Lesothosaurus* (Fig. 17.7) has a small predentary bone that is loosely joined to the tips of the two mandibles. It is generally accepted that the predentary bone of ornithischians supported a horny beak. The first premaxillary tooth is set back from the tip of the bone, and the general structure of this area suggests that a small upper beak may have been present. The premaxillary teeth are recurved and quite different from the serrated and laterally flattened maxillary and lower jaw teeth. A lower horny beak fitted inside the small upper beak and premaxillary teeth. This presumably formed a more effective cropping mechanism than anterior teeth in both jaws. The maxillary and dentary teeth are robust, and the bases of the crowns are fairly wide lateromedially. On the basis of wear on isolated teeth, Thulborn (1971) claimed that maxillary and mandibular teeth alternated with one another, that is, a lower tooth met two upper teeth and vice versa, and in so doing produced two oblique wear facets

Figure 17.7. Skull of the Liassic ornithischian dinosaur *Lesothosaurus*.

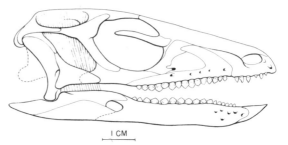

| CM

on each tooth. An undescribed maxilla (Fig. 17.8) and dentary of *Lesothosaurus* from southern Africa, with a complete complement of teeth, confirms Thulborn's observation that this form occluded its teeth. The occlusal pattern is similar to that of the herbivorous lizard *Uromastyx* (Robinson 1976).

The skull of *Lesothosaurus* is lightly built, and although the volume of adductor musculature relative to tooth surface was only slightly greater than in prosauropods, they must have lacked sufficient adductor force to break down tough plant material. However, *Lesothosaurus* does represent an advance over *Massospondylus* in that the maxillary and dentary teeth occluded. They were, therefore, better able to break down plant food than the contemporaneous *Massospondylus*.

Heterodontosaurus (see Fig. 17.11) (Crompton and Charig 1962; Charig and Crompton 1974, in press; Santa Luca, Crompton, and Charig 1976; Santa Luca 1980) is characterized by a deep skull, a large predentary, reduced anterior premaxillary teeth, and a premaxilla that was almost certainly covered at the tip by a well-developed horny beak with a sharp cutting edge, which worked against the sharp cutting edge on a lower beak that was supported by the predentary bone (Fig. 17.9). The lower beak was larger than the upper, and the former fitted snugly between the three premaxillary teeth on either side. Wear facets on the premaxillary teeth confirm this close relationship. The maxillary and dentary teeth are columnar, with prominent ridges on the external and internal surfaces of the upper and lower teeth, respectively. The enamel is thicker on the surface supporting the ridges and extremely thin on the opposite surface (Fig. 17.10). A distinct feature of *Heterodontosaurus*, and also characteristic of other ornithopods, is that the crowns of the upper curve inward and those of the lowers curve outward. There is a superficial resemblance between the unworn teeth of *Heterodontosaurus* and those of *Lesothosaurus*. However, the teeth of *Heterodontosaurus* are taller, and the outer surface of the upper and inner surface of the teeth has thick enamel. In addition, the upper and lower teeth curve toward each other (Fig. 17.10).

Fully erupted teeth of *Heterodontosaurus* are truncated by flat, oblique wear facets that are roughly on the same level as the facets on the adjoining teeth, so that a single continuous wear surface is present along both the maxillary and dentary tooth rows. The facet on the upper teeth faces downward and inward, and that on the lower teeth faces upward and outward. As *Heterodontosaurus* teeth erupt, the tip of the crown grows into the occlusal plane and is truncated until a wear facet covers the entire anteroposterior and lateromedial extent of the crown. Because the crown is curved, the most proximal part of the enamel surface and the shear facet meet at an angle that varies along the tooth row and in the course of eruption from 65° to 90°. At the beginning of occlusion, the enameled inner edge of the continuous wear surface of the lower dentition meets the outer enameled edge of the continuous wear surface of the upper dentition. The cutting action is analogous to that of a paper cutter where one straight edge shears past another straight edge. Wear

Figure 17.8. Medial view of maxillary dentition of *Lesothosaurus* to illustrate tooth replacement and double wear facets. a.w.f., anterior wear facet; g.d.l., groove for the dental lamina; p.r.t., pits for replacing teeth; p.w.f., posterior wear facet; rep 8, 10, replacement teeth 8 and 10.

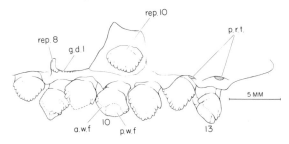

Figure 17.9. Skull of the Liassic ornithischian *Heterodontosaurus*.

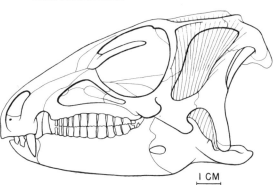

Figure 17.10. Comparison of occlusion in *Lesothosaurus* and *Heterodontosaurus*.

LESOTHOSAURUS

HETERODONTOSAURUS

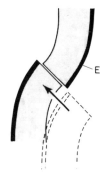

facets are wider on the posterior teeth than on the anterior teeth. The broad facets could only have been formed if the lower teeth moved medially relative to the uppers during occlusion. Several other genera that are considered heterodontosaurids have been described from southern Africa (see Hopson 1975), but they lack the characteristic features of the *Heterodontosaurus* dentition (Charig and Crompton in press).

In mammals, during opening and the first stage of jaw closing, the hemimandible on the side that will be involved in occlusion (working-side mandible) moves laterally, and during occlusion moves medially. Consequently, when the tip of the jaw is viewed from in front, it subscribes an elliptical orbit (Hiiemae 1978; Hiiemae and Crompton 1985). In contrast, in *Heterodontosaurus* and all other ornithischians the movement of the tip of the lower jaw was restricted by the premaxillary teeth and beaks to a vertical plane. However, occlusion between the maxillary and dentary teeth was bilateral in *Heterodontosaurus*, and the wear facets indicate that the dentary teeth moved medially relative to the maxillary teeth. There are two ways in which this could have been achieved, either by moving the mandibular rows inward relative to stationary uppers (Fig. 17.11A,C) or by moving the maxillary rows outward relative to direct vertical movements of the lowers (Fig. 17.11B). Weishampel (1983, 1984a) has proposed the latter mechanism for ornithopods (excluding *Heterodontosaurus*). The firm junction of the palatal regions of the maxillary and premaxillary bones in *Heterodontosaurus* rules out laterally directed movement of the maxilla and associated elements. For *Heterodontosaurus*, Weishampel (1984a) has suggested that the upper surface of each hemimandible rotated inward around its longitudinal axis during occlusion. The planar wear facets and the transversely widened quadrate–articular joint do not support the view that rotation alone was the only mechanism for moving the lower molars medially relative to the uppers.

A smooth ball-and-socket contact was present between both dentaries and the predentary in *Heterodontosaurus* (Fig. 17.11C), suggesting that these elements could move relative to one another. Transverse movements of the lower molariform teeth relative to the upper could be achieved if, as the jaws closed, both mandibles moved in a medial direction as a result of rotation around their contact points with the predentary. This implies that either the quadrate–articular joint permitted transverse translation or the quadrates rotated medially about their contact with the opisthotic and squamosal. The ventral surface of the quadrate condyle slopes inward and upward, and the joint apparently was not designed to prevent mediolateral movement of the ar-

ticular. If each mandible were to rotate around its predentary contact in the manner described above, medially directed movement would increase with increasing distance away from the rotation point. This would account for the wider facets on the more posterior teeth.

The masticatory apparatus of *Heterodontosaurus* stands in sharp contrast to that of *Massospondylus* and *Lesothosaurus*. In *Heterodontosaurus*, the dentition and horny beaks form an efficient and powerful cutting mechanism. Muscular cheeks were probably present, which suggests some intraoral manipulation and progressive breakdown of food. The deep mandible and skull and the structure of the lateral temporal opening indicate a high ratio of adductor muscle mass to tooth and beak surfaces.

The presence of a predentary bone in *Heterodontosaurus* appears to be a functional requirement

Figure 17.11. Alternative mechanisms of obtaining transverse movement of lower teeth relative to upper teeth in ornithopod dinosaurs. In **A** and **B** movement as seen in a transverse section is illustrated; in **C** movement as seen in a ventral view is illustrated.

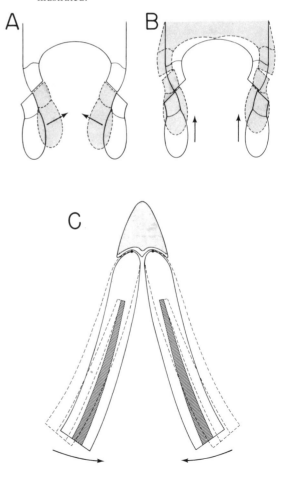

of bilateral occlusion, and it is probable that this bone and bilateral occlusion developed synchronously. The ancestral ornithischian probably had a horny beak supported by both dentaries, that is, similar to the situation assumed to be present in *Massospondylus*. In *Massospondylus*, the upper and lower teeth do not contact one another, and consequently the mandible would not have had to move medially during biting. In contrast, in *Lesothosaurus* the cheek teeth do occlude, and, because of the tapering profile of the teeth (Fig. 17.9), the lower jaws must both have been forced inward during occlusion. Although this movement was probably minimal, it would have, in the absence of a predentary bone, caused movement of the anterior tips of the dentaries relative to the horny beak. In birds, a horny beak is separated from the bone supporting it by a thin layer of skin (Lucas and Stetterheim 1972). Growth of the horn takes place at the basal epidermal layer lying external to the dermis. Capillaries that arise from arterioles within the dermis invade the basal layer to supply the nutrients necessary for growth. Part of the blood supply of the dermis is via vascular canals that perforate the bone below the horn. Extensive movement between bone and horn would disrupt the nutrient supply to the growing horn. This could be avoided if the lower beak was supported by a separate ossification, the predentary, which had its own blood supply distinct from that of the dentary. A joint developed between this bone and both dentaries.

Lesothosaurus represents an early stage in the development of the predentary, and in this form medial movement of the mandibles during occlusion was small. In *Heterodontosaurus*, with the development of transversely widened wear facets, the amount of medial movement during occlusion was far greater. Coupled with this is the development of a larger predentary and well-defined matching articulating surfaces at the junction of the dentaries and predentary. It is possible that a synovial joint was present in this position.

The basic features of the masticatory apparatus in *Heterodontosaurus* (thick enamel on one side of the teeth and curvature of upper and lower teeth toward one another) are retained in all later ornithopods. We consider these features of the masticatory apparatus diagnostic for ornithopods and do not accept Santa Luca's (1984) argument that an obturator process is the definitive feature for characterizing ornithopods (see Charig and Crompton in press). In all the ornithopods the dentary–predentary junction appears to have been loose, and it is, therefore, possible that occlusion involved medial movement of the lower jaw rather than lateral movement of the maxillae (Weishampel 1983, 1984a). However, irrespective of the precise mechanism of transverse movements of the lower jaw relative to the maxilla during bilateral occlusion, all the ornithopods possessed large anterior beaks, a massive dentition, and a large mass of adductor muscles, and they were, therefore, able to deal with resistant plant material and possibly able to manipulate it for repeated chews within the oral cavity. On the other hand, the stegosaurs, ankylosaurs, fabrosaurids, and *Scelidosaurus* (although they occluded their teeth) retained the rather fragile dentitions and small muscle volume to biting area ratio.

Conclusions

The Carnian–Norian transition is characterized by a major change in the masticatory mechanism of the larger terrestrial herbivores. In the Carnian, these herbivores (traversodonts, anomodonts, rhynchosaurs) were characterized by the ability to generate powerful bite forces, and the ratio of jaw adductors to cutting and crushing surfaces of the beaks and/or teeth was large. In contrast, in the Norian the masticatory apparatus of the dominant herbivores was weak and unspecialized. The prosauropod dinosaurs had fragile jaws and skulls, slender bladelike teeth that did not occlude, and, compared with the above-mentioned Carnian forms, a greatly decreased mass of adductor muscles relative to skull length. Some Norian aetosaurs also appear to have been herbivores, and members of this group may have preceded the prosauropods in becoming herbivores without developing a powerful masticatory apparatus.

In the Liassic, prosauropods continued to flourish as the dominant herbivores, but other forms that could deal with tougher vegetation became a significant part of the fauna. The masticatory apparatus of fabrosaurids and *Scelidosaurus* represents only a slight advance over that of prosauropods, but heterodontosaurids and tritylodontids are characterized by adaptations for the intraoral manipulation and breakdown of resistant plant food.

If it can be established that *Pisanosaurus* is a heterodontosaurid, it means that the earliest ornithischian dinosaurs were capable of dealing with tough vegetation as were the contemporary Carnian therapsids and rhynchosaurs. However, *Pisanosaurus* did not form a major part of the fauna of the Carnian (being known only from a single poorly preserved specimen), and other heterodontosaurids are unknown in Norian deposits. Both tritylodontids and heterodontosaurids were characterized by a relatively large mass of jaw-closing musculature and complex lower jaw movements that are rivaled only by modern herbivorous mammals. Tritylodontids do not appear directly to replace traversodonts; there are only fragmentary remains of tritylodontids from Norian deposits, but they flourished during Liassic times. It appears that a powerful masticatory mechanism was a basic necessity for most terrestrial herbivores during Carnian times, but this requirement seems to have disappeared during the Norian and was only partially restored during the Liassic and Late Mesozoic. Dutuit (1972), Galton (1984), and Hopson (1984) have claimed on the basis of isolated teeth that prosauropods and fabrosaurids extend back to the Carnian. However, as in the case of heterodontosaurids, these groups do not form a significant element in the known fauna until the Norian in the case of the prosauropods and the Liassic in the case of the fabrosaurids and heterodontosaurids.

Why was there a shift from a Carnian fauna dominated by terrestrial herbivores with jaws, teeth, and beaks adapted to cut and break down tough vegetation to a Norian one dominated by forms that lacked any marked cranial adaptations for dealing with tough vegetation?

It will be difficult to determine the possible reasons for the transition from powerful- to weak-jawed herbivores across the Carnian–Norian border until precise correlations of strata containing terrestrial vertebrate herbivores have been determined. There is a broad temporal overlap between carnivorous therapsids and archosaurs. This is also true for prosauropod, ornithischian, and sauropod dinosaurs. This suggests a gradual replacement of carnivorous therapsids by archosaurs and of prosauropods by ornithischian and sauropod dinosaurs. It is, however, difficult to document a gradual replacement of anomodonts, rhynchosaurs, and traversodonts by prosauropods, although isolated anomodonts [*Placerias* (Camp and Welles 1956) and Soeberg trackway (Olsen and Galton 1984)] and traversodonts [*Scalenodontoides* (Crompton and Ellenberger 1957; Hopson 1984)] are known to have coexisted with early Norian prosauropods. However, except for these few examples, the transition is well defined. Carnian deposits (for example, Ischigualasto, Elgin, Manda, Santa Maria) are dominated by traversodonts, rhynchosaurs, and anomodonts. Norian deposits (for example, Middle Keuper, basal Elliot, Upper Los Colorados) are dominated by prosauropods Therefore, if a competitive replacement between prosauropods and the older rhynchosaurs, traversodonts, and anomodonts did in fact occur, it may have taken place in locations and/or times for which we have no fossil record. If these forms were in direct competition, other adaptive features of the prosauropods may have been more important than the specialized masticatory apparatus of the Carnian herbivores. It is possible, for example, that prosauropods and fabrosaurids did not require a powerful masticatory mechanism to break down food because, as Charig (1984) suggested, major advances in the skeleton (for example, parasa-

gittal gait and a mesotarsal joint) may have increased the foraging range of these dinosaurs. This may have enabled them to exploit widely dispersed plants, parts of the vegetation that did not require a powerful masticatory mechanism, or vegetation that could not be reached by the squat, quadrupedal dicynodonts, traversodonts, and rhynchosaurs. Ash (Chapter 2) states that no sudden change in the floral composition occurred during Late Triassic times. On the other hand, prosauropod success may have been due to the development in this group of a new way of breaking down fibrous food. Their stomach gastroliths may have been able to triturate plant material. Attridge discovered a cluster of stomach stones within the rib cage of a Zimbabwean *Massospondylus* skeleton (Andras 1982). Sauropods may also have had a gastric mill, as Janensch (1929) has described gastroliths in association with several sauropod skeletons from Tendaguru (Upper Jurassic). Stomach stones that help crocodiles in breaking down mice have been observed with cinefluoroscopy. A combination of improved gait and a gastric mill may account for prosauropod success even in the absence of floral change.

Prosauropods, on the other hand, may have coexisted with a different vegetation than did the Carnian herbivores. Benton (1983) has suggested that the *Dicroidium* flora dominated the lowland environments of the Southern Hemisphere during most of the Triassic, that is, up to the end of the Norian, whereas the Northern Hemisphere flora lacked seed ferns and was dominated by cycadophytes, ferns, and conifers. The *Dicroidium* flora, according to Benton, did not survive the end of the Norian and was replaced by a worldwide flora dominated by conifers and bennettitaleans. He suggests that dinosaurs became ecologically important in the north in places where rhynchosaurs and the *Dicroidium* flora were absent (Germany and North America). The extinction of both rhynchosaurs and the *Dicroidium* flora and the worldwide extension of a coniferous flora thereafter permitted the new dinosaurs to migrate south and radiate even more extensively. This suggests that prosauropods may have evolved with a different flora than that to which the anomodonts, rhynchosaurs, and traversodonts were adapted, and that the dinosaurs later migrated south only after the *Dicroidium* flora and associated fauna had become extinct. In this view, the dinosaurs were not directly involved in the extinction of the rhynchosaurs, traversodonts, and anomodonts and did not compete directly with them. This interpretation assumes an initial northern radiation of prosauropods. However, according to the stratigraphic correlation suggested by Olsen and Galton (1977, 1984) the earliest undoubted Norian prosauropods of Europe and southern Africa were contemporaneous. If footprint data are considered in addition to skeletal remains, the early to mid-Norian dinosaur fauna was greatly diversified in southern Africa, North America, and Europe. This makes it difficult to associate the radiation of prosauropods with specific changes in the flora.

Charig (1980, 1984) does not consider direct competition between therapsids, rhynchosaur, and dinosaur herbivores but rather competition between archosaurs as a whole and nonarchosaurs (that is, rhynchosaurs and therapsids), with the carnivorous archosaurs exterminating the rhynchosaurs, anomodonts, and traversodonts. He considers this to have been a rather gradual process, but, if extinction of anomodonts, traversodonts, and rhynchosaurs was caused by this process and was completed toward the end of the Carnian, the prosauropods may have diversified in the absence of other competitive terrestrial vertebrate herbivores.

The virtual disappearance of the rhynchosaurs, anomodonts, and traversodonts at the end of the Carnian may coincide with a major mass extinction of marine invertebrates during the Norian (Kummel 1970; Raup and Sepkoski 1982, 1984; Raup 1984). Raup and Sepkoski (1982, 1984) draw a distinction between mass extinctions and background extinctions. Raup (1984) states that "The conventional wisdom on this point is that interspecific competition is important in normal background extinction, but less so in mass extinction where an external physical or chemical stress forces the event." Raup and Sepkoski (1984) suggest that mass extinctions were short-term events that may have been caused by extraterrestrial forces. Raup (1984) states that extinction was "a discontinuous process wherein species are not at risk of extinction most of the time, but that the probability of extinction during certain short intervals in their history is very high." The abrupt change in the vertebrate herbivore fauna and flora (as reflected by palynological evidence) in late Carnian or early Norian times is suggestive of a mass extinction of several groups because, in common with other mass extinctions of marine faunas, the extinctions of terrestrial vertebrates at the end of the Carnian or early Norian were selective. Rhynchosaurs and virtually all anomodonts and traversodonts disappeared, but aetosaurs and carnivorous archosaurs (thecodontians plus dinosaurs) did not undergo major changes across this boundary. This suggests that aetosaurs and carnivorous archosaurs were more resistant than rhynchosaurs, anomodonts, and traversodonts to an environmental perturbation. It is possible that these latter groups were effective, efficient, competitive herbivores and that (with the possible exception of *Pisanosaurus*) their presence blocked the evolution of dinosaurian herbivores. If a major environmental perturbation

caused by external factors such as meteorite impacts (Hallam 1984; Raup 1984; Raup and Sepkoski 1984) led to the extinction of the Carnian herbivores, archosaurs (in the virtual absence of other terrestrial vertebrate herbivores) may have evolved an "adequate" masticatory apparatus in order not only to survive, but also to flourish. The effect of this mass extinction hypothesis and a slightly modified version of Charig's (1984) "overkill" hypothesis implies that prosauropods radiated in the absence of other vertebrate terrestrial herbivores, and, consequently, competition from other terrestrial herbivores may not have been a factor in the initial radiation of prosauropods.

The dramatic changes in the terrestrial fauna at the end of the Triassic coincide with major stratigraphic boundaries; therefore, the abrupt faunal change may be more apparent than real. Fossiliferous strata that were laid down during the extinction period are necessary to determine the degree of "overlap" between Carnian and Norian vertebrate terrestrial faunas.

It is also not possible at present to establish whether the marine and terrestrial "mass" extinction events were synchronous. Nevertheless, there is an increasing body of literature suggesting periodic and selective extinctions of marine life that were caused by short-lived perturbations of the environment. The enigmatic, dramatic, and apparent reversal in the efficiency of masticatory mechanisms of vertebrate terrestrial herbivores during the Late Triassic is difficult to explain in terms of competitive interactions between terrestrial herbivores. It is, therefore, worth considering the view that short-lived changes in the environment led to the rapid decline of the rhynchosaurs, traversodonts, and anomodonts. In the Liassic, herbivores with masticatory mechanisms more diverse and, in some cases, more powerful than those of the prosauropods and aetosaurs arose. However, the prosauropods had established themselves and coexisted (at least during the Liassic) with the more advanced forms. Fabrosaurids flourished in the Liassic and continued through to the Jurassic–Cretaceous boundary.

Acknowledgments

Most of the material described in this chapter was collected with the aid of NSF Grant DEB78–01327 and National Geographic grants. We are grateful to the Navajo Tribal Council for permission to conduct paleontological field work on the lands of the Navajo Nation. We thank W. W. Amaral for most of the preparation, L. L. Meszoly for rendering the figures, L. Maloney for secretarial assistance, and A. J. Charig and P. M. Galton for their critical comments on the manuscript.

References

Andras, G. 1982. Kavics-leletek a Bakony hgysegi kozepsojura pelagikus kepzodmenyekbol. Foldtani Kozlony, *Bull. Hungarian Geol. Soc.* 112:373–81.

Attridge, J., A. W. Crompton, and F. A. Jenkins. 1985. Common Liassic southern African prosauropod *Massospondylus* discovered in North America. *J. Vert. Paleontol.* 5:128–132.

Bakker, R. T. 1977. Tetrapod mass extinctions – (a model of the regulation of speciation rates and immigration by cyles of topographic diversity. *In* Hallam, A. (ed.), *Patterns as Illustrated by the Fossil Record* (New York: Elsevier), pp. 439–68.

Barghusen, H. R. 1968. The lower jaw of cynodonts (Reptilia:Therapsida) and the evolutionary origin of mammal-like adductor jaw musculature. *Postilla*, 116:1–49.

Benton, M. J. 1979. Ecological succession among late Palaeozoic and Mesozoic tetrapods. *Paleogeogr., Palaeoclimat., Palaeoecol.* 26:127–50.

1983. Dinosaur success in the Triassic: a new competitive model. *Quart. Rev. Biol.* 58:29–51.

Bonaparte, J. F. 1963a. La Familia Traversodontidae. *Acta Geol. Lilloana.* 4:63–94.

1963b. Descripción de *Ischignathus sudamericanus*, nuevo cinodonte gonfodonte del Triásico Medio Superior de San Juan, Argentina. *Acta Geol. Lilloana.* 4:111–28.

1971. Los tetrapodos del sector superior de la Formación Los Colorados, La Rioja, Argentina (Triásico Superior). *Opera Lilloana* 22:1–183.

1976. *Pisanosaurus mertii*: Casamiquela and the origin of the Ornithischia. *J. Paleontol.* 50:808–20.

1978. El Mesozoica de America del Sur y sus tetrapodos. *Opera Lilloana.* 26:1–596.

1982. Faunal replacement in the Triassic of South America. *J. Vert. Paleontol.* 28:362–71.

Camp, C. L., and S. P. Welles. 1956. Triassic dicynodont reptiles. Part I. The North American genus *Placerias*. *Mem. Univ. Calif.* 13:255–304.

Casamiquela, R. M. 1960. Noticia preliminar sobre dos nuevos estagonolepoideos argentinos. *Ameghiniana*. 2:3–9.

1961. Dos nuevos estagonolepoideos argentinos. *Rev. Asoc. Geol. Argentina*. 16:143–203.

Charig, A. J. 1979. *A New Look at the Dinosaurs* (London: Heinemann).

1980. Differentiation of lineages among Mesozoic tetrapods. *Mem. Soc. Geol. Fr. N.S.*, No. 239:207–10.

1982. Problems in dinosaur phylogeny: a reasoned approach to their attempted resolution. *Geobios, Mem. Spec.* 6:113–26.

1984. Competition between therapsids and archosaurs during the Triassic period: A review and synthesis of current theories. *Symp. Zool. Soc. London* 52:597–628.

Charig, A. J., and A. W. Crompton. 1974. The alleged synonymy of *Lycorhinus* and *Heterodontosaurus*. *Ann. S. Afr. Mus.* 64:167–89.

in press. The Lower Jurassic ornithischian *Heterodontosaurus tucki*: skull, dentition and relationships. *Ann. S. Afr. Mus.*

Charig, A. J., J. Attridge, and A. W. Crompton. 1965. On the origin of the Sauropoda and classification of the Saurischia. *Proc. Linn. Soc. London* 176:197–221.

Chatterjee, S. 1980. The evolution of Rhynchosaurs. *Mem. Soc. Geol. Fr. Fr. N.S.* 139:57–65.

1982. A new cynodont reptile from the Triassic of India. *J. Paleontol.* 56:203–14.

Clark, J., and J. A. Hopson, 1985. Distinctive mammal-like reptile from Mexico and its bearing on the phylogeny of the Tritylodontidae. *Nature.* (London) 315:398–400.

Colbert, E. H. 1961. *Dinosaurs: Their Discovery and Their World* (New York: E. P. Dutton).

1971. Tetrapods and continents. *Quart. Rev. Biol.* 46:250–69.

1980. *Evolution of the Vertebrates*, 3rd ed. (New York: Wiley).

1981. A primitive ornithischian dinosaur from the Kayenta formation of Arizona. *Mus. North. Ariz. Bull.* 53:1–61.

Cooper, M. R. 1981. The prosauropod dinosaur *Massospondylus carinatus* Owen from Zimbabwe: its biology, mode of life and phylogenetic significance. *Occ. Papers, Natl. Mus. Rhod. (Zimb.) B, Nat. Sci.* 6:689–840.

1982. A mid-Permian to earliest Jurassic tetrapod biostratigraphy and its significance. *Arnoldia Zimbabwe* 9:77–104.

Cox, C. B. 1965. New Triassic dicynodonts from South America, their origins and relationships. *Phil. Trans. Roy. Soc. B* 248:457–516.

1967. Changes in terrestrial vertebrate faunas during the Mesozoic. *In* Harland, W.B., C.H. Holland, M.R. House, N.F. Hughs, A.M. Reynolds, M.J.S. Rudwick, G.E. Satterthwaite, L.B.H. Tarlo, and E.C. Willey (eds.), *The Fossil Record* (Geological Society of London, London), pp. 77–89.

Crompton, A. W. 1968. The enigma of the evolution of mammals. *Optima.* 18:137–51.

1972a. Postcanine occlusion in cynodonts and tritylodontids. *Bull. Brit. Mus. Nat. Hist. (Geol.).* 21:27–71.

1972b. The evolution of the jaw articulation in cynodonts. *In* Joysey, K.A. and T.S. Kemp (eds.), *Studies in Vertebrate Evolution* (Edinburgh: Oliver & Boyd), pp. 23–53.

1974. The dentitions and relationships of the southern African Triassic mammals *Erythrotherium parringtoni* and *Megazostrodon rudnerae*. *Bull. Brit. Mus. Nat. Hist. (Geol.)* 24:400–37.

Crompton, A. W., and A. J. Charig. 1962. A new ornithischian from the Upper Triassic of South Africa. *Nature (London)* 196:1074–7.

Crompton, A. W., and F. Ellenberger. 1957. On a new cynodont from the Molteno beds and the origin of the tritylodontids. *Ann. S. Afr. Mus.* 44:1–14.

Crompton, A. W., and N. Hotton. 1967. Functional morphology of the masticatory apparatus of two dicynodonts (Reptilia : Therapsida). *Postilla* 109:1–51.

Crompton, A. W., and W. L. Hylander. 1986. Changes in the mandibular function following the acquisition of a dentary-squamosal articulation. *In* Hotton, N., P. D. Maclean, J. J. Roth, and F. C. Roth (eds.), *The Ecology and Biology of Mammal-like Reptiles* (Washington, D.C.: Smithsonian Press).

Cruickshank, A.R.I. 1978. Feeding adaptations in Triassic dicynodonts. *Palaeontol. Afr.* 21:121–32.

Dutuit, J. M. 1972. Decouvert d'un Dinosaure ornithischien dans le Trias superieur de l'Atlas. *Compt. Rend. Acad. Sci. Paris (D)* 275:2841–44.

Galton, P. M. 1976. Prosauropod dinosaurs (Reptilia : Saurischia) of North America. *Postilla.* 169:1–98.

1978. Fabrosauridae, the basal family of ornithischian dinosaurs (Reptilia : Ornithopoda). *Palaontol. Z.* 52:138–59.

1984. Cranial anatomy of the prosauropod dinosaur *Plateosaurus* from the Knollenmergel (Middle Keuper, Upper Triassic) of Germany. I. Two complete skulls from Trossingen/Wurt. with comment on the diet. *Geol. Palaeontol.* 18:139–71.

1985a. The diet of prosauropod dinosaurs from the late Triassic and early Jurassic. *Lethaia* 18:105–23.

1985b. Notes on the Melanorosauridae, a family of large prosauropod dinosaurs (Saurischia : Sauropodomorphy). *Geobios.* 18:671–76.

Gow, C. E. 1975. A new heterodontosaurid from the Redbeds of South Africa showing clear evidence of tooth replacement. *Zool. J. Linn. Soc.* 57:335–9.

Hallam, A. 1984. The causes of mass extinctions. *Nature (London).* 308:686.

Hiiemae, K. M. 1978. Mammalian mastication: a review of the activity of jaw muscles and the movements they produce in chewing. *In* Butler, P.M., and K. A. Joysey (eds.), *Studies in the Development, Structure and Function of Teeth* (London: Academic Press), pp. 360–98.

Hiiemae, K. M., and A. W. Crompton. 1985. Mastication, food transport and swallowing. *In* M. Hildebrand, et al., *Functional Vertebrate Morphology* (Cambridge, Massachusetts: Harvard University Press), pp. 262–91.

Hopson, J. A. 1964. The braincase of the advanced mammal-like reptile *Bienotherium. Postilla.* 83:1–30.

1971. Postcanine replacement in the gomphodont cynodont *Diademodon. Zool. J. Linn. Soc. Suppl.* 50 (1):1–21.

1975. On the generic separation of the ornithischian dinosaurs *Lycorhinus* and *Heterodontosaurus* from the Stormberg Series (Upper Triassic) of South Africa. *S. Afr. J. Sci.* 71:302–5.

1984. Late Triassic traversodont cynodonts from Nova Scotia and southern Africa. *Palaeontol. Afr.* 25:181–201.

Janensch, W. 1929. Material und Formengehalt der Sauropoden in der Ausbeute der Tendaguru - Expedition. *Palaeantographica, Suppl.* 7(2):1–34.

Kemp, T. S. 1982. *Mammal-Like Reptiles and the Origin of Mammals* (New York: Academic Press).

Kermack, D. 1984. New prosauropod material from

South Wales. *Zool. J. Linn. Soc.* 82:101–17.

Kitching, J. W., and M. A. Raath. 1984. Fossils from the Elliot and Clarens Formations (Karoo sequence) of the Northeastern Cape, Orange Free State and Lesotho, and a suggested biozonation based on tetrapods. *Palaeontol. Afr.* 25:111–25.

Kummel, B. 1970. *History of the Earth* (San Francisco: Freeman).

Lucas, A. M., and P. R. Stettenheim. 1972. *Avian Anatomy: Integument. Part II.* Agriculture Handbook 362 (Washington, D.C., U.S. Dept. Agriculture).

Olsen, P. E., and P. M. Galton. 1977. Triassic–Jurassic tetrapod extinctions: are they real? *Science.* 197:983–6.

1984. A review of the reptile and amphibian assemblages from the Stormberg Group of southern Africa with special emphasis on the footprints and the age of the Stormberg. *Palaeontol. Afr.* 25:87–110.

Ostrom, J. H. 1969. Terrestrial vertebrates as indicators of Mesozoic climates. *Proc. North Am. Paleontol. Conven.*, pp. 347–76.

Raup, D. M. 1984. Evolutionary radiations and extinctions. *In* Holland, A. D. and A. F. Trendal (eds.), *Patterns of Change in Earth Evolution* (New York: Springer-Verlag), pp. 5–14.

Raup, D. M., and J. J. Sepkoski. 1982. Mass extinction in the marine fossil record. *Science.* 215:1501–2.

1984. Periodicity of extinctions in the geologic past. *Proc. Nat. Acad. Sci. U.S.A.* 81:801–5.

Robinson, P. L. 1976. How *Sphenodon* and *Uromastyx* grew their teeth and used them. *Morpho. Biol. Reptiles. Linn. Soc. Symp. Ser.* 3:43–64.

Romer, A. S. 1966. *Vertebrate Paleontology*, 3rd ed. (Chicago: University of Chicago Press).

Romer, A. S., and L. I. Price. 1944. *Stahleckeria lenzii,* a giant Triassic Brazilian dicynodont. *Bull. Mus. Comp. Zool. Harv.* 43:465–90.

Santa Luca, A. P. 1980. The postcranial skeleton of *Heterodontosaurus tucki* (Reptilia : Ornithischia) from the Stormberg of South Africa. *Ann. S. Afr. Mus.* 79:159–211.

1984. Postcranial remains of Fabrosauridae (Reptilia : Ornithischia) from the Stormberg of southern Africa. *Palaeontol. Afr.* 25:151–80.

Santa Luca, A. P., A. W. Crompton, and A. J. Charig. 1976. A complete skeleton of the Late Triassic ornithischian *Heterodontosaurus tucki. Nature (London)* 264:324–8.

Sill, W. D. 1971. Functional morphology of the rhynchosaur skull. *Forma Functio.* 4:303–18.

Smith, K. K. 1984. The use of the tongue and hyoid apparatus during feeding in lizards (*Ctenosaura similis* and *Tupinambis nigropunctatus*). *J. Zool.* 201:115–43.

Sues, H.-D. 1983. Tritylodontidae from the Kayenta Formation (Early Jurassic) of northeastern Arizona. Ph.D. thesis, Harvard University.

1984. Inferences concerning feeding and locomotion in the Tritylodontidae (Synapsida). *3rd Symp. Mesozoic Terr. Ecosystems, Short Pap.*, pp. 231–6.

1985. The relationships of the Tritylodontidae (Synapsida). *Zool. J. Linn. Soc.* 85:205–17.

Throckmorton, G. S. 1976. Oral food processing in two herbivorous lizards, *Iguana iguana* (Iguanidae) and *Uromastyx aegypticus* (Agamidae). *J. Morphol.* 48:363–90.

Thulborn, R. A. 1970. The skull of *Fabrosaurus australis*, a Triassic ornithischian dinosaur. *Palaeontology.* 13:414–32.

1971. Tooth wear and jaw action in the Triassic ornithischian dinosaur *Fabrosaurus. J. Zool.* 164:165–79.

1974. A new heterodontosaurid dinosaur (Reptilia: Ornithischia) from the Upper Triassic Red Beds of Lesotho. *Zool. J. Linn. Soc.* 55:151–75.

Van Heerden, J. 1979. The morphology and taxonomy of *Euskelosaurus* (Reptilia : Saurischia, Late Triassic) from South Africa. *Navors. Nas. Mus. Bloemfontein.* 4:21–81.

Weishampel, D. B. 1983. Hadrosaurid jaw mechanics. *Acta Palaeontol. Polonica* 28:271–80.

1984a. Evolution of jaw mechanisms in ornithopod dinosaurs. *Advan. Anat. Embryol. Cell Biol.* 87:1–110.

1984b. Interactions between Mesozoic plants and vertebrates: fructification and seed predation. *N. Jahrb. Geol. Palaont. Abh.* 167:228–50.

Young, C. C. 1951. The Lufeng saurischian fauna in China. *Paleontol. Sinica*, 13:1–96.

18 On Triassic and Jurassic mammals

WILLIAM A. CLEMENS

In the years since the publication of *Mesozoic Mammals: The First Two-thirds of Mammalian History* (Lillegraven, Kielan-Jaworowska, and Clemens 1979), the understanding of the Triassic and Jurassic history of the Mammalia has been significantly advanced. This is not just the result of finding new material, although such discoveries have introduced us to new species of mammals and extended the temporal and biogeographic ranges of some previously known forms. Two other contributions have expanded our knowledge of these early mammals.

First, revisions and refinements in the determination of the ages of fossil localities continue to increase the resolution of our perception of the tempo and pattern of mammalian evolution. Particularly, recognition that many local faunas once referred to the Rhaetian are of Early Jurassic age has begun to fill in a major gap in the history of mammalian evolution in Laurasia.

Also, cladistic analyses of mammals and "mammal-like reptiles," the nonmammalian synapsids, have resulted in several wide-ranging proposals to restructure their classification. For example, McKenna's (1975) and Prothero's (1981) analyses of major groups of mammals and Kemp's (1982) studies primarily of the nonmammalian synapsids have brought into question the criteria for diagnosis of the Mammalia and its content.

Rather than adopting one of the new, possibly ephemeral classifications, this chapter will follow Kemp (1982), who employed Romer's (1966) influential classification in recognizing a paraphyletic group, Synapsida ("mammal-like reptiles"), and a descendant group, the Mammalia, which might be polyphyletic in its origins. Although questioning the utility of some of the diagnostic characters he employed, I shall also accept Kemp's (1982, p. 295) designation of the last common ancestor of morganucodontids and *Kuehneotherium* as the first mam-

mal. This classification is familiar and is chosen to facilitate communication. However, it does not highlight the great antiquity of the last common ancestor of modern reptiles and mammals, which probably was a member of a Carboniferous fauna.

In the first of the following sections, some major changes in our understanding of the stratigraphic framework of the fossil record are reviewed. Then, announcements of discoveries of Triassic and Jurassic mammals or systematic analyses of these groups that have appeared since publication of *Mesozoic Mammals* (Lillegraven et al. 1979) are summarized. Next, major questions concerning diagnosis and content of the Mammalia are addressed with the purpose of highlighting current discussions rather than offering answers. Finally, our knowledge of the biogeographic pattern and tempo of mammalian evolution in the Late Triassic and Jurassic is summarized, and its implications for interpreting the evolution of this group during the Cretaceous and early Cenozoic are explored.

Stratigraphic framework

A major advance in understanding the evolutionary transition from synapsids to mammals is being facilitated by increased refinement of determination of the ages and correlations of local faunas that until recently were considered to be of Late Triassic (Rhaetian) or Early Jurassic age (Clemens et al. 1979). Current research shows that the so-called "Rhaetic fissure fillings" of England and Wales, which have produced a wealth of specimens of early mammals and advanced synapsids, were deposited at various times during this interval. Most of the fillings yielding *Morganucodon* and *Kuehneotherium* are now thought to be of Liassic (Early Jurassic), probably Sinemurian, age (Kermack, Mussett, and Rigney 1981). Research conducted by Sigogneau-Russell (1983c and references cited) and her associates strengthens the interpretation that a sequence of "Rhaetian" faunas can be recognized in continental Europe. The site at Saint-Nicolas-de-Port probably is older than the classic "Rhaetic bone beds" of Germany and Switzerland. Apparently, many of these European local faunas can be ordered in at least a four-unit scale: Norian, earlier and later Rhaetian, and Liassic. It must be noted, however, that the utility of the "Rhaetic" as a stratigraphic unit is being challenged [note Hallam (1981) and references cited], and this stratigraphic scale probably will undergo significant revision.

A second major change in the stratigraphic framework stems from a more refined correlation of units on other continents that have yielded advanced synapsids and mammals with the European Late Triassic–Early Jurassic section. The Kayenta Formation of North America, the upper part of the Stormberg Group of southern Africa, the Kota Formation of India, and the Lufeng beds of China now are thought to be broadly correlative units and of Early Jurassic age. In contrast, the Cumnock Formation of the Newark Supergroup, the source of *Microconodon* and *Dromatherium*, is considered of Carnian (Late Triassic) age (see Olsen and Galton 1977; Olsen, Hubert, and Mertz 1981; Puffer et al. 1981; Olsen, McCune, and Thomson 1982).

Additions to the fossil record

Since the publication of *Mesozoic Mammals* (Lillegraven et al. 1979), the pace of discovery and analysis of Late Triassic and Jurassic mammals has been increasing rapidly. The following are revisions of or additions to summaries given in the second chapter of that book (Clemens et al. 1979). Each of the entries begins with the name of the collecting locality or fossiliferous lithostratigraphic unit, the current best estimate of the age of the local fauna(s), and its geographic location.

Late Triassic

Emborough fissure filling
Norian; England; Fraser, Walkden and Stewart (1985) described the first record of the therian *Kuehneotherium* in Norian deposits.

Medernach
Middle Norian; Luxembourg: Wouters, Lepage, and Coupatez (1983) discovered therapsid-like teeth in a bonebed in the Steinmergel-Gruppe. On the basis of analysis of dental morphology, Hahn, Lepage, and Wouters (1984) allied the synapsid from Metternach, *Pseudotriconodon wildi*, with *Dromatherium* and *Microconodon* (North America, Cumnock Formation, Carnian) and *Therioherpeton* (South America, Santa Maria Formation, latest Middle or earliest Late Jurassic) in the Family Dromatheriidae.

Saint-Nicolas-de-Port
Earlier Rhaetian; France: As the result of research by Sigogneau-Russell (1978, 1983a, 1983b, 1983c) and associates (Sigogneau-Russell, Cappetta, and Taquet 1979), a number of mammals and other vertebrates now are known to have been members of this local fauna. The most abundantly represented are haramiyids, and some of the teeth are of new morphological types. Several isolated teeth, once thought to be parts of the dentition of a *Paulchoffatia*-like multituberculate, are referred to a new family (see Chapter 8). The Morganucodontidae is represented by *Brachyzostrodon coupatezi* (Sigogneau-Russell 1983c) and the Kuehneotheriidae by *Woutersia mirabilis* (Sigogneau-Russell 1983a).

Habay-la-Vieille
Rhaetian; Belgium: Wouters, Sigogneau-Russell, and Lepage (1984) described a tooth referable to *Haramiya*.

Rhaetic bonebeds of Württemberg
Possibly later Rhaetian; West Germany: A review of the collections made by E. von Heune and others (Clemens 1980) led to the recognition of *Tricuspes tubingensis* as a valid taxon, possibly of mammalian affinities. Also, all the teeth of haramiyids from these bonebeds complete enough to be identified at a generic level are referable to *Thomasia*. [A catalog of the taxa of multituberculates, including the haramiyids, has been compiled by Hahn and Hahn (1983).]

Hallau bonebed
Possibly later Rhaetian; Switzerland: A review of the samples of this local fauna (Clemens 1980) led to the discovery of a tooth resembling the type of *Tricuspes tubingensis* and recognition of both haramiyid form genera, *Thomasia* (including cf. *T. antigua* and *T. anglica*) and *Haramiya* (*H. moorei*). The Morganucodontidae is represented by a new species of *Morganucodon*, *M. peyeri*. Two new genera, probably referable to the Mammalia, *Helvetiodon schutzi* and *Hallautherium schalchi*, were recognized. Additional nonmammalian vertebrate fossils were subsequently described by Kindlimann (1984). Interestingly, this collection, like the older samples of the Hallau bonebed, does not contain any specimens representing the tritylodonts.

Khuabberg
Late Triassic; northern Namibia: Hopson and Reif (1981) demonstrated that the type specimen of *Archaeodon reuningi*, which had been identified as the tooth of a plagiaulacoid multituberculate or a tritylodont, is a chalcedony infilling of a vesicle in a block of lava.

Early Jurassic
Fissure fillings near Bridgend,
Glamorgan

Early Sinemurian; Great Britain: Kermack, Mussett, and Rigney (1973, 1981), in two extensive monographs, provided detailed studies of the cranial anatomy of *Morganucodon watsoni* based on large collections from Glamorgan and a skull of *M. oehleri* from China. Parrington (1978) added additional anatomical data, and Evans (1984) argued that she could find no evidence of epipubic bones in this early mammal. Unfortunately, the literature on this mammal has been muddled through reference as either *Morganucodon* or *Eozostrodon*. A suggested solution to this problem (Clemens 1979) is to limit the content of *Eozostrodon parvus* to the poorly known morganucodontid now represented only in Holwell Quarry local fauna in England. *Morganucodon* is accepted as the appropriate generic reference for several species of morganucodontids now known in local faunas in Europe, China, and the United States.

The dentition of *Kuehneotherium* was the subject of a detailed study by Mills (1984). A comparison of the samples of teeth of this early mammal from fissure fillings at Pant and Pontalun quarries, Wales, revealed differences in root structure and pattern of wear that suggest marked differences in ontogeny and function of their dentitions.

Kota Formation

Liassic; India: Datta, Yadagiri, and Rao (1978) greatly expanded the biogeographic scope of our knowledge of Mesozoic mammals with the discovery of a microvertebrate fauna in the Kota Formation of peninsular India. A kuehneotheriid, *Kotatherium haldanei*, was the first mammal of this fauna to be described (Datta 1981). Subsequently, two kuehneotheriid or kuehneotheriid-like forms, *Trishulotherium kotaensis* and *Indotherium pranhitai*, also known only from isolated teeth, were described by Yadagiri (1984).

Lufeng Faunas

Probably Liassic; Yunnan Province, People's Republic of China: In the past decade, knowledge of the faunas of the Lufeng Formation has greatly increased. Sigogneau-Russell and Sun (1981) summarized the paleontological record of the unit and argued that the stratigraphically higher fauna of the Lower Lufeng Formation, which has produced the mammals discussed below, is probably of Liassic age. Sun et al. (1985) presented a listing of the members of the Lufeng saurischian fauna. All the Lufeng mammals described to date appear to be referable to a very broadly defined Order Triconodonta, a grouping clearly inviting additional study and refinement.

In addition to *Sinoconodon rigneyi* Patterson and Olson 1961, two additional species have been recognized: *S. parringtoni* Young 1982 and *S. yangi* Zhang and Cui 1983. Zhang and Cui argued that the shift of accessory bones from the lower jaw to the middle ear region, not the earlier acquisition of a dentary–squamosal articulation, should be taken as the diagnostic character of the Mammalia. Consequently, Zhang and Cui (1983) placed *Sinoconodon* and the Sinoconodontidae among the synapsid Cynodontia. In contrast, Jenkins (1984, pp. 42–3) noted that Crompton's and Sun's studies of new material of *Sinoconodon* indicate that a dentary–squamosal articulation was developed, and these authors regard this a clear justification to refer this genus to the Mammalia.

In a posthumously published study, Young (1982) described *Lufengoconodon changchiawaensis*. Zhang (1984) pointed out morphological similarities of this new genus to both sinoconodontids and morganucodontids. Judgment on its phylogenetic affinities was reserved until a thorough study of the sample, which includes both cranial and postcranial material, is completed.

The Morganucodontidae is represented by *Morganucodon oehleri* (note Kermack et al. 1973, 1981) and *"Eozostrodon" heikuopengensis*. The morphological differences separating the two species are not great (see Zhang 1984). If the latter species proves valid, reference to *Morganucodon* probably is warranted.

Luzhang

?Early Jurassic; Sichuan Province, People's Republic of China: Chow and Rich (1984b) reported the discovery of a fragment of a triangular mammalian tooth.

Kayenta Formation

Early Jurassic; United States: Jenkins, Crompton, and Downs (1983) discovered representatives of three groups of mammals. The morganucodontid is referred to *Morganucodon* sp. *Dinnetherium nezorum* is referred to the Order Triconodonta, Family *incertae sedis*. A jaw of this mammal documents the evolution of an angular region as a neomorphic process (also note Jenkins 1984). Finally, one isolated tooth appears to record the presence of a haramiyid.

Middle Jurassic
Great Estuarine Group, Isle of Skye

Bathonian; Great Britain: In a review of the geology and paleontology of the Kilmalaug Formation or "Ostracod Limestones" of the Isle of Skye, Savage

(1984) noted that the fauna includes both the docodont *Borealestes serendipitus* and a new genus and species of pantothere that resembles forms known from the Stonesfield Slate and Guimarota local faunas.

Forest Marble; Kirtlington local fauna
Bathonian, Oxfordshire, Great Britain: Fossils found in the Forest Marble are thought to be approximately equivalent in age to those from the Kilmalaug Formation of the Isle of Skye (Savage 1984) and slightly younger than the fauna of the Stonesfield Slate, the only other units in Great Britain from which mid-Jurassic mammalian local faunas have been described.

The available sample of the Kirtlington local fauna documents a diversity of mammalian lineages. In his study of this local fauna, Freeman (1979) recognized a new morganucodontid, *Wareolestes rex*. The Multituberculata is represented by a molar, and fragments of molars document the presence of a member of the Tritylodontia. The docodont found at Kirtlington was referred provisionally to *Borealestes*. A new genus and species, *Cyrtlatherium canei*, was referred to the Kuehneotheriidae. The Peramuridae is represented by *Palaeoxonodon ooliticus* that might be closely related to *Amphitherium*, which is known from the Stonesfield Slate and questionably recognized at Kirtlington. Fragmentary teeth record the presence of dryolestids.

Shishugou Formation
Middle or Late Jurassic; Uygar Autonomous Region, People's Republic of China: A new triconodont, *Klamelia zhaopengi*, was referred to the Family Amphilestidae by Chow and Rich (1984a). They pointed out close affinities with the Mongolian amphilestids *Gobiconodon* and *Guchinodon* (note Trofimov 1978, 1981) and an undescribed amphilestid from the Cloverly Formation of the United States (see Jenkins and Crompton 1979, pp. 78, 84–5). In another report, Chow and Rich (1984b) noted the discovery of a fragmentary mammalian dentary in Early Cretaceous deposits in the same general area.

Late Jurassic
Guimarota coal mine fauna
Kimmeridgian; Portugal; Krusat (1980) presented a detailed analysis of the dental and cranial morphology of *Haldanodon exspectatus*. Henkel and Krusat (1980) described a partial skeleton of this docodont.

Shilong-zhai
Late Jurassic; Sichuan Province, People's Republic of China: *Shuotherium dongi* Chow and Rich (1982) is based on a lower jaw containing six postcanine teeth and the alveolus for a seventh. The molariform teeth are characterized by a curious, bladelike "pseudotalonid" anterior to a typical therian trigonid. Chow and Rich established a new therian order, Shuotheridia, for the species, and thus provided ample warning of the limitations of our knowledge of the diversification of therians prior to the Cretaceous.

The type locality of *Shuotherium* is in the Upper Shaximiao Formation that yields fossil vertebrates typical of the *Mamenchisaurus* fauna. Dong (1980) suggested that this fauna is of later Jurassic age, broadly correlative with those of the Morrison Formation (North America), Purbeck Beds (England), and the deposits at Tendagaru (Tanzania).

Como Bluff, Quarry 9 local fauna
Late Jurassic; Wyoming, United States: Quarry 9 was one of the collecting sites in the Morrison Formation worked in the late 1870s and the 1880s by crews employed by O. C. Marsh. Prothero (1981) discussed a new sample from this quarry, which was obtained in the summers of 1968–70. This collection includes specimens of a triconodontid; a docodont, *Docodon victor*; plagiaulacid multituberculates; and a large suite of dryolestids.

In his analysis of the dryolestids Prothero (1981): (1) described *Comotherium richi*, n. gen. et sp.; (2) argued that the previously recognized two species of *Melanodon* Simpson and two species of *Herpetairus* could be distinguished; and (3) demonstrated that *Malthacolestes osborni* Simpson is a junior synonym of *Melanodon oweni* Simpson. He also presented a wide-ranging cladistic analysis and dichotomous classification of nontribosphenic therians.

Fruita fauna
Late Jurassic; Colorado, United States: Rasmussen and Callison (1981) recognized a new species of *Priacodon*, *P. fruitaensis*, in the local fauna found in the Salt Wash Member of the Morrison Formation. They indicated that the local fauna includes at least one more species of triconodont as well as several species of multituberculates and dryolestids.

What is a mammal?
In his influential textbook, Romer (1966) recognized the great antiquity, dating back to the Carboniferous, of the separation of lineages leading to modern reptiles and birds, on one hand, and modern mammals, on the other, but did not formally express this dichotomy in his classification. Instead the basal members of the latter lineage, the synapsids or "mammal-like reptiles," were treated as the "earliest to appear of known reptilian groups" and described as "taking a leading role in the archaic reptilian radiation" (Romer 1966, p. 173). The

Mammalia was recognized as a possibly polyphyletic group including descendants of these synapsids. Romer usually employed the evolution of a dentary–squamosal articulation between the jaws and skull as the diagnostic character of the Mammalia (note Romer 1966, p. 185), a convention that had been discussed at length by Simpson (1960).

The limitations of the then available fossil record abetted maintenance of a major, class-level separation between synapsids and mammals. Specimens of the most advanced synapsids and earliest mammals were fragmentary and rare. Romer (1966, p. 185) wrote of a poorly documented and little understood evolutionary "no-man's-land." An additional obstruction to detailed analyses of the relationships of members of these two groups stemmed from the fact that many of the advanced synapsids were known from well-preserved partial or complete skeletons, while, at best, Late Triassic and Jurassic mammals were documented by fragmentary jaws and isolated teeth. Most studies of synapsids emphasized cranial and postcranial anatomy, but gave short shrift to their dental morphology and thereby lost the opportunity for detailed comparative studies.

In the past decade or so, long-prevailing views concerning the definition and classification of the Mammalia have been challenged, particularly by those workers committed to cladistic analyses of characters and the development of dichotomous classifications. In part this has been the result of discovery and analysis of cranial and postcranial material of Late Triassic and Jurassic mammals [for example, Kermack et al. (1973, 1981) and Jenkins and Parrington (1976)]. Also, paleontologists, many of whom are experienced in analyses of mammalian dentitions, began to study the dental morphology of synapsids and reptiles. Examples of the latter are Crompton's (1972) analysis of the occlusion of postcanine dentitions of cynodonts and tritylodonts, Lees and Mills's (1983) analysis of the synapsid *Pattsia*, and the review by Hahn et al. (1984) of the dromatheriids that includes detailed comparisons of the teeth of these synapsids with similar triconodont-like teeth of the prolacertilians *Tanystropheus* and *Macrocnemus* and the pterosaur *Eudimorphodon*.

Second, studies such as those of Kemp (1982, 1983), Kermack and Kermack (1984), and Jenkins (1984), which deal comprehensively with the evolution of advanced synapsids and primitive mammals, appear to herald a major change in methodologies and goals of interpretation and classification of synapsid and mammalian relationships. Current research is challenging the conventional criteria for diagnosis of the Mammalia and perceptions of the evolutionary interrelationships of Triassic and Jurassic synapsids and mammals. Kemp (1982,

p. 295) reviewed the problems posed by the recognition of a variety of mammal-like characters in the cranial and postcranial skeletons of some of the most advanced synapsids and advanced reasons for including tritylodonts and trithelodonts in the Mammalia.

Finally, the prevalent interpretation a decade ago of the origin of the Mammalia called for an early basic division between therian and nontherian (or atherian) lineages (Crompton and Jenkins 1979). This dichotomy was substantiated on the basis of apparent differences in structure of the braincase and in patterns of symmetry in occlusal relationships of the cheek teeth. Presley (1981 and references cited) has advanced persuasive embryological evidence challenging the supposed significance of the purported differences in structure of the braincase. Kemp (1983) argued that the linear (anteroposterior) occlusal pattern of the cusps of the cheek teeth of "nontherian" mammals is simply a retention of the primitive synapsid pattern. The unity of the "nontherian" group is challenged, and the evolutionary relationships of monotremes to other groups of mammals are open to reinterpretation.

In summary, the stage is being set for a major revision of the classification of tetrapods through removal of the Synapsida from the Reptilia and its unification with the Mammalia. Such a change in the classification will revitalize the sauropsid–theropsid division recognized by Goodrich (1916) in the early years of this century. The utility of employing the evolution of a dentary–squamosal articulation as the diagnostic character of the Mammalia, seriously challenged by discovery of the complex articulations in *Diarthrognathus* and *Morganucodon*, has been weakened further. Finally, the interpretation of a basic dichotomy separating mammals into therian and nontherian groups probably should be abandoned.

Mesozoic mammalian biogeography and evolution

In spite of recognition that many local faunas formerly assigned a Late Triassic–Early Jurassic age are probably Jurassic correlatives, the evolutionary roots of the Mammalia as here defined remain firmly planted in the Triassic. The oldest record of the group is still a haramiyid in the *Plateosaurus* beds near Halberstadt (German Democratic Republic). Discovery of *Kuehneotherium* in Norian deposits of England and the diverse mammalian fauna from Saint-Nicolas-de-Port, which includes at least haramiyids, a morganucodontid, and a new kuehneotheriid, testify to the breadth of the radiation of mammals in the Late Triassic.

Assignment of an Early Jurassic age to local faunas in Europe, Asia, Africa, and North America begins to fill what was once perceived as a major gap in the fossil record of vertebrate evolution (Padian and Clemens, 1985). The composition of these faunas now confirms a suggestion, based on the presence of advanced synapsids in Triassic faunas in the same areas, that mammals were widely dispersed through Laurasia and Gondwana before fragmentation of the supercontinents. The Laurasian fauna included haramiyids, morganucodontids, and kuehneotheriids. The kuehneotheriids found in the Kota Formation of peninsular India show that at least one of these families was represented in the Gondwanan fauna. Given the small size of the sample of this local fauna, it would not be surprising if other groups of mammals are discovered as collecting and analysis continues.

Until recently, our knowledge of mammalian diversity during the Middle Jurassic was limited to a relatively small sample of mammals from the Stonesfield Slate of England. Discoveries of mammals in the approximately contemporaneous Forest Marble and Kilmalaug formations of Great Britain have greatly increased the mammalian faunal list. Together they document the survival of morganucodontids and kuehneotheriids beyond the Early Jurassic. Additionally, they contain the first records of nonharamiyid multituberculates, docodontids, amphilestids, peramurids, and dryolestids. Thus, by the Middle Jurassic most of the mammalian families that are known to be represented in Late Jurassic and earlier Cretaceous faunas of the Northern Hemisphere had made their appearance.

As for the Middle Jurassic, knowledge of the mammalian faunas of the Late Jurassic is essentially limited to the Northern Hemisphere; an edentulous dentary of a eupantothere from the Tendaguru (Tanzania) remains the only record from the Southern Hemisphere. The mammalian faunas of the Morrison Formation of North America and Purbeck Beds of England show broad similarity with five of the ten families known to occur in both areas. However, resemblance decreases at the generic and specific levels. Although separated by a much shorter distance, the approximately contemporaneous mammalian faunas of Porto Pinheiro and Guimarota coal mine, Portugal, have noticeably different compositions. The former is similar in composition to the Purbeck local fauna, while the latter lacks common Late Jurassic mammals such as symmetrodonts and triconodonts, but includes representatives of paulchoffatiid multituberculates and pantotheres unknown in other local faunas. These differences might reflect evolution of a distinct coal swamp fauna. Records of mammals from China include an amphilestid (of either Middle or Late Jurassic age) and a remarkable new type of mammal, *Shuotherium*, with a "pseudo-tribosphenic" dentition. The fossil record

of other Late Jurassic terrestrial vertebrates includes samples of faunas of most southern continents. The typical biogeographic pattern, well illustrated by the sauropods, for example, is one of widespread distribution of major groups with regional differentiation at the generic or specific levels (Colbert 1981).

Recognition that by the end of the Jurassic mammals not only were inhabitants of both Laurasia and Gondwana but also had evolved some locally distinct faunas is promoting changes in interpretations of their evolution during the Cretaceous and origins of the mammalian-dominated Cenozoic faunas. For many years biogeographic analyses of patterns of mammalian evolution were dominated by the views of W. D. Matthew (1915), who argued that the center of origin of Cenozoic mammalian radiations was in the Northern Hemisphere, and that as new groups arose, the relicts of earlier radiations were pushed or drifted southward. Hershkovitz (1968, p. 316) dubbed this pattern the "Sherwin–Williams effect," a reference to this paint company's advertisement depicting a can of paint being poured on the North Pole of the globe and dripping southward. He argued that this biogeographic pattern probably erroneously describes the evolutionary origins of many mammalian groups found in the first Cenozoic faunas of South America and other southern continents.

Frequently, analyses involving the "Sherwin–Williams effect" started with the assumption that the lack of fossil records in the Southern Hemisphere documented an absence of mammalian members of the Jurassic and Cretaceous faunas in these areas, or that no compelling evidence required speculation about the possibility of the presence of mammals in the Southern Hemisphere at this time. Thus, research tended to be focused on the question of when, during the Late Cretaceous or earliest Tertiary, the ancestors of various groups of mammals dispersed from the Northern Hemisphere into the southern continents. Although Matthew's thesis appears to describe accurately the evolutionary histories of various groups of mammals that originated in the Cenozoic, several workers have joined Hershkovitz and argued that it has been misapplied in analyses of Mesozoic mammalian evolution (see Marshall 1980; Hoffstetter 1981; Marshall, De Muizon, and Sigé 1983; and Woodburne 1984 and references cited).

In part, these challenges are based on changes in understanding of the physical evolution of continents and ocean basins during the Cretaceous. In the chapter on paleogeography in *Mesozoic Mammals*, Lillegraven, Kraus, and Bown (1979) cited geologic studies indicating that the Early Cretaceous (ca. 135–100 MYA) was a time of continued fragmentation and separation of the continents. Subsequent research strengthens this interpretation.

Particularly toward the end of the Early Cretaceous, a combination of the effects of movement of continents, growth of the Tethyan seaway, and variable flooding by epicontinental seas began to fragment terrestrial areas complexly. The geologic framework was appropriate for increased regional (continental) differentiation of the mammalian faunas of Laurasia and Gondwana during the Cretaceous, yet set the stage for intermittent dispersal from one area to another.

Paleontological data play an even stronger role in these challenges. For example, pantotheres, a group currently suspected to include the ancestors of marsupials and placentals, are known to have been present in Laurasia (North America and Europe) and Gondwana (Africa) in the Late Jurassic. They, or their descendants, very probably were members of the Early Cretaceous faunas of many, if not most, of the fragments of these supercontinents.

In the Northern Hemisphere, occurrences of the triconodont *Gobiconodon* in Mongolia and North America (Jenkins 1984) and the broad similarities of the therians of eutherian–metatherian grade (probably descendants of the pantotheres) in Eurasia and North America document some faunal interchange in the Early Cretaceous. During the middle of the Cretaceous, origin, diversification, and dispersal of the angiosperms probably were basic causal factors in a major remodeling of the terrestrial faunas of, at least, the Northern Hemisphere. The dinosaurian fauna was significantly changed by the rapid diversification of the hadrosaurids and ceratopsids. Many of the mammalian lineages that were abundantly represented in the Late Jurassic faunas (triconodonts and symmetrodonts, for example) dwindled and did not survive the end of the period. In contrast, there was a radiation of advanced multituberculates as well as early eutherians, marsupials, and other groups apparently derived from pantotherian ancestors. The absence of marsupials in Asia and dissimilarities of Asian and North American eutherians and multituberculates, for example, suggest increasingly effective barriers to mammalian dispersal (note Kielan-Jaworowska 1982, 1984; Nessov 1984).

Except for discoveries of a few species in latest Cretaceous strata of South America, the mammalian fossil record of the southern continents is a blank. However, records of other groups of Cretaceous terrestrial vertebrates and the interrelationships of the oldest Cenozoic mammals of these continents hint at the course of their Mesozoic history. In his analysis of dinosaurian faunas, Charig (1979) suggested that the Cretaceous was a time of evolution of endemic dinosaurian faunas in both hemispheres that were enriched by limited intercontinental dispersals. Colbert (1981) was more specific and argued that the isolation of many modern continents did not occur until late in the period. Bonaparte (1984) analyzed the differences and similarities between the reptilian faunas of North and South America and concluded that some faunal interchange took place during the Late Cretaceous. Although primitive marsupials and condylarths dispersed between the Americas at about this time, other groups did not. Absence of South American representatives of taxonomically diverse and abundant North American mammalian groups, including ptilodontoid and taeniolabidoid multituberculates and palaeoryctoid proteutherians, for example, strongly suggests that movement of vertebrates between these continents was limited by environmental filters or the result of chance dispersal events.

Although far from demonstrating the historical events, these biogeographic patterns support the working hypothesis that pantotheres or their descendants probably were members of the Late Jurassic faunas of Gondwana. Even if they were not, distribution patterns of groups, such as the hadrosaurids and pachycephalosaurids, which originated in the Cretaceous, suggest that there were opportunities for interchange of some mammalian lineages between Northern and Southern Hemispheres during the Cretaceous (Sues 1980).

Finally, the members of the oldest known mammalian faunas of the southern continents provide definite hints of their long residency in these areas. The marsupial-dominated fauna of Australia long has been recognized as probably having a history stretching well back into the Cretaceous. The research of Woodburne and Zinsmeister (1982, and see Woodburne 1984) on the evolution of the Southern Hemisphere biota, focusing both on marsupials and high latitude southern marine faunas, adds support to this interpretation. Marshall et al. (1983) likewise argued that the discovery of the very derived, Late Cretaceous, South American marsupial *Roberthoffstetteria nationalgeographica* points to a long history of the group on this continent. Finally, as emphasized in McKenna's (1975) analysis of mammalian interrelationships, the edentates (armadillos, sloths, and the American anteaters) seem to be distant relatives of all other placental orders. Rather than focusing on the hypothesis that edentates are descendants of some group of Northern Hemisphere placentals that dispersed to South America late in the Cretaceous, equal attention must now be given to the hypothesis that edentates are a relic of an endemic Cretaceous evolutionary radiation of primitive placentals.

In summary, in spite of the essential lack of direct documentation of the presence of mammals in the southern continents throughout the Cretaceous, a variety of indirect evidence suggests that these areas were an important arena for mammalian evolution.

Conclusions

Recent discoveries, changes in age assignments of some local faunas, and new systematic analyses have modified perceptions of the evolution of advanced synapsids and mammals during the Late Triassic and Early Jurassic. Certainly one of the major products of this recent research is recognition of the need to put increased emphasis on the relationships of the synapsids and mammals and the great antiquity of the last common ancestor that they shared with the members of the Reptilia. In the Late Triassic, haramiyids, morganucodontids, and kuehneotheriids were members of at least the European fauna. More widespread sampling documents an increased taxonomic diversity in the Early Jurassic faunas of Laurasia and Gondwana. By the Middle Jurassic, most of the families that were represented in Laurasian Late Jurassic and Early Cretaceous faunas had differentiated. During the Cretaceous, mammals probably had a worldwide distribution, and endemic groups evolved as the Laurasian and Gondwanan supercontinents underwent major fragmentation. Rather than being entirely of Northern Hemisphere origin, some members of southern Cenozoic faunas (edentates, for example) might have originated in that hemisphere.

Acknowledgments

In preparation of this chapter, I have benefited from the help of many colleagues who have provided information about their studies of Triassic and Jurassic mammals. Also, discussions of problems of mammalian classification with many colleagues, particularly Mr. Tim Rowe, Ms. Nancy Simmons, Mr. Luo Zhexi, and other graduate students at Berkeley, have been most rewarding. *Nota bene*: They do not support all the conclusions presented here; however, I gratefully acknowledge their sometimes contentious but always constructive contributions to the refining of my interpretations of the available data. Mr. Luo Zhexi alerted me to and translated several papers reporting research carried on in the People's Republic of China. In similar fashion, Mr. George Shkurkin helped in bringing me up to date on pertinent publications of Russian colleagues. Some aspects of the research reported here were supported by National Science Foundation Grant BSR 81–19217 and the Annie M. Alexander Endowment, Museum of Paleontology, University of California Berkeley. All this assistance is acknowledged with my sincere thanks.

References

Bonaparte, J. F. 1984. Late Cretaceous faunal interchange of terrestrial vertebrates between the Americas. *In* Reif, W.-E., and F. Westphal (eds.), *Third Symposium on Mesozoic Terrestrial*

Ecosystems (Tübingen: ATTEMPTO), pp. 19–24.

Charig, A. 1979. *A New Look at the Dinosaurs* (London: William Heinemann Ltd.), pp. 1–160.

Chow, M. and T. H. V. Rich. 1982. *Shuotherium dongi*, n. gen. and sp., a therian with pseudo-tribosphenic molars from the Jurassic of Sichuan, China. *Austral. Mammal* 5:127–42.

—— 1984a. A new triconodontan (Mammalia) from the Jurassic of China. *J. Vert. Paleontol.* 3:226–31.

—— 1984b. New Mesozoic mammal sites from China. *Alcheringa* 8:304.

Clemens, W. A. 1979. A problem in morganucodontid taxonomy (Mammalia). *Zool. J. Linn. Soc. London* 66:1–14.

—— 1980. Rhaeto-Liassic mammals from Switzerland and West Germany. *Zitteliana* 5:51–92.

Clemens, W. A., J. A. Lillegraven, E. H. Lindsay, and G. G. Simpson. 1979. Where. when, and what — A survey of known Mesozoic mammal distribution. *In* Lillegraven, J. A., Z. Kielan-Jaworowska, and W. A. Clemens (eds.), *Mesozoic Mammals: The First Two-Thirds of Mammalian History* (Berkeley: University of California Press), pp. 7–58.

Colbert, E. H. 1981. The distribution of tetrapods and the break-up of Gondwana. *In* Cresswell, M. M. and P. Vellas (eds.), *Gondwana Five: Selected Papers and Abstracts of Papers Presented at the Fifth International Gondwana Symposium* (Rotterdam: A. A. Balkema), pp. 278–82.

Crompton, A. W. 1972. Postcanine occlusion in cynodonts and tritylodonts. *Bull. Brit. Mus. (Nat. Hist.).* 21:27–71.

Crompton, A. W., and J. A. Jenkins Jr. 1979. Origin of mammals. *In* Lillegraven, J. A., Z. Kielan-Jaworowska, and W. A. Clemens (eds.), *Mesozoic Mammals: The First Two-Thirds of Mammalian History* (Berkeley: University of California Press), pp. 59–73.

Datta, P. M. 1981. The first Jurassic mammal from India. *Zool. J. Linn. Soc. London* 73:307–12.

Datta, P. M., P. Yadagiri, and B. R. J. Rao. 1978. Discovery of Early Jurassic micromammals from Upper Gondwana sequence of Pranhita Godavari Valley, India. *J. Geol. Soc. India.* 19:64–8.

Dong, Z. 1980. China's dinosaurian faunas and faunal succession. *J. Strat., Acta Strat. Sinica.* 4:256–63 [in Chinese].

Evans, S. E. 1984. On the question of epipubic bones in morganucodontids. *J. Paleontol.* 58:1339.

Fraser, N. C., G. M. Walkden, and V. Stewart. 1985. The first pre-Rhaetic therian mammals. *Nature (London)* 314:161–2.

Freeman, E. F. 1979. A Middle Jurassic mammal bed from Oxfordshire. *Palaeontology* 22:135–66.

Goodrich, E. S. 1916. On the classification of the Reptilia. *Proc. Roy. Soc. London, Ser. B* 89:261–76.

Hahn, G., and R. Hahn. 1983. Multituberculata. *In* F. Westphal (ed.), *Fossilium Catalogus, I: Animalia* (Amsterdam: Kugler,), pp. 1–409.

Hahn, G., J.-C. Lepage, and G. Wouters. 1984. Cynodontier-Zähne aus der Ober-Trias von Medernach, Grossherzogtum Luxemburg. *Bull. Soc. Belg. Geol.* 93:357–73.

Hallam, A. 1981. The end-Triassic bivalve extinction event. *Palaeogeog., Palaeoclim., Palaeoecol.* 35:1–44.

Henkel, S., and G. Krusat. 1980. Die Fossil-lagerstätte in der Kohlengrube Guimarota (Portugal) und der erste Fuld eines Docodontiden-skelettes. *Berliner Geowiss. Abh. (A)* 20:209–14.

Hershkovitz, P. 1968. The recent mammals of the neotropical region: a zoogeographic and ecological review. In Keast, A., F. C. Erk, and B. Glass (eds.), *Evolution, Mammals, and Southern Continents* (Albany: State University of New York Press), pp. 311–431.

Hoffstetter, R. 1981. Historia biogeografica de los mamíferos terrestres sudamericanos: problemas y enseñanzas. *Acta Geol. Hisp.* 16:71–88.

Hopson, J. A., and W.-E. Reif. 1981. The status of *Archaeodon reuningi* von Huene, a supposed late Triassic mammal from southern Africa. *N. Jahrb. Geol. Palaont. Mh.* 1981:307–10.

Jenkins, F. A. Jr. 1984. A survey of mammalian origins. In Broadhead, T. W. (ed.), *Mammals, Notes for a Short Course* (University of Tennessee, Dept. of Geological Science, Knoxville, Tennessee), pp. 32–47.

Jenkins, F. A., Jr., and Crompton, A. W. 1979. Triconodonta. In Lillegraven, J. A., Z. Kielan-Jaworowska, and W. A. Clemens (eds.), *Mesozoic Mammals: the First Two-Thirds of Mammalian History* (Berkeley: University of California Press), pp. 74–90.

Jenkins, F. A., and R. Parrington. 1976. The postcranial skeletons of the Triassic mammals *Eozostrodon, Megazostrodon,* and *Erythrotherium. Roy Soc. London, Phil. Trans. Ser. B* 273:387–431.

Jenkins, F. A., Jr., A. W. Crompton, and W. R. Downs. 1983. Mesozoic mammals from Arizona: new evidence on mammalian evolution. *Science* 222:1233–5.

Kemp, T. S. 1982. *Mammal-Like Reptiles and the Origin of Mammals* (London: Academic Press).
1983. The relationships of mammals. *Zool. J. Linn. Soc. London* 77:353–84.

Kermack, D. E., and K. A. Kermack. 1984. *The Evolution of Mammalian Characters* (London: Croom Helm).

Kermack, K. A., F. Mussett, and H. W. Rigney. 1973. The lower jaw of *Morganucodon. Zool. J. Linn. Soc. London* 53:87–175.
1981. The skull of *Morganucodon. Zool. J. Linn. Soc. London* 71:1–158.

Kielan-Jaworowska, Z. 1982. Marsupial-placental dichotomy and paleogeography of Cretaceous theria. In Gallitelli, E. M. (ed.), *Paleontology, Essential of Historical Geology: Proc. First Intern. Meet. on "Paleontology, Essential of Historical Geology"* (Modena: S.T.E.M. Mucchi), pp. 376–83.
1984. Evolution of the Therian mammals in the Late Cretaceous of Asia. Part VII. Synopsis. *Paleont. Polonica.* 46:173–83.

Kindlimann, R. 1984. Ein bisher unerkannt gebliebener Zahn eines synapsiden Reptils aus dem Rät von Hallau (Kanton Schaffhausen, Schweiz). *Mit. Naturforsch. Gesell. Schaffhausen.* 32:3–11.

Krusat, G. 1980. Contribuição para o conhecimento da fauna do Kimeridgiano da mina de lignito Guimarota (Leiria, Portugal). IV Parte. *Haldanodon exspectatus* Kuhne & Krusat 1972. (Mammalia, Docodonta). *Mem. Serv. Geol. Portugal.* 27:1–79.

Lees, P. M., and R. Mills. 1983. A quasi-mammal from Lesotho. *Acta Palaeont. Polonica.* 28:171–80.

Lillegraven, J. A., Z. Kielan-Jaworowska, and W. A. Clemens (eds.), 1979. *Mesozoic Mammals: The First Two-Thirds of Mammalian History.* (Berkeley: University of California Press).

Lillegraven, J. A., M. J. Kraus, and T. M. Bown. 1979. *Paleogeography of the World of the Mesozoic. In* Lillegraven, J. A., Z. Kielan-Jaworowska, and W. A. Clemens (eds.), *Mesozoic Mammals: The First Two-thirds of Mammalian History* (Berkeley: University of California Press), pp. 277–308.

Marshall, L. G. 1980. Marsupial paleobiogeography. In Jacobs, L. L. (ed.), *Aspects of Vertebrate History* (Flagstaff, Arizona: Museum of Northern Arizona Press), pp. 345–86.

Marshall, L. G., C. De Muizon, and B. Sige. 1983. Late Cretaceous mammals (Marsupialia) from Bolivia. *Geobios* 16:739–45.

Matthew, W. D. 1915. Climate and evolution. *Ann. N. Y. Acad. Sci.* 24:171–318.

McKenna, M. C. 1975. Toward a phylogenetic classification of the Mammalia. In Luckett, W. P. (ed.), *Phylogeny of the Primates: a multidisciplinary approach.* (New York: Plenum), pp. 21–46.

Mills, J. R. E. 1984. The molar dentition of a Welsh pantothere. *Zool. J. Linn. Soc. London* 82:189–205.

Nessov, L. A. 1984. Concerning some discoveries of the remains of mammals from the Cretaceous deposits of middle Asia. *Vestnik Zoologii.* 2:60–5 [in Russian].

Olsen, P. E., and P. M. Galton. 1977. Triassic–Jurassic tetrapod extinctions: are they real? *Science* 197:983–6.

Olsen, P. E., J. F. Hubert, and K. A. Mertz. 1981. Comment and reply on "Eolian dune field of Late Triassic age, Fundy Basin, Nova Scotia." *Geology* 9:557–9.

Olsen, P. E., A. R. McCune, and K. S. Thomson. 1982. Correlation of the early Mesozoic Newark Supergroup by vertebrates, principally fishes. *Am. J. Sci.* 282:1–44.

Padian, K. and W. A. Clemens. 1985. Terrestrial vertebrate diversity: episodes and insights. In Valentine, J. W. (ed.), *Factors in Phanerozoic Diversity* (Princeton: Princeton University Press), pp. 41–96.

Parrington, R. 1978. A further account of the Triassic mammals. *Phil. Trans. Roy. Soc. London* 282B:177–204.

Patterson, B., and E. C. Olson. 1961. A triconodontid mammal from the Triassic of Yunnan. In Vandebroek, G. (ed.), *International Colloquium on the Evolution of Lower and Nonspecialized Mammals* (Kon. Vlaamse Acad. Wetensh., Lett. Schone Kunsten Belge, Brussels, Belgium), pp. 129–91.

Presley, R. 1981. Alisphenoid equivalents in placentals, marsupials, monotremes, and fossils. *Nature* (*London*) 294:668–70.

Prothero, D. R. 1981. New Jurassic mammals from Como Bluff, Wyoming, and the interrelationships of non-tribosphenic Theria. *Bull. Am. Mus. Nat. Hist.* 167:277–326.

Puffer, J. H., D. O. Hurtubise, F. J. Geiger, and P. Lechler. 1981. Chemical composition and stratigraphic correlation of Mesozoic basalt units of the Newark Basin, New Jersey, and the Hartford Basin, Connecticut: Summary. *Geol. Soc. Am. Bull.* 92:155–9.

Rasmussen, T. E., and G. Callison. 1981. A new species of triconodont mammal from the Upper Jurassic of Colorado. *J. Paleont.* 55:628–34.

Romer, A. S. 1966. *Vertebrate Paleontology*, 3rd ed. (Chicago: University of Chicago Press).

Savage, R. J. G. 1984. Mid-Jurassic mammals from Scotland. *In* Reif, W.-E., and F. Westphal (eds.), *Third Symposium on Mesozoic Terrestrial Ecosystems* (Tübingen: ATTEMPTO), pp. 211–13.

Sigogneau-Russell, D. 1978. Découverte de Mammifères rhétiens (Trias supérieur) dans l'est de la France. *Compt. Rend. Acad. Sci. Paris* 287:991–3.

1983a. A new therian mammal from the Rhaetic locality of Saint-Nicolas-de-Port (France). *Zool. J. Linn. Soc. London* 78:175–86.

1983b. Caractéristiques de la faune mammalienne du Rhétien de Saint-Nicolas-de-Port (Meurthe-et-Moselle). *Bull. Inf. Géol. Bas. Paris* 20:51–3.

1983c. Nouveaux taxons de mammifères rhétiens. *Acta Paleont. Polonica* 28:233–49.

Sigogneau-Russell, D., and Sun, A. 1981. A brief review of Chinese synapsids. *Geobios* 14:275–9.

Sigogneau-Russell, D., H. Cappetta, and P. Taquet. 1979. Le gisement rhétien de Saint-Nicolas-de-Port et ses conditions de depot. *7e Reunion Annuelle des Sciences de la Terre, Lyon. (Abst.)*, p. 429.

Simpson, G. G. 1960. Diagnosis of the Classes Reptilia and Mammalia. *Evolution.* 14:388–91.

Suess, H.-D. 1980. A pachycephalosaurid dinosaur from the Upper Cretaceous of Madagascar and its paleobiological implications. *J. Paleontol.* 54: 954–62.

Sun, A., G. Cui, Y. Li, and X. Wu 1985. A verified list of Lufeng saurischian fauna. *Vert. Palas.* 23:1–12 [in Chinese with English abstract].

Trofimov, B. A. 1978. Pervye trikonodonty (Mammalia, Triconodolta) iz Mongolii. *Akad. Nauk SSSR, Dokl.* 243:213–16 [in Russian].

Trofimov, B. S. 1981. *The first Triconodonts (Mammalia, Triconodonta) from Mongolia* (Sbripta Publ. Co., Silver Spring, Maryland) [*Akad. Nauk SSSR, Dokl.* 243:219–22, translation of Trofimov 1978].

Woodburne, M. O. 1984. Families of marsupials: relationships, evolution and biogeography. *In* Broadhead, T. W. (ed.), *Mammals, Notes for a Short Course* (Knoxville: University of Tennessee), pp. 48–71.

Woodburne, M. O., and W. J. Zinsmeister. 1982. Fossil land mammal from Antarctica. *Science* 218:284–6.

Wouters, G., J.-C. LePage, and P. Coupatez. 1983. Note préliminaire sur des dents d'aspect thérapside du Keuper supérieur du Grand-Duché de Luxembourg. *Bull. Soc. Belge Géol.* 92:63–4.

Wouters, G., D. Sigogneau-Russell, and J.-C. LePage. 1984. Découverte d'une dent d'Haramiyide (Mammalia) dans des niveaux rhetiens de la Gaume (en Lorraine belge). *Bull. Soc. Belge Géol.* 93:351–5.

Yadagiri, P. 1984. New symmetrodonts from Kota Formation (Early Jurassic) India. *J. Geol. Soc. India* 25:514–21.

Young, C. C. (Yang, Z.). 1982. Two primitive mammals from Lufeng, Yunnan. *Selected Works of Yang Zhungjian* (Beijing, People's Republic of China: Science Press), pp. 21–4.

Zhang, F. 1984. Fossil record of Mesozoic mammals of China. *Vert. Palas.* 22:29–38 [in Chinese with English abstract].

Zhang, F., and Cui, G. 1983. New material and new understanding of *Sinoconodon. Vert. Palas.* 21:32–41 [in Chinese with English abstract].

IV Early Jurassic vertebrate taxa and faunas

19 The early radiation and phylogenetic relationships of the Jurassic sauropod dinosaurs, based on vertebral anatomy

JOSE F. BONAPARTE

Introduction

In the past fifteen years, several discoveries and studies of prosauropods (Galton 1973, 1976; Bonaparte 1971, 1978) and primitive sauropods (Raath 1972; Jain et al. 1975, 1977; Ogier 1975; Berman and McIntosh 1978) have given a new perspective to further attempts to interpret the origins, systematics, and evolution of Jurassic sauropods. The discoveries of the Liassic *Barapasaurus tagorei* (Jain et al. 1975, 1977) and of the Callovian *Volkheimeria chubutensis* and *Patagosaurus fariasi* (Bonaparte 1979) have substantially improved the possibilities of building a reasonable picture of the systematics of the group. In addition, the discoveries of *Vulcanodon karibaensis* Raath (1972) in Zimbabwe and of undescribed advanced prosauropods in the Late Triassic of Argentina help to understand possible links between prosauropods and sauropods.

This attempt to update the rather static picture of the evolution and systematics of the Sauropoda is based on a comparative analysis of the presacral vertebrae of advanced prosauropods, several genera of Early and Middle Jurassic sauropods, and the best known Late Jurassic sauropods of North America and East Africa.

It appears that the anatomical and evolutionary characters of the vertebral column of sauropods are more variable and, therefore, more suitable to evaluate than are those characters from the appendicular skeleton. Actually, I hope that the still fresh sentence by the late Professor A. S. Romer (1968, p. 138) may soon be overcome: "It will be a long time, if ever, before we obtain a valid, comprehensive picture of sauropod classification and phylogeny."

I feel sure that an appropriate interpretation of the evolution of these enormous beasts, which wandered all around the world during most of the Mesozoic, may provide important evidence of the capabilities of land vertebrates to produce different lines of evolution and varied adaptive types within the realm of oversized quadrupeds.

Vertebral links between prosauropods and sauropods

The general type of dorsal vertebrae in Triassic archosaurs, and in particular in Late Triassic prosauropods, have centra without pleurocoels and a rather low neural arch, with the zygapophyses not far above the neural canal and with a laterally flat neural spine (see Fig. 19.3A). On the other hand, the dorsal vertebrae of the Early and Middle Jurassic sauropods, *Barapasaurus*, *Cetiosaurus*, and *Patagosaurus*, have a tall neural arch, with the zygapophyses far above the neural canal. A neural cavity just above the neural canal and below the zygapophyses, provided with lateral openings, has evolved. The neural spine is no longer short and laterally flat, but has become quite tall (Fig. 19.3C) and developed four divergent laminae. However, in the dorsal centra, true pleurocoels were not developed. As shown below, these very different types of dorsal vertebrae are phylogenetically related, with the first (Triassic prosauropod) type probably ancestral to the second (primitive sauropod) type.

The biomechanical and evolutionary reasons for such strong structural modifications of the ancestral prosauropod form into the primitive sauropod form of the dorsal vertebrae are not understood in detail at present. However, it is considered that the huge size developed by the Late Triassic prosauropods required strong anatomical modifications in the whole skeleton. According to the available fossil record, the anatomical modifications of adaptive features were achieved first in the cervicals and the appendicular bones (see following discussion of

Vulcanodon) and proceeded more slowly in the dorsal and sacral sections of the vertebral column.

The cervicals and anterior dorsals of some prosauropods exhibit significant modifications that approach the Early and Middle Jurassic sauropod vertebrae. In the cervicals (Fig. 19.1A), these modifications include the proportions and progressive slenderness toward the skull; the asymmetrical ventral curvature of the anterior cervical centra in lateral view; the median ventral keel in the anterior half; the pendant diapophyses; the low, axially elongate neural spine coalescent with the postzygapophyses; the depressed morphology ventral and anterior to the postzygapophyses; and the laminae running from the diapophyses toward the front and the rear. All of these characters represent a set of derived features in the cervicals of advanced prosauropods approaching the anatomical structure of Jurassic sauropods (Fig. 19.1B,C).

In the first dorsal vertebrae of advanced prosauropods (Fig. 19.2A), we can see again several sauropod-like characters: tall but axially short neural arches; high and inclined zygapophyses; high and axially short neural spines; well-developed suprapostzygapophyseal laminae forming a deep posterior axial depression between them; and the anterior side of the neural spine free of laminae, long, and transversely convex. In posterior view, large infrapostzygapophyseal cavities are present, and, in the anterior view, small infraprezygapophyseal cavities are present. The diapophyses project ventrolaterally, and a posterior lamina connects them to the postzygapophyses (Fig. 19.2). Most of these characters are very much as in the third or fourth dorsal of the cetiosaurid *Patagosaurus fariasi* (Fig.19.2B).

The posterior dorsals of the prosauropods (Fig. 19.3A) do not show any evidence of morphological approach to the corresponding vertebrae of Jurassic sauropods (Fig. 19.3B,C). They are low, with laterally compressed centra and low neural

Figure 19.1. **A**, Posterior cervicals of an advanced prosauropod from the Los Colorados Formation of Argentina. **B**, The cetiosaurid *Patagosaurus fariasi*. **C**, The diplodocid *Diplodocus*. Not to scale.

Figure 19.2. Anterior dorsals of the taxa represented in Figure 19.1.

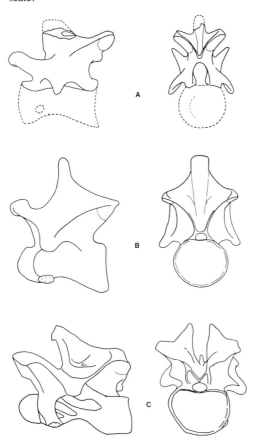

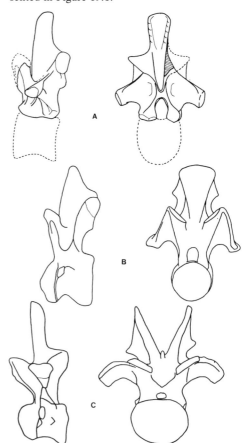

arches. The zygapophyses are near the neural canal, and bear well–developed hypantra and hyposphenes, and the neural spine is low and laterally flat. The transverse processes bear short infradiapophyseal, parapophyseal–diapophyseal, and prezygapophyseal–diapophyseal laminae. Most of these characters are present not only in prosauropods, but also in other Triassic archosaurs, such as ornithosuchids and rauisuchids.

It seems reasonable to me to conclude that the evolution of sauropod characters in the vertebrae of advanced prosauropods proceeded first in the cervicals and anterior dorsals, while the central and posterior dorsals remained anatomically more conservative. When comparisons are made to Middle and Late Jurassic sauropod vertebrae, it is easy to realize that the posterior dorsals of some families (Cetiosauridae, Diplodocidae, Camarasauridae, and Brachiosauridae) exhibit more generalized, primitive features than do the central and anterior dorsals, where several derived characters have replaced them. Phylogenetically, the information from the

Figure 19.3. **A**, Posterior dorsals of the prosauropod *Plateosaurus*. **B**, The cetiosaurid *Volkheimeria chubutensis*. **C**, The cetiosaurid *Patagosaurus fariasi*. Not to scale.

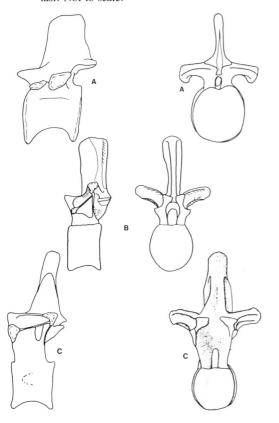

vertebral anatomy of prosauropods and primitive sauropods leaves no room for doubt (Cruickshank 1975) that they are very closely related groups, as several authors have stated.

The basis of this study is the analysis and interpretation of the vertebral morphology of Jurassic sauropods. It is assumed that the peculiar and varied osteological features of the vertebrae represent a diagnostic and reliable set of characters that, if properly interpreted, may give results that, as far as I know, are not possible to obtain through the study of appendicular bones.

This chapter suggests that before the evolution of the advanced, well–known sauropods from the Late Jurassic Morrison Formation of North America and the Tendaguru Beds of Tanzania, three previous, primitive stages of the evolution of the sauropods may be recognized. The earliest of them is probably represented by *Vulcanodon karibaensis* Raath (1972) from the Triassic–Jurassic boundary or possibly the Liassic; the next by the Middle Jurassic species of "*Bothriospondylus* sp." (Ogier 1975) from Madagascar, and by *Volkheimeria chubutensis* Bonaparte (1979) from Patagonia; and the third stage by the typical cetiosaurs *Barapasaurus tagorei* Jain et al. (1975), *Patagosaurus fariasi*, and both species of *Cetiosaurus* from England.

Vulcanodon karibaensis Raath

This important species from the Triassic–Jurassic boundary (or Liassic) of Zimbabwe was described by Raath in 1972. Raath's interpretation of this species is not clear because he recognized affinities with sauropods of Late Jurassic and Cretaceous age (Raath, 1972, p. 36), but at the same time admitted that *Vulcanodon karibaensis* was provided with carnivorous teeth (Raath, 1972, p. 28), and concluded that the described species may be interpreted as a prosauropod. Such a complex assumption, although in accordance with the prevailing knowledge of the 1970s (cf. Charig, Attridge, and Crompton 1965), does not agree with what is known at present of the general anatomy of the Prosauropoda and Sauropoda. The discovery of a carnivorous (rauisuchid or theropod) tooth among the remains of *Vulcanodon* is a rather common finding in most of the excavations of semiarticulated sauropodomorphs. The tooth should be considered to belong not to the sauropodomorphs, but to a predator from one of the groups of carnivorous archosaurs.

The most significant features of *Vulcanodon* are of the sauropod type. These include the pubes (Raath 1972, Fig. 5), which are anatomically even more advanced than in some primitive sauropods, such as *Barapasaurus* (Jain et al. 1977, Fig. 97B). They have proximolateral projections that make them proportionally wider in the proximal region than in primitive sauropods, and even more different from the subparallel, flat pubes of the Prosauropoda. The ischia exhibit an interesting sauropod feature in their proximal area, the reduction of the medioventral projection of the symphyseal laminae, which reduces the space between the sacrum and the puboischiadic laminae.

Another significant, definitive sauropod feature is seen in the elongate, strong metacarpals, which are nearly as long as the metatarsals. This size ratio discards the possible assignment of *Vulcanodon* to the Prosauropoda, and strongly suggests that it is a sauropod.

Finally, another decisive bone suggesting sauropod affinities is the astragalus. In the Prosauropoda, it is almost rectangular in dorsal view, with a wide internal side; but in *Vulcanodon* it is roughly triangular, with a wide external side and a pointed, narrower internal side, as it is in most Jurassic and Cretaceous sauropods.

The incomplete sacral and caudal vertebrae of *Vulcanodon*, as far as can be seen in the figures and photographs given by Raath, and following his own interpretation of them, do resemble the prosauropod much more than the sauropod condition. The relatively elongate and low centra with lateral, wide depressions very much resemble the prosauropod condition (Huene 1907–8; Bonaparte 1971; Galton 1976). In the Liassic *Barapasaurus* and the Bathonian and Callovian *Cetiosaurus* and *Patagosaurus*, the first caudal centra are axially short and dorsoventrally high, with some indication of slight procoely. So, as far as the anterior caudals of *Vulcanodon* are concerned, they suggest that this part of the vertebral column and probably the posterior section of the sacrum are of the prosauropod type.

Although the absence of any dorsal vertebrae strongly limits a definite systematic assessment of the family level of *Vulcanodon karibaensis*, I consider that the evidence of its anterior caudals of prosauropod type, together with several sauropod characters in the pubis, ischium, tarsus, and metacarpals, represent a unique set of primitive and derived characters that may support the proposal to distinguish a new family for this primitive sauropod. As shown below, the implications of the anatomy of *Vulcanodon* for the knowledge of the early evolution of the sauropods are significant.

The anterior caudals of *Vulcanodon* differ in many respects from those of *Cetiosaurus*, *Barapasaurus*, *Patagosaurus*, and other Middle Jurassic sauropods, such as "*Bothriospondylus* sp." and *Volkheimeria chubutensis*, but, at the same time, have many affinities with the prosauropod type. These characteristics strongly suggest that the sauropod characters of the pelvis and limbs were fully developed at a stage when the caudal and sacral section of the vertebral column were still nearer to the prosauropod than to the primitive sauropod condition. The *Vulcanodon* stage appears to be more primitive than the two further stages of primitive sauropods represented within the Cetiosauridae.

The Cetiosauridae

The Cetiosauridae Lydekker (1888) represent a primitive family, perhaps the basal stock of the Sauropoda, as stated by Steel (1970) and several authors previously. The cetiosaurids are recorded mainly from the Early and Middle Jurassic.

Some of the Late Jurassic genera referred to the Cetiosauridae or Cetiosaurinae are probably not part of this family. "*Elosaurus parvus*" Peterson and Gilmore (1902) is actually a juvenile, considered by McIntosh (1981) to be an immature specimen of the genus *Apatosaurus*. *Haplocanthosaurus priscus* Hatcher (1906), generally considered to belong to the Cetiosaurinae (Steel 1970) or Cetiosauridae (McIntosh 1981), has a very specialized vertebral anatomy, quite different from *Cetiosaurus*, *Barapasaurus*, and *Patagosaurus*, and it is considered here nearer to the Dicraeosauridae than to any other Jurassic sauropod family.

In this chapter, we consider the cetiosaurs a family of the Sauropoda, as did McIntosh (1981), and the constituent genera afford good anatomical basis for a confident assignment to the family. They are *Cetiosaurus*, *Barapasaurus*, *Amygdalodon*, *Lapparentosaurus* nov., *Patagosaurus*, *Volkheimeria*, and possibly *Rhoetosaurus* Longman (1926). Within these genera, the structure of the neural arches of the dorsal and sacral vertebrae suggests that two successive stages are present. The first is less specialized: Its neural spines are laterally flat, and the neural arch below the zygapophyses is relatively low. This stage is represented by the genera *Lapparentosaurus* nov. (discussed in the following pages) and *Volkheimeria* (Fig. 19.3B). The second stage, with elongated, tetraradiate neural spines and high neural arches (Fig. 19.3C), is represented by *Cetiosaurus*, *Barapasaurus*, *Amygdalodon*, *Patagosaurus*, and possibly *Rhoetosaurus*. The two stages of specialization of vertebral anatomy present in the cited genera are considered, not without doubt, to belong to the same family on the grounds that several other characters of the centra and neural arches, as well as most of the postcranial skeleton, are common to both stages.

"Bothriospondylus sp." Ogier and Volkheimeria chubutensis Bonaparte

The generic name *Bothriospondylus* was proposed by Owen (1875) for the species *B. suffosus*, based on four dorsal centra from the Kimmeridgian of England. Later, Lydekker (1895) used the same generic name for several vertebral centra from the "Bathonian?" of the Narinda Bay, northwestern Madagascar, on which he based the new species *Bothriospondylus madagascariensis*. In 1907, Thevenin published several incomplete remains of sauropods from the Narinda Bay, near Analabava, from beds considered Bathonian, and referred them to *B. madagascariensis*.

The generic identification of the Bathonian sauropod remains from Madagascar with the vertebral centra of *Bothriospondylus* from the Kimmeridgian of England has no serious basis. The presence of deep pleurocoels, the only character used by Lydekker (1895) to recognize the same genus for the vertebral centra of Madagascar, has no systematic value at the generic level for elements at very different positions in the vertebral column. To avoid misleading interpretations of the systematics and of the stratigraphic and geographic paleodistribution of Jurassic sauropods, and because the Malagasy species recently described by Ogier (1975) is so distinct, we cannot accept the generic identification of Lydekker (1895). So, the Bathonian sauropod remains from Madagascar studied by Lydekker (1895) and by Thevenin (1907), and an almost complete juvenile specimen studied by Ogier (1975), are considered here to represent a new genus, *Lapparentosaurus*, with the species name *L. madagasmadagascariensis*.

Infraorder Sauropoda
Family Cetiosauridae

Lapparentosaurus nov. gen.

Holotype. Two associated neural arches of posterior dorsals (Fig. 19.4). Collection: Muséum Nationale d'Histoire Naturelle, Paris, MAA. 91–92.

Locality and age. South of the Kamoro River, near Andranomamy, northern Madagascar; Bathonian (Facies Isalo III).

Diagnosis. Cetiosaurid with rather flat neural spine, without divergent laminae present in *Cetiosaurus*, *Barapasaurus*, and *Patagosaurus*.

Lapparentosaurus madagascariensis nov. sp.

Diagnosis. As for the genus.

Comment. The type specimen is part of a collection of at least five incomplete specimens, made in five small hills very near one another. The collected material, described by Ogier (1975), affords good information about the general anatomy of this primitive cetiosaur.

Lapparentosaurus is a member of the Cetiosauridae, according to the general morphology of well–known cetiosaurids, such as *Barapasaurus ta-*

gorei from the Liassic of India (Jain et al. 1975, 1977), *Patagosaurus fariasi* (Bonaparte 1979, pers. obs.) and *Cetiosaurus* itself, but with primitive development of the neural arches of the posterior dorsal and sacral vertebrae. The neural spine is relatively low, laterally flat, and without the divergent laminae present in the above cited genera. The relative distance between the zygapophyses and the contact with the centrum is short, with poor dorsoventral development of the neural arch (see Fig. 19.4).

Volkheimeria chubutensis Bonaparte (1979) is a cetiosaurid from the Callovian of Patagonia, which is represented by one incomplete specimen with well-preserved dorsal and sacral vertebrae, plus some complete bones of the girdles and limbs. The morphology of the neural arches of the posterior dorsal and sacral vertebrae is basically of the same type as in *Lapparentosaurus*, with simple, laterally flat neural spines and with rather low neural arches in the area between the zygapophyses and the centra, but with laterally open neural cavities. It is probable that both *Lapparentosaurus* and *Volkheimeria* rep-

Figure 19.4. *Lapparentosaurus madagascariensis* nov. sp. Two posterior dorsal neural arches in anterior, posterior, and lateral views, with broken diapophyses, from the Bathonian of Madagascar. After Ogier (1975).

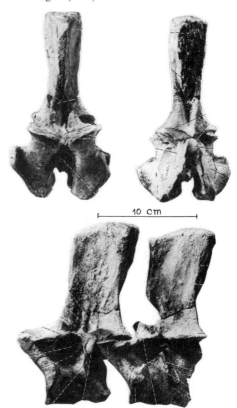

10 cm

resent a more primitive stage in the specialization of the neural arch than do *Barapasaurus*, *Cetiosaurus*, and *Patagosaurus*. These last three genera have a neural arch that is tall and a neural spine that is not flatter but is made of four divergent laminae. As shown before, it is assumed that in the evolution of primitive sauropods the relative height of the whole neural arch increased, and the neural spine overcame the laterally flat condition present in prosauropods to become tetraradiate in advanced cetiosaurids.

Summing up, the primitive cetiosaurids *Lapparentosaurus madagascariensis* and *Volkheimeria chubutensis* represent an early stage in the development of sauropod vertebrae, probably more specialized than in *Vulcanodon karibaensis* but more primitive than in typical cetiosaurids such as *Cetiosaurus*.

The cetiosaurids Barapasaurus, Cetiosaurus, and Patagosaurus

At present, not a single dorsal vertebrae of *Cetiosaurus* has been properly figured. However, the one illustrated by Phillips (1871, Fig. 86), although distorted, exhibits the typical features of the central dorsals of Cetiosauridae: the neural spine with large lateral depressions, the wide anterior concavity below the prezygapophyses and between the parapophyses, and the elongated neural arch with a significant distance between the zygapophyses and the neural canal.

The genus *Cetiosaurus* requires review and updating, especially now that some additional well-preserved remains tentatively referred to the genus have been recorded. However, the vertebral morphology of Owen's genus *Cetiosaurus*, and what is known of significant postcranial bones, such as the scapula, humerus, pubis, and ischium, agree very well with corresponding pieces of *Barapasaurus* and *Patagosaurus*; therefore, I conclude that these genera belong to the same family.

Barapasaurus tagorei, from the Liassic beds of the Pranhita–Godavari Valley, India, was briefly diagnosed by Jain et al. (1975) based on several incomplete specimens that afforded good information about most of the postcranial skeleton. In 1977, Jain et al. described and figured a good part of the postcranium, with valuable descriptions and figures of dorsal vertebrae. They recognized the presence of the neural cavity, inside the neural arch just above the neural canal, in the dorsal vertebrae of *Barapasaurus*.

The *Barapasaurus* vertebrae include typical sauropod cervicals, opisthocoelous, with elongated pleurocoels, pendant diapophyses, rather low and not bifurcated neural spines, and axially elongated and wide laminae connecting the neural spine to both pre– and postzygapophyses and connecting the latter with the diapophyses.

The transition between the cervical and dorsal types occurs in the anterior dorsals, which are opisthocoelous, axially short and high, and with deep lateral depressions (pseudopleurocoels) in the centra. The zygapophyses are in a high position; the neural spine is high, axially very short, and connected to the postzygapophyses only by the posterior laminae. Strong diapophyses have connecting laminae to both pre– and postzygapophyses.

The posterior dorsals are proportionally high and amphiplatyan, with modest lateral depressions in the centra. The zygapophyses are in a high position, and an internal cavity inside the neural arch is present from above the roof of the neural canal up to near the level of the zygapophyses, enclosed in bone all around except for a lateral opening below each transverse process. The anterior face of the neural arch between the zygapophyses and the top of the centrum is very concave, laterally bordered by both parapophyseal projections. The opposite side, in the rear of the neural arch, is roughly convex. The neural spine is high, with four dorsoventral depressions, and those on the lateral sides are more pronounced than those in front. These depressions are made by four divergent laminae running dorsoventrally. The general morphology of the *Barapasaurus* vertebrae is very much like that of *Patagosaurus*, with some slight differences that are probably not greater in the *Cetiosaurus* vertebrae.

Patagosaurus fariasi Bonaparte (1979), from the Callovian beds of Patagonia, Argentina, is based on several incomplete skeletons representing most of the postcranium, some teeth, a premaxilla, and

Figure 19.5. *Patagosaurus fariasi*. Central dorsals in lateral and anterior views. Total height 73 cm.

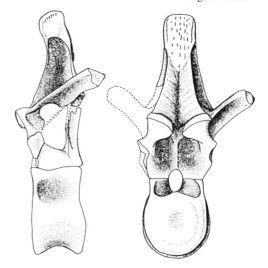

two maxillae. The vertebrae (Figs. 19.3C, 19.5) are very much as in *Barapasaurus*, but the opisthocoelia extend more toward the middle dorsals, the neural spine is proportionally higher, the pubes are proximally wider, and the sacrum bears an enormous expansion of the neural canal in the second, third, and fourth sacrals.

These three genera, *Cetiosaurus, Barapasaurus*, and *Patagosaurus*, show the same general level of organization of the vertebral column. Although they have almost the same general osseous anatomy as *Lapparentosaurus* and *Volkheimeria* discussed above, a noticeable difference is seen in the neural spine, and it leads us to assume that these two genera are not as specialized in the development of the neural spine as are *Cetiosaurus, Barapasaurus*, and *Patagosaurus*. The tetralaminate neural spine of the latter three genera represents a set of derived characters that is not present in *Lapparentosaurus* and *Volkheimeria*.

It is considered that the vertebral anatomy of the Cetiosauridae, and in particular that of the advanced three genera, is probably ancestral to the vertebral anatomy present in Diplodocidae, Camarasauridae, and Brachiosauridae, and with some doubts to the Dicraeosauridae and *Haplocanthosaurus*.

The Diplodocidae

This family was partially reviewed by Berman and McIntosh (1978) in their study of *Apatosaurus*. As far as the scope of this chapter permits, I consider that the Jurassic sauropods that can be properly placed in the Diplodocidae are *Diplodocus, Barosaurus*, and *Apatosaurus*. Other genera, such as *Cetiosauriscus* and *Mamenchisaurus*, included by Berman and McIntosh within the family, may be correctly assigned, but as far as I know we do not have the evidence from the dorsal vertebrae, which are, as stated by Berman and McIntosh (1978; p. 33), "very diagnostic among the sauropods." Finally, the tentative inclusion of *Dicraeosaurus* in the Diplodocidae by Berman and McIntosh appears untenable because the well-known morphology of the dorsal vertebrae of *Dicraeosaurus* (Janensch 1929) is quite different from that of the genera *Diplodocus, Barosaurus*, and *Apatosaurus*, as shown in this chapter.

In its vertebral anatomy, *Diplodocus* Hatcher (1901) is probably the least specialized genus of the family, except for the strong development of the pleurocoel system (Fig. 19.6). *Barosaurus* developed highly derived characters in the elongated cervicals and in the cervicalization of the anterior dorsals (Berman and McIntosh 1978). *Apatosaurus* developed a set of derived characters in the para- and diapophyses of the cervicals and in the capitulum and tuberculum of the cervical ribs, which resulted

in a sort of horizontal symmetry (seen from the front) of bifurcated neural spines above and well-developed parapophyses and capitula below.

The cervical vertebrae of *Diplodocus* very much resemble the vertebrae of the Cetiosauridae, except that *Diplodocus* has more pronounced pleurocoels and a system of trabeculae, features that can easily be understood as derived from those of the Cetiosauridae.

The bifurcation of the neural spines of sauropods is a derived character, and its functional reason is not easy to understand (see Janensch 1929, 1950). However, the bifurcation itself did not conceal the main character states of the vertebrae, and it appears to be a feature with no clear phylogenetic meaning.

The dorsal vertebrae of *Diplodocus* (Fig. 19.6) resemble, in a significant way, the vertebrae of the cetiosaurid *Patagosaurus*. They are proportionally tall and axially short, with a long distance between the zygapophyses and the base of the neural arch. The portion of the neural arch below the transverse processes is very much as in *Patagosaurus*, and the system of laminae connecting the diapophyses with different parts of the neural arch is basically the same. Most of the dorsal vertebrae of *Diplodocus* retain parapophyseal–diapophyseal laminae, which in all other Late Jurassic sauropods have vanished. The cavity of the neural arch just above the neural canal, which is present at least in the posterior dorsals of *Diplodocus*, shows great similarity to the same cavity of *Patagosaurus* (Fig. 19.7A,C).

Finally, the structure and morphology of the neural spines of the dorsal vertebrae of *Diplodocus*, in lateral view, are basically the same as in *Patagosaurus*, with similar laminae involved in their structure, up to the sixth dorsal (see Fig. 19.6). Posteriorly *Diplodocus* has a lateral lamina running up

Figure 19.6. *Diplodocus carnegii*. Sixth to ninth dorsal vertebrae in lateral view. After Hatcher (1901).

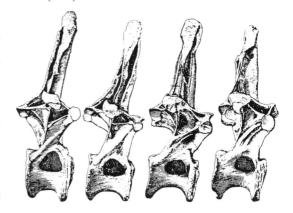

the neural spine, the supradiapophyseal lamina, which in *Patagosaurus* is just beginning to develop, so it is quite small.

Summing up, it is considered that *Diplodocus* and the Diplodocidae probably evolved from the Cetiosauridae. The derived characters developed in the vertebrae of *Diplodocus* did not depart very much from its forebears, and several primitive characters are still present. The supposed relationships of the genera of the Diplodocidae are shown in Figure 19.8.

Figure 19.7. Schematic sections of the neural arch just above the neural canal (nc), to show morphological changes. **A**, The cetiosaurid *Patagosaurus fariasi*. **B**^{a}, anterior and **B**^{b}, posterior dorsals of the camarasaurid *Camarasaurus* sp. **C**, Posterior dorsal of the diplodocid *Diplodocus carnegii*. **D**, Posterior dorsal of *Haplocanthosaurus* sp. Not to scale.

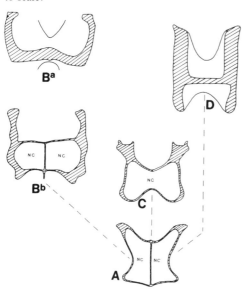

The Camarasauridae

In this chapter, it is considered that the Camarasauridae are represented only by the Late Jurassic genus *Camarasaurus*. The vertebral column of this genus developed several derived characters that suggest a strong departure from the assumed ancestral family, the Cetiosauridae. In general terms, the presacral vertebrae of *Camarasaurus* (Osborn and Mook 1921) are low, some of them very low, and particularly wide because of the development of large, strong transverse processes. This is easily seen from the sixth cervical backward, with some variants up to the last dorsal. In addition, the opisthocoelia are strongly developed in *Camarasaurus* and extend to the last dorsal. The neural pedicels of the centra are strongly developed (Fig. 19.9) and frequently enclose the neural canal completely. In most of the dorsals, the anterior face of the neural arch, below the prezygapophyses, shows the concavity present in the Cetiosauridae but modified into a deep fossa, with angular lateral walls (Fig. 19.9). On the opposite side of the neural arch, that is to say in the back, the convex area present in the cetiosaurids has become a concave, depressed region with a median vertical ridge (see Fig. 19.7B^b). These modifications of the neural arch strongly affected the neural cavity above the neural canal. In the anterior and middle dorsals, the cavities were probably obliterated (Fig. 19.7B^a), or they migrated dorsally, and in the posterior dorsals they were significantly reduced.

The neural spine is extremely reduced dorsoventrally, and the spines are bifurcated up to the sixth and seventh dorsals. The neural spine of the dorsals, although reduced, is rather generalized in structure, without development of axial spinal laminae (Fig. 19.9).

Several characters of the *Camarasaurus* dorsal vertebrae suggest that the cetiosaurid type of ver-

Figure 19.8. Possible phylogenetic relationships among the Jurassic Diplodocidae based on vertebral structure.

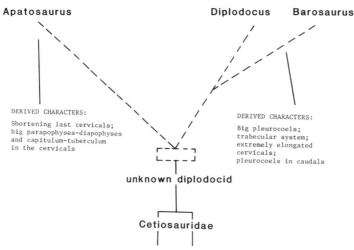

tebrae may be the ancestral model from which they were derived. Compared to the ancestral condition, *Camarasaurus* is a very unusually modified type, which diverged markedly from its ancestors. As shown below, several characters of the dorsal vertebrae of *Camarasaurus* suggest some kind of relationship with the Brachiosauridae.

The Brachiosauridae

This family represents an extreme adaptive type, huge, with enormous cervicals, and with the forelimbs much longer than the hindlimbs. *Brachiosaurus*, recorded in the Late Jurassic of North America (Riggs 1904) and East Africa (Janensch 1950), appears to be the only Jurassic genus that we can properly refer to the family.

The cervical vertebrae are hypertrophied and cavernous in structure. Cervical and dorsal neural spines are not bifurcated. The dorsal vertebrae have proportionally large, elongated centra, and rather low but axially elongated neural arches. The opisthocoelia are well developed up to the last dorsal in *B. brancai* and up to the penultimate dorsal in *B. altithorax* (Fig. 19.10). The neural spines of the sacrals and last dorsals are short, but gradually increase in

height toward the front. The spine has an axially large base and is made of dorsal projections of the suprapre- and suprapostzygapophyseal laminae (Fig. 19.10). In addition, the supradiapophyseal laminae are present in each lateral side of the spine. Finally, incipient prespinal and postspinal laminae are also present to reinforce the neural spine.

An evolutionary relationship between the Camarasauridae and the Brachiosauridae is very probable, based on morphological comparisons of their vertebrae, and there is no anatomical problem with the possibility that the cetiosaurids included the ancestors of the Brachiosauridae.

Although the dorsal vertebrae of *Brachiosaurus* represent a peculiar adaptive type, some features deserve comparison with the *Camarasaurus* vertebrae:

1. The opisthocoelia are developed to a similar extent in both genera.
2. The neural spines are proportionally low, with modest development of the supradiapophyseal laminae.
3. The neural spines of the last dorsals show poor development of the axial spinal laminae frequently developed in other sauropods.

It is difficult to ascertain the phylogenetic meaning, if any, of these few affinities between the dorsal vertebrae of *Brachiosaurus* and *Camarasaurus*. In addition, the similar dentition of both genera suggests that if convergence cannot be demonstrated, some type of relationship between the two families may be assumed.

Figure 19.9. *Camarasaurus* Cope. Last dorsal of *Camarasaurus supremus* in anterior and lateral views, after Osborn and Mook (1921); and posterior dorsal of *Camarasaurus grandis* in anterior view, after Ostrom and McIntosh (1966).

Figure 19.10. *Brachiosaurus* Riggs. Last dorsal vertebra in posterior, anterior, and lateral views. After Janensch (1950).

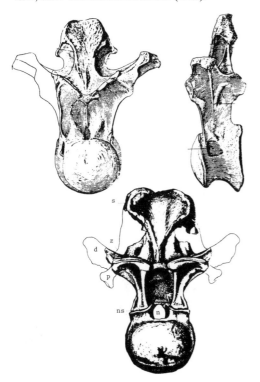

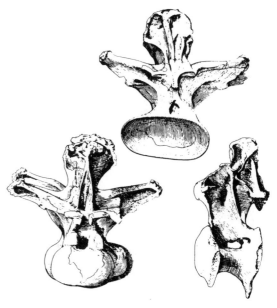

As summarized in the phylogenetic chart of Figure 19.13, it seems to me that the Brachiosauridae and Camarasauridae might be considered closer to each other than either of them is to the Diplodocidae, Dicraeosauridae, or *Haplocanthosaurus*.

The Dicraeosauridae

Huene (1956) recognized this family, placing in it only the genus *Dicraeosaurus* from the Late Jurassic of East Africa, described by Janensch (1929). The distinct anatomy of the vertebrae of this genus supports the erection of a family of its own. In lateral view, the cervical vertebrae of *Dicraeosaurus* exhibit very elongate neural spines that in anterior view are deeply bifurcated. Apart from the differences of the neural spines, however, the cervical vertebrae are similar to those of cetiosaurids, with less pronounced lateral cavities in the centra.

Along the dorsal vertebrae, the lateral depressions of the centra are very shallow (Fig. 19.11), as in the last cervicals. I consider that this feature may be a derived character for *Dicraeosaurus*. The neural arches of the dorsals have upwardly projecting trans-

verse processes, as in *Haplocanthosaurus*, with a modest development of the infradiapophyseal laminae. The parapophyseal–diapophyseal laminae become well developed and replace the prezygapophyseal–diapophyseal laminae in most of the dorsals. This feature is again a derived character for both *Dicraeosaurus* and *Haplocanthosaurus* (Figs. 19.11, 19.12).

The lower half of the neural arch is transversely reduced in the posterior dorsals of *Dicraeosaurus*. A significant anteroposterior reduction has also developed in this part of the neural arch, making deep anterior and posterior depressions laterally bordered by bone (Fig. 19.11). This morphology is also present in *Haplocanthosaurus* Hatcher (1906), and a transverse section of this part of the neural arch of the latter genus is shown in Figure 19.7. Such reduction is quite different from the morphology of this part of the neural arch in the Cetiosauridae, Diplodocidae, Camarasauridae, and Brachiosauridae, as shown in Figure 19.7, and probably represents a common trend in *Dicraeosaurus* and *Haplocanthosaurus*. The neural spines of *Dicraeosaurus* (Fig. 19.11) show a strong departure from the ancestral type present in cetiosaurids, and their specialized condition is in accord with the advanced morphology of the centra and the neural arch. In the last dorsals, where the spines are not affected by bifurcation, it is possible to see that the supraprezygapophyseal laminae have a modest participation in the spine. Most of the medial axial laminae are made of anterior and posterior spinal laminae, running most of the length of the spines except in the

Figure 19.11. *Dicraeosaurus* Janensch. Last dorsal vertebrae in lateral and anterior views. After Janensch (1929).

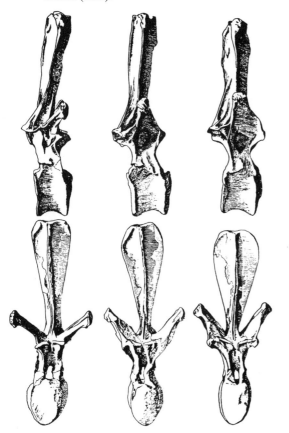

Figure 19.12. *Haplocanthosaurus* Hatcher. Last dorsal vertebrae in lateral, anterior, and posterior views. After Hatcher (1906).

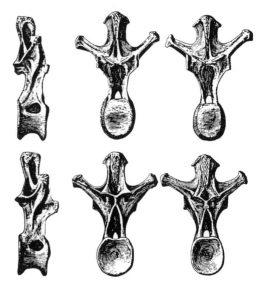

lower part. This type of neural spine, although morphologically different from that of *Haplocanthosaurus*, is formed from the same components: spinal laminae, supradiapophyseal laminae, and (to a much lesser degree) suprazygapophyseal "laminae."

The phylogenetic relationships of *Dicraeosaurus* are difficult to trace. However, I think that the general anatomy of the cervicals, elongated neural spines aside, is the same as in the Cetiosauridae, which may be a good indication of its probable origin. It seems reasonable to assume that *Dicraeosaurus* followed an evolutionary trend to develop long, bifurcated neural spines along most of the vertebral column and to reduce the size and depth of the pseudopleurocoels.

The similarities of some aspects of the vertebral anatomy of *Dicraeosaurus* and *Haplocanthosaurus*, such as the elongation of the neural arches below the zygapophyses, the upwardly projecting transverse processes, the reduced transverse width of the lower part of the neural arch, the similar section across this part of the neural arch, and the dominance of the axial spinal laminae over the suprazygapophyseal laminae, are significant, and these similarities suggest that these genera may be part of a lineage that produced two distinct adaptive types.

On this basis, it is suggested that the Dicraeosauridae may include *Haplocanthosaurus* as a genus that exhibits several common features with *Dicraeosaurus*, but, at the same time, reveals significant differences in the shape (not structure) of the neural spine.

Figure 19.13. Phylogenetic chart of some Jurassic sauropods based on the vertebral anatomy. The presacral vertebrae of *Vulcanodon* are not known. The teeth of *Haplocanthosaurus* are not known.

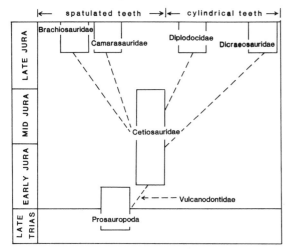

Conclusions

1. It is suggested that the origin of the Sauropoda from the advanced Prosauropoda of the Late Triassic is supported by the complex anatomy of the cervicals and anterior dorsals of Melanorosauridae from the Los Colorados Formation of Argentina.

2. It is recognized that the specialization of the vertebral column in advanced prosauropods and primitive (if not all) sauropods proceeded from the anterior cervicals backward, and that the last dorsals, although affected by the morphological changes, are the most conservative of the presacral vertebrae.

3. It is suggested that the first stage of the evolution of the Sauropoda is represented by the primitive family Vulcanodontidae (containing only *Vulcanodon*), which exhibits sauropod features in the appendicular bones and prosauropod features in the anterior caudal and posterior sacral vertebrae.

4. The Cetiosauridae is probably not the earliest stage of development of the Sauropoda. They exhibit a rather uniform evolutionary stage in the presacral vertebral anatomy, except in the neural spines. The genera *Lapparentosaurus* and *Volkheimeria* have laterally flat neural spines. The genera *Cetiosaurus*, *Barapasaurus*, and *Patagosaurus* have tetraradiate neural spines formed of laminae, suggesting a more advanced condition for this part of the vertebrae.

5. It is concluded that the Cetiosauridae, ranging from the Liassic up to the Middle Jurassic, is the ancestral stock from which radiated the families Camarasauridae and Brachiosauridae, retaining the spatulate type of teeth of the Cetiosauridae (Fig. 19.13); the Diplodocidae, with not so strong a modification in the dorsal vertebrae but acquiring the cylindrical type of teeth; and in a divergent evolutionary direction, the Dicraeosauridae (tentatively including *Haplocanthosaurus*), with strongly derived characters in the dorsal vertebrae and, at least in the Dicraeosauridae, with the cylindrical type of teeth.

Acknowledgments

This study was begun at the Carnegie Museum of Natural History through a grant to visit that institution in 1982, and completed at the Field Museum of Natural History of Chicago after a grant from the Visiting Scientists Program of that Museum in 1984. The field work was made possible by grants from the National Geographic Society, the Consejo Nacional de Investigaciones Científicas y Técnicas, and the Museo Argentino de Ciencias Naturales de Buenos Aires. Very fruitful discussions were held with Dr. John McIntosh. Dr. Kevin Padian revised much of the

English and made a number of editorial suggestions on the manuscript. Dr. Mary Dawson (Carnegie Museum) permitted me to use untouchable specimens in order to draw sections of neural arches. Dr. Leonard Ginsburg of Paris, with open generosity, permitted me to name a sauropod specimen he collected in Madagascar. My friends Dr. H. Jaeger of the Naturkunde Museum of East Berlin and Dr. A. Charig of the British Museum (Natural History) permitted to see and study everything I asked for. My crew in Argentina, M. Vince, J. Powell, J. C. Leal, O. Gutierrez, J. L. Gomez, and G. Rougier, were always active in field and laboratory work.

References

Bonaparte, J. F. 1971. Los tetrapodos del sector superior de la Formación Los Colorados, La Rioja, Argentina (Triásico Superior). *Opera Lilloana* 22:1–183.

1978. *Coloradia brevis* n.g. et n.sp. (Saurischia, Prosauropoda) dinosaurio Plateosauridae de la Formación Los Colorodos, Triásico Superior de La Rioja, Argentina. *Ameghiniana* 15:3–4.

1979. Dinosaurs: A Jurassic assemblage from Patagonia. *Science* 205:1377–9.

Berman, D. S., and J. S. McIntosh. 1978. Skull and relationships of the Upper Jurassic sauropod *Apatosaurus* (Reptilia, Saurischia). *Bull. Carnegie Mus. Nat. Hist.* 8:1–35.

Charig, A. J., J. Attridge, and A. W. Crompton. 1965. On the origin of the sauropods and the classification of the Saurischia. *Proc. Linn. Soc. London* 176:197–221.

Cruickshank, A. R. I. 1975. The origin of sauropod dinosaurs. *S. Afr. J. Sci.* 71:89–90.

Galton, P. M. 1973. On the anatomy and relationships of *Efraasia diagnostica* (Huene) n. gen., a prosauropod dinosaur (Reptilia: Saurischia) from the Upper Triassic of Germany. *Palaeont. Z.* 47(3/4):229–55.

1976. Prosauropod dinosaurs (Reptilia, Saurischia) of North America. *Postilla* 169:1–98.

Hatcher, J. B. 1901. *Diplodocus* Marsh, its osteology, taxonomy and probable habits, with a restoration of the skeleton. *Mem. Carnegie Mus.* 1:1–64.

1906. Osteology of *Haplocanthosaurus*, with a description of a new species, and remarks on the probable habits of the Sauropoda and the age and origin of the *Atlantosaurus* beds. *Mem. Carnegie Mus.* 2(1):1–75.

Huene, F. von. 1907–8. Die Dinosaurier des europaischen Triasformations. *Geol. Palaeont. Abh., Suppl.* 1:1–419.

1956. *Palaeontologie und Phylogenie der neideren Tetrapoden* (Jena: Gustav Fischer).

Jain, S. L., T. S. Kutty, T. K. Roy Chowdhury, and S. Chatterjee. 1975. A sauropod dinosaur from the Lower Jurassic Kota Formation of India. *Proc. Roy. Soc. London Ser. B* 188(1091):221–8.

1977. Some characteristics of *Barapasaurus tagorei*, a sauropod dinosaur from the Lower Jurassic of Deccan, India. *4th Gondwana Symp., Calcutta*, pp. 204–20.

Janensch, W. 1929. Die Wirbelsaule der Gattung *Dicraeosaurus*. *Palaeontographica, Suppl.* VII:39–133.

1950. Die Wirbelsaule von *Brachiosaurus brancai*. *Palaeontographica* 1950, Suppl. 7(1/3/2):27–103.

Longman, H. A. 1926. A giant dinosaur from Durham Downs, Queensland. *Mem. Queensland Mus. (Brisbane)* 8:183–94.

Lydekker, R. 1888. Note on a new Wealden Iguanodont and other dinosaurs. *Quart. J. Geol. Soc. London* 44:46–61.

1895. On bones of a sauropod dinosaur from Madagascar. *Quart. J. Geol. Soc. London* 51:329–36.

McIntosh, J. S. 1981. Annotated catalog of the dinosaurs (Reptilia, Archosauria) in the collections of the Carnegie Museum of Natural History. *Bull. Carnegie Mus. Nat. Hist.* 18:1–67.

Ogier, A. 1975. Etude de nouveaux ossements de *Bothriospondylus* (Sauropode) d'un gisement du Bathonien de Madagascar. Doctoral Thesis. (3e.cycle), Université Paris VI.

Ostrom, J. H., and J. S. McIntosh. 1966. *Marsh's Dinosaurs: The Collections from Como Bluff* (New Haven: Yale University Press).

Osborn, H. F., and C. C. Mook. 1921. *Camarasaurus, Amphicoelias* and other sauropods of Cope. *Memoirs Am. Mus. Nat. Hist. N.S.* 3(3):251–387.

Owen, R. 1875. A monograph on the fossil Reptilia of the Mesozoic formations. Part II: *Bothriospondylus, Cetiosaurus. Palaeontograph. Soc. (London)*, pp. 15–93, pl. 3–10.

Peterson, O. A., and C. W. Gilmore. 1902. *Elosaurus parvus*, a new genus and species of the Sauropoda. *Ann. Carnegie Mus.* 1:490–9.

Phillips, J. 1871. Geology of the Oxford and the valley of the Thames (Oxford: Clarendon Press).

Raath, M. A. 1972. Fossil vertebrate studies in Rhodesia: a new dinosaur (Reptilia, Saurischia) from near the Trias–Jurassic boundary. *Arnoldia (Rhodesia)* 5(30):1–37.

Riggs, E. S. 1904. Structure and relationships of opisthocoelian Dinosaurs. Part II. The Brachiosauridae. *Publ. Field Columbian Mus., Geol. Ser.* 2:229–47.

Romer, A. S. 1968. *Notes and Comments on Vertebrate Paleontology* (Chicago: University of Chicago Press).

Steel, R. 1970. Saurischia. *In* Kuhn, O. (ed.), *Encyclopedia of Paleoherpetology, Part 14* (Stuttgart: Gustav Fischer).

Thevenin, A. 1907. Dinosauriens de Madagascar. *Ann. Paleont.* 2:121–36.

20 Earliest records of *Batrachopus* from the southwestern United States, and a revision of some Early Mesozoic crocodylomorph ichnogenera

PAUL E. OLSEN AND KEVIN PADIAN

Introduction

During the field season of 1983, a field party from the University of California investigated the faunas of the Kayenta and Moenave formations in northeastern Arizona, on lands of the Navajo and Kaibab–Paiute nations. This fieldwork was a continuation of paleontological, sedimentologic, and biostratigraphic reconnaissance of the region begun in 1981 and supported by grants from the National Geographic Society and the Museum of Paleontology of the University of California (UCMP). Work to date has been summarized by Clark and Fastovsky (Chapter 23) and by Padian et al. (1982). The tracks described here were discovered by J. M. Clark, and collected by Clark, K. Padian, S. M. Gatesy, and E. Cobabe. Upon arrival at Berkeley, the sandstone slabs bearing the footprints were washed and cleaned with a soft brush to remove dirt; latex molds of the best preserved individual tracks and trackways were made by Kyoko Kishi.

The purpose of this chapter is to describe these tracks, the first southwestern records of the crocodylomorph ichnogenus *Batrachopus*. To do so, however, it has proved necessary to revise the ichnogenus *Batrachopus* and other named ichnogenera, mostly from the Newark Supergroup of eastern North America. We also review the stratigraphic distribution of trackways comparable to *Batrachopus*, and we conclude that the basis of first records suggests correlation of the Dinosaur Canyon Member of the Moenave Formation with the Early Jurassic horizons of the Newark Supergroup.

The trackways described here were collected from two different areas in the Dinosaur Canyon Member of the Moenave Formation. The first locality, V85012, is some 500 m east of the type locality of *Protosuchus richardsoni* (UCMP locality V3828) near "Protosuchus Pillar," worked by Barnum Brown and field crews from the American Museum of Natural History (AMNH) in 1931 and 1934. As Colbert and Mook (1951) noted, eight *Protosuchus* specimens came from the "Protosuchus Pillar" area in Dinosaur Canyon during those years; seven are AMNH specimens, and the eighth is a UCMP specimen (36717) discovered by Dr. S. P. Welles (locality V4120). From this same region, less than half a kilometer south–southeast of Protosuchus Pillar, six protosuchian skeletons and the first set of footprints mentioned above were discovered within 100 m of each other by James M. Clark in 1983. The tracks came from the base of a cross-bedded orange–red sandstone bed, 1.5–3.0 m thick, with blue–white spheres and burrows; the crocodile skeletons were higher in this bed. Clark (UCMP field notes, 1983) determined that the type specimen of *Protosuchus* is from the next higher sandstone horizon, similar to the lower bed but without cross bedding, and separated from it by one meter of shaly sandstone, brown with green–white spheres and tunnels. Clark concluded, and we agree, that all the *Protosuchus* localities in this area are much higher in the Moenave Formation than Colbert and Mook (1951) realized: Locally, the Moenave is a graded sandstone terrace, above which the talus and cliffs of Dinosaur Canyon rise. The cliffs are primarily formed of Kayenta rocks, with a cap of Navajo Sandstone, and the Moenave–Kayenta contact is probably within fifteen meters of the highest occurrence of *Protosuchus*.

The second footprint locality, V84239, is about two miles (3.5 km) south–southeast of the Landmark, north of Tanahakaad Wash, Coconino County, Arizona, near the base of the Adeii Eechii Cliffs. [A third record of *Batrachopus*, uncollected, occurs in the Dinosaur Canyon Member of the

Moenave Formation in Tanahakaad Wash, in association with other skeletons of *Protosuchus* (J. M. Clark pers. comm.).]

The footprints from both localities are preserved in slabs of coarse orange-red sandstone, badly weathered and friable, up to 2 cm thick, interbedded between layers of finer orange-red mudstone that have been easily washed from the footprint-bearing layers. In the first locality surveyed (V85012), about 0.4 m^2 of track-bearing sandstone was collected; from the second locality (V84239), about 1.8 m^2 were collected. We took measurements and made composite drawings of the manus and pes from latex molds of the footprints. When we compared our molds, drawings, and composites to specimens, photographs, and drawings of similar ichnites, we were able to assign the new tracks to the ichnogenus *Batrachopus* E. Hitchcock 1845, which was first recognized in the Early Mesozoic Newark Supergroup of the Connecticut Valley.

As noted above, the purpose of this chapter is to describe the Moenave footprints and to justify their assignment to *Batrachopus*. We also discuss their biostratigraphic significance, especially with respect to the age of the Moenave Formation. In the course of our study, however, it became clear that the ichnogenus *Batrachopus* required substantial revision and comment before any tracks could be referred to it.

The standard reference for Connecticut Valley footprints is Lull (1915, revised 1953). Unfortunately, several pervasive problems with Professor Lull's work must be frankly discussed, specifically those dealing with diagnosis and reconstruction of the trackways that constitute most of the paleovertebrate evidence of the Connecticut Valley. In many cases, including the ichnogenus *Batrachopus*, Lull incorrectly identified the type specimens designated by Edward Hitchcock in the mid 1800s, and often incorrectly recognized nominal priority. Lull's inferences about the possible trackmakers of most tracks, including many badly preserved ones, have generally stood up very well. On the other hand, his drawings of the trackways are often not reliable. They are less drawings of specimens than idealizations of footprint forms. In contemporary ichnological work, composites of manus and pes are normally made by comparing prints in a series, reversing and superimposing tracings of successive tracks to ascertain their consistent features. It is clear that Lull did not do this, and often (e.g., Lull 1953, fig. 54) merely repeated reversed drawings of single impressions to simulate trackways. His reconstructions of trackways suffer because he did not faithfully reproduce three important components: the distances between manus and pes, the distances between successive left and right prints, and the

orientation of manus and pes prints with respect to each other and to the direction of movement. Finally, although he did recognize the importance of ontogenetic growth to the relative proportions of footprints (e.g., Lull 1953, pp. 295–307), it did not seem to affect his acceptance of many of Hitchcock's form genera as valid, regardless of their similarities to slightly larger tracks or of their unique features resulting from substrate differences or poor preservation. To begin with, therefore, we will revise the ichnogenus *Batrachopus*.

Figure 20.1. Trackways of *Batrachopus deweyi*. Scale is 3 cm. **A**, Neotype trackway of *Batrachopus deweyi* (A.C. 26/5 and 26/6; locality unknown) showing the impression of digit V in the first two pedal impressions. **B**, Type trackway of *B. "gracilis"* (A.C. 42/3; Turners Falls Sandstone of Massachusetts).

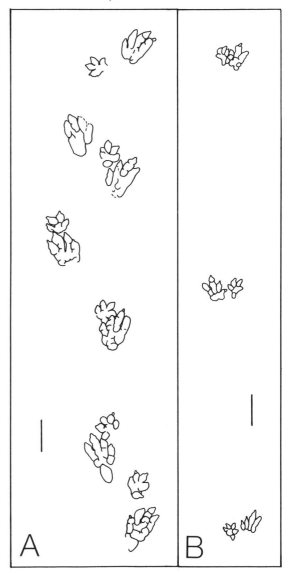

Systematic paleontology
Batrachopus E. Hitchcock 1845

E. Hitchcock 1845, p. 25
Sillimanius E. Hitchcock 1845, p. 24 (in part)
Palamopus E. Hitchcock 1845, p. 24
Anisopus E. Hitchcock 1848, p. 226
Cheirotheroides E. Hitchcock 1858, p. 130
Arachnichnus E. Hitchcock 1858, p. 117
Shepardia E. Hitchcock 1858, p. 131
Macropterna E. Hitchcock 1858, p. 24
Exocampe E. Hitchcock 1858, p. 142
Chelonoides E. Hitchcock 1858, p. 140
Sustenodactylus E. Hitchcock 1858, p. 116
Orthodactylus E. Hitchcock 1858, p. 114
Antipus E. Hitchcock 1858, p. 115
Comptichnus E. Hitchcock 1865, p. 9
Anisichnus C. H. Hitchcock 1871, p. xxi
Parabatrachopus Lull 1942, p. 421

Type species: Batrachopus deweyi
Included species: deweyi, parvulus, dispar

Emended diagnosis:
Small quadrupedal archosaurian ichnite. The manus has five toes and is usually rotated so that digit II points forward, digit IV points laterally, and digit V points posteriorly. The pes is functionally tetradactyl and digitigrade. Digit V of the pes, when impressed, is reduced to an oval pad posterior to and nearly in a line with digit III. Digit III of the pes is longest and digit I is shortest. (See Figs. 20.1, 20.2.) The pes length, from the base of digit I to the tip of digit IV, ranges between approximately 2 and 8 cm.

Geologic range:
?Latest Triassic (but see below), Early Jurassic, Newark Supergroup of eastern North America: McCoy Brook Formation of Fundy Basin; Turners Falls Sandstone of Deerfield Basin; Shuttle Meadow, East Berlin, and Portland formations of Hartford Basin; Feltville, Towaco, Boonton, and uppermost meter of Passaic formations, Newark Basin (Olsen 1980a–c, 1981, 1983; Olsen and Baird 1982; Olsen

Figure 20.2. Type trackways of species synonymous with *Batrachopus deweyi*. Scale is 3 cm. **A,** Two parallel trackways that make up the type of *B. "gracilior"* (A.C. 46/3; ?Turners Falls ss.; locality unrecorded). **B,** Type trackway of *B. "bellus"* (A.C. 26/21; Turners Falls ss.). **C,** Two intersecting trackways that make up the type of *"Chirotheroides pilulatus"* (A.C. 34/37; ?Turners Falls ss.; locality unrecorded).

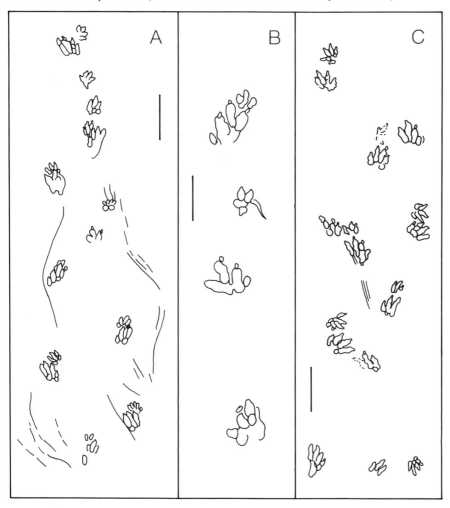

and Galton 1984). Infra-Lias (?Hettangian) of Vendée, France; Upper Stormberg Group (Early Jurassic) of southern Africa; La Cantera Formation (Early Cretaceous) of Argentina (see remarks and references in the section entitled Biogeographic and Stratigraphic Implications below).

Batrachopus deweyi
(E. Hitchcock 1843) E. Hitchcock
1845, emend.

Objective synonyms

> *Sauroidichnites deweyi* E. Hitchcock 1843, p. 261, Plate 11, Fig. 9
> *Batrachopus deweyanus* E. Hitchcock 1845, p. 25
> *Anisopus deweyanus* E. Hitchcock 1848, p. 226, Plate 16, Figs. 5, 6; E. Hitchcock 1858, p. 60, Plate 9, Fig. 3; Plate 41, Fig. 2; Plate 43, Fig. 1–2; Plate 53, Fig. 8; Plate 58, Fig. 11
> Lull 1904a, p. 483; Lull 1915, p. 175, Fig. 32; Lull 1953, p. 237, Figs. 104–5

Subjective synonyms

> *Anisopus gracilis* E. Hitchcock 1845; 1848, p. 228, Plate 16, Figs. 3–4; E. Hitchcock 1858, p. 61, Plate 9, Fig. 4; Plate 35, Fig. 5; Plate 36, Fig. 1; Plate 43, Figs. 3–5; Plate 58, Fig. 9
> *Anisichnus gracilis* C. H. Hitchcock 1889, p. 119
> *Batrachopus gracilis* Lull 1904a, p. 484; Lull 1904b, p. 381, Figure ("Probable footprint of *Stegomosuchus longipes*"); Lull 1915, p. 177, Fig. 34; Lull 1953, p. 238, Fig. 107
> *Anisopus gracilior* E. Hitchcock 1863, p. 54; E. H. Hitchcock 1865, p. 6, Plate 1, Fig. 3
> *Anisichnus gracilior* C. H. Hitchcock 1889, p. 119
> *Batrachopus gracilior* Lull 1904a, p. 484; Lull 1915, p. 177, Fig. 35; Lull 1953, p. 239, Fig. 108 (wrong specimen citation listed in Fig. 108, but correct in text)
> *Apatichnus bellus* E. Hitchcock 1858, p. 101, Plate 17, Fig. 6; Plate 35, Fig. 8; Plate 45, Fig. 6
> *Batrachopus bellus* Lull 1904a, p. 485; Lull 1915, p. 178; Lull 1953, pp. 239–40
> *Cheirotheroides pilulatus* E. Hitchcock 1858, p. 130, Plate 23, Fig. 4; Plate 36, Fig. 6; Plate 54, Fig. 3
> Lull 1904a, p. 485; Lull 1915, p. 170, Fig. 37; 1953, pp. 240–1, Figs. 110–11
> *Arachnichnus dehiscens* E. Hitchcock 1858, p. 117, Plate 20, Figs. 12–13; Plate 37, Fig. 2; E. Hitchcock 1865, p. 24, Plate 17, Fig. 2
> Lull 1904a, p. 539; Lull 1915, p. 261, Fig. 122; Lull 1953, pp. 265–6, Figs. 140–1

Tentative subjective synonyms

> ?*Comptichnus obesus* E. Hitchcock 1865, p. 9, Plate 5, Fig. 4; Plate 18, Fig. 6; Lull 1904a, p. 538; Lull 1915, p. 260, Fig. 121; Lull 1953, pp. 241–2, Figs. 112–13. Holotype A.C. 55/5

> ?*Shepardia palmipes* E. Hitchcock 1858, p. 131, Plate 24, Fig. 2; Lull 1904a, p. 538; Lull 1915, p. 260, Fig. 120; Lull 1953, p. 243, Fig. 114. Holotype A.C. 33/47
> ?*Palamopus palmatus* E. Hitchcock 1841, p. 483, Plate 34, Figs. 15–16. Holotype A.C. 27/3
> ?*Palamopus gracilipes* E. Hitchcock 1858, p. 129, Plate 23, Fig. 6; Plate 34, Fig. 1. Holotype A.C. 35/23
> ?*Palamopus rogersi* E. Hitchcock 1841, p. 496, Plate 45, Fig. 41 (in part). Holotype A.C. 36/52
> ?*Exocampe arcta* E. Hitchcock 1858, p. 142, Plate 25, Figs. 5, 6, 10; Plate 49, Fig. 5. Holotype A.C. 35/24
> ?*Exocampe ornata* E. Hitchcock 1858, p. 143, Plate 25, Fig. 11; Plate 48, Figs. 1, 6. Holotype A.C. 39/69
> ?*Exocampe minima* E. Hitchcock 1865, p. 11, Plate 18, Fig. 3. Holotype A.C. 55/4
> ?*Chelonoides incedens* E. Hitchcock 1858, p. 140, Plate 31, Fig. 3. Holotype A.C. 6/1
> ?*Sustenodactylus curvatus* E. Hitchcock 1858, p. 116, Plate 20, Fig. 11; Plate 34, Fig. 3. Holotype A.C. 34/43
> ?*Orthodactylus floriferus* E. Hitchcock 1858, p. 114, Plate 20, Fig. 7; Plate 45, Fig. 2. Holotype A.C. 6/1
> ?*Orthodactylus intro-vergens* E. Hitchcock 1858, p. 114, Plate 20, Fig. 8; Plate 51, Fig. 1. Holotype A.C. 34/32
> ?*Orthodactylus linearis* E. Hitchcock 1858, p. 115, Plate 20, Fig. 9; Plate 48, Fig. 4. Holotype A.C. 27/15
> ?*Antipus flexiloquus* E. Hitchcock 1858, p. 115, Plate 20, Fig. 10. Holotype A.C. 41/52

Neotype

A.C. 26/5 and 26/6 (counterparts) as given by Lull (1904a) (see discussion, below, for details). Locality data unrecorded, but matrix looks like East Berlin Formation, Hartford Basin, at Mount Tom, Massachusetts.

Emended diagnosis

Batrachopus in which the complete manus impression is about 75 percent of the length of the pes, including the metatarsophalangeal pads. Pedal digits IV and II are subequal in length, and the distal phalangeal pad of digit I is approximately opposite the crease between the two most proximal phalangeal pads of digit II. The pes length, from the base of digit I to the tip of digit IV, ranges between approximately 2 and 6 cm.

Geologic range

?Latest Triassic (but see below), Early Jurassic, Newark Supergroup: Portland, East Berlin, and Shuttle Meadow formations of the Hartford Basin, Turners Falls Sandstone of the Deerfield Basin, and Feltville, Towaco, Boonton, and uppermost Passaic formations of the Newark Basin (see references listed for the genus, above).

Discussion

The first Connecticut Valley ichnite recognized as quad-rupedal was named *Sauroidichnites deweyi* by E. Hitchcock in 1843 (Plate XI, Fig. 9). The slab figured shows a partial trackway and an isolated manus–pes set. No locality was given for this specimen, and no number was given to it. No slab matching its description is listed in E. Hitchcock's (1865) catalog of the Amherst collection; the figured spec-imen is not mentioned in the 1843 paper. Our attempts to locate the specimen at Amherst have failed. Lull (1904a, 1915, 1953) cited a clearly different specimen (A.C. 26/5 and 26/6) as the type. This specimen also lacks locality data, but Hitchcock did include it in *deweyi* in 1858. It is clear from E. Hitchcock's (1843) lithograph that digit I of the pes is relatively longer than in *B. parvulus* (see below), and in all ways it seems to belong to the same ichnospecies as the slab that Lull designated as the type. Because the type specimen appears to be lost, but the species is ap-parently still valid, we accept as the neotype the specimen that Lull believed was the holotype.

The neotype (A.C. 26/5 and 26/6) of *deweyi* is an excellent trackway with seven manus–pes sets that show all the characters typical of the genus and species (see Fig. 20.1A); it falls in the upper part of the known size range. The two most posterior manus–pes impressions show a well–defined pad for digit V, which we have occasionally observed in other specimens referrable to the genus.

The species *deweyi* was originally included in the higher group name *Sauroidichnites*. Hitchcock used higher names, such as *Ornithichnites* and *Sauroidichnites*, not as generic names of actual specimens, but rather as classifi-catory ideals (Lull, 1915, pp. 171–2). Therefore, they do not have priority over *Batrachopus* (see also Baird 1957). In 1845, E. Hitchcock applied generic names to the spec-imens themselves, with a clear table of synonymy. Included in this list was the new genus *Batrachopus*, which included the single species *deweyanus* (a new spelling of *deweyi*). The date of establishment for *Batrachopus* is thus 1845. In the same list, Hitchcock named two other genera, *Silli-manius* and *Palamopus*, which we regard as subjective syn-onyms of *Batrachopus*. They are poor specimens that cannot be diagnosed to the specific level, and thus are inappropriate as the basis of a genus.

Batrachopus gracilis (Fig. 20.1B) was also named by E. Hitchcock in 1845, but he did not figure or describe it until 1848. The type specimen (A.C. 42/3) consists of a long row of clear manus–pes sets. Digit I of the pes is definitely longer than in *B. parvulus*, and its most distal pad lies opposite the crease between the first and second phalangeal pads of digit II (Fig. 20.3). The manus is about 75 percent of the pes in length. Apart from its smaller size, longer pace, and the lack of impressions of digit V in the pes, this form is identical to *B. deweyi*, and we therefore synonymize the two.

The type slabs of *Batrachopus gracilior*, *B. bellus*, *Arachnichnus dehiscens*, and *Cheirotheroides pilulatus* con-sist of trackways that show no appreciable differences from *Batrachopus deweyi* other than size, when all of the manus–pes sets are considered and their composites compared (Figs. 20.1, 20.2, 20.4). We thus consider these taxa sub-jective synonyms of *B. deweyi*. The differences among them as illustrated by Lull (1904a, 1915, 1953) result from

inaccurate renderings of isolated manus–pes sets that do not represent entire trackways.

The ichnotaxa listed above as "tentative subjective synonyms," as well as their objective synonyms (which Lull listed in 1953), cannot be distinguished from *Batrachopus deweyi*; unfortunately, the specimens are too poorly pre-served for us to tell if they share all of the characters of the genus and species *B. deweyi*.

Batrachopus parvulus (E. Hitchcock 1841) E. Hitchcock 1845, emend.

Ornithichnites parvulus E. Hitchcock 1841, Plate 39, Fig. 26.

Holotype

20/4 A.C., Amherst College; collected from the sidewalks of Middletown (E. Hitchcock, 1865) or Middlefield (Lull, 1915, 1953), Connecticut, by Dr. Joseph Barratt around 1835. (Middlefield was separated from Middletown in 1886; the latter is more likely to have been paved with large flagstones in 1835). Judging from town records, the slab, which is of the same lithology as typical middle Portland Formation sandstone, probably was excavated from the old Portland quarry across the Connecticut River. This slab also bears the holotype of *Sauropus barrattii*. No referred specimens.

Emended diagnosis

Batrachopus distinguished by a very short pedal digit I, with its most distal pad lying about opposite the most prox-imal phalangeal pad on digit II. Digits IV and II are about equal in their forward projection. The manus is about 71 percent of the length of the pes.

Geological range

Early Jurassic, Middle Portland Formation, Hartford Basin, Newark Supergroup of eastern North America.

Discussion

What E. Hitchcock called *Ornithichnites parvulus* in 1841 was the first named ichnospecies of what he later called *Batrachopus*. The trackway consists of two successive manus–pes impressions on the same slab as the type of *Sauropus barrattii* (see Chapter 6). In 1843, when E. Hitch-cock named *Sauroidichnites deweyi*, he recognized that *par-vulus* might be the same sort of track. In 1845, when he named *Batrachopus*, he did not list *parvulus* as one of the species. In 1858, however, he explicitly synonymized *par-vulus* with *deweyi*, and never mentioned *parvulus* after-ward; in fact, in the 1865 Supplement to the *Ichnology*, E. Hitchcock described the type specimen of *parvulus* as *An-isopus gracilis*.

Examination of the type slab (Fig. 20.3) suggests that *parvulus* is a determinate species, distinguished prin-cipally by its shorter digit I. It comes from the middle Portland Formation of the Hartford Basin, whereas all the other type specimens of *Batrachopus* come from either the Turners Falls Sandstone of the Deerfield Basin or the East Berlin Formation of the Hartford Basin.

Batrachopus dispar Lull 1904

Lull 1904a, p. 483, Fig. 2; Lull 1915, p. 176, Fig. 33; Lull 1953, p. 237–8, Fig. 106.

Holotype

21/7 A.C., Amherst College, from the Turners Falls Sandstone of Lily Pond, Gill, Massachusetts, Deerfield Basin, Newark Supergroup. No referred specimens.

Emended diagnosis

Large *Batrachopus*, supposedly distinguished by a relatively small manus impression, but the manus is too incomplete for comparison.

Discussion

The type specimen of this species unfortunately has only one well–defined manus impression, and it is very lightly impressed: it shows only digits II, III, and IV (not four toes as shown by Lull 1904a, 1915, 1953) (Fig. 20.5). The digits impressed are, however, as long proportionally as in *B. deweyi*. Digits II and IV are subequal. The apparent difference in manus size could result from the incomplete manus impression in *B. dispar*. The pedes are identical, except for size. The pes of *B. dispar* is roughly 60 percent longer than that of *B. deweyi* and is the largest *Batrachopus* in the Amherst collection. Because of the size difference and the lack of complete information on the manus, we provisionally retain *dispar* as a valid species pending the description of better material.

Description of the Moenave Formation specimens

The trackways collected from the Moenave Formation (UCMP 130583–130596) are preserved in slabs of coarse orange–red sandstone about 1.0–1.5 cm thick. Footprints are abundant on the slabs and are usually impressed to a depth of 3–5 mm. All the footprints appear to be of the same type, though they range in size from about 20 to 60 mm pedal length. Some tracks are poorly preserved and yield

Figure 20.3. Type of *Batrachopus parvulus* (A.C. 20/4; Portland Formation, sidewalk, Middletown, Connecticut), a natural cast. **A**, Entire type slab, including the type of *Sauropus barrattii* (the larger, five-toed track), the type of *Batrachopus parvulus* (on right), and several badly worn ?grallatorid tracks and possible tail drag marks. Scale is 10 cm. **B**, Detail of *Batrachopus parvulus* trackway in **A**. Scale is 4 cm.

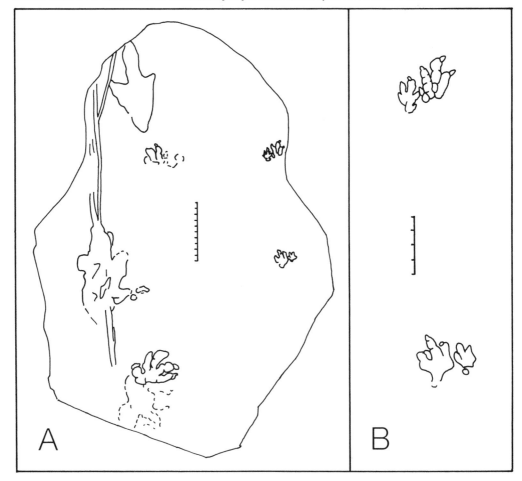

little information; these are frequently the deeper tracks, which suggests high water content and poor substrate competence at the time the tracks were made. Manus impressions are variably present and are usually less distinct than pedal impressions.

All of the relatively clear footprints show the same diagnostic characters of *Batrachopus deweyi*. The line drawings of specimens in Figures 20.6 and 20.7, the composite restoration in Figure 20.8, and the photographs in Figure 20.9 obviate long discussion of characteristics. In all the manus–pes sets in which the digits can be discerned, the manus is outwardly rotated and there is no impression of pedal digit V in all but the deepest tracks. These features are typical of *Batrachopus*. The manus is about 75 percent of the length of the pes, and the distal pad of pedal digit I is approximately opposite the crease between the two most proximal phalangeal pads of digit II, as in *Batrachopus deweyi*.

Comparison of Batrachopus to other ichnotaxa

As Baird (1954, 1957) noted, *Batrachopus* illustrates the culmination of the large-scale trend, visible through the early Mesozoic, of the reduction in the significance of digit V in the dominant quad-

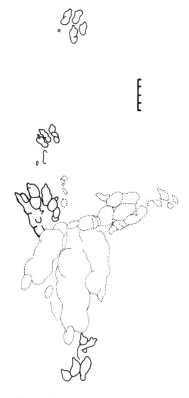

Figure 20.5. Type slab of *Batrachopus dispar* (A.C. 21/7; Turners Falls ss., Massachusetts). Extraneous tracks, mostly *Grallator spp.*, in dotted lines. Scale is 4 cm.

Figure 20.4. Composites of trackways of *Batrachopus deweyi* and *B. dispar*. Scale is 1 cm. All drawn as right manus–pes sets. **A**, Composite of slabs A.C. 26/5 and 26/6 (see Fig. 20.1A), type of *Batrachopus deweyi*, showing impression of digit V. **B**, Composite of type trackway of *B. "gracilis"*, A.C. 42/3 (see Fig. 20.1B). **C**, Composite of slab A.C. 46/3 (Fig. 20.2A), type of *B. "gracilior"*. **D**, Composite of slab A.C. 26/21 (Fig. 20.2B), type of *B. "bellus"*. **E**, Composite of slab A.C. 34/37 (Fig. 20.2C), type of "*Cheirotheroides pilulatus*". **F**, Best manus–pes set of slab A.C. 21/7 (Fig. 4), *B. dispar*.

Figure 20.6. *Batrachopus deweyi* from the Dinosaur Canyon Member, Moenave Formation, northeastern Arizona. Scale is 3 cm. UCMP specimen numbers 130587 (**A**), 130594 (**B**), 130586 (**C**), 130588–91 (**D**), 130592–3 (**E**), and 130596 (**F**).

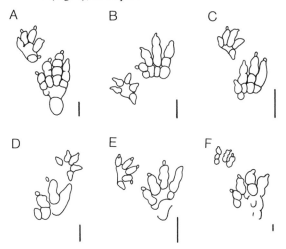

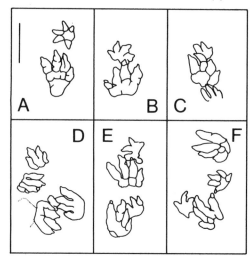

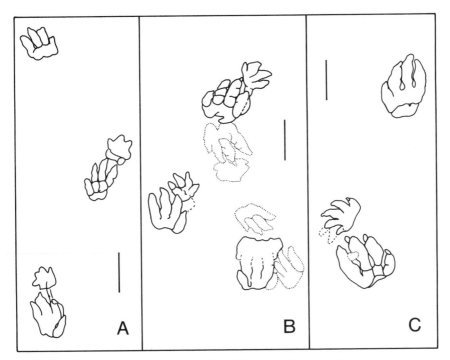

Figure 20.7. Trackways of *Batrachopus deweyi* from the Dinosaur Canyon Member, Moenave Formation, northeastern Arizona. Scale is 3 cm. UCMP specimen numbers 130595 (**A**), 130585 (**B**), and 130584 (**C**) (also shown in Fig. 20.11).

rupedal, nondinosaurian archosaur ichnofaunas (Fig. 20.10). In *Batrachopus*, digit V is only very rarely impressed, and when it is, only an oval pad posterior to digit III is present. *Otozoum* (Fig. 20.10) is considered a scaled-up, graviportal version of its contemporary *Batrachopus* (Baird 1954, 1957). *Brachychirotherium*, characteristic of the Late Triassic horizons of the Newark and of the Early to Late horizons of the German Keuper, has a slightly larger pad for digit V that projects slightly laterally and is present in almost all trackways. *Chirotherium*, dominant in the Early and Middle Triassic, has the least reduced digit V, and in most tracks it projects strongly laterally and is usually recurved (Fig. 20.10).

Along with the loss of a functional digit V in the pes, the manus of *Batrachopus* shows a greater outward rotation than in any other nondinosaurian archosaurs. The manus in *Chirotherium* and *Brachychirotherium* are partially rotated outward, with digit V pointing laterally in most trackways. In *Batrachopus*, however, digit V of the manus usually points backward. A similar orientation is seen in the manus of the related crocodylomorph track *Pteraichnus* (Padian and Olsen 1984) and in the unrelated dinosaur track *Anomoepus*.

Figure 20.8. Composites of *Batrachopus deweyi* material. Scale is 1 cm. All drawn as right manus–pes sets. **A**, Composite of Moenave material, based on material drawn in Figs. 20.6 and 20.7. **B**, Composite of Newark Supergroup material, based on material drawn in Figure 20.1–20.4. Note that the impression of pedal digit V, as shown here, occurs only rarely.

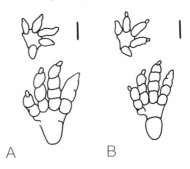

A   B

To avoid confusion, we note that we are using the terms "crocodilian" and "crocodylomorph" in their osteotaxonomic sense. Crocodilians (= Crocodylia) include the traditional grades Protosuchia, Mesosuchia, and Eusuchia (Romer 1966); Crocodylomorpha includes Crocodylia plus their closest relatives, the sphenosuchids and pedeticosaurs *sensu lato* (genera include, for example, *Sphenosuchus, Pedeticosaurus, Saltoposuchus, Hesperosuchus, Pseudhesperosuchus, Terrestrisuchus*). The sphenosuchids and pedeticosaurs are small, lightly built, terrestrial, possibly bipedal forms. We will suggest below that *Batrachopus* could have been made by a crocodilian *sensu stricto*, but we are not sure about

the other crocodylomorphs mentioned above, because their pedal structures are poorly known. Therefore, the presence of noncrocodilian crocodylomorph bones in Late Triassic rocks does not necessarily imply that *Batrachopus* would have been present there as well.

Possible trackmakers of Batrachopus

A pedal digit V reduced as much as it is in *Batrachopus* is seen in several Mesozoic archosaurs, including *Lagosuchus* (Romer 1972; Bonaparte 1975), *Lagerpeton* (Romer 1972), crocodiles, and

Figure 20.9. Photograph of UCMP 130584, *Batrachopus deweyi*, from the Moenave Formation, Dinosaur Canyon Member, Arizona. Diameter of lens cap is 54 mm.

Figure 20.10. Comparison of ichnites similar to *Batrachopus*, all drawn as right manus–pes sets. Scale is 1 cm.
A, *Chirotherium barthi*, from the Moenkopi Formation of Arizona (after Baird, 1957), Early Triassic.
B, *Brachychirotherium parvum*, from the Passaic Formation of New Jersey (after Baird, 1957), Late Triassic.
C, *Batrachopus deweyi*, from the Newark Supergroup. **D**, *Alligator ?mississipiensis*, drawn from lithograph in Dean (1861; Plate 21), with manus and pes positioned according to trackways of *Caiman* sp. in Padian and Olsen (1984). **E**, *Otozoum moodii*, from the Portland Formation of Massachusetts, drawn reversed from A.C. 15/14.

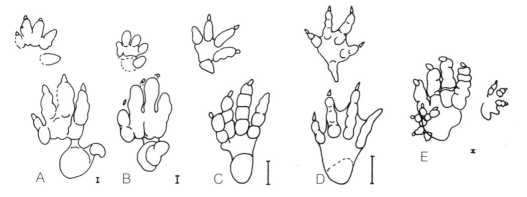

dinosaurs. The basic form of the rest of the pes of *Lagosuchus* and *Lagerpeton* is very different from the form of the reconstructed pes of *Batrachopus*, but the possibility cannot be eliminated that a very primitive, as yet unknown dinosaur could have made *Batrachopus* tracks. This is because *Batrachopus* retains a number of features plesiomorphic to all archosaurs, which most dinosaurs modified, and which today are found only in crocodiles: for example, quadrupedality, five digits in both the manus and pes, and four long toes with the third toe longest.

As reconstructed from the trackways, the bones of the manus and pes of *Batrachopus* cannot be distinguished from those of crocodilians. Moreover, the strongly out-turned, five-fingered manus is definitely present in the tracks of modern crocodiles (Dean 1861; Schaeffer 1941; Padian and Olsen 1984). [The five-fingered manus is also out-turned in *Anomoepus* (Lull 1953, p. 194, Fig. 61), usually considered an ornithischian track; however, the pes is functionally tridactyl in the latter, which proves that the two trackmakers were very different.] The osteology of the manus and pes of *Batrachopus*, using the rules worked out by Peabody (1948) and Baird (1954, 1957) strongly resembles those of both modern and Early Mesozoic crocodiles (Fig. 20.11). Finally, in both the Newark Supergroup and the Glen Canyon Group, crocodile skeletons in the same size range as *Batrachopus* occur in the same strata (*Protosuchus* in the Moenave Formation of the Glen Canyon Group and *Stegomosuchus* in the Newark). In fact, J. M. Clark discovered six protosuchian skeletons less than 100 m from the *Batrachopus* trackways collected from locality V85012.

The inferred association of *Batrachopus* with crocodilian skeletons has an interesting history. Lull (1904a) suggested that *B.* "*gracilis*" (our *B. deweyi*) is probably the trackway of *Stegomosuchus longipes*, a small armored reptile known from a partial, incompletely preserved skeleton discovered in the Portland Formation near Longmeadow, Connecticut. At first, the skeleton was referred to the small aetosaur *Stegomus*, of which one species, *S. arcuatus*, was known from a much lower horizon in the New Haven Arkose of New Haven, Connecticut. Differences of the carapaces and limbs of the two specimens convinced Lull to erect a separate genus, and he did so noting the long foot of the new form, which lacked a fifth free digit. He therefore suggested (Lull 1904b, p. 381) the possible association of *Stegomosuchus* with *Batrachopus*, regarding the former (and hence the trackmaker of the latter) as a quadrupedal "pseudosuchian" allied to the aetosaurs but distinctly different. As it turns out, Lull's view of the association appears to be a good inference, but *Stegomosuchus*, instead of aetosauroid, is properly regarded as a crocodile (Walker 1968). This supports our assessment on independent grounds that *Batrachopus* is the footprint of a crocodylomorph.

It is important to note, however, that as strong as the osteological resemblance is between the reconstructed *Batrachopus* feet and those of protosuchid and modern crocodiles, the same resemblance could be shared with many noncrocodilian crocodylomorphs, for which complete pedes have not been preserved. Unfortunately, the foot is not well-known in "paracrocodiles" (Walker 1970) in gen-

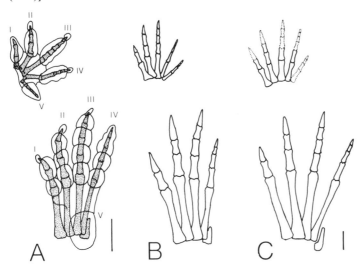

Figure 20.11. Comparison of reconstructed pes of (**A**) *Batrachopus deweyi* with (**B**) manus and pes of *Alligator* sp. [from Romer (1956)], and with (**C**) manus and pes of *Protosuchus richardsoni* [from Colbert and Mook (1951)]. Scale is 1 cm. All drawn as right manus–pes sets.

eral. *Terrestrisuchus*, from the Late Triassic Carboniferous limestone fissure fills of Cowbridge, Glamorgan, Wales (Crush 1984), is an exception because it appears to have a digit V with two small phalanges. *Batrachopus* shows no evidence of phalanges on pedal digit V, although *Chirotherium* and *Brachychirotherium* do, so we can presume to eliminate *Terrestrisuchus* and similar forms from consideration as the trackmaker of *Batrachopus*. We conclude that the trackmaker was most probably a true crocodilian or a crocodylomorph with a pedal digit V reduced to the state seen in crocodilians.

Stratigraphic implications

Batrachopus was originally discovered in a sandstone slab probably from the East Berlin Formation of the Hartford Basin, Newark Supergroup, and has since been found in nine additional formations in the Hartford and four other basins. In every case, *Batrachopus* is restricted to horizons directly below the oldest basalt and sediments interbedded and above it. These horizons, long thought to be Triassic in age, have been reassigned to the Lower Jurassic on the basis of radiometric dating of the basalt flows and by biostratigraphic correlation of palynomorphic and vertebrate fossils (Cornet, Traverse, and McDonald 1973; Olsen and Galton 1977; Cornet 1977; Olsen 1980a–c; Cornet and Olsen 1985).

Footprints from three other regions have been referred to *Batrachopus*. The stratigraphically oldest form so ascribed is *B. varians*, from the Middle Triassic of France (Demathieu 1970). The pes is proportionally very similar to *Batrachopus*, specifically *B. deweyi*, but the manus points straight ahead and is not of the *Batrachopus* type, which Demathieu, working only with Lull's figures, could not have realized. The existing material of *B. varians* suggests to us a brachychirotheriid rather than a true *Batrachopus*.

Another supposedly Middle Triassic form is *Batrachopus* (*Parabatrachopus*) *argentina* (Lull 1942), from what Lull listed as the Paganzo Beds III of the Sierra de la Quijadas, Argentina. Lull also described a species of *Grallator*, as *Anchisauripus australis*, from the same beds. However, the basin exposed in the Sierra de la Quijadas is the San Luis Basin (see Anderson and Anderson 1970), in which the Paganzo Beds proper do not occur. Instead, Stipanicic (1956) listed the track-bearing strata as Division III of the San-Luisense interval, and noted similarities between this sequence and the Botucatu Group of Brazil, which contains a lower sequence of possible Early Jurassic age (Cordani, Kawashita, and Filho 1978) and an upper sequence of lavas and interbedded sediments of Early Cretaceous age. However, the stratigraphy of the *Batrachopus*-bearing beds has since been revised (Criado Roque, Mombru, and Moreno 1981; Bonaparte 1981), and these beds are now recognized as from the La Cantera Formation (Flores 1969) of the Gigante Group. The La Cantera Formation has produced a Neocomian pollen assemblage and has, therefore, been reassigned to the Early Cretaceous (Yrigoyen 1975). The La Cantera footprint faunule is completely separate from the Triassic, *Chirotherium*-bearing Los Rastros beds of the Ischigualasto–Ville Union Basin, with which it is often listed (e.g., Haubold 1971).

Other occurrences of *Batrachopus* were reported from the Infralias (probably Hettangian) of Veillon, Vendée, France. Lapparent and Montanet (1967) named these *Batrachopus gilberti*. These appear to be true *Batrachopus* and are indistinguishable from *B. deweyi*. Similar tracks occur in the Upper Stormberg Group of southern Africa and were named *Plateotetrapodiscus rugosus*, *Suchopus bakoenaorum*, *Molapopentapodiscus pilosus*, and *Synaptichnium motutongense* (Ellenberger 1970, 1974). Although the available material fits within the range of variation known in *Batrachopus* trackways, it has no diagnostic features; definite assignment to that genus must await the discovery of better material (Olsen and Galton 1984). The Upper Stormberg is thought to be of Early Jurassic age on the basis of associated reptile and mammal remains, tetrapod ichnofossils (Olsen and Galton 1977, 1984), and some limited palynological information and radiometric dates (Aldiss, Benson, and Rundel 1984).

The diagnostic features of the *Batrachopus* trackways and their absence in earlier strata form an interesting pattern. Virtually all *Batrachopus* reported from eastern North America, Europe, and South Africa occur in strata of Early Jurassic age, where dates can be assigned or correlated on other grounds. The one exception is a faunule with *Batrachopus* from the uppermost meter of the Passaic Formation, Newark Basin, Newark Supergroup. Olsen and Galton (1977) and Olsen (1980a,b, 1983) considered the age of this assemblage Late Norian (Rhaetian of earlier authors). However, strata in the same position in other parts of the Newark Basin have produced good Early Jurassic palynoflorules, in which the transition to Late Triassic palynoflorules occurs some 10–20 m lower (Cornet 1977; Cornet and Olsen 1985). The Passaic *Batrachopus* assemblage may, therefore, also be Early Jurassic (Hettangian) in age. This faunule overlies and is completely distinct from the older *Brachychirotherium–Apatopus* footprint assemblage characteristic of the rest of the Passaic Formation (Chapter 6), and in all other biostratigraphic particulars is of typical "Jurassic" aspect.

We conclude from the above that, apart from the possibility that *Batrachopus* may yet be found in horizons of latest Triassic age, its known distribution is Hettangian to Neocomian (Early Jurassic to Early Cretaceous). We qualify this generalization for several reasons. Bones of crocodylomorphs are known from sediments of Norian and younger age, although at present records of crocodylomorphs with clearly reduced fifth pedal digits are restricted to Hettangian and younger sediments. Protosuchian crocodiles are known from the Los Colorados Formation of Argentina (Bonaparte 1971), which we regard as of Norian age. Known protosuchians lack phalanges on the fifth digit, so it can be presumed that *Batrachopus*-type footprints could be found in horizons slightly earlier that their known range. The Los Colorados fauna is a good admixture of typical Norian and Liassic faunal types, and could well be transitional. At present, it provides the only datum suggesting potential extension of the known range of inferred trackmakers of *Batrachopus*, that is, crocodylomorphs with reduced pedal digit V.

Age of the Moenave Formation

The Moenave Formation has long eluded a well-founded assessment of age; various proponents have argued for either a Late Triassic or an Early Jurassic age based on isolated factors (reviews in Harshbarger, Repenning, and Irwin 1957; Pipiringos and O'Sullivan 1978; Peterson and Pipiringos 1979). The Glen Canyon Group, which includes the Wingate, Moenave, Kayenta, and Navajo formations, has yielded relatively few fossils of any kind until recently, and there are as yet no radiometric age determinations for any part of the Glen Canyon Group. However, recent palynological investigations of the Whitmore Point Member of the Moenave Formation have provided indications that the Glen Canyon Group, with the exception of the basal Rock Point Member of the Wingate Formation (see below), is probably entirely Jurassic in age. Based on a series of comparisons of Whitmore Point pollen to pollen from the Newark Supergroup (detailed in Peterson and Pipiringos 1979, pp. 31–3), Bruce Cornet and his colleagues have correlated the Moenave locality with the upper–lower to lower–middle part of the Portland Formation of the Newark Supergroup, which is late Sinemurian to early Pliensbachian in age (Peterson, Cornet, and Turner-Peterson 1977). Cornet based these determinations both on the predominance of striate *Corollina*, the principal palynomorphic indicator of Liassic horizons in Europe, and on the presence of forms referable to *Corollina itunensis*, *Chasmatosporites apertus*, and *Callialasporites*, known only from Liassic and younger strata (the first species only from middle Liassic and younger rocks). Peterson and Pipiringos agreed with Cornet's conclusion that the entire Glen Canyon Group above the Rock Point

Member of the Wingate Formation was Liassic in age (middle Sinemurian to late Toarcian). However, it should be mentioned that the Whitmore Point Member, from which pollen samples were taken, is stratigraphically just above the Dinosaur Canyon Member, from which the *Protosuchus* and *Batrachopus* were collected (Harshbarger et al. 1957).

Unfortunately, repeated attempts to find additional palynofloras have failed. In 1981, Carol Hotton of the UCMP field party collected twenty-eight pollen samples and tested nine others from the Kayenta Formation of northeastern Arizona, all of which proved barren (Chapter 23; Padian et al. 1982). Two samples from the Whitmore Point Member of the Moenave Formation were tested, but only the one from the locality sampled by Cornet proved fossiliferous. The predominance of striate *Corollina* suggested a Lower Jurassic age. In the sample she studied, Hotton did not find *Corollina itunensis* or *Callialasporites*, but also found no taxa characteristic of Late Triassic European horizons. Although further testing is always desirable, we accept the Liassic determination of Peterson, Cornet, and Turner-Peterson, in view of the high correspondence of the Moenave and Portland palynofloras.

Vertebrate fossils from the Moenave have also been sparse; Clark and Fastovsky (Chapter 23) review the fauna. The crocodile *Protosuchus*, one candidate for the *Batrachopus* trackmaker, has been the most common vertebrate fossil collected. Eight specimens were recovered from the Dinosaur Canyon Member of the Moenave Formation by field parties from the American Museum of Natural History and the Museum of Paleontology of the University of California between 1931 and 1941. In 1983 the UCMP field party, as mentioned above, found six other skeletons less than half a mile from the original *Protosuchus* locality. These, like the other specimens, were found near the top of the Moenave Formation, close to its contact with the Kayenta Formation. Considerably lower in the Whitmore Point Member, at the same pollen locality discussed above, J. M. Clark and members of the UCMP field party collected bone scraps of fishes and reptiles in 1981 and 1983. These included the first records of turtles (a partial shell) and theropod dinosaurs (a vertebra) from the Moenave Formation. The subholostean fish *Semionotus* and the conchostracan *Cyzicus* have also been collected from the Whitmore Point Member (discussed in Olsen, McCune, and Thomson 1982), and several species referable to *Semionotus* (including *Lepidotes*) have come from the Springdale Sandstone or upper Dinosaur Canyon Member farther north (review in Harshbarger et al. 1957). S. P. Welles and other UCMP investigators have collected similar fish remains from the Moenave Formation near Kanab, Utah.

None of the paleovertebrate taxa recovered from the Moenave Formation is diagnostic of Triassic or Jurassic horizons. Therefore, the age of the Moenave should not be based only on vertebrate correlations. However, given the abundance of Moenave crocodiles, and the absence of Late Triassic crocodiles or crocodiloid tracks anywhere in the world except for the questionably dated Los Colorados Formation of South America (Cornet and Olsen 1985), we suggest that the *Batrachopus*-bearing horizons may be correlative and of Early Jurassic age (Peterson et al., 1977). This inference seems to be supported by a strong correspondence between Moenave and Portland palynofloras. In addition, there is a marked universal nonconformity near the base of the Glen Canyon Group, separating the Rock Point Member of the Wingate Formation (which intergrades with the uppermost Chinle Formation) from the Lukachukai Member of the Wingate [which intergrades with overlying Kayenta and Navajo Formations in an apparently continuous sequence of deposition (Pipiringos and O'Sullivan 1978)]. The Rock Point Member has yielded typical Late Triassic Chinle vertebrates, a fauna much different from those of overlying Glen Canyon sediments. Although there is no evidence either way for the age of the earliest Moenave and Wingate deposits, the possibility exists that this unconformity discussed by Pipiringos and O'Sullivan separates the available Triassic and Jurassic records in the southwestern United States. At this point, all available evidence points to an Early Jurassic age for the Glen Canyon Group proper, excluding the Rock Point Member, and including the vertebrate–, footprint–, and pollen–bearing horizons discussed here.

Note added in proof

Fr. Giuseppe Leonardi, the eminent Brazilian paleoichnologist, has recently informed us that *Batrachopus* has been discovered in the Prado region of southwestern Colombia, in sediments dated by ammonites as late Norian (below the traditionally recognized "Rhaetian") in age. This information corresponds to our prediction in this paper that footprints of crocodylomorphs, such as *Batrachopus*, would be discovered in sediments from the Late Triassic (Norian), from which age skeletal material of crocodylomorphs has been known for some years. At the same time, if this new record is correctly identified, it extends the stratigraphic range of *Batrachopus*, which by itself can no longer be regarded as evidence for a maximal Jurassic age. The footprints described to us by Fr. Leonardi have not yet been published; it will be interesting to see the degree to which they conform to known records of *Batrachopus* from sediments of Jurassic and later age.

Acknowledgments

Our best thanks go to J. M. Clark, who found the specimens described here and was responsible for coordinating the field effort in the Moenave and Kayenta formations that resulted in their discovery. Steve Gatesy, Emily Cobabe, R. A. Long, and Kyoko Kishi assisted in the collection, transportation, and preparation of the specimens. We thank J. M. Clark, Dr. Hartmut Haubold, Nicholas McDonald, Dr. W. A. S. Sarjeant, and also Dr. Donald Baird, dean of ichnologists, for helpful criticism of the manuscript and for aid in tracking down elusive specimens and citations. Fieldwork was supported by grants to K. P. and colleagues from the National Geographic Society (2327–81 and 2484–82) and the Museum of Paleontology, University of California. P. E. O. was supported by a fellowship from the Miller Institute for Basic Research in Science, 1983–4.

References

Aldiss, D. T., J. M. Benson, and C. C. Rundel. 1984. Early Jurassic pillow lavas and palynomorphs of eastern Botswana. *Nature (London)* 310:302–4.

Anderson, H. M., and J. M. Anderson. 1970. A preliminary review of the biostratigraphy of the uppermost Permian, Triassic, and lowermost Jurassic of Gondwanaland. *Palaeontol. Afr. 13* (Suppl.): 1–22.

Baird, D. 1954. *Chirotherium lulli*, a pseudosuchian reptile from New Jersey. *Mus. Comp. Zool. (Harvard Univ.), Bull.* 1:163–92.

1957. Triassic reptile footprint faunules from Milford, New Jersey. *Mus. Comp. Zool. (Harvard Univ.), Bull.* 117: 449–520.

Bonaparte, J. F. 1971. Los tetrapodos del sector Superior de la Formación Los Colorados, La Rioja, Argentina (Triásico Superior): I Parte. *Opera Lilloana* 22: 1–183.

1975. Nuevos materiales de *Lagosuchus talampeyensis* Romer (Thecodontia-Pseudosuchia) y su significado en el origen de los Saurischia. Chañarense Inferior, Triásico Medio de Argentina. *Acta Geol. Lilloana*. 13: 5–90.

1981. Los fosiles mesozoicos. *In* M. Yrigoyen (ed.), *Geología y Recursos Naturales de la Provincia de San Luis. Relatorio del VII Congreso Geológico Argentino*. (Buenos Aires: Asociación Geológica Argentina), pp. 97–9.

Colbert, E. H., and C. C. Mook. 1951. The ancestral crocodile *Protosuchus*. *Am. Mus. Hist., Bull.* 97: 143–82.

Cordani, U. G., K. Kawashita, and A. T. Filho. 1978. Application of the rubidium–strontium method to shales and related rocks. *Am. Assoc. Petrol. Geol., Stud. Geol.* 6: 93–117.

Cornet, B. 1977. The palynostratigraphy and age of the Newark Supergroup. Ph.D. thesis, Department of Geosciences, University of Pennsylvania.

Cornet, B., A. Traverse, and N. G. McDonald. 1973. Fossil spores, pollen, fishes from Connecticut indicate Early Jurassic age for part of the Newark

Group. *Science* 182: 1243–7.

Cornet, B., and P. E. Olsen. 1985. A summary of the biostratigraphy of the Newark Supergroup of eastern North America, with comments on early Mesozoic provinciality. *III Congr. Latin-Am. Paleontología. México. Memoria*, pp. 67–81.

Criado Roque, P., C. A. Mombru, and J. Moreno. 1981. Sedimentitas mesozoicas. *In* M. Yrigoyen (ed.), *Geología y Recursos Naturales de la Provincia de San Luis. Relatorio del VII Congreso Geológico Argentino.* (Buenos Aires: Asociación Geológica Argentina), pp. 79–96.

Crush, P. J. 1984. A Late Upper Triassic sphenosuchid crocodilian from Wales. *Palaeontology* 27: 131–57.

Dean, J. 1861. *Ichnographs from the Sandstone of the Connecticut River* (Boston: Little, Brown).

Demathieu, G. 1970. *Les Empreintes de Pas de Vertébrés du Trias de la Bordure Nord-Est du Massif Central. Cahiers de Paléontologie* (Paris: Editions CNRS).

Ellenberger, P. 1970. Les niveaux paléontologiques de première apparition des Mammifères primordiaux en Afrique du Sud, et leur ichnologie. Etablissement de zones stratigraphiques détaillées dans le Stormberg du Lesotho (Afrique du Sud) (Trias supérieur à Jurassique). *Proc. Pap., 2nd Gondwana Symp. CSIR Pretoria, S. Afr.* 1970: 343–70.

1974. Contribution à la classification des pistes de vertés du Trias. Les types du Stormberg d'Afrique du Sud, (Ilème partie: le Stormberg Supérieur — I. Le biome de la zone B/1 ou niveau de Moyeni: ses biocenoses). *Palaeovert., Mém. Extraord., Montpellier*, pp. 1–155.

Flores, M. A. 1969. El Bolsón de Las Salinas en la Provincia de San Luis. *Actas IV J. Geol. Argent.* 1:311–27.

Harshbarger, J. W., C. A. Repenning, and J. H. Irwin. 1957. Stratigraphy of the uppermost Triassic and the Jurassic rocks of the Navajo country [Colorado Plateau]. *U.S. Geol. Surv. Prof. Pap.* 291: 1–74.

Haubold, H. 1971. Ichnia Amphibiorum et Reptiliorum fossilium. *Handbuch der Paläoherpetologie*, Teil 18 (Gustav Fischer, Stuttgart).

Hitchcock, C. H. 1871. Account and complete list of the Ichnozoa of the Connecticut Valley. *Wallings and Gray's Official Topographical Atlas of Massachusetts* (Boston: Wallings and Gray), pp. XX–XXI.

1889. Recent progress in ichnology. *Proc. Boston Soc. Nat. Hist.* 24: 117–27.

Hitchcock, E. 1841. *Final Report on the Geology of Massachusetts. Amherst and Northampton*, Pt. III, pp. 301–714.

1843. Description of five new species of fossil footmarks, from the red sandstone of the valley of the Connecticut River. *Am. Assoc. Geol. Natural., Trans.* 1843:254–64.

1845. An attempt to name, classify, and describe the animals that made the fossil footmarks of New England. *Proc. 6th Mtg, Am. Assoc. Geol. and Naturalists, New Haven, Conn.*, pp. 23–5.

1848. An attempt to discriminate and describe the animals that made the fossil footmarks of the United States, and especially New England. *Mem. Am. Acad. Arts Sci.* (2)3: 129–256.

1858. *Ichnology of New England. A Report on the Sandstone of the Connecticut Valley, Especially Its Fossil Footmarks* (Boston: William White).

1863. New facts and conclusions respecting the fossil footmarks of the Connecticut Valley. *Am. J. Sci.* (2) 36: 46–57.

1865. *Supplement to the Ichnology of New England* (Boston: Wright and Potter).

Lapparent, A.-F., and C. Montenat. 1967. Les empreintes des pas de reptiles de l'Infralias du Veillon. *Mem. Soc. Geol. France.* 107: 1–44.

Lull, R. S. 1904a. Fossil footprints of the Jura-Trias of North America. *Boston Soc. Nat. Hist. Mem.* 5:461–557.

1904b. Notes on the probable footprints of *Stegomus longipes*. *Am. J. Sci.* (4)17: 381–2.

1915. Triassic life of the Connecticut Valley. *Conn. State Geol. Nat. Hist. Surv., Bull.* 24: 1–285.

1942. Triassic footprints from Argentina. *Am. J. Sci.* (5)(240): 421–5.

1953. Triassic life of the Connecticut Valley. *Conn. State Geol. Nat. Hist. Surv., Bull.* 81: 1–331.

Olsen, P. E. 1980a. A comparison of the vertebrate assemblages from the Newark and Hartford Basins (early Mesozoic, Newark Supergroup) of Eastern North America. *In* Jacobs, L. L. (ed.), *Aspects of Vertebrate History: Essays in Honor of Edwin Harris Colbert* (Flagstaff, Arizona: Museum of North Arizona Press), pp. 35–53.

1980b. Triassic and Jurassic Formations of the Newark Basin. *In:* Manspeizer, W. (ed.), *Field Studies in New Jersey Geology and Guide to Field Trips, 52nd Ann. Mtg. New York State Geol. Assoc.*, Newark College of Arts and Sciences, Newark, Rutgers University, pp. 2–39.

1980c. Fossil great lakes of the Newark Supergroup in New Jersey. *In* Manspeizer, W. (ed.), *Field Studies in New Jersey Geology and Guide to Field Trips, 52nd Ann. Mtg. New York State Geol. Assoc.*, Newark College of Arts and Sciences, Newark, Rutgers University, pp. 352–98.

1981. Comment on "Eolian dune field of Late Triassic age, Fundy Basin, Nova Scotia." *Geology.* 9: 557–61.

1983. Relationship between biostratigraphic subdivisions and igneous activity in the Newark Supergroup. Southeastern Sect., *Geol. Soc. Am., Abstr. Prog.* 15(2): 71.

Olsen, P. E., and D. Baird. 1982. Early Jurassic vertebrate assemblages from the McCoy Brook Formation of the Fundy Group (Newark Supergroup, Nova Scotia, Canada). *Geol. Soc. Am., Abstr. Prog.* 14(1–2): 70.

Olsen, P. E., and P. M. Galton. 1977. Triassic-Jurassic terapod extinctions: are they real? *Science 197:* 983–6.

1984. A review of the reptile and amphibian assemblages from the Stormberg of southern Africa, with special emphasis on the footprints and the age of the Stormberg. S. H. Haughton Memorial Volume. *Palaeontol. Afr.* 25: 87–110.

Olsen, P. E., A. R. McCune, and K. S. Thomson. 1982. Correlation of the early Mesozoic Newark Supergroup by vertebrates, principally fishes. *Am. J. Sci.* (5)282: 1–44.

Padian, K., J. M. Clark, D. E. Foster, and C. Hotton. 1982. Preliminary biostratigraphic–sedimentologic exploration of the Kayenta Formation of Arizona. *National Geographic Society Research Final Reports* 1982, in press.

Padian, K., and P. E. Olsen. 1984. The fossil trackway *Pteraichnus*: not pterosaurian, but crocodilian. *J. Paleontol.* 58: 178–84.

Peabody, F. E. 1948. Reptile and amphibian trackways from the lower Triassic Moenkopi Formation of Arizona and Utah. *Bull. Dept. Geol. Sci., Univ. Calif., Berkeley* 27: 295–468.

Peterson, F., B. Cornet, and E. C. Turner-Peterson. 1977. New data bearing on the stratigraphy and age of the Glen Canyon Group (Triassic and Jurassic) in southern Utah and northern Arizona. *Geol. Soc. Am. Abst. Prog.* 9(6):755.

Peterson, F. and G. N. Pipiringos. 1979. Stratigraphic relations of the Navajo Sandstone to Middle Jurassic formations, southern Utah and northern Arizona. *U.S. Geol. Surv. Prof. Pap.* 1035-B: 1–43.

Pipiringos, G. N., and R. B. O'Sullivan. 1978. Principal unconformities in Triassic and Jurassic rocks, Western Interior, United States – a preliminary survey. *U.S. Geol. Surv. Prof. Pap.* 1035-A: 1–29.

Romer, A. S. 1956. *Osteology of the Reptiles.* (University of Chicago Press, Chicago), pp. 1–722.

1966. *Vertebrate Paleontology*, 3rd ed. (Chicago: University of Chicago Press), pp. 1–468.

1972. The Chañares (Argentina) reptile fauna. XV. Further remains of the thecodonts *Lagerpeton* and *Lagosuchus*. *Mus. Comp. Zool. (Harvard Univ.) Breviora.* 394: 1–7.

Schaeffer, B. 1941. The morphological and functional evolution of the tarsus in amphibians and reptiles. *Am. Mus. Nat. Hist. Bull.* 78: 395–472.

Stipanicic, P. N. 1956. El sistema Triásico en la Argentina. *Cong. Geol. Internat., Sec. II, El Mesozoico del Aemisferio Occidental y sus Correlaciones Mundiales* (Durango, Mexico: Editorial Stylo), pp. 73–111.

Walker, A. D. 1968. *Protosuchus, Proterochampsa*, and the origin of phytosaurs and crocodiles. *Geol. Mag.* 105: 1–14.

1970. A revision of the Jurassic reptile *Hallopus victor* (Marsh), with remarks on the classification of crocodiles. *Phil. Trans. Roy. Soc. London B* 257: 323–72.

Yrigoyen, M. R. 1975. La edad Cretácica del Grupo Gigante (San Luis) y su relación con cuencas circunvecinas. *Actas I Congr. Argent. Paleont. Bioestr.* 2: 29–56.

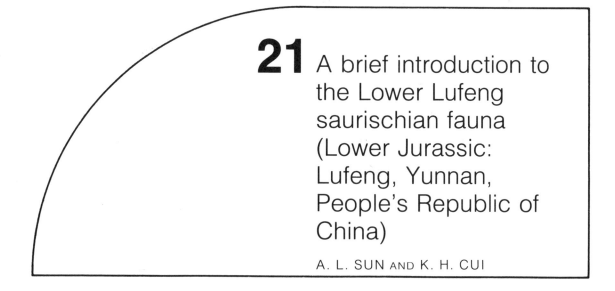

21 A brief introduction to the Lower Lufeng saurischian fauna (Lower Jurassic: Lufeng, Yunnan, People's Republic of China)

A. L. SUN AND K. H. CUI

Vertebrate remains were first collected from Lufeng, Yunnan Province (People's Republic of China) in 1938 by M. N. Bien. C.-C. Young and others joined the fieldwork immediately, and they were rewarded with a considerable collection of fossil reptiles. The fossiliferous beds were named the Lufeng Series (Bien 1940; Young 1940). In the following years, Young published a number of papers on various dinosaurs and mammal-like reptiles. In his monograph of 1951, ten genera were recorded.

New material and information have greatly increased since then. Rigney's collections, gathered by Oehler in 1948–9, were studied by Patterson and Olson (1961), Hopson (1964), Simmons (1965), Kermack, Mussett, and Rigney (1973, 1981), Carroll and Galton (1977), and others. These collections include remains of mammals, tritylodonts, dinosaurs, crocodylomorphs, and lizards. In China, the majority of the new material was collected by members of the Institute of Vertebrate Paleontology and Paleoanthropology (IVPP) and some material was collected by workers from the Geological Museum, the Ministry of Geology, and other institutions.

Because most of the specimens discovered in recent years have not yet been fully described, this chapter will give a very preliminary synopsis of the fauna.

The division of the Lufeng Series into the Upper and Lower Lufeng Series is still used, except the term "Series" has been changed to "Formation." The fossil record from the Upper Lufeng Formation is poor: Only isolated fish spines, teeth, and scales and fragmentary turtle shells are known. All these materials came from the Variegated Beds (Fig. 21.1). In contrast, the Lower Lufeng Formation is rich in vertebrate remains. Lithologically, it is divisible into two members, lower "Dull Purplish Beds" and upper "Dark Red Beds" (Fig. 21.2).

So far, the known fauna of the Dull Purplish Beds is dominated by saurischians and the tritylodont *Bienotherium*. The saurischian dinosaurs are perfectly identical to those from the overlying Dark Red Beds: They include the prosauropods *Lufengosaurus* and *Gyposaurus*, the coelurosaur *Lukousaurus*, and the carnosaur *Sinosaurus*. No new forms have been recorded.

Among the tritylodonts, most can be referred to *Bienotherium*. They are of moderate size, with an average skull length of 130 mm. A small form, *Bienotherium minor*, is now believed to pertain to the genus *Lufengia* (Hopson and Kitching 1972).

Dianosaurus petilus (Young 1982) is represented by a posterior part of a small skull. It has a pair of temporal openings high on the skull roof and a deep jugal arch beneath. The orbit is similar in size to the temporal fossa. Young mentioned a possible relationship to araeoscelids and trilophosaurs, but its systematic position is still waiting to be clarified.

The 20-m-thick Massive Green Sandstone, which was taken as a line of demarcation between the Dull Purplish Beds and the Dark Red Beds by Bien (1940) and Young (1951) is, in fact, not present everywhere; but the color of the strata changes markedly. The Dark Red Beds are dominated by dark red clays, shales, and sandstones. Most of the small skulls were contained in those nodules spreading over the slopes of the sediments.

Up to the present, except for a giant form of *Bienotherium* (Chow 1962) and the type skull of *Dianzhongia* (Cui 1981), all the other tritylodonts from the Dark Red Beds are small animals with skull lengths of 40–50 mm.

Yunnania is characterized by a cusp formula of 2–3–2, which is distinguished from the 2–3–3 formula of *Lufengia* and *Bienotherium*. The lower jaw, which Young referred to *Oligokyphus sinensis* (V4009), is most likely not *Oligokyphus* itself, because there are only four cusps on each tooth instead of six; but *Oligokyphus*-like teeth do occur in the strata. The lower jaw identified by Young as *Lufengia delicata* (V4008) possesses six cusps arranged in two rows.

Among the other Therapsida, *Kunminia* is the only problematic taxon. Because the only specimen of *Kunminia* was lost during World War II and no new material has since come to light, its taxonomic position is not definite. According to Young (1947b), *Dromatherium* is the only animal that could be compared with *Kunminia* in the small size, the shape of the jaw, and the divided roots and accessory cusps of the posterior upper cheek teeth. Nevertheless, he concluded that *Kunminia* holds quite an isolated position in the Ictidosauria. Hopson and Kitching (1972) considered that the small skull of *Kunminia* may be the same as that of *Morganucodon oehleri*. We are not inclined to overemphasize its *Morganucodon* features because the posterior part of the dentary shows distinct dissimilarities to *Morganucodon*, and there is also no angular process. The anterior cheek teeth at the left side of their figure are peculiar and completely different from those of *Morganucodon*.

The Dark Red Beds are rich in mammals. So far, over twenty skulls have been obtained, and all pertain to the genera *Morganucodon* and *Sinoconodon*. Postcranial bones are comparatively rare.

Three additional taxa of Crocodylomorpha have been added since the description of *Platyo-gnathus*, including *Dibothrosuchus*, *Strigosuchus*, and a Crocodylomorpha *incertae sedis* (Simmons 1965). In the spring of 1984, a party from the IVPP excavated a beautiful crocodylomorph skull together with the anterior part of a skeleton. According to X. C. Wu, it is a quite typical sphenosuchid. Simmons, based on the material previously collected, had noted the sphenosuchid-like appearance of the skull of *Platyognathus*. Wu further noticed that apart from the scutes, Simmons's specimens are closer to our new sphenosuchid than to Young's type of *Platyognathus*.

New discoveries have been made in addition to the previously recorded forms of protosuchians. Besides the materials of *Microchampsa* introduced by Simmons, a skull with its accompanying trunk was found in the Dark Red Beds. It is now under preparation. *Dianosuchus*, based on another small skull, was referred to the Protosuchia (Young 1982). It has surface sculpture similar to that of crocodiles, but the characters of the posterior part of the skull are not yet ascertained.

Pachysuchus imperfectus is represented by a skull element, the part around the nasals, and the middle portion of the maxilla. Young had referred it to *Lufengosaurus magnus* (Young 1947a), but transferred it to the Parasuchia in 1951. The reason he gave is that "the sudden elevation of the skull at the part before the nasal openings looks very much the same as that of *Mystriosuchus planirostris* H. v. Meyer" (Young 1951). Owing to poor documenta-

Figure 21.1. Map of Lufeng Basin with fossiliferous sites of Dull Purplish Beds (●) and Dark Red Beds (▲).

Figure 21.2. A generalized columnar section of the Lufeng basin. Modified from Young (1951).

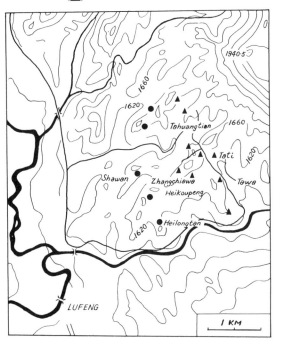

Division	Column	Thickness (m)	Lithology
Shihmen series		100 ±	Capping sandstone
Lufeng Formation — Upper		80 ±	Wine-red Beds
		120 ±	Variegated Beds
		70 ±	Greenish–yellow banded Beds
		5–10	Red sandstone
		150 ±	Dark Red Beds
Lufeng Formation — Lower		570 ±	Dull Purplish Beds
		5–20	Basal breccia
Metamorphic series		?	Metamorphic series

Table 21.1 *Vertebrate taxa of the Lufeng saurischian fauna*

	R[a]	P[a]
Amphibia		
Labyrinthodontia indet.	×	
Reptilia		
Chelonia		
Proganochelyidae indet.	×	
Crocodylomorpha		
Platyognathus hsuii Young 1944	×	
Dibothrosuchus elaphros Simmons 1965	×	
Strigosuchus licinus Simmons 1965	×	
Crocodylomorpha *incertae sedis*	×	
Dibothrosuchus sp.	×	
Protosuchia		
Microchampsa scutata Young 1951	×	
Dianosuchus changchiawaensis Young 1982	×	
?Protorosauria		
Dianosaurus petilus Young 1982		×
Lepidosauria		
Fulengia youngi Carroll and Galton 1977	×	
?Phytosauria		
Pachysuchus imperfectus Young 1951	×	
Saurischia		
Lufengosaurus huenei Young 1941	×	×
Lufengosaurus magnus Young 1941	×	×
Yunnanosaurus huangi Young 1941	×	×
Yunnanosaurus robustus Young 1941	×	×
Gyposaurus sinensis Young 1941	×	×
Sinosaurus triassica Young 1948	×	×
Lukousaurus yini Young 1948	×	×
Ornithischia		
Tatisaurus oehleri Simmons 1965	×	
Tawasaurus minor Young 1982	×	
Dianchungosaurus lufengensis Young 1982	×	
Therapsida		
Bienotherium yunnanense Young 1940		×
Bienotherium elegans Young 1947		×
Bienotherium minor Young 1947		×
Bienotherium magnus Chow 1962	×	
Kunminia minima Young 1947	×	
Lufengia delicata Chow and Hu 1959	×	
?Oligokyphus sinensis Young 1974	×	
Yunnania brevirostre Cui 1976	×	
Dianzhongia longirostrata Cui 1981	×	
Mammalia		
Morganucodon (*Eozostrodon*) *heikoupengensis* Young 1978	×	
Morganucodon oehleri Rigney 1963	×	
Sinocondon rigneyi Patterson and Olson 1961	×	
Sinoconodon parringtoni Young 1982	×	
Sinoconodon changchiawaensis Young 1982	×	
Sinoconodon yangi Zhang and Cui 1983	×	

Note: [a]R, Dark Red Beds; P, Dull Purplish Beds.

tion and the subsequent loss of the specimen, its generic identification and affinity with the Parasuchia may have to remain uncertain.

New findings of saurischian dinosaurs are rare. A sauropod skeleton has been excavated from the neighboring basin of Wudin. The horizon was thought to be equivalent to that of the Lower Lufeng Formation of the Lufeng Basin (Chao pers. comm.). However, up to the present, there is no evidence of sauropods from Lufeng. Consequently, we are not able to be sure whether these two fossiliferous beds are chronologically correlative.

As for the other dinosaurian groups, Simmons demonstrated the presence of the ornithischian *Tatisaurus*. Later, Young assigned two more genera to this group: *Tawasaurus* and *Dianchungosaurus*. The former is known from the snout region of a skull with complete lower jaws. Nevertheless, the identification of *Tawasaurus* as an ornithischian is questioned because there is no sign of the predentary bone (Z. M. Dong pers. comm.).

A number of small vertebrae represent the presence of labyrinthodont amphibians (Sun 1962). According to M. Chow, who collected the specimen, it came from Heikoupeng together with a partial skull of *Lufengia*. Well-preserved turtle shells, most likely of Proganochelyidae (F. Zhang pers. comm.), are also known. *Fulengia youngi* is the only representative of the Lepidosauria thus far known in this assemblage (Carroll and Galton 1977).

By comparison to the Dark Red Beds, the apparent monotony of the fossil assemblage in the Dull Purplish Beds is considered, at the present moment, the possible result of insufficient collecting. Perhaps stimulated by the frequent presence of fossils of small mammals, almost all field parties in recent years have concentrated their attention on the Dark Red Beds, looking for those small nodules that contain mammalian fossils. We are not sure if various forms present in the overlying beds also occurred in the underlying Dull Purplish Beds.

Because there are no radiometric records of these sediments, the dating relies upon paleontological evidence. The Rhaetic age (Upper Triassic) of the Lower Lufeng proposed by Bien and Young on the basis of vertebrate fossils has been held for a long time. The view was challenged only when various forms of invertebrates (ostracods, pelecypods, gastropods, and so forth) were collected during 1966–71 by the "Yunnan Red Beds Expedition" led by the Paleontological Institute of Nanking. In their book (1975, in Chinese), the Lower Lufeng Formation was dated as Lower Jurassic, which is accepted by the present authors.

A current listing of the known fossil vertebrate taxa from the Lower Lufeng Formation is given in Table 21.1.

References

Bien, M. N. 1940. Discovery of Triassic saurischian and primitive mammalian remains at Lufeng, Yunnan. *Bull. Geol. Soc. China* 20(3–4):225–34.

Carroll, R. L., and P. M. Galton. 1977. Modern lizard from the upper Triassic of China. *Nature (London)* 266:252–5.

Chow, M. 1962. A tritylodontid specimen from Lufeng, Yunnan. *Vert. Paleontol.* 6(4):365–7.

Chow, M., and C. C. Hu. 1959. A new tritylodontid from Lufeng, Yunnan. *Vert. Paleontol.*, 3(1):9–12.

Cui, G. 1976. *Yunnania*, A new tritylodont genus from Lufeng, Yunnan. *Vert. Paleontol.* 14(2):85–90.

  1981. A new genus of Tritylodontoidea. *Vert. Paleontol.*, 19(1):5–10.

Hopson, J. A. 1964. The braincase of the advanced mammal-like reptile *Bienotherium*. *Postilla* 87:1–30.

Hopson, J. A., and J. W. Kitching. 1972. A revised classification of cynodonts (Reptilia, Therapsida). *Paleontol. Afr.* 14, 71–85.

Kermack, K. A., F. Mussett, and H. W. Rigney. 1973. The lower jaw of *Morganucodon*. *Linn. Soc. Lond. Zool. J.* 53:87–175.

  1981. The skull of *Morganucodon*. *Zool. J. Linn. Soc.* 71:1–158.

Paleontological Institute of Nanking et al. (sponsors) 1975. Collected papers of the on-the-spot conference on the Mesozoic–Cenozoic redbeds of southern China. Untitled, in Chinese.

Patterson, B., and E. C. Olson. 1961. A triconodontid mammal from the Triassic of Yunnan. *Intern. Coll. Evolution of Lower and Nonspecialized Mammals*, 129–91. *Brussels*, pp. 129–91.

Rigney, H. W. 1963. A specimen of *Morganucodon* from Yunnan. *Nature (London)* 197:1122–3.

Simmons, D. J. 1965. The Non-therapsid reptiles of the Lufeng Basin, Yunnan, China. *Fieldiana: Geol.*, 15(1):1–96.

Sun, A. L. 1962. Discovery of neorhachitomous vertebrae from Lufeng, Yunnan. *Vert. Paleontol.* 6(1):109–10.

Young, C. C. 1940. Preliminary note on the Lufeng Vertebrate Fossils. *Bull. Geol. Soc. China*, 20(3–4):235–40.

  1947a. On *Lufengosaurus magnus* Young (sp. nov.) and additional finds of *L. huenei* Young. *Pal. Sinica* 12:1–53

  1947b. Mammal-like reptiles from Lufeng, Yunnan, China. *Proc. Zool. Soc. Lond.* 117:537–97.

  1951. The Lufeng saurischian fauna in China. *Paleontol. Sin., N.S. C*, No. 13, pp. 1–96.

  1974. New discoveries on therapsids from Lufeng, Yunnan. *Vert. Paleontol.*, 12(2):111–14.

  1978. New materials of *Eozostrodon*. *Vert. Paleontol.* 16(1):1–3.

  1982. *Selected works of Yang Zhungjian (Young Chung - Chien)* (Science Press, Beijing).

Zhang, F., and G. Cui. 1983. New material and new understanding of *Sinoconodon*. *Vert. Paleontol.* 21(1):32–41.

22 Relationships and biostratigraphic significance of the Tritylodontidae (Synapsida) from the Kayenta Formation of northeastern Arizona

HANS-DIETER SUES

Introduction

The Late Triassic and Early Jurassic represent a crucial period in the evolutionary history of the Synapsida, especially because the earliest mammals make their appearance in the fossil record. The stratigraphically oldest find confirmed so far is a haramiyid molariform tooth from the Knollenmergel (km4, Norian) of Halberstadt (East Germany), which was fortuitously discovered during the preparation of a skeleton of the prosauropod dinosaur *Plateosaurus* (Hahn 1973). Along with the earliest mammals, two other groups of advanced synapsids, each with a host of mammalian characters, occur: the Tritylodontidae and the Tritheledontidae. These apparent instances of extensive parallel evolution have been the subject of seemingly endless, largely semantic discussions about a possibly polyphyletic origin of the Mammalia. Until quite recently, both the Tritylodontidae and the Tritheledontidae were poorly understood in anatomical terms.

The Tritylodontidae are particularly distinguished by quadrilateral multicusp cheek teeth that meet in precise occlusion. Tritylodontids are very common in Early Jurassic terrestrial tetrapod assemblages and are presumed to have been herbivorous. The nominal genus, *Tritylodon* Owen 1884, and related forms were generally considered to be mammals until the 1940s, when Kühne (1943) demonstrated the existence of a quadrate–articular contact in *Oligokyphus* ("*Tritylodon*"). [Since that time this articulation has been demonstrated to coexist with a dentary–squamosal joint in the most primitive known mammals, the Morganucodontidae (Kermack, Mussett, and Rigney 1973, 1981).] The recent discovery of numerous, excellently preserved skeletal remains referable to three taxa of Tritylodontidae from the "Silty Facies" of the Kayenta Formation in northeastern Arizona has permitted a detailed anatomical survey of these advanced synapsids (Sues 1983, in prep.). The genera represented are *Kayentatherium* D. M. Kermack 1982, *Oligokyphus* Hennig 1922, and a new form to be described elsewhere (Sues 1986).

Relationships

Following Hopson and Kitching (1972), the Tritylodontidae are placed within the Tritylodontoidea Simpson 1928 *emend*. Hopson and Kitching 1972 and are considered the sister group of the Traversodontidae (Sues 1985b). The Tritylodontidae are a strictly monophyletic taxon, particularly characterized by the following craniodental synapomorphies (Hopson and Kitching 1972; Sues 1983) (Fig. 22.1, Node 1):

1. Cheek teeth multiple-rooted, quadrilateral, with three (upper) or two (lower) anteroposterior rows of crescentic cusps
2. First lower and second upper incisors much enlarged
3. First lower incisor procumbent
4. Canines absent
5. Postorbital (and postorbital bar) and prefrontal absent
6. Articulation between quadrate and anterior paroccipital process (crista parotica) *without* contact between quadrate and squamosal (with possible exception of *Oligokyphus*)

These characters were established by outgroup comparison with the Diademodontidae and the Traversodontidae (see also Sues 1985b).

Most genera of Tritylodontidae are still poorly known, despite the availability of a substantial number of specimens. The group appears to be remarkably uniform in its pattern of structural organization. The pattern of cusps on the postcanine teeth has been the principal criterion for classification, commencing with the studies by Hennig (1922) and Simpson (1928). Relatively few other differences have been identified, but even some of these are subject to considerable ontogenetic variation or cannot be determined for most genera (J.A. Hopson pers. comm.).

Two genera, *Stereognathus* Charlesworth 1855 and *Oligokyphus* Hennig 1922 are clearly distinguished by their respective cusp patterns. *Stereognathus*, known only from the Middle Jurassic (Bathonian) of the British Isles, has upper postcanine teeth with two buccal, two median, and two lingual cusps (Simpson 1928, Fig. 4B; Waldman and Savage 1972, Fig. 1). *Oligokyphus*, from the Lower Jurassic of England, southern Germany, and Arizona (Kühne 1956; Sues 1985a), has upper postcanine teeth with three buccal, four median, and four lingual cusps (Fig. 22.2A) and lower teeth with three cusps in both the buccal and lingual rows (Fig. 22.2B). [The former feature is apparently also developed in juvenile specimens of *Tritylodon* (J.A. Hopson pers. comm.).] Furthermore, crown length exceeds crown width in the postcanine teeth of *Oligokyphus*; the reverse is true in all other Tritylodontidae (Savage 1971, p. 82).

The other tritylodontid genera described so far have two buccal, three median, and, with one unambiguous exception, three lingual cusps on the upper cheek teeth. (Owing to wear and the small size of the anteriormost cusps, cusp formulas given by different authors are often at variance.) They include *Bienotheroides* Young 1982, *Bienotherium* Young 1940, *Kayentatherium* D. M. Kermack 1982, *Lufengia* Chow and Hu 1959, (adult) *Tritylodon* Owen 1884, an unnamed genus from Holwell, England (Savage 1971), and a new genus from the Kayenta

Formation (Sues 1986). Following Hopson and Kitching (1972), I have treated *Likhoelia* Ginsburg 1961 and *Tritylodontoideus* Fourie 1962 as subjective junior synonyms of *Tritylodon*; this decision is supported by the range in size documented for *Kayentatherium* (Sues 1983). Cui (1976, 1981) has described two additional genera, *Yunnania* Cui 1976 and *Dianzhongia* Cui 1981, from the Lower Lufeng Series of Yunnan, China. There is little evidence to justify these taxonomic decisions. "*Yunnania*" [a preoccupied generic name (Sun 1984)] might be synonymous with *Lufengia* from the same strata because both have relatively long crests on the cusps of the upper postcanine teeth. One of the diagnostic characters for "*Dianzhongia*" given by Cui, the presence of only two lingual cusps on the upper cheek teeth, is plainly contradicted by his own drawing (Cui 1981, Fig. 3), which clearly shows *three* cusps, much as in *Tritylodon* and related forms.

A preliminary hypothesis of interrelationships for the better known taxa of Tritylodontidae is presented in the form of a cladogram in Figure 22.1. A more definitive assessment must await publication of Hopson's monographic revision of the genera *Bienotherium* and *Lufengia*, based on specimens collected by American missionaries from the Lower Lufeng Series of Yunnan in the late 1940s.

Oligokyphus has a number of autapomorphic traits, including the ratio of crown length to crown width for its cheek teeth, the almost complete bony enclosure formed by the dorsal border of the squamosal for the external auditory meatus, and the monotreme-like disposition of the proximal femoral trochanters (Kühne 1956). The presences of (albeit tiny) second and third lower incisors and of two lacrimal foramina are plesiomorphic characters as is apparent from comparison with other Tritylodontoidea *sensu* Hopson and Kitching (1972). Further-

Figure 22.1. Cladogram depicting a hypothesis of interrelationships of selected genera of Tritylodontidae. Characters defining nodes 1 to 4 are discussed in the text. The phylogenetic positions of *Bienotherium* and *Kayentatherium* cannot be resolved further at present. *Dinnebitodon* is the generic name proposed for the new tritylodontid from the Kayenta Formation (Sues 1986).

Figure 22.2. Postcanine teeth referable to *Oligokyphus* Hennig 1922, from the Kayenta Formation of northeastern Arizona. **A**, Left upper postcanine tooth (MCZ 8846), occlusal view. **B**, Slightly worn left lower postcanine tooth (MCZ 8851), occlusal view. Scale bars each equal 1 mm; ant, anterior.

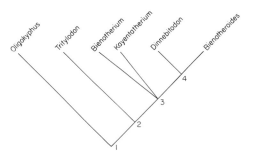

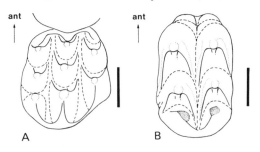

more, the long posterodorsal process of the quadrate may have still contacted the squamosal in *Oligokyphus* (Kühne 1956, Fig. 9); this contact is absent in other Tritylodontidae.

Other Tritylodontidae, with the exception of juvenile *Tritylodon* (3–4–4) and of *Stereognathus*, have upper postcanine teeth with two buccal, three median, and two or three lingual cusps and lower postcanine teeth with two cusps and an accessory posterior cuspule in each row (see Fig. 22.1, Node 2). Chow and Hu's (1959) description of *Lufengia* is insufficient for a more precise phylogenetic assessment of this genus.

Bienotheroides, *Bienotherium*, and *Kayentatherium* (see Fig. 22.1, Node 3) share the absence of the postincisive constriction of the snout commonly found in other cynodonts and in both *Oligokyphus* and *Tritylodon*. Subadult specimens of *Bienotherium* show a slight constriction (J.A. Hopson pers. comm.). *Bienotherium*, from the Lower Lufeng Series of Yunnan, China, is quite similar to *Kayentatherium*, but differs from the latter in the presence of two or three upper incisors, rather than one, and the presence of a prominent angular process of the dentary, on level with the lower tooth row, in adult specimens (Young 1947). I cannot accept the two derived characters in the upper cheek teeth of *Kayentatherium* mentioned by D. M. Kermack (1982). First, the substantially higher ratio of crown width to crown length in this genus (1.52, compared to 1.24 in *Tritylodon* and 1.29 in *Bienotherium*) is undoubtedly the result of the distortion of the upper postcanine teeth in the holotype of *K. wellesi* (UCMP 83671). I have obtained ratios of 1.04 to 1.14 on a sample of well-preserved teeth in my material, more closely comparable to ratios listed for other Tritylodontidae by D. M. Kermack (1982; Table 1). Second, D. M. Kermack claimed that the upper postcanine teeth of *Kayentatherium* have two principal cusps in each row plus an anterior accessory cuspule in the median row. Slight additional cleaning of the left fifth upper postcanine in UCMP 83671 revealed the presence of a small cusp anterior to the two larger lingual cusps (Fig. 22.3B), much as in *Tritylodon* and *Bienotherium*. Both this anterolingual and the anteromedian cusps are small (see Fig. 22.3C), but this is also the case in both *Bienotherium* and *Tritylodon* where D. M. Kermack accepts them as true cusps.

Bienotheroides, from the Middle of Upper Jurassic of Sichuan (Sun 1984), apparently forms a clade with *Stereognathus* and with a newly described genus from the ?Jurassic La Boca Formation of Mexico (Clark and Hopson, 1985). The new tritylodontid from the Kayenta Formation (Sues 1986) also appears to belong here (see Fig. 22.1, Node 4). The

skull of these forms, exemplified by that of *Bienotheroides* (Sun 1984; Figs. 1–4), is noteworthy for

1.   The exclusion of the maxilla from the side of the face by the enlarged premaxilla and lacrimal

2.   Complete overlap of the maxilla laterally by the jugal (which extends down to the level of the upper tooth row)

3.   The contact between the premaxilla and palatine on the palate, excluding the maxilla from participation in the formation of the secondary bony palate.

The last of these features is foreshadowed to some extent in immature *Kayentatherium* (e.g., MCZ 8811) where the premaxilla extends backward between the anteriormost upper postcanine teeth. In the large adult specimen MCZ 8812, however, the suture between premaxilla and maxilla on the palate lies well anterior to the upper cheek teeth, much as in *Bienotherium* and *Tritylodon*.

The tritylodontid material from the uppermost portion of the Kayenta Formation at Comb Ridge, Arizona, mentioned by Lewis, Irwin, and Wilson (1961, p. 1437), still awaits full description, but my review of these specimens revealed no differences

Figure 22.3. *Kayentatherium wellesi* D. M. Kermack 1982. **A**, Incomplete skull of an immature specimen (UCMP 83671, holotype), restored in left lateral view. Scale bar equals 1 cm. **B**, Oblique buccal view of the left fifth upper postcanine tooth of UCMP 83671. Arrow points to anterolingual cusp. Scale bar equals 2 mm. **C**, Occlusal view of the right third upper postcanine tooth of MCZ 8811. Hatching denotes broken surface. Scale bar equals 2 mm; ant, anterior; buc, buccal.

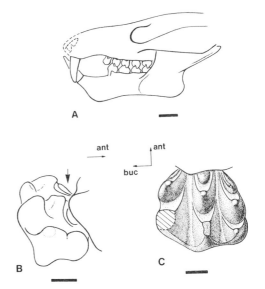

precluding reference to *Kayentatherium wellesi*. A particularly well-preserved skull of an immature individual (USNM 317203) is illustrated in Figure 22.4.

Biostratigraphic significance

Until recently, the Kayenta Formation, along with much or all of the Glen Canyon Group of the southwestern United States, was considered Late Triassic in age by most authors. Galton (1971) reviewed much of the earlier relevant literature. The principal dissenting opinion was that of Welles (1954) who maintained, on the basis of his work on the theropod dinosaur *Dilophosaurus* ("*Megalosaurus*") *wetherilli*, that the Kayenta Formation was of Early or even Middle Jurassic age. He based this estimate primarily on the relative limb proportions in this theropod, but Galton (1971) demonstrated the limitations of this approach. (In fact, the limb proportions of *Dilophosaurus* are most similar to those calculated for "*Halticosaurus*" *liliensterni* from the German Keuper!) Galton (1971) described fragmentary skeletal remains of the prosauropod dinosaur *Ammosaurus* from the Navajo Sandstone, the lower part of which intertongues with the Kayenta Formation (Lewis et al. 1961, p. 1439), and Galton reiterated a Late Triassic date for this formation. The congeneric *Ammosaurus major* occurs in the upper portion of the Portland Formation (Newark Supergroup) of the Connecticut Valley, then considered Late Triassic in age. Walker (1968, p. 6) noted the similarity between the crocodile *Protosuchus* from the Moenave Formation of the Glen Canyon Group and *Stegomosuchus* from the Longmeadow Sandstone of the Portland Formation.

It is extremely difficult to establish the ages of the continental sediments of the Early Mesozoic in both eastern and western North America relative to the largely marine classic sequences in Europe. Vertebrate and palynofloral assemblages offer the only reasonably reliable data for intercontinental stratigraphic correlation at the present time.

Establishing the position of the Triassic–Jurassic boundary within the Glen Canyon Group and elsewhere is very important because of an alleged major crisis in the terrestrial biota at that transition (Olsen and Galton 1977; Chapter 25). Cornet, Traverse, and McDonald (1973) redated the lower part of the Portland Formation in the Connecticut Valley as Lower Liassic ("not older than Sinemurian") on the basis of the palynoflora. Building on this work, Olsen and Galton (1977) concluded that "all of the Glen Canyon Group of the southwestern United States ... [is] Early Jurassic." According to their tabulation, deposition of the sediments forming the Glen Canyon Group commenced in Hettangian times and continued well into the Pliensbachian. [Olsen and Galton do not include the phytosaur-bearing Rock Point Member of the Wingate Formation as part of the Glen Canyon Group. According to Pipiringos and O'Sullivan (1978, p. A19), the Rock Point Member does not intertongue with the overlying Lukachukai Member of the Wingate, as had been reported earlier, but is separated from it by an unconformity (J–0).] A palynoflora from the lower part of the Whitmore Point Member of the Moenave Formation has been correlated by Cornet (quoted in Olsen and Galton 1977) with the Lower Liassic on the basis of the extreme abundance of species of *Corollina* [which constitutes about 95–99 percent of the entire sample (Cornet, quoted in Peterson and Pipiringos 1979, p. B31)]. Cornet suggests correlation of the Whitmore Point Member with the "upper-lower to lower-middle part" of the Portland Formation, which he considers late Sinemurian to early Pliensbachian (*Corollina torosus* palynoflora). The Kayenta Formation, which overlies and intertongues with the Moenave Formation, would then be no older than late Sinemurian. The Connecticut prosauropods are from the upper portion of the Portland Formation that Cornet (quoted in Olsen 1980, p. 49) considers Toarcian or even younger in age. This possibly suggests a similar age for the *Ammosaurus*-bearing Navajo Sandstone, which is unconformably overlain by early Bajocian sediments of the San Rafael Group (Pipiringos and O'Sullivan 1978, p. A20).

Numerous specimens referable to the small tritylodontid *Oligokyphus* have been recovered from the "Silty Facies" of the Kayenta Formation in recent years and are of considerable biostratigraphic interest (Sues 1985a). *Oligokyphus* was first recorded on the basis of two isolated teeth from the "Rhaeto-Liassic" bonebeds of Baden-Württemberg, southern Germany (Hennig 1922; Simpson 1928).

Figure 22.4. *Kayentatherium wellesi* D. M. Kermack 1982. Skull of an immature specimen (USNM 317203), in right lateral view. Slightly restored. Scale bar equals 1 cm; d, dentary; ep, epipterygoid; j, jugal; m, maxilla; pm, premaxilla; q, quadrate; qj, quadratojugal; sm, septomaxilla; sq, squamosal.

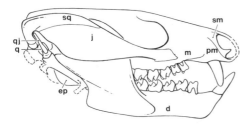

The locality for the type of *O. triserialis* is not known with certainty. The lower postcanine tooth, called "*O. biserialis*" (Hennig 1922, Plate II, Fig. 11), is almost certainly from the bonebed of the Olgahain near Bebenhausen (Hennig 1922, p. 225). This locality has since also yielded upper teeth referable to *O. triserialis* (unpublished material collected by O. H. Schindewolf in 1948). After a thorough review of the stratigraphic position of various "Rhaeto-Liassic" bonebeds from southern Germany, Clemens (1980, p. 59) concluded that the Olgahain bonebed is definitely younger than uppermost Middle Keuper (Knollenmergel, Norian), but probably no younger than Hettangian. Abundant but disarticulated remains referable to *Oligokyphus* from Windsor Hill Quarry ("Mendip 14") in Somerset, England, the subject of an exemplary monograph by Kühne (1956), were collected from fissure fillings in Carboniferous limestones. These fissures apparently opened beneath the sea and contain remains of marine invertebrates in association with the tritylodontid bones and teeth. The belemnites and brachiopods, particularly *Squamirhynchia fronto*, did permit a precise dating of the *Oligokyphus*-bearing deposits as Charmouthian (Lias Gamma), probably no older than Pliensbachian (Kühne 1956, p. 9). Only one of the *Morganucodon*-bearing Welsh "Rhaeto-Liassic" assemblages, the fauna from fissure 4 at Pant Quarry, contains *Oligokyphus* (K. A. Kermack pers. comm.). These assemblages are considered Hettangian to early Sinemurian in age on the basis of unspecified evidence (Kermack et al. 1981).

The European records of *Oligokyphus* thus range in time from the Hettangian to the Pliensbachian. The alleged occurrence of *Oligokyphus* in the Lower Lufeng Series of Yunnan, China, is based on an incomplete right dentary (Young 1974, Plate 2), probably referable to an immature *Bienotherium yunnanense* (J. A. Hopson pers. comm.). The diversified fauna of terrestrial vertebrates from the upper portion of the Lower Lufeng Series has been variously considered Rhaetic (Young 1951) or Early Jurassic (Cui 1976) in age.

D. M. Kermack (1982) considered the dental structure of another Kayenta tritylodontid, *Kayentatherium*, more derived than that of *Tritylodon* or *Bienotherium*, and, therefore, she concurred with Welles's (1954) age estimate for the Kayenta Formation. As mentioned above, her arguments are not valid, and, therefore, *Kayentatherium* offers no evidence for or against any particular biostratigraphic assessment of the Kayenta Formation. The third Kayenta tritylodontid, to be formally described elsewhere (Sues 1986), apparently belongs to the *Bienotheroides* clade, of which *Stereognathus* definitely and *Bienotheroides* at least are Middle Jurassic in age. Therefore, its presence favors a late Early Jurassic dating of the Kayenta Formation.

I conclude then that the biostratigraphic evidence afforded by the presence of *Oligokyphus* and the new tritylodontid genus is fully compatible with the palynological evidence (from the underlying Moenave Formation) in support of an Early Jurassic age for the Kayenta Formation, which is probably no older than late Sinemurian. This is significant in view of the now well-established resemblances among the tetrapod assemblages from the Kayenta Formation, the upper part of the Lower Lufeng Series of Yunnan, and the Upper Stormberg Group (Clarens and upper Elliot formations) of southern Africa. These similarities suggest that the faunas in question are broadly contemporaneous and that the Triassic–Jurassic transition was a period of widespread faunal interchange between the various regions of Pangea (Olsen and Galton 1977, 1984; Sues 1985a).

Conclusions

The Kayenta Formation, long considered Late Triassic in age, has yielded three species of Tritylodontidae: *Kayentatherium wellesi* D. M. Kermack 1982, *Oligokyphus* sp., and a new genus and species to be described elsewhere. *Kayentatherium* appears to be most closely related to *Bienotherium* Young 1940, from the Lower Lufeng Series of Yunnan, China. Certain allegedly derived characters of the upper cheek teeth are shown to be nonexistent and, therefore, are of no particular relevance to discussions concerning the age of the Kayenta Formation. *Oligokyphus* has a known stratigraphic range from the Hettangian to the Pliensbachian in Europe. Its presence, along with that of the new tritylodontid (a presumed member of the *Bienotheroides* clade), in the Kayenta Formation is thus compatible with other lines of evidence in support of an Early Jurassic age of this formation.

Acknowledgments

I am most grateful to F. A. Jenkins, Jr. for the opportunity to study the Kayenta Tritylodontidae. K. Padian arranged for the loan of the holotype of *K. wellesi* and T. M. Bown put USNM 317203 at my disposal. I thank J. A. Hopson and P. E. Olsen for valuable suggestions and comments. Financial support for part of this work was provided by The Geological Society of America and the Anderson Foundation of Harvard University. Institutional acronyms preceding specimen numbers: MCZ, Museum of Comparative Zoology, Harvard University; UCMP, University of California Museum of Paleontology, Berkeley; USNM, National Museum of Natural History, Washington, D.C. (currently in the collections of the United States Geological Survey, Denver, Colorado).

References

Chow, M., and C. Hu. 1959. A new tritylodontid from Lufeng, Yunnan. *Vert. Palasiat.* 3: 9–12.

Clark, J. M., and J. A. Hopson. 1985. A distinctive mammal-like reptile from Mexico and its bearing on the phylogeny of the Tritylodontidae. *Nature (London)* 315: 398–400.

Clemens, W. A. 1980. Rhaeto–Liassic mammals from Switzerland and West Germany. *Zitteliana* 5: 51–92.

Cornet, B., A. Traverse, and N. G. McDonald. 1973. Fossil spores, pollen, and fishes from Connecticut indicate Early Jurassic age for part of the Newark Group. *Science* 182: 1243–8.

Cui, G. 1976. *Yunnania*, a new tritylodont genus from Lufeng, Yunnan. *Vert. Palasiat.* 14: 85–90 [in Chinese].

1981. A new genus of Tritylodontidae. *Vert. Palasiat.* 19: 5–10 [In Chinese].

Galton, P. M. 1971. The prosauropod dinosaur *Ammosaurus*, the crocodile *Protosuchus*, and their bearing on the age of the Navajo Sandstone of northeastern Arizona. *J. Paleontol.* 45: 781–95.

Hahn, G. 1973. Neue Zähne von Haramiyiden aus der deutschen Ober-Trias und ihre Beziehungen zu den Multituberculaten. *Palaeontographica Abt. A* 142: 1–15.

Hennig, E. 1922. Die Säugerzähne des württembergischen Rhät-Lias-Bonebeds. *N. Jahrb. Min. Geol. Paläont., Beilage-Bd.* 46: 181–267.

Hopson, J. A., and J. W. Kitching. 1972. A revised classification of cynodonts (Reptilia, Therapsida). *Palaeontol. Afr.* 14: 71–85.

Kermack, D. M. 1982. A new tritylodontid from the Kayenta Formation of Arizona. *Zool. J. Linn. Soc.* 76: 1–17.

Kermack, K. A., F. Mussett, and H. W. Rigney. 1973. The lower jaw of *Morganucodon*. *Zool. J. Linn. Soc.* 53: 87–175.

1981. The skull of *Morganucodon*. *Zool. J. Linn. Soc.* 71: 1–158.

Kühne, W. G. 1943. The dentary of *Tritylodon* and the systematic position of the Tritylodontidae. *Ann. Mag. Nat. Hist.* (11)10: 589–601.

1956. *The Liassic Therapsid Oligokyphus* London: Trustees of the British Museum, pp. 1–149.

Lewis, G. E., J. H. Irwin, and R. F. Wilson. 1961. Age of the Glen Canyon Group (Triassic and Jurassic) on the Colorado Plateau. *Geol. Soc. Am. Bull.* 72: 1437–40.

Olsen, P. E. 1980. A comparison of the vertebrate assemblages from the Newark and Hartford Basins (Early Mesozoic, Newark Supergroup) of eastern North America. *In* Jacobs, L. L. (ed.), *Aspects of Vertebrate History* (Flagstaff, Arizona: Museum of Northern Arizona Press,), pp. 35–53.

Olsen, P. E., and P. M. Galton. 1977. Triassic–Jurassic tetrapod extinctions: are they real? *Science* 197: 983–6.

1984. A review of the reptile and amphibian assemblages from the Stormberg of southern Africa, with special emphasis on the footprints and the age of the Stormberg. *Palaeontol. Afr.* 25: 87–110.

Peterson, F., and G. N. Pipiringos. 1979. Stratigraphic relations of the Navajo Sandstone to Middle Jurassic Formations, southern Utah and northern Arizona. *U. S. Geol. Surv. Prof. Pap.* 1035-B: B1–B43.

Pipiringos, G. N., and R. B. O'Sullivan. 1978. Principal unconformities in Triassic and Jurassic rocks, western interior United States–a preliminary survey. *U.S. Geol. Surv. Prof. Pap.* 1035-A: A1–A29.

Savage, R. J. G. 1971. Tritylodontid incertae sedis. *Proc. Bristol Nat. Hist. Soc.* 32: 80–83.

Simpson, G. G. 1928. *A Catalogue of the Mesozoic Mammalia in the Geological Department of the British Museum* (London: British Museum [Natural History]).

Sues, H.-D. 1983. Advanced mammal-like reptiles from the Early Jurassic of Arizona. Ph.D. dissertation, Harvard University.

1985a. First record of the tritylodontid *Oligokyphus* (Synapsida) from the Lower Jurassic of western North America. *J. Vert. Paleontol.* 5(4): 328–39.

1985b. The relationships of the Tritylodontidae (Synapsida). *Zool. J. Linn. Soc.* 85: 209–17.

1986. *Dinnebitodon amarali*, a new tritylodontid (Synapsida) from the Lower Jurassic of western North America. *J. Paleontol.* 60: 758–62.

Sun, A.-L. 1984. Skull morphology of the tritylodont genus *Bienotheroides* of Sichuan. *Scient. Sin. B* 27: 970–84.

Waldman, M., and R. J. G. Savage. 1972. The first Jurassic mammal from Scotland. *J. Geol. Soc. London* 128: 119–25.

Walker, A. D. 1968. *Protosuchus*, *Proterochampsa*, and the origin of phytosaurs and crocodiles. *Geol. Mag.* 105: 1–14.

Welles, S. P. 1954. New Jurassic dinosaur from the Kayenta Formation of Arizona. *Geol. Soc. Am. Bull.* 65: 591–8.

Young, C.-C. 1947. Mammal-like reptiles from Lufeng, Yunnan, China. *Proc. Zool. Soc. London* 117: 537–97.

1951. The Lufeng saurischian fauna in China. *Palaeont. Sin. N.S. C* 13: 1–96.

1974. New Materials of Therapsida from Lufeng, Yunnan. *Vert. Palasiat.* 12: 111–16 [in Chinese].

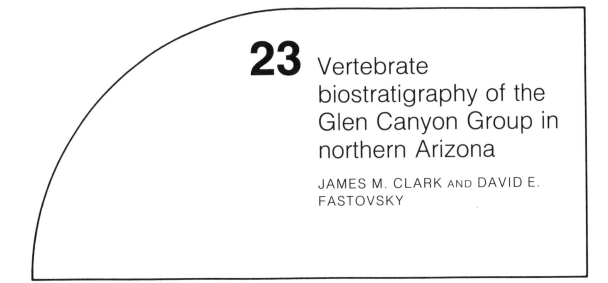

23 Vertebrate biostratigraphy of the Glen Canyon Group in northern Arizona

JAMES M. CLARK AND DAVID E. FASTOVSKY

Introduction

The nonmarine Glen Canyon Group in northern Arizona has the best known and most diverse assemblage of vertebrate fossils from the latest Triassic or Early Jurassic in North America. Harshbarger, Repenning, and Irwin (1957) reviewed the stratigraphy and paleontology of this group, but since then a period of intense collecting has markedly increased both the number of specimens and diversity of the taxa. The purpose of this chapter is to review the stratigraphic distributions of the vertebrate fossils within the group, especially those of the Moenave and Kayenta formations, and to interpret the implications of these distributions for the biostratigraphy of the group.

The Glen Canyon Group extends throughout much of the Colorado Plateau region of Arizona, Utah, Colorado and New Mexico (Baker, Dane, and Reeside 1936; Imlay 1980). Vertebrate fossils are known mainly from northern Arizona, especially from the Adeii Eechii (Painted) Cliffs south of Tuba City, Arizona (Fig. 23.1). The only published records of fossils found outside Arizona are semionotid fish from Southern Utah (Hesse 1935; Wilson 1967) and tetrapod footprints in Colorado and Utah (Bunker 1950; Faul and Roberts 1951; Stokes 1978, among others).

Lithostratigraphic units

A detailed study of the lithostratigraphic boundaries of the Glen Canyon Group and of the units within it is outside the scope of this chapter. The following summary of the lithostratigraphy of the Glen Canyon Group is based mainly upon the published record, especially Harshbarger et al. (1957), Cooley et al. (1969), and Peterson and Pipiringos (1979).

The Glen Canyon Group has been divided into four formations that are reported to intertongue to varying degrees. In ascending order they are the Wingate, Moenave, Kayenta, and Navajo formations (Fig. 23.2). The Wingate Sandstone, Kayenta Formation, and Navajo Sandstone extend throughout the Colorado Plateau, but the Moenave Formation is restricted to northern Arizona and southern Utah.

Fluvio-lacustrine siltstones and mudstones of the Late Triassic Chinle Formation underlie the Glen Canyon Group throughout nearly all of its extent. There is a difference of opinion in the literature concerning the boundary between the Glen Canyon Group and the Chinle Formation. Throughout much of their mutual extent there is a clear erosional unconformity between them, but they are reported to intertongue with one another in northeastern Arizona. Harshbarger et al. (1957) maintained that the uppermost part of the Chinle, the "A" Unit of Gregory (1917), intertongues with the Wingate Sandstone, so they included the Chinle "A" in the Wingate Sandstone as the Rock Point Member. Peterson and Pipiringos (1979), however, found evidence for a regional erosional unconformity above the Rock Point Member [the J–0 unconformity of Pipiringos and O'Sullivan (1978)] in this same area. We have not studied this contact, and because it does not directly bear upon our biostratigraphic conclusions, we defer to the results of the most recent study (Fig. 23.2).

The thick sandstones of the Wingate Sandstone (here restricted to the Lukachukai "Member") intertongue with the Moenave Formation where they occur together in the southern part of the Adeii Eechii Cliffs (Harshbarger et al. 1957). The only published record of fossils from the Wingate Formation are dinosaur footprints reported from Colorado (Bunker 1950).

The Moenave Formation has been divided into three members: In ascending order they are the Dinosaur Canyon, Whitmore Point, and Springdale Sandstone members. The Dinosaur Canyon Member (Colbert and Mook 1951) is the only member present along the Adeii Eechii Cliffs (Fig. 23.3), but elsewhere in northern Arizona and southern Utah it is overlain by the Springdale Sandstone Member. The Whitmore Point Member lies between these two members in a small area along the western Arizona–Utah border (Wilson 1967). All three members are reported to grade into one another (Wilson 1967), and the Springdale Sandstone is reported to grade into the Kayenta Formation in most areas (Harshbarger et al. 1957; Wilson 1967). The Dinosaur Canyon Member is reported to grade into the Kayenta Formation in the northern part of the Adeii Eechii Cliffs (Harshbarger et al. 1957), but our observations suggest that this is ambiguous (as we discuss in the section on Tuba City–Moenkopi Wash).

The Kayenta Formation has been divided into two laterally contiguous facies differing in the relative proportions of siltstones versus sandstones. Throughout most of its extent the Kayenta contains a high proportion of pale red cross-stratified sandstones; this part of the formation is termed the "typical facies" (Harshbarger et al. 1957). In the southern part of the Echo Cliffs a few kilometers north of Moenkopi Wash, there is a higher proportion of siltstones and mudstones that extends southward into the Adeii Eechii Cliffs; this is the "silty facies" of Harshbarger et al. (1957). The division between the two facies is probably an oversimplification because the "typical facies" is actually quite variable. Both facies are known to intertongue with the overlying Navajo Sandstone (Harshbarger et al. 1957; J.M. Clark and D.E. Fastovsky pers. obs.).

The Navajo Sandstone is among the thickest formations of the Colorado Plateau; it reaches 677 m in Zion National Park (Wilson 1965). It thins considerably to the south (it is about 50 m thick along the Adeii Eechii Cliffs) and is absent in the region around Chinle, Arizona and to the southeast (Harshbarger et al. 1957). The Navajo Sandstone was thought to intertongue with the overlying Middle Jurassic Carmel Formation of the San Rafael Group (Phoenix 1963), but Peterson and Pipiringos (1979) could not trace the reported Navajo tongues into the Navajo Sandstone. They show that the eolian sandstone that intertongues with the Carmel is separated from the Navajo by a regional unconformity [the J-2 unconformity of Pipiringos and O'Sullivan (1978)], and they named the eolian strata above this surface the Page Sandstone.

In southwestern Utah, a unit lying between the Navajo Sandstone and the Carmel Formation has been designated the Temple Cap Sandstone by Peterson and Pipiringos (1979). According to these authors the sandstones, siltstones, and gypsum lenses of this formation rest unconformably on the Navajo Sandstone [the J-1 unconformity of Pipiringos and O'Sullivan (1978)] and are overlain unconformably by the Carmel Formation [the J-2 unconformity of Pipiringos and O'Sullivan (1978)].

Figure 23.1. Map of Arizona and southern Utah showing localities mentioned in the text.

Figure 23.2. Idealized relationships of lithostratigraphic units discussed in the text. The vertical axis represents only the sequence of deposition; time and thickness are not to scale. Strippled areas are depositional hiatuses inferred from unconformities.

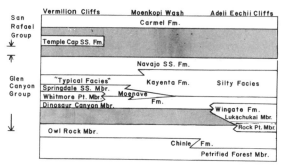

History of vertebrate paleontological work in the Glen Canyon Group

Collection of fossil vertebrates from the Glen Canyon Group has been sporadic, and only in recent years have large areas been systematically prospected. Here we briefly summarize the historical sequence in which important specimens have been collected from the formations comprising the Glen Canyon Group.

The first recorded fossil vertebrates collected from the group were fish collected by a United States Geological Survey party led by C. D. Walcott in 1879 (Cross 1908; Eastman 1917). The specimens were collected from the Whitmore Point Member of the Moenave Formation near Kanab, Utah. [These beds were considered to be part of the Chinle Formation until the work of Wilson (1967).] Many specimens have been collected from the original locality [by the Museum of Paleontology of the University

Figure 23.3. Detailed geologic map of Adeii Eechii Cliffs, modified from Cooley et al. (1969). The stippled region is the Wingate Formation (Lukachukai Member only). The ruled area is the Kayenta Formation. The space between them is the Moenave Formation. The space to the east of the Kayenta Formation is the Navajo Formation. Abbreviations: A, Moenkopi Wash; B, Moenkopi Point; C, Dinosaur Canyon; D, The Landmark; E, Tanahakaad Wash; F, Gold Springs; G, Willow Springs; H, Rock Head.

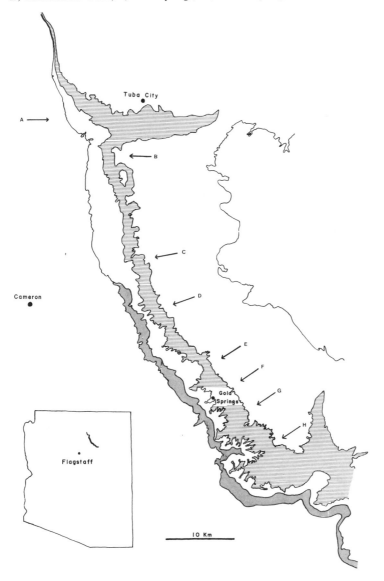

of California (UCMP) for example (Camp 1930)], and other Moenave localities with fossil fish have been reported in the Whitmore Point Member along the Vermilion Cliffs near Fredonia, Arizona (Wilson 1967) and in the Springdale Sandstone Member in Zion Canyon, Utah (Hesse 1935).

The first tetrapod fossils collected from the Glen Canyon Group were the type skeleton and several referred specimens of the crocodile (in the sense of Langston 1973) *Protosuchus richardsoni* from the Dinosaur Canyon Member of the Moenave Formation. These specimens were collected by Barnum Brown of the American Museum of Natural History in 1931 from Dinosaur Canyon due east of Cameron, Arizona (Brown 1933; Colbert and Mook 1951). Another specimen was collected from the same area by S. P. Welles of the UCMP in 1941, but there were no significant efforts to collect more fossils from the Dinosaur Canyon Member until the recent work of Jenkins described below (Crompton and Smith 1980) and our own work.

The first fossil discovered in the Navajo Sandstone was found in 1933 in Segi (or Tsegi) Canyon near Shonto, Arizona. The specimen, a partial skeleton of a small theropod dinosaur, was described by Camp (1936) as *Segisaurus halli*. Fragmentary "prosauropod" dinosaur specimens were collected in the same area in the 1930s by the Museum of Northern Arizona (MNA) and the UCMP (Brady 1935, 1936; Galton 1971, 1976), and a "prosauropod" dinosaur trackway was collected by the MNA in 1958 from near the head of Segi Canyon (Baird 1980). In 1968, a UCMP party found a fragmentary specimen of a small crocodilian in the same area (Galton 1971). The only other fossils reported from the Navajo since then are trackways in Utah (Stokes 1978).

The first significant fossil vertebrates found in the silty facies of the Kayenta were collected from Moenkopi Wash near Tuba City, Arizona in 1942 by a UCMP party. The most important specimens collected from this locality are the type and a referred specimen of *Dilophosaurus wetherilli* (Welles 1954, 1970, 1984).

A locality with abundant remains of tritylodontid therapsids in the upper part of the typical facies of the Kayenta was discovered in 1952 by B. C. Hoy near Kayenta, Arizona (Harshbarger et al. 1957; Lewis 1958). A large collection was made by G. E. Lewis of the United States Geological Survey, and smaller collections were later made by the Field Museum in Chicago (W. Turnbull pers. comm.) and the American Museum of Natural History in New York (E. Gaffney pers. comm.).

In 1964 and 1968, UCMP parties returned to the Tuba City area to collect a third large theropod dinosaur near the locality of the first two (Welles 1970, 1984). During the 1968 trip, two important

localities were discovered: a site with abundant crocodilian remains in the silty facies several kilometers south of the theropod localities (UCMP locality V6899), and a locality yielding a skull and fragmentary skeleton of a tritylodontid therapsid in the "typical facies" near Many Farms, Arizona. The latter specimen became the type of *Kayentatherium wellesi* D.M. Kermack 1982. Casts were also made at this time of theropod dinosaur trackways in the Kayenta Formation in Moenkopi Wash (Welles 1971).

From the late 1960s until the present, personnel of the Geology Department of the Museum of Northern Arizona have periodically collected from the silty facies of the Kayenta Formation, especially in the Moenkopi Wash and Rock Head areas. Among the most important of their finds is the type skeleton of the ornithischian dinosaur *Scutellosaurus lawleri* Colbert 1981, which was collected in 1971 near Rock Head.

A major collecting effort in the silty facies of the Kayenta was initiated in 1977 by parties from the Harvard University Museum of Comparative Zoology (MCZ) under the direction of F. A. Jenkins, Jr. (Jenkins, Crompton, and Downs 1983). From 1977 to 1983 these parties systematically examined the extensive exposures of the silty facies along the Adeii Eechii Cliffs from Dinosaur Canyon southward to Rock Head and made large collections from numerous localities (Crompton and Smith 1980; Jenkins et al. 1983; Padian 1984; Attridge, Crompton, and Jenkins 1985). During the course of this work, a specimen of *Protosuchus* was also found in the Dinosaur Canyon Member of the Moenave Formation (Crompton and Smith 1980). Most significantly, a locality with an abundance of small vertebrates, including some associated skeletons, was found in the silty facies at Gold Springs. The large collection from this locality includes the earliest known mammals in the Western Hemisphere (Jenkins et al. 1983).

During the summers of 1981 and 1983, our parties, sponsored by the UCMP, collected a representative sample of fossils from the Kayenta and Moenave Formations. We made small collections from the Dinosaur Canyon Member of the Moenave Formation along the Adeii Eechii Cliffs and from the Whitmore Point Member near Fredonia, Arizona. We made a more extensive collection from the silty facies of the Kayenta along the Adeii Eechii Cliffs, and we screen-washed one site in the silty facies, in the Dinosaur Canyon area, for microvertebrates.

Stratigraphic distribution of fossil vertebrates

The development of a refined stratigraphy within the Kayenta and Moenave Formations is hampered by the difficulty of establishing time lines. The Kayenta and Moenave formations in the Adeii Eechii Cliffs are largely the products of fluvial processes (see below for a preliminary report on the sedimentology of one site in the silty facies). Because the fluvial deposits are lenticular, individual strata within the Moenave and Kayenta cannot be traced laterally for more than a few kilometers. Furthermore, without examining the lithofacies in detail, it is not possible to distinguish whether the deposits within a local section are derived from a single meandering stream or from multiple sources at significantly different times. Thus, sequences of fluvial deposits can produce a fossil record giving the illusion of abrupt evolution when, in fact, many small sedimentary hiatuses are simply not recognized. This has profound consequences for any hope of documenting faunal changes within these formations.

Although a moderately diverse fauna has been collected from the excellent exposures of the Kayenta and Moenave formations, vertebrate fossils are rare in relation to the thickness and lateral extent of the formations. Furthermore, very little of the recently collected material has been fully studied. We have, therefore, been conservative in the inferences we have drawn from the available data.

We have relocated the localities of most of the fossils collected from the Glen Canyon Group and have determined their positions within sections measured in most of the important collecting areas. We summarize below these stratigraphic relationships for each local area with important fossil localities. The columnar sections presented for these areas are composites, and the relationships between them cannot form the basis for a valid chronostratigraphy.

The locality of one important fossil remains enigmatic at present. Olsen and Galton (1977) identified the ichnotaxon *Anomoepus* among tracings made by S. P. Welles and considered it diagnostic of a Jurassic age (see discussion below). These tracings had been made at a number of localities in the Glen Canyon Group of Utah and Arizona (S. P. Welles pers. comm.). The tracings could not be located, however, so the locality (or localities) from which the *Anomoepus* footprints came remains unknown.

Lewis tritylodont locality

The extremely productive quarry in the "typical facies" of the Kayenta near Kayenta, Arizona (Lewis 1958) is the only known locality in this region (see Fig. 23.1). It lies near the top of the Kayenta Formation about 3 m beneath the Navajo Sandstone and immediately above a tongue of the latter (Fig. 23.4). Only two taxa are known from this locality: the tritylodontid therapsid *Kayentatherium wellesi*

(Chapter 22) and a single specimen of a primitive crocodilian. Our observations of the crocodilian, identified as *Protosuchus richardsoni* by Lewis (1958) and housed in the collections of the United States Geological Survey in Denver, suggest that it is not complete enough to identify at the generic level.

Kayentatherium type locality

The type specimen of the tritylodont *Kayentatherium wellesi* D. M. Kermack 1982 was collected from the "typical facies" of the Kayenta near the residence of Mr. Sam Benale (Fig. 23.1) (not Bernale as in D. M. Kermack 1982) approximately 9 km west of Many Farms, Arizona (UCMP locality V6897). Two other specimens described by D. M. Kermack (1982) were found in close proximity to the type specimen. The fossils were found approximately 10 m above the base of the formation, which is approximately 30 m thick in this area (S. P. Welles field notes, August 12, 1968).

Segi Canyon–Shonto area

Fossil vertebrates have been collected from both the "typical facies" of the Kayenta and the Navajo Sandstone in the area near Shonto, Arizona (approximately 30 km northwest of Tuba City). The single locality of the Kayenta Formation in this area

Figure 23.4. Generalized composite columnar section measured at the Lewis Tritylodont Locality (see text), Navajo County, Arizona.

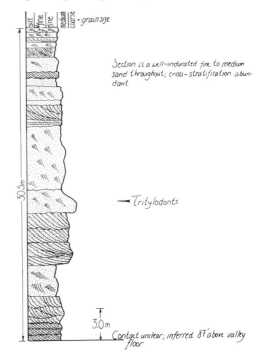

(UCMP V5690) is in a limestone conglomerate at the base of the formation in the mouth of Segi Canyon. The specimens are all isolated bones and bone fragments, mostly from fishes.

This is the only area from which fossil bones have been reported in the Navajo Sandstone, which is approximately 300 m thick in this region (Harshbarger et al. 1957, p. 22). "Prosauropod" dinosaur remains have been found at two sites: An articulated pes (MNA G2 7233) was found approximately 25–30 m below the top of the formation northwest of Segi Canyon (Brady 1935), and an articulated manus (UCMP 82961) was found near the Betatakin Ruins at an unknown stratigraphic level (Galton 1971, 1976) (UCMP locality V6910). These specimens were referred to the genus *Ammosaurus* by Galton (1971, 1976), but this genus may be a synonym of *Massospondylus* (Olsen and Galton 1984). The type specimen of the theropod dinosaur *Segisaurus halli* was found in Segi Canyon approximately 30 m below the top of the formation (Camp 1936). A specimen of a primitive crocodilian consisting of a series of dorsal scutes and a partial pes [UCMP 61229, not 61299 as listed by Galton (1971)] was collected in Segi Canyon from what was described as the "*Segisaurus* level" (from description of locality V69146 in UCMP files). Galton (1971) identified the specimen as "*Protosuchus* sp.," but it is also similar to the postcrania of several undescribed new crocodilians from the Kayenta Formation (see Appendix) and cannot be identified at the generic level.

The type specimen of the ichnotaxon *Navahopus falcipollex* Baird 1980 was found in the upper part of the Navajo Sandstone far to the west of these specimens (approximately 45 km west of the Shonto Trading Post) near Copper Mine Trading Post (MNA locality 226). It has been identified as the trackway of a "prosauropod" dinosaur (Baird 1980).

Vermilion Cliffs

Fossil fish are extremely common in the Whitmore Point Member of the Moenave Formation along the western part of the Vermilion Cliffs of northern Arizona and southern Utah (Wilson 1967). These thinly bedded silts and muds were deposited under lacustrine conditions (Wilson 1967). The major collections have been from near Kanab, Utah, including the specimens described in detail in the literature. Three taxa have been identified: *Lepidotes walcotti* Eastman 1917, *Semionotus* cf. *gigas* (Hesse 1935), and *S. kanabensis* Schaeffer and Dunkle 1950. The exact stratigraphic position of these specimens within the member was not recorded by these authors. We have collected fossils from the type section of the Whitmore Point Member (UCMP locality V85009; field numbers JMC 83–31 to 33) near Fredonia, Arizona (see Fig. 23.1) that include

the fragmentary remains of perhaps several large tetrapods, as well as several fish specimens and coprolites. All specimens came from a horizon very near the top of the member. Pollen samples were collected from the same section by C. Hotton and are discussed later in this chapter. The pollen samples are from a horizon 2–3 m beneath the level from which the fossil vertebrates were collected.

Tuba City–Moenkopi Wash

The type (UCMP 37302) and a referred specimen (UCMP 37303) of the theropod dinosaur *Dilophosaurus wetherilli* Welles (1954, 1970, 1984) were found in close proximity (UCMP locality V4214) in the silty facies of the Kayenta in Moenkopi Wash near Tuba City, Arizona (see Fig. 23.3). A third theropod skeleton (UCMP 77270) (Welles 1970) was found nearby (UCMP locality V6468), but has not yet been described. Welles (1984) believes that it may represent a separate genus. Welles (1971) published a description of a measured section in this area that locates the stratigraphic positions, in ascending order:

1. The type locality of the theropod ichnotaxon *Dilophosauripus williamsi* (UCMP locality V67239)
2. The localities of the three theropod specimens 26 ft (9 m) above that
3. The type locality of the theropod ichnotaxa *Kayentopus hopii* and *Hopiichnus shingi* (UCMP locality V6898) 367 ft (121 m) above that

In his section, Welles placed the contact with the Moenave Formation at 27 ft (9 m) below the *Dilophosauripus* locality, but these beds are not certainly within the Moenave Formation. The Moenave–Kayenta contact identified by Harshbarger et al. (1957, section 4) is 31 m below this locality, and this is in our opinion more likely to be correct. Because the dip of the beds appears to change over this distance and much of the exposure is covered by alluvium, this is only a reasonable approximation. Fossils have also been found in a limestone conglomerate near the base of the Kayenta in this region (UCMP locality V82302).

The contact between the Kayenta Formation and the Dinosaur Canyon Member of the Moenave Formation in this area is not as clearly defined as in other localities of the silty facies of the Kayenta. The same contact is marked elsewhere in the Adeii Eechii Cliffs by an extensive limestone conglomerate with chert pebbles, but no such laterally continuous unit has been identified between Moenkopi Wash and Dinosaur Canyon. The presence of beds in the lower part of the Kayenta lithologically similar to the Dinosaur Canyon Member further confuses recognition of the contact. The two formations may be conformable, but there is no evidence for inter-tonguing of Dinosaur Canyon beds with the lower Kayenta beds. The contact in this region, although locally reflecting differences in lithology, is not demonstrably conformable or unconformable.

Moenkopi Point

Immediately south of Moenkopi Wash a prominent landmark known as Moenkopi Point marks the northern edge of the Adeii Eechii Cliffs (see Fig. 23.3). A locality with numerous crocodilian specimens (UCMP locality V6899) lies approximately 25 m above the base of the Kayenta formation within beds that are lithologically similar to sandstones of the Dinosaur Canyon Member of the Moenave (Fig. 23.5). The crocodilians represent two taxa that are being described by J. M. Clark; one of these appears to be similar to *Edentosuchus tienshanensis* Young 1973 from supposed Early Cretaceous beds in China (see Appendix). Two tritylodont specimens (possibly representing only one animal) – a partial mandible (V6899/130857) and a partial rostrum (V6899/130858) – are also known from the locality, but are insufficient to identify at the generic level. Coprolites and fish scales are common at the locality, and a theropod scapula referred to *Dilophosaurus* (K. Padian pers. comm.) and dinosaur footprints (S. P. Welles pers. comm.) were found

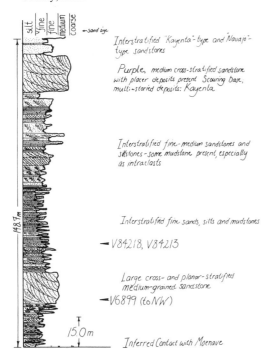

Figure 23.5. Generalized composite columnar section measured at Moenkopi Point, Coconino County, Arizona.

nearby at the same level. A small collection was made from beds higher in the formation nearer the base of Moenkopi Point, but only one specimen, a crocodilian similar to the *Edentosuchus*-like form (UCMP 130082 from locality V84218), is identifiable at a low taxonomic level (see Appendix). A few fragments of a turtle were also found nearby (V84243/130859); it is comparable to the taxon found in abundance to the south, but the fragments are insufficient for certain identification.

Dinosaur Canyon

The type and several referred specimens of *Protosuchus richardsoni* (Brown 1933, 1934; Colbert and Mook 1951) were collected from beds near the top of the Dinosaur Canyon Member of the Moenave Formation in Dinosaur Canyon due East of Cameron, Arizona (see Fig. 23.3). We collected four more specimens of this taxon in 1983 from an area near the type locality (UCMP locality V85012; JMC 83–25 to 30). Based upon photographs of the type locality in Colbert and Mook (1951, Plate 9) and field notes of S. P. Welles (June 27 to 29, 1941), our observations place the stratigraphic level of the new specimens at a horizon immediately below the level that yielded the specimens described by Colbert and Mook (1951). Footprints identified as *Batrachopus* (UCMP 130583–130596) and described by Olsen and Padian (Chapter 20) were found in a thin bed of coarse sandstone at the bottom of the beds containing the newly collected specimens. Colbert and Mook (1951, Fig. 2) placed the stratigraphic level of the type specimen in the middle of the Dinosaur Canyon beds, but our observations place it very close to the top of these beds. Within the 202 ft (62 m) section of the Dinosaur Canyon Member (Cooley, Akers, and Stevens 1964), the level of the type specimen lies 7–12 m below a white sandstone at the top of the Moenave, and the level of our new locality lies 2–4 m beneath the level of the type. F. E. Peabody located dinosaur footprints in this area and recorded that they are "below Brown's loc. level by *at least* 50 ft (16 m)" [Peabody 1941 field notes, (UCMP), p.245 (his emphasis)].

The type specimen of the crocodilian *Eopneumatosuchus colberti* Crompton and Smith 1980 is from a locality in the silty facies of the Kayenta 5.3 km north of the *Protosuchus* localities (MCZ field number 79a/7). It lies in blue silts at the base of the upper portion of the silty facies, approximately 3 m above the top of Kayenta beds with sandstones of Dinosaur Canyon lithology. Screen-washing and surface collecting of this and another site nearby at the same level (UCMP locality V82374) has produced a small fauna, including a pterosaur bone (UCMP 128227) (Padian 1984); nearly all specimens are isolated bones. A detailed study of these specimens has not yet been attempted, but the fauna includes remains of theropod dinosaurs, rhynchocephalians, turtles, tritylodonts, pterosaurs, and crocodilians. A mandible that may be referable to the *Edentosuchus*-like crocodilian (MCZ 8816) (see Appendix) was found in beds immediately overlying the *Eopneumatosuchus* locality (MCZ field number 79a/6).

The Landmark

Several specimens from both the Moenave Formation and the silty facies of the Kayenta have been collected near The Landmark, a prominent point on the Adeii Eechii Cliffs 8.2 km south of the *Protosuchus* type locality (see Fig. 23.3). We collected a single specimen that may be a lepidosaur from a sandstone in the Dinosaur Canyon Member of the Moenave Formation in this area approximately 30 m below the Kayenta–Moenave contact (UCMP locality V84240). Trackways identified as *Batrachopus* (UCMP 130586–130596) (Chapter 20) were found in the middle of the Dinosaur Canyon Member in the same area (UCMP locality V84239). Three tritylodont specimens were found by an MCZ party in blue silt beds of the Kayenta silty facies at the base of The Landmark, and one (MCZ 8842) was identified as *Kayentatherium wellesi* by H.-D. Sues (pers. comm.). The specimens were found in the middle third of the formation, but the Kayenta section has not been measured in this area.

Tonahakaad Wash

Two specimens of *Protosuchus richardsoni* have been collected from orange sandstones of the Dinosaur Canyon Member of the Moenave Formation near the southernmost tributary of Tonahakaad Wash and the northernmost part of the Red Rock Cliffs (see Fig. 23.3). One specimen (MCZ 6727 from locality 79/2a), which has been given a preliminary description by Crompton and Smith (1980), was found approximately 7 m below a yellow-orange high-angle cross-bedded sandstone (possibly a tongue of the Wingate Formation), which lies immediately below the Kayenta Formation. A second specimen (V84246/130860) was collected nearby from beds approximately 5 m below the level of the first (UCMP locality V84245). Trackways similar to those identified as *Batrachopus* (Chapter 20) were observed at approximately the same level as the UCMP *Protosuchus* specimen but were not collected. Crompton and Smith suggested that the relatively small MCZ *Protosuchus* specimen "should perhaps be placed in a separate species or new genus" (Crompton and Smith 1980, p.196), but comparison with the new specimens and with the type specimen of *Protosuchus richardsoni* suggests that the differences may be ontogenetic.

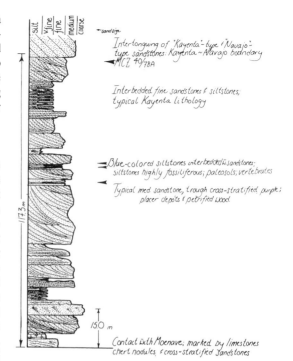

Figure 23.6. Generalized composite columnar section measured at Gold Springs, Coconino County, Arizona.

Gold Springs

The most productive area for fossil vertebrates in the silty facies of the Kayenta has been the Gold Springs area, which lies at the top of the northernmost extent of the Red Rock Cliffs (see Fig. 23.3). A section measured through the Kayenta Formation in this areas (Fig. 23.6) locates the fossil-producing horizons: the beds around the most productive levels are detailed in Figure 23.9. The vast majority of the fossils in this area were collected from two horizons, known as the "Upper and Lower Blue" layers, by Harvard field parties. The only locality in the Glen Canyon Group that has produced articulated microvertebrates (Jenkins et al. 1983) lies in the "Upper Blue" layer, and the paratype of *Scutellosaurus lawleri* Colbert 1981 is from the same layer (MNA locality 291). We made a small collection of turtles, tritylodontids, and ornithischian dinosaurs from a third "Blue" horizon, below the Lower Blue (UCMP locality V85013; field number JMC 81–19 to 25).

Willow Springs

We obtained a large collection from the Willow Springs area immediately south of Gold Springs (see Fig. 23.3). Most of the specimens were collected from a single blue silt bed, but several of the others were found in sandstone channels immediately above or below this one (Fig. 23.7). The blue silt

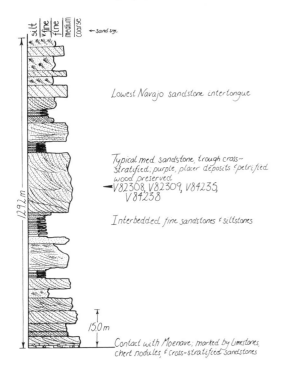

Figure 23.7. Generalized composite columnar section measured at Willow Springs, Coconino County, Arizona.

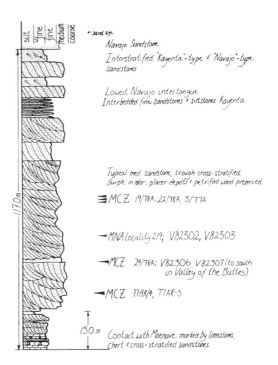

Figure 23.8. Generalized composite columnar section measured at Rock Head, Coconino County, Arizona.

bed was traced northward and is at approximately the same level as the "Lower Blue" layer of Gold Springs. Two Harvard localities in this area (MCZ field numbers 38 and 41/78a) are from the same horizon. The fauna is nearly identical to that of Gold Springs, including abundant turtles and tritylodonts and, more rarely, ornithischian and theropod dinosaurs.

Rock Head–Valley of the Buttes–Sand Mesa

The fauna from the southwesternmost exposures of the silty facies of the Kayenta near the prominent landmark Rock Head (see Fig. 23.3) is generally similar to that of the Gold Springs and Willow Springs area, but differs in several respects. A measured section is presented in Figure 23.8. Among the fossils found here are a skull of the "prosauropod" dinosaur *Massospondylus* (MCZ field number 77AR/5) (Attridge et al. 1985) and the type specimen of the pterosaur *Rhamphinion jenkinsi* (MNA V4500, MCZ field number 23/78a) (Padian 1984). These are the only specimens of these genera from the Kayenta, but other "prosauropod" and pterosaur specimens from elsewhere in the formation are not complete enough to be identified at the generic level. A nearly complete skeleton of an undescribed taxon of theropod dinosaur (MCZ field

number 3/77a) (T. Rowe pers. comm.) was found in a channel sandstone near the top of the formation, and the type locality of *Scutellosaurus lawleri* Colbert 1981 (MNA locality 219) is in a siltstone in the middle part of the formation. The turtle that is the most common element of the fauna in Gold Springs and Willow Springs is not known from the Rock Head area, although we collected a number of specimens from the Valley of the Buttes to the north of Rock Head (UCMP localities V82306, V82307). Tritylodonts are known from five specimens at Rock Head, and one (MCZ 8811; field number 22/78a) has been identified as *Kayentatherium wellesi* (H.-D. Sues pers. comm.). Isolated, water-worn bones of large animals are common in the channel sandstones at the base of Rock Head (UCMP locality V82303 for example).

Sedimentology of the Gold Springs mammal locality

With the exception of the fossiliferous blue-colored silt horizons, the sedimentary regime at Gold Springs is like that seen throughout the Adeii Eechii Cliffs – especially at Willow Springs and Rock Head – and thus we have examined it more closely to exemplify this entire region. The 28-m section through the mammal locality (see Fig. 23.9) consists of a series of four stacked medium sandstone deposits interstratified with finer sediments, which we have interpreted as channel deposits. Channel deposits in the silty facies of the Kayenta are commonly recognizable as large, purple, cross-stratified, placer-bearing, medium to fine sandstone deposits.

Petrographically, Kayenta sandstones contain approximately equivalent ratios of quartz and feldspar and less common lithic fragments (40 : 45 : 15 percent, respectively). Occasionally, however, the proportion of lithic fragments increases significantly (up to 30 percent) with respect to the other constituents. The sandstones tend to be moderately sorted, with subangular to subrounded grains that have low sphericity. They are matrix-lean, consisting of about 90 percent framework grains. The quartz is generally plain or undulose, although an occasional foliated or sutured fragment has been observed. The feldspars include abundant microcline, plagioclase, and orthoclase. The lithic fragments tend to be primarily of volcanic origin; we commonly observed intersertal and hyalophitic textures in the clasts. A few, rare metamorphic lithics were seen. Trace accessory minerals include zircon, magnetite, epidote, clinozoisite, and sphene. Celadonite is present, and it clearly is the alteration product of volcanic precursors. Illite is also present as a product of the alteration of volcanic grains in these rocks. Kayenta sandstones are cemented by hematite and calcite, the latter being dominant, and extensively eroded framework components are present in places. The hematite may be suggestive of deposition in an oxidizing environment. Several generations of clay cements are present, and relict textures have been observed. In summary, then, Kayenta sandstones are semimature units eroded from a terrain containing volcanic constituents.

The channels shown in Figure 23.9 may be migrations of the same river through time. If so, then microstratigraphic faunal succession would not be expected within this section (i.e., between the "Blue" layers) because very little time may be represented. At least two of the channels were traced over an area of about 3 km.

Paleocurrent studies show a general northwest trend (Fig. 23.10). This is in agreement with the northwest direction of paleostream flow found in Kayenta sandstones of this area by Poole (1961) and contrasts with the southwest and west directions reported by Poole (1961) for the Kayenta Formation in southern Utah and northernmost Arizona. This suggests that there were different source areas for the deposits in the two regions.

Most of the stratification is small-scale trough cross stratification, although some planar and tabular stratification was observed. Small-scale ripple deposits, including climbing ripples, are common, especially in finer sediments. The channel sandstones commonly contain petrified wood fragments that reveal evidence of water abrasion. The channel cross sections observed in this region are fairly small – about 12 m wide and 2 m deep. Lateral accretion deposits were observed in association with the channel sediments. The sedimentary structures, sediment size, and channel size are in accord with those expected in low flow intensity, sediment-rich streams of modest size (Harms, Southard, and Walker 1982).

Interstratified with the channel sandstones are fine-grained sediments, mostly siltstones, but locally some mudstones. Weathered surfaces are commonly green or gray to black, and hence some have been described as "blue layers." The green color associated with these layers is caused by the weathering and drying of an olive (5Y 5/4; Goddard et al. 1951, rock color chart) silty mudstone. Preliminary X-ray diffraction analysis shows the dominant clay species is an illite–smectite mixed layer clay. Horizons that

Figure 23.9. Columnar section of outcrop at the mammal quarry of Jenkins et al. (1983), Gold Springs, Coconino County, Arizona. Scale is in centimeters.

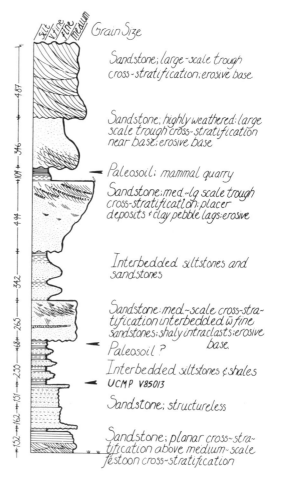

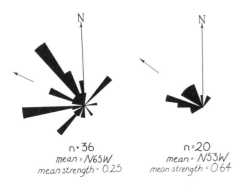

Figure 23.10. Rose diagrams of cross-bedding dip vectors measured in two channel deposits, Gold Springs, Coconino County, Arizona.

show the remnants of roots are present in association with the finer sediments, and differentiation of soil horizons is locally possible.

Our preliminary reconstruction of the depositional environment of the Kayenta silty facies in the Adeii Eechii Cliffs area suggests a floodplain drained by modest-sized, low-energy, sediment-rich streams. Our geographically restricted study does not preclude the possibility of larger rivers nearby. The association of point bars, fine sedimentary structures indicative of low energy (small- and large-scale ripples showing trough cross stratification), and soil formation suggest a morphology close to classic meandering models (Allen 1970). Water was probably abundant, and the land supported a floodplain fauna, indicated especially by the presence of aquatic turtles in the mudstones (J. H. Hutchison pers. comm.). The large volume of petrified wood we observed suggests that the floodplain was rich in plant life.

Radiometric date from the Temple Cap Sandstone

We sampled two bentonites in the Temple Cap Sandstone at Gunlock Lake near Gunlock, Utah (see Fig. 23.1) in order to obtain a radiometric date giving a youngest possible age for the Glen Canyon Group. They have an age of 166 ± 3.62 MY (with a 2σ error) when dated by the potassium–argon (K–Ar) method (R. Drake pers. comm.). Bentonite at this same locality, possibly one of the two we sampled, was dated by Marvin, Wright, and Walthall [(1965), sample 1; discussed by Odin and Obradovich (1982) as "NDS 102"], but a K–Ar date differed significantly from a rubidium–strontium date of the same sample. Their K–Ar date was 165 ± 8.25 MY, and the Rb–Sr date was 85 ± 50 MY; these dates were recalculated by Odin and Obradovich (1982) using a different constant, resulting in dates of 169 ± 8 MY and 88 ± 50 MY, respectively. Both sets of authors concluded that alteration of the bio-

tite accounted for the discrepancy. Unlike their samples, ours was treated with hydrofluoric acid to cleanse it of any alterations.

Discussion
Biostratigraphic correlations within the Glen Canyon Group

The stratigraphic distributions outlined above lead to several tentative conclusions. Contrary to conventional understanding (Lewis, Irwin, and Wilson 1961; Galton 1971), the fauna of the Moenave Formation, though poorly known, is distinct from that of the Kayenta Formation and Navajo Sandstone. *Protosuchus* is known only from the Moenave Formation, whereas crocodilian taxa from the Kayenta are demonstrably more derived than *Protosuchus* (see Appendix). Two of these taxa are from the lowermost beds of the Kayenta (UCMP locality V6899), and *Protosuchus* is known from the highest beds in the Dinosaur Canyon Member of the Moenave Formation 13 km to the south. Thus, there is evidence to suggest that there may be a depositional hiatus between the Dinosaur Canyon Member of the Moenave and the Kayenta Formation in the northern part of the Adeii Eechii Cliffs. Our observations of the geology of this contact do not contradict this possibility. It should be noted, though, that the fauna of the Elliot Formation of South Africa, which is very similar to that of the Kayenta Formation, includes crocodilians that are probably congeneric with *Protosuchus* [*Lesothosuchus* Whetstone and Whybrow 1983 and *Baroqueosuchus* Busbey and Gow 1984 are probably junior synonyms of *Protosuchus* (J. Clark pers. obs.)].

The presence of *Kayentatherium wellesi* at two separate localities in the "typical facies" of the Kayenta and throughout the silty facies from The Landmark southward (Chapter 22) suggests that most of the formation was deposited during the same biostratigraphic interval. Within the silty facies, the localities in the southern part of the Adeii Eechii Cliffs (The Landmark, Gold Springs, Willow Springs, and Rock Head) appear to be homotaxial: They share not only *Kayentatherium* but also, in most areas, the ornithischian dinosaur *Scutellosaurus* and a new taxon of cryptodiran turtle that is currently being studied (J. H. Hutchison and E. Gaffney pers. comm.). The absence of these forms from the northern part of the silty facies may be due to sampling because there are undiagnostic tritylodont and turtle specimens from Moenkopi Point and the *Eopneumatosuchus* locality, but the absence of diagnostic specimens prevents correlating these beds with the remainder of the formation. Thus, it is possible that the taxa found in the northern part of the silty facies (especially *Dilophosaurus* and three undescribed crocodilians) may not be of the same age as those from elsewhere in the formation. This is important when considering the fauna of the Kayenta Formation as a unit, as in discussions of the age of the formation, because there is no evidence at present that there is a single fauna.

The fauna of the Navajo Sandstone is insufficiently known to compare critically with that of the Kayenta Formation, but two points are worthy of note. First, the "prosauropod" dinosaur *Ammosaurus* reported from this formation (Galton 1971) may be a synonym of *Massospondylus* (Olsen and Galton 1984), which is known from the Kayenta Formation (Attridge et al. 1985). Second, the Lewis tritylodont locality, which contains the typical Kayenta tritylodontid, is almost within the Navajo Sandstone and lies above a tongue of the same. The fossil evidence at present is insufficient to determine if the vertebrate fossil localities in the Navajo Sandstone are significantly younger than those in the Kayenta Formation.

Age of the Glen Canyon Group

The age of the Glen Canyon Group and its constituent formations has been the subject of much dispute. The dispute has centered upon whether the fauna of the Kayenta Formation is Late Triassic or Early Jurassic in age. Lewis et al. (1961) and Colbert (1981) summarize the arguments for considering the age to be Late Triassic, and the arguments for an Early Jurassic age are posited by Welles (1954), Olsen and Galton (1977, 1984), Peterson and Pipiringos (1979), Olsen, McCune, and Thomson (1982), D. M. Kermack (1982), Olsen and Sues (Chapter 25), and Olsen and Padian (Chapter 20). There seems to be little question that the Kayenta fauna is similar to and, therefore, roughly equivalent in age to the faunas of the upper part of the Elliot Formation and the Clarens Formation in South Africa (Olsen and Galton 1984, i.e., the *Massospondylus* range zone of Kitching and Raath 1984), and the upper part of the Lower Lufeng Formation in China (Sigogneau-Russell and Sun 1981; Chapter 21), so the debate extends to these formations as well. Olsen and Galton (1984) provide a summary of much of this evidence; therefore, we will restrict our comments to points not covered in their paper and to differences of fact or interpretations.

The dispute has developed from the difficulty of correlating nonmarine faunas with the marine faunas upon which the standard biostratigraphic time scale is based. There is, at present, no direct evidence of marine invertebrates from any of the formations in question (the Glen Canyon Group, the Lower Lufeng Formation, and the Elliot and Clarens formations). Pollen found along with marine invertebrates has enabled palynologists to establish a pollen chronology for this time period, and pollen is much more likely to be found in these beds than are marine invertebrates. Radiometrically dated beds interstratified with marine sediments offer a second possible means of indirectly correlating nonmarine beds with the marine record.

The age of the Glen Canyon Group can be bracketed between the ages of the Chinle Formation beneath it and the San Rafael Group above it. Pollen samples from the Petrified Forest Member of the Chinle Formation 135 km to the southeast of the Adeii Eechii Cliffs have been correlated with the late Carnian (early Late Triassic) marine invertebrate age (Ash 1980), and samples from the Cameron area beneath the Adeii Eechii Cliffs are of a similar age (S. Ash pers. comm.). Plant remains are poorly known in the higher members of the Chinle Formation, but the vertebrate fauna of the Owl Rock Member is similar to that of the uppermost part of the Petrified Forest Member (R. Long pers. comm.). The vertebrate fauna of the Rock Point Member is poorly known, but isolated teeth from east central Arizona have been identified as belonging to phytosaurs (Harshbarger et al. 1957). Thus, it is possible that the Glen Canyon Group could be as old as the early Norian (middle Late Triassic) stage.

The Carmel Formation, which lies unconformably above the Navajo Sandstone, has an invertebrate marine fauna in southern Utah that correlates with those from the late part of the Bajocian (early Middle Jurassic) stage (Imlay 1980). Thus, the Glen Canyon Group could be as young as middle Bajocian. A somewhat older minimum age is suggested by lithological correlations of the unfossiliferous Temple Cap Sandstone, which lies between the Carmel and Navajo Sandstone Formations in southern Utah. It has been correlated with the lithologically similar Gypsum Springs Member of the Twin Creek Limestone Formation of northern Utah (Peterson and Pipiringos 1979), which is stratigraphically equivalent to the Carmel Formation. The Gypsum Springs Member lies unconformably above the Nugget Sandstone that is equivalent to the Navajo Sandstone. The Gypsum Springs Member has an early Bajocian invertebrate fauna (Imlay 1980).

The 166 MY date from the bentonites we sampled in the Temple Cap Sandstone is anomalously young compared to the time scales of Harland et al. (1982) and Kennedy and Odin (1982). These time scales would place this date within either the Callovian or Bathonian stage, which are both younger than the Bajocian invertebrates of the Carmel Formation overlying the Temple Cap Sandstone. However, the chronometry of these time periods is very poorly documented (cf. Harland et al. 1982, Fig. 3.4f; Kennedy and Odin 1982, Table 10). Therefore, little confidence can be placed at present on correlations of the chronometric time scale with the marine stages for the Middle Jurassic. Because the radiometric date appears to be younger than the marine invertebrates of the Carmel Formation, it is not considered useful in bracketing the upper age limit of the Glen Canyon group.

Nonmarine invertebrates are known from the Glen Canyon Group, but they are so poorly known that they offer little for age determination. Lewis et al. (1961) and Colbert (1981) suggested that invertebrates collected around the turn of the century by H. E. Gregory and described by Yen (1951) provide evidence of a Late Triassic age for the Kayenta Formation, but we question the usefulness of these fossils. The specimens *may* be from the Kayenta Formation, but their locality is virtually unknown [see the discussion by Harshbarger et al. (1957, p.28)], and no further specimens of these taxa have been collected during the intensive work of the last decade. Furthermore, the Late Triassic age is based not upon correlations with marine faunas but upon the presence of one of the two species in the Late Triassic Chinle Formation (Yen 1951). Considering the general paucity of Early and Middle Jurassic nonmarine invertebrate faunas, the poor understanding of their age significance, and the poor evidence we have of their stratigraphic distributions, this evidence is particularly weak. Other invertebrates collected subsequently from the silty facies of the Kayenta and the Whitmore Point Member of the Moenave Formation likewise have not proved to be of biostratigraphic use (J. Hanley pers. comm.).

Welles has long held that the Kayenta Formation may be as young as Middle Jurassic because the features of the theropod dinosaur *Dilophosaurus wetherilli* are more advanced than those of typical Triassic forms (Welles 1954). Welles (1984) now finds that this taxon cannot be demonstrably allied with theropods from unquestionable Jurassic beds; therefore, it offers no unique evidence for a Jurassic age of the Kayenta Formation.

In her description of *Kayentatherium*, D.M. Kermack (1982) argued that its advanced tooth morphology suggested an Early or Middle Jurassic age for the Kayenta Formation. Several problems with this study are addressed by Sues (Chapter 22); further comment here is unnecessary.

A pollen sample from the Whitmore Point Member of the Moenave Formation was given a preliminary description by Bruce Cornet (in Peterson and Pipiringos 1979, p. 33). Based upon a trend in the dominance of the palynomorph *Corollina torosus* within the Newark Supergroup (Cornet and Traverse 1975), Cornet suggested a correlation with the late Sinemurian to early Pliensbachian portion of the Portland Formation of the Newark. However, a pollen sample from the same locality examined by C. Hotton (pers. comm.) failed to contain several Ju-

rassic palynomorphs identified by Cornet. A detailed examination of this flora is clearly necessary, and until then the biostratigraphic evidence it provides must be considered tentative. Over eighty pollen samples were taken from the Kayenta silty facies by Hotton as part of our project, but the sediments proved to be too oxidized, and no pollen was recovered. Pollen samples in the Kayenta Formation silty facies would be of considerable interest but are not expected to be forthcoming in the near future.

Attempts have been made to establish the age of the Glen Canyon Group based upon biostratigraphic correlations with the Late Triassic to Early Jurassic Newark Supergroup of eastern North America (Galton 1971; Olsen and Galton 1977; Olsen et al. 1982). The Newark sequence is well dated with numerous pollen records (Cornet and Traverse 1975) and radiometrically dated basalts. Trackways and fish are the only well-represented vertebrate fossils in the Jurassic part of the Supergroup. The "prosauropod" dinosaurs of the Early Jurassic Portland Formation (Galton 1976) may provide biostratigraphic evidence, but the taxonomy of "prosauropods" is currently in a state of flux. The abundant footprint faunas of the Newark do provide some evidence for an Early Jurassic age; two ichnotaxa, *Batrachopus* and *Anomoepus*, are known only from the Jurassic portion of the Newark and are also known from the Glen Canyon Group (Olsen and Galton 1977; Chapter 20). *Batrachopus* is known from the Moenave Formation, but the locality of the *Anomoepus* tracks is currently unknown. The stratigraphic distribution of *Anomoepus* offers some evidence that some part of the Glen Canyon Group is Early Jurassic, but without locality data no further conclusions can be drawn from it. If the identification of *Batrachopus* as a protosuchian crocodile (Chapter 20) is correct, however, then the evidence it offers for a Jurassic age is weakened by the presence of protosuchians in the Triassic. A protosuchian crocodile that is nearly identical to *Protosuchus* in most features *(Hemiprotosuchus leali)* is known from the Los Colorados Formation of Argentina, and the fauna of this formation found in direct association with it (J. Bonaparte pers. comm.) includes several taxa [aetosaurid, ornithosuchid, and rauisuchid "thecodonts" (Bonaparte 1971)] that are known only from the Late Triassic. Studies of fossil fish in the Newark Supergroup suggest that *Semionotus kanabensis* from the Whitmore Point Member of the Moenave is most similar to forms in the Newark that are found well into the Early Jurassic (Olsen et al. 1982). However, the study of *S. kanabensis* was only preliminary, and a more detailed study has not yet been completed.

Vertebrate faunas from fissures in Wales are similar to the vertebrate fauna of the southern part of the silty facies of the Kayenta in sharing the tritylodont *Oligokyphus* and the mammal *Morganucodon* (Jenkins et al. 1983). Pacey (unpublished dissertation cited in K. E. Kermack, Mussett, and Rigney 1981) has found evidence for correlating some of the fissure faunas with the Sinemurian stage of the Early Jurassic. Unfortunately, stratigraphy within fissures is often untrustworthy because older fossils can be mixed with younger ones, and no details of these correlations have been published. The discovery of a typical element of the supposed Early Jurassic fauna in a Late Triassic fissure (Fraser, Walkden, and Stewart 1985) further complicates this evidence.

Records of an aetosaur and "small, armored thecodonts" in the Kayenta have been noted by Colbert (1981, pp. 52 and 56). The group "Thecodontia" as used by Colbert (including aetosaurs) is not otherwise known from beds younger than the Triassic, and the presence of this group in the Kayenta was used as evidence for a Late Triassic age. However, the aetosaur specimen consists only of an isolated osteoderm, which is not diagnostic of any known aetosaur taxon (R. Long pers. comm.). The small "thecodonts" mentioned by Colbert are, in fact, protosuchian crocodilians (currently being studied by J. Clark).

To summarize, there is some evidence to suggest that at least some parts of the Glen Canyon Group along the Vermilion and Adeii Eechii Cliffs are Early Jurassic in age. This evidence includes the presence of semionotid fish, *Anomoepus* trackways, and possibly *Corollina*-dominated pollen samples. This is a highly tentative conclusion, however, and much more evidence relevant to this question needs to be found. More importantly, it should be realized that determining which side of the Triassic–Jurassic boundary these faunas are from is not as important as is determining their utility for studying faunal changes during the early part of the Mesozoic. The difficulties we found in making stratigraphic correlations within these beds due to fluvial sedimentation and a relatively sparse fossil record suggest that the Glen Canyon Group may not offer strong paleobiological evidence relevant to these faunal changes at the present time.

Note added in proof. Morales (1986) has recently noted the presence of additional dinosaur tracks in the Kayenta Formation.

Appendix: crocodilians of the Glen Canyon Group

In order to substantiate our observation that the crocodilians of the Kayenta Formation are more derived than is *Protosuchus* from the Moenave Formation, we present a brief description of the derived features of *Eopneumatosuchus colberti* and three undescribed crocodilian taxa from the Kayenta Formation.

Eopneumatosuchus colberti

Material. Holotype, MNA P1. 2460; probably also several isolated crocodilian elements from the type locality and a nearby locality.

Localities. MCZ field number 79a/7 (= UCMP V82347) near Dinosaur Canyon.

Comments. Hecht and Tarsitano (1983) have recently questioned the allocation of this taxon to the Suborder Protosuchia of the Crocodylia by Crompton and Smith (1980). They cite five reasons for excluding it from the Protosuchia and consider some of these to preclude assignment to the Crocodilia. However, contrary to their statements, there is no evidence that the quadrate, which is not preserved, was (1) "more vertical in position than in all Crocodilia" and (2) not sutured to the laterosphenoid. Indeed, the quadrate was probably less vertical than in *Protosuchus*, and the quadrate is the only element that could have articulated with the suture on the posterior edge of the laterosphenoid because the prootic clearly was not exposed on the outside of the braincase. (3) The cranioquadrate canal is *not* enclosed within the exoccipital, and this was described by Crompton and Smith (1980, p.201). (4) The complex pneumatic spaces within the basicranium are not unique to this taxon; in fact, they are not unusual among protosuchians. Similar, although less elongated, pneumatic spaces are found in *Protosuchus* (UCMP V84246/130860), *Hemiprotosuchus* (Bonaparte 1971), and the Kayenta specimens referred here to *Edentosuchus*. (5) Finally, although the supratemporal fenestrae are much larger than those of any protosuchian, they are not unusual among crocodilians. The large size of the supratemporal fenestrae suggests that *Eopneumatosuchus*, which is known only from the posterior part of the skull, is a longirostrine crocodilian; this suggests a possible relationship with the Early Jurassic teleosaurs. There are no known features of *Eopneumatosuchus* that are more primitive than the conditions found in *Protosuchus*. Reference to "flattened" teeth in this form (Crush 1984, p.153) is without foundation because the type does not preserve the dentition.

?Edentosuchus undescribed species 1

Material. UCMP 125358, 125359, 97638.

Localities. UCMP locality V6899 near Moenkopi Point.

Comments. This taxon is similar to *Edentosuchus tienshanensis* Young (1973) from the "Tugalo Group," supposed Early Cretaceous beds in China. Direct comparisons have not been made with the Chinese material, which consists of a partial skull, a pair of mandibles, and a fragmentary postcranial skeleton. Based upon the published figures, similarities include small size, bulbous teeth, a long, narrow mandibular symphysis, and an extremely short tooth row. The new form seems to differ from the figured type material in that the teeth are transversely broader than they are long.

This taxon retains many primitive crocodilian features found in *Protosuchus*, but it is advanced beyond *Protosuchus* in that the palatines form a primitive secondary palate composed of medially directed shelves that do not meet along the midline. Furthermore, there is a median depression in the pterygoids immediately posterior to the palatines that represents the probable position of the internal choana. A similar structure was described for *Orthosuchus* (Nash 1975) from the Elliot Formation of South Africa, but the position of the choana in this form suggested by Nash is posterior to that of the most advanced crocodilians, and there is no depression in the position seen in the *Edentosuchus*-like form. The posterior maxillary and dentary teeth of this form are peculiar in having two distinct cusps set transversely.

?Edentosuchus undescribed species 2

Material. UCMP 130082 is the only known specimen.

Locality. UCMP locality V84218 near Moenkopi Point.

Comments. This single specimen consists of a partial skeleton, including a skull with nearly complete mandibles and maxillae containing a complete dentition. The dentition differs from that of the first species in that the posterior teeth are more circular in shape and have numerous small cusps rather than two large ones. Because this specimen is significantly larger than the specimens of the first species (the skull is approximately twice as long), this may be an ontogenetic difference, but that degree of ontogenetic change would be remarkable in a crocodilian.

Undescribed new genus

Material. UCMP 97639 and 97640.

Locality. UCMP locality V6899 near Moenkopi Point.

Comments. This new form is poorly known, but is certainly more advanced than *Protosuchus* in that (1) the opisthotic is broadly sutured to the quadrate (Busbey and Gow 1984), (2) the quadrate lacks large fenestrae opening onto its dorsal surface, (3) the mandible has a posteriorly projecting retroarticular process, and (4) the preorbital region is broader. This form lacks the bulbous teeth of the *Edentosuchus*-like form and appears to lack palatine shelves as well.

Acknowledgments

The fieldwork necessary for this project was only possible through the generosity of the Navajo people, who allowed us to study the rocks in areas that are holy to them. We especially thank the Nez family of Gold Springs, the Manygoat family of Willow Springs, and Tom Lee of Moenkopi Point for allowing us on their lands. The invaluable advice of Cayce Boone and the permission of the Cameron, Tuba City, and Coalmine Mesa chapter houses are also greatly appreciated. The Kaibab–Paiute Tribe was similarly generous in allowing us access to Whitmore Point. K. Padian (University of California, Berkeley, Paleontology Department) guided the project along and was indefatigable in working out numerous problems over the years; we owe him a great deal. Assistance in the field was rendered by Cole Abel, Emily Cobabe, Steve Gatesy, Carol Hotton, C. Kellner, and K. Padian. C. Hotton collected and processed innumerable Kayenta samples in a valiant but fruitless effort to find palynomorphs. Will Downs screenwashed the sediments from V82374 and eased us into Kaoyenta fieldwork. Farish Jenkins, Jr. and Chuck Schaff of Harvard generously made available field locality data for their collections, and they gave welcome advice on many aspects of the project. The K–Ar dating of the bentonite from Gunlock was determined by R. Drake of the University of California, Berkeley, Geology and Geophysics Department. Sam Welles (University of California, Berkeley, Museum of Paleontology) was an inspiration to the project from its inception and provided field data for the many specimens he collected. Ned Colbert and Mike Morales of the Museum of Northern Arizona arranged our weekend stays at the Museum of Northern Arizona and were helpful in many other ways. For access to specimens in their care we thank T. Bown (United States Geological Survey, Denver), E. Gaffney (American Museum, New York), J. H. Hutchison (University of California Museum of Paleontology), F. Jenkins and C. Schaff (Museum of Comparative Zoology), and M. Morales (Museum of Northern Arizona). F. Peterson and K. Padian made many useful comments on an earlier draft of this chapter. Funding for the project was provided by National Geographic Society Grants 2327–81 and 2484–82 to K. Padian. We would also like to thank S. Ash, R. Cifelli, W. A. Clemens, Jr., R. Dott, J. Hanley, J. Hopson, J. H. Hutchison, R. Long, P. Olsen, F. Peterson, T. Rowe, and H.-D. Sues for their help and advice. The observations and conclusions of this study are, of course, our own, and we take full responsibility for their veracity. Authorship of this chapter is alphabetical: no seniority is implied.

References

Allen, J. R. L. 1970. A quantitative model of grain size and sedimentary structures in lateral deposits. *Geol. J.* 7:129–46.

Ash, S. R. 1980. Upper Triassic floral zones of North America. *In* D. L. Dilcher and T. N. Taylor (eds.), *Biostratigraphy of Fossil Plants* (Stroudsburg, Pennsylvania: Dowden, Hutchinson, and Ross), pp. 153–70.

Attridge, J., A. W. Crompton, and F. A. Jenkins, Jr. 1985. Southern African Liassic prosauropod *Massospondylus* discovered in North America. *J. Vert. Paleontol.* 5:128–32.

Baird, D. 1980. A prosauropod dinosaur trackway from the Navajo Sandstone (lower Jurassic) of Arizona. *In* Jacobs, L. (ed.), *Aspects of Vertebrate History*. (Flagstaff, Arizona: Museum of Northern Arizona Press), pp. 219–30.

Baker, A. A., C. H. Dane, and J. B. Reeside, Jr. 1936. Correlation of the Jurassic formations of parts of Utah, Arizona, New Mexico, and Colorado. *Prof. Pap. U.S. Geol. Surv.* 183:1–66.

Bonaparte, J. F. 1971. Los tetrapodos del sector superior de la Formación Los Colorados, La Rioja, Argentina. *Opera Lilloana* 22:1–183.

Brady, L. F. 1935. Preliminary note on the occurrence of a primitive theropod in the Navajo. *Am. J. Sci.* 30:210–15.

   1936. A note concerning the fragmentary remains of a small theropod recovered from the Navajo Sandstone in northern Arizona. *Am. J. Sci.* 31:150.

Brown, B. B. 1933. An ancestral crocodile. *Am. Mus. Novitates* 683:1–4.

   1934. A change of names. *Science* 79:80.

Bunker, C. M. 1950. Theropod saurischian footprint discovery in the Wingate (Triassic) Formation. *J. Paleontol.* 31(5):973.

Busbey, A. B., III, and C. E. Gow. 1984. A new protosuchian crocodile from the upper Triassic Elliot Formation of South Africa. *Palaeontol. Afr.* 25:127–49.

Camp, C. L. 1930. A study of the phytosaurs, with description of new material from western North America. *Mem. Univ. Calif.* 10:1–161.

    1936. A new type of small bipedal dinosaur from the Navajo Sandstone of Arizona. *Bull. Univ. Calif. Dept. Geol. Sci.* 24:39–56.

Colbert, E. H. 1981. A primitive ornithischian dinosaur from the Kayenta Formation of Arizona. *Bull. Mus. North. Ariz.* 53:1–61.

Colbert, E. H., and C. C. Mook. 1951. The ancestral crocodilian *Protosuchus. Bull. Am. Mus. Nat. Hist.* 97:147–82.

Cooley, M. E., J. P. Akers, and P. R. Stevens. 1964. Geohydrologic data in the Navajo and Hopi Indian Reservations, Arizona, New Mexico, and Utah. Part 3, selected lithologic logs, drillers' logs, and stratigraphic sections. *Water Res. Rept. Ariz. State Land Dept.* 12C:1–157.

Cooley, M. E., J. W. Harshbarger, J. P. Akers, and W. F. Hardt. 1969. Regional hydrogeology of the Navajo and Hopi Indian Reservations, Arizona, New Mexico and Utah. *Prof. Pap. U.S. Geol. Surv.* 521A:1–61.

Cornet, B., and A. Traverse. 1975. Palynological contributions to the chronology and stratigraphy of the Hartford Basin in Connecticut and Massachusetts. *Geosci. and Man* 11:1–33.

Crompton, A. W., and K. K. Smith. 1980. A new genus and species of crocodilian from the Kayenta Formation (Late Triassic?) of Northern Arizona. *In* Jacobs, L. (ed.), *Aspects of Vertebrate History* (Flagstaff, Arizona: Museum of Northern Arizona Press), pp. 193–217.

Cross, W. 1908. The Triassic portion of the Shinarump group, Powell. *J. Geol.* 16:97–123.

Crush, P. J. 1984. A late Upper Triassic sphenosuchid crocodilian from Wales. *Palaeontology* 27(1):131–57.

Eastman, C. R. 1917. Fossil fishes in the collections of the United States National Museum. *Proc. U.S. Natl. Mus.* 52:235–304.

Faul, H., and W. A. Roberts. 1951. New fossil footprints from the Navajo(?) Sandstone of Colorado. *J. Paleontol.* 25:266–74.

Fraser, N. C., G. M. Walkden, and V. Stewart. 1985. The first pre-Rhaetic therian mammal. *Nature (London)* 314:161–3.

Galton, P. 1971. The prosauropod dinosaur *Ammosaurus*, the crocodile *Protosuchus*, and their bearing on the age of the Navajo Sandstone of northeastern Arizona. *J. Paleontol.* 45:781–95.

    1976. Prosauropod dinosaurs (Reptilia : Saurischia) of North America. *Postilla Yale Peabody Mus.* 169: 1–98.

Goddard, E. N. (Chairman). 1951. *Rock-Color Chart*

(New York: Geology Society of America), pp.1–6.

Gregory, H. E. 1917. Geology of the Navajo country – a reconnaissance of parts of Arizona, New Mexico, and Utah. *Prof. Pap. U.S. Geol. Surv.* 93:1–161.

Harland, W. B., A. V. Cox, P. G. Llewellyn, C. A. G. Pickton, A. G. Smith, and R. Walters. 1982. *A Geologic Time Scale* (New York: Cambridge University Press), pp. 1–131.

Harms, J. C., J. B. Southard, and R. G. Walker. 1982. Structures and sequences in clastic rocks, *Society of Economic Paleontology and Minerology Short Course No. 9*, Chaps. 2–3.

Harshbarger, J. W., C. A. Repenning, and J. H. Irwin. 1957. Stratigraphy of the uppermost Triassic and the Jurassic rocks of the Navajo country. *Prof. Pap. U.S. Geol. Surv.* 291: 1–74.

Hecht, M. K., and S. F. Tarsitano. 1983. On the cranial morphology of the Protosuchia, Notosuchia and Eusuchia. *N. Jahrb. Geol. Pal. Mh.* 1983(11):657–68.

Hesse, C. J. 1935. *Semionotus* cf. *gigas* from the Triassic of Zion Park, Utah. *Am. J. Sci.* 29:526–31.

Imlay, R. W. 1980. Jurassic paleobiogeography of the conterminous United States in its continental setting. *Prof. Pap. U.S. Geol. Surv.* 1062:1–134.

Jenkins, F. A., Jr., A. W. Crompton, and W. R. Downs. 1983. Mesozoic mammals from Arizona: new evidence on mammalian evolution. *Science* 222:1233–5.

Kennedy, W. J., and G. S. Odin. 1982. The Jurassic and Cretaceous time scale in 1981. *In* Odin, G. S. (ed.), *Numerical Dating in Stratigraphy* (New York: Wiley), pp.557–92.

Kermack, D. M. 1982. A new tritylodontid from the Kayenta formation of Arizona. *Zool. J. Linn. Soc.* 76:1–17.

Kermack, K. E., F. Mussett, and H. W. Rigney. 1981. The skull of *Morganucodon. Zool. J. Linn. Soc.* 71:1–158.

Kitching, J. W., and M. A. Raath. 1984. Fossils from the Elliot and Clarens Formations (Karoo sequence) of the northeastern Cape, Orange Free State and Lesotho, and a suggested biozonation based on tetrapods. *Palaeontol. Afr.* 25:111–25.

Langston, W. 1973. The crocodilian skull in historical perspective. *In* Gans, C., and T. S. Parsons (eds.), *Biology of the Reptilia*, Vol. 4 (New York: Academic Press,), pp.263–84.

Lewis, G. E. 1958. American Triassic mammal-like vertebrates. *Bull. Geol. Soc. Am.* 69:1735.

Lewis, G. E., J. H. Irwin, and R. F. Wilson. 1961. Age of the Glen Canyon Group on the Colorado Plateau. *Bull. Geol. Soc. Am.* 72:1437–40.

Marvin, R. F., J. C. Wright, and F. G. Walthall. 1965. K–Ar and Rb–Sr ages of biotite from the Middle Jurassic part of the Carmel Formation. *Prof. Pap. U.S. Geol. Surv.* 525B:104–7.

Morales, M. 1986. Dinosaur tracks in the Lower Jurassic Kayenta Formation near Tuba City, Arizona. *In* Lockley, M. (ed.), *Dinosaur Trackways* (Denver: University of Colorado Department of Geology, Special Publication No. 1), pp. 14–16.

Nash, D. 1975. The morphology and relationships of a crocodilian, *Orthosuchus stormbergi*, from the upper Triassic of Lesotho. *Ann. South Afr. Mus.* 67:227–329.

Odin, G. S., and J. D. Obradovich. 1982. NDS 102. *In* Odin, G. S. (ed.), *Numerical Dating in Stratigraphy* (New York: Wiley), pp. 766–7.

Olsen, P. E., and P. Galton. 1977. Triassic–Jurassic tetrapod extinctions: are they real? *Science* 197:983–6.

    1984. A review of the reptile and amphibian assemblages from the Stormberg of Southern Africa with special emphasis on the footprints and the age of the Stormberg. *Palaeontol. Afr.* 25:87–110.

Olsen, P. E., A. R. McCune, and K. S. Thomson. 1982. Correlation of the early Mesozoic Newark Supergroup by vertebrates, principally fishes. *Am. J. Sci.* 282:1–44.

Padian, K. 1984. Pterosaur remains from the Kayenta Formation (?early Jurassic) of Arizona. *Paleontology* 27(2):407–13.

Peterson, F., and G. N. Pipiringos. 1979. Stratigraphic relations of the Navajo Sandstone to Middle Jurassic formations, southern Utah and northern Arizona. *Prof. Pap. U.S. Geol. Surv.* 1035B:1–43.

Phoenix, D. A. 1963. Geology of the Lees Ferry area, Coconino County, Arizona. *Bull. U.S. Geol. Surv.* 1137:1–86.

Pipiringos, G. N., and R. B. O'Sullivan. 1978. Principal unconformities in Triassic and Jurassic rocks, western interior United States – a preliminary survey. *Prof. Pap. U. S. Geol. Surv.* 1035A:1–29.

Poole, F. G. 1961. Stream directions in Triassic rocks of the Colorado Plateau. *Prof. Pap. U.S. Geol. Surv.* 424C(199):139–41.

Schaeffer, B., and D. Dunkle. 1950. A semionotid fish from the Chinle Formation, with consideration of its relationships. *Am. Mus. Novitates.* 1457:1–29.

Sigogneau-Russell, D., and A.-L. Sun. 1981. A brief review of Chinese synapsids. *Geobios* 14(2):275–9.

Stokes, W. L. 1978. Animal tracks in the Navajo-Nugget Sandstone. *Contrib. Geol., Univ. Wyoming.* 16(2):103–7.

Welles, S. P. 1954. New Jurassic dinosaur from the Kayenta Formation of Arizona. *Bull. Geol. Soc. Am.* 65:591–8.

    1970. *Dilophosaurus* (Reptilia: Saurischia), a new name for a dinosaur. *J. Paleontol.* 44(5):989.

    1971. Dinosaur footprints from the Kayenta Formation of northern Arizona. *Plateau* 44(1):27–38.

    1984. *Dilophosaurus wetherilli* (Dinosauria, Theropoda) osteology and comparisons. *Palaeontographica Abt. A.* 185(4–6):85–180.

Whetstone, K., and P. Whybrow. 1983. A "cursorial" crocodilian from the Triassic of Lesotho (Basutoland), southern Africa. *Occ. Pap. Mus. Nat. Hist. Univ. Kansas* 106:1–37.

Wilson, R. F. 1965. Triassic and Jurassic strata of southwestern Utah. Geology and Resources of South-central Utah – Resources for Power. *Utah Geol. Soc. Guidebook Geol. Utah* 19:31–46.

    1967. Whitmore Point, a new member of the Moenave Formation in Utah and Arizona. *Plateau* 40(1):29–40.

Yen, T.-C. 1951. Some freshwater gastropods from northern Arizona. *Am. J. Sci.* 249:671–5.

Young. C.-C. 1973. *Edentosuchus tienshanensis* from the Early Cretaceous of the Dzungar Basin in Sinkiang, China. *Mem. Inst. Vert. Paleontol. Peking* 11:37–44 [in Chinese].

V Macroevolutionary patterns of the Triassic–Jurassic transition

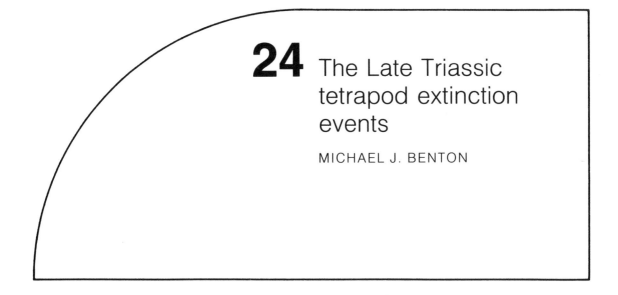

24 The Late Triassic tetrapod extinction events

MICHAEL J. BENTON

Introduction

The Triassic period (245–208 MYA) is widely recognized as having been crucial in the evolution of the tetrapods. During that period, many of the typical Late Paleozoic tetrapods – labyrinthodonts, procolophonids, and mammal-like reptiles – disappeared, or declined markedly, and new groups – dinosaurs, crocodiles, pterosaurs, lepidosaurs, testudinates, and mammals – appeared for the first time. The changeover has been described in terms of a competitively based replacement, a sudden "mass extinction," or something between these two extremes. Recently, there has been some controversy over both the pattern of the replacement and the mechanisms that may have produced that pattern. Colbert (1958a,b, 1966), Bakker (1977), Olson (1982), Wild (1982), Tucker and Benton (1982), and Benton (1983a,b, 1984a, 1985a) have noted a mass extinction among tetrapods in the Late Triassic, whereas Charig (1979, 1980, 1984), Bonaparte (1982), and others have seen the replacement as having been gradual (i.e., lasting for 25–30 MY). Furthermore, most authors have argued that the replacement was largely competitively based, whether or not they recognized a mass extinction, whereas Benton (1983a,b, 1984a, 1985a,b) has argued strongly that there is no evidence for competition.

In this chapter, I intend to review briefly the evidence that has been presented for and against the competitive models for the radiation of the dinosaurs and other "new" tetrapods in the Late Triassic. I argue that "competition" is a much overused and abused concept in macroevolution. I shall present preliminary data on the changeover as it is recorded in the Keuper sediments of southwestern Germany, and then review other evidence for worldwide tetrapod extinctions in the Late Triassic.

Mass extinction or competition in the Late Triassic?

The idea of a mass extinction of tetrapods near or at the end of the Triassic is not new. Colbert (1949, 1958a,b, 1966, 1969) described the extinction of a whole range of groups, such as labyrinthodonts, procolophonids, "protorosaurs" (= prolacertiform diapsids), nothosaurs, placodonts, rhynchosaurs, "thecodontians," and most mammal-like reptiles, at the Triassic–Jurassic boundary. These animals were replaced by new groups, such as the lissamphibians, turtles, ichthyosaurs, plesiosaurs, lepidosaurs, crocodiles, dinosaurs, and mammals. These extinctions and replacements have been noted by many authors. Explanations for these events have recently shifted toward a greater emphasis on competition.

Colbert (1949) initially recognized that the origin or radiation of several of the "new" groups in the Jurassic did not result from competition. The phytosaurs "were a highly successful and very dominant group of reptiles in the final phases of Triassic history" (Colbert 1949), but they died out for unknown reasons. The crocodiles, which were already present as small terrestrial carnivores, radiated into various aquatic niches after the extinction of the phytosaurs; therefore, competition was not responsible. However, in a later account, Colbert (1958a) concluded that some of the groups disappeared at the end of the Triassic because of competition with newly evolved forms. Thus, "eosuchians" (early "lizard-like" forms) "crowded out" the procolophonids, and the thecodontians may have outcompeted the mammal-like reptiles. Colbert had special problems in accounting for the extinction of the thecodontians: "they were well adapted to their environment, and they were widely distributed over several continents in great numbers" (Colbert 1949). He could not explain how the small early dinosaurs and crocodiles could possibly have "competed" with the thecodontians and the phytosaurs. However, by 1969, Colbert argued that the thecodontians were eliminated by competitive pressure from their descendants, the dinosaurs, and likewise, that the mammal-like reptiles "vanished because of the highly progressive nature of their descendants [the mammals]. They evolved themselves into oblivion" (Colbert 1969, pp. 166–7).

The fact that many tetrapod groups died out at about the same time has suggested to several authors that an environmental change of some kind may have been involved. Colbert (1958b) noted that environments were changing in the Late Triassic to become generally more arid, and he hinted that this might have had something to do with the extinctions. A number of authors accepted this view and sought to link it with the competition-based theories for dinosaur success by arguing that different aspects of the physiology of dinosaurs gave them great advantages over the mammal-like reptiles and thecodontians in the new, more arid conditions. Some of these explanations of dinosaur superiority include: endothermy and nakedness (for heat loss) (Cox 1967; Crompton 1968), uricotely (for water-retention) and ectothermy (Robinson 1971; Hotton 1980), improved locomotor ability (Bakker 1968, 1971; Charig 1972, 1980, 1984), endothermy (Bakker 1971, 1972, 1975, 1980), or inertial homeothermy (Spotila et al. 1973; Benton 1979; Spotila 1980). The list of "explanations" for the success of the dinosaurs could fill several pages and, if nothing else, it demonstrates the ingenuity of paleontologists (and nonpaleontologists) in making up mechanisms to explain "competitive replacements."

The hypothesis that the origin of the dinosaurs resulted from their successful competition with all comers has been challenged. Tucker and Benton (1982) and Benton (1983a,b, 1984a) presented evidence that several groups of dominant terrestrial reptiles (dicynodonts, diademodonts, rhynchosaurs, as well as most cynodonts and thecodontians) died out at the same time (middle Norian, Late Triassic) and that the dinosaurs radiated only *after* that extinction event. They argued that there was no need for competitive scenarios to explain the success of the dinosaurs: The dinosaurs took their chance and radiated opportunistically into empty ecospace in the Late Triassic, just as the mammals did in the Paleocene.

There are thus two diametrically opposed hypotheses to explain the "success" of the dinosaurs: the competition-based scenarios and the mass-extinction scenario. These should be testable insofar as we can discern different patterns in the fossil record: Benton (1983a) gave sets of criteria that might allow this to be done. However, the mechanisms underlying the scenarios cannot be tested. Scenario making in paleontology is several steps removed from hard facts – it deals in probabilities, assumptions and guesswork – and it is heavily colored by personal viewpoints. This seems very clear in the attempts that paleontologists make to explain major events in the history of life, such as the Triassic tetrapod replacements. The very terminology that has been used in describing events in the Triassic is based on the assumption that large-scale competition was taking place. For example, Colbert (1958a) distinguished Paleozoic "holdovers," groups such as the labyrinthodonts, procolophonids, and the mammal-like reptiles, from "progressive" forms that arose during the Triassic, such as the lissamphibians, thecodontians, dinosaurs, and mammals. These two kinds of tetrapods have also been termed "palaeotetrapods" and "neotetrapods," respectively by Charig (1979, 1980, 1984). These authors, and many others, have assumed that a progressive group could always beat a holdover group – that the order of appearance of taxa in geological time is directly proportional to their competitive ability. The simple a posteriori observation that group A appears later in time than group B, and seems to have occupied a similar adaptive zone, is taken to prove that group A outcompeted group B. In other words, it is assumed that evolution leads to all round improvement through time in a regular machine-like way; but, who is to say that present-day mammals would "out-compete" their Pleistocene forebears in all conditions? The fact that present-day mammals live 1 MY after Pleistocene mammals does not mean that they are better and much improved creatures.

Competition in macroevolution

The role of competition in macroevolution is of particular relevance here. Many evolutionary biologists have assumed that competition is a major force in evolution. However, there is little evidence for this assumption.

A large body of recent research in community ecology has cast doubt on the all-encompassing role of competition. There is no question that competition can be shown to occur between members of the same species or of two similar species. However, it seems far from clear to many ecologists that competition actually shapes the majority of ecological communities or causes long-term evolutionary changes in species distributions or adaptations. Other factors, such as environmental fluctuations, predation, or a combination of several biotic and abiotic aspects of the ecology of a species (individualistic response) may be more important (see Connell 1980; Simberloff 1983; Strong et al. 1984; Price, Slobodchikoff, and Gaud 1984).

In macroevolutionary studies, "competition"

has been used to describe interactions between families, orders, classes, or even phyla. There are numerous problems associated with this view:

1. Confusion of pattern and mechanism. Most of the classic examples of long-term "competitive" replacements have been based on a particular kind of pattern that was observed from the fossil record. This has been called the "double-wedge" pattern by Gould and Calloway (1980): One taxon decreases in abundance through time while the other increases – correlated waxing and waning. Of course, such a pattern does not in any way prove that it was produced by competition: the two taxa might not have been interacting at all (e.g., the terrestrial flowering plants were radiating at the same time as the marine ichthyosaurs were declining), or the two taxa might have been responding differently to a new kind of predation or to one or more environmental changes (biotic and/or abiotic). "Competition" describes a mechanism, not a pattern.

2. Oversimplification. Competitive scenarios usually boil down to explanations of major faunal replacements in terms of simple key adaptations that supposedly gave their possessors great advantages. One common scenario states that the key adaptation of later archosaurs was their semierect or erect gait, and this is supposed to have been sufficient for them to vanquish the diverse mammal-like reptiles, rhynchosaurs, and the rest. This hypothesis must be a biological oversimplification and, like many such hypotheses, it does not bear close scrutiny. The rhynchosaurs, and many Middle to Late Triassic mammal-like reptiles, had semierect gait, just like many of their supposed betters, the thecodontians. However, erect gait was not the sole preserve of the dinosaurs. Most Late Triassic thecodontians were also erect: Ornithosuchidae, Rauisuchidae, Poposauridae, and Stagonolepididae, as well as some early crocodylomorphs (Saltoposuchidae) and the pterosaurs (Bonaparte 1984; Parrish 1984; Benton 1984b; Padian 1983).

3. Lack of evidence. Competition cannot be assumed as the mechanism that has produced most extinctions and mass extinctions in the history of life. The probabilities of other explanations must be assessed in any particular case.

4. Incorrect scaling of concepts. There are three points to my critique here. First, it may be wholly inappropriate to apply the terms of individual and species interactions to interactions between larger taxonomic entities. Biologists often try to describe macroevolutionary phenomena that lasted for millions of years ("geological time") in the language of present-day community ecology, which applies to events that take place over days, weeks, or years at most ("ecological time"). It seems likely that major evolutionary events involve mechanisms, such as the

causes of mass extinctions, that are quite distinct from the day-to-day events going on down in the woods.

The second problem of scaling concerns the identification of key adaptations to explain the success of whole groups. Such adaptations (e.g., endothermy or erect gait) might have been selectively advantageous to the first species that possessed them, but it is hard to see how adaptations of these kinds have relevance for higher taxa, which include many and various species. How could a particular key adaptation prove to be advantageous to all of the species within one taxon and in all situations?

The third problem of scaling of competition in macroevolution concerns the duration of selection pressures. Most so-called competitive replacements lasted over millions of generations (typical examples lasted 2–30 MY), and it is hard to see how a differential selection pressure could have been maintained for so long. The advantage, when reduced to the level of the individual organism (because we are considering natural selection), would have been so miniscule as to be indistinguishable from stochastic effects.

A detailed example: the Keuper of southwestern West Germany

The Late Triassic in the southwest of the Federal Republic of Germany (Baden–Württemberg) is represented by a succession of terrestrial sediments, the Keuper, which has yielded abundant tetrapods at various levels. I studied this succession because the vertebrate-bearing beds are better dated than most other Late Triassic terrestrial sequences. I tried to assess the patterns of faunal replacement in this single case study. and to examine the timing and nature of the radiation of the dinosaurs in particular. The present account is preliminary: some of the results are summarized in Benton (1984c).

The oldest dinosaurs

Most authors now accept that the first dinosaurs appeared in the Late Triassic. However, many general accounts published in the 1970s draw the different dinosaur lineages well back into the Middle Triassic. The records of Middle Triassic dinosaurs have arisen from three problem areas: (1) thecodontian and indeterminate remains described as those of dinosaurs, (2) imprecise definition of what a "dinosaur" is, and (3) incorrectly dated geological formations that contain dinosaurs.

Doubtful early dinosaurs

As to the first problem, a large number of remains of doubtful early dinosaurs have been recorded from the Middle, and even the Lower, Triassic of Germany and elsewhere. For example, Huene (1914,

1932) noted ten named dinosaurs from the German Muschelkalk. These have subsequently turned out to be prolacertiforms, unidentifiable archosaurs, or even ?placodonts (Wild 1973; Benton 1984c). One of the key groups of supposed early dinosaurs has been the Teratosauridae from the Middle and Late Triassic of Europe, as well as the Late Triassic and Early Jurassic of South Africa and China. However, these specimens turn out to be an assemblage of skulls and teeth of rauisuchid thecodontians (*Teratosaurus*) or Archosauria inc. sed., together with the skeletons of prosauropods (Walker 1964; Charig, Attridge, and Crompton 1965; Galton 1973; Benton 1984b,c).

The Dinosauria

The second factor that may have led to the identification of Middle Triassic dinosaurs concerns the definition of a dinosaur. Until recently, and with only a few exceptions (e.g., Bakker and Galton 1974; Bonaparte 1976), the dinosaurs were thought to have evolved as several separate lineages that derived from ancestors in the Middle Triassic or earlier. However, a remarkable consensus of opinion that the dinosaurs are a monophyletic group has now been reached by several workers who have independently carried out cladistic analyses of the archosaurs (Benton 1984c; Gauthier 1984; Norman 1984; Padian 1984; Parrish 1984; Paul 1984; Sereno 1984; Benton 1984d). The closest sister groups of the Dinosauria are *Lagosuchus*, the Ornithosuchidae, and, controversially, the Pterosauria (Padian 1984; Gauthier 1984). A monophyletic Dinosauria tends to move the origins of the group upward, possibly to the very top of the Ladinian.

Stratigraphy

The third problem that gave rise to extensive records of Middle Triassic dinosaurs was one of stratigraphy. Until the mid 1970s, many authors, especially Romer (e.g., 1970), assigned all beds that contained rhynchosaurs to the Middle Triassic; these included the Santa Maria Formation of Brazil, the Ischigualasto Formation of Argentina, the Maleri Formation of India, and the Lossiemouth Sandstone Formation of Scotland. These are now all firmly dated in the Upper Triassic, either as late Carnian (Bonaparte 1978; Chapter 25), or as early Norian (Anderson and Cruickshank 1978; Tucker and Benton 1982; Benton and Walker 1985).

In this chapter, I accept the stratigraphic assignments given by Anderson and Cruickshank (1978) and Tucker and Benton (1982) for the Early and Middle Triassic reptile beds, and the assignments given by Olsen and Galton (1977, 1984) and Olsen, McCune, and Thomson (1982) for the Late Triassic and Early Jurassic formations. The majority of horizons that Anderson and Cruickshank (1978)

placed in the early–middle Norian are reassigned to the middle–late Carnian (e.g., Santa Maria, Ischigualasto, Maleri, Lossiemouth, Argana, Lockatong, Stockton, Wolfville, Popo Agie). The Chinle and Dockum Formations are assigned wholly to the late Carnian by Olsen et al. (1982) on the basis of stratigraphic evidence from the fossil fish and palynology (Dunay and Fisher 1974, 1979; Gottesfeld 1980). I accept a late Carnian assignment for the lower portions of the Dockum and Chinle. However, some reptile-bearing upper units of both formations may still belong in the early–middle Norian (Chapters 10 and 11), and I assign them there in this chapter.

The oldest dinosaurs, then, are known from the middle to upper Carnian interval (Upper Triassic) from several places around the world: *Staurikosaurus* (Santa Maria Formation, Brazil); *Herrerasaurus, Ischisaurus*, and *Pisanosaurus* (Ischigualasto Formation, Argentina); *Saltopus* (Lossiemouth Sandstone Formation, Scotland); unnamed forms (Maleri Formation, India); *Coelophysis* (Chinle Formation, Dockum Formation, western United States); *Azendohsaurus* (Argana Formation, Morocco); and unnamed "fabrosaurids" (Weishampel and Weishampel 1982) from the Wolfville Formation of Nova Scotia, the Chinle Formation of Arizona, the Dockum Formation of Texas, the Chatham Group of North Carolina, and the New Oxford Formation of Pennsylvania. Many of the unnamed "fabrosaurids" are represented only by odd teeth and other fragments, and some may turn out not to be dinosaurs. The earliest dinosaurs from Germany occur in the Unterer Stubensandstein (early to middle Norian): a plateosaurid from Ochsenbach near Heilbronn, and a plateosaurid from Ebersbach a.d. Fils near Stuttgart.

Stratigraphy of the Upper Triassic in Germany

Fossil tetrapods have been collected from at least seventy localities in the Keuper of Baden–Württemberg in a strip that runs from Heilbronn and Schwäbisch Hall in the northeast to Donaueschingen in the southwest (Fig. 24.1). The most abundant finds come from the general areas of Stuttgart and Tübingen, and the most important museum collections are housed in those two cities. The Keuper deposits continue northeast into Franconia and Thuringia, and southwest into Switzerland and into Lorraine and Luxembourg, where further similar reptile finds have been made, but are less abundant than in Baden–Württemberg.

The lithostratigraphy of the Keuper of Baden–Württemberg is well established on the basis of detailed field observations throughout the whole region (Brenner 1973, 1978a,b), and standard sections have been drawn up (Brenner and Villinger 1981; Gwinner 1981). A typical section (Fig. 24.2) shows a sequence of largely terrestrial sediments. The marine Muschelkalk (Middle Triassic, not shown) passes up through the transitional marine and brackish Lettenkeuper (Unterer Keuper) into the terrestrial Mittlerer Keuper. The Rät (Oberer Keuper) represents a return to marine conditions. During the critical Mittlerer Keuper episode, and especially in the Stubensandstein, when dinosaurs appeared in Germany, there were no major environmental changes (Brenner 1973, 1979).

It has proved difficult so far to correlate the almost entirely terrestrial German Keuper with standard marine ammonite zones of the southern Alps. The ammonite zones of the Late Triassic (Tozer 1967, 1974, 1979) have been tentatively correlated with provisional standard palynological zones for terrestrial sediments (Visscher, Schuur-

man, and Van Erve 1980; Visscher and Brugman 1981; Anderson 1981) (see Fig. 24.3). Palynomorphs are well known from several horizons in the Keuper, but until a worldwide standard is agreed, it will be hard to date individual beds precisely. There appear to be two ways of interpreting the German Keuper, and these are shown in Figure 24.3. According to interpretation (A), the Rote Wand and Kieselsandstein are early Norian (Geiger and Hopping 1968; Fisher 1972; Fisher and Bujak 1975; Dunay and Fisher 1979; Dockter et al. 1980; Anderson 1981; Schröder 1982), while according to interpretation (B), those two horizons are late Carnian (Kozur

Figure 24.1. Localities in southwestern West Germany that have yielded Late Triassic tetrapods. Main rivers and towns are marked for orientation, and each fossil locality is coded by the stratigraphic horizon that yielded the tetrapod(s) (see explanatory box on the figure). The data came from the collections of Keuper tetrapods in Stuttgart and Tübingen, and from the paleontological and geological literature. A base map of central Europe is also given. Abbreviations: AUS, Austria; BRD, West Germany; CZ, Czechoslovakia; DDR, East Germany; FR, France; N, Netherlands; SW, Switzerland.

Figure 24.2. A summary lithostratigraphic section through the Keuper of southwestern West Germany. The rock units are named on the left, and the current stratigraphic terms are given on the right, with the standard abbreviations. Tetrapods have come from numerous horizons within this sequence (shown by arrows). Based on Brenner and Villinger (1981).

			Oberer Keuper ko
Rät			
	Knollenmergel	km5	
4. Stubensandstein	Oberer		
3. Stubensandstein		Stuben-	
2. Stubensandstein	Mittlerer	sandstein	
		km4	Mittlerer Keuper km
1. Stubensandstein	Unterer		
Obere Bunte Mergel			
Kieselsandstein	Kieselsandstein	km3	
Untere Bunte Mergel	Rote Wand		
	Schilfsandstein	km2	
	Gipskeuper	km1	
Lettenkeuper			Unterer Keuper ku

Figure 24.3. Biostratigraphy of the Upper Triassic. The ammonite zones (* = Tozer 1979, 1974, 1979) are tentatively correlated with provisional palynological zones (** = Visscher et al. 1980, Visscher and Brugman 1981, Anderson 1981). See the text for explanations of the alternative stratigraphic and chronometric assignments.

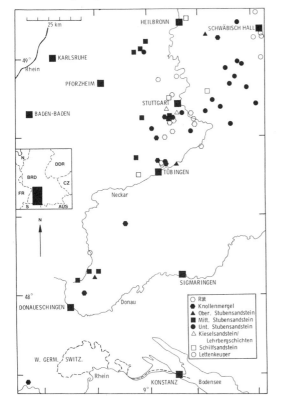

Ammonite zones *			Palynological zones **	Zonations of German Keuper (A)	(B)	Dates (Ma) (1)	(2)	(3)
(Rhaetian)		Crickmayi Amoenum Cordilleranus	Rhaetian/ Sevatian	Rät	Rät	213	204	208
Norian	U	Columbianus Rutherfordi Magnus	U. Norian (Alaunian)	Knollenmergel Stubensandstein	Knollenmergel Stubensandstein			
	M							
	L	Dawsoni Kerri	L. Norian (Lacian)	Kieselsandstein Rote Wand		225	220	225
Carnian	U	Macrolobatus Welleri Dilleri	Tuvalian	Schilfsandstein	Kieselsandstein Rote Wand			
	L	Nanseni Obesum	Julian Cordevolian	Gipskeuper	Schilfsandstein Gipskeuper	231	229	230
Ladinian	U	Sutherlandi Maclearni Meginae	Langobardian	Lettenkeuper	Lettenkeuper			

1975; Gall, Durand, and Muller 1977; Olsen et al. 1982). Magnetostratigraphic evidence indicates that the Schilfsandstein is latest Carnian in age (Hahn 1984), and this favors interpretation (A). For the purposes of the rest of this study, interpretation (A) is followed: If (B) had been selected, my conclusions would be little changed.

A final problem with the stratigraphy concerns the numerical ages of the various formations. Three recently published time scales (Odin 1982; Harland et al. 1982; Palmer 1983) offer different age dates for the stage boundaries (1, 2, and 3, respectively, in Fig. 24.3). I have arbitrarily selected Palmer's (1983) dates, as the more recent compilation. Again, my conclusions would not be materially affected by the use of either of the other two time scales.

Distribution of tetrapods in the southwestern West German Keuper

Tetrapod remains have been collected from numerous localities (Fig. 24.1) and horizons (Fig. 24.2) in the Baden–Württemberg Keuper. The largest collections of these are preserved in the Staatliches Museum für Naturkunde Stuttgart and the Institut für Geologie und Paläontologie der Universität Tübingen, with smaller collections in the museums in München, Erlangen, Göttingen, Berlin, and London. I examined the specimens in all of these collections, and took notes of locality, stratigraphic horizon, material present, and collecting data for each specimen. I attempted to track down each locality precisely with the help of topographic and geological maps and to establish the precise stratigraphic horizon that had yielded reptiles by the use of the collecting data, the general literature, and geological maps and memoirs. This work was done in the excellent libraries attached to the museums in Stuttgart and Tübingen. I then made up a sheet for each locality, listing all of the tetrapod specimens that had been collected, and estimated the minimum numbers of individual animals present at each. For this, I attempted to count only skulls, or only left femora of a particular species. Full details of localities, stratigraphy and species counts will be published elsewhere.

For the present general study, I wanted to assess the approximate changes in the relative composition of the faunas through time. I added up the numbers of individuals from each locality that had come from a particular horizon, such as the Schilfsandstein or the Unterer Stubensandstein, and converted these to percentages of the totals. To my knowledge, no tetrapods have been found in the Gipskeuper or the Rote Wand, and only very few specimens are known from the Kieselsandstein or Bunte Mergel (Lehrbergschichten). I did not make

estimates of tetrapod distributions in the Rät bonebeds. Furthermore, the figures for the Lettenkeuper were based on data for the Kupferzell excavation (Wild 1980). Both the Lettenkeuper and the Rät represent very different environments of deposition from the remaining Mittlerer Keuper – both show clear marine–brackish water influence. The significant faunal changes that are of interest here occurred in the more typical floodplain sandstone deposits of the Mittlerer Keuper, and the faunal changes are not associated with any obvious environmental change.

The results of the preliminary study (Table 24.1 and Fig. 24.4) show that the earlier faunas were heavily dominated by amphibians (Lettenkeuper, Schilfsandstein). The gap in the record between the Schilfsandstein and the Stubensandstein is unfortunate, because by early Stubensandstein times, the amphibians were much rarer (ca. 10 percent of the fauna). The commonest animals were early turtles (*Proganochelys and Proterochersis*). Thecodontians (rauisuchids and phytosaurs) were present, as well as a couple of plateosaurid dinosaurs, the first dinosaurs from Germany (ca. 7 percent of the fauna).

By middle Stubensandstein times, the proportion of dinosaurs had risen to 21 percent, and several genera are known: the prosauropods *Plateosaurus, Sellosaurus, Thecodontosaurus*, and *Efraa-*

Figure 24.4. Faunal changes through southwestern German Keuper (Late Triassic). The time scale for stage boundaries is from Palmer (1983), and the Keuper formations are spaced arbitrarily through the whole time span. Nonmarine tetrapod faunas are known from several horizons, but only those from the Lettenkeuper, Schilfsandstein, Stubensandstein, and Knollenmergel were abundant enough to give data on proportional composition. The total number of specimens of each genus that had been found in each formation was assessed (see text for details), and approximate percentages were calculated for each (Table 24.1). Labyrinthodonts dominated early faunas, but the dinosaurs radiated rapidly in Stubensandstein times. Abbreviations: DINO, dinosaurs; LABY, labyrinthodonts; PHYT, phytosaurs; TEST, testudinates; THEC, thecodontians; VAR, various (odd diapsids, ?cynodonts).

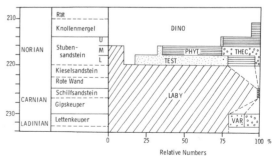

Table 24.1 *Typical nonmarine tetrapod faunas from the Keuper southwestern West Germany (FRG).*

		No.	%
Lettenkeuper (Wild 1980)[a]			
LABY	*Plagiosternum*		70
	Mastodonsaurus, etc.		10
DIAP	*Tanystropheus*, etc.		6
THEC	rauisuchid; indet.		10
THER	? cynodonts		4
Schilfsandstein			
LABY	*Metoposaurus*	8	12
	Cyclotosaurus	50+	?85
THEC	*Belodon*	1 ⎫	
	Dyoplax	1 ⎬	3
	indet.	2 ⎭	
Unterer Stubensandstein			
LABY	*Cyclotosaurus*	2 ⎫	10
	Plagiosaurus	1 ⎭	
TEST	*Proganochelys/Proterochersis*	18	60
THEC	*Teratosaurus*	1	3
	phytosaur	6	20
PROS	"plateosaurid"	2	7
Mittlerer Stubensandstein			
LABY	*Cyclotosaurus*	10 ⎫	
	Plagiosternum	4 ⎬	10
	Metoposaurus	3 ⎭	
TEST	*Proganochelys*	3	2
THEC	*Aetosaurus*	31	19
	Teratosaurus	14	8
	phytosaurs	65	39
PROS	*Plateosaurus/Sellosaurus*	22 ⎫	
	Thecodontosaurus	4 ⎬	18
	Efraasia	3 ⎭	
COEL	*Procompsognathus*	2 ⎫	3
	Halticosaurus	3 ⎭	
CROC	*Saltoposuchus*	2	1[a]
Oberer Stubensandstein			
TEST	*Proganochelys*	3	6
THEC	phytosaurs	9	17
PROS	*Plateosaurus*	41	77
Knollenmergel[a]			
THEC	phytosaur	1	5
PROS	*Plateosaurus*	19	95

Note: The data are based mainly on the collections in Stuttgart and Tübingen. Abbreviations: COEL, Coelurosauria; CROC, Crocodylia; DIAP, Diapsida (odd groups); LABY, Labyrintho-dontia; PROS, Prosauropoda; TEST, Testudines; THEC, "thecodontians"; THER, Theropoda. *Note:* [a]Estimated percentages.

sia, and the theropods *Procompsognathus* and *Halticosaurus*. Phytosaurs, rauisuchids, and aetosaurs were also fairly abundant. The late Stubensandstein faunas consist almost entirely of the dinosaur *Plateosaurus* (ca. 77 percent of the fauna) with rarer turtles and phytosaurs. By Knollenmergel times, the proportion of dinosaurs had risen to 95 percent.

These data from the German Keuper show an increase in the proportion of dinosaurs present in typical terrestrial faunas from 7 to 95 percent in a time span of 8 MY or so, and the main increase, from 21 to 77 percent, occurred in as little as 2–4 MY (Mittlerer–Oberer Stubensandstein, Middle Norian). However, the data cannot simply be taken as defining the shape of the adaptive radiation of the dinosaurs. As mentioned earlier, there are clearly collection and preservation biases that cannot be precisely assessed, but the changes in proportions (7–95 percent) are probably large enough to have some biological meaning. More important is the fact that the generic diversity of dinosaurs declined from the Mittlerer Stubensandstein to the Knollenmergel; there were five or six genera of dinosaurs in the former, and probably only one in the latter. The "radiation" from 21 to 95 percent was all *Plateosaurus*. A second problem is that *Plateosaurus* is already quite an advanced dinosaur, and dinosaurs had appeared earlier in other parts of the world (see above). The timing of events suggests that what we are seeing in Germany is the result of an immigration of dinosaurs from elsewhere, and the sample of dinosaurs from the Knollenmergel is probably restricted by preservation bias.

The absence of mammal-like reptiles and rhynchosaurs in Germany makes it hard to compare events there with the typical Gondwana pattern. The oldest known dinosaurs are middle or late Carnian in age (see above) – the dinosaurs in North and South America arose at least 8–10 MY before they reached Germany. The oldest European dinosaur, *Saltopus* (if it is a dinosaur), also dates from the late Carnian. The earliest clearly identifiable true dinosaurs were podokesaurids (e.g., *Coelophysis*), and this group is known from the Mittlerer Stubensandstein *(Halticosaurus, Procompsognathus)*. A first for Germany, however, was the appearance of prosauropods, which were diverse in the Mittlerer Stubensandstein (*Plateosaurus, Sellosaurus, Efraasia, Thecodontosaurus*), and then spread to other parts of Europe and to South America in the late Norian, and to South Africa, North America, and Southeast Asia in the Early Jurassic.

The early evolution of dinosaurs is still obscure in many respects, but there is evidence now that the small- to medium-sized podokesaurids and others radiated (i.e., became diverse and widespread) to some extent in the middle and late Carnian (225–8

MY), and then much more so in the middle Norian (215–21 MY). The prosauropods became established and radiated in the middle Norian, as seen in Germany, and then spread worldwide in the late Norian, and especially in the Early Jurassic. The middle Norian radiation of podokesaurids and of prosauropods occurred some 4–5 MY *after* the extinction of a range of thecodontian and mammal-like reptile groups at the end of the Carnian, and the Early Jurassic radiations again occurred *after* the extinction of the last thecodontians (see below).

Evidence for mass extinctions of nonmarine tetrapods in the Late Triassic

Several authors (e.g., Colbert 1949, 1958b; Bakker 1977; Olson 1982; Tucker and Benton 1982; Wild 1982; Benton 1983a,b, 1984a) have noted a mass extinction of nonmarine tetrapods in the Late Triassic, or at the Triassic–Jurassic boundary. There are many problems associated with identifying mass extinctions among fossil vertebrates (the patchy record and stratigraphic uncertainty), but several kinds of evidence point to the occurrence of *two* such events in the Late Triassic.

The term "mass extinction" can have different meanings. In this chapter, I use the term only as a descriptor of a *pattern* that may be observed in the fossil record, and I define a mass extinction as a genuine fall in diversity that occurred during a stratigraphic stage. This definition calls for the observation of a drop in the number of taxa present when one stratigraphic stage is compared with its forerunner, and the strong belief that the drop is not simply the result of a gap in the fossil record (produced by preservation failure or collection failure). The restriction of time to a stratigraphic stage is simply a pragmatic measure of the maximum resolution that is possible for most terrestrial vertebrate deposits – we may believe that the mass extinction occurred "overnight," but usually proof of that is lacking. I do not imply that a mass extinction need be caused by an excessively high extinction rate [cf. Raup and Sepkoski's (1982) statistical test] because a marked fall in diversity may equally be caused by a low origination rate. In other words, the definition of a mass extinction in terms of rates would imply the assumption of one kind of *mechanism* as having been involved. I believe that the study of rates, and of other aspects of mass extinctions, such as biological and physical causes, must be secondary to the observation of a mass extinction pattern.

Family diversity analysis

A plot of the numbers of families of nonmarine tetrapods present through time was produced (Fig. 24.5). This study is concerned with the fossil record of nonmarine tetrapods: this is taken to include terrestrial, freshwater and flying forms, but excludes fully marine families. A listing of all of the families of terrestrial tetrapods was made using the most recent taxonomic reviews, as well as secondary literature, to the end of 1984 [details of the main sources of data are given in Benton (1985a)]. A small number of extinct families was excluded from the analysis because they were based on single, often incomplete, specimens. This left a total of 730 families, of which 469 are extinct. The range in geological time for each family was determined from the most recent available literature, and this was resolved to the level of the stratigraphic stage (duration, 2–19 MY; mean duration, 6 MY). The time scale selected was that of Palmer (1983), which is based on several recent compilations, including Harland et al. (1982), Odin (1982), and Snelling (in press). The term "Rhaetian" refers to a stage at the very top of the Triassic that is often included in the Norian now (Tozer 1979; Hallam 1981).

The graph of tetrapod family diversity versus time shows several declines, including one in the Late Triassic (Fig. 24.5). Total family diversity fell from 25 in the Norian to 22 in the "Rhaetian" and

17 in the Hettangian. The Norian–"Rhaetian" drop in diversity represents a loss of 28 percent of families that were present at the start (cf. the Maastrichtian drop of 14 percent).

Extinction, origination, and diversification rates were then calculated from the data on non-marine tetrapod family diversity. Total extinction (R_e) and total origination (R_s) rates were calculated as the number of families that disappeared or appeared, respectively, during a stratigraphic stage, divided by the estimated duration of that stage (Δt):

$$R_e = \frac{E}{\Delta t} \quad \text{and} \quad R_s = \frac{S}{\Delta t}$$

where E is the number of extinctions and S is the number of originations. Per taxon extinction (r_e) and origination (r_s) rates were calculated by dividing the total rates by the end-of-stage family diversity D (Sepkoski 1978):

$$r_e = \frac{1}{D} \cdot \frac{E}{\Delta t} \quad \text{and} \quad r_s = \frac{1}{D} \cdot \frac{S}{\Delta t}$$

The per taxon rates can be seen as the "probability of origin" or the "risk of extinction." The diversi-

Figure 24.5. Diversity (total number of families recorded in a stage) through time for families of terrestrial tetrapods. The upper curve shows total diversity through time, and six apparent mass extinctions are indicated by drops in diversity (see the text for explanations). These are numbered 1–6, and the relative magnitude of each drop is given in terms of the percentage of families that disappeared. Three assemblages of families succeeded each other through time; I, labyrinthodont amphibians, anapsids, mammal-like reptiles; II, early diapsids, dinosaurs, pterosaurs; III, the "modern groups": frogs, salamanders, lizards, snakes, turtles, crocodiles, birds, and mammals. These assemblages (I–III) are shown for illustrative purposes only: the data are not robust enough for a factor analysis (cf. Sepkoski 1981). Abbreviations: CARB., Carboniferous; DEV., Devonian; TRIAS., Triassic.

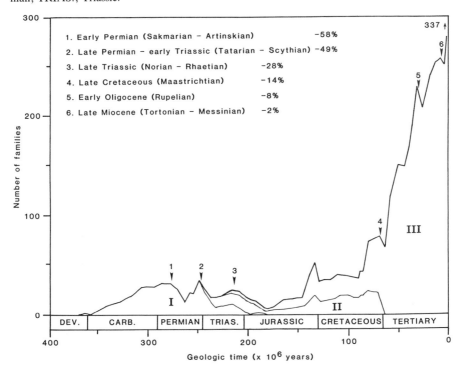

fication rate (r_d) was calculated as the difference between the per taxon origination and the per taxon extinction rate (Sepkoski 1978):

$$r_d = r_s - r_e$$

In this example, the per taxon family extinction and origination rates were not very high in the Norian, but they were high in the "Rhaetian" (Fig. 24.6). The diversification rate was negative in the "Rhaetian," although not quite as low as in the terminal Cretaceous (Maastrichtian) (Fig. 24.6) (Benton 1985a). The very low diversification rate in the

Pliensbachian, and the high rates from the Bajocian to the Kimmeridgian, may be an artifact of the poor fossil record of terrestrial tetrapods during parts of the Early and Middle Jurassic.

The record of turnover of nonmarine tetrapod families may be studied in more detail. Many of the important tetrapod-bearing formations have been tentatively assigned to particular parts of the five (or six) standard Triassic stages. For example, Anderson and Cruickshank (1978) were able to determine twenty tetrapod "substages" in the Triassic. I have modified their compilation of data for the Late Trias-

Figure 24.6. The pattern of origination, extinction, and diversification of terrestrial tetrapod families between the Late Permian and the Early Eocene. Per taxon rates are plotted on the vertical axis, and time (in millions of years) on the horizontal axis. The time scale is from Palmer (1983). The high extinction rate and low diversification rate in the Early Jurassic are probably the result of the poor fossil record of parts of the Early and Middle Jurassic, rather than being true rates. Abbreviations of stratigraphic stages: Aa, Aalenian; Alb, Albian; An, Anisian; Ap, Aptian; Baj, Bajocian; Be, Berriasian; Br, Barremian; Bth, Bathonian; Ce, Cenomanian; Cl, Callovian; Cmp, Campanian; Co, Coniacian; Cr, Carnian; Da, Danian; Ha, Hauterivian; He, Hettangian; Km, Kimmeridgian; Kz, Kazanian; L, Ladinian; Ma, Maastrichtian; Nor, Norian; Oxf, Oxfordian; Pl, Pliensbachian; Po, Portlandian; R, Rhaetian; Sa, Santonian; Sc, Scythian; Si, Sinemurian; Ta, Tatarian; Th, Thanetian; To, Toarcian; Tu, Turonian; U, Ufimian; Vlg, Valanginian; Yp, Ypresian.

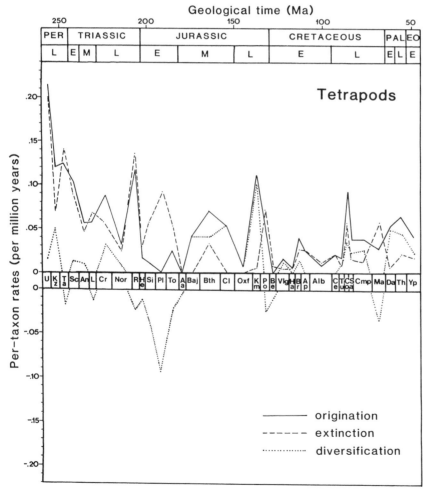

sic according to the new dating scheme of Olsen and Galton (1977, 1984). I plotted family diversity and total origination and extinction rates for families of Triassic nonmarine tetrapods (Fig. 24.7). Family diversity fluctuated throughout the Triassic, with particularly noticeable drops at the end of the Scythian (cf. "substages" 5 and 6) and in the Late Triassic (substages 15–16 and 20–Jurassic 1). Extinction rates were high in the early and late Scythian (substages 1 and 5), the early Anisian (substage 6), the late Carnian (substage 15), and the "Rhaetian" (substage 20). The high Scythian and Anisian rates may be partly explained as artifacts of a subsequent poor fossil record (i.e., some of the recorded extinctions might have occurred later, during the gap in the record). The high rate in the late Carnian, which was associated with a marked drop in diversity, however, cannot be explained in that way, although the "Rhaetian" extinction might be partly an artifact of a poorer Hettangian record (cf. Table 24.3 and Fig. 24.9 below).

Figure 24.7. The turnover of nonmarine tetrapod families in the Triassic. The data on family distributions, and the tetrapod "substages" 1–20 are taken from Anderson and Cruickshank (1978), with stratigraphic modifications according to Olsen and Galton (1977, 1984, see the text). Family diversity and total extinction and origination rates are shown from the early Scythian to the early Hettangian (lowermost Jurassic). The time scale is from Palmer (1983). Abbreviations: J, Jurassic; Sc, Scythian; An, Anisian; L, Ladinian; Cr, Carnian; "Rh," "Rhaetian"; He, Hettangian; J, Jurassic.

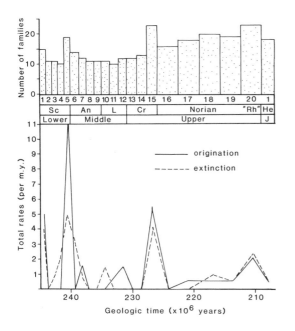

It must be noted that, although the extinction rates in the Late Triassic, in general, and in the late Carnian and "Rhaetian" in particular, are high, they are not detected as excessively high outliers from a linear regression analysis, using the techniques of Raup and Sepkoski (1982). There are, however, statistical and theoretical problems with this technique (Quinn 1983; Raup, Sepkoski, and Stigler 1983), and one further problem is that it takes no account of the effects of a depressed origination rate that can mimic the effects of a mass extinction by reducing overall taxonomic diversity. Total origination rates were low in the "substages" following the late Carnian and the "Rhaetian" (Fig. 24.7). It should be noted, in this context, that the marked drop in the diversity of nonmarine tetrapods at the end of the Cretaceous period (Fig. 24.5) is not associated with a statistically high total extinction rate (using the methods of Raup and Sepkoski 1982). Several key groups had their last known representatives then (e.g., dinosaurs and pterosaurs), but most of the fall in diversity was caused by a low origination rate (Fig. 24.6) (Benton 1985a).

Species diversity analysis

A more detailed study of the distribution in time of all Late Triassic nonmarine tetrapods was carried out. A list of all species of amphibians and reptiles from the Carnian, Norian, and Rhaetian was compiled from the most recent available sources. Where possible, a monographic review of a whole family was used, and more recently described taxa were appended from other sources. Species that are based on single fragmentary and questionable specimens were omitted. The main references for the species data are given in Table 24.2.

Stratigraphic assignments were taken from Olsen and Galton (1977, 1984), as explained above. The data are shown in Figure 24.8, which takes the form of a spindle diagram of families in which the width of the spindle is directly related to the number of species present in each time interval. The time scale is that of Palmer (1983).

This more detailed overview of the occurrence of tetrapod species in the Late Triassic shows that there was not a single time during which an overall and decimating mass extinction occurred. However, there are several intervals during which species diversity levels fell markedly. These may be further elucidated by studying the data in Figure 24.8.

It is evident from Figure 24.8 that the fossil record is incomplete in places: Every dashed line represents a time interval during which fossils have not been found. However, the group is assumed to have been present because it is known from below and above that particular time interval. Thus, for example, family No. 1 (the Capitosauridae) is not recorded from the early or middle Carnian or the lower middle Norian, but it is known from above and below each of these time intervals. Furthermore, it is evident that certain time intervals (e.g., the early Carnian and the early Norian) have poorer fossil records overall than the others. A measure of the completeness of the record of nonmarine tetrapods was calculated by taking the ratio of families recorded to the families apparently present for each of the time intervals used in the Late Triassic (Table 24.3). The apparent number of species per time interval was then divided by the completeness ratio in order to give an indication of the probable number of species that would have been collected if the fossil record were equally complete (or incomplete) throughout the Late Triassic.

The unadjusted species counts and the adjusted species counts (Table 24.3 and Figure 24.9) show three drops against an overall rising trend through time. These drops, in order of magnitude, represent losses of 66 percent (late Carnian to early

Norian), 43 percent (early middle to late middle Norian), and 38 percent (late middle to late Norian) of all species present (from the adjusted species counts). The small drop from the "Rhaetian" to the Hettangian, seen in the unadjusted species counts, is removed when the completeness ratio is applied. These figures are high and the first one certainly approaches the figures calculated for marine species in the major terminal Permian extinction event [77–96 percent loss of species (Sepkoski 1982)].

The analysis of family diversity in the Late Triassic suggests that there were two main extinction events, one at the end of the late Carnian, and one at the end of the "Rhaetian," thus at the Triassic–Jurassic boundary. In terms of magnitude, the first of these was the greater. The analysis of species diversity confirms the first event, and its impact, but there is no clear drop in species numbers at the Triassic–Jurassic boundary. In fact, two possible declines are identified in the middle and late Norian.

Table 24.2 *Main sources of data on the species of Late Triassic nonmarine tetrapods*

	Taxon	Source
1–5	Labyrinthodontia	Carroll and Winer 1977
6	Procolophonidae	Colbert 1960
7	Proganochelyidae	Gaffney and Meeker 1983
8–11	Various diapsids	Benton 1985b
12	Proterochampsidae	Sues 1976
13	Aetosauridae	Krebs 1976
14	Rauisuchidae	Bonaparte 1981
15	Poposauridae	Chatterjee 1985
16	Phytosauridae	Westphal 1976, Buffetaut and Ingavat 1982
17	Ornithosuchidae	Bonaparte 1975
18	Scleromochlidae	Benton and Walker 1985
19	Eudimorphodontidae	Wild 1978
20	Dimorphodontidae	Wild 1978
21	Saltoposuchidae	Bonaparte 1978
22	Sphenosuchidae	Bonaparte 1978
23	Protosuchidae	Nash 1975
24	Podokesauridae	Steel 1970
25	Anchisauridae	Galton 1976
26	Plateosauridae	Huene 1932, Galton 1976
27	Melanorosauridae	Bonaparte 1978
28	Fabrosauridae	Colbert 1981
29	Heterodontosauridae	Hopson 1975
30	Sphenodontidae	Benton 1985b
31	Kannemeyeriidae	Keyser and Cruickshank 1979
32	Traversodontidae	Bonaparte 1978, Chatterjee 1982
33	Chiniquodontidae	Bonaparte 1978
34	Tritylodontidae	Kermack 1982
35	Tritheledontidae	Chatterjee 1983
36–38	Mammals	Lillegraven et al. 1979

Note: See Figure 24.8. Many more references were used, but they could not all be listed.

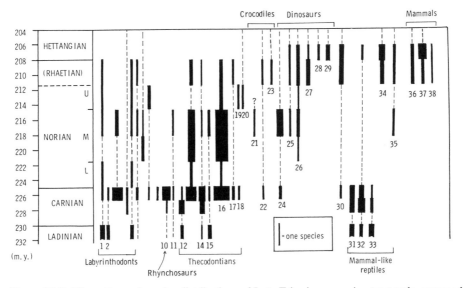

Figure 24.8. The pattern of species distributions of Late Triassic nonmarine tetrapods, arranged by families. The time scale is based on Palmer (1983). Each vertical shape represents one tetrapod family, and the width of the shape shows the number of species present worldwide through time. Certain families that have been based on single, or indeterminate, specimens are omitted, as are marine forms. The main sources of data are given in Table 24.2; stratigraphic distributions are mainly from Olsen and Galton (1977, 1984). The families are: 1, Capitosauridae; 2, Mastodonsauridae; 3, Metoposauridae; 4, Chigutisauridae; 5, Plagiosauridae; 6, Procolophonidae; 7, Proganochelyidae; 8, Kuehneosauridae; 9, Trilophosauridae; 10, Rhynchosauridae; 11, Tanystropheidae; 12, Proterochampsidae; 13, Aetosauridae; 14, Rauisuchidae; 15, Poposauridae; 16, Phytosauridae; 17, Ornithosuchidae; 18, Scleromochlidae; 19, Eudimorphodontidae; 20, Dimorphodontidae; 21, Saltoposuchidae; 22, Sphenosuchidae; 23, Protosuchidae; 24, Podokesauridae; 25, Anchisauridae; 26, Plateosauridae; 27, Melanorosauridae; 28, Fabrosauridae; 29, Heterodontosauridae; 30, Sphenodontidae; 31, Kannemeyeriidae; 32, Traversodontidae; 33, Chiniquodontidae; 34, Tritylodontidae; 35, Tritheledontidae; 36, Haramiyidae; 37, Morganucodontidae; 38, Kuehneotheriidae.

Table 24.3 *Completeness of the Late Triassic nonmarine tetrapod fossil record*

	No. of families		Completeness ratio $\left(\dfrac{\text{recorded}}{\text{apparent}}\right)$	No. of species	
	Apparent	Recorded		Recorded	Actual
Hettangian	18	12	0.667	22	33
"Rhaetian"	23	18	0.783	26	33
Late Norian	19	7	0.368	9	24
Middle Norian (L.)	20	15	0.750	29	39
Middle Norian (E.)	18	4	0.222	15	68
Early Norian	16	4	0.250	5	20
Late Carnian	23	22	0.957	56	59
Middle Carnian	13	7	0.538	17	32
Early Carnian	12	0	0	0	?

Note: The number of families recorded per stratigraphic unit (see Fig. 24.8) is compared with the apparent number of families (i.e., those that are recorded, plus those that occur below and above the unit in question). The ratio of recorded to apparent numbers gives a measure of how complete the record is for each stratigraphic unit. This completeness ratio was used to establish the approximate numbers of tetrapod species that would have been recorded per stratigraphic unit, assuming that the fossil record was equally complete (or incomplete) throughout.

Relative abundances

Benton (1983a) presented a study of the relative abundances of the major groups of nonmarine tetrapods through the Late Permian and the Triassic. This analysis was done mainly to test the commonly held view that the dinosaurs arose and radiated as the culmination of a series of drawn-out "competitive" processes in which the thecodontians gradually took over from the mammal-like reptiles, and the dinosaurs then successfully competed with the thecodontians. In fact, it became clear that the thecodontians never convincingly "beat" the carnivorous mammal-like reptiles, the cynognathoid cynodonts. Various mammal-like reptiles, thecodontians, rhynchosaurs, and other groups, died out at the end of the Carnian. The dinosaurs, which were already present as rare faunal elements, apparently radiated in the early and middle Norian. Several thecodontian lineages lived on right to the end of the Triassic, and the dinosaurs (especially the prosauropods) spread worldwide in the Early Jurassic. Benton (1983a) regarded the late Carnian extinction event (mistakenly dated there as middle Norian) as crucial to the "opportunistic" radiation of the dinosaurs, and thus the central event in shaping terrestrial vertebrate faunas for the rest of the Mesozoic.

Studies of changes in relative abundance through time are, of course, prone to various sources of error. [The assumption is made that the diversity of specimens in museums represents, in some way, the diversity of the living assemblage, and allowance has to be made for collection and preservation bias

Figure 24.9. The diversity of species of nonmarine tetrapods in the Late Triassic. The recorded totals are taken from Figure 24.8, and the "corrected" totals include a supplement to take account of the different levels of incompleteness of the fossil record in each "substage" (data from Table 24.3). Declines in diversity occurred after the Upper Carnian, the lower Middle Norian, and the upper Middle Norian.

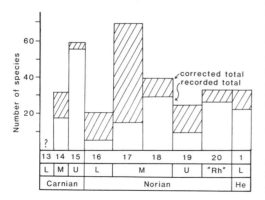

(see Benton 1983a)]. However these sources of bias can give a general impression of an extra dimension that is missing from simple taxonomic plots. They add a measure of the relative importance of different groups in typical ecosystems through time. Thus, for example, the rhynchosaurs were never represented by more than a couple of species worldwide at any one time, and yet where they occur, they regularly make up 40–70 percent of all tetrapods collected in a fauna. In contrast, the various groups of thecodontians may be represented by fifteen to twenty species at any one time (ca. 40 percent of all species of tetrapods worldwide), and yet each species could be based on only a handful of specimens (ca. 5–10 percent of all individual tetrapods). Both approaches are worthwhile. In the present example, both the taxic and the relative abundance studies suggest that the end-Carnian event was more significant in its impact on communities.

Discussion

The Late Triassic tetrapod extinctions have been linked to an increasing aridity that was observed in reptile-bearing beds in various parts of the world (e.g., Colbert 1958b; Cox 1967; Robinson 1971). Tucker and Benton (1982) and Benton (1983a,b) elaborated this hypothesis, noting an increase in hot subtropical arid to semiarid conditions, as indicated by tetrapod-bearing sediments, in the latest Triassic of most parts of the world. Linked with these climatic changes, they noted that there were also major floral replacements, in the Norian and "Rhaetian." The *Dicroidium* flora of Gondwanaland was replaced by a worldwide conifer–bennettitalean flora in the "Rhaetian" and Early Jurassic. Tucker and Benton (1982) suggested that the climatic and floral changes could have led to the extinction of various tetrapod groups.

Bakker (1977) presented an alternative theory, linking the Late Triassic (and other) tetrapod extinctions to marine regressions and reduced orogenic activity. According to this theory, the Late Triassic marine regression exposed great areas of lowlands around all continents, and the low level of mountain building reduced the geographic–topographic diversity. This meant that fewer habitats were available on land, reproductive barriers to speciation were removed, and speciation rates fell. The overall effect was a reduction in terrestrial tetrapod diversity. In support of this view is the fact that the Late Triassic extinction event was associated with a reduced family origination rate as much as with an elevated extinction rate.

There were mass extinctions among other groups in the Late Triassic. The marine extinction events were particularly severe: It is thought that nearly all species of ammonites and bivalves died out at the end of the Triassic, and only a few scraped through into the Jurassic to establish new radiations (Hallam 1981). Several major groups of brachiopods, nautiloids, and gastropods, and the last of the conodonts also disappeared at the end of the Triassic. The Norian marine extinction event was comparable in magnitude overall to the Maastrichtian, Frasnian, and Ashgillian events (Raup and Sepkoski 1982, 1984; Sepkoski 1982) – over 20 percent of approximately 300 extant families of marine invertebrates and vertebrates were eliminated.

The timing of the Late Triassic marine extinction event has not been given precisely for all groups. The bivalves declined in diversity from a Carnian–early Norian peak, and they were affected by a major extinction event at the end of the "Rhaetian"–Sevatian (Hallam 1981). Similarly, the main group of Triassic ammonites, the ceratites, reached their peak of diversity in the Carnian and declined thereafter. The last genera disappeared at the end of the "Rhaetian," but it is not clear how extreme this event was (Kennedy 1977; Hallam 1981). The mass extinctions of brachiopods and conodonts appear to have occurred at the end of the "Rhaetian" (Hallam 1981). Bakker (1977) and Benton (1983a) placed the extinction event of nonmarine tetrapods rather earlier, in the middle Norian; however, it seems, from the evidence given in this chapter, that there were two events, the earlier or major one at the end of the Carnian, and the later or lesser one at the end of the Triassic ("Rhaetian").

One final comment concerning the extinctions of marine vertebrates: These have been tied in with the terminal Triassic invertebrate extinctions (e.g., Raup and Sepkoski 1982; Sepkoski 1982). However, as far as we know, most of the Late Triassic marine reptile families died out in the Carnian (Thalattosauridae, Nothosauridae, Cymatosauridae, Henodontidae, Shastasauridae); others are uncertain (Simosauridae), and only one died out at the Triassic–Jurassic boundary (Placochelyidae) (Anderson and Cruickshank 1978).

The precise dating and nature of the Late Triassic extinction event(s) are uncertain, then. Several kinds of explanations have been given for the extinctions of all groups at the end of the Triassic: widespread marine regression followed by an anoxic event (Hallam 1981, 1984), temperature changes (Stanley 1984), extraterrestrial impact, or other extraterrestrial event (McLaren 1983; Raup and Sepkoski 1984). Most of these hypotheses are tied to the idea that major extinctions, of which the Late Triassic event was one, have occurred in regular cycles through time. Much more information is needed on the pattern of the Late Triassic extinction (and most of the others) before the proposed mechanisms can be adequately considered.

Acknowledgments

I thank Dr. Rupert Wild (Stuttgart/Ludwigsburg) and Dr. Frank Westphal (Tübingen) for their assistance and great kindness when I was in Germany. I thank the President and Fellows of Trinity College, Oxford, for financial support during this work, and the Royal Society for supporting my trip to Germany. Three reviewers (Peter Dodson, Paul Olsen, and Kevin Padian) offered extremely helpful comments for which I am very grateful.

References

Anderson, J. M. 1981. World Permo-Triassic correlations: their biostratigraphic basis. *In* Cresswell, M. M., and P. Vella (eds.), *Gondwana Five* (Rotterdam: Balkema), pp. 3–10.

Anderson, J. M., and A. R. I. Cruickshank. 1978. The biostratigraphy of the Permian and the Triassic. Part 5. A review of the classification and distribution of Permo-Triassic tetrapods. *Palaeontol. Afr.* 21: 15–44.

Bakker, R. T. 1968. The superiority of dinosaurs. *Discovery* 3(1): 11–22.

1971. Dinosaur physiology and the origin of the mammals. *Evolution* 25: 636–58.

1972. Anatomical and ecological evidence of endothermy in dinosaurs. *Nature (London)* 238: 81–5.

1975. Experimental and fossil evidence for the evolution of tetrapod bioenergetics. *In* Gates, D. M., and R. B. Schmerl (eds.), *Perspectives of Biophysical Ecology* (New York: Springer Verlag), pp. 365–99.

1977. Tetrapod mass extinctions – a model of the regulation of speciation rates and immigration by cycles of topographic diversity. *In* Hallam, A. (ed.), *Patterns of Evolution as Illustrated by the Fossil Record* (Amsterdam: Elsevier), pp. 439–68.

1980. Dinosaur heresy – dinosaur renaissance: why we need endothermic archosaurs for a comprehensive theory of bioenergetic evolution. *In* Thomas, R. D. K., and E. C. Olson (eds.), *A Cold Look at the Warm-Blooded Dinosaurs* (Boulder, Colorado: Westview Press), pp. 351–462.

Bakker, R. T., and P. M. Galton. 1974. Dinosaur monophyly and a new class of vertebrates. *Nature (London)* 248: 168–72.

Benton, M. J. 1979. Ectothermy and the success of the dinosaurs. *Evolution* 33: 983–97.

1983a. Dinosaur success in the Triassic: a noncompetitive ecological model. *Quart. Rev. Biol.* 58: 29–55.

1983b. The age of the rhynchosaur. *New Scient.* 98: 9–13.

1984a. Dinosaur's lucky break. *Nat. Hist.* 93(6): 54–9.

1984b. Rauisuchians and the success of dinosaurs. *Nature (London)* 310: 101.

1984c. Fossil reptiles of the German late Triassic and the origin of the dinosaurs. *In* Reif, W.-E., and F. Westphal (eds.), *Third Symposium on Mesozoic Terrestrial Ecosystems, Tübingen 1984, Short*

Papers (Tübingen: ATTEMPTO), pp. 13–18.

1984d. Consensus on archosaurs. *Nature (London)* 312: 599.

1985a. Patterns in the diversification of Mesozoic nonmarine tetrapods, and problems in historical diversity analysis. *Spec. Pap. Palaeontol.* 33: 185–202.

1985b. Classification and phylogeny of the diapsid reptiles. *Zool. J. Linn. Soc.* 84: 97–164.

Benton, M. J., and A. D. Walker. 1985. The palaeoecology, taphonomy and dating of the Permo-Triassic reptiles from Elgin, north-east Scotland. *Palaeontology* 28: 207–34.

Bonaparte, J. F. 1975. The family Ornithosuchidae. *Coll. Intern. C.N.R.S.* 218: 485–502.

1976. *Pisanosaurus mertii* Casamiquela and the origin of the Ornithischia. *J. Paleontol.* 50: 808–20.

1978. El Mesozoico de America del Sur y sus tetrapodos *Opera Lilloana* 26: 1–596.

1981. Descripción de *"Fasolasuchus tenax"* y su significado en la sistematica y evolución de los Thecodontia. *Rev. Mus. Argent. Cienc. Nat., Palaeontol.* 3: 55–101.

1982. Faunal replacement in the Triassic of South America. *J. Vert. Paleontol.* 21: 362–71.

1984. Locomotion in rauisuchid thecodonts. *J. Vert. Paleontol.* 3: 210–18.

Brenner, K. 1973. Stratigraphie und Paläogeographie des Oberen Mittelkeupers in Südwest-Deutschland. *Arb. Inst. Geol. Paläontol. Univ. Stuttgart, N.F.* 68: 101–222.

1978a. Profile aus dem Oberen Mittelkeuper Südwest-Deutschlands. *Arb. Inst. Geol. Paläontol. Univ. Stuttgart, N.F.* 72: 103–203.

1978b. Sammlung und Revision der bis 1978 veröffentlichten Profile aus dem Oberen Mittelkeuper Südwest-Deutschlands. *Arb. Inst. Geol. Paläontol. Univ. Stuttgart, N.F.* 72: 205–39.

1979. Paläogeographische Raumbilder Südwest-Deutschlands für die Ablagerungszeit von Kiesel- und Stubensandstein. *J. Ber. Mitt. Oberrhein. Geol. Ver., N.F.* 61: 331–35.

Brenner, K., and E. Villinger. 1981. Stratigraphie und Nomenklatur des südwestdeutschen Sandsteinkeupers. *Jh. Geol. Landesamt Baden-Württemberg* 23: 45–86.

Buffetaut, E., and R. Ingavat. 1982. Phytosaur remains (Reptilia, Thecodontia) from the Upper Triassic of North-Eastern Thailand. *Géobios* 15: 7–17.

Carroll, R. L., and L. Winer. 1977. Appendix to accompany Chapter 13. Patterns of amphibian evolution: an extended example of the incompleteness of the fossil record. *In* Hallam, A. (ed.), *Patterns of Evolution as Illustrated by the Fossil Record* (Amsterdam: Elsevier), pp. 405–37.

Charig, A. J. 1972. The evolution of the archosaur pelvis and hindlimb: an explanation in functional terms. *In* Joysey, K. A., and T. S. Kemp (eds.), *Studies in Vertebrate Evolution* (Edinburgh: Oliver and Boyd), pp. 121–55.

1979. *A New Look at the Dinosaurs* (London: Heinemann).

1980. Differentiation of lineages among Mesozoic tetrapods. *Mem. Soc. Géol. France* 139: 207–10.

1984. Competition between therapsids and archosaurs during the Triassic period: a review and synthesis of current theories. *Symp. Zool. Soc. Lond.* 52: 597–628.

Charig, A. J., J. Attridge, and A. W. Crompton. 1965. On the origin of the sauropods and the classification of the Saurischia. *Proc. Linn. Soc. Lond.* 176: 197–221.

Chatterjee, S. K. 1982. A new cynodont reptile from the Triassic of India. *J. Paleontol.* 56: 203–14.

1983. An ictidosaur fossil from North America. *Science* 220: 1151–3.

1985. The poposaurid *Postosuchus* from the Dockum of Texas. *Phil. Trans. R. Soc. Lond. B* 309: 395–460.

Colbert, E. H. 1949. Progressive adaptations as seen in the fossil record. *In* Jepsen, G. L., E. Mayr, and G. G. Simpson (eds.), *Genetics, Paleontology, and Evolution* (Princeton, New Jersey: Princeton Univ. Press), pp. 390–402.

1958a. The beginning of the age of dinosaurs. *In* Westoll, T. S. (ed.), *Studies on Fossil Vertebrates* (London: Athlone Press), pp. 39–58.

1958b. Tetrapod extinctions at the end of the Triassic period. *Proc. Nat. Acad. Sci. U.S.A.* 44: 973–7.

1960. A new Triassic procolophonid from Pennsylvania. *Am. Mus. Novitates* 2022: 1–19.

1966. *The Age of Reptiles* (New York: W.W. Norton).

1969. *Evolution of the Vertebrates* (New York: Wiley).

1981. A primitive ornithischian dinosaur from the Kayenta Formation of Arizona. *Bull. Mus. N. Arizona* 53: 1–61.

Connell, J. H. 1980. Diversity and the coevolution of competitors, or the ghost of competition past. *Oikos* 35: 131–8.

Cox, C. B. 1967. Changes in terrestrial vertebrate faunas during the Mesozoic. *In* Harland, W. B. et al. (eds.), *The Fossil Record* (London: Geology Society), pp. 77–89.

Crompton, W. A. 1968. The enigma of the evolution of mammals. *Optima* 18: 137–51.

Dockter, J., P. Puff, G. Seidel, and H. Kozur. 1980. Zur Triasgliederung und Symbolgebung in der DDR. *Z. Geol. Wiss* 8: 951–63.

Dunay, R. E., and M. J. Fisher. 1974. Late Triassic palynofloras of North America and their European correlatives. *Rev. Palaeobot. Palynol.* 17: 179–86.

1979. Palynology of the Dockum Group (Upper Triassic), Texas, U.S.A. *Rev. Palaeobot. Palynol.* 28: 61–92.

Fisher, M. J. 1972. The Triassic palynofloral succession in England. *Geosci. Man.* 4: 101–9.

Fisher, M. J., and J. Bujak. 1975. Upper Triassic palynofloras from Arctic Canada. *Geosci. Man.* 11: 87–94.

Gaffney, E. S., and L. J. Meeker. 1983. Skull morphology of the oldest turtles: a preliminary description of *Proganochelys*. *J. Vert. Paleontol.* 3: 25–8.

Gall, J.-C., M. Durand, and E. Muller. 1977. Le Trias

de part d'autre du Rhin. Corrélations entre les marges et le centre du bassin germanique. *Bull. Bur. Rech. Géol. Minières* 3(2): 193–204.

Galton, P. M. 1973. On the anatomy and relationships of *Efraasia diagnostica* (Huene) n. gen., a prosauropod dinosaur (Reptilia : Saurischia) from the Upper Triassic of Germany. *Paläontol. Z.* 47: 229–55.

1976. Prosauropod dinosaurs (Reptilia : Saurischia) of North America. *Postilla* 169: 1–98.

Gauthier, J. A. 1984. A cladistic analysis of the higher systematic categories of the Diapsida, Ph.D. dissertation, University of California, Berkeley.

Geiger, M. E., and C. A. Hopping. 1968. Triassic stratigraphy of the southern North Sea Basin. *Phil. Trans. Roy. Soc. Lond. B* 254: 1–36.

Gottesfield, A. S. 1980. Upper Triassic palynofloras of the western United States. *IV Int. Palynol. Conf. Lucknow (1976–77)* 2: 295–308.

Gould, S. J., and C. B. Calloway. 1980. Clams and brachiopods – ships that pass in the night. *Paleobiology* 6: 383–96.

Gwinner, M. P. 1981. Eine einheitliche Gliederung des Keupers (Germanische Trias) in Süddeutschland. *N. Jb. Geol. Paläontol., Mh.* 1980: 229–34.

Hahn, G. 1984. Paläomagnetische Untersuchungen im Schilfsandstein (Trias, Km 2) Westeuropas. *Geol. Rundsch.* 73: 499–516.

Hallam, A. 1981. The end-Triassic bivalve extinction event. *Palaeogeogr., Palaeoclimatol., Palaeoecol.* 35: 1–44.

1984. Pre-Quaternary sea-level changes. *Ann. Rev. Earth Planet. Sci.* 12: 205–43.

Harland, W. B., A. V. Cox, P. G. Llewellyn, C. A. G. Pickton, A. G. Smith, and R. Walters. 1982. *A Geological Time Scale* (Cambridge: Cambridge University Press).

Hopson, J. A. 1975. On the generic separation of the ornithischian dinosaurs *Lycorhinus* and *Heterodontosaurus* from the Stormberg Series (Upper Triassic) of South Africa. *S. Afr. J. Sci.* 71: 302–5.

Hotton, N., III 1980. An alternative to dinosaur endothermy: the happy wanderers. *In* Thomas, R. D. K., and E. C. Olson (eds.), *A Cold Look at the Warm-Blooded Dinosaurs* (Boulder, Colorado: Westview Press), pp. 311–50.

Huene, F. von. 1914. Saurischia et Ornithischia triadica. *Foss. Cat* 4: 1–21.

1932. Die fossile Reptil-Ordnung Saurischia, ihre Entwicklung und Geschichte. *Monogr. Geol. Palaontol.* 4(1):1–361.

Kennedy, W. T. 1977. Ammonite evolution. *In* Hallam, A. (ed.), *Patterns of Evolution as Illustrated by the Fossil Record* (Amsterdam: Elsevier), pp. 251–304.

Kermack, D. M. 1982. A new tritylodontid from the Kayenta Formation of Arizona. *Zool. J. Linn. Soc.* 76: 1–17.

Keyser, A. W., and A. R. I. Cruickshank. 1979. The origins and classification of Triassic dicynodonts. *Trans. Geol. Soc. S. Afr.* 82: 81–108.

Kozur, H. 1975. Probleme der Triasgliederung und Parallelisierung der germanischen und tethyalen

Trias. Teil II: Anschluss der germanischen Trias an die internationale Triasgliederung. *Freiberger ForschHft. (C)* 3: 58–65.

Krebs, B. 1976. Pseudosuchia. *Handb. Paläoherpetol.* 13: 40–98.

Lillegraven, J. A., Z. Kielan-Jaworowska, and W. A. Clemens. 1979. *Mesozoic Mammals, The First Two-thirds of Mammalian History* (Berkeley, California: University of California Press).

McLaren, D. J. 1983. Bolides and biostratigraphy. *Bull. Geol. Soc. Am.* 94: 313–24.

Nash, D. S. 1975. The morphology and relationships of a crocodilian, *Orthosuchus stormbergi*, from the Upper Triassic of Lesotho. *Ann. S. Afr. Mus.* 67: 227–329.

Norman, D. B. 1984. A systematic reappraisal of the reptile order Ornithischia. *In* Reif, W.-E. and F. Westphal (eds.), *Third Symposium on Mesozoic Terrestrial Ecosystems, Tubingen 1984, Short Papers* (Tübingen: ATTEMPTO), pp. 157–62.

Odin, G. S. 1982. *Numerical Dating in Stratigraphy*. (Chichester: Wiley).

Olsen, P. E., and P. M. Galton. 1977. Triassic–Jurassic extinctions: are they real? *Science* 197: 983–6.

1984. A review of the reptile and amphibian assemblages from the Stormberg of Southern Africa with special emphasis on the footprints and the age of the Stormberg. *Palaeontol. Afr.* 25: 87–110.

Olsen, P. E., A. McCune, and K. S. Thomson. 1982. Correlation of the Early Mesozoic Newark Supergroup by vertebrates, principally fishes. *Am. J. Sci.* 282: 1–44.

Olson, E. C. 1982. Extinctions of Permian and Triassic nonmarine vertebrates. *Spec. Pap. Geol. Soc. Am.* 190: 501–11.

Padian, K. 1983. A functional analysis of flying and walking in pterosaurs. *Paleobiology* 9(3): 218–39.

1984. The origin of pterosaurs. *In* Reif, W.-E., and F. Westphal (eds.), *Third Symposium on Mesozoic Terrestrial Ecosystems, Tübingen 1984, Short Papers* (Tübingen: ATTEMPTO), pp. 163–8.

Palmer, A. R. 1983. The Decade of North American Geology 1983 Time Scale. *Geology* 11: 503–4.

Parrish, J. M. 1984. Locomotor grades in the Thecodontia. *In* Reif, W.-E., and F. Westphal (eds.), *Third Symposium on Mesozoic Terrestrial Ecosystems, Tübingen 1984, Short Papers* (Tübingen: ATTEMPTO), pp. 169–73.

Paul, G. S. 1984. The archosaurs: a phylogenetic study. *In* Reif, W.-E. and F. Westphal (eds.), *Third Symposium on Mesozoic Terrestial Ecosystems, Tübingen 1984, Short Papers* (Tübingen: ATTEMPTO), pp. 175–80.

Price, P. W., C. N. Slobodchikoff, and W. S. Gaud. 1984. *A New Ecology* (New York: Wiley Interscience).

Quinn, J. F. 1983. Mass extinctions in the fossil record. *Science* 219: 1239–40.

Raup, D. M., and J. J. Sepkoski, Jr. 1982. Mass extinctions in the marine fossil record. *Science* 215: 1501–3.

1984. Periodicity of extinctions in the geologic past.

Proc. Nat. Acad. Sci. U.S.A. 81: 801–5.

Raup, D. M., J. J. Sepkoski, Jr., and S. M. Stigler. 1983. Mass extinctions in the fossil record. *Science* 219: 1240–1.

Robinson, P. L. 1971. A problem of faunal replacement on Permo-Triassic continents. *Palaeontology* 14: 131–53.

Romer, A. S. 1970. The Triassic faunal succession and the Gondwanaland problem. *Gondwana Stratigraphy. IUGS Symposium Buenos Aires 1967* (Paris: UNESCO), pp. 375–400.

Schröder, B. 1982. Entwicklung des Sedimentbeckens und Stratigraphie der klassischen Germanischen Trias. *Geol. Rundsch.* 71: 783–94.

Sepkoski, J. J., Jr. 1978. A kinetic model of Phanerozoic taxonomic diversity I. Analysis of marine orders. *Paleobiology* 4: 223–51.

1981. A factor analytic description of the Phanerozoic marine fossil record. *Paleobiology* 7: 36–53.

1982. Mass extinctions in the Phanerozoic oceans: a review. *Spec. Pap. Geol. Soc. Am.* 190: 283–9.

Sereno, P. C. 1984. The phylogeny of the Ornithischia, a reappraisal. *In* Reif, W.-E., and F. Westphal (eds.), *Third Symposium on Mesozoic Terrestrial Ecosystems, Tübingen 1984, Short Papers* (Tübingen: ATTEMPTO), pp. 219–26.

Simberloff, D. 1983. Competition theory, hypothesis testing, and other community ecological buzzwords. *Am. Nat.* 122: 626–35.

Snelling, N. J. (ed.) (in press). Geochronology and the Geological Record. *Geol. Soc. London.*

Spotila, J. R. 1980. Constraints of body size and environment on the temperature regulation of dinosaurs. *In:* Thomas, R. D. K., and E. C. Olson (eds.), *A Cold Look at the Warm-blooded Dinosaurs* (Boulder, Colorado: Westview Press), pp. 233–52.

Spotila, J. R., P. W. Lommen, G. S. Bakken, and D. M. Gates. 1973. A mathematical model for body temperatures of large reptiles: implications for dinosaur ecology. *Am. Nat.* 107: 391–404.

Stanley, S. M. 1984. Temperature and biotic crises in the marine realm. *Geology* 12: 205–8.

Steel, R. 1970. Saurischia. *Handb. Paläoherpetol.* 14: 1–87.

Strong, D. R., D. Simberloff, L. G. Abele, and A. B. Thistle (eds.). 1984. *Ecological Communities: Conceptual Issues and the Evidence* (Princeton, New Jersey: Princeton University Press).

Sues, H.-D. 1976. Thecodontia incertae sedis: Proterochampsidae. *Handb. Paläoherpetrol.* 13: 121–6.

Tozer, E. T. 1967. A standard for Triassic time. *Bull. Can. Geol. Surv.* 156: 1–103.

1974. Definitions and limits of Triassic stages and substages: suggestions prompted by comparisons between North America and the Alpine–Mediterranean region. *Österr. Akad. Wiss. Schriftenr. Erdwiss. Komm.* 2: 195–206.

1979. Latest Triassic ammonoid faunas and biochronology, western Canada. *Geol. Surv. Canada, Pap.* 79, 127–35.

Tucker, M. E., and M. J. Benton. 1982. Triassic environments, climates and reptile evolution. *Palaeogeogr., Palaeoclimatol., Palaeoecol.* 40: 361–79.

Visscher, H., and W. A. Brugman. 1981. Ranges of selected palynomorphs in the Alpine Triassic of Europe. *Rev. Palaeobot. Palynol.* 34: 115–28.

Visscher, H., W. M. L. Schuurman, and A. W. Van Erve. 1980. Aspects of a palynological characterization of Late Triassic and Early Jurassic "Standard" units of chronostratigraphical classification in Europe. *IV Int. Palynol. Conf., Lucknow (1976–77)* 2: 281–7.

Walker, A. D. 1964. Triassic reptiles from the Elgin area: Ornithosuchus and the origin of carnosaurs. *Phil. Trans. Roy. Soc. Lond.* 248: 53–134.

Weishampel, D. B., and J. B. Weishampel. 1982. Annotated localities of ornithopod dinosaurs: implications to Mesozoic paleobiogeography. *Mosasaur* 1: 43–87.

Westphal, F. 1976. Phytosauria. *Handb. Paläoherpetol.* 13: 99–120.

Wild, R. 1973. Die Triasfauna der Tessiner Kalkalpen. XXIII *Tanystropheus longobardicus* (Bassani) (Neue Ergebnisse). *Schweiz. Paläontol. Abh.* 95: 1–162.

1978. Die Flugsaurier (Reptilia, Pterosauria) aus der oberen Trias von Cene bei Bergamo, Italien. *Boll. Soc. Paleontol. Ital.* 17: 176–256.

1980. Die Triasfauna der Tessiner Kalkalpen. XXIV. Neue Funde von *Tanystropheus* (Reptilia, Squamata). *Schweiz. Paläontol. Abh.* 102: 1–43.

1982. Die Evolution der Reptilien in der Triaszeit. *Geol. Rundsch.* 71: 725–39.

25 Correlation of continental Late Triassic and Early Jurassic sediments, and patterns of the Triassic–Jurassic tetrapod transition

PAUL E. OLSEN AND
HANS-DIETER SUES

Introduction

The Late Triassic–Early Jurassic boundary is frequently cited as one of the thirteen or so episodes of major extinctions that punctuate Phanerozoic history (Colbert 1958; Newell 1967; Hallam 1981; Raup and Sepkoski 1982, 1984). These times of apparent decimation stand out as one class of the great events in the history of life.

Renewed interest in the pattern of mass extinctions through time has stimulated novel and comprehensive attempts to relate these patterns to other terrestrial and extraterrestrial phenomena (see Chapter 24). The Triassic–Jurassic boundary takes on special significance in this light. First, the faunal transitions have been cited as even greater in magnitude than those of the Cretaceous or the Permian (Colbert 1958; Hallam 1981; see also Chapter 24). Second, like the Cretaceous-Tertiary boundary, the Triassic–Jurassic boundary heralded a new, long-lasting regime of dominant animals, the dinosaurs. Third, but unlike the Cretaceous, a definite bolide impact structure is known in the Late Triassic. The 70 km Manicouagan crater in Québec, Canada has been dated at 210 ± 4 MY (Grieve 1982), which is within the margin of error of the currently accepted dates for the Triassic–Jurassic boundary. Despite considerable uncertainty, scenarios of asteroid impact have already been proposed to explain the Triassic–Jurassic extinctions, much as they have been for the Cretaceous–Tertiary extinctions (Raup and Sepkoski 1984; Rampino and Stothers 1984); but as such attempts at explanation proceed, the pattern itself must be continually examined. Experts on the patterns of each of the supposed mass extinction events must define the terms and taxa involved, and ask whether each "event" is real or artificial.

Olsen and Galton (1977) previously asked this question of the Triassic–Jurassic extinctions. The apparent answer was that the supposed mass extinctions in the tetrapod record were largely an artifact of incorrect or questionable biostratigraphic correlations. On reexamining the problem, we have come to realize that the kinds of patterns revealed by looking at the change in taxonomic composition through time also profoundly depend on the taxonomic levels and the sampling intervals examined. We address those problems in this chapter. We have now found that there does indeed appear to be some sort of extinction event, but it cannot be examined at the usual coarse levels of resolution. It requires new fine-scaled documentation of specific faunal and floral transitions.

Stratigraphic correlation of geographically disjunct rocks and assemblages predetermines our perception of patterns of diversity, extinctions, and originations. This poses an especially difficult problem for the Early Mesozoic because there are virtually no unquestioned Early Jurassic continental vertebrate assemblages. Correlations are of such paramount importance to any study of change during the Early Mesozoic that we devote the first part of this chapter to a summary description of the rationale for correlating various continental sequences with those of the Late Triassic and Early Jurassic type areas of Europe.

The second part of this chapter details the taxonomic changes through the Early Mesozoic based on these correlations. We examine the skeletal record of continental tetrapods by looking at the global record at the family and stage levels. Unfortunately, the family and stage levels are too coarse a level of analysis for these kinds of questions. We try to circumvent this problem by examining two subsets of the world data, data from the Newark Supergroup (Froelich and Olsen 1984) and data from the European Early Mesozoic.

The unique periodic lacustrine cycles of the

Newark allow us to look at chronometric sampling intervals of two million years as well as the individual stage lengths. Although the data are reliable at the generic level, we use these data at the level of the family rather than the genus or species because too many species are from single localities for the compilations of generic distributions to be meaningful. In addition, it can be argued that when looking at one geographic area, sampling can exert a severe bias, especially in the Newark, where osseous remains are not common and there is a bias toward lacustrine taxa. Therefore, we also look at vertebrate ichnotaxa because they are sampled in the same kind of depositional environment through the Newark, and they are extremely abundant and therefore not as subject to the problems of small sample size that plague bony remains. Pollen and spore taxa are examined as an independent check on the diversity patterns. They are reliable at a much finer taxonomic level than bones or ichnotaxa, and they do not suffer from small sample size.

We examine the European Early Mesozoic only at a stage level, because there is as yet no way to calibrate the section independently at a finer stratigraphic level. Relative dates are not reliable, and chronostratigraphic measures are elusive. For the most part, absolute dating has not been done. As we do for the global data, we also restrict our analysis of the European fossil vertebrates to the family level because we are unsure of potential synonymies at lower taxonomic levels within the assemblages themselves. Finally, we compare taxonomic rates from the global record of tetrapods to those of the marine invertebrate record of the Early Mesozoic, at the stage and family level. Of interest here is the comparison among the different patterns. We treat all of these data, including the global data, with consistent methods of calculating the average number of taxa, normalized origination and extinction rates per million years, and probabilities of extinction and origination.

In the third section of this chapter, we identify which taxa in particular are responsible for the observed patterns. We go on to examine physical and biological changes through the Early Mesozoic and comment on events that might be synchronous with (and therefore perhaps related to) faunal changes that stand above background levels.

The rationale for correlation of continental Early Mesozoic tetrapod assemblages

A dramatic change in facies marks the transition between Triassic and Jurassic systems in the Germanic Basin of Central Europe. The continental and paralic Germanic facies of the Upper Triassic, the Keuper, give way to the fully marine Lias of the Lower Jurassic (Gall, Durand, and Muller 1977; Chapter 1). Terrestrial tetrapod remains are fairly common in the Keuper, but are all but absent in the Lias. As a consequence, it is impossible to compare Triassic and Jurassic tetrapod assemblages directly in the type area of the Early Mesozoic. The main goal of this chapter is to make just such a comparison, and thus it is necessary to correlate, by whatever means available, other continental beds with those of the European Early Mesozoic. Unfortunately, correlation of principally continental beds with the marine Jurassic beds has proved very difficult. Obviously, whatever sections we choose as correlative with those of the European Lias determine our view of the transition in tetrapods. This view is necessarily indirect, and therefore our conclusions based on cumulative faunal lists of the world are somewhat uncertain (as stressed by Colbert in Chapter 1).

Correlation problems within the type areas in Europe

Colbert (Chapter 1) reviews the origins of the main divisions of the Triassic and Jurassic within Europe. For our purposes, it is necessary to say a little more about these divisions and outline the crucial problems of correlating the type areas of the Triassic and Jurassic systems with the type areas of the standard marine stages.

The type area of the Triassic is the Germanic Basin of Central Europe. The earliest Mesozoic in the Germanic Basin consists of three vertically segregated facies: (1) a lower continental and paralic sequence, the Buntsandstein; (2) a middle marine sequence, the Muschelkalk; and (3) an upper continental and paralic sequence, the Keuper. These are lithological divisions.

The wholly marine sequences of the Alps provide the type areas for the stages of the Late Triassic; their history is reviewed by J. T. Gregory (in prep.). The stages in the Alpine Triassic are recognized principally by marine invertebrate zones, especially ammonites. These stages are time-stratigraphic, not lithological, units, and there are problems in correlating the type areas of the stages of the Late Triassic. Specifically, the youngest of the stages of the Triassic, the Rhaetian, contains only one ammonite zone and is now generally included as the uppermost division of the Norian (Tozer 1974, 1979; Hallam 1981; Pearson 1970). Accordingly, we do not recognize the Rhaetian as a separate stage. On the other hand, correlation of the Alpine zones with the Muschelkalk of the Germanic Basin has been fairly straightforward, with the Anisian and Ladinian stages mostly represented by the Muschelkalk. However, the upper and lower boundaries of the Muschelkalk have proved diachronous, as might be expected. The upper part of the Middle and the Upper Buntsandstein and the Lower Keuper are included in the Early Anisian and Late Ladinian, respectively. The rest of the Buntsandstein is apparently equivalent to the Scythian of the Alpine section (Gall et al. 1977).

Correlation of the different lithostratigraphic divisions of the Keuper with the Alpine stages is weak. At present, there are two different correlation schemes for the Keuper and Alpine Triassic (reviewed by Benton, Chapter 24). The main difficulties with the correlation center on the Upper Gipskeuper [the Kieselsandstein, Bunte Mergel, and Rote Wand = km3 of the standard stratigraphic scheme (Laemmlen 1958; Gwinner, 1980)]. Perhaps because the data are scarce, palynologists and invertebrate paleontologists differ on the correlation. The paleomagnetic data of Hahn (1984) probably demonstrate that at least some Keuper units (such as the Schilfsandstein = km2) are time-transgressive. However, these data cannot yet bear on stage level correlations because the magnetic stratigraphy of the Triassic is known at far too coarse a level; alternative stage level correlations to other magnetozones cannot be excluded. For the purposes of this chapter we accept the correlations of Kozur (1975) and Brinkmann (1960) (see Fig. 25.1) that place the km3 beds within the Carnian, because this agrees with placement of the Haupt Unconformity at the basal Norian and at the base of km4 beds (Steinmergelkeuper) in the Keuper (Schroeder 1982).

The type intervals of the Jurassic System are exposed in the Jura Mountains of southeastern France, the rocks of which are marine and highly fossiliferous. The type areas of the various stages of the Jurassic occur widely through much of Europe. Even correlation of stage boundaries outside the type areas has been reasonably straightforward and was worked out by the middle of the nineteenth century. Correlation is principally by Oppel ammonite zones, and the main works on the detailed correlation stand as classics in biostratigraphy (Oppel 1856–8; Arkell 1933). In marine rocks, the stage boundaries of the Jurassic have been successfully extended essentially worldwide (Tozer 1979).

Correlation of the European Early Mesozoic outside the type areas
Lithostratigraphic correlation
Because the type Triassic of the Germanic Basic consists of roughly two-thirds continental rocks, correlation to other continental areas is aided by the shared suites of continental fossils. Unfortunately, the exposed continental sequences of the Germanic Basin are very thin compared to the total amount of time they must represent. This is especially true for the Keuper, which has an exposed

thickness of around 500 m (Rutte 1957; Brinkmann 1960). When correlated with the Alpine section, the sequence covers the interval from late Ladinian to the Triassic–Jurassic boundary. Measured against the radiometric scale of Palmer (1984), this interval covers roughly 24 MY. This is equivalent to a net sedimentation rate of 0.02 mm/year. In contrast, the correlative parts of the exposed portions of most other Late Triassic continental deposits, such as the Chinle Formation in Arizona, United States, and the Newark Supergroup of eastern North America would have much faster net sedimentation rates (500 m/5 MY = 0.10 mm/year for the Chinle and 6600 m/20 MY = 0.31 mm/year for the entire Late Triassic part of the Newark Basin of the Newark Supergroup). The relative thinness of the Keuper and the very slow average sedimentation rates compared to other Triassic continental sequences are very important, because in principally fluvial and "deltaic" sequences, such as the Keuper, sedimentation rates are not constant at all. In general, as the mean sedimentation rate drops, the variance in sedimentation rates measured over shorter intervals increases, and the completeness of the record decreases (Sadler 1981; Sadler and Dingus 1982; Schindel 1982; Retallack 1984).

The intervals of the Keuper that are composed largely of sandstone, such as the crucial and richly fossiliferous Schilfsandstein (km2) and the Blasensandstein (km3), are fluvial and represent one or a few sedimentation packages, each deposited very rapidly. This strongly suggests that, in comparison to other continental areas, the sedimentation rate fluctuated to extremes (viewed over a short time interval), and large intervals of time are represented by few or no sediments in the Keuper. As previously mentioned, Hahn's paleomagnetic data from the Schilfsandstein indicate at least that sandstone unit is time-transgressive. Thus, fossil assemblages from vertically adjacent Keuper units may be separated by comparatively large hiatuses, and fossils from the same lithologic units need not be contemporaneous.

In addition, there is a regionally recognized unconformity above the Upper Gipskeuper (km3 = Kieselsandstein and Rote Wand), the so-called Haupt Unconformity (Schroeder 1982). In Eastern Europe, all of the pre-km4 sediments are truncated by this slightly angular unconformity so that (for example) in northern Poland, km4 sediments rest on pre-Triassic rocks (Schroeder 1982). The strongly episodic sedimentation rates, coupled with a low total accumulation rate and the presence of the basal km4 unconformity, make correlation with the vastly thicker and more complete continental Mesozoic sequences in other areas difficult. Applying Alpine stage designations to the correlative continental areas is, therefore, even more difficult.

Correlation outside the Germanic Basin of km5, the youngest Germanic Basin Triassic (Rhaetian of older works), is extremely uncertain, because correlation of these mixed marine, paralic, and continental rocks with the Alpine Late Norian (Rhaetian) is itself uncertain. This is all the more confusing because at least some of the classic German "Rhaetian" horizons have proved to be earliest Jurassic (Achilles 1981). We designate these Triassic beds only as latest Norian.

Palynological correlation

Pollen and spore assemblages provide perhaps the most direct means of correlation among the Germanic Basin Triassic, the Alpine Triassic, and continental rocks of other areas. Correlation by pollen and spores in the Triassic largely depends on the recognition of taxa with ranges limited to a portion of the system in Europe. Of the several hundred presumably valid morphospecies that have been described, only a fraction have ranges limited to less than two stages (Fisher and Dunay 1981). The upper and lower limits of these ranges represent the basic data for correlation with other areas. Triassic palynoflorules from Antarctica, Australia, India, southern Africa, and South America are radically different from all more northern assemblages, regardless of the details of correlation (Anderson and Anderson 1970; Dolby and Balme 1976). The megafossil florules from the same southern areas are equally different from the northern florules, which makes paleobotanical correlation of these areas with those of the Germanic Basin and the Alpine Triassic all the more uncertain. Fortunately, there are several areas where marine rocks interfinger with plant- and palynomorph-bearing sections, and these can be tied to the European section (Retallack 1977, 1979).

For the Early and Middle Triassic, we accept the correlations of Anderson and Anderson (1970) and Anderson and Cruickshank (1978), which are based principally on pollen and spore and megafossil plant assemblages. It is necessary to revise the correlations of the Late Triassic and Early Jurassic, however.

The first problem is to differentiate those vertebrate assemblages of Carnian and Norian age. In Anderson and Anderson (1970) and Anderson and Cruickshank (1978) (followed by Benton in Chapter 24), most of the Late Triassic assemblages of the world are considered Norian or early Norian in age. Recent work suggests that most of these are Carnian assemblages.

Recent work on the exposed portions of the Chinle Formation and Dockum Group of the western North American Triassic has failed to locate any palynoflorules that indicate an age younger than latest Carnian–early Norian or older than middle Car-

nian (Ash et al. 1978; Dunay and Fisher 1979; Ash 1980; Chapter 9). We believe that the supposedly Norian assemblages discussed by Chatterjee (Chapter 10) are also Carnian. Those portions of the Newark Supergroup that until recently have produced the bulk of the vertebrates have likewise proved to be middle and late Carnian in age (Hope and Patterson 1970; Cornet 1977a,b; Cornet and Olsen 1985). Younger, much less fossiliferous sequences make up the bulk of the Newark, however.

The Maleri Formation of India and correlative formations were first thought to be Norian in age (Chatterjee 1980), but recent palynological work by Kumaran and Maheswari (1980) on the Tiki Formation has suggested a Carnian age. On the basis of the presence of the rhynchosaur *Hyperodapedon huxleyi* (Ghosh and Mitra 1970; Chatterjee 1980; Cooper 1981; Benton 1983) in both formations, the Tiki Formation correlates with the Maleri. Likewise, the vertebrate-rich Ischigualasto Formation of Argentina is best viewed as Carnian, not Norian, on the basis of megafossil plants and palynomorph assemblages (Bonaparte 1982).

The Early Jurassic formations of Europe present the greatest challenge to correlation. Where lateral correlations can be established with certainty, the rocks are almost entirely marine; where continental rocks are present, their correlation to the stage level is very tentative. Terrestrial vertebrates are virtually absent, and the pollen and spore assemblages are not as well studied as those of the Late Triassic. Nonetheless, pollen and spore assemblages provide the best link with continental areas outside the European marine Jurassic.

Unfortunately, palynological characterization of the Triassic–Jurassic transition is not well established, even in Europe. The transition between the Late Triassic of the Germanic Basin and the European Liassic is marked by change in the relative abundance of taxa, but by surprisingly few extinctions or originations of pollen and spore taxa marking the boundary (Morbey 1975; Schuurman 1979; Fisher and Dunay 1981; Visscher and Brugman 1981). In many areas, this change in relative abundance consists of a dramatic increase in the percentage of the conifer palynotaxon *Corollina*, especially *C. meyeriana* (Visscher, Schuurman, and Van Erve 1980; Alvin 1982). However, even in Europe, *Corollina* does not consistently increase through the boundary at various localities (Cornet 1977a). In fact, the European section appears to straddle an Early Jurassic boundary between a northern palynological province, in which *Corollina* may be absent or only a minor part of palynoflorules, and a southern province, in which *Corollina* is strongly dominant (Hughes 1973; Pederson and Lund 1980; Alvin 1982; Olsen and Galton 1984; Cornet and Olsen 1985). The later Early Jurassic is characterized by a number of diagnostic taxa, but they are fewer than the characteristic taxa of divisions of the Triassic. The recognition of the Triassic–Jurassic boundary by the extreme dominance of *Corollina* probably reflects relative abundance of the *Corollina* producers, the cheirolepidaceous conifers (Cornet 1977a; Alvin 1982) that replaced the older, more diverse plant assemblage.

The only region in which both adequate palynological data and adequate terrestrial vertebrate assemblage are in direct association is the Newark Supergroup of eastern North America. Here, the upper Newark Supergroup has produced abundant palynoflorules, which, based on all the available evidence, correlate with those of the European Lias. *Corollina meyeriana* becomes strongly dominant a few meters below the oldest extrusive basalts in all Newark basins, and this is where the Triassic–Jurassic boundary has been placed by Cornet (1977a) and Cornet and Olsen (1985). The overlying extrusive basalt flows consistently give K–Ar and ^{40}Ar/^{39}Ar dates on the younger side of the Triassic–Jurassic boundary (Armstrong and Besancon 1970; Cornet, Traverse, and McDonald 1973; Armstrong 1982; McHone and Butler 1984; Seideman et al. 1984). In strata higher in the Newark section, palynomorph taxa characteristic of the Early Jurassic occur, just as they do in Europe (Cornet 1977a).

Unfortunately, the taxonomic diversity within Newark Supergroup palynoflorules of Jurassic age is very low. Cornet (1977a) has been forced to divide the more than 2000 m of post-Triassic Newark Supergroup into three palynologically defined zones, which are largely based on the relative proportion of three *Corollina* species. These are the *Corollina meyeriana* zone, *C. torosus* zone, and the *C. murphii* zone, which he correlates with the Hettangian + Sinemurian, Pliensbachian, and Toarcian, respectively. Whereas it seems certain that at least most of these strata are Early Jurassic, precise correlation with the European Jurassic stages must be regarded as uncertain, even though these are not conflicting data.

The age assessment of three other classic areas with diverse continental vertebrates has recently shifted from the Late Triassic to the Early Jurassic, again principally on the basis of floral remains. The Moenave Formation of the lower part of the Glen Canyon Group of the southwestern United States has produced a palynoflorule dominated by *C. torosus* (Olsen and Galton 1977; Peterson, Cornet, and Turner-Peterson, 1977; Peterson and Pipiringos 1979; Chapters 20, 22, and 23). This indicates that most of the Glen Canyon Group, including the vertebrate-rich Kayenta Formation, is Jurassic in age.

Likewise, beds in the Tuli Basin of Zimbabwe and Botswana equivalent to the Clarens Formation of the Karoo Basin have produced a palynoflora dominated by *Corollina intrareticulatus*, known otherwise only from Late Sinemurian to Tithonian age sediments of Argentina (Volkheimer 1971; Aldiss, Benson, and Rundel 1984). Interbedded basalt flows and the older flows of the Drakensberg Basalts from the Karoo Basin of South Africa give K–Ar dates of 169–193 MY (Fitch and Miller, 1971; Cleverly, 1979; Bristow and Saggerson, 1983; Aldiss et al., 1984). These basalts appear to be somewhat younger than those of the Newark Supergroup and fall well within the Early Jurassic in all the current radiometric scales (Busbey and Gow 1984). Finally, the "Dark Red Beds" and possibly the underlying "Dull Purplish Beds" of the Lower Lufeng of Yunnan, China appear to be Early Jurassic on the basis of floral remains (Cui 1976; Sigogneau-Russell and Sun 1981; Chapter 21).

Two more deposits that must be mentioned are the Evergreen Formation and the Marburg Sandstone of Australia. Dated as late Early Jurassic on the basis of palynoflorules dominated by *Corollina classoides* and *Tsugaepollenites* spp. (de Jersey and Paten 1964; Reiser and Williams 1969), these units have produced the youngest known labyrinthodont amphibians, *Siderops* and *Austropelor* (Family Chigutisauridae) (Warren and Hutchinson 1983). At least one group of labyrinthodonts, it seems, survived the Triassic (Chapter 1).

Correlation by vertebrates

Although correlation by pollen and spores provides the best link between the Early Mesozoic of Europe and areas of wholly continental deposition, the vertebrate evidence is in good agreement. For example, the portion of the Newark Supergroup that is assigned to the Late Triassic on the basis of pollen and spores contains reptiles and amphibians characteristic of the Middle and Upper Keuper. In addition, many vertebrate taxa characteristic of the Jurassic portion of the Newark are absent from the German Keuper (Olsen and Galton 1977, 1984; Cornet and Olsen 1985). The osseous remains of phytosaurs and their referred footprints, for instance, extend to the very top of the palynologically defined Late Triassic in the Newark Basin, and they extend to the very top of the Keuper (Rhaet = km5) as well. No phytosaurs have ever been found in the part of the Newark thought to be Jurassic on the basis of pollen and spores, and phytosaurs have never been found in the disputed or undisputed Jurassic sediments anywhere else in the world. Phytosaurs are absent from the Glen Canyon Group (Olsen and Galton 1977) [the phytosaur-bearing Rock Point Member is now assigned to the Chinle Formation (Pipiringos and

O'Sullivan 1978)], the Upper Stormberg Group of southern Africa, and also the Lower Lufeng [the one possible specimen is now lost and regarded as questionable (Chapter 21)]. The tetrapod assemblages from these units, now correlated with the Early Jurassic, are distinctly different from the classic Late Triassic assemblages (Olsen and Galton 1977, 1984; Attridge, Crompton, and Jenkins 1985).

It should be stressed that not all authors agree with an Early Jurassic age for many of these continental strata (Colbert 1958; Busbey and Gow 1984; Chapter 1), and it must be admitted that these assignments are by no means certain [see reviews by Clark and Fastovsky (Chapter 23) and Sues (Chapter 22)]. However, there is considerable published evidence independent of the vertebrates that supports an Early Jurassic age for these strata. A counterargument presenting positive evidence to the contrary has yet to be presented, in our view.

Both Newark Supergroup and Chinle–Dockum assemblages dated as late Carnian by pollen and spore assemblages yield distinctive vertebrate assemblages dominated by the advanced phytosaur *Rutiodon* and the labyrinthodont amphibian *Metoposaurus* (Olsen, McCune, and Thomson 1982). Such an assemblage is not present in the Germanic Keuper. Instead, the youngest *Metoposaurus* assemblages in km3 contain *Paleorhinus*-type phytosaurs [*Parasuchus* of Chatterjee (Chapter 10)]. The *Paleorhinus*-type phytosaurs also occur with *Metoposaurus* in the Chinle Formation and the Dockum Group in beds with middle Carnian pollen and spores, and these beds occasionally contain *Rutiodon* as well. The hypothesis that the *Rutiodon–Metoposaurus* assemblage is represented by the post-km3 unconformity in Europe is used here, but must be tested by additional work.

The Lossiemouth beds of Scotland present an additional problem because there are no independent means to assess the age of the vertebrates. Walker (1961) assigned a Norian age based on the close relationship of the aetosaur *Stagonolepis* to *Aetosaurus* from the German Stubensandstein, of Norian age. Recently, however, Baird and Olsen (1983) have reported the Lossiemouth procolophonid *Leptopleuron* (*Telerpeton*) from the late Carnian Wolfville Formation of the Newark Supergroup. The Nova Scotian form may even be conspecific with that from Scotland. In addition, the Lossiemouth beds have produced the type species of the rhynchosaur *Hyperodapedon*, thus suggesting correlation with the Maleri and Tiki formations. Rhynchosaurs are as yet unknown in beds regarded on other grounds as younger than Carnian. In addition, *Stagonolepis*-like aetosaurs occur in both Carnian and Norian beds in the Newark Supergroup (Olsen et al. 1982; Baird and Olsen 1983) and Ar-

gentina (Bonaparte 1982). We therefore prefer, on balance, to regard the Lossiemouth beds as Carnian (Fig. 25.1).

Two important continental assemblages have even more uncertain ages than those discussed thus far: the fissure-fills of Great Britain and the Los Colorados Formation of Argentina. The classic fissure fillings of Great Britain have been traditionally divided into two suites (Robinson 1957): a supposedly Norian suite dominated by sauropsids, especially *Kuehneosaurus* and other reptiles, and a Rhaeto-Liassic suite dominated by mammals and tritylodonts. One of the latter fissure fillings yields the tritylodont *Oligokyphus* and marine invertebrates indicating a Pliensbachian age (Kuehne 1956). One fissure contains both *Oligokyphus* and the mammal *Morganucodon*, and other fissure fills containing *Morganucodon* have also produced a "Rhaeto-Liassic" plant assemblage (Pacey 1978; Marshall and Whiteside 1980) dominated by cheirolepidaceous conifers. Recently, all these "Rhaeto-Liassic" fissures have been regarded by a number of authors as more definitely Early Jurassic, probably early Sinemurian (Evans 1980; Kermack, Mussett, and Rigney 1981). The mammal *Kuehneotherium* has been considered characteristic of the younger fissure fills, but Fraser and Walkden (1984) and Fraser, Walkden, and Stewart (1985) reported this mammal in association with *Kuehneosaurus*, thus blurring the faunal distinction between the two sets of fissures. The two sets of fissures also share very similar sphenodontids (Pacey 1978; Fraser and Walkden 1983). Thus, it is unclear what range of ages can be assigned to these assemblages, although at least some are regarded as Early Jurassic in age and some just as certainly as late Norian in age (Marshall and Whiteside 1980). For the purposes of our range chart (Fig. 25.2), we regard all the Jurassic fissures listed by Kermack, Mussett, and Rigney (1973), except the Neptunian dike described by Kuehne (1956), as Hettangian, and all the others as Norian.

The La Esquina local fauna of the Los Colorados Formation of Argentina poses a major problem because the assemblage contains elements typical of both Late Triassic and Early Jurassic faunas, and there are no other associated forms of age-correlative data. Specifically, it is the only assemblage known to contain tritheledonts, tritylodonts, protosuchid crocodiles, and sphenosuchids together with stagonolepidids and rauisuchids (Bonaparte 1982). It has not been documented in the published literature that this assemblage represents a single faunal horizon. If it does not represent a mixture of Late Triassic and Early Jurassic horizons, this fauna may be truly transitional, and, in the absence of contrary evidence, we list it here as Norian.

The correlations of the early Mesozoic areas

discussed above with the European type area are shown in Figure 25.1. The ranges of families of tetrapods through the Triassic and Early Jurassic based on these correlations are shown in Figure 25.2. We cannot overemphasize the importance of precise correlations to the pattern of originations that we discuss in the following section.

Patterns of diversity, origination, and extinctions in the Early Mesozoic

Every method that expresses some metric of diversity, extinctions, or originations over an interval of time has its own advantages, disadvantages, and implicit assumptions. In fact, choice of the metric may determine the kind of pattern that the analysis of raw data reveals. The main assumption of most

simple metrics is that there is an even distribution of taxa within an interval. In most cases, this is exactly what we do not know because we have selected the shortest reliably correlated interval of time as a unit of resolution.

A metric of diversity is particularly sensitive to the distribution of originations and extinctions through the time interval considered. For example, it is not really useful simply to give the total number of taxa present in the interval, because some of the taxa go extinct or have their origin during that interval. What we really want is the average number of taxa per million years per interval. To know this accurately we would have to know the actual duration of taxa at the million year scale, which we do not. However, if we assume that on the average, taxa go extinct or originate at the middle of the stage,

Figure 25.1. Correlation of Early Mesozoic deposits considered in this chapter with the Standard Ages and with the standard section for the Triassic of the Germanic Basin. Germanic Basin correlation from Brinkmann (1960), Gall, Durand, and Muller (1977), and Brenner and Villinger (1981). Radiometric scale from Palmer (1984). Explanation of abbreviations follows. Germanic Basin Section: su–so, mu–mo, ku, km1–km4, and ko are standard abbreviations for subdivisions of Germanic Triassic. Newark Supergroup: PT, Portland Formation (Hartford Basin) faunule; M, McCoy Brook Formation faunule; P, Passaic Formation (Newark Basin)–New Haven Arkose (Hartford Basin) faunules; L, Lockatong Formation (Newark Basin)–New Oxford Formation (Gettysburg Basin)–Cow Branch Formation (Dan River Basin) faunules; PR, Pekin Formation (Deep River Basin)–Richmond Basin–Taylorsville Basin faunules.

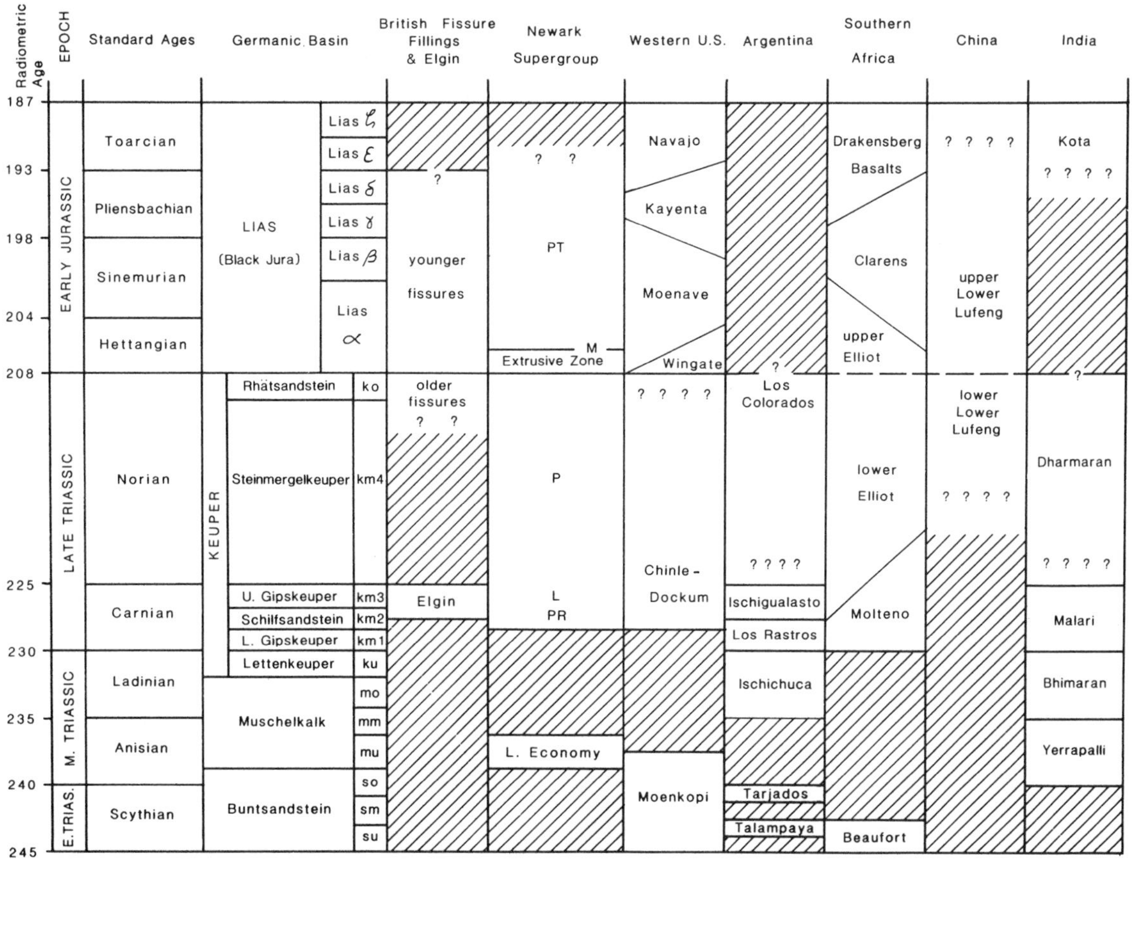

```
                                                         stages
        Family              S     A     L     C     N     H     S     P     T     Aal     B
 1  Trematosauridae         XXXXX
 2  Uranocentrodontidae     XXXXX
 3  Benthosuchidae          XXXXX
 4  Indobrachyopidae        XXXXX
 5  Rytidosteidae           XXXXX
 6  Dissorophidae           XXXXX
 7  Lystrosauridae          XXXXX
 8  Myosauridae             XXXXX
 9  Cynognathidae           XXXXX
10  Diademodontidae         XXXXXXXXXX
11  Erythrosuchidae        +XXXXXXXXXX
12  Brachyopidae            XXXXXXXXXX
13  Ctenosauriscidae              XXXXX
14  Shansiodontidae               XXXXX
15  Proterosuchidae        +XXXXXXXXXXXXXXX
16  Kannemeyeriidae         XXXXXXXXXXXXXXXXXXXXXXXXX
17  Traversodontidae        XXXXXXXXXXXXXXXXXXXXXXXXX
18  Stahleckeriidae               XXXXXXXXXXXXXXXX
19  Rhynchosauridae               XXXXXXXXXXXXXXXX
20  Chiniquodontidae                    XXXXXXXXXX
21  Metoposauridae                      XXXXXXXXXX
22  Proterochampsidae       S     A     XXXXXXXXXX    N     H     Sin   P     T     Ael     B
23  Lagosuchidae                              XXXXX
24  Erpetosuchidae                            XXXXX
25  Scleromochlidae                           XXXXX
26  Capitosauridae                      XXXXXXXXXX
27  Procolophonidae        +XXXXXXXXXXXXXXXXXXXXXXXXXXXXXX
28  Mastodonosauridae             XXXXXXXXXXXXXXXXXXXX
29  Rauisuchidae                  XXXXXXXXXXXXXXXXXXXX
30  Trilophosauridae              XXXXXXXXXXXXXXXXXXXX
31  Tanystropheidae               XXXXXXXXXXXXXXXXXXXX
32  Plagiosauridae                      XXXXXXXXXXXXXXX
33  Ornithosuchidae                     XXXXXXXXXX
34  Stagonolepididae                    XXXXXXXXXX
35  Phytosauridae                       XXXXXXXXXX
36  Kuehneosauridae                     XXXXXXXXXX
37  Drepanosauridae                           XXXXX
38  Endennasauridae                           XXXXX
39  Prolacertidae           XXXXXXXXXXXXXXXXXXXXXXXXXXXXXXXXXXXXXX
40  Proganochelyidae                          XXXXXXXXXX
41  Kuehneotheriidae                          XXXXXXXXXX
42  Haramiyidae                               XXXXXXXXXX
43  Tritheledontidae        S     A     L     C     XXXXXXXXXX    Sin   P     T     Ael     B
44  Gephyrosauridae                                 XXXXXXXXXX
45  Melanorosauridae                          XXXXXXXXXXXXXXXX
46  Heterodontosauridae                             XXXXXXXXXXXXXXXXXX
47  "Scelidosauridae"                               XXXXXXXXXX
48  Anchisauridae                             XXXXXXXXXXXXXXXXXXXXXXXXXXXX
49  Chigutisauridae                     XXXXXXXXXXXXXXXXXXXXXXXXXX
50  Procompsognathidae                  XXXXXXXXXXXXXXXXXXXXXXXXXXXXXXXXXXXXXXXXXXXXXXX+
51  Fabrosauridae                       XXXXXXXXXXXXXXXXXXXXXXXXXXXXXXXXXXXXXXXXXXXXXXX+
52  Sphenodontidae                      XXXXXXXXXXXXXXXXXXXXXXXXXXXXXXXXXXXXXXXXXXXXXXX+
53  Stegomosuchidae                           XXXXXXXXXXXXXXXXXXXXXXXXXXXXXXXXXXXXXXXXX+
54  Sphenosuchidae                            XXXXXXXXXXXXXXXXXXXXXXXXXXXXXXXXXXXXXXXXX+
55  "Dimorphodontidae"                        XXXXXXXXXXXXXXXXXXXXXXXXXXXXXXXXXXXXXXXXX+
56  "Eudimorphodontidae"                      XXXXXXXXXXXXXXXXXXXXXXXXXXXXXXXXXXXXXXXXX+
57  Morganucodontidae                         XXXXXXXXXXXXXXXXXXXXXXXXXXXXXXXXXXXXXXXXX+
58  Tritylodontidae                           XXXXXXXXXXXXXXXXXXXXXXXXXXXXXXXXXXXXXXXXX+
59  Megalosauridae                                  XXXXXXXXXXXXXXXXXXXXXXXXXXXXXXXXXXX+
60  Casichelyidae                                         XXXXXXXXXXXXXXXXXXXXXXXXXXX+
61  Cetiosauridae                                         XXXXXXXXXXXXXXXXXXXXXXXXXXX+
```

Figure 25.2. Cumulative distribution of tetrapods through the Triassic and Early Jurassic, based on the correlations in Figure 25.1. Distributional data from Anderson and Anderson (1970) and Anderson and Cruickshank (1978) for the Early and Middle Triassic and from Kitching and Raath (1984), Olsen, McCune, and Thomson (1982), and Olsen and Galton (1977, 1984) for the Late Triassic and Early Jurassic.

Notes on the families. Please note here that (1) "family" is a very arbitrarily erected hierarchical category; (2) "families" are not strictly comparable among groups; (3) they are often paraphyletic. Our justification for using them is that they often form cohesive morphologic, taxonomic, and stratigraphic units. This cannot, unfortunately, be said of most lower-level groups. We include the Poposauridae within the Rauisuchidae. *Vulcanodon* is included in Melanorosauridae, although Cooper (1984) suggests it should belong in a new family along with *Barapasaurus*, which we include here within the Cetiosauridae. We give the age of *Vulcanodon* as Sinemurian, which is the oldest range for the pollen taxa found in inter-Drakensberg Basalt sediments (Aldiss, Benson, and Rundel 1984), although we recognize it could well be younger. The oldest ornithischians (from the Carnian Ischigualasto Formation and the Newark Supergroup) are included in the "fabrosaurs," although, in our opinion, they cannot really be assigned to a family. *Scelidosaurus* from the English Lias and *Scutellosaurus* are almost certainly thyreophorans, but they cannot be placed in either the Ankylosauria or Stegosauria. Therefore, we place them in the group "Scelidosauridae." Fabrosaurs are known from the Early Cretaceous in the form of *Echinodon*. We include the Saltoposuchidae (including *Terrestrisuchus*) in the Sphenosuchidae. We also regard *Hallopus* from the Morrison Formation as a possible sphenosuchid. Stegomo-

the average diversity per million years turns out to be exactly equal to the average number of taxa at risk per interval as defined by Van Valen (1984). This average number of taxa per million years per interval is equal to the number of originations plus the number of extinctions per interval divided by two, with that value added to the total number of taxa that survive the previous interval and survive through the interval under consideration. The measures of diversity, extinction, origination, and turnover that we use are given in Table 25.1.

We define originations and extinctions as first appearances and last appearances, respectively, within a selected interval. Origination and extinction rates present a problem similar to that of diversity. If originations and extinctions in an interval are distributed through its length, then normalization to the length of an interval is appropriate. This method thus gives the "density" of originations or extinctions during that interval as if they occurred randomly through it. However, if the originations and extinctions are concentrated in a short part of the stage, normalization will mask the origination or extinction "event." Furthermore, large differences in interval length could make curves of different data sets artificially look similar. For these reasons, we also calculate the probability of extinction per taxon per time interval and the proportion of origination. These are calculated according to the method used by Hoffman and Ghiold (1985); the probability of

extinction is equal to the number of taxa that have their last appearances during an interval divided by the *total* number of taxa at risk during that interval. The latter is equal to the number of taxa surviving from the previous interval plus the originations during that interval. This is preferable to the similar method of Van Valen (1984) in which the number of extinctions is divided by the *average* number of taxa at risk, because the latter method often results in probabilities greater than one. The proportion of originations is equal to the number of first appearances within an interval divided by the number of taxa entering the stage.

Normalizing raw origination and extinction rate to numbers per million years makes the choice of calibrated scales particularly important, because the lengths of the stages of the Triassic and Early Jurassic vary greatly. For example, the Norian is two to five times the length of the average Mesozoic stage, depending on which time scale is used. Fortunately, most modern scales reflect reasonably similar relative durations for the stages of the Triassic and Early Jurassic [Armstrong 1982; Odin and Letolle 1982; Harland 1982 (Norian added to Rhaetian); Palmer 1984]. We have chosen to use Palmer's (1984) scale because it is the most recent and seems to reflect best the relative duration of the stages as represented by the relative thickness of marine and continental sequences assigned to each stage. We recognize that this approach exacerbates the prob-

Table 25.1 *Measures of diversity, extinctions, originations, and turnover used*

T	= total number of taxa
E	= number of extinctions
O	= number of originations
C	= number of taxa continuing (taxa that both enter from a previous interval and survive into the next interval)
E_n	= number of taxa entering an interval from the previous interval
D	= duration of an interval in millions of years

Average number of taxa $= (E + O)/2 + C$

Extinction rate = normalized extinction rate $= E/D$

Origination rate = normalized extinction rate $= O/D$

Turnover rate $= (E + O)/D$

P_e = probabilistic extinction rate $= E/T$

P_o = probabilistic origination rate $= O/E_n$

P_t = probabilistic turnover rate $= (E + O)/E_n$

Caption to Fig. 25.2 (*cont.*)

suchidae = Protosuchidae; a representative of this family may occur in the Morrison Formation and does occur in the Early Cretaceous of Mongolia (J. Clark pers. comm.). We regard *Dyoplax* as indeterminate. All of the British fissure fillings that seem to be Jurassic we place within the Hettangian, except for the one Neptunian dike containing *Oligokyphus* (Kuehne 1956), but we recognize that they could be younger.

lems already outlined with correlating the continental assemblages with the marine stages, because most of the isotopically dated sections must themselves be correlated with the marine stages through sometimes convoluted and uncertain paths.

We also calculate measures of turnover for all the sets of data. The normalized turnover rate is equal to the sum of originations plus extinctions divided by the duration of the stage. This is equivalent to the "density" of total change. The probabilistic turnover rate is equal to the total number of originations and extinctions divided by the number of taxa entering an interval. This can be greater than one because a taxon can both originate and become extinct within a single interval.

The global tetrapod data

For the global tetrapod data (Figs. 25.2, 25.3, and Table 25.2), we feel justified in working only at the levels of family and stage. At present, identification of taxa at the genus and species level is problematic among different geographic areas and formations. Some taxa are surely synonymous with others elsewhere, whereas others thought to be synonymous are not: Determinations vary with individual workers, much material has not been restudied for years, and it is difficult for individual workers to study firsthand all of the necessary material in collections around the world. (Except as noted, we accept the generic and specific identification of other authors.) On the other hand, we recognize the prob-

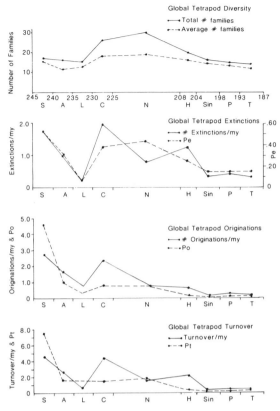

Figure 25.3 Taxic curves for global tetrapod distribution at the stage level. Data derived from Table 25.2 and Figure 25.2.

Table 25.2 *Diversity, extinction, origination, and turnover data for global tetrapods*[a]

Stage	S	A	L	C	N	H	Sin	P	T
No. of families	17	16	15	26	30	20	16	15	14
No. of extinctions	9	5	1	10	13	5	2	2	2
No. of originations	14	8	4	12	14	3	1	1	1
No. continuing	3	5	10	7	5	12	13	12	10
No. entering	3	8	11	14	16	17	15	14	13
Av. no. of families	14.50	11.50	12.50	18.00	18.50	16.00	12.5	13.50	11.50
Duration of stage (MY)	5	5	5	5	17	4	6	5	6
Extinction rate	1.80	1.00	0.20	2.00	0.76	1.25	0.33	0.40	0.33
Origination rate	2.80	1.60	0.80	2.40	0.78	0.75	0.17	0.20	0.17
Turnover rate	4.60	2.60	0.80	4.40	1.58	2.10	0.50	0.60	0.50
P_e	0.53	0.31	0.07	0.38	0.43	0.25	0.13	0.13	0.14
P_o	4.67	1.00	0.36	0.86	0.88	0.18	0.07	0.07	0.08
P_t	7.67	1.63	0.45	1.57	1.93	0.47	0.20	0.21	0.23

[a]Abbreviations: S, Scythian; A, Anisian; L, Ladinian; C, Carnian; N, Norian; H, Hettangian; Sin, Sinemurian; P, Pliensbachian; T, Toarcian.

lems inherent in using the family as the main level of analysis, such as pseudo-extinction (Padian 1984); however, we believe that it is unrealistic to use finer taxonomic levels at this time. We also cannot examine rates on a worldwide basis at a level finer than a stage, because that is the finest level at which the worldwide correlations are reliable; as the discussion above indicates, even accepting correlation at that level may be overly optimistic.

Perhaps the most striking aspect of the global data on tetrapod families (Figs. 25.2 and 25.3) is the lack of any dramatic change in diversity from the Carnian through the Early Jurassic. Normalized extinction rates show a dramatic decline in the Norian, a rise in the Hettangian, and a drop through the rest of the Early Jurassic. The curve of probability of extinction shows a slight increase from the Carnian to the Norian and a drop into the Early Jurassic. Similar trends are seen in the origination and turnover curves. The drop in extinctions, originations, and turnover from the highs in the Scythian may reflect the very poor Ladinian record. Clearly, these data show no evidence of a major extinction in the Norian. Taken at face value, the normalized extinction rates show a dramatic peak in the Carnian, but this is matched by the Carnian origination rates. There is also a decrease in turnover into the Jurassic. These patterns are very different from those discussed by Benton (Chapter 24) and Colbert (1958), who cover the same time interval. The differences are a direct consequence of our stratigraphic revi-

sions described above, especially the recognition of distinctive Early Jurassic tetrapod assemblages. Blurring the differences between late Norian and Hettangian assemblages only makes changes in diversity and origination and extinction rates less significant across the Triassic–Jurassic boundary, which must perforce be examined at the *stage* level.

The Germanic Basin

The Lias of the Germanic Basin is marine, and this makes any direct comparison of tetrapods over the Triassic–Jurassic boundary clearly spurious. However, if we list as present in the Early Jurassic those taxa that occur in unquestionably Middle Jurassic and younger beds elsewhere and occur in the Germanic Triassic, we can partially circumvent this problem. Corrected in this manner, we see a pattern in diversity comparable to that seen in the global data. Normalized extinction, origination, and turnover rates are rather uniform from the Carnian through the Hettangian (Figs. 25.4, 25.5 and Table 25.3). However, unlike the global data, there is a Norian peak in both the normalized extinction rate and in the probability of extinction. The peak in Hettangian normalized extinction rates corresponds to that seen in the global data. Considering the problems of the vast differences in facies between the Germanic Triassic and Liassic, we do not know if these Norian maxima justify notice as a major extinction.

Figure 25.4 Tetrapods from the Germanic Basin. Taxa in the Jurassic are based on the overlying Jurassic and Cretaceous in Europe (see text for explanation).

```
                                       stages
      family              S |  A    L    C  | N  |  H  |  S  |  P  |  T  |
 1  Procolophonidae         |XXXXX|            |
 2  Ctenosauriscidae        |XXXXX|            |
 3  Macronemidae            |XXXXXXXXXXXXXXX|
 4  Mastodonsauridae        |XXXXXXXXXXXXXXX|
 5  Metoposauridae          | XXXXXXXXXX|
 6  Rauisuchidae            |XXXXXXXXXXXXXXXXXXXXXXXX|
 7  Capitosauridae          |XXXXXXXXXXXXXXXXXXXXXXXX|
 8  Tanystropheidae         |XXXXXXXXXXXXXXXXXXXXXXXX|
 9  Plagiosauridae          | XXXXXXXXXXXXXXXXX|
10  Stagonolepididae                   |XXXXXXXXXX|
11  Phytosauridae                      |XXXXXXXXXX|
12  Anchisauridae                      |XXXXX|
13  Proganochelyidae                   |XXXXX|
14  Sphenosuchidae                     |XXXXX|
15  Drepanosauridae                    |XXXXX|
16  Endennasauridae                    |XXXXX|
17  Kuehneotheriidae                        |XXXXXXXXXX|
18  Haramiyidae                             |XXXXXXXXXX|
19  "Scelidosauridae"                            |      |XXXXX|
20  Procompsognathidae                      |XXXXXXXXXXXXXXXXXXXXXXXXXXXXXX+
21  Sphenodontidae                          |XXXXXXXXXXXXXXXXXXXXXXXXXXXXXX+
22  Morganucodontidae                       |XXXXXXXXXXXXXXXXXXXXXXXXXXXXXX+
23  "Dimorphodontidae"                      |XXXXXXXXXXXXXXXXXXXXXXXXXXXXXX+
24  "other Pterosauria"                     |XXXXXXXXXXXXXXXXXXXXXXXXXXXXXX+
25  Megalosauridae                               |XXXXXXXXXXXXXXXXXXXXXXXXX+
26  Tritylodontidae                              |XXXXXXXXXXXXXXXXXXXXXXXXX+
27  ?Cetiosauridae                                                  |XXXXX+
```

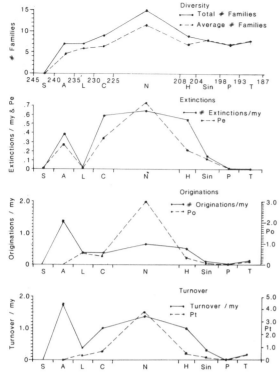

Figure 25.5 Taxic curves for the Germanic Basin based on Table 25.3 and Figure 25.4. Note that Anisian levels are almost certainly an artifact.

Newark Supergroup tetrapods

The Newark Supergroup diversity, origination, and extinction data, viewed at the stage level, are comparable to those of the world (Figs. 25.6, 25.7 and Table 25.4). Comparisons must be limited, however, to the Carnian through Toarcian because, apart from a single Anisian assemblage, older and younger rocks are not known in the Newark (Olsen, McCune, and Thomson, 1982). Like the global data, the Newark extinction rate and probability of extinction for the Norian are lower than for the Carnian. At the stage level, the Newark data, like those of the world, show no evidence of a major extinction at the Triassic–Jurassic boundary. In fact, the Newark data seem to show either a major extinction in the Carnian or only a drop in extinction rates into the Jurassic.

The bulk of the Newark, quite unlike all other described Early Mesozoic deposits, consists of repetitive and periodic (in thickness) sedimentary cycles (Van Houten 1969; Olsen, 1984a,b; Olsen and Imbrie in prep; Chapter 6). These cycles were formed by the rise, fall, and evaporation of very large lakes and appear to have been responses to climatic changes controlled by astronomical variation in the earth's orbit. These pervasive cycles allow the estimation of time between successive fossil assemblages and allow a time scale to be applied to individual Newark Supergroup sections at a ±10,000 year scale (discussed in more detail in Chapter 6). However,

Table 25.3 *Diversity, extinction, origination, and turnover data for Germanic Basin[a]*

Stages	S	A	L	C	N	H	Sin	P	T
No. of families	0	7	7	9	15	9	8	7	8
No. of extinctions	0	2	0	3	11	2	1	0	0
No. of originations	0	7	2	2	12	2	1	0	1
No. continuing	0	0	5	4	0	5	7	7	7
No. entering	0	0	5	7	6	7	7	7	7
Av. no. of families	0	4.50	16.00	6.50	11.50	7.00	8.00	7.00	7.50
Duration of stage (MY)	5	5	5	5	17	4	6	5	6
Extinction rate	0	0.40	0	0.60	0.65	0.50	0.17	0	0
Origination rate	0	1.40	0.40	0.40	0.71	0.50	0.17	0	0
Turnover rate	0	2.80	0.40	1.00	1.36	1.00	0.34	0	0.17
P_e	0	0.29	0	0.33	0.73	0.22	0.13	0	0
P_o	0	0	0.40	0.29	2.00	0.29	0.14	0	0.14
P_t	0	0	0.40	0.71	3.83	0.57	0.29	0	0.17

[a]For abbreviations see Table 25.2.

correlation between basins, with some notable exceptions, cannot yet be resolved any better than at a two-million-year level. Therefore, although we have applied a time scale for the Newark sections divided into millions of years, we actually examine them over two-million-year intervals (Fig. 25.8 and Table 25.5). We have used the palynologically placed Triassic–Jurassic boundary in the Newark to fit Palmer's (1984) isotopic scale to the lacustrine cycle-based Newark time scale (see also Chapter 6). The stage boundaries as defined in the Newark on paleontological grounds fall very close to the iso-

Figure 25.6 Distribution of skeletal remains of tetrapods from the Newark Supergroup based on correlations presented in Cornet and Olsen (1985) and faunal lists in Olsen (1980a, in press).

```
                                        2 million year interval
                            228 224 220 216 212 208 204 200 196 192
     family           325    230|226|222|218|214|210|206|202|198|194|190
                                 - |- |- |- |- |- |- |- |- |- |- |
Capitosauridae         X     - |XXX|- |- |- |- |- |- |- |- |- |- |
Rauisuchidae           X        XXX|
Kannemeyeriidae        X        XXX|
Doswelliidae                    XXX|
Chiniquodontidae                XXX|
Tanystropheidae        X        XXXXX
Traversodontidae       X        XXXXX
Metoposauridae                  XXXXX
Kuehneosauridae                   | |X|
Stagonolepididae                XXXXXXXXX|
Procolophonidae        X        XXXXXXXXXXXXXXXXXXXXXXX
Phytosauridae                    |XXXXXXXXXXXXXXXXXXXX|
"Fabrosauridae"                   |XXXXXXXXXXXXXXXXXXXXX
Sphenodontidae                             |XXX|
Procompsognathidae                          |X|
Trithelodontidae                            |X|
Stegomosuchidae                             XXXXXXXXXXXXXX|
Anchisauridae                               XXXXXXXXXXXXXXXXX
                      325    230|26 | 22 | 18 | 14 | 10 |06 |02  98   94 | 90
                      |A|            C |          N         |H  | S | P | T
                                           stages
```

Table 25.4 *Diversity, extinction, origination, and turnover data for Newark tetrapods at the stage level*[a]

Stages	C	N	H	Sin	P	T
No. of families	13	5	6	2	2	1
No. of extinctions	9	3	4	0	1	0?
No. of originations	6	1	4	0	0	0?
No. continuing	1	1	0	2	1	1?
No. entering	6?	4	2	2	2	1?
Average no. of taxa	8.5?	3	4	2	1.5	1?
Duration of state	5	17	4	6	5	6
Extinction rate	1.8	0.2	1	0	0.2	0?
Origination rate	1.2	0.1	1	0	0	0?
Turnover rate	2.0	0.3	2	0	0.2	0?
P_e	0.69	0.60	0.67	0	0.50	0?
P_o	1.00?	0.25	0.50	0	0	0?
P_t	2.50	1.00	4.00	0	0.50	0?

[a]For abbreviations see Table 25.2.

topically defined stage boundaries; any of the current isotopic scales work equally well, however. This is discussed in detail in Olsen (1984a).

Viewed at the two-million-year level rather than the stage level, Newark tetrapod diversity clearly peaked between 230 and 225 MY and between 210 and 204 MY (Fig. 25.8). These intervals correspond to the late Carnian and the late Hettangian. Extinctions are gathered in two similarly placed, well-defined peaks. The general pattern resembles that for the stage level, but the extinctions appear to be concentrated in the Carnian and at the Triassic–Jurassic boundary. Either the extinctions are really concentrated where they appear to be or

the pattern is an artifact of poor sampling in intervals between the two peaks. In this case we can argue the latter. For example, tetrapod bones are exceedingly rare throughout the Newark Jurassic. The diversity and extinction peaks in the early Hettangian are due to the recent discovery of a rich early Hettangian bone assemblage in Nova Scotia (Olsen and Baird 1982); only a few of the vertebrates in this tetrapod assemblage are known elsewhere in the Newark, but more of these Nova Scotian taxa are known from Middle and Late Jurassic rocks in other regions. If taxa known to persist through the Hettangian (specifically sphenodontids, stegomosuchids/protosuchids, "procompsognathids," and "fabrosaurids") are inserted through the rest of the Newark Jurassic (even though they are not recorded), the Hettangian diversity and extinction peaks between 208 and 206 MY disappear, and only the late Norian peaks (between 210 and 208 MY) remain.

Figure 25.7 Taxic curves for skeletal remains of tetrapods from the Newark Supergroup at the stage level based on the data in Figure 25.6 and Table 25.4. Question marks show estimates that cannot be directly calculated from the data in figures and tables. They represent minimum values. Abbreviations as in Table 25.1 and Figure 25.3 except as follows: Ave. # Fam., average number of families; Ext./my, extinctions/MY; Orig./my, originations/MY; To./my, Turnover/MY.

Figure 25.8 Taxic curves for skeletal remains of tetrapods from the Newark Supergroup sampled at the two-million-year level based on the data in Figure 25.6 and Table 25.5. Abbreviations as in Figure 25.7.

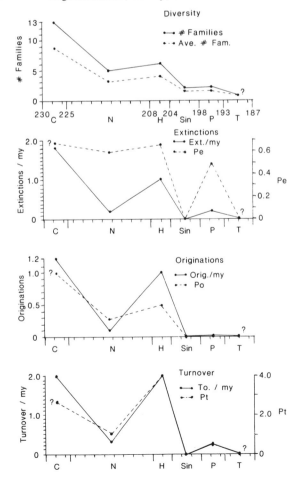

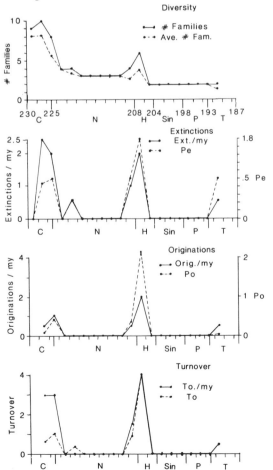

Table 25.5 *Diversity, extinction, origination, and turnover data for the tetrapods of the Newark Supergroup during a two-million-year interval*

Time interval (MY)	228	226	224	222	220	218	216	214	212	210	208	206	204	202	200	198	196	194	192
No. of families	10	8	4	4	3	3	3	3	3	4	6	2	2	2	2	2	2	2	
No. of extinctions	5	4	0	1	0	0	0	0	0	2	4	0	0	0	0	0	0	1	
No. of originations	1	2	0	0	0	0	0	0	0	1	4	0	0	0	0	0	0	0	
No. continuing	5	3	4	3	3	3	3	3	3	1	0	2	2	2	2	2	2	1	
No. entering	10	6	4	4	3	3	3	3	3	3	2	2	2	2	2	2	2	2	
Average no. of families	8	5.5	4	3.5	3	3	3	3	3	2.5	4	2	2	2	2	2	2	1.5	
Extinction rate	2.5	2	0	0.5	0	0	0	0	0	1	2	0	0	0	0	0	0	0.5	
Origination rate	0.5	1	0	0	0	0	0	0	0	0.5	2	0	0	0	0	0	0	0	
Turnover rate	3	3	0	0	0	0	0	0	0	1.5	4	0	0	0	0	0	0	0.5	
P_e	0.46	0.50	0.50	0.25	0	0	0	0	0	0.50	1	0	0	0	0	0	0	0.50	
P_o	0.10	0.33	0.33	0	0	0	0	0	0	0.33	2.0	0	0	0	0	0	0	0	
P_t	0.60	1.0	1.0	0.25	0	0	0	0	0	1.0	4.0	0	0	0	0	0	0	0.50	

Newark tetrapod ichnotaxa

A good proxy of tetrapod diversity is reflected in the abundant tetrapod ichnotaxa from the Newark. Ichnofossil assemblages are directly tied to one of the repetitive facies of the lacustrine cycles, and, therefore, a similar environment is sampled through the entire Newark. At individual sites, tracks can be very common, so both sample size and distribution are much less problematic for Newark tracks than for bones. Because the ichnotaxa generally reflect larger cursorial tetrapods, they should be an excellent indicator of major tetrapod extinctions. Comparison among pedal skeletons of tetrapods suggests that ichnogenera and even ichnospecies correspond more or less to families of tetrapods based on skeletal remains (Chapter 20).

Viewed at the stage level, ichnotaxic diversity drops into the Jurassic (Figs. 25.9, 25.10, and Table 25.6). The probability of extinction is highest in the Norian and drops dramatically into the Jurassic. However, as in all other stage level comparisons, this Norian peak is not seen in the normalized extinction curve. The same data viewed over two-million-year intervals (Fig. 25.11 and Table 25.7) show a strong rise of diversity into the middle Norian, with a drop afterward. Extinctions show three peaks: one in the Carnian between 228 and 226 MY, one in the middle Norian between 216 and 214 MY, and one in the latest Norian between 210 and 208 MY. There are no extinctions after the Hettangian in the footprint data. The first and last extinction peaks correspond exactly to the peaks seen in the skeletal data; the mid-Norian peak in the footprint data does not.

In the actual sections, the first footprint faunules of Jurassic aspect ("Connecticut Valley" aspect) occur in a stratigraphic interval just above the palynologically placed Triassic–Jurassic boundary (Chapter 6). Although the tracks and palynoflorules come from different localities in the same basin (Newark Basin), they are almost certainly correlative within 400,000 years, based on an estimate of possible error in sedimentation rates in sedimentary cycles (Olsen 1984a; Chapter 6). The palynofloral boundary comes from within 60 m of the oldest lava flow in the basin and the oldest Jurassic-aspect track assemblage appears within the last meter below the basalt. Unfortunately, the 1,000-m (about 8-MY) interval between the palynologically fixed Triassic–Jurassic boundary and the youngest well-known Norian assemblage is very poorly sampled for footprints. All other intervals in the Newark, except the oldest Newark Carnian and Anisian age strata, have been relatively densely sampled in either the Newark or Hartford Basins. Thus, it is possible that this 1,000-m gap is wholly responsible for the mid-Norian peak in the extinctions of ichnotaxa, which could have occurred at any time within the unsampled interval.

Overview of tetrapod data

When the tetrapod data of the world are viewed at the stage and family level, there does not appear to be any sign of a major set of extinctions at the Triassic–Jurassic boundary, but there is a peak of extinctions in the Carnian. The same pattern holds for the Newark Supergroup tetrapod data, for the

Figure 25.9 Distribution of vertebrate ichnotaxa in the Newark Supergroup.

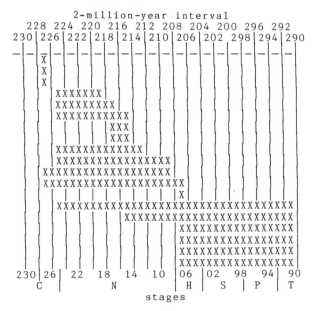

Carnian to Early Jurassic data of the Germanic Basin, and for the Newark ichnotaxa data, all at the stage level. However, when the Newark track and skeletal data are viewed on a two-million-year sampling interval, the two data sets show strong concentrations of extinctions in the Carnian and in the latest Norian. The way in which the high Norian extinctions disappear in the skeletal data and the high Carnian extinctions disappear in the track data when both are subjected to sampling at the stage interval demonstrates that, no matter what the real pattern of extinctions might be, the stage level is an inappropriate level of analysis for seeking major concentrations of extinctions. It tends to decrease the significance of highly concentrated intervals of extinction and turn several small sets of extinctions spread over the stage into a major extinction event. This is true no matter what metric of diversity and extinction is used.

If we extend this line of reasoning to the global data, it seems clear that although at face value there is no evidence of a major tetrapod extinction at the end of the Triassic, the stage level of analysis may mask the real pattern.

Newark pollen and spores

The pollen and spore record of the Newark is known at a finer level than any other category of fossils, largely due to the work of Bruce Cornet and his associates (Cornet et al. 1973; Cornet and Traverse 1975; Cornet 1977a,b; Cornet and Olsen 1985). Pollen and spore records for the Newark Supergroup, correlated to the nearest two million years, are given in Cornet and Olsen (1985). These records provide a completely independent data set to compare with the Newark track and skeletal records. The Newark palynomorph record detailed by Cornet and Olsen consists of 121 "species" sampled from more than a hundred localities. Palynomorphs have the advantages of usually being very abundant in individual samples and of having relatively high morphological and hence high taxonomic diversity. Assemblages occur in all nonmetamorphosed gray and black fine clastics of the Newark, and species counts and identifications based on separate samples are highly reproducible. Their disadvantage is that their taxonomy is a parataxonomy such as that for footprints; they are organ taxa, and a palynomorph species or genus does not necessarily correspond to an equivalent taxonomic level in whole organisms. A pollen cone of one conifer might produce two or three easily recognized pollen species, or several conifer species might produce only a single morphotype of pollen grain. Nonetheless, if used consistently, the available record is legitimately compared within and between formations, and as a record of plant diversity, palynomorphs are unsurpassed in the Newark.

We must qualify these statements, however, by noting that Cornet and Olsen's (1985) work is only a first attempt to correlate the palynological records of the various basins of the Newark Supergroup and provide a cumulative range chart for the entire supergroup at such a fine level. We believe that this record will be subject to change, but, at the present, it is still more refined than all the other classes of data.

At the stage level, pollen and spore diversity and extinction patterns show a resemblance to the global tetrapod data and a very strong resemblance to the Newark tetrapod and footprint data (Figs. 25.12, 25.13, and Tables 25.8 and 25.9). Especially noteworthy is the high diversity in the Carnian and Norian and the high probability of extinction in Norian in both the Newark track and pollen records.

As was true for the Newark track and skeletal data, the diversity and extinction curves viewed on

Figure 25.10 Taxic curves for footprints from the Newark Supergroup at the stage level based on the data in Figure 25.9 and Table 25.6. Abbreviations as in Figure 25.7.

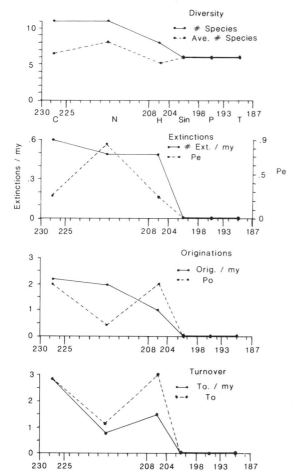

Table 25.6 *Diversity, extinction, origination, and turnover data for Newark ichnotaxa at the stage level*[a]

	C	N	H	S	P	T
No. of forms	11	11	8	6	6	6
No. of extinctions	3	9	2	0	0	0
No. of originations	8	3	4	0	0	0
No. continuing	1	2	2	6	6	6
No. entering	4	8	2	6	6	6
Average no. of ichnotaxa	6.5	8	5	6	6	6
Duration of stage	5	17	4	6	5	6
Extinction rate	0.6	0.5	0.5	0	0	0
Origination rate	2.2	0.2	1	0	0	0
Turnover rate	2.8	0.7	1.5	0	0	0
P_e	0.27	0.82	0.25	0	0	0
P_o	2.00	0.38	2.00	0	0	0
P_t	2.75	1.30	3.00	0	0	0

[a]For abbreviations see Table 25.2.

a two-million-year level look rather different than they do at the stage level (Fig. 25.14). In particular, the Carnian high in diversity is all contained in the first two million years of the Newark record, and the rest of the Carnian through Norian record of diversity is relatively flat, with one peak in the late Norian (214–212 MYA). The Jurassic record is of uniformly low diversity. It is of some interest that the Carnian high in palynomorph diversity, extinctions, and originations falls very close to those seen in the Newark skeletal and footprint curves. Likewise, the terminal Norian peak in pollen and spore extinctions is also apparent in the Newark footprint and skeletal assemblages. The terminal Norian extinction peak corresponds to the palynologically defined Triassic–Jurassic boundary within the Newark. However, like the Newark skeletal data, the palynomorph extinction curves show no peak in the mid-Norian such as we see in the curves for the footprints. We avoid consideration of the apparent periodicity exhibited by the data in Figure 25.14 at this time.

The correspondence between the Newark palynomorph, skeletal, and track extinction curves suggest that they may be causally related. Taken at face value, there really is a terminal Triassic extinction event and perhaps also one in the Carnian. We consider these patterns robust only where the taxonomic data can be examined on finer levels than the stage, for at the stage level these patterns become much less clear. The absence of a peak in extinctions in the Norian in all three data sets when examined at the stage level is an artifact.

Figure 25.11 Taxic curves for tetrapod footprints from the Newark Supergroup sampled at the two-million-year level based on the data in Figure 25.9 and Table 25.7. Abbreviations as in Figure 25.7.

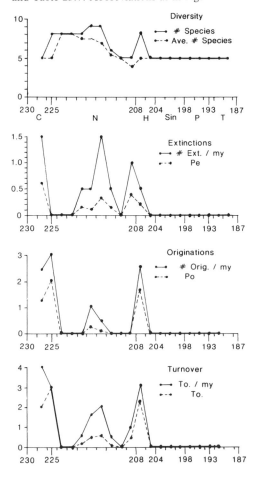

Table 25.7 *Diversity, extinction, origination, and turnover data for Newark ichnotaxa at the two-million-year level*

Time interval (MY)	228	226	224	222	220	218	216	214	212	210	208	206	204	202	200	198	196	194	192	190
No. of forms	5	8	8	8	8	9	9	6	5	5	8	8	5	5	5	5	5	5	5	5
No. of extinctions	3	0	0	0	1	1	3	1	0	2	2	0	0	0	0	0	0	0	0	0
No. of originations	5	6	0	0	0	2	1	0	0	0	5	5	5	0	0	0	0	0	0	0
No. continuing	1	2	8	8	7	6	5	5	5	3	2	2	5	5	5	5	5	5	5	5
No. entering	4	2	8	8	8	7	8	6	5	5	3	3	5	5	5	5	5	5	5	5
Average no. of families	5	5	8	8	7.5	7.5	7	5.5	5	4	5	5	5	5	5	5	5	5	5	5
Extinction rate	1.5	0	0	0	0.5	0.5	1.5	0.5	0	1	0.5	0.5	0	0	0	0	0	0	0	0
Origination rate	2.5	3	0	0	0	1	0.5	0	0	0	2.5	2.5	0	0	0	0	0	0	0	0
Turnover rate	4.0	3	0	0	0.5	1.5	2.0	0.5	0	1	3.0	3.0	0	0	0	0	0	0	0	0
P_e	0.60	0	0	0	0.13	0.11	0.30	0.17	0	0.40	0.25	0	0	0	0	0	0	0	0	0
P_o	1.25	2.00	0	0	0	0.29	0.13	0	0	0	1.67	0	0	0	0	0	0	0	0	0
P_t	2.00	3.00	0	0	0.13	0.43	0.50	0.17	0	0.40	2.33	0	0	0	0	0	0	0	0	0

Species list (rows 1–69):

1 "Triangulatisporites" maximus
2 Zebrasporites corneolus
3 Aratrisporites saturnii
4 Polycingulatisporites mooniensis
5 Osmundacidites senectus
6 Aratrisporites fimbriatus
7 Calamospora nathorstii
8 Striatoabieites aytugii
9 Apiculatisporites laviverrucosus
10 Pityosporites devolvens
11 Pityosporites inclusus
12 Camerosporites secatus
13 Triadispora cf. T. aurea
14 Paracirculina scurrilis
15 Duplicisporites granulatus
16 Cyclotriletes oligogranifer
17 Lycospora imperialis
18 Lagenella martinii
19 Cyclogranisporites oppressus
20 Tigrisporites dubius
21 Convolutispora affluens
22 Pityosporites neomundanus
23 Raistrickia crassiornata
34 Triletes cf. T. verrucatus
35 Leschikisporis aduncus
36 Raistrickia grovensis
37 Triletes subtriangularis
38 "Tuberculatosporites" hebes
39 "Placopollis raymondii"
40 Lunatisporites acutus
41 Microcachryidites doubingeri
42 Tetrad type 39
43 Neoraistrickia americana
44 Retisulcites sp. 126
45 Acathotriletes varius
46 Adivisporites dispertitus
47 Vallasporites sp. 68
48 Cycadopites sp. 103
49 Camerozonosporites rudis
50 Plicatisaccus badius
51 Camerosporites pseudoverrucatus
52 Gunthoerisporites cancellosus
03 Microcachryidites sp. 143
54 Triadispora cf. T. obscura
55 Retinonocolpites sp. 173
56 Triadispora modesta
57 Triadispora verrucata
58 Triadispora sp. 165
59 Alisporites cf. A. perlucidus
60 Spiritisporites spirabilis
61 Colpectopollis sp. 142
62 Camerosporites verrucosus
63 Distaverrusporites sp. 167
64 Foveolatitriletes sp. 235
65 Lycopodiumsporites cf. L. semimurus
66 Verrucosisporites morulae
67 Conbaculatisporites mesozoicus
68 Osmandacidites cf. O. alpinus
69 Carnisporites granulatus

Top chart axis labels:
2 million year interval
Radiometric dates: 190 90 | 94 | 08 | 02 | 208 06 | 10 | 14 | 18 | 22 | 228 26 | 230

Bottom chart axis labels:
Radiometric dates / Stages
190 193 | 198 | 204 208 | 225 | 228 230
T P S H N C

Figure 25.12 Distribution of palynomorph taxa in the Newark Supergroup, based on Cornet and Olsen, 1985. **A**, Palynomorph distribution at the stage level. **B**, Palynomorph distribution at the two-million-year level.

Marine invertebrates

The end of the Triassic is one of the commonly cited intervals in which a mass extinction of invertebrates is supposed to have occurred. Raup and Sepkoski (1982, 1984) summarized the marine data and showed a very important extinction in the Norian. They also showed major extinctions in the Carnian and in the Rhaetian. Raup and Sepkoski (1982), however, normalized the extinction rates to time by dividing the number of extinctions in the stage by its length in millions of years. Neither the post–Paleozoic stage lengths used nor their sources were given by Raup and Sepkoski. The source used, however, seems to have been Armstrong (1978), in which the Carnian, Norian, and Rhaetian are given equal durations of 5 MY. As pointed out by Odin and Letolle (1982) and others, this is unreasonable, as

Figure 25.13 Taxic curves for pollen and spores from the Newark Supergroup at the stage level based on the data in figure 25.12 and Table 25.8. Question marks show estimates that cannot be directly calculated from the data in figures and tables. They represent minimum values. Abbreviations as in Figure 25.7.

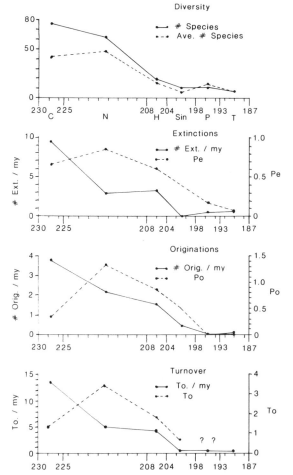

discussed in the first part of this chapter. The very thin Rhaetian record, the very thick Norian record, and the presence of only one ammonite zone in the Rhaetian compared to six in the Norian and five in the Carnian (Tozer 1967) all militate against arbitrarily using equal stage lengths. Using Raup and Sepkoski's (1982) data (from their graph) and normalizing it to Palmer's (1984) time scale, the Norian has an extinction rate of about four families per million years, which is not significant by their own standard. The Carnian extinction is just barely significant by their measure (95 percent confidence level).

On the other hand, Raup and Sepkoski (1984) use percent extinction, which is equal to the probability of extinction [as defined by Hoffman and Ghiold (1985)] multiplied by 100. Raup and Sepkoski (1984) used the Harland (1982) time scale, which still recognizes the Rhaetian as a stage and gives 7 MY to the Carnian, 6 MY to the Norian, and 6 MY to the Rhaetian. It is not important to their major points that the Norian and Rhaetian are given equal lengths by Harland. However, if we combine the Norian and Rhaetian extinctions, the result is an extinction that appears more important than that of the Maastrichtian. This, however, represents a deliberate loss of resolution, as the 2-MY analysis shows. If we consider the Norian and Rhaetian data separately, we cannot know if the extinctions are concentrated at a short interval within a stage or are distributed though it. The Norian seems to be a long stage, and the Rhaetian (if recognized) would be a short one; therefore, if the extinctions were randomly distributed through the stages, the extinctions would definitely be more concentrated in the Rhaetian than in the Norian. According to Hallam (1981), who combines the Rhaetian with the Norian, there were a large number of extinctions distributed through the Norian with, however, a major concentration at the end of the stage, just at the Triassic–Jurassic boundary. If this is so, then the peak in the Norian (without the Rhaetian) extinctions of Raup and Sepkoski (1984) must be spread through that interval, and the "Rhaetian" (i.e., our terminal Norian) and Carnian concentrations of extinctions are the most significant of the Triassic. Viewed in this way, the marine invertebrate data agree strongly with our curves for Newark pollen, tetrapod skeletons, and footprints.

Causes

Two levels of cause can be addressed here: (1) taxa responsible for the observed patterns; (2) processes responsible for taxonomic changes. The first can be derived from an examination of the data that make up the curves. However, at this time, we can only speculate about the second because to imply processes we need to correlate events external to the taxonomic curves in time. Such correlation is even more tentative than the stratigraphic correlation that underlies the global continental tetrapod taxic curves.

The extinction peak that occurs at the end of the Carnian reflects the last known occurrences of a variety of synapsids (the Stahleckeriidae, Kannemeyeriidae, Traversodontidae, Chiniquodontidae), labyrinthodont amphibians (Metoposauridae, Capitosauridae), and nondinosaurian archosaurs and archosauromorphs (Rhynchosauridae, Erpetosuchidae, Scleromochlidae, Proterochampsidae). Norian assemblages differ from Carnian assemblages not only in lacking these taxa, but also by having the first definite records of pterosaurs, Sphenosuchidae (including *Terrestrisuchus* and *Saltoposuchus*), Protosuchidae, Melanorosauridae, Anchisauridae (including Plateosauridae), Proganochelyidae, Tritylodontidae, Morganucodontidae, Kuehneotheriidae, Haramiyidae, and Tritheledontidae. The latter mammals and therapsids are known only from the youngest Norian (what was called Rhaetian).

Unfortunately, early Norian vertebrate assemblages are very poorly known, and, therefore, it is difficult to place much faith in the peak of Carnian extinctions. Furthermore, it is not at all clear whether these extinctions really were concentrated at a single peak within the Carnian.

Despite these problems, it is possible to show that the extinctions that characterize the middle Carnian in the pollen record predate the vertebrate extinctions by perhaps 2 MY. The very diverse vertebrate assemblages of the Petrified Forest Member of the Chinle Formation (Chapter 12) and the correlative Lockatong and Wolfville vertebrate assemblages of the Newark Supergroup include almost all of the typically Carnian families. These assemblages are associated with late Carnian palynoflorules that have a relatively low diversity and are definitely younger than the highly diverse middle Carnian assemblages. No period of major palynofloral extinctions seems to characterize the end of the late Carnian vertebrate assemblages. Unlike the Triassic–Jurassic boundary assemblages, the floral and faunal extinctions do not appear synchronous, and there is no need to look for a common cause.

At the Triassic–Jurassic boundary, the Plagiosauridae, Mastodonsauridae, Procolophonidae, Kuehneosauridae [although see Estes (1983)], Trilophosauridae, Tanystropheidae, Phytosauridae, Rauisuchidae (including Poposauridae), Ornithosuchidae, and Stagonolepididae became extinct. This is the same number of families that became extinct at the end of the Carnian. In contrast to the Carnian–Norian extinctions, however, the Hettangian is distinguished only by the appearance of the Tritheledontidae and Heterodontosauridae (not including the problematic South American *Pisanosaurus*), very rare taxa with very poorly defined times of origin. The Tritheledontidae occur definitely only in the Upper Stormberg Group of Africa and the Newark Supergroup of Nova Scotia, Canada, and the Heterodontosauridae are known only from the Stormberg Group and the Kayenta Formation of the southwestern United States. These are listed as Hettangian through Toarcian only because they are dat-

Table 25.8 *Diversity, extinction, origination, and turnover data for palynomorphs of the Newark Supergroup*[a]

	C	N	H	S	P	T
No. of species	75	61	20	11	11	9
No. of extinctions	48	50	12	0	2	3?
No. of originations	19?	35	9	3	0	0?
No. continuing	12?	4	6	5	11	6?
No. entering	56?	26	11	6	0	3?
Average no. of spp.	45.5?	46.5	16.5	6.5	12	7.5?
Duration of stage	5	17	4	6	5	6
Extinction rate	9.6	2.9	3	0	0.4	0.5?
Origination rate	3.8?	2.1	1.5	0.5	0	0?
Turnover rate	13.4	5.0	4.5	0.5	0.4	0.5?
P_e		0.64	0.82	0.60	0	0.17	0.40?
P_o		0.34?	1.35	0.82	0.50	0	0?
P_t		1.20?	3.27	1.91	0.50	U	1.00?

[a]For abbreviations see Table 25.2.

Table 25.9 *Diversity, extinction, origination, and turnover data for Newark Supergroup palynomorphs*

Time interval (MY)	228	226	224	222	220	218	216	214	212	210	208	206	204	202	200	198	196	194	192	190
No. species	62	29	27	35	34	30	31	45	31	33	20	8	8	11	11	11	11	11	11	11
No. of extinctions	36	2	0	2	4	1	0	14	7	22	12	0	0	0	0	0	0	0	0	11
No. of originations	7	3	0	8	1	0	2	12	0	9	9	0	0	3	0	0	0	0	2	2
No. continuing	19	24	27	25	29	29	29	21	24	10	6	8	8	8	11	11	11	11	11	0
No. entering	55	26	27	27	33	30	29	31	31	34	14	8	8	8	11	11	11	11	11	11
Average no. spp.	40.5	26.5	27	30	31.5	29.5	30	44	27.5	25.5	16.5	8	8	9.5	11	11	11	11	11	10
Extinction rate	18	1	0	1	2	0.5	0	7	3.5	11	6	0	0	0	0	0	0	0	0	1
Origination rate	4	1.5	0	4	0.5	0	1	6	0	4.5	4.5	0	0	1.5	0	0	0	0	0	0
Turnover rate	22	2.5	0	5	2.5	0.5	1	13	3.5	15.5	10.5	0	0	1.5	0	0	0	0	0	1
P_e	0.58	0.07	0	0.06	0.12	0.03	0	0.31	0.23	0.67	0.60	0	0	0	0	0	0	0	0	0.78
P_o	0.13	0.12	0	0.03	0.03	0	0.06	0.39	0	0.26	0.64	0	0	0.37	0	0	0	0	0	0
P_t	0.78	0.19	0	0.37	0.15	0.03	0.07	0.84	0.23	0.91	1.50	0	0	0.38	0	0	0	0	0	0.18

able only to within the "Early Jurassic." As far as the tetrapod skeletal record goes, the Early Jurassic, quite unlike the Norian, is distinguished *only by a lack of taxa* characteristic of the previous stage (Norian).

Of considerable interest is certain evidence that suggests that the taxa that become extinct at the end of the Norian may have persisted right up to the Triassic–Jurassic boundary, rather than becoming extinct over a longer span. The fissure fillings of Great Britain seem to include assemblages in which "typical" Triassic elements (such as procolophonids) are mixed with others (such as mammals) found in fissures known to be Jurassic on the basis of floral remains and invertebrates. The same sort of mixture occurs in the St. Nicolas-de-Port assemblage (Sigogneau–Russell, Cappetta, and Taquet 1979; Sig-

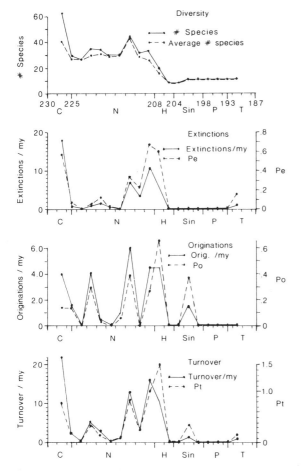

Figure 25.14 Taxic curves for pollen and spores from the Newark Supergroup sampled at the two-million-year level based on the data in Figure 25.12 and Table 25.9. Abbreviations as in Figure 25.7.

ogneau-Russell 1983). In addition, a mixed assemblage appears to occur in the Los Colorados Formation of Argentina. The very tentative picture that emerges is one in which the Triassic–Jurassic boundary is characterized by the extinction of at least ten families over what is beginning to look like a very short period of time, perhaps less than 500,000 years. This certainly counts as a mass extinction, at least at the family level. How many genera and species were included within these families at their times of extinction is very much an open question.

The Newark Supergroup ichnological picture is somewhat muddled. All of the ichnotaxa that became extinct at the close of the Triassic are thought to have been produced by families that became extinct at the very end of the Triassic [ichnogenus *Apatopus* = "skeletal" family Phytosauridae (but see Chapter 4); *Brachychirotherium* = ?Stagonolepididae + Rauisuchidae; *Procolophonichnium* = Procolophonidae]. The ichnotaxic extinctions appear to occur within 1,000 m (about 4 MY) of the Triassic–Jurassic boundary, but those 1,000 m are unsampled. Sampling this gap could show that

1. The extinctions remain concentrated at the mid-Norian.
2. They are spread out over the interval.
3. They combined with the latest Norian extinction.

If the last is true, the latest Norian extinctions in the footprint data would show a nearly complete turnover in composition, with only the theropod dinosaur tracks continuing unabated. Conversely, the latest Norian extinctions and originations could be spread through this gap, and this could eliminate any concentration of turnover or extinction.

Three ichnogenera characterize the Newark Jurassic (*Batrachopus, Anomoepus,* and *Otozoum*), but only *Batrachopus* is definitely known from the earliest Jurassic beds. Whether these forms occur in the 1,000-m unsampled interval below remains to be seen. If the Newark ichnotaxa characteristic of the Norian, such as *Procolophonichnium* sp., *Rhynchosauroides brunswickii, Chirotherium lulli, Brachychirotherium* spp., and *Atreipus* (Chapter 6), disappear just below the palynologically defined Triassic–Jurassic boundary *and* if *Batrachopus* occurs below the boundary, the ichnological situation will closely resemble what we see in the skeletal record. In the Newark, however, it will be possible to document the amount of time that the ichnological transition takes and to demonstrate its relationship to the palynological transition. These questions are answerable within the Newark and clearly deserve attention.

Should the Newark Supergroup ichnological picture prove to be as we have outlined above, the timing of the extinctions of skeletal taxa and ichnotaxa and the rise to dominance of the *Corollina* producers would seem close indeed. A change in floral biogeographic patterns can also be tentatively associated with this Triassic–Jurassic transition. Apparently, the rise to dominance of the cheirolepidaceous conifers (and bennettitalians) and *Corollina* was marked by the extinction of the distinctive Gondwana macrofloral associations dominated by *Dicroidium* and *Thinnfeldia* and the Ipswich–Onslow microfloral assemblage. Unfortunately, it is beyond the current level of biostratigraphic resolution to know if this biogeographic change was really synchronous with the Triassic–Jurassic boundary. It is very tempting to correlate, in a tentative fashion, the homogenization of the world floras with the vertebrate extinctions at the boundary. This homogenization was maintained over a long interval in the Jurassic of what seem to be very low turnover rates, compared to the Triassic, in both tetrapods at a familial level and microflora at a specific level.

If the terminal Norian invertebrate extinctions prove to be synchronous with the tetrapod extinctions and the floral transition (and there is no evidence that they are not), and if these changes can be shown to occur in a relatively short period of time, then the magnitude of these events would indeed compare with the largest of the major Phanerozoic extinction events as Benton (Chapter 24) and Colbert (1958) have suggested.

If there really is a mass extinction at the Triassic–Jurassic boundary that involves plants, tetrapods, and invertebrates, it is appropriate to seek a common cause and perhaps to look at hypotheses relating this event to others in the Phanerozoic record. Hallam (1981) has hypothesized a causal link between a major terminal Norian regression and a Hettangian marine anoxic event. Bakker (1977) has proffered a major regression as the cause of Triassic extinctions as well. On the other hand, because there is a very large impact structure (Manicouagan) known to be of an age that is close to the the Triassic–Jurassic boundary, it is also tempting to associate the extinctions with a bolide impact, as Alvarez et al. (1980) have proposed for the Cretaceous. These hypotheses warrant detailed analysis because all seem plausible mechanisms to explain the observed patterns. However, the bolide impact and the anoxic event model are not mutually exclusive, because the former could cause the latter. At least in the case of the Manicouagan impact, we can look for conclusive evidence in the Newark Supergroup sections; the Fundy Basin was only 500 km away from the impact site in eastern Canada.

However, some of the basic questions till remain unanswered:

1. Over what time period do these extinctions occur?
2. What is the magnitude of the tetrapod extinctions at lower taxonomic levels?
3. How do the Triassic–Jurassic extinctions compare in detail with the "background" rates of extinction through the Early Mezozoic at the same level of resolution?
4. How well do the floral, tetrapod, and invertebrate extinctions correlate within single sections and over larger, even global areas?

The answers to these questions are not yet available and cannot be sought at the standard level of resolution of stage and family. The next step is to look for specific rock units that have characteristics appropriate to these specific questions.

Conclusions
Methodological conclusions

It is crucial to recognize that stratigraphic correlation predetermines the patterns of taxonomic curves on a scale larger than the single section. In this particular study, differences in correlation of Early Mesozoic continental assemblages of the world produce the major differences between our extinction curves and those of Benton (Chapter 24). Without some level of certainty in these correlations, studies of worldwide diversity, extinctions, and originations cannot be expected to yield consistent or meaningful results.

The interval over which worldwide patterns are generally examined is that of the stage. At the stage level, it is impossible to distinguish long intervals of uniform but somewhat high extinction from unified extinction events. An extinction event must be defined in terms of time, but the lengths (4–17 MY) of early Mesozoic stages are too great to qualify as a single "event." From our examination of the Newark data at the stage and two-million-year level, we conclude that the stage level is simply too coarse. As a consequence, it is probably inappropriate to put much faith in taxic curves compiled at the stage level.

Different metrics of diversity, extinction, and origination can yield different results when applied to the same data. A suite of metrics is always desirable because each has its particular assumptions of the distribution of extinctions or originations through the interval. We rarely know which assumptions are valid.

At this time, for Early Mesozoic continental rocks, the most reliable level of analysis is clearly the basin or the series of tightly correlated basins in which there are ways to apply a chronostratigraphic scale with a level of resolution finer than that of the stage. A uniform sampling interval of at most two million years is desirable. At present only the Newark Supergroup meets these requirements.

Diversity and periods of major extinctions in the Triassic and Early Jurassic

All of the curves of diversity suggest higher average diversity in the Late Triassic than in the Early Jurassic. This is visible at all levels of analysis in the marine invertebrates and terrestrial tetrapods of the world, as well as in subsets of those records and in Newark Supergroup pollen and spores. Particularly high levels of diversity appear to typify the Carnian.

Of the more detailed analyses of the Newark Supergroup and the marine invertebrates as studied by Hallam (1981), there seem to be two particularly important periods of extinctions: one in the Carnian and one in the late Norian. The latter seems to coincide with the Triassic–Jurassic boundary. These events are not discernible in the global tetrapod record and can only be seen in data sets compiled at a level finer than that of the stage.

The coincidence among the Newark footprint, skeletal, and pollen and spore records and the record of the marine invertebrates at a fine level of resolution seems to reflect a major extinction event at or very near the Triassic–Jurassic boundary. It also suggests a period of (perhaps less concentrated) extinctions in the Carnian. The Carnian is probably better considered a period of very rapid turnover, because originations are also high at the same time. However, the terminal Norian extinction event was not matched by high origination rates, and the net result was a drop in diversity. In contrast to the Triassic, the Early Jurassic was characterized by very low turnover rates.

The Carnian tetrapod extinctions do not seem to have been synchronous with the floral changes, but the late Norian tetrapod extinctions may have been. The latter also may correlate with the major marine invertebrate extinctions. Unfortunately, at this time, it is not possible to be more precise about either the potential correspondence among these extinctions or their relation to external, perhaps causal, events.

Acknowledgments

We would first like to thank all of our friends and colleagues who made available their information on taxonomic distributions. We especially thank Bruce Cornet for unpublished palynomorph data and Kevin Padian and Rupert Wild for reviewing and updating the taxonomic lists. Antoni Hoffman and Kevin Padian supplied many comments and suggestions that substantially improved the manuscript. We thank Mark Anders for pointing out to us the possible correspondence between the Manicouagan impact structure and the Triassic–Jurassic Boundary. We also sincerely thank Kevin Padian and the staff of the Paleontology Department at the University of California at Berkeley for their invitation to participate in the Society of Vertebrate Paleontology Triassic–Jurassic Symposium and their hospitality during the meeting. Finally, research for this work by P.O. was supported by a fellowship from the Miller Institute for Basic Research in Science at the University of California at Berkeley for 1983–4.

References

Achilles, H. 1981. Die raetische und liassische Mikroflora Frankens. *Palaeontographica, Abt. B.* 197: 1–86.

Aldiss, D. T., J. M. Benson, and C. C. Rundel. 1984. Early Jurassic pillow lavas and palynomorphs in the Karoo of eastern Botswana, *Nature (London)* 310: 302–4.

Alvarez, L. W., W. Alvarez, F. Asaro, and H. V. Michel. 1980. Extraterrestrial cause for the Cretaceous–Tertiary Extinction. *Science* 208: 1095–108.

Alvin, K. L. 1982. Cheirolepidaceae. Biology, structure, and paleoecology. *Rev. Palaeobot. Palynol.* 37: 71–98.

Anderson, H. M., and J. M. Anderson. 1970. A preliminary review of the biostratigraphy of the uppermost Permian, Triassic, and lowermost Jurassic of Gondwanaland. *Palaeontol. Afr.* 13 (Suppl.): 1–22.

Anderson, J. M., and A. R. I. Cruickshank. 1978. The biostratigraphy of the Permian and Triassic. Part 5. A review of the classification and distribution of Permo-Triassic tetrapods. *Palaeontol. Afr.* 21: 15–44.

Arkell, W. J. 1933. *The Jurassic System in Great Britain* (Oxford: Clarendon Press).

Armstrong, R. L. 1978. Pre-Cenozoic Phanerozoic time scale – computer file of critical dates and consequences of new and in-progress decay-constant revisions. *In* Cohee, G. V., M. F. Glaessner, and H. Hedberg (eds.), *Contributions to the Geologic Time Scale. Amer. Assoc. Petrol. Geol., Stud. Geol.* No. 6.

1982. Late Triassic–Early Jurassic time-scale calibration in British Columbia, Canada. In Odin, G.S., (ed.), *Numerical Dating in Stratigraphy* (New York: Wiley), pp. 509–14.

Armstrong, W. J., and J. Besancon. 1970. A Triassic time scale dilemma: K-Ar dating of Upper Triassic igneous rocks, eastern U.S.A. and Canada and post-Triassic plutons, western Idaho, U.S.A. *Eclogae Geol.Helv.* 63: 15–28.

Ash, S. R. 1980. Upper Triassic Floral zones of North America. *In* Dilcher, D. L., and Taylor, T. N. (eds.), *Biostratigraphy of Fossil Plants* (Stroudsburg, Pennsylvania: Dowden, Hutchinson & Ross), pp. 153–70.

Ash, S. R., W. E. Dean, J. F. Stone, P. Tasch, D. J. Weber, and G. C. Lawler. 1978. Paleoecology of Lake Ciniza. *In* Ash, S. R., (ed.), *Geology, Pa-*

leontology, and Paleoecology of a Late Triassic Lake, Western New Mexico. *Brigham Young Univers. Geol. Ser.* 25, Pt. 2, pp. 89–95.

Attridge, J., A. W. Crompton, and F. A. Jenkins, Jr. 1985. The southern African Liassic prosauropod *Massospondylus* discovered in North America. *J. Vert. Paleontol.* 5: 128–32.

Baird, D., and P. E. Olsen. 1983. Late Triassic herpetofauna from the Wolfville Fm. of the Minas Basin (Fundy Basin) Nova Scotia, Can. *Geol. Soc. Amer., Abst. Prog,* 15: 122.

Bakker, R. T. 1977. Tetrapod mass extinctions – a model of the regulation of species rates and immigration by cycles of topographic diversity. *In* A. Hallam (ed.), *Patterns of Evolution as Illustrated by the Fossil Record.* (Amsterdam: Elsevier), pp. 439–68.

Benton, M. J. 1983. The Triassic reptile *Hyperodapedon* from Elgin: functional morphology and relationships. *Phil. Trans. Roy. Soc. Lond. B.* 302: 605–717.

Bonaparte, J. F. 1982. Faunal replacement in the Triassic of South America. *J. Vert. Paleontol.* 2: 362–71.

Brenner, K., and E. Villinger. 1981. Stratigraphie and Nomenklatur des Suedwestdeutschen Sandsteinkeupers. *Jh. Geol. Landesamt Baden-Wuerttemberg.* 23: 45–86.

Brinkmann, R. 1960. *Geologic Evolution of Europe* (trans. from German by J. E. Sanders) New York: Hafner Publishing.

Bristow, J. W., and E. P. Saggerson. 1983. A review of Karoo Vulcanicity in southern Africa. *Bull. Vulcan.* 46:135–59.

Busbey, A. B., III and C. Gow. 1984. A new protosuchian crocodile from the Upper Triassic Elliot Formation of South Africa. *Palaeontol. Afr.* 23: 127–149.

Chatterjee, S. 1980. The evolution of rhynchosaurs. *Mém. Soc. Géol. Fr., N.S.* 139: 57–65.

Cleverly, R. W. 1979. The volcanic geology of the Lebombo monocline in Swaziland. *Trans. Geol. Soc. S. Afr.* 82: 343–8.

Colbert, E. H. 1958. Tetrapod extinctions at the end of the Triassic Period. *Proc. Natl. Acad. Sci. U.S.A.* 44: 973–7.

Cooper, M. R. 1981. A mid-Permian to earliest Jurassic tetrapod biostratigraphy and its significance. *Arnoldia Zimbabwe* 9: 77–104.

1984. A reassessment of *Vulcanodon karibensis* Raath (Dinosauria : Saurischia) and the origin of the Sauropoda. *Palaeontol. Afr.* 25: 203–31.

Cornet, B. 1977a. The palynostratigraphy and age of the Newark Supergroup. Ph.D. thesis, Department of Geosciences, Pennsylvania State University.

1977b. Preliminary investigation of two Late Triassic conifers from York County, Pennsylvania. *In* Romans, R. C. (ed.), *Geobotany* (New York: Plenum), pp. 165–72.

Cornet, B., and P. E. Olsen. 1985. A summary of the biostratigraphy of the Newark Supergroup of eastern North America, with comments on early Me-

sozoic provinciality. *III. Congr. Latino-Amer. Paleontol. Mexico. Memoria,* pp. 67–81.

Cornet, B., and A. Traverse. 1975. Palynological contribution to the chronology and stratigraphy of the Hartford Basin in Connecticut and Massachusetts. *Geosci. Man* 11:1–33.

Cornet, B., A. Traverse, and N. G. McDonald. 1973. Fossil spores, pollen, fishes from Connecticut indicate Early Jurassic age for part of the Newark Group. *Science* 182: 1243–7.

Cui, G. 1976. [*Yunnania,* a new tritylodont genus from Lufeng, Yunnan.] *Vert. Palasiat.* 14: 85–90.

De Jersey, N. J., and R. J. Paten. 1964. Jurassic spores and pollen from the Surat Basin. *Publs. Geol. Surv., Queensland.* 322: 1–23.

Dolby, J. H., and B. E. Balme. 1976. Triassic palynology of the Carnarvon Basin, Western Australia. *Rev. Palaeobot. Palynol.* 22: 105–68.

Dunay, R. E., and M. J. Fisher. 1979. Palynology of the Dockum Group (Upper Triassic), Texas, U.S.A. *Rev. Palaeobot. Palynol.* 28: 61–92.

Estes, R. 1983. Sauria terrestria, Amphisbaenia. *Handbuch der Palaeoherpetologie* (Stuttgart: Gustav Fischer Verlag), Pt. 10A, pp. 1–249.

Evans, S. E. 1980. The skull of a new eosuchian reptile from the Lower Jurassic of South Wales. *Zool. J. Linn. Soc.,* 73: 81–116.

Fisher, M. J., and R. E. Dunay. 1981. Palynology and the Triassic–Jurassic boundary. *Rev. Palaeobot. Palynol.* 34: 129–35.

Fitch, F. J., and J. A. Miller. 1971. Potassium–Argon radioages of Karroo volcanics rocks from Lesotho. *Bull. Vulcan.* 35: 1–8.

Fraser, N. C., and G. M. Walkden. 1983. The ecology of a Late Triassic reptile assemblage from Gloucestershire, England. *Palaeogeog. Palaeoclimat. Palaeoecol.* 42: 341–65.

1984. Two Late Triassic terrestrial ecosystems from South West England. *In* Reif, W.-E., and Westphal, F. (eds.), *Third Symposium on Mesozoic Terrestrial Ecosystems, Tuebingen, 1984* (Tübingen: ATTEMPTO), pp. 87–92.

Fraser, N. C., G. M. Walkden, and V. Stewart. 1985. The first pre-Rhaetic therian mammal. *Nature (London)* 314: 161–3.

Froelich, A. J., and P. E. Olsen. 1984. Newark Supergroup, a revision of the Newark Group in eastern North America. *U.S. Geol. Surv. Bull.* 1537A: A55–8.

Gall, J.-C., M. Durand, and E. Muller. 1977. Le Trias de part d'autre du Rhin. Correlations entre les marges et le centre du bassin germanique. *Bur. Recher. Géol. Géophys. Minieres, 2nd sér. Sec. IV,* No. 3: 193–204.

Ghosh, P. K., and N. D. Mitra. 1970. A review of recent progress in the studies of the Gondwanas of India. *Proc. Pap. 2nd Gondwana Symp., Cape Town. 1970* pp. 29–48.

Grieve, R. A. F. 1982. The record of impact on Earth: Implications for a major Cretaceous/Tertiary impact event. *Geol. Soc. Amer., Spec. Pap.* 190: 25–37.

Gwinner, M. P. 1980. Eine einheitliche Gliederung des Keuper (Germanische Trias) in Sueddeutschland. *N. Jb. Geol. Palaeont. Mh.* 4: 229–34.

Hahn, G. 1984. Palaeomagnetische Untersuchungen im Schilfsandstein (Trias, Km 2). *Geol. Rundsch.* 73: 499–516.

Hallam A. 1981. The end-Triassic bivalve extinction event. *Palaeogeogr. Palaeoclim. Palaeoecol.* 35: 1–44.

Harland, W. B. 1982. *A Geologic Time Scale* (New York: Cambridge University Press).

Hoffman, A., and J. Ghiold. 1985. Randomness in the pattern of "mass extinctions" and "waves of origination." *Geol. Mag.* 122: 1–4.

Hope, R. C., and O. F. Patterson III. 1969. Triassic flora from the Deep River Basin, North Carolina. *N.C. Dept. Conserv. Dev., Spec. Publ.* 2:1–22.

1970. *Pekinopteris auriculata*: A new plant from the North Carolina Triassic. *J. Paleontol.* 44: 1137–9.

Hughes, N. F., 1973. Mesozoic and Tertiary distributions and problems of land plant evolution. *In* Hughes, N. F. (ed.), *Organisms and Continents through Time. Special Paper in Palaeontology* (London: Palaeontological Association), pp. 188–98.

Kermack, K. A., F. Mussett, and H. W. Rigney. 1981. The skull of *Morganucodon. Zool. J. Linn. Soc.* 71; 1–158.

Kitching, J. W., and M. A. Raath. 1984. Fossils from the Elliot and Clarens formations (Karoo Sequence) of the Northeastern Cape, Orange Free State and Lesotho, and a suggested biozonation based on tetrapods. *Palaeontol. Afr.* 25: 111–25.

Kozur, H. 1975. Probleme der Triasgliederung und Parallelisierung der germanischen und tethyalen Trias. Teil II: Anschluss der germanischen Trias an die internationale Triasgliederung. *Freiberger Forchungsch. (C)* 3: 58–65.

Kuehne, W. G. 1956. *The Liassic Therapsid Oligokyphus.* (London: Trustees British Museum).

Kumaran, K. P. N., and H. K. Maheswari 1980. Upper Triassic sporae dispersae from the Tiki Formation 2: Miospores from the Janar Nala Section, South Gondwana Basin, India. *Palaeontographica Abt. B* 173: 26–84.

Laemmlen, M. 1958. Keuper. Léxique stratigraphique international. *CNRS. 1 Eur. Pt. 5, Allemagne*, 2: 1–235.

Marshall, J. E. A., and D. I. Whiteside. 1980. Marine influence in the Triassic "uplands." *Nature (London)* 287: 627–8.

McHone, J. G., and J. R. Butler. 1984. Mesozoic igneous provinces of New England and the opening of the North Atlantic Ocean. *Geol. Soc. Am. Bull.* 95: 757–65.

Morbey, S. J. 1975. The palynostratigraphy of the Rhaetian stage, Upper Triassic, in the Kendelbachgraben, Austria. *Palaeontographica B.* 152: 1–75.

Newell, N. D. 1967. Revolutions in the history of life. *In* Albritton, C. C., Jr. (ed.) *Uniformity and Simplicity. Geol. Soc. Amer. Spec. Pap.* 89: 62–92.

Odin, G. S., and R. Letolle. 1982. The Triassic time scale in 1981; *In:* Odin, G. S., (ed.), *Numerical Dating in Stratigraphy* (New York: Wiley), pp. 523–36.

Olsen, P. E. 1980a. A comparison of the vertebrate assemblages from the Newark and Hartford Basins (Early Mesozoic, Newark Supergroup) of eastern North America. *In* Jacobs, L. L. (ed.), *Aspects of Vertebrate History: Essays in Honor of Edwin Harris Colbert* (Flagstaff, Arizona: Museum of Northern Arizona Press), pp. 35–53.

1980b. Triassic and Jurassic Formations of the Newark Basin. *In* Manspeizer, W. (ed.), *Field Studies in New Jersey Geology and Guide to Field Trips.* 52nd Annual Mtg. New York State Geol. Assoc., Newark College of Arts and Sciences, Rutgers University, Newark, pp. 2–39.

1980c. Fossil great lakes of the Newark Supergroup in New Jersey. In Manspeizer, W. (ed.), *Field Studies in New Jersey Geology and Guide to Field Trips*, 52nd Annual Mtg. New York State Geol. Assoc. Newark College of Arts and Sciences, Rutgers University, Newark, pp. 352–98.

1984a. Comparative paleolimnology of the Newark Supergroup: a study of ecosystem evolution. Ph.D. Thesis, Biology Department, Yale University.

1984b. Periodicity of lake-level cycles in the Late Triassic Lockatong Formation of the Newark Basin (Newark Supergroup, New Jersey and Pennsylvania). *Milankovitch and Climate. NATO Symposium* (Dordrecht: D. Reidel Publishing.), pp. 129–46.

Olsen, P. E. in press. Stratigraphic review, faunal and floral assemblages, and summary of paleoecology. *In* Manspeizer, W., J. K. Costain, J. Z. deBoer, A. J. Froelich, J. G. McHone, P. E. Olsen, D. C. Prowell, J. H. Puffer (eds.), *Post-Paleozoic Activity in the Appalachians* (Geological Society of America).

Olsen, P. E., and D. Baird. 1982. Early Jurassic vertebrate assemblages from the McCoy Brook Fm. of the Fundy Group (Newark Supergroup, Nova Scotia, Can.) *Geol. Soc. Am., Abst. Prog.* 14(2):70.

Olsen, P. E., and P. M. Galton. 1977. Triassic–Jurassic tetrapod extinctions: are they real? *Science.* 197: 983–6.

1984. A review of the reptile and amphibian assemblages from the Stormberg of southern Africa, with special emphasis on the footprints and the age of the Stormberg. *Palaeontol. Afr.* 25: 87–110.

Olsen, P. E., A. R. McCune, and K. S. Thomson. 1982. Correlation of the Early Mesozoic Newark Supergroup by Vertebrates, principally fishes. *Am. J. Sci.* 282: 1–44.

Oppel, A. 1856-8. *Die Juraformation Englands, Frankreichs und des suedwestlichen Deutschlands.* (Stuttgart: Ebner and Seubert).

Pacey, D. E. 1978. On a tetrapod assemblage from a Mesozoic fissure-filling in South Wales. Ph.D. thesis, Department of Geology, University of London.

Padian, K. 1984. The possible influence of sudden events on biological radiations and extinctions (group report). *In* Holland, H. D., and A. F. Trendall (eds.), *Patterns of Change in Earth Evolution. Dahlem Konferenzen 1984* (Berlin: Springer, Verlag), pp. 77–102.

Palmer, A. R. 1984. The decade of North American Geology 1983 time scale. *Geology* 11: 503–4.

Pearson, A. B. 1970. Problems of Rhaetian stratigraphy with special reference to the lower boundary of the stage. *Quart. J. Geol. Soc. Lond.* 126: 125–50.

Pederson, K. R., and J. J. Lund. 1980. Palynology of the plant-bearing Rhaetian to Hettangian Kap Stewart Formation, Scorsby Sund, East Greenland. *Rev. Palaeobot. Palynol.* 31: 1–69.

Peterson, F., and G. N. Pipiringos. 1979. Stratigraphic relations of the Navajo Sandstone to Middle Jurassic Formations, southern Utah and northern Arizona. *U.S. Geol. Surv. Prof. Pap.* 1035B: B1–B43.

Peterson, F., B. Cornet, C. E. Turner-Peterson. 1977. New data bearing on the stratigraphy and age of the Glen Canyon Group (Triassic and Jurassic) in southern Utah and northern Arizona. *Geol. Soc. Am., Abst. Prog.* 9(6): 755.

Pipiringos, G. N., and R. B. O'Sullivan. 1978. Principal unconformities in Triassic and Jurassic rocks, western interior, United States – a preliminary survey. *U.S.Geol. Surv. Prof. Pap.* 1035A: A1–A29.

Rampino, M. R., and R. B. Stothers. 1984. Geological rhythms and cometary impacts. *Science* 226: 1427–31.

Raup, D. M., and J. J. Sepkoski, Jr. 1982. Mass extinctions and the marine fossil record. *Science* 215: 1501–3.

1984. Periodicity of extinctions in the geological past. *Proc. Natl. Acad. Sci., U.S.A.* 81: 801–5.

Reiser, R. F., and A. J. Williams. 1969. Palynology of the Lower Jurassic sediments of the northern Surat Basin, Queensland. *Publs. Geol. Surv., Queensland* 339: 1–24.

Retallack, G. 1977. Reconstructing Triassic vegetation of eastern Australia: a new approach for the biostratigraphy of Gondwanaland. *Alcheringa.* 1: 247–77.

1979. Middle Triassic coastal outwash plain deposits in Tank Gully, Canterbury, New Zealand. *J. Roy. Soc. New Zealand* 9: 397–414.

1984. Completeness of the rock and fossil record: some estimates using fossil soils. *Paleobiology* 10: 59–78.

Robinson, P. L. 1957. The Mesozoic fissures of the Bristol Channel area and their vertebrate faunas. *J. Linn. Soc., Lond.* 43: 260–82.

Rutte, E. 1957. *Einfuehrung in die Geologie von Unterfranken.* Würzburg.

Sadler, P. M. 1981. Sediment accumulation rates and the completeness of stratigraphic sections. *J. Geol.* 89: 569–84.

Sadler, P. M., and L. W. Dingus. 1982. Expected completeness of sedimentary sections: estimating a time-scale dependent, limiting factor in the resolution of the fossil record. *Proc. 3rd North Am. Paleontol. Conv.* 2: 461–4.

Schindel, D. E. 1982. Resolution analysis: a new approach to the gaps in the fossil record. *Paleobiology* 8: 340–53.

Schroeder, B. 1982. Entwicklung des Sedimentbeckens und Stratigraphie der klassischen Germanischen Trias. *Geol. Rundsch.* 71: 783–94.

Schuurman, W. M. L. 1979. Aspects of Late Triassic palynology. 3. Palynology of the latest Triassic and earliest Jurassic deposits of the Northern Limestone Alps in Austria and southern Germany, with special reference to a palynological characterization of the Rhaetian Stage in Europe. *Rev. Palaeobot. Palynol.* 27: 53–75.

Seideman, D. E., Masterson, W. D., Dowling, and K. K. Turekian. 1984. K–Ar dates and ^{40}Ar/^{39}Ar age spectra for Mesozoic basalt flows of the Hartford Basin, Connecticut, and the Newark Basin, New Jersey. *Geol. Soc. Am. Bull.* 95: 594–8.

Sigogneau-Russell, D. 1983. Nouveaux taxons de mammifères Rhétiens. *Acta Palaeontol. Polon.* 28: 233–49.

Sigogneau-Russell, D., H. Cappetta, and P. Taquet. 1979. Le Gisement Rhétien de Saint-Nicolas-de-Port et ses conditions de dépôt. *7th Réunion Ann. Sci. Terre, Lyon.* 1979: 429.

Sigogneau-Russell, D., and A.-L. Sun. 1981. A brief review of Chinese synapsids. *Géobios* 14: 275–9.

Tschudy, R. H. 1984. Palynological evidence for change in continental floras at the Cretaceous–Tertiary boundary. *In* Berggren, W. A., and J. A. Van Couvering (eds.), *Catastrophes and Earth History.* (Princeton, New Jersey: Princeton University Press), pp. 315–37.

Tozer, E. T. 1967. A standard for Triassic time. *Bull. Geol. Surv. Can.* 156: 1–103.

1974. Definitions and limits of Triassic stages and substages: suggestions prompted by comparisons between North America and the Alpine–Mediterranean region. *In* Zapfe, H., (ed.), *The Stratigraphy of the Alpine–Mediterranean Triassic.* Schriftenreihe Erdwiss. Komm., 2 (Berlin: Springer-Verlag), pp. 195–206.

1979. Latest Triassic ammonoid faunas and biochronology, Western Canada. *Geol. Sur. Can., Pap.* 79–1B: 127–35.

Van Houten, F. B. 1969. Late Triassic Newark Group, north central New Jersey and adjacent Pennsylvania and New York. *In* Subitsky S., (ed.), *Geology of Selected Areas in New Jersey and Eastern Pennsylvania and Guidebook of Excursions* (New Brunswick, New Jersey: Rutgers University Press), pp. 314–47.

Van Valen, L. M. 1984. A resetting of Phanerozoic community evolution. *Nature (London)* 307: 50–2.

Visscher, H., and W. A. Brugman. 1981. Ranges of selected palynomorphs in the Alpine Triassic of Europe. *Rev. Palaeobot. Palynol.* 34: 115–28.

Visscher, H., W. M. L. Schuurman, and A. W. Van Erve. 1980. Aspects of a palynological characterization of Late Triassic and Early Jurassic "Standard" units of chronostratigraphical classification in Europe. *IV Int. Palynol. Conf., Lucknow (1976–*

1977) 2: 281–7.

Volkheimer, W. 1971. Zur stratigraphischen Verbreitung von Sporen und Pollen im Unter-und Mitteljura des Neuquen–Beckens (Argentinien). *Münster. Forsch. Geol. Palaeont.* 20/21: 297–321.

Walker, A. D. 1961. Triassic reptiles from the Elgin area: *Stagonolepis, Dasygnathus* and their allies.

Phil. Trans. Roy. Soc. Lond., B 244(709): 103–204.

Warren, A., and M. N. Hutchinson. 1983. The last labyrinthodont? A new brachyopoid (Amphibia, Temnospondyli) from the Early Jurassic Evergreen Formation of Queensland, Australia. *Phil. Trans. Roy. Soc. Lond., B.* 303: 1–62.

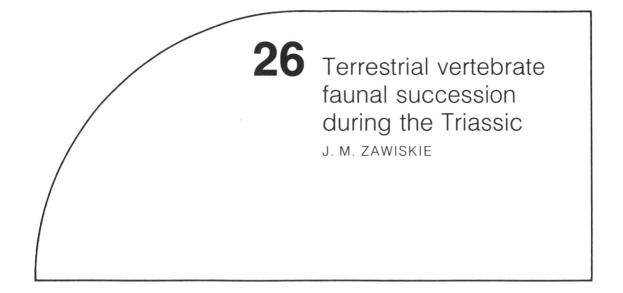

26 Terrestrial vertebrate faunal succession during the Triassic

J. M. ZAWISKIE

Introduction

During the interval spanning the Late Permian through the latest Triassic, nonmammalian therapsids were stratigraphically succeeded by archosaurs in the continental fossil assemblages of Pangaea. This pattern of faunal succession has recently been reviewed and reinterpreted by Benton (1983a). Previous hypotheses called for archosaur (i.e., dinosaur or thecodontian) "superiority" with respect to locomotion (e.g., Ostrom 1969; Bakker 1971; Charig 1972, 1980) or physiology (e.g., Bakker 1971, 1975, 1980; Robinson 1971; Halstead 1975; Benton 1979; Hotton 1980), and a consequent selective advantage in interspecific competition with therapsids. Benton (1983a, p. 30) characterized these as "untestable speculations." Instead, he preferred a noncompetitive opportunistic radiation of dinosaurs. He presumed this radiation to have occurred by chance in empty ecological space following a relatively sudden mass extinction of dicynodonts, rhynchosaurs, and "thecodontians" (nondinosaurian archosaurs). An extrinsic cause for the extinctions might have been a well-defined, progressive climatic deterioration and a shift to more arid conditions, with a concomitant floral turnover near the end of the Triassic (Tucker and Benton 1982).

By contrast, Bonaparte (1982) outlined the paleontologic evidence from South America and provided a more conventional competitive ecological model. He cited archosaurian specializations for improved locomotion together with their capacity for size increase to explain the transition to archosaur-dominated faunas during the Late Triassic. Bonaparte concluded that therapsids declined in response to direct interspecific competition with archosaurs. More recently, Charig (1984) reviewed the literature on the faunal turnover and also argued for archosaur competitive superiority for the same reasons cited

by Bonaparte (1982). Benton (1984a), however, was not persuaded and stated that there was no evidence of sustained competition between supposedly inferior mammal-like reptiles and superior archosaurs regardless of the adaptation considered. He later referred to the initial dinosaur radiation as a "lucky break" (Benton 1984b, p. 42).

It is significant that community neoecology is presently in an introspective phase (see reviews in Price, Slobodchikoff, and Gaud 1984), and that the typical textbook emphasis placed on interspecific competition (e.g., Odum 1971; Pianka 1974; Ricklefs 1979) as the major organizing force in extant communities is now being seriously questioned. In light of the complex variables and problems encountered by neoecologists studying modern faunas, Benton's reassessment of the fossil record and criticism of competitive paleoecological models for the Triassic faunal succession are valid. However, there are problems with completely accepting Benton's chance and opportunism scenario. The objectives of this chapter are to offer a critique of Benton's analysis and an alternative hypothesis of differential archosaur survival in changing adaptive zones without necessarily invoking direct interspecific competition.

As demonstrated by Bakker (1977), freshwater aquatic tetrapods and those less than 10 kg (adult weight) do not show clear patterns of mass extinction. Therefore, this analysis is primarily concerned with the larger, terrestrial to semiterrestrial vertebrates. The taxonomic groups considered include the nonmammalian therapsids, rhynchosaurs, early archosaurs ("thecodontians"), and early saurischian and ornithischian dinosaurs. It shall be argued that the first wave of Late Triassic extinction within the specified size categories [? late Carnian (Chapter 25)] displays "taxonness" (Schopf 1977): specifically, the differential survival of archosaurs relative to nonmammalian therapsids and rhynchosaurs. I suggest that the differential survival of archosaurs and the initial radiation of dinosaurs was related to the development of novel herbivore-feeding strategy and locomotor adaptations that were favorable in habitats that expanded during the Late Triassic as a result of extrinsic geologic events. I will attempt to support this interpretation by presenting faunal succession within the context of the ecostratigraphic paradigm (Hoffman 1981). However, prior to the discussion of specific aspects of Late Triassic faunal change, I will first briefly review the concepts of ecostratigraphy and community types.

Community types and ecostratigraphy

Community types

Bakker (1977) recognized seven "dynasties" or cycles of vertebrate diversification and extinction in several macrohabitats in Permo-Triassic and later Mesozoic continental rocks. Within this framework, a number of more specific "empires" have been named (Anderson and Cruickshank 1978; Benton 1983a). I prefer to abandon these anthropomorphic terms and replace them with "community types" (Whittaker 1970; Boucot 1975) as defined by Hoffman (1981, p. 2):

a group of communities making part of a single bioprovince, resembling each other in their ecologic structure, and limited by the same environmental factors. The taxonomic composition may vary among particular communities but the replacement species are expected to be ecologically very close to each other. This concept refers thus to the permanence of ecologic structure, given a constant environmental framework.

I want to emphasize that these associations of fossil assemblages are an epiphenomenon of overlaps in the distributional patterns of organisms. They are controlled largely by environmental, depositional, and taphonomic factors, and the actual community representation should be regarded as suspect (Hoffman 1979, 1981). Table 26.1 outlines the four community types under consideration. Benton's Rhynchosaur–Diademodontid community type is modified to restrict its range to Middle and Upper Triassic rocks.

Ecostratigraphy

Hoffman (1981) has provided the clearest statement and criticism of the central paradigm of ecostratigraphy. It is a multidisciplinary approach concerned with changes in the fossil record at the community type, taxocene, or species group level. Its basic tenet is that the fossil record of certain community types shows a punctuated rather than a gradual pattern of change in evolutionary time. The discontinuities between successive community types are attributed to extrinsic geologic events that caused major environmental disturbances, altered niche patterns, and at least temporarily elevated rates of evolution. Most extrinsic geologic events are regional in scale, for example, tectonic elimination of habitats or local transgressive–regressive cycles. Therefore, the time planes between successive paleocommunity types are obviously not significant on a worldwide scale. However, as Hoffman stated, the goal of ecostratigraphy is not the recognition of worldwide time planes or stage boundaries (i.e., the goal of chronostratigraphy), but rather to gain information about "remarkable events" in the course of geobiologic history.

It can be expected that the discrete patterns of community type change would be further accentuated in continental rocks, in which there is commonly a complex interfingering of facies combined with small- and large-scale unconformities (Retallack 1978). Further bias is introduced by the restriction of rock accumulation (and hence fossils) to basinal areas, and outcrop availability by erosional truncation. Despite these problems, ecostratigraphy may contribute to a better understanding of Triassic and Early Jurassic faunal change.

The opportunistic model

I agree with Benton's conclusion that dinosaurs radiated into empty ecological space, and that there is no way to prove that interspecific competition was involved in the succession of nonmammalian therapsids and rhynchosaurs by archosaurs in terrestrial habitats during the Triassic. However, I contend that archosaurs were better able than most therapsids to become adapted to changing adaptive zones, and that this is a case of differential survival

Table 26.1 *Community types discussed in text with approximate distribution in time and space*

Community type	Distribution
Lystrosaurid–Kannemeyeriid	Pangaea (early to late Early Triassic)
Rhynchosaur–Traversodontid	Gondwana (middle to early Late Triassic)
Phytosaur–Metoposaur	Laurasia (early to late Late Triassic)
Prosauropod	Pangaea (mid-Late Triassic to Early Jurassic)

Source: Modified from Anderson and Cruickshank (1978) and Benton (1983a).

through an evolutionary bottleneck and not a lucky break.

Benton (1983a) indicated that the first "prosauropod-dominated" faunas were "dominated by dinosaurs as medium to large herbivores and carnivores," and that the rise of dinosaurs was rapid following the extinction of nonmammalian therapsids, rhynchosaurs, and early archosaurs. However, the lumping of the nonmammalian therapsid and rhynchosaur extinction with that of the archosaurs is not accurate. In fact, the first wave of Late Triassic extinction (? late Carnian) did not significantly affect the early archosaurs. For example, footprint evidence suggests an overlap of large dinosaurs and early archosaurs in a palynologically dated Rhaetic correlate in the Newark Supergroup (Olsen 1980). Furthermore, it is likely that rauisuchians were present with early sauropodomorphs ("prosauropods") in the Germanic Trias, as their probable trackways are common, and the material of *Teratosaurus,* from the Stubensandstein, is now generally regarded as rauisuchian (Benedetto 1973; Dawley 1979; Bonaparte 1981; Chapter 24). In Gondwana, rauisuchids, ornithosuchids, and aetosaurs were still important elements in the early sauropodomorph faunas of South America (Bakker 1977; Bonaparte 1982). The numerous teeth reported from the Triassic sauropodomorph-bearing Red Beds (lower Elliot Formation) of South Africa may also represent rauisuchians (Dawley 1979) rather than carnivorous prosauropods (Charig, Attridge, and Crompton 1965). Therefore, it seems that there was taxonness associated with the first Late Triassic extinctions, with archosaurs differentially surviving relative to nonmammalian therapsids and rhynchosaurs.

While it is certainly possible that this pattern is stochastic, the apparent taxonness and possible causal relationship to extrinsic Late Triassic events also allow for a deterministic explanation. In the following section, I shall argue that early archosaurs and dinosaurs that coexisted with the nonmammalian therapsids and rhynchosaurs had already become adapted for environments that would later become widespread; but first, several points regarding Benton's assessment of paleocommunity composition and faunal dominance should be made.

Community composition and dominance

Benton (1983a, p. 33) recognized that "the fossilized sample rarely corresponds to a true ecological community and that preservational bias clearly tends to emphasize medium-to-large waterside animals in Triassic faunas." He suggested that collecting practices compensate for this bias and presented relative abundance and faunal dominance on the basis of known and estimated numbers of skeletal specimens. However, he provided little or no detail on the precise stratigraphic, taphonomic, or lithologic context of the fossil assemblages except for three broad lithofacies associations (Tucker and Benton 1982; Benton 1983a). It is doubtful that collecting practices can compensate for sample bias introduced by facies control. This problem was emphasized by Bakker (1977), who suggested that we typically get a better representation of lacustrine or swamp vertebrates, as opposed to those living in seasonally well-drained floodplains. In short, Benton's inferences on relative abundance are interesting, but if taken at face value may be misleading, unless put in a detailed facies context. In the absence of further sedimentologic and taphonomic data, it is difficult to decide whether rare taxa or erratics, such as early dinosaurs, are occasional immigrants from extrabasinal, more upland communities or unobtrusive elements of the basinal fauna (Olson 1971; Hoffman 1979). This problem is clearly demonstrated by ichnologic studies. For instance, in the Newark Supergroup (Baird 1980, p. 229): "footprints numbering in the tens of thousands belong to numerous species of *Anchisauripus, Grallator,* and *Eubrontes,* ichnogenera which are reliably correlated with the coelurosaur–carnosaur group of dinosaurs. Yet these same formations have yielded only two partial skeletons of coelurosaurs, both assigned to the genus *Coelophysis* (Colbert and Baird 1958; Colbert 1964)." Therefore, the presence of even a single dinosaur skeleton in a fossil assemblage must be considered an important paleoecological datum.

Community type succession and extrinsic factors

The transition between faunal assemblages with high abundance and taxonomic diversity of nonmammalian therapsids and rhynchosaurs to archosaur-dominated assemblages is best represented in Gondwana, and to a lesser extent in the southwestern United States. In these regions the rhynchosaur–traversodontid and phytosaur–metoposaur community types, respectively, were stratigraphically succeeded by the prosauropod community type during the middle to late Late Triassic. The objective of the following sections is to relate extrinsic events and environmental reorganization causally to community type replacement and the differential survival of archosaurs.

Paleoclimatic setting

A monsoonal climatic pattern has been hypothesized for Pangaea during the Late Paleozoic (Robinson 1973; J. T. Parrish et al. 1982). This cyclic circulation is thought to have become progressively stronger throughout the Triassic, with a major arid phase near the end of the period as Pangaea drifted northward and subtropical dry belts expanded (Tucker and Benton 1982; J. M. Parrish et al. 1986). This increase in aridity was supposedly sharpest in the southwestern United States and in Gondwana (Tucker and Benton 1982). In both of these regions, nonmammalian therapsids, rhynchosaurs, and early archosaurs were components of terrestrial ecosystems prior to the postulated arid phase. Tucker and Benton (1982) and Benton (1983a) suggested a causal relationship between the progressive aridification in Gondwana and the demise of the *Dicroidium* lowland flora and its associated rhynchosaur–traversodontid community type. Benton (1983a), however, denied the possibility that archosaurs differentially survived during the environmental reorganization, although he stated that the new habitats were "unsuitable for rhynchosaurs and synapsids."

The paleogeographic distribution of nonmammalian therapsids provides additional circumstantial evidence that they may have been vulnerable in the dry habitats that spread during the Late Triassic. According to J.M. Parrish et al. (1986): "During their time of highest diversity, therapsids were generally distributed at middle and high latitudes on the eastern side of Pangaea, adjacent to the warm Tethyan seaway and away from the very warm, dry equatorial region." However, they caution that this distribution may simply be an artifact of rock outcrop availability. The Germanic terrestrial faunas lack nonmammalian therapsids and were archosaur-dominated throughout the Triassic, perhaps as a result of the persistence of predominantly hot and dry, low rainfall conditions (Tucker and Benton 1982).

Community type succession

Several authors (Anderson and Cruickshank 1978; Bakker 1977; Kemp, 1982; Benton 1983a) have provided detailed summaries of the pattern of relative abundance and taxonomic diversity of the component taxa of the Triassic community types, and Tucker and Benton (1982) summarized their general environmental setting.

Nonmammalian therapsids gained ecological dominance in lowland basins during the Late Permian, and this trend continued in the Early Triassic lystrosaurid–kannemeyeriid community type. The herbivores were squat quadrupedal forms (dicynodonts, diademodontids) that fed no higher than a meter above the ground, and probably were most common in aquatic waterside and thickly vegetated terrestrial settings (Dingle, Siesser, and Newton 1983). This herbivore complex supported a mixed association of therapsid and archosaurian predators. The therapsid predators (galesaurids, cynognathids) were small- to medium-sized quadrupeds. Quadrupedal archosaurs (proterosuchids, erythrosuchids) were large- to medium-sized terrestrial to semiaquatic carnivores, but the small lightly built bipedal insectivores (euparkeriids) were probably disassociated from the herbivore complex.

During the range of this community type in the Karoo Basin of southern Africa (*Lystrosaurus* and *Cynognathus* assemblage zones, early to late Early Triassic), environmental changes occurred, and archosaurs demonstrated the ability to become adapted to hot, dry climates and more open habitats. The *Lystrosaurus* assemblage zone fauna developed in a warm temperate lowland with seasonal floods (Tucker and Benton 1982) in lacustrine and anastomosed fluvial environments, distal to sandy braidplain facies (Dingle et al. 1983; Hiller and Stavrakis 1984). The basinal flora comprised mixed *Glossopteris–Dicroidium,* while extensive coniferous forests may have occupied the basin periphery (Dingle et al. 1983). The transition to the *Cynognathus* assemblage zone records a shift to a hotter climate that was increasingly arid with reduced plant cover and ephemeral streams (Tucker and Benton 1982; Dingle et al. 1983; Hiller and Stavrakis 1984). Hiller and Stavrakis (1984) have related this local climatic shift to progressive uplift within the Cape Fold Belt to the south. Archosaurs are rare in the collections from the wetter, more vegetated *Lystrosaurus* assemblage zone, being represented only by the aquatic to semiterrestrial proterosuchids. However, with the shift to drier, open habitats, the more terrestrially adapted erythrosuchids and euparkeriids abruptly appeared, and early archosaurs account for nearly 14 percent of all known specimens (Benton, 1984a), even though they were exclusively carnivores.

No continuous stratigraphic sequence with well sampled faunas records the transition between the lystrosaurid–kannemeyeriid and rhynchosaur–Traversodontid community types. In South Africa, a major interval of uplift occurred in the Cape Fold Belt in post-*Cynognathus* zone times that also caused uplift of adjacent basinal areas and the development of a major unconformity, spanning most of the Middle Triassic (Dingle et al. 1983). The raising of land surfaces along the Gondwanide Fold Belt may have been instrumental in the rise to dominance of the *Dicroidium* flora (Retallack 1978).

The environmental framework of the rhynchosaur–traversodontid community type (Middle to early Late Triassic) needs further documentation; however, the climate was apparently seasonal (monsoonal) with dry sandy to savannah-like interchannel areas (Benton 1983a). Vegetation and rainfall varied cyclically from abundant to sparse, and hot, dry deserts existed in middle to low latitudes (Tucker and Benton 1982).

The Middle Triassic Manda Formation in East Africa provides a sample of an early phase of the rhynchosaur–traversodontid community type, and the South American sequence fills in the late Middle to early Late Triassic history. Nonmammalian therapsid herbivores and carnivores maintained their ecological dominance in lowland basins with the widespread *Dicroidium* flora. However, dicynodonts were largely replaced by rhynchosaurs and traversodontids with dental specialization for handling tougher vegetation. Benton (1983a) considered this an example of differential survival, although dicynodonts had not yet become fully extinct. All members of the herbivore complex were squat quadrupeds that fed no more than a meter above the ground. Medium to large quadrupedal chiniquodontids were the primary therapsid carnivores, well adapted for feeding on members of the herbivore complex. The numerical abundance of rhynchosaurs and the nonmammalian therapsids probably reflects their ecological dominance in areas closer to water and hence centers of sediment dispersal (levee and proximal floodplain) with higher preservation potential.

Early archosaurs became taxonomically important in this community type (Bonaparte 1982). Rauisuchians (Fig. 26.1) were the largest terrestrial predators and showed advances in locomotor capability (Bakker 1971; Bonaparte 1984). Lagosuchids and ornithosuchids were small- to medium-sized, light, facultative bipeds closely related to dinosaurs. Gracile, terrestrially adapted crocodylomorphs, bipedal ornithosuchids and saurischians, and armored herbivorous stagonolepids with vertical limb posture are also found in these fossil assemblages [see Bonaparte (1982) for detailed stratigraphic treatment].

Bonaparte (1982) emphasized that the archosaurian taxa show morphological improvements in locomotor capacity that would certainly have been advantageous in the cyclically open, dry savannah-like interchannel subenvironments. In fact, the numerical rarity of some archosaurian taxa (e.g., saur-

ischian and ornithischian dinosaurs) may be the result of their occasional emigration from subenvironments with poor preservation potential, or even from separate extrabasinal or basin margin habitats.

According to Bonaparte (1982) the major climatic deterioration in South America began in the upper Ischigualasto Formation (early Late Triassic). Only the archosaurian component and small traversodontids of the rhynchosaur–traversodontid community type survived during the environmental reorganization and transition to the archosaur-dominated prosauropod community type in the overlying Los Colorados Formation.

In the Karoo Basin of southern Africa, a significant stratigraphic hiatus preceded deposition of the Stormberg Group (Molteno, Elliot, and Clarens Formations). The archosaur-dominated prosauropod community type (middle Late Triassic to Early Jurassic) appeared in the Elliot Formation and continued with modification into the overlying Clarens Formation. The Elliot Formation records the southerly migration of distal floodplain facies toward the mountains bordering the basin and the establishment of dry conditions (Dingle et al. 1983). Major stream channels were separated by broad, monotonous tracts of floodplain, subjected to sharply seasonal rainfall and long periods of subaerial exposure (Tankard et al. 1982). Aridification is thought to have intensified upward in the Elliot Formation (late Late Triassic) with an eventual shift to playa lakes, ephemeral streams, and eolian dune fields in the overlying Clarens Formation (Early Jurassic).

Dinosaurs dominate this community type in South Africa. Sauropodomorphs such as the large, high-browsing plateosaurids (Chapter 16) and lighter-limbed anchisaurids and light bipedal orni-

Figure 26.1 Rauisuchids confront juvenile herrerasaurids and ornithosuchids feeding on a rhynchosaur carcass during the Late Triassic in Argentina (Ischigualasto Formation, rhynchosaur–traversodontid community type). Reconstruction by Gregory Paul.

thischians comprised the terrestrial herbivore complex. The teeth of large carnivores (?rauisuchians) occur in the plateosaurid-dominated lower Elliot Formation, but otherwise large carnivores are not known. Lightly built bipedal coelurosaurs and crocodylomorphs, also gracile facultative bipeds, were the medium-sized carnivores. The stratigraphic tendency for an upward increase in the percentage of highly mobile, light-limbed forms from the lower Elliot into the Clarens Formation has long been interpreted as a result of the need for improved mobility with progressive aridification (Haughton 1924, 1953).

In Laurasia, the *Dicroidium* flora was absent, but some elements of the rhynchosaur–traversodontid community type are present in the terrestrial to semiterrestrial component of the largely aquatic phytosaur–metoposaur community type. Much new information is currently being gained from the faunal composition of this community type in Upper Triassic formations in the southwestern United States. Squat, low-feeding dicynodonts, trilophosaurids, and armored early archosaurs with vertical limb posture (aetosaurs) were the main elements of the herbivore complex along with rare rhynchosaurs (Chatterjee 1985) and ornithischian dinosaurs (Jacobs and Murry 1980). Terrestrial carnivores were exclusively archosaurian (rauisuchians, ornithosuchids, coelurosaurs).

As in Gondwana, Middle and Late Triassic tectonism resulted in a mountain belt that separated the Chinle basin from the ameliorating effects of the ocean and a monsoonal climate prevailed (Blakey and Gubitosa 1983). The Late Triassic to Early Jurassic shift to more open savannah-like and desert habitats, containing the prosauropod community type that occurred in Gondwana, also occurred in the southwestern United States (Tucker and Benton 1982; Blakey and Gubitosa 1983; Blodgett 1984).

It has been suggested that an unconformity separates the Upper Triassic Chinle Formation (phytosaur–metoposaur community type) from the overlying Glen Canyon Group, with the prosauropod community type (Pipiringos and O'Sullivan 1978). However, Blakey and Gubitosa (1983) and Blodgett (1984) consider the contact conformable. If so, the various members of the Chinle and basal Glen Canyon Group may in part represent laterally equivalent environments. Therefore, the faunas they contain in vertical succession may give an indication of the lateral intergrading of community types that was present over the broad latitudinal zones spanning the subhumid to arid belts.

The Germanic terrestrial faunas, as noted earlier, were archosaur-dominated throughout the Triassic. The climate was consistently "hot and dry with little rainfall," and "desert conditions persisted until the latest Triassic" (Tucker and Benton 1982,

p. 370). As Benton (1984c and Chapter 24) documented, there was a steady increase in dinosaur abundance over a ten-million-year interval, so presumably they had become adapted to the drier continental interior conditions that would later expand into lower latitudes. Bonaparte (1981) and Tucker and Benton (1982) both suggested that the South American sauropodomorphs may have been immigrants from Europe. Therefore, instead of an opportunistic radiation we are probably dealing with the differential survival and spread of a community type that had already become adapted to environmental conditions that were expanding at the end of the Triassic as a result of Pangaea's northward drift (Tucker and Benton 1982; J.T. Parrish et al. 1982).

Differential archosaur survival

In the succession from the lystrosaurid–kannemeyeriid to the rhynchosaur–traversodontid/phytosaur–metoposaur to the prosauropod community types, a shift occurred from faunas dominated by nonmammalian therapsids and rhynchosaurs to archosaur-dominated faunas. The environmental and floral reorganization accompanying the radiation of dinosaurs no doubt occurred in a diachronous, stepwise fashion from the Late Triassic to the Early Jurrasic. During this time, the squat, quadrupedal, low-feeding therapsids and rhynchosaurs became extinct along with the coexisting therapsid predators. As Benton's (1983a, Fig. 2) own figures show, ornithischians, coelurosaurs, and sauropodomorphs were already present as erratics or "unobtrusive" elements in the rhynchosaur–traversodontid/phytosaur–metoposaur community types. Their low numerical representation probably reflects their preference for the expanding, cyclically dry, savannah-like habitats and basin margin uplands or different latitudinal and climatic belts. They may not have had "ecological access" (Simpson 1953) to the habitats occupied by the mixed nonmammalian therapsid–rhynchosaur–early archosaur faunas.

Was the differential survival of archosaurs an accident during the environmental reorganization? As already stressed, the herbivores that became extinct in the first Late Triassic extinction (nonmammalian therapsids, rhynchosaurs) were heavy, squat forms with stout barrel-like bodies that fed on low vegetation. Their demise may have been coupled with floral change and a shift to more open habitats. Although the rhynchosaurs attained semierect posture (Benton 1983b), they certainly were not highly mobile or cursorial, nor were the nonmammalian therapsids. They would have been easy prey away from densely vegetated areas. Conversely, the early ornithischians were light bipeds and among the most cursorial dinosaurs (Coombs 1978) and had dentitions specialized for the tough, fibrous, or waxy plant material (Chapter 17) that would be expected in dry habitats. Early sauropodomorphs were medium-to-large facultative bipeds or quadrupeds of advanced mediportal grade (Coombs 1978) and developed an evolutionary novelty: They were the first high browsers in the vertebrate record (Chapter 16). Therefore, they had become adapted to exploiting different vegetation that may not have been affected to the same degree as the lower foliage. Aetosaurs possessed erect posture and also developed an evolutionary novelty: armor that would have imparted protection from predation as habitats became progressively more open.

The therapsid carnivores were probably heavily dependent on the rhynchosaur and nonmammalian therapsid herbivore complex and did not have the locomotor capacity to run down ornithischians in savannah-like settings or the size to shift to sauropodomorph prey. Rauisuchians, on the other hand, were large and agile, and at least some forms were facultative bipeds (Chatterjee 1985). The light, bipedal, cursorial coelurosaurs (Coombs 1978) were independent of the rhynchosaur and nonmammalian therapsid herbivore complex, except perhaps as carrion feeders, and were well suited for open habitats. The same reasoning holds for the light facultative bipedal ornithosuchids. Early crocodylomorphs were also gracile, terrestrially adapted, and at least facultative bipeds.

Therefore, it seems that archosaurs displayed a pattern of differential survival, relative to the nonmammalian therapsids and rhynchosaurs during the environmental deterioration, and that there was taxonness in the first Late Triassic extinction. This pattern resulted from archosaurian adaptations for open habitats that had already been acquired prior to the extinction event, and from their exploitation of high foliage. The later extinction of rauisuchians and ornithosuchids in the latest Triassic is more problematic and may have been the result of stochastic processes.

Summary

Ecological dominance of nonmammalian therapsid herbivores was established in lowland basins during the Late Permian, and this pattern continued into the Early Triassic Lystrosaurid–Kannemeyeriid community type. Coexisting therapsid and archosaurian predators were supported by this herbivore complex. The environmental framework of this community type in Gondwana was altered during the late Early and Middle Triassic by progressive uplift within the Gondwanian Orogen, and a northward drift into lower latitudes. The uplift isolated the vertebrate-bearing foreland basins from the Panthalassic ocean. The associated uplift of basinal areas may have been a factor in the spread of the *Dicro-*

idium flora, and the mountains bordering the basin may have provided subhabitats with diverse coniferous floras.

The rhynchosaur–traversodontid community type (Middle to early Late Triassic) became widespread in lowland areas in Gondwana in conjunction with the *Dicroidium* flora. Nonmammalian therapsids were the dominant herbivores and a new archosauromorph group, the rhynchosaurs, was added to the herbivore complex. Traversodontids and rhynchosaurs partially replaced dicynodonts during the later stages of this community type [differential survival (Benton, 1983a)], and armored archosaurian herbivores (stagonolepids) appeared. The nonmammalian therapsid–rhynchosaur assemblage comprised a group of heavy, squat quadrupeds that fed on low vegetation. Coexisting therapsid carnivores (chiniquodontids) were similar to the herbivores in size and locomotor capacity. I suggest that this complex was probably closely tied to waterside and closed, densely vegetated subhabitats proximal to waterways. During the stratigraphic range of this community type, climatic patterns became more strongly cyclic or monsoonal, with a sharp and sometimes oscillating latitudinal climatic zonation across Pangaea. This caused the expansion of open, savannah-like subhabitats that were cyclically subjected to long dry spells and sporadically changing patterns of vegetation. Rare, herbivorous dinosaurs in these faunas show adaptations for cursorial locomotion and mastication of tough, waxy vegetation (early ornithischians) and high browsing (early sauropodomorphs). These adaptations may have been developed as a result of selection pressure in the broad, periodically dry savannah-like habitats, and in basin margin or other subhabitats with high foliage. Similarly, early archosaurian predators showed increasingly efficient locomotor capacity (vertical limb posture, facultative bipedality), and coelurosaurs were bipedal cursorial predators dissociated from the therapsid–rhynchosaur herbivore complex.

In Laurasia, a similar environmental setting and inferred ecological structure is represented in the terrestrial component of the dominantly aquatic phytosaur–metoposaur community type (early to late Late Triassic). It differs in lacking *Dicroidium* and in the lower diversity of nonmammalian therapsids and numbers of rhynchosaurs.

The environmental framework of the rhynchosaur–traversodontid and phytosaur–metoposaur community types was disturbed by continued progressive aridification and floral change. During the first Late Triassic extinction (? late Carnian), the low-feeding nonmammalian therapsid herbivores and rhynchosaurs and associated therapsid carnivores became extinct. Conversely, early archosaurs and dinosaurs differentially survived into the Prosauropod community type. This taxonness is attributed to the development of locomotor adaptations and novel feeding strategy acquired by archosaurian taxa prior to the expansion of the arid to semiarid climatic belts and spread of coniferous flora. The later extinction of rauisuchians and ornithosuchids during the late Late Triassic may have resulted from stochastic processes. The stepwise nature of the Late Triassic vertebrate extinctions (Chapter 25), and their association with major environmental reorganization, obviate the need to invoke extraterrestrial mechanisms (Chapter 17).

Future research

Further data are now needed on the taxonomic composition and relative abundance of individual taxa in Triassic fossil vertebrate assemblages. However, these data must be placed within the construct of detailed basin and facies analysis (e.g., Blakey and Gubitosa 1983) and taphonomic studies of fossil-bearing beds. Analysis of paleosols may provide greater resolution regarding paleoclimatic change, sedimentation rates, and the density and type of floodplain vegetation (e.g., Retallack 1983; Blodgett 1984). Comparisons must be made between the type of plant assemblages found near lowland waterways and those in more distal floodplain settings with those from inferred upland settings. A stronger link between paleoclimate and possible floral and faunal changes is also needed. The community types considered here should be further subdivided and analyzed on a more local, regional scale. All facies should be carefully searched for skeletal and ichnologic evidence, instead of concentrating on just those beds that are most productive. Hopefully, palynomorphs may provide a stronger link between marine and nonmarine biostratigraphy. This is needed to give an independent means for correlating faunas on a Pangaean scale and to assess the connection, if any, between paleoclimatic variation and eustatic events.

Such a multidisciplinary approach, within the context of the ecostratigraphic paradigm and modern neoecology, may reveal the extent of environmental control on community type composition and provide a stronger data base for theories regarding faunal dynamics during the Triassic and Early Jurassic. I hope that the model presented here will stimulate more fieldwork and closer cooperation between vertebrate paleontologists, paleobotanists, neoecologists, and sedimentologists.

Acknowledgments

I extend my sincere gratitude to the late John W. Cosgriff for introducing me to the subject of Triassic vertebrate paleontology and his moral support and kindness over the years that I had the privilege to know him. Kevin Padian and Mike Parrish reviewed the manuscript, and their comments and criticism have greatly improved it. Patricia Mullen and Pat Sheehy typed several drafts of the text, and Greg Paul drew Figure 26.1. Some of the ideas presented here were formulated during my tenure as a visiting research student with Nicholas Hotton III at the National Museum of Natural History. Lastly, I thank Michael J. Benton for reviewing the manuscript and his refreshing reassessment of Triassic and Jurassic faunal change.

References

Anderson, J. M., and A. R. I. Cruickshank. 1978. The biostratigraphy of the Permian and Triassic: Part 5. A review of the classification and distribution of Permo-Triassic tetrapods. *Palaeontol.. Afr.* 21: 15–44.

Baird, D. 1980. A prosauropod dinosaur trackway from the Navajo Sandstone (Lower Jurassic). *In* Jacobs, L. L. (ed.), *Aspects of Vertebrate History* (Flagstaff, Arizona: Museum of Northern Arizona Press), pp. 219–30.

Bakker, R. T. 1971. Dinosaur physiology and the origin of mammals. *Evolution.* 25: 636–58.

　1975. Dinosaur renaissance. *Sci. Am.* 232(4): 58–78.

　1977. Tetrapod mass extinctions – a model of the regulation of speciation rates and immigration by cycles of topographic diversity. *In* Hallam, A. (ed.), *Patterns of Evolution as Illustrated by the Fossil Record* (Amsterdam: Elsevier), pp. 439–68.

　1980. Dinosaur heresy – dinosaur renaissance: why we need endothermic archosaurs for a comprehensive theory of bioenergetic evolution. *In* Thomas., R. D. K., and E. C. Olson (eds.), *A Cold Look at the Warm-Blooded Dinosaurs* (Boulder, Colorado: Westview Press), pp. 351–462.

Benedetto, J. L. 1973. Herrerasauridae, neuva familia de saurisquios Triásicos. *Ameghiniana.* X: 89–102.

Benton, M. J. 1979. Ectothermy and the success of dinosaurs. *Evolution* 33: 983–97.

　1983a. Dinosaur success in the Triassic: a noncompetitive ecological model. *Quart. Rev. Biol.* 58: 29–55.

　1983b. The Triassic reptile *Hyperodapedon* from Elgin: Functional morphology and relationships. *Phil. Trans. Roy. Soc. Lond. B* 302: 605–17.

　1984a. The relationships and early evolution of the Diapsida. *Symp. Zool Soc. Lond.* 52: 575–96.

　1984b. Dinosaurs' lucky break. *Natural History* 6: 54–9.

　1984c. Fossil reptiles of the German late Triassic and the origin of the dinosaurs. *In* Reif, W. E., and F. Westphal (eds.), *Third Symposium on Mesozoic Terrestrial Ecosystems, Short Papers* (Tubingen: ATTEMPTO), pp. 13–18.

Blakey, R. C., and R. Gubitosa. 1983. Late Triassic paleogeography and depositional history of the Chinle Formation, southern Utah and northern Arizona. *In* Reynolds, M. W., and E. D. Dolly (eds.), *Mesozoic Paleogeography of the West-Cen-*

tral United States: Rocky Mountain Sec. SEPM, Rocky Mtn. Paleogeography Symp. 2: 51–76.

Blodgett, R. H. 1984. Nonmarine depositional environments and paleosol development in the upper Triassic Dolores Formation, southwestern Colorado. *In* Brew, Douglas C. (ed.), *Field Trip Guidebook: Durango, Colorado, Fort Lewis College, 37th Ann. Mtg. Rocky Mtn. Sec. Geol. Soc. Am.*, pp. 46–92.

Bonaparte, J. F. 1981. Discripción de *Fasolasuchus tenax* y su significado en la sistematica y evolución de los thecodontia. Revista Museo Argentino de Ciencias Naturales. *Paleontología* 3(2): 55–101.

1982. Faunal replacement in the Triassic of South America. *J. Vert. Paleontol.* 2: 362–71.

1984. Locomotion in rauisuchid thecodonts. *J. Vert. Paleontol.* 3(4): 210–18.

Boucot, A. J. 1975. *Evolution and Extinction Rate Controls* (Amsterdam: Elsevier).

Charig, A. J. 1972. The evolution of the archosaur pelvis and hindlimb: an explanation in functional terms. *In* Joysey, K. A., and T. S. Kemp (eds.), *Studies in Vertebrate Evolution* (Edinburgh: Oliver and Boyd), pp. 121–55.

1980. Differentiation of lineages among Mesozoic tetrapods. *Mem. Soc. Geol. Fr.* 139: 207–10.

1984. Competition between therapsids and archosaurs during the Triassic Period: A review and synthesis of current theories. *Symp. Zool. Soc. Lond.* 52: 597–628.

Charig, A. J., J. Attridge, and A. W. Crompton. 1965. On the origin of the sauropods and the classification of the Saurischia. *Proc. Linn. Soc. London* 176(2): 197–221.

Chatterjee, S. 1985. *Postosuchus*, a new thecodontian reptile from the Triassic of Texas and the origin of tyrannosaurs. *Phil. Trans. Roy. Soc. London B* 309: 395–460.

Colbert, E. H. 1984. The Triassic dinosaur genera *Podokesaurus* and *Coelophysis*. *Am. Mus. Nov.* 2168: 1–12.

Colbert, E. H., and D. Baird. 1958. Coelurosaur bone casts from the Connecticut Valley Triassic. *Am. Mus. Nov.* 1901: 1–11.

Coombs, W. P., Jr. 1978. Theoretical aspects of cursorial adaptations in dinosaurs. *Quart. Rev. Biol.* 53: 393–417.

Dawley, R. M. 1979. *Heptasuchus clarki*, a new rauisuchid thecodont from the upper Triassic of Wyoming. M. Sc. thesis. Wayne State University, Detroit, Michigan.

Dingle, R. V., W. G. Siesser, and A. R. Newton. 1983. *Mesozoic and Tertiary Geology of Southern Africa* (Rotterdam: A. A. Balkema).

Halstead, L. B. 1975. *The Evolution and Ecology of the Dinosaurs* (London: Peter Lowe).

Haughton, S. H. 1924. The fauna and stratigraphy of the Stormberg Series. *S. Afr. Mus. Ann.* 12: 1–323.

1953. Gondwanaland and the distribution of early reptiles. *Trans. Proc. Geol. Soc. S. Afr.* 56 (Annex): 1–30.

Hiller, N., and N. Stavrakis. 1984. Permo-Triassic fluvial systems in the southeastern Karoo Basin, South

Africa. *Palaeogeog., Palaeoclim., Palaeoecol.* 45(1): 1–23.

Hoffman, A. 1979. Community paleoecology as an epiphenomenal science. *Paleobiology* 5: 357–79.

1981. The ecostratigraphic paradigm. *Lethaia* 14: 1–7.

Hotton, N., III. 1980. An alternative to dinosaur endothermy: the happy wanderers. *In* Thomas, R. D. K., and E. C. Olson (eds.), *A Cold Look at Warm-Blooded Dinosaurs* (Boulder, Colorado: Westview Press), pp. 311–45.

Jacobs, L. L., and P. A. Murry. 1980. The vertebrate community of the Triassic Chinle Formation near St. Johns, Arizona. *In* Jacobs, L. L. (ed.), *Aspects of Vertebrate History* (Flagstaff, Arizona: Museum of Northern Arizona Press), pp. 55–72.

Kemp, T. S. 1982. *Mammal-like Reptiles and the Origin of Mammals* (London: Academic Press)

Odum, E. P. 1971. *Fundamentals of Ecology*, 3rd ed. (Philadelphia, Saunders).

Olsen, P. E. 1980. A comparison of the vertebrate assemblages from the Newark and Hartford Basins (early Mesozoic, Newark Supergroup) of eastern North America. *In* Jacobs, L. L. (ed.), *Aspects of Vertebrate History, Essays in Honor of Edwin Harris Colbert* (Flagstaff, Arizona: Northern Arizona Press), pp. 35–53.

Olson, E. C. 1971. *Vertebrate Paleozoology* (New York: Wiley).

Ostrom, J. H. 1969. Terrestrial vertebrates as indicators of Mesozoic climates. *Proc. N. Am. Paleontol. Conv. D*, pp. 347–76.

Parrish, J. M., J. T. Parrish, and A. M. Ziegler. 1986. Permo-Triassic paleogeography and paleoclimatology and implications for therapsid distributions. *In* Hotton, N., P. D. Maclean, J. J. Roth, and F. C. Roth (eds.). *Ecology and Biology of Mammal-like Reptiles* (Washington, D.C.: Smithsonian Press).

Parrish, J. T., A. M. Ziegler, and C. A. Scotese. 1982. Rainfall patterns and the distribution of coals and evaporites in the Mesozoic and Cenozoic. *Palaeogeogr., Palaeoclimatol., Palaeoecol.* 40: 67–101.

Pianka, E. R. 1974. *Evolutionary Ecology*. (New York: Harper and Row).

Pipiringos, G. N., and R. B. O'Sullivan. 1978. Principal unconformities in Triassic and Jurassic rocks, western interior United States – a preliminary survey. *U.S. Geol. Survey Prof. Paper.* 1035-A: 29.

Price, P. W., C. N. Slobodchikoff, and W. S. Gaud. (eds.). 1984. *A New Ecology: Novel Approaches to Interactive Systems* (New York: Wiley).

Retallack, G. J. 1978. Floral ecostratigraphy in practice. *Lethaia* 11: 81–3.

1983. Late Eocene and Oligocene paleosols from Badlands National Park, South Dakota. *Geol. Soc. Amer. Spec. Pap.* 193: 1–82.

Ricklefs, R. E. 1979. *Ecology*, 2nd ed. (Newton, Massachusetts: Chiron Press).

Robinson, P. L. 1971. A problem of faunal replacement on Permo-Triassic continents. *Paleontology* 14: 131–53.

1973. Paleoclimatology and continental drift. *In* Tarling, D. H., and S. K. Runcorn (eds.), *Implications of Continental Drift to the Earth Sciences*

(London: Academic Press), pp. 451–76.

Schopf, T. J. M. 1977. Evolving paleontologic views on deterministic and stochastic approaches. *Paleobiology* 3: 337–52.

Simpson, G. C. 1953. *The Major Features of Evolution* (New York: Simon and Schuster).

Tankard, A. J., et al. 1982. *Crustal Evolution of Southern Africa: 3.8 Billion Years of Earth History* (New York: Springer-Verlag).

Tucker, M. E., and M. J. Benton. 1982. Triassic environments, climates and reptile evolution. *Palaeogeogr., Palaeoclimat., Palaeoecol.* 40: 361–79.

Whittaker, R. H. 1970. *Communities and Ecosystems* (London: Macmillan).

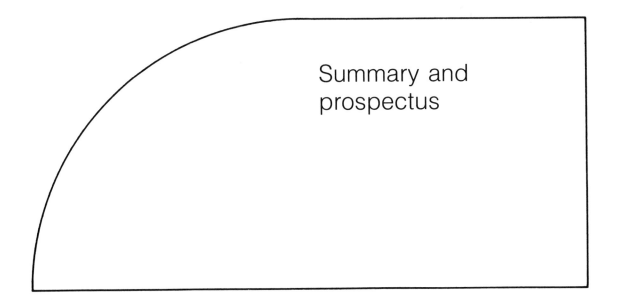

Summary and prospectus

The objective of this book has been to gather together a collection of chapters that would exemplify recent research being carried out on Late Triassic and Early Jurassic taxa and faunas and to summarize the present state of major evolutionary questions that occupy the interest of paleontologists working on the Triassic–Jurassic boundary. It may be presumptuous to attempt to organize in any one way the results and viewpoints presented here; however, from the present context of research, thought, and discussion, a series of themes for future work seems to emerge. The following considerations are by no means exhaustive.

Taxonomic diversity of the Late Triassic and the Early Jurassic

The basic paleontological data for asking any kinds of evolutionary questions are always the fossils themselves. It is evident from the contributions in this volume that the knowledge of Late Triassic and Early Jurassic vertebrates, especially in North America, has increased in the past decade at a prodigious rate. On the Triassic side, Chatterjee's explorations in the Dockum Formation of Texas have unearthed several new forms of dinosaur, pterosaur, and other archosaurs. The possibility that the horizons of these newest animals are of Norian age (rather than Carnian, as the entire Dockum has traditionally been regarded) may significantly extend the known temporal range of faunas of this vast formation, which is clearly far from exhausted paleontologically, as the efforts of Murry and Parrish and Carpenter also demonstrate. The same can be said for the Chinle Formation, of which the Petrified Forest Member shows the greatest promise for both taxonomic expansion and biostratigraphic correlation based on vertebrates and pollen. As Long and Padian report, there are now perhaps as many as

two dozen additional taxa to be added to the five known by Charles Camp in 1930. Most notable in the Petrified Forest horizons is the fine-scaled stratigraphic distribution of species of phytosaurs and aetosaurs, which provide an excellent opportunity to test Camp's hypothesis of evolutionary change and biostratigraphic correlation (Long and Ballew 1985). Even the Newark Supergroup of eastern North America, with its great lateral and temporal extents, has proved to be exceptionally fertile ground in the last decade: witness the new saurischian dinosaurs, early mammal-like therapsids, and several new kinds of footprints, in addition to literally thousands of new specimens of fishes (Olsen 1980; Olsen, McCune, and Thomson 1982).

As the chapters in this volume show, Late Triassic vertebrate diversity comprised a number of groups that were radiating quickly; many of these are confined stratigraphically to the Late Triassic. Metoposaurian and latiscopid amphibians shared the ponds and waterways with a variety of sharks and nonteleostean bony fishes, as well as phytosaurs and small aquatic archosauromorphs (Chapter 9). Large therapsids such as *Placerias* were not diverse, but could have been locally numerous, as the Placerias Quarry at St. Johns, Arizona, suggests; no other animals of the late Carnian are known to have been so large and to have exploited the ecological resources used by these hippopotomus-like beasts (Chapter 7). On land, unusual reptilian groups, such as the hypsognathid procolophonids, the rhynchosaurs, and the trilophosaurs, each produced minor radiations during this interval of time, as did the archosaurian aetosaurs, rauisuchians, poposaurs, and ornithosuchids. The first sphenodontids and perhaps the first pterosaurs appear in the late Carnian, along with several taxa traditionally regarded as closely related to the first true pterosaurs (*Scler-*

omochlus) and dinosaurs (*Lagosuchus* and *Lagerpeton*). The first crocodylomorphs show up in the Norian, as do the first unquestionable pterosaurs and the chelonians; all of these groups radiated quickly and continued quite successfully through the Mesozoic. Other unusual groups, such as proterochampsians, cerritosaurs, gomphodont cynodonts, and the problematic *Doswellia*, demonstrate both the apparent speed of Late Triassic evolutionary change and its frequently ephemeral success. In general, as Chapters 15, 24, and 25 show, the records of bones and footprints appear to agree on the picture of diversity, and despite differences in the composition of individual local faunas, it seems safe to conclude that dinosaurs and many of the taxa that would later take over the land were low in both abundance and diversity during the Late Triassic.

The recent increase in knowledge of these faunas has resulted from three approaches: new explorations, new analyses of previous collections, and new techniques. For example, screen-washing of bulk sediment samples, a technique pioneered for Tertiary mammals by Claude Hibbard, has been used by Murry, Downs, Jacobs, and others to recover microvertebrates in both the Chinle and Dockum formations and has recently been applied to the Newark Supergroup. It has also had spectacular results in some of the European deposits, as the work of Sigogneau-Russell and colleagues demonstrates. In some cases, this technique has more than doubled the known vertebrate diversity and has provided a much fuller picture of community elements that are rarely sampled by traditional approaches (Jacobs and Murry 1980; Tannenbaum 1983). In the Kayenta Formation of the Glen Canyon Group, it was responsible for the discovery of the oldest mammals in North America, as well as nonmammalian therapsids, small sphenodontids, lizards, crocodiles, and many other microvertebrates (Jenkins, Crompton, and Downs 1983).

The present studies, based on both osseous remains and footprints, tend to support the impression that dinosaurian diversity in the Late Triassic was generally low, despite local records both high in diversity and in relative abundance of dinosaurian specimens. The Carnian seems to have the earliest definite remains of dinosaurs, and these were reviewed in the Introduction to this volume. Many horizons around the world have been identified to be of Carnian age, often more on the basis of the vertebrate taxa that they contain than on any independent criterion. Although by no means infallible, pollen and spores may provide such a criterion, but the possibility must remain open that vertebrates and plants evolved and were preserved differentially through time and space (Chapter 9).

During the Norian, a much longer stage than

the Carnian, it is clear that the saurischian dinosaurs diverged into several theropod and sauropodomorph types, although the latter apparently did not reach the "sauropod grade" of gigantic size and columnar stance. The theropods remained relatively small. Ornithischians appear to have remained rather generalized, although in the Early Jurassic both they and the saurischians began to diversify. The exact taxonomic diversity of dinosaurs is difficult to assess because many taxa have been named on the basis of very partial specimens, often ranging in size and probable ontogenetic stage. A conservative list might include something like *Coelophysis*, *Procompsognathus*, and possibly the problematic *Halticosaurus* among the theropods and the anchisaurid sauropodomorphs *Plateosaurus* and *Euskelosaurus*. It is unfortunate that the poorly preserved *Pisanosaurus* was for so long the most complete representative of Triassic ornithischians, but many new records have recently surfaced (see the Introduction). Other dinosaurs, such as *Herrarasaurus*, *Ischisaurus*, and *Staurikosaurus*, are difficult to place within either the Saurischia or the Ornithischia and probably stand outside the common ancestor of these two orders (Gauthier 1984). In view of the very high diversity of many nondinosaurian groups known from the Late Triassic, it seems reasonable to take at face value the evidently low diversity of the contemporaneous dinosaurs.

Across the division of the Triassic and Jurassic marked in the Newark Supergroup and extended to other horizons worldwide (Olsen and Galton 1977, 1974; Chapter 25), a distinct change in vertebrate composition occurs. First, the overall taxonomic diversity is much lower, and, second, very few new major groups seem to have appeared. Nearly all taxonomic lineages of terrestrial vertebrates either disappeared or suffered drastic reductions. Among the amphibians, only a single lineage of the large labyrinthodonts appears to have survived into the Jurassic, although the Lissamphibia (modern amphibians), now beginning to be known from microvertebrate assemblages, obviously persisted. The sphenodontids survived, and therefore so must have the lepidosaurian lineage leading to the Squamata (lizards and snakes), although it remains difficult to determine at what point true squamates appeared. Among the therapsids only cynodonts remained; the largest were the tritylodonts, which persisted to the mid-Jurassic, and the trithelodonts, but there were also kuehneotheres, haramiyids, morganucodontids, and diarthrognathids. All but the last group were present in the Norian. Proganochelyid chelonians did not survive, but casichelyid turtles, which appeared in the Norian, continued on. Two lineages of pterosaurs were already present in the Norian, and they persisted and diversified in the Early Jurassic. The first crocodylomorphs – sphenosuchids and their relatives – were present in the Norian, but true crocodiles may not have evolved before the time represented by the Moenave Formation, perhaps as late as Pliensbachian time (middle Early Jurassic). By the Toarcian, crocodiles had invaded the seas, and their skeletons are found in the marine deposits of the Upper Lias of Europe.

Then there were the dinosaurs. Small, generalized theropods included the dog-sized *Syntarsus* and *Segisaurus*; a few, such as *Dilophosaurus*, grew as large as 5 m in overall length. The less specialized "prosauropod" sauropodomorphs continued to diversify; Anchisauridae and Melanorosauridae were joined by several forms that are clearly true sauropods or very close to them, such as *Vulcanodon*, *Barapasaurus*, and *Ohmdenosaurus* (see Chapter 19). If Triassic ornithischians are generalized and nondescript, the same cannot be said for their Early Jurassic counterparts; a host of them appeared, and they diversified into many of the major lineages of the later groups. *Scutellosaurus* (Colbert 1981) and *Scelidosaurus* represent the earliest Thyreophora (the group including Ankylosauria and Stegosauria), while *Heterodontosaurus* and other "fabrosaurids" appear to be more closely related to the "cheeked" lineages of Ceratopsia, Pachycephalosauria, and Hadrosauria (Gauthier 1984). Distinct members of any of these five major ornithischian groups are unknown, however; only the two major divisions are demarked.

In general, then, despite the successes of the dinosaurs, pterosaurs, crocodiles, lissamphibians, lepidosaurs, chelonians, tritylodonts, and mammals, a great part of the world's terrestrial vertebrate fauna was lost at the end of the Triassic, and diversity at both higher and lower taxonomic levels was clearly lower in the Early Jurassic than in the Carnian and Norian, when these groups and many others flourished. This conclusion appears to be inescapable on the basis of the available evidence of terrestrial vertebrates. On the other hand, although artifacts of preservation or sampling cannot be separated from the empirical data with reliability, the fishes do not seem to have shown substantial taxonomic change across the Triassic–Jurassic boundary. If anything, there is a slight decrease in diversity, but Late Triassic *Lagerstaetten* may be responsible for the apparent trend (Chapter 13).

The confidence with which paleontologists can document particular macroevolutionary patterns depends on two things: understanding the evolutionary relationships of the groups involved and fitting their records into a reliable framework of stratigraphy and time. Fossils by themselves do not tell time, at least in an absolute sense. For example, phytosaurs are only known from sediments regarded on other grounds to be of Triassic age; but there is no reason why they could not have persisted into the Jurassic in some places (Chapter 1 gives a similar example for the labyrinthodont amphibians). Biostratigraphic hypotheses risk circularity unless we use many kinds of fossils and other lines of evidence to erect and test such hypotheses. With some idea of the *dramatis personae*, then, we may move to the question of establishing the ages on which these hypotheses are based.

Biostratigraphic resolution of Late Triassic and Early Jurassic horizons

Perhaps the major question in this volume, and one that has been debated for most of the past decade, is whether the reassignment of many supposed "Late Triassic" beds to the Early Jurassic is warranted. The history and rationale for the change was explained briefly in the Introduction and has been reiterated in Chapter 25. Yet at least two other chapters in this volume call the reassignment into question, and on slightly different grounds. Colbert does not dispute the suggested correlation of several geographically disparate formations on the basis of their vertebrates, but suggests that it is quite arbitrary where the "boundary" between the Triassic and Jurassic is fixed. Clark and Fastovsky question the correlation of geographically disparate beds for two reasons: First, absolute ages are poorly known, and, second, correlations on the basis of family level taxa, none of which is restricted to Jurassic beds, are of too poor a level of resolution to merit confidence. Although they do not propose that these beds are of Triassic age, they see little evidence for assigning them a Jurassic age.

Colbert has taken the theme of the whole history of the concepts of "Triassic" and "Jurassic," and has shown how these ideas evolved from the type section of European marine sediments to apply to terrestrial horizons that contained vertebrates similar to those of beds underlying and interfingering with the European marine sections. The prosauropod and small theropod dinosaurs of the Newark Supergroup were sufficiently similar to their Triassic European counterparts to warrant correlation, and by extension so were the faunas of the Glen Canyon Group, the Upper Stormberg Series of South Africa, and the Lufeng beds of China; but how are the dates of any of these horizons to be determined? As Colbert points out, for decades geologists have debated whether the "Rhaetian" should be placed at the uppermost Triassic or the lowermost Jurassic, and now some prefer to discard it entirely or to subsume it within the Norian. Without a "golden spike" demarking the Triassic–Jurassic boundary in the type area – which, like many type areas, is by no means the most complete geologic section available – and without a clear correlation between marine and terrestrial exposures, in some very important ways the entire question of the Triassic–Jurassic boundary becomes insoluble, or at least very arbitrary.

Clark and Fastovsky do not question the Jurassic assignment of part of the Newark Supergroup, but they do question its extension to the Glen Canyon Group of the American Southwest because most of the taxa used to establish the biostratigraphic correlation are either questionably known or not diagnostic of Jurassic horizons. Protosuchian crocodilians, such as appear in the Moenave Formation and in the Jurassic portion of the Newark, are also known from the Los Colorados beds of Argentina, which also preserve classic Triassic "marker taxa." Most geographic areas do not share vertebrate taxa at the generic level, and "family" level correlations are not sufficiently fine-scaled. Pollen and spores are poorly preserved in the Glen Canyon Group, and the few radiometric dates are too high in the section to be relevant to the question at hand.

These and other thoughtful criticisms make it clear that proponents of an Early Jurassic age need to specify what is "Jurassic" about the faunas being reassigned. As mentioned before, there are few terrestrial vertebrate data from the type sections of the European Early Jurassic; age assignments for the terrestrial beds of other continents have had to be made through palynological and radiometric criteria. Several considerations should be investigated.

1. As Chapters 24 and 25 make clear, the Jurassic is not characterized by the appearance of any new vertebrate taxa (at least at the family level), but by the loss of taxa at many hierarchical levels. It is of considerable interest that none of the extensive and often exhaustive exploration of supposed Jurassic horizons has yielded any specimens of what have traditionally been called "thecodonts" – nondinosaurian, noncrocodilian, nonpterosaurian, nonavian archosaurs such as phytosaurs, aetosaurs, rauisuchians, ornithosuchids, and lagosuchids – that are so characteristic of Late Triassic faunas. Certainly there are lithological changes worldwide through this interval, perhaps implying ecological or climatic changes (Tucker and Benton 1982), but is the probability realistic that the worldwide absence of "thecodonts" and other classic Triassic forms in many different environments can be completely ascribed to preservational bias? If so, then we must wonder why the taxa that survive the Triassic extinctions, such as the dinosaurs, turtles, and mammals, have been found (often in great abundance) in Late Triassic environments that were clearly very different from those of the presumed Jurassic horizons above. This argument may amount to an endorsement of negative evidence, but in fact this is nothing new for paleontology. We often characterize faunas and ages by the absence of certain taxa, often with circular results, as Camp (1952) observed on the traditional, a priori exclusion of dinosaurs from Tertiary sediments. This brings us to the second point.

2. Vertebrate fossils alone are probably not sufficiently preserved and studied to allow fine-scale correlation of disjunct geological formations. At the present state of knowledge, no single taxon can serve as a reliable marker taxon: A comparison of all available taxonomic lineages should be investigated to reinforce biostratigraphic correlations suggested by single taxa. (As a rule, no single taxon should ever be used to demarcate biostratigraphic boundaries.) Our best efforts should involve palynomorphic and radiometric data, and all lines of evidence must be tested against each other. Sometimes the results will be inconclusive or too grossly scaled to be of real use, but each method can provide upper and lower age ranges that may be compared against each other.

3. It is always tempting to resolve paleontological disputes with the rallying cry for more fossils and more research. Who would argue against gathering more evidence? Yet it is also important to open the evidence we have to new investigations. Clark and Fastovsky (Chapter 23) and Olsen and Sues (Chapter 25) call for renewed efforts on the systematic relationships of the vertebrate groups in question. Refined resolution of stratigraphic ranges is the most important potential outcome of this kind of analysis. For example, the primitive sauropodomorphs known as "prosauropods" are found in both Late Triassic and Early Jurassic formations. As a group, then, the "Prosauropoda" is biostratigraphically unenlightening, for our purposes; but a subset of this group, for example, the Vulcanodontidae (Chapter 17) or certain tritylodontids (Chapter 22), may only appear in the Jurassic. Therefore, phylogenetic analysis of Triassic–Jurassic taxa is of primary importance in resolving biostratigraphic questions. Monophyletically restricted subgroups of larger taxonomic groups are more likely to have restricted, diagnostic ranges; however, it is important to stress the need to use only monophyletic groups in such analyses.

4. Independent methods of establishing chronology have been of tremendous importance to paleontology. Yet radiometric dating is not a panacea for all stratigraphic problems. One difficulty is that the igneous sediments amenable to these techniques are not always found precisely where we would like them. Therefore, when they appear and can be dated, we are more likely to gain only a minimum or maximum age for a pile of rocks that we would like to subdivide much more finely. A second problem is that as we move back in time, the error mar-

gins associated with radiometric dating necessarily become larger, so that at a dstance of 200 MY the "slop" either way is likely to be as high as 8–10 MY – more than the average length of a stage. Even magnetic stratigraphy becomes more difficult in the more distant past because there is greater chance of overprinting, contamination, or losing a reliable signal. However, some kind of nonbiological data is always desirable, especially in cases such as the Los Colorados Formation, where an anomaly to a general pattern occurs. Through the continued use of nonbiological dating techniques, we may learn more about both the timing of the history of life and the degree of resolution of the methods themselves.

Systematic, functional, and ecological considerations

Much of this volume is devoted to macro-evolutionary questions of change in faunas through time and space. These form general themes to which much evidence here and in the future may be addressed; but beyond the evidence of large-scale taxonomic and biostratigraphic patterns, there are the biological considerations of the organisms themselves. The large-scale patterns, unless they were produced by random processes and are therefore artifactual, can only make sense if biological insights reveal the underlying causes.

Functional considerations of extinct vertebrates have often been used to explain patterns of adaptation, radiation, or replacement. On a large scale, Benton (Chapter 24) argues that the dinosaurs merely got a "lucky break" and showed no real competitive replacement of other taxa. Zawiskie (Chapter 26) agrees with the difficulty of identifying the role of competition in the fossil record, but argues that the dinosaurs really were better able to take advantage of climatic and community changes at the end of the Triassic. Charig (1984) and Bakker (1977, 1980), among other authors, have based their arguments on biological grounds, some of which themes are borne out here.

The principal foci of functional studies of fossil animals have historically been feeding and locomotion. In this volume, two chapters address herbivory in dinosaurs. Galton provides a phylogenetic survey of herbivorous adaptations, discussing the presence of cheeks, slicing teeth, mechanical leverage of the jaws, and other features related to processing plant food. It seems as though the sauropodomorphs took an early lead in abundance and diversity of herbivorous types, while the ornithischians kept a low profile until the Jurassic. Galton suggests that ornithischian and even mammalian herbivores were much better adapted to their diets, and agrees with Crompton and Attridge (Chapter 17) that competition was probably not a factor in the early success of the "prosauropods." After the Early Jurassic, the giant sauropods replaced the early sauropodomorphs, and these diversified in ways much different those of the varied ornithischian types. Crompton and Attridge contrast a generalized model of dinosaurian herbivory with one for Triassic rhynchosaurs, anomodonts, and cynodonts. They further note a puzzling change from a Carnian fauna of herbivores with adaptations for cutting and breaking down rough vegetation to a Norian fauna with no obvious adaptations. As Ash (Chapter 2) notes, there were few, if any, major botanical changes over this interval, and Crompton and Attridge conclude that another, perhaps completely unrelated causal agent was responsible for the very sudden demise and replacement of the Carnian forms. This hypothesis is quite viable, although again the possibility of preservational bias against many of the "Carnian" herbivores in the Norian, which is generally low in herbivores, must be considered (J.M. Parrish pers. comm.).

The locomotion of dinosaurs, with their "fully erect" postures and parasagittal gaits, has long figured in explanations of their success. Welles (Chapter 3) gives a thoughtful reconsideration of the problem of the pelvis and hindlimb in dinosaurs. Acknowledging the importance of the mesotarsal ankle as an innovation, Welles suggests that the emphasis on stance and gait has been misplaced. Usually "analyses treat the leg as a pendulum swinging freely from the acetabulum," when in fact "the problem should consider the foot as the fulcrum . . . and the limb is a pillar, a lever, rocking in a parasagittal plane upon this base" (Welles, Chapter 3). He goes on to consider that when a biped walks, it is the role of each limb to support the whole body when the other is in the air. The pelvic musculature does this through generation of forward momentum and by shifting the body during locomotion. Padian (Chapter 5) notes that the functional anatomy of the hindlimb points to the conclusion that even the earliest theropod dinosaurs stood and walked like their descendants the birds in all major particulars, including a more or less horizontal femur and a tibia and fibula that swung in a parasagittal plane. The "advantage" traditionally presumed for the "fully erect" stance of the dinosaurs is that they were able to walk parasagittally, while other forms were "semierect" or "sprawling." As Gauthier (1984) notes, however, the truth is that the dinosaurs were *restricted* to fully erect posture; and as Parrish (in prep. and Chapter 4) and Bonaparte (Chapter 19) have suggested for aetosaurs, phytosaurs, and rauisuchians, there is more than one way to achieve "erect" posture. The joints of nondinosaurian archosaurs often permit more potential excursion than dinosaurian joints do, and this may lead us to the conclusion that there was something sloppy about their locomotory patterns. However, as anyone who has dissected a crocodilian pelvis knows, the role of cartilage, tendons, and ligaments in shaping joint surfaces and limb excursions should not be underestimated simply because it is not obvious from the bony skeleton alone. Sometimes, the absence of obvious restrictions of joint morphology on stance and limb excursions require us to consider evidence from the entire limb (including areas for muscle attachment, shapes of bony processes, torsion of bone shafts) to arrive at conclusions about posture and locomotion (Chapter 7).

Footprints are a major source of information about the stance and locomotory patterns of extinct vertebrates. As Haubold (Chapter 15) notes, the relationship between bones and footprints is often tenuous, and it is often difficult to match tracks with trackmakers. Tracks can stand on their own as biostratigraphic data (Chapters 6, 15, 20, and 25), but it is especially satisfying to gain insight about locomotion from footprints whose makers can be identified with some reliability. The feet of crocodilian skeletons from the Moenave Formation can be easily fit to the tracks of *Batrachopus* in the same sediments (Chapter 20), suggesting that crocodilian locomotory patterns have changed little in nearly 200 MY. *Atreipus*, a new dinosaurian ichnogenus named by Olsen and Baird (Chapter 6), implies a trackmaker with a tridactyl foot that might fit easily in either the Theropoda or the Ornithischia, but its manus impression does not easily accord with either group. Either there is a type of dinosaur out there for which we have no osseous remains, or we must reconsider structural and functional assumptions about known groups, one of which must have been responsible for *Atreipus*. This method of working structural and functional considerations back and forth between bones and tracks also surfaces in Parrish's (Chapter 4) suggestion that *Apatopus*, identified as a phytosaurian trackway by Baird (1957), does not seem to have been made by the foot of any known phytosaur, which implies further work on the known structure of phytosaurs as well as consideration of what other animal might have made the tracks.

Prospectus

Several major unanswered questions support paths of future research. Most paleontological problems, including those surrounding the Triassic–Jurassic boundary, could be ameliorated by more fossils, higher resolution of stratigraphic and chronometric determinations, and thorough phylogenetic analyses of the taxa involved. Several particular problems may be mentioned briefly.

1. "Jurassic" faunas need more extensive study, diagnosis, and differentiation from those of the Triassic. Finer-level taxonomic analysis may help to restrict the perceived chronologic ranges of certain taxa; associations of taxa at several hierarchical levels and across broad taxonomic lines may provide more useful characterizations (Chapters 24 and 25).

2. The question of evolutionary change must be tested against potential change in the preservation of certain environments through time (Chapters 24 and 25). For example, was there really a warming, drying trend in the Late Triassic and Early Jurassic? Is it reflected in the paleoecology of many sediment types, or are only certain types preserved through this interval? If the latter is true, then the possibility cannot be eliminated that the pace of evolution may have been quite different, and the apparent pattern is due more to environmental preservation than to evolutionary change.

3. What is the tempo and mode of evolutionary events in the Late Triassic and Early Jurassic? This question depends on the durations of time reflected by individual formations and members, as well as the completeness of the depositional record. Relative thicknesses of rock do not reliably indicate duration of deposition, and, of course, the length of time represented by a geological unconformity could be from a few weeks to millions of years. This is especially difficult to assess when rock types change substantially through a given sequence, as for example with the Triassic–Jurassic exposures of the southwestern United States. We simply have no measure of the amount of time represented by each of the Chinle, Dockum, Wingate, Moenave, Kayenta, and Navajo formations, although we have a rough idea from palynological and radiometric methods of the total span of time involved. The understanding of large-scale evolutionary transitions depends in part on the understanding of the time and pace of the event. This question is as important for the beginnings of the dinosaurs as it is for the extinction of many of them at the end of the Cretaceous.

4. Does any biological factor unify many of the groups that replaced others or that were replaced in the Late Triassic? The chapters in this volume suggest that feeding patterns do not seem to explain differential success in many herbivorous lineages, and that the locomotory picture is not simple. Bakker (1980) suggested factors of physiological advantage for the dinosaurs and their relatives, and this idea has been hotly debated (see other papers in Thomas and Olson 1980). With increased taxonomic and stratigraphic resolution, can biological resolution of these factors emerge? What aspects might be amenable to further testing?

5. If we now have a Lower Jurassic record, is there hope that our knowledge of the Middle Jurassic terrestrial faunas will become more complete? Mid-

dle period records are poor for the entire Mesozoic (Padian and Clemens 1985), but recent discoveries on several continents, including South America (J. F. Bonaparte pers. comm.), China (Sigogneau-Russell and Sun 1981), and North Africa (P. Taquet pers. comm.) suggest that some of the gaps may be filled in the not too distant future. From what is known of the Early Jurassic, it seems likely that the Middle Jurassic will reveal some very interesting events in vertebrate history, including the radiation of stegosaurs and sauropod dinosaurs, the disappearance of the last prosauropod dinosaurs and non-mammalian therapsids, and perhaps even the appearance of the first birds. It would be of great interest to see whether the wonderfully diverse faunal components of the Late Jurassic radiated quickly at that time, built slowly from the Early Jurassic paucity of types, or surged suddenly in the Middle Jurassic; but this is another frontier about which only time, as usual, will tell.

References

Baird, D. 1957. Triassic reptile footprint faunules from Milford, New Jersey. *Mus. Comp. Zool. (Harvard Univ.) Bull.* 117: 449–520.

Bakker, R. T. 1977. Tetrapod mass extinctions – a model of the regulation of speciation rates and immigration by cycles of topographic diversity. *In* Hallam, A. (ed.), *Patterns of Evolution as Illustrated by the Fossil Record* (Amsterdam: Elsevier Scientific), pp. 439–68.

   1980. Dinosaur heresy – Dinosaur renaissance: why we need endothermic archosaurs for a comprehensive theory of bioenergentic evolution. *In* Thomas, R. D. K., and E. C. Olson (eds.), *A Cold Look at the Warm-Blooded Dinosaurs.* AAAS Selected Symposium 28 (Boulder, Colorado: Westview Press), pp. 351–462.

Camp, C. L. 1930. A study of the phytosaurs, with descriptions of new material from western North America. *Mem. Univ. Calif.* 10: 1–174.

   1952. Geologic boundaries in relation to faunal changes and diastrophism. *J. Paleontol.* 26: 353–8.

Charig, A. J. 1984. Competition between therapsids and archosaurs during the Triassic Period: a review and synthesis of current theories. *Symp. Zool. Soc. Lond.* 52: 597–628.

Colbert, E. H. 1981. A primitive ornithischian dinosaur from the Kayenta Formation of Arizona. *Mus. N. Ariz. Bull.* 53: 1–61.

Gauthier, J. A. 1984. A cladistic analysis of the higher systematic categories of the Diapsida. Ph.D. dissertation, Department of Paleontology, University of California, Berkeley.

Jacobs, L. L., and P. A. Murry. 1980. The vertebrate community of the Triassic Chinle Formation near St. Johns, Arizona. *In* Jacobs, L. L. (ed.), *Aspects of Vertebrate History: Essays in Honor of Edwin Harris Colbert* (Flagstaff, Arizona: Museum of North Arizona Press), pp. 55–72.

Jenkins, F. A., Jr., A. W. Crompton, and W. R. Downs. 1983. Mesozoic mammals from Arizona: new evidence of mammalian evolution. *Science* 222: 1233–5.

Long, R. A., and K. L. Ballew. 1985. Aetosaur dermal armor from the Late Triassic of Southwestern North America, with special reference to material from the Chinle Formation of Petrified Forest National Park. *Bull. Mus. N. Ariz.* 54: 35–68.

Olsen, P. E. 1980. A comparison of vertebrate assemblages from the Newark and Hartford Basins (Early Mesozoic, Newark Supergroup) of Eastern North America. *In* Jacobs, L. L. (ed.), *Aspects of Vertebrate History: Essays in Honor of Edwin Harris Colbert* (Flagstaff, Arizona: Museum of North Arizona Press), pp. 39–54.

Olsen, P. E., and P. M. Galton. 1977. Triassic–Jurassic tetrapod extinctions: are they real? *Science* 197: 983–6.

   1984. A review of the reptilian and amphibian assemblages from the Stormberg Group of South Africa, with special emphasis on the footprints and the age of the Stormberg. *Palaeontol Afr.* 25: 87–110.

Olsen, P. E., A. R. McCune, and K. S. Thomson. 1982. Correlation of the early Mesozoic Newark Supergroup (Eastern North America) by vertebrates, principally fishes. *Am. J. Sci.* 282: 1–44.

Padian, K., and W. A. Clemens. 1985. Terrestrial vertebrate diversity: episodes and insights. *In* Valentine, J. W. (ed.), *Phanerozoic Diversity Factors* (Princeton, New Jersey: Princeton University Press), pp. 41–96.

Sigogneau-Russell, D., and A.-L. Sun. 1981. A brief review of Chinese Synapsids. *Geobios* 14 (2): 275–9.

Tannenbaum, F. A. 1983. The microvertebrate fauna of the Placerias and Downs Quarries, Chinle Formation (Upper Triassic), near St. Johns, Arizona. M.S. thesis, University of California, Department of Paleontology.

Thomas, R. D. K., and E. C. Olson (eds.), 1980. *A Cold Look at the Warm-Blooded Dinosaurs*, AAAS Selected Symposium 28 (Boulder, Colorado: Westview Press).

Tucker, M. E., and M. J. Benton. 1982. Triassic environments, climates and reptile evolution. *Palaeogeogr., Palaeoclimat., Palaeoecol.* 40: 361–79.

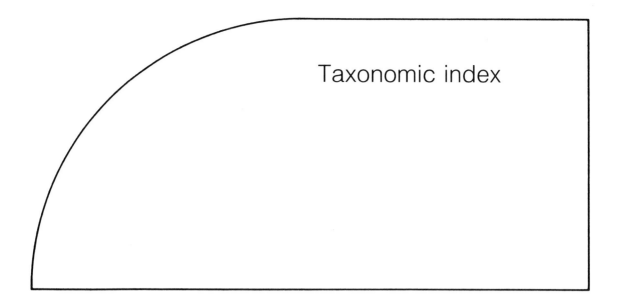

Taxonomic index

This index refers only to osteological taxa; ichnological taxa are given in a separate index, immediately following. Invertebrate taxa are not listed. Paleobotanical and palynological taxa and issues are not indexed; they are discussed and illustrated on pages 20–9, 73–4, 80–3, 131, 232, 270, 282, 296–8, 307, 316, 322, 324–5, 337–8, 340–1, 344–8, and 356–7.

Osteological taxa below the level of class are listed, as are the classes Mammalia and Aves because they appear infrequently. Species names are omitted. Taxa are listed according to authors' usages, which may vary. Cross references to some taxonomic higher categories that may be similar or synonymous are provided. The same term may be used with different meanings and different memberships by different authors. In the text, traditional names that apply to unnatural (nonmonophyletic) groups are often placed by authors in quotation marks, which are omitted in this index.

The letter "f" following a page number refers to a figure on that page; "t" refers to a table.

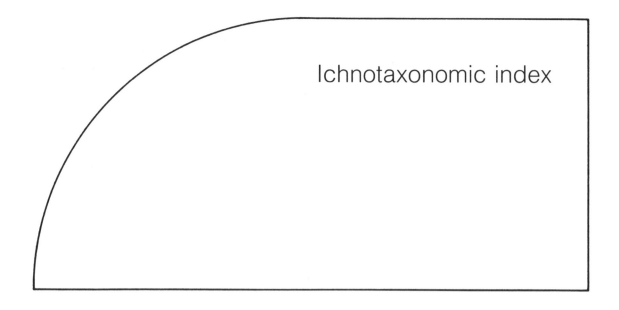

Ichnotaxonomic index

Genera and higher taxa are included here, except names for osteological taxa that are used in an osteological sense. For example, the ichnogenus *Batrachopus* is normally placed in the osteological Order Crocodylia; the latter name is not listed here.

Conversely, the names Melanorosauria and Sauropoda, which are also osteological names, are used as ichnotaxa on the text pages given. As with the osteological index, taxa are listed according to authors' usages, which may vary.